Intermediate Algebra

eText Reference

Kirk Trigsted

University of Idaho

Randy Gallaher

Lewis & Clark Community College

Kevin Bodden

Lewis & Clark Community College

PEARSON

Boston   Columbus   Hoboken   Indianapolis   New York   San Francisco
Amsterdam   Cape Town   Dubai   London   Madrid   Milan   Munich   Paris   Montreal   Toronto
Delhi   Mexico   São Paulo   Sydney   Hong Kong   Seoul   Singapore   Taipei   Tokyo

Editorial Director: Christine Hoag
Editor in Chief: Michael Hirsch
Senior Acquisitions Editor: Dawn Giovanniello
Editorial Assistant: Megan Tripp
Program Management Team Lead: Karen Wernholm
Program Manager: Beth Kaufman
Project Management Team Lead: Christina Lepre
Project Manager: Ron Hampton
Associate Design Director: Andrea Nix
Art Director: Heather Scott
Math Media Producer: Aimee Thorne
Software Development: TestGen, Marty Wright; MathXL, Rebecca Williams
Marketing Manager: Rachel Ross
Associate Marketing Manager: Alicia Frankel
Procurement Specialist: Carol Melville

Credits: Chapter 7: Blend Images/Pete Saloutos/Getty Images; Screenshots from Texas Instruments, Courtesy of Texas Instruments; **Chapters 9–11:** Screenshots from Texas Instruments, Courtesy of Texas Instruments.

The author and publisher have used their best efforts in preparing this book. These efforts include the development, research, and testing of the theories and programs to determine their effectiveness. The author and publisher make no warranty of any kind, expressed or implied, with regard to these programs or the documentation contained in this book. The author and publisher shall not be liable in any event for incidental or consequential damages in connection with, or arising out of, the furnishing, performance, or use of these programs.

1   2   3   4   5   6—RRD–H—19   18   17   16   15

PEARSON

ISBN-13: 978-0-321-99041-9
ISBN-10:   0-321-99041-2

Contents

Preface

Introduction

This *Intermediate Algebra eText Reference* contains the pages of Kirk Trigsted, Randy Gallaher, and Kevin Bodden's *Intermediate Algebra* eText in a portable, bound format. The structure of the *eText Reference* helps students organize their notes by providing them with space to summarize the videos and animations. Students can also use it to review the eText's material anytime, anywhere.

A Note to Students

This eText was created for you! Unlike a traditional text, we wanted to create content that gives you, the reader, the ability to be an active participant in your learning. The eText was specifically designed to be read online. The eText pages have large, readable fonts and were designed without the need to scroll. Throughout the material, we have carefully placed thousands of hyperlinks to definitions, previous chapters, interactive videos, animations, and other important content. Many of the videos and animations allow you to actively participate and interact. Take some time to "click around" and get comfortable with the navigation of the eText and explore its many features.

Before you attempt each homework assignment, read the appropriate section(s) of the eText. At the beginning of each section, you will encounter a feature called Things to Know. This feature includes all the prerequisite objectives from previous sections that you will need to successfully learn the material presented in the new section. If you do not have a basic understanding of these objectives, click on the desired hyperlinks and rework through the objectives, taking advantage of any videos or animations.

An additional feature of the eText is the inclusion of an audio icon. By clicking on this icon, you can listen to the text as you read. While you read through the pages of the eText, use the margins of this *Intermediate Algebra eText Reference* to take notes, summarize key points, and list helpful tips and reminders. An additional option, if made available by your instructor, is to use the *Guided Notebook* to guide your note taking as you work through the eText. The *Guided Notebook* provides more specific direction on how to proceed through the material while providing more space for note taking.

Try testing yourself by working through the corresponding You Try It exercises. Remember, you learn math by doing math! The more time you spend working through the videos, animations, and exercises, the more you will understand. If your instructor assigns homework in MyMathLab or MathXL, rework the exercises until you get them right. Be sure to go back and read the eText at any time while you are working on your homework. This text caters to your educational needs. We hope you enjoy the experience!

A Note to Instructors

Today's students have grown up in a technological world where everything is "clickable." We have taught with MyMathLab for many years and have experienced firsthand how fewer and fewer students have been referring to their traditional textbooks, opting instead to use electronic resources. As the use of technology plays an ever increasing role in how we are teaching our students, it seems only natural to have a textbook that mirrors the way our students are learning. We are excited to have written a text that was conceived from the ground up to be used as an online, interactive tool. Unlike traditional printed textbooks, this eText was specifically designed for students to read online while working in MyMathLab. Therefore, we wrote this text entirely from an online perspective, with MyMathLab and its existing functionality specifically in mind. Every hyperlink, video, and animation has been strategically integrated within the context of each page to engage the student and maximize his or her learning experience. All of the interactive media was designed so students can actively participate while they learn math. Many of the interactive videos and animations require student interaction, giving specific feedback for incorrect responses.

We are proponents of students learning terms and definitions. Therefore, we have created hyperlinks throughout the text to the definitions of important mathematical terms. We have also inserted a tremendous amount of "just-in-time review" throughout the text by creating links to prerequisite topics. Students have the ability to reference these review materials with just a click of the mouse. You will see that the exercise sets are concise and nonrepetitive. Since MyMathLab will be used as the main assessment engine, there is no need for a repetitive exercise set in the hardcopy version of the text. Every exercise is available for you to assign within MyMathLab or MathXL. For the first time, instructors can assess reading assignments. We have created five conceptual reading assessment questions for every section of the eText, giving the student specific feedback for both correct and incorrect responses. Each feedback message directs them back to the appropriate location for review. Our hope is that every student that uses the eText will have a positive learning experience.

Acknowledgments

First of all, we want to thank our wives for their loving support, encouragement, and sacrifices. Thanks to our children for being patient when we needed to work and for reminding us when we needed to take breaks. We could not have completed this project without generous understanding and forgiveness from each of you.

Writing this textbook has been one of the most difficult and rewarding experiences of our lives. We truly could not have accomplished our goal without the support and contributions of so many wonderful, talented people. These extraordinary talents deserve every accolade we can give them. From Pearson, we would like to thank Dawn Giovanniello, our editor and friend, for believing in us; Chris Hoag for her continued support; Heather Scott and Andrea Nix for their design brilliance; Rachel Ross and Alicia Frankel for their marketing expertise; Aimee Thorne and Ruth Berry for taking care of all the details; Eileen Moore, Rebecca Williams, and the rest of the MathXL development team for an amazing job; Andrea Stefanowicz of Lumina Datamatics; Ron Hampton, our project manager, for all his support.

Along the way, we had the help of so many people from around the world and from many different walks of life. The contributions of Elaine Page cannot be measured. Her detailed reviews and numerous suggestions have improved this text considerably. Our conversations were invaluable throughout the entire writing process—thank you for everything! We appreciate everyone else who made this book a reality: Alice Champlin and Anthony T. J. Kilian at Magnitude Entertainment for creating quality videos and animations; Helen Medley for reading the entire eText and catching the little things; Sue Glascoe for creating the PowerPoints; and Kim Johnson for her work on the *Guided Notebook*.

The following list attempts to recognize all of the reviewers. Please accept our deepest apologies if we have inadvertently omitted anyone. We have benefited greatly from your honest feedback, constructive criticism, and thoughtful suggestions. Each of you has helped us create a resource we truly believe will be effective in helping students learn mathematics. We are deeply grateful and we genuinely thank you all from the bottom of our hearts.

—Randy Gallaher
Kevin Bodden

Reviewers

CHAPTER R
Review Chapter

CHAPTER R CONTENTS

R.1 Sets of Numbers

OBJECTIVES

1 Identify Sets
2 Classify Real Numbers
3 Plot Real Numbers on a Number Line
4 Use Inequality Symbols to Order Real Numbers
5 Compute the Absolute Value of a Real Number

OBJECTIVE 1 IDENTIFY SETS

Let's review some definitions involving sets.

A **set** is a collection of objects. Each object in a set is called an **element** or a **member** of the set. Typically, capital letters are used to name sets, and braces { } group the list of elements in a set. For example, the seven days of the week form a set. To name this set D, write

$$D = \{\text{Sunday, Monday, Tuesday, Wednesday, Thursday, Friday, Saturday}\}.$$

The **element symbol** ($\in$) indicates that an object is an element of a set. For example, Friday $\in D$ means "Friday is an element of the set D." A slash through an element symbol forms the **not-an-element symbol** ($\notin$), which indicates that an object is not an element of a set. For example, January $\notin D$ means "January is not an element of the set D."

A set that contains no elements is called an **empty set** or **null set** and is denoted by a pair of **empty braces** { } or the **null symbol** $\varnothing$. For example, suppose A represents the set of all days of the week that begin with the letter B. Since there are no such days, $A = \{ \} = \varnothing$.

 The symbol $\{\varnothing\}$ is not the empty set. It is the set containing the element $\varnothing$. Be sure to use the symbols { } or $\varnothing$ to represent the empty set, not $\{\varnothing\}$.

Some sets are **finite**, meaning that the set has a fixed number of **elements**. For example, set D contains the 7 days of the week. Set D is written using the **roster method**, which displays the elements in a list.

(eText Screens R.1-1–R.1-24)

Some sets are **infinite**, meaning that they have an unlimited number of elements. For example, the set of **natural numbers**, or **counting numbers**, denoted by $\mathbb{N}$, is the set

$$\mathbb{N} = \{1, 2, 3, 4, 5, \dots\}.$$

The **ellipsis** ($\dots$) indicates that the list continues forever. Note that 0 is not included in the set of natural numbers.

Sets also can be written in **set-builder notation**, which includes a general element of the set (usually a letter that represents the numbers in the set), a vertical bar (|) to indicate "such that," and any conditions that must be satisfied by each element of the set. For example, if set $S = \{0, 1, 2, 3, 4, 5, 6, 7, 8, 9\}$, then S is written in set-builder notation as

$$S = \{x \mid x \text{ is a single-digit number}\}.$$

This statement reads as "S equals the set of all x such that x is a single-digit number."

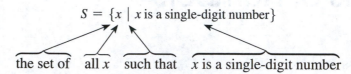

$$S = \{x \mid x \text{ is a single-digit number}\}$$

the set of   all x such that x is a single-digit number

My video summary ▶ **Example 1 Using the Symbols $\in$ and $\notin$**

Fill in the blank with the correct symbol, $\in$ or $\notin$.

a. 100 _____ $\mathbb{N}$ **b.** $\dfrac{2}{3}$ _____ $\mathbb{N}$ **c.** 11 _____ $\{x \mid x \text{ is an even number}\}$

Solutions

Read through the following, or watch this **video** for solutions to all three parts.

a. $\mathbb{N}$ indicates the set of **natural numbers**. If we write out the list of natural numbers, we will eventually reach 100. So, $100 \in \mathbb{N}$.

b. No fractions are included within the set of natural numbers, so $\dfrac{2}{3} \notin \mathbb{N}$

c. 11 is an **odd number**, not even, so $11 \notin \{x \mid x \text{ is an even number}\}$.

You Try It Work through this You Try It problem.

Work Exercises 1–6 in this eText or in the MyMathLab Study Plan.

My video summary ▶ **Example 2 Listing the Elements of a Set**

Using the **roster method**, list the elements in the set described.

a. Months with exactly 30 days

b. Natural numbers between 4 and 8, inclusive

c. Natural numbers greater than 10

d. Months with 35 days

Solutions

a. Table 1 shows the numbers of days in each month. The months April, June, September, and November each have exactly 30 days, so the set is $\{April, June, September, November\}$.

Table 1 Days per Month

Month	Days	Month	Days
January	31	July	31
February	28 or 29	August	31
March	31	September	30
April	30	October	31
May	31	November	30
June	30	December	31

b. The natural numbers *between* 4 and 8 are 5, 6, and 7. The word *inclusive* means that 4 and 8 are also included in the set. So, the set is {4, 5, 6, 7, 8}.

c.–d. Try listing the elements of these sets on your own. View the **answers**, or watch this **video** for complete solutions to all four parts.

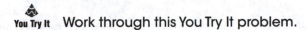 **You Try It** Work through this **You Try It** problem.

Work Exercises 7–10 in this eText or in the MyMathLab **Study Plan.**

Two sets A and B are **equal sets**, denoted as $A = B$, if they contain the exact same elements. For example, if $A = \{1, 2, 3\}$ and $B = \{3, 2, 1\}$, then $A = B$ because both sets contain the same three elements, just listed in a different order.

If every element of set A is also an element of set B, then A is a **subset** of B, denoted as $A \subseteq B$. If every element of A is an element of B, but A and B are not equal sets, then A is a **proper subset** of B, denoted as $A \subset B$. For example, if $C = \{1, 2, 3, 4, 5, 6\}$ and $D = \{2, 4, 6\}$, then $D \subset C$ because every element in D is also in C.

OBJECTIVE 2 CLASSIFY REAL NUMBERS

Now let's formalize some definitions of specific **sets of numbers**.

Definition Natural Numbers

A **natural number**, or **counting number**, is an element of the set $\mathbb{N} = \{1, 2, 3, 4, 5, \ldots\}$.

Adding the element 0 to the set of natural numbers gives the set of *whole numbers*, denoted by the symbol $\mathbb{W}$.

Definition Whole Numbers

A **whole number** is an element of the set $\mathbb{W} = \{0, 1, 2, 3, 4, 5, \ldots\}$.

TIP Every natural number is also a whole number, so $\mathbb{N} \subset \mathbb{W}$.

Adding the **negative numbers** $-1, -2, -3, -4, \ldots$ to the set of whole numbers gives the set of *integers*, denoted by $\mathbb{Z}$.

Definition   Integers

An **integer** is an element of the set $\mathbb{Z} = \{\ldots, -4, -3, -2, -1, 0, 1, 2, 3, 4, \ldots\}$.

 Every whole number is also an integer, so $\mathbb{N} \subset \mathbb{W} \subset \mathbb{Z}$.

The integers are still "whole" in the sense that they are not fractions. If fractions are included, the result is the set of *rational numbers*, denoted by $\mathbb{Q}$.

Definition   Rational Numbers

A **rational number** is an element of the set $\mathbb{Q} = \left\{\dfrac{p}{q} \middle| p \text{ and } q \text{ are integers, } q \neq 0\right\}$.

A rational number can be written as the **quotient** of two **integers** p and q as long as $q \neq 0$. $\dfrac{p}{q}$ is called the **fraction form** for the rational number. p is the **numerator**, and q is the **denominator**. Examples of rational numbers include $\dfrac{3}{4}$, $-\dfrac{7}{8}$, and $\dfrac{13}{11}$.

 Every integer is also a rational number because the integer can be written with a denominator of 1. For example, $3 = \dfrac{3}{1}$, $0 = \dfrac{0}{1}$, and $-8 = \dfrac{-8}{1}$. So, $\mathbb{N} \subset \mathbb{W} \subset \mathbb{Z} \subset \mathbb{Q}$.

Rational numbers are often written in **decimal form**. When a **fraction** is written as a **decimal**, the result is either a **terminating decimal**, such as $\dfrac{3}{4} = 0.75$ and $-\dfrac{7}{8} = -0.875$, or a **repeating decimal**, such as $\dfrac{5}{6} = 0.8333\ldots = 0.8\overline{3}$ and $\dfrac{13}{11} = 1.181818\ldots = 1.\overline{18}$.

 Every rational number can be expressed as a terminating or a repeating decimal. Conversely, every number that can be written as a terminating or a repeating decimal is a rational number.

Numbers that cannot be written as **terminating** or **repeating decimals** form the set of *irrational numbers*, denoted by $\mathbb{I}$.

Definition   Irrational Numbers

An **irrational number** is an element of the **set**

$$\mathbb{I} = \{x \mid x \text{ is a number that is not rational}\}.$$

An irrational number cannot be written as the **quotient** of two **integers**, and its **deci-mal** form does not repeat or terminate. **Square roots** and other numbers expressed with **radicals** are often irrational numbers. For example, $\sqrt{3} = 1.7320508\ldots$

and $\sqrt[3]{5} = 1.7099759\ldots$ have decimal forms that continue indefinitely without a repeating pattern, so each is an irrational number. Another example of an irrational number is $\pi = 3.1415926\ldots.$

 Do not automatically assume that a number expressed with a radical is an irrational number. Some numbers expressed with radicals are rational. For example, $\sqrt{9} = 3, \sqrt{2.25} = 1.5$ and $\sqrt[3]{8} = 2$ are all rational numbers.

If the set of **rational numbers** is combined with the set of **irrational numbers**, the result is the set of *real numbers*, denoted by $\mathbb{R}$.

Definition   **Real Numbers**

A **real number** is an element of the set

$$\mathbb{R} = \{x \mid x \text{ is a rational number or an irrational number}\}.$$

Real numbers are used in everyday activities such as computing salaries, making purchases, and measuring distances. As we progress through algebra, we will occasionally encounter numbers that are not real. For example, $\sqrt{-4}$ is not a real number because there is no real number that can be multiplied by itself to equal -4. We will define such *non-real complex numbers* later in this eText. Barring those, every number you can think of is a real number.

Figure 1 gives a visual of the set of **real numbers** that shows the following relationships:

- Every real number is either **rational** or **irrational**;
- All **natural numbers** are whole numbers;
- All whole numbers are **integers**; and
- All integers are rational numbers.

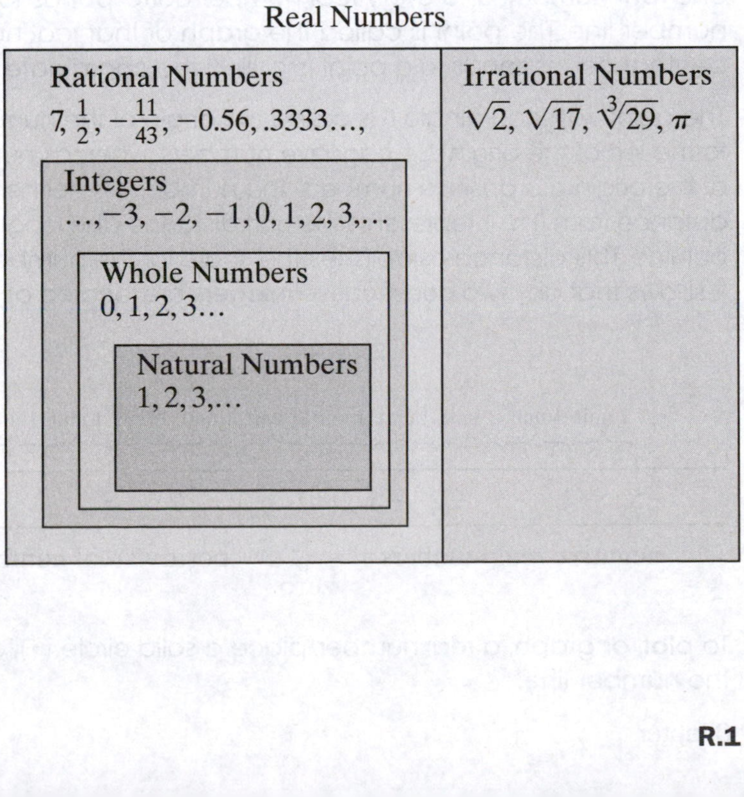

Real Numbers

Rational Numbers	Irrational Numbers
$7, \frac{1}{2}, -\frac{11}{43}, -0.56, .3333\ldots,$	$\sqrt{2}, \sqrt{17}, \sqrt[3]{29}, \pi$

Integers
$\ldots, -3, -2, -1, 0, 1, 2, 3, \ldots$

Whole Numbers
$0, 1, 2, 3 \ldots$

Natural Numbers
$1, 2, 3, \ldots$

Figure 1
The Real Numbers

Example 3 Classifying Real Numbers

Classify each real number as a natural number, whole number, integer, rational number, and/or irrational number. Each number may belong to more than one set.

a. 8 b. -4.8 c. $\sqrt{10}$ d. -7

e. $-\dfrac{4}{7}$ f. $\sqrt{25}$ g. 0 h. $3.\overline{45}$

Solutions

a. Notice that 8 is one of the natural (or counting) numbers. From **Figure 1**, it is a natural number, a whole number, an integer, and a rational number.

b. Because -4.8 is **negative**, it cannot be a natural number or a whole number. Also, because it contains a fractional portion (.8), it cannot be an integer. Notice that -4.8 is a **terminating decimal**. Therefore -4.8 is a rational number.

c. $\sqrt{10} = 3.1622776\ldots$ is not a terminating or **repeating decimal**, so it cannot be a rational number. Therefore, $\sqrt{10}$ is an irrational number.

 *My video summary* d.–h. Try classifying these numbers on your own. Check your **answers**, or watch this **video** for the complete solutions to parts d–h.

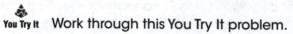 **You Try It** Work through this **You Try It** problem.

Work Exercises 11–24 in this eText or in the MyMathLab Study Plan.

⚠ Before classifying a **real number**, check to see if it can be **simplified**.

OBJECTIVE 3   PLOT REAL NUMBERS ON A NUMBER LINE

Let's review some facts about number lines.

The **real number line**, simply known as a number line, is a **graph** that represents the set of all **real numbers**. Every **point** on the number line corresponds to exactly one real number, and every real number corresponds to exactly one point on the number line. The point is called the **graph** of that real number, and the real number that corresponds to a point is called the **coordinate** of that point.

The point with coordinate 0 is called the **origin** of the number line. Numbers located to the left of the origin are **negative numbers**, whereas numbers located to the right of the origin are **positive numbers**. The number 0 is neither negative nor positive. The distance from 0 to 1 represents the **unit distance** (the length of one unit) for the number line. This distance establishes the **scale** for the number line. For example, Figure 2 shows that any two **consecutive integers** are located at a scale of one unit apart.

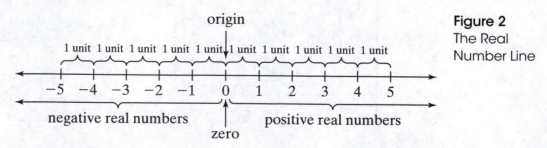

Figure 2
The Real Number Line

To **plot**, or **graph**, a real number, place a solid circle (•), or point, at its location on the number line.

My video summary ▶ **Example 4** Plotting Real Numbers on the Number Line

Plot the following set of numbers on the number line.

$$\left\{-4, 3, \frac{1}{2}, 0, -2.25\right\}$$

Solution

These five **real numbers** are plotted on the number line in Figure 3. Watch this **video** for a complete solution.

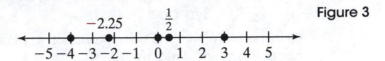

Figure 3

![You Try It] **Work through this You Try It problem.**

Work Exercises 25–28 in this eText or in the MyMathLab Study Plan.

OBJECTIVE 4   USE INEQUALITY SYMBOLS TO ORDER REAL NUMBERS

To find the **order** of two **real numbers** x and y means to determine if the first number x is smaller than, larger than, or equal to the second number y.

Order of Real Numbers

1. If x is located to the left of y on the **real number line**, shown in Figure 4(a), then x **is less than** y, denoted as $x < y$.

2. If x is located to the right of y on the **real number line**, shown in Figure 4(b), then x **is greater than** y, denoted as $x > y$.

3. If x and y are located at the same position on the **real number line**, shown in Figure 4(c), then x **is equal to** y, denoted as $x = y$.

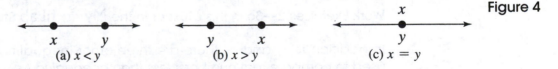

Figure 4

The symbol for equality ($=$) is the **equal sign**. Symbols for less than ($<$) and greater than ($>$) are called **inequality symbols**. These symbols are used to state the order of real numbers. For example, since -5 is located to the left of -2 on the **number line**, as shown in Figure 5, write "-5 *is less than* -2" as $-5 < -2$. Write "-2 *is greater than* -5" as $-2 > -5$.

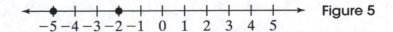

Figure 5

Figure 6 shows that 0.5 and $\frac{1}{2}$ share the same position on the number line. Write "0.5 *equals* $\frac{1}{2}$" as $0.5 = \frac{1}{2}$.

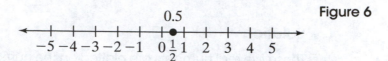

Figure 6

My video summary ▶ **Example 5 Ordering Real Numbers Using Equality and Inequality Symbols**

Fill in the blank with the correct symbol, $<$, $>$, or $=$, to make a true statement about the order of **real numbers**.

a. -1 ____ 0 **b.** -2.3 ____ -4.5 **c.** $-\frac{9}{4}$ ____ -2.25 **d.** $\frac{7}{8}$ ____ $\frac{5}{6}$

Solutions

Read through the following, or watch this **video** for solutions to all four parts.

a. View this **graph** to see -1 and 0 plotted on the real number line. Note -1 is to the left of 0 on the number line, so $-1 < 0$.

b. View this **graph** to see -2.3 and -4.5 plotted on the real number line. Note -2.3 is to the right of -4.5 on the number line, so $-2.3 > -4.5$.

c. View this **graph** to see $-\frac{9}{4}$ and -2.25 plotted on the number line. The two numbers share the same position, so $-\frac{9}{4} = -2.25$.

d. To compare $\frac{7}{8}$ and $\frac{5}{6}$, first it is helpful to convert both **fractions** to **decimals**: $\frac{7}{8} = 0.875$ and $\frac{5}{6} = 0.8\overline{3}$. On the **number line**, 0.875 is to the right of $0.8\overline{3}$, so $\frac{7}{8} > \frac{5}{6}$. View this **graph** to see $\frac{7}{8}$ and $\frac{5}{6}$ plotted on the number line.

You Try It Work through this **You Try It** problem.

Work Exercises 29–36 in this eText or in the MyMathLab Study Plan.

In addition to $<$ and $>$, there are three other **inequality symbols** that can be used to compare real numbers: *less than or equal to* ($\leq$), *greater than or equal to* ($\geq$), and *not equal to* ($\neq$). Notice that the $\leq$ and $\geq$ inequality symbols allow for equality as part of the comparison. For this reason, inequalities of the forms $x \leq y$ and $x \geq y$ are called **non-strict inequalities**. Similarly, inequalities of the forms $x < y, x > y$, and $x \neq y$ are called **strict inequalities** because they do not allow for equality.

For non-strict inequalities, remember that the inequality contains two parts ($<$ or $>$, and equality). In order for the inequality to be true, only one of the two parts must be a true statement.

$$a \leq b \begin{cases} \text{True if } a < b & (4 \leq 8 \text{ is true because } 4 < 8 \text{ is true}) \\ \text{True if } a = b & \left(3.5 \leq \dfrac{7}{2} \text{ is true because } 3.5 = \dfrac{7}{2} \text{ is true}\right) \end{cases}$$

$$a \geq b \begin{cases} \text{True if } a > b & (9 \geq 6 \text{ is true because } 9 > 6 \text{ is true}) \\ \text{True if } a = b & \left(2.5 \geq \dfrac{5}{2} \text{ is true because } 2.5 = \dfrac{5}{2} \text{ is true}\right) \end{cases}$$

Example 6 Using Inequality Symbols

Determine if each statement is true or false.

a. $\dfrac{7}{10} \leq 0.7$

b. $-8 \geq 4$

c. $-2 \geq -4$

d. $\dfrac{7}{3} \leq 1.\overline{3}$

e. $-\dfrac{9}{4} \neq -2.75$

Solutions

a. In order for the statement to be true, we need either $\dfrac{7}{10} < 0.7$ or $\dfrac{7}{10} = 0.7$ to be true. The comparison may be easier if we first convert the **fraction to a decimal**. In decimal form, $\dfrac{7}{10} = 0.7$ So, the statement $\dfrac{7}{10} \leq 0.7$ is true because $\dfrac{7}{10} = 0.7$.

b. In order for the statement to be true, we need either $-8 > 4$ or $-8 = 4$ to be true. Because -8 lies to the left of 4 on the **real number line**, we have $-8 < 4$. So, the statement $-8 \geq 4$ is false.

My video summary ▶ c.–e. Try to answer these problems on your own. Check your **answers**, or watch this **video** for complete solutions to parts c–e.

You Try It Work through this You Try It problem.

Work Exercises 37–40 in this eText or in the MyMathLab Study Plan.

OBJECTIVE 5 COMPUTE THE ABSOLUTE VALUE OF A REAL NUMBER

Recall that the distance from 0 to 1 on a real number line is the **unit distance** for the number line. See **Figure 2**. The concept of unit distance is related to the concept of *absolute value*.

> **Definition** Absolute Value
>
> The **absolute value** of a real number a, denoted as $|a|$, is the distance from 0 to a on the real number line.

The notation $|4|$ reads as "the absolute value of four," whereas $|-4|$ reads as "the absolute value of negative four." Since both 4 and -4 are 4 units from 0 on the number line, both absolute values equal 4. So, $|4| = 4$ and $|-4| = 4$. See Figure 7.

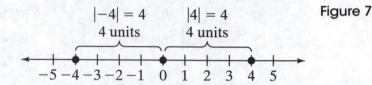

Figure 7

$|-4| = 4$ $|4| = 4$
4 units   4 units

$$-5 \; -4 \; -3 \; -2 \; -1 \; \; 0 \; \; 1 \; \; 2 \; \; 3 \; \; 4 \; \; 5$$

 TIP Because distance cannot be negative, the absolute value of a real number will always be **non-negative**.

 ▶ **Example 7 Computing Absolute Values**

Compute each **absolute value**.

a. $|2|$ b. $|-1.5|$ c. $\left|-\dfrac{7}{2}\right|$ d. $|0|$

Solutions Try to find each absolute value on your own. Check your **answers**, or watch this **video** for a complete solution.

You Try It Work through this **You Try It** problem.

Work Exercises **41–46** in this eText or in the MyMathLab Study Plan.

R.1 Exercises

In Exercises 1–6, fill in the blank with the correct symbol, $\in$ or $\notin$.

1. e _____ $\{a, e, i, o, u\}$ 2. 4.5 _____ $\mathbb{N}$

3. 25 _____ $\{3, 6, 9, \ldots, 45\}$ 4. 75 _____ $\{5, 10, 15, 20, \ldots\}$

5. 13 _____ $\{x \,|\, x \text{ is an odd number}\}$ 6. 6.4 _____ $\{x \,|\, x \text{ is a number smaller than 7}\}$

In Exercises 7–10, use the roster method to list the elements in each set.

7. Natural numbers between 3 and 9, inclusive

8. Multiples of 4 that are greater than 1 and less than 30

9. Natural numbers greater than 20

10. Natural numbers less than 0.5

In Exercises 11–22, classify each number as a natural number, whole number, integer, rational number, irrational, and/or real number. Each number may belong to more than one set.

11. -15 12. 9.5 13. $-\sqrt{11}$ 14. $\dfrac{2}{5}$

15. $-\dfrac{24}{3}$ 16. 18   17. $4.\overline{35}$ 18. $\sqrt{36}$

19. 259   20. $-11{,}401$ 21. $\dfrac{43}{7}$ 22. $\sqrt{7}$

In Exercises 23–24, given each set of real numbers, list the numbers that are (a) natural numbers, (b) whole numbers, (c) integers, (d) rational numbers, and/or (e) irrational numbers.

23. $\left\{ -17, -\sqrt{5}, -\dfrac{25}{29}, 0, 0.331, 1, \dfrac{\pi}{2} \right\}$

24. $\left\{ -11, -3.\overline{2135}, -\dfrac{3}{9}, \dfrac{\sqrt{7}}{2}, 21.1 \right\}$

In Exercises 25–28, plot each set of numbers on the real number line.

25. $\left\{ -1, 1.75, -\dfrac{9}{4}, 5 \right\}$ 26. $\left\{ \dfrac{2}{3}, 3.8, 0, -\dfrac{5}{2} \right\}$

27. $\left\{ -3, 2.5, -\dfrac{7}{4}, \sqrt{10} \right\}$ 28. $\left\{ \dfrac{3}{4}, -4.\overline{2}, 0, \sqrt{21}, -2 \right\}$

In Exercises 29–36, fill in the blank with the correct symbol $<$, $>$, or $=$.

29. $0 \underline{\quad} -7$ 30. $12.5 \underline{\quad} 12.37$ 31. $-9 \underline{\quad} -2$ 32. $8.5 \underline{\quad} 8.\overline{5}$

33. $-2.6 \underline{\quad} -3.2$ 34. $-\dfrac{7}{2} \underline{\quad} -3.5$ 35. $\dfrac{9}{5} \underline{\quad} \dfrac{7}{4}$ 36. $-\dfrac{1}{2} \underline{\quad} -\dfrac{2}{5}$

In Exercises 37–40, determine if each statement is true or false.

37. $-13 \le -12$ 38. $\dfrac{15}{4} \ge 3.75$ 39. $5 \ne -5$ 40. $6.2 < 6.20$

In Exercises 41–46, compute the absolute value.

41. $|-18|$ 42. $|34|$ 43. $\left| \dfrac{3}{8} \right|$ 44. $|-9.23|$

45. $|0.374|$ 46. $\left| -\dfrac{13}{6} \right|$

R.2 Order of Operations and Properties of Real Numbers

THINGS TO KNOW

Before working through this section, be sure you are familiar with the following concepts:

| | VIDEO | ANIMATION | INTERACTIVE |

1. Compute the Absolute Value of a Real Number (Section R.1, Objective 5)

OBJECTIVES

1 Perform Operations on Real Numbers

2 Simplify Numeric Expressions Containing Exponents and Radicals

3 Identify and Use the Properties of Real Numbers

4 Use Order of Operations to Evaluate Numeric Expressions

....

OBJECTIVE 1   PERFORM OPERATIONS ON REAL NUMBERS

The four **basic operations** of arithmetic are *addition, subtraction, multiplication,* and *division.* Let's review the rules for each operation.

ADDITION OF REAL NUMBERS

The result of adding two **real numbers** is called the **sum** of the numbers. The numbers being added are called **terms**, or **addends**. How to add terms depends on their **signs (positive or negative)**.

Adding Two Real Numbers

1. If the signs of the two **terms** are the same (both positive or both negative), add their **absolute values** and use the common sign as the sign of the **sum**.

2. If the signs of the two **terms** are different (one positive and one negative), subtract the smaller absolute value from the larger absolute value and use the sign of the number with the larger absolute value as the sign of the sum.

My video summary ▶ **Example 1 Adding Two Real Numbers**

Add.

a. $-9 + (-5)$ b. $2 + (-8)$ c. $-\dfrac{5}{6} + \dfrac{7}{8}$ d. $-3.4 + 2.8$

Solutions

a. Both -9 and -5 are **negative**, so the sum is negative. Add the **absolute values**, $|-9| = 9$ and $|-5| = 5$, and keep the negative sign in the result. $9 + 5 = 14$, so $-9 + (-5) = -14$.

b. The signs are different: $|2| = 2$ and $|-8| = 8$. Since -8 has a larger absolute value than 2, the sum is negative. Subtract 2 from 8 and write a negative sign in the result. $8 - 2 = 6$, so $2 + (-8) = -6$.

c. The fractions have different signs: $\left|-\dfrac{5}{6}\right| = \dfrac{5}{6}$ and $\left|\dfrac{7}{8}\right| = \dfrac{7}{8}$. $\dfrac{7}{8}$ has a larger absolute value, so the sum is **positive**. Try to finish this problem by subtracting $\dfrac{5}{6}$ from $\dfrac{7}{8}$ and writing a positive result. Remember to find a **common denominator**. Check your **answer**, or watch this **video** for complete solutions to all four points.

d. Try to work this problem on your own. Check your **answer**, or watch this **video** for the complete solutions to all four parts.

You Try It Work through this You Try It problem.

Work Exercises 1–8 in this eText or in the MyMathLab Study Plan.

OPPOSITES OR ADDITIVE INVERSES

What if the two **terms** being added have the same **absolute value** but different signs? For example, consider $4 + (-4)$. Both 4 and -4 have the same absolute value, 4. Because they have different signs, subtract their absolute values. Since $4 - 4 = 0$, the sum is $4 + (-4) = 0$, which is neither positive nor negative. Numbers such as 4 and -4 are called *opposites*, or *additive inverses*.

Definition   Opposites (or Additive Inverses)

Two numbers are **opposites**, or **additive inverses**, if their **sum** is zero.

When two numbers are opposites, each number is the opposite of the other. For example, -8 is the opposite of 8, and 8 is the opposite of -8 (because $8 + (-8) = 0$). A negative sign $(-)$ denotes "opposite of." So, -8 reads as "the opposite of eight," and $-(-8)$ reads as "the opposite of negative eight." This means $-(-8) = 8$. In general, if a is a number, then $-(-a) = a$. Zero is its own opposite.

Finding the Opposite of a Real Number

To find the opposite of a real number, change its **sign**.

My video summary ▶ **Example 2 Finding Opposites of Real Numbers**

Find the **opposite** of each number.

a. 17 b. -8.4 c. $\dfrac{2}{9}$ d. 0

Solutions

Watch this **video**, or read through the following for solutions to all four parts.

a. The opposite of 17 is -17.

b. 8.4 is the opposite of -8.4 because $-(-8.4) = 8.4$.

c. The opposite of $\frac{2}{9}$ is $-\frac{2}{9}$.

d. 0 is the opposite of 0.

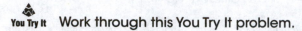 **You Try It** Work through this You Try It problem.

Work Exercises 9–12 in this eText or in the MyMathLab Study Plan.

SUBTRACTION OF REAL NUMBERS

The result of subtracting two **real numbers** is called the **difference** of the numbers. The number being subtracted is called the **subtrahend**, and the number being subtracted from is called the **minuend**.

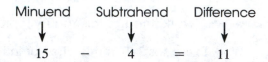

$$
\begin{array}{ccc}
\text{Minuend} & \text{Subtrahend} & \text{Difference} \\
\downarrow & \downarrow & \downarrow \\
15 & - \quad 4 & = \quad 11
\end{array}
$$

Subtraction is defined in terms of **addition**. To subtract a number, add its **opposite**.

Subtracting Two Real Numbers

If a and b represent two real numbers, then

$$a - b = a + (-b).$$

My video summary ⓥ **Example 3 Subtracting Two Real Numbers**

Subtract.

a. $4 - 12$ **b.** $-2 - 7$ **c.** $\frac{1}{4} - \left(-\frac{2}{3}\right)$ **d.** $-2.1 - (-4.8)$

Solutions

For each subtraction, add the **opposite**.

a. Change to add the opposite of 12: $4 - 12 = 4 + (-12)$

Now use the appropriate **addition rule**. The signs are different. $|4| = 4$ and $|-12| = 12$. Since -12 has the larger **absolute value**, this sum is **negative**. Subtract 4 from 12 and write a negative sign in the result. $12 - 4 = 8$, so $4 + (-12) = -8$. The difference is $4 - 12 = -8$.

b. Change to add the opposite of 7: $-2 - 7 = -2 + (-7)$

Both -2 and -7 are negative, so the sum is negative also. $|-2| = 2$ and $|-7| = 7$. Since $2 + 7 = 9$, the result is $-2 + (-7) = -9$. The difference is $-2 - 7 = -9$.

c. Change to add the opposite of $-\frac{2}{3}$: $\frac{1}{4} - \left(-\frac{2}{3}\right) = \frac{1}{4} + \frac{2}{3}$

Try to finish this problem on your own. Remember to find a **common denominator**. Check your **answer**, or watch this **video** for complete solutions to all four parts.

d. Try to work this problem on your own. Check your **answer**, or watch this **video** for complete solutions to all four parts.

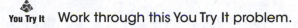 **You Try It** Work through this You Try It problem.

Work Exercises 13–20 in this eText or in the MyMathLab Study Plan.

MULTIPLICATION OF REAL NUMBERS

The result of multiplying two **real numbers** is called the **product** of the numbers. The numbers being multiplied are called **factors**. The **sign** of a product depends on the signs of the factors.

Multiplying Two Real Numbers

Multiply the **absolute values** of the two **factors** to get the absolute value of the **product**. Determine the sign of the product using the following rules:

1. If the signs of the two factors are the same, then the product is positive.
2. If the signs of the two factors are different, then the product is negative.

✎ *My video summary* ▶ ### Example 4 Multiplying Two Real Numbers

Multiply.

a. $(-9)(-3)$ **b.** $6(-7)$ **c.** $-\dfrac{2}{3}\left(\dfrac{5}{4}\right)$ **d.** $(-5.6)(-3.25)$

Solutions

a. $|-9| = 9$ and $|-3| = 3$. The absolute value of the product is $9 \cdot 3 = 27$. Since -9 and -3 are both negative, the product is positive. So, $(-9)(-3) = 27$.

b. One factor is **positive** and the other is **negative**, so the **product** is negative. Since $6 \cdot 7 = 42$, the product is $6(-7) = -42$.

c. The factors have different **signs**, so the product is negative. Try to finish this problem by multiplying $\dfrac{2}{3}$ and $\dfrac{5}{4}$, and writing a negative sign in the result.

Remember to write the **fraction** in **lowest terms**. Check your **answer**, or watch this **video** for complete solutions to all four parts.

d. Try to work this problem on your own. Check your **answer**, or watch this **video** for complete solutions to all four parts.

You Try It Work through this You Try It problem.

Work Exercises 21–28 in this eText or in the MyMathLab Study Plan.

RECIPROCALS OR MULTIPLICATIVE INVERSES

Consider the product $-5\left(-\dfrac{1}{5}\right) = 1$. Numbers such as -5 and $-\dfrac{1}{5}$ are called *reciprocals* or *multiplicative inverses*.

> **Definition** Reciprocals (or Multiplicative Inverses)
>
> Two numbers are **reciprocals**, or **multiplicative inverses**, if their product is 1.

Every nonzero **real number** b has a reciprocal, denoted as $\dfrac{1}{b}$. When two numbers are reciprocals, each is the reciprocal of the other. For example, $-\dfrac{1}{5}$ is the reciprocal of -5, and -5 is the reciprocal of $-\dfrac{1}{5}$.

⚠ Zero does not have a reciprocal because there is no number that can be multiplied by 0 to result in 1.

> **Finding the Reciprocal of a Nonzero Real Number**
>
> Find the reciprocal of a nonzero real number as follows:
>
> 1. If the number is written as a **fraction**, interchange the **numerator** and **denominator** and simplify.
> 2. If the number is not written as a fraction, write it in the denominator of a fraction with 1 as the numerator and simplify.

My video summary ▶ **Example 5 Finding Reciprocals**

Find the reciprocal of each number.

a. -8 b. $\dfrac{5}{6}$ c. 0.4 d. $-\dfrac{1}{7}$ e. $\sqrt{2}$

Solutions

Read the following, or watch this **video** for solutions to all five parts.

a. Since -8 is not written as a fraction, think of it as $\dfrac{-8}{1}$. Now, interchange the numerator -8 and the denominator 1 to get $\dfrac{1}{-8}$ or $-\dfrac{1}{8}$. The reciprocal of -8 is $-\dfrac{1}{8}$.

b. Since $\dfrac{5}{6}$ is a fraction, interchange the numerator 5 and the denominator 6. The reciprocal of $\dfrac{5}{6}$ is $\dfrac{6}{5}$.

c. The reciprocal of the decimal number 0.4 is $\dfrac{1}{0.4} = \dfrac{1}{0.4}\cdot\dfrac{10}{10} = \dfrac{10}{4} = \dfrac{5}{2}$ or 2.5.   *Multiply by 1* ↓

d. The reciprocal of the fraction $-\dfrac{1}{7}$ is $-\dfrac{7}{1} = -7$.

e. The reciprocal of $\sqrt{2}$ is $\dfrac{1}{\sqrt{2}}$. We will learn to simplify numbers of this form later in this eText.

You Try It Work through this You Try It problem.

Work Exercises 29–32 in this eText or in the MyMathLab Study Plan.

TIP The reciprocal of a number will have the same sign as the number. So, the reciprocal of a **positive number** is positive, and the reciprocal of a **negative number** is negative.

DIVISION OF REAL NUMBERS

The result of dividing two **real numbers** is called the **quotient** of the numbers. The number being divided is called the **dividend**, and the number being divided by is called the **divisor**.

Dividend Divisor Quotient

$$20 \div 4 = 5$$

Dividend $\longrightarrow \dfrac{20}{4} = 5 \longleftarrow$ Quotient
Divisor $\longrightarrow$

To divide by a number, multiply by its reciprocal.

Dividing Two Real Numbers Using Reciprocals

If a and b represent two real numbers and $b \neq 0$, then

$$a \div b = \frac{a}{b} = a \cdot \frac{1}{b}.$$

Because division is defined in terms of multiplication, the rules for finding the **sign** of a quotient are the same as the rules for the sign of a **product**.

Dividing Two Real Numbers

Divide the absolute values of the numbers to obtain the absolute value of the quotient. Determine the sign of the quotient by the following rules:

1. If the signs of the two numbers are the same, then their quotient is positive.
2. If the signs of the two numbers are different, then their quotient is negative.

My video summary ▶ **Example 6 Dividing Two Real Numbers**

Divide.

a. $-50 \div 10$ **b.** $\dfrac{-30}{-6}$ **c.** $\dfrac{1}{4} \div \left(-\dfrac{2}{3}\right)$ **d.** $-3 \div \left(-\dfrac{1}{5}\right)$

Solutions

a. $|-50| = 50$ and $|10| = 10$. The absolute value of the **quotient** is $50 \div 10 = 5$. Since the signs of -50 and 10 are different, the quotient is negative. So, $-50 \div 10 = -5$.

R.2 Order of Operations and Properties of Real Numbers **R-17**

b. Both numbers are negative, so the quotient is positive. Since $30 \div 6 = 5$, the quotient is $\dfrac{-30}{-6} = 5$.

c. To divide by a **fraction**, we multiply by its **reciprocal**. So, $\dfrac{1}{4} \div \left(-\dfrac{2}{3}\right) = \dfrac{1}{4} \cdot \left(-\dfrac{3}{2}\right)$.
Try to finish this problem on your own. Check your **answer**, or watch this **video** for the complete solutions to all four parts.

d. Try to work this problem on your own. Check your **answer**, or watch this **video** for the complete solutions to all four parts.

You Try It   Work through this **You Try It** problem.

Work Exercises 33–40 in this eText or in the MyMathLab **Study Plan.**

OBJECTIVE 2   SIMPLIFY NUMERIC EXPRESSIONS CONTAINING EXPONENTS AND RADICALS

Consider the **product** $2 \cdot 2 \cdot 2 \cdot 2 \cdot 2$. The **factor** 2 is repeated five times. We can write this product as the *exponential expression* 2^5. The superscript number 5 indicates that the factor 2 is repeated five times. The number 5 is called an *exponent*, or *power*, and the number 2 is called the *base*.

$$\underbrace{2 \cdot 2 \cdot 2 \cdot 2 \cdot 2}_{5 \text{ factors of } 2} = 2^5 \leftarrow \text{Exponent}$$
$$\uparrow$$
$$\text{Base}$$

We read 2^5 as "2 raised to the fifth power."

Definition   Exponential Expression

If a is a real number and n is a natural number, then the **exponential expression** a^n represents the product of n factors of a.

$$a^n = \underbrace{a \cdot a \cdot a \cdot \ldots \cdot a}_{n \text{ factors of } a}$$

a is the **base** of the expression, and n is the **exponent** or **power**.

We read a^n as "a raised to the nth power" or "a to the nth." An exponent of 2 is usually read as "squared," whereas an exponent of 3 is read as "cubed." For example, 5^2 is "five squared" and 5^3 is "five cubed."

An **exponent** of 1 is usually not written. For example, $2^1 = 2$.

We **simplify** numeric **exponential expressions** by multiplying the **factors**. For example, $2^5 = 2 \cdot 2 \cdot 2 \cdot 2 \cdot 2 = 32$. When a negative sign is involved, we must see if the negative sign is part of the **base**. For example, consider $(-5)^2$. The parentheses around -5 indicate that the negative sign is part of the base. So, $(-5)^2 = (-5)(-5) = 25$. In the expression -5^2, however, the negative sign is not part of the base. Instead, we need to find the **opposite** of 5^2. So, $-5^2 = -(5 \cdot 5) = -25$.

My video summary ▶ **Example 7 Simplifying Numeric Expressions Containing Exponents**

Simplify each exponential expression.

a. 5^3 **b.** $(-3)^4$ **c.** -3^4 **d.** $(-2)^5$ **e.** $\left(\dfrac{3}{4}\right)^2$

Solutions

a. The exponential expression 5^3 means to multiply three factors of 5, so $5^3 = 5 \cdot 5 \cdot 5 = 125$.

b. The parentheses around the -3 indicate that the base is -3, and the negative sign is part of the base. So, $(-3)^4 = (-3)(-3)(-3)(-3) = 81$.

c. Since there are no parentheses around the -3, the negative sign is not part of the base. Instead we find the opposite of 3^4. So, $-3^4 = -(3 \cdot 3 \cdot 3 \cdot 3) = -81$.

d.–e. Try to work these problems on your own. Check your answers to parts d and e, or watch this video for complete solutions to all five parts.

 You Try It Work through this You Try It problem.

Work Exercises 41–48 in this eText or in the MyMathLab Study Plan.

⚠ A common error that occurs when simplifying **exponential expressions** is to multiply the **base** by the **exponent**. For example, $10^2 = 10 \cdot 10 = 100$, not $10 \cdot 2 = 20$.

A **radical expression** is an expression that contains a **radical sign** $\sqrt{}$. For example, $\sqrt{16}$ and $\sqrt[3]{8}$ are all **radical expressions**. The expression $\sqrt{16}$ represents the *principal*, or *positive*, *square root* of 16. The **principal square root** of a number a is a **positive number** b whose **square** is a. So, $\sqrt{16} = 4$ because 4 is positive and $4^2 = 16$.

The expression $\sqrt[3]{8}$ represents the *cube root* of 8. The **cube root** of a number a is a number b (positive or negative) whose **cube** is a. So, $\sqrt[3]{8} = 2$ because $2^3 = 8$.

My video summary ▶ **Example 8 Simplifying Numeric Expressions Containing Radicals**

Simplify each radical expression.

a. $\sqrt{25}$ **b.** $-\sqrt{100}$ **c.** $\sqrt{-4}$

d. $\sqrt[3]{64}$ **e.** $\sqrt[3]{-8}$ **f.** $\sqrt{\dfrac{1}{9}}$

Solutions

a. For $\sqrt{25}$, we must find a **positive number** whose square is 25. The number 5 is positive and $5^2 = 25$, so $\sqrt{25} = 5$.

b. The negative sign in front of the radical indicates that the **square root** is negative. $-\sqrt{100} = -10$ because $10^2 = 100$.

c. For $\sqrt{-4}$, we need to find a positive number whose square is -4. When a real number is squared, the result is always non-negative, so $\sqrt{-4}$ is not a real number.

d. For $\sqrt[3]{64}$, find a number whose cube is 64. $4^3 = 64$, so $\sqrt[3]{64} = 4$.

e.–f. Try to simplify these radical expressions on your own. Check your **answers** for parts e and f, or watch this **video** for complete solutions to all six parts.

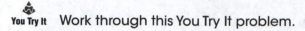

You Try It Work through this You Try It problem.

Work Exercises 49–56 in this eText or in the MyMathLab Study Plan.

OBJECTIVE 3 IDENTIFY AND USE THE PROPERTIES OF REAL NUMBERS

Now we identify useful properties of **real numbers**. These properties help with simplifying both **numeric expressions** and **algebraic expressions**.

Let's begin with the *commutative properties*. The **commutative property of addition** states that the *order* of the **terms** in addition does not affect the **sum**. For example, $3 + 5 = 8$ and $5 + 3 = 8$, so $3 + 5 = 5 + 3$.

Commutative Property of Addition

If a and b are real numbers, then $a + b = b + a$.

The **commutative property of multiplication** states that the *order* of the **factors** in multiplication does not affect the **product**. For example, $(-3)(7) = -21$ and $7(-3) = -21$, so $(-3)(7) = 7(-3)$.

Commutative Property of Multiplication

If a and b are real numbers, then $a \cdot b = b \cdot a$.

 Subtraction and division do not have commutative properties: $a - b \neq b - a$; $a \div b \neq b \div a$. See **why.**

Next, let's look at the *associative properties*. The **associative property of addition** states that regrouping **terms** does not affect the **sum**. For example, $(3 + 5) + 4 = 8 + 4 = 12$ and $3 + (5 + 4) = 3 + 9 = 12$, so $(3 + 5) + 4 = 3 + (5 + 4)$.

Associative Property of Addition

If $a, b,$ and c are real numbers, then $(a + b) + c = a + (b + c)$.

The **associative property of multiplication** states that regrouping **factors** does not affect the **product**. For example, $(-2 \cdot 5) \cdot 3 = (-10) \cdot 3 = -30$ and $-2 \cdot (5 \cdot 3) = -2 \cdot (15) = -30$, so $(-2 \cdot 5) \cdot 3 = -2 \cdot (5 \cdot 3)$.

Associative Property of Multiplication

If $a, b,$ and c are real numbers, then $(a \cdot b) \cdot c = a \cdot (b \cdot c)$.

 Subtraction and division do not have associative properties: $(a - b) - c \neq a - (b - c)$; $(a \div b) \div c \neq a \div (b \div c)$. See **why.**

My video summary ▶ **Example 9 Using the Commutative and Associative Properties**

Use the given property to complete each statement.

a. Associative property of multiplication: $5(7y) = $ _____

b. Associative property of addition: $(4x + 2) + 7 = $ _____

c. Commutative property of addition: $8(3x + 2) = $ _____

d. Commutative property of multiplication: $x(y + 4) = $ _____

Solutions

Watch this video, or read the following for solutions to all four parts.

a. The associative property of multiplication states that the **factors** can be regrouped. So, $5(7y) = (5 \cdot 7)y$.

b. By the associative property of addition, $(4x + 2) + 7 = 4x + (2 + 7)$.

c. The commutative property of addition allows us to change the order of the **terms** being added. So, $8(3x + 2) = 8(2 + 3x)$.

d. By the commutative property of multiplication, $x(y + 4) = (y + 4)x$.

You Try It Work through this You Try It problem.

Work Exercises 57–60 in this eText or in the MyMathLab Study Plan.

The **distributive property** states that multiplication distributes over addition (or subtraction). For example, compare $5(3 + 4) = 5(7) = 35$ with $(5 \cdot 3) + (5 \cdot 4) = 15 + 20 = 35$. Since both expressions simplify to 35, they are equivalent: $5(3 + 4) = (5 \cdot 3) + (5 \cdot 4)$.

Distributive Property

If a, b, and c are real numbers, then $a(b + c) = ab + ac$.

Because multiplication is **commutative**, we can also write the distributive property as

$$(b + c)a = ba + ca.$$

The distributive property also applies to subtraction.

$$a(b - c) = ab - ac$$

The distributive property extends to sums (or differences) involving more than two terms.

$$a(b + c + d) = ab + ac + ad$$

My video summary ▶ **Example 10 Using the Distributive Property**

Use the distributive property to multiply.

a. $4(x + 9)$ b. $(x - 4) \cdot 2$ c. $-5(x + y - 4)$

Solutions

a. Begin with the original expression: 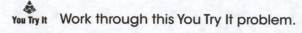 $4(x + 9)$

Apply the distributive property: $= 4 \cdot x + 4 \cdot 9$

Find each product: $= 4x + 36$

b.–c. Try to complete these problems on your own. Check your **answers** to parts b and c, or watch this **video** for complete solutions to all three parts.

You Try It Work through this You Try It problem.

Work Exercises 61–66 in this eText or in the MyMathLab Study Plan.

We complete our discussion of the properties of **real numbers** with *identity* and *inverse properties*.

If 0 is added to any real number, then the **sum** is the original number. For example, $5 + 0 = 5$ and $0 + 2 = 2$. So, the *identity* of a number is not changed by adding the number plus 0. Since 0 is the only real number for which this is true, 0 is called the **identity element for addition**, or the **additive identity**.

Identity Property of Addition

If a is a real number, then $a + 0 = a$ and $0 + a = a$.

Similarly, if 1 is multiplied by a real number, then the **product** is the original number. For example, $7 \cdot 1 = 7$ and $1 \cdot 3 = 3$. So, the *identity* of a number is not changed by multiplying the number and 1. Since 1 is the only real number for which this is true, 1 is called the **identity element for multiplication**, or the **multiplicative identity**.

Identity Property of Multiplication

If a is a real number, then $a \cdot 1 = a$ and $1 \cdot a = a$.

The **opposite** of a number is also called the **additive inverse** of the number. The **reciprocal** of a number is also called the **multiplicative inverse** of the number. The **inverse property of addition** states that every real number has a **unique** additive inverse.

Inverse Property of Addition

If a is a **real** number, then there is a unique real number $-a$ such that

$$a + (-a) = -a + a = 0.$$

The **inverse property of multiplication** states that every nonzero real number has a unique multiplicative inverse.

Inverse Property of Multiplication

If a is a nonzero real number, then there is a unique real number $\frac{1}{a}$, such that

$$a \cdot \frac{1}{a} = \frac{1}{a} \cdot a = 1.$$

My video summary ▶ **Example 11 Identifying Properties of Real Numbers**

Identify the property of real numbers demonstrated in each statement.

a. $(2x + 0) - 4 = 2x - 4$

b. $\frac{1}{6}(6x) = \left(\frac{1}{6} \cdot 6\right)x$

c. $(-13 + 13) + 7 = 0 + 7$

d. $2x - 8 = 2(x - 4)$

Solutions

Try to identify each property on your own. Check your **answers**, or watch this video for complete solutions.

 Work through this **You Try It** problem.

Work Exercises 67–72 in this eText or in the MyMathLab Study Plan.

OBJECTIVE 4   USE ORDER OF OPERATIONS TO EVALUATE NUMERIC EXPRESSIONS

The **numeric expression** $2 + 3 \cdot 4$ contains two **operations**: addition and multiplication. Depending on which operation is **performed** first, a different answer will result.

Add first, then multiply: $2 + 3 \cdot 4 = 5 \cdot 4 = 20$

Multiply first, then add: $2 + 3 \cdot 4 = 2 + 12 = 14$

⟩Different results

For this reason, mathematicians have agreed on a set **order of operations**, which determines the priority in which operations should be performed. For example, we perform multiplication before addition from left to right. So the correct **evaluation** for our numeric expression above is 14, not 20.

Order of Operations

1. **Parentheses (or other grouping symbols)** Evaluate operations within parentheses (or other grouping symbols) first, starting with the innermost set and working out.

2. **Exponents and Radicals** Work from left to right evaluating any exponential or radical expressions as they occur.

3. **Multiplication and Division** Work from left to right and perform any multiplication or division operations as they occur.

4. **Addition and Subtraction** Work from left to right and perform any addition or subtraction operations as they occur.

Example 12 Using Order of Operations to Evaluate Numeric Expressions

Simplify each expression using the order of operations.

a. $50 \div 5 \cdot 2$

b. $18 - 6(4) + \dfrac{12}{3}$

c. $7 - 4^2 + 6 \div 3(8 - 10)^2$

Solutions

a. The two **operations** in this problem are **division** and **multiplication**, which have equal priority. So, work from left to right and perform the division and multiplication in order.

$$\begin{aligned}
\text{Begin with the original expression:} \quad & 50 \div 5 \cdot 2 \\
\text{Divide } 50 \div 5: \quad &= 10 \cdot 2 \\
\text{Multiply:} \quad &= 20
\end{aligned}$$

b. There are four operations in this problem: **subtraction**, multiplication, **addition**, and division. Working from left to right, perform the multiplication and division first, in that order. Then perform the addition and subtraction from left to right.

$$\begin{aligned}
\text{Begin with the original expression:} \quad & 18 - 6(4) + \frac{12}{3} \\
\text{Multiply } 6(4); \text{ divide } \frac{12}{3}: \quad &= 18 - 24 + 4 \\
\text{Subtract } 18 - 24: \quad &= -6 + 4 \\
\text{Add:} \quad &= -2
\end{aligned}$$

 My video summary ▶ **c.**

$$\begin{aligned}
\text{Begin with the original expression:} \quad & 7 - 4^2 + 6 \div 3(8 - 10)^2 \\
\text{Simplify within the parentheses first:} \quad &= 7 - 4^2 + 6 \div 3(-2)^2 \\
\text{Evaluate each exponential expression:} \quad &= 7 - 16 + 6 \div 3 \cdot 4
\end{aligned}$$

There are four operations remaining in this expression. Try to finish simplifying this problem on your own. Check your **answer**, or watch this **video** for a complete solution to part c.

You Try It Work through this **You Try It** problem.

Work Exercises **73–82** in this eText or in the MyMathLab **Study Plan.**

Recall that fraction bars −, **absolute value** symbols | |, and **radicals** √ can serve as **grouping symbols**. Simplify expressions separately in the **numerator** and **denominator** of a fraction before dividing. Likewise, simplify expressions within absolute value symbols before evaluating the absolute value. Also, simplify expressions beneath the radical sign before evaluating the radical expression.

⚠ Though not specified in the **order of operations**, absolute values should be evaluated in the same priority with exponents and radicals. Of course, any expression within the absolute value bars must be simplified first.

Example 13 Using Order of Operations with Special Grouping Symbols

Simplify each expression.

a. $\dfrac{8(-5) + 4^2}{(-3)^2 - 5}$

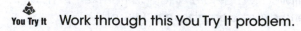 **b.** $48 \div \sqrt{6 \cdot 7 + 3(-2)} \div |11 - 3^2|$

Solutions

a. Begin by simplifying the **numerator** and **denominator** separately before dividing.

Begin with the original expression:
$$\frac{8(-5) + 4^2}{(-3)^2 - 5}$$

Evaluate the exponents separately in the numerator and denominator:
$$= \frac{8(-5) + 16}{9 - 5}$$

Multiply in the numerator:
$$= \frac{-40 + 16}{9 - 5}$$

Add and subtract separately in the numerator and denominator:
$$= \frac{-24}{4}$$

Divide:
$$= -6$$

My video summary ⏵ **b.** First, simplify beneath the radical and within the **absolute value** bars. Try to do this on your own. Check your answer, or watch this video for a complete solution to part b.

🔺 **You Try It** Work through this **You Try It** problem.

Work Exercises 83–88 in this eText or in the MyMathLab Study Plan.

When numeric expressions contain **nested grouping symbols** (grouping symbols within grouping symbols), begin with the innermost set of grouping symbols and work outward.

Example 14 Using Order of Operations with Nested Grouping Symbols

Simplify each expression.

a. $[2^3 - 3(5 - 7)^2] \div 6 - 9$

b. $\dfrac{|-5^2 + 2^3| - 10}{4^2 - 6 \cdot 5}$

Solutions

a. Begin with the original expression:
$$[2^3 - 3\underbrace{(5 - 7)}_{\text{Innermost grouping 1st}}^2] \div 6 - 9$$

Evaluate $(5 - 7)$:
$$= [2^3 - \underbrace{3(-2)^2}_{\text{Outermost grouping 2nd}}] \div 6 - 9$$

Evaluate exponents 2^3 and $(-2)^2$:
$$= [8 - 3(4)] \div 6 - 9$$

Multiply $3(4)$:
$$= [8 - 12] \div 6 - 9$$

Subtract $8 - 12$:
$$= (-4) \div 6 - 9$$

Divide $-4 \div 6$:
$$= -\frac{2}{3} - 9 \leftarrow \boxed{\frac{-4}{6} = \frac{-2 \cdot 2}{3 \cdot 2} = \frac{-2}{3} = -\frac{2}{3}}$$

Subtract $-\dfrac{2}{3} - 9$:
$$= -\frac{29}{3} \leftarrow \boxed{-\frac{2}{3} - 9 = -\frac{2}{3} - \frac{27}{3} = -\frac{2}{3} + \left(-\frac{27}{3}\right) = -\frac{29}{3}}$$

My video summary ⏵ **b.** First, simplify within the absolute value bars. Try to do this problem on your own, then check your answer. Watch this video for a complete solution.

🔺 **You Try It** Work through this **You Try It** problem.

Work Exercises 89–94 in this eText or in the MyMathLab Study Plan.

R.2 Exercises

In Exercises 1–8, add.

1. $-7 + 10$

2. $-13 + (-9)$

3. $7 + (-9)$

4. $-12 + 12$

5. $-\dfrac{4}{3} + \left(-\dfrac{7}{3}\right)$

6. $-\dfrac{8}{9} + \dfrac{5}{6}$

7. $5.1 + (-7.4)$

8. $-2.78 + (-1.25)$

In Exercises 9–12, find the opposite of each real number.

9. -13

10. 52

11. $-\dfrac{4}{15}$

12. -9.34

In Exercises 13–20, subtract.

13. $8 - 11$

14. $-8 - 14$

15. $5 - (-6)$

16. $-7 - (-3)$

17. $\dfrac{7}{3} - \dfrac{3}{7}$

18. $-\dfrac{5}{8} - \left(-\dfrac{3}{4}\right)$

19. $7.5 - 10.1$

20. $-2.1 - (-3.2)$

In Exercises 21–28, multiply.

21. $(-12)(-6)$

22. $8(-4)$

23. $9 \cdot 17$

24. $27 \cdot 0$

25. $(-7)\left(\dfrac{5}{14}\right)$

26. $\left(-\dfrac{5}{12}\right)\left(-\dfrac{3}{10}\right)$

27. $-7.2(6.75)$

28. $(-2.3)(-10.5)$

In Exercises 29–32, find the reciprocal of each real number.

29. 12

30. $-\dfrac{6}{11}$

31. $-\dfrac{1}{9}$

32. 0.8

In Exercises 33–40, divide.

33. $200 \div (-8)$

34. $\dfrac{-75}{-5}$

35. $\dfrac{0}{-8}$

36. $\dfrac{1}{3} \div \dfrac{1}{4}$

37. $-\dfrac{9}{16} \div \dfrac{5}{12}$

38. $-15 \div \dfrac{5}{2}$

39. $\dfrac{-0.6}{-4}$

40. $-3.9 \div (-0.2)$

In Exercises 41–48, simplify each exponential expression.

41. 8^3

42. -5^4

43. $(-4)^3$

44. $(-3)^4$

45. $\left(\dfrac{3}{2}\right)^4$

46. $\left(-\dfrac{5}{8}\right)^2$

47. $(0.2)^5$

48. $-(-0.1)^4$

In Exercises 49–56, simplify each radical expression.

49. $\sqrt{169}$

50. $\sqrt{\dfrac{1}{25}}$

51. $\sqrt{0.64}$

52. $\sqrt{\dfrac{4}{81}}$

53. $\sqrt[3]{216}$

54. $\sqrt[3]{-125}$

55. $\sqrt[3]{-1000}$

56. $-\sqrt{121}$

In Exercises 57–60, use the given property to complete each statement.

57. Commutative property of addition: $ab + c =$ _____

58. Commutative property of multiplication: $6(v + 4) =$ _____

59. Associative property of addition: $4 + (a + 11) =$ _____

60. Associative property of multiplication: $45 \cdot (y \cdot z) =$ _____

In Exercises 61–66, use the distributive property to multiply.

61. $5(x + 8)$ 62. $4(2y - 5)$ 63. $(x - 6) \cdot 3$ 64. $0.2(1.5 + 3.4x)$

65. $-7(m - n + 5)$ 66. $x(x + y - 2)$

In Exercises 67–72, identify the property of real numbers demonstrated in each statement.

67. $x + 14 + (-14) = x$

68. $0 + 9 = 9$

69. $(5 \cdot 1)x^2 = 5x^2$

70. $\left(\dfrac{3}{4} \cdot \dfrac{4}{3}\right)y = 1y$

71. $(5 + x) - 5 = (x + 5) - 5$

72. $(2x + 4) + (-4) = 2x + (4 + (-4))$

In Exercises 73–94, simplify each expression using the order of operations.

73. $4(2 - 5)^3$

74. $20 - 4 \cdot 9 + 12$

75. $1.8 - 4.2 \div 6 - 1.2$

76. $100 \div 25 \cdot 4$

77. $7 - (4 \cdot 6 - 26)$

78. $10^2 + 20^2 \div 5^2$

79. $2^2 - 5 \cdot 2 + (2^2 - 5) \cdot 2$

80. $(40 \div 2)^2 - 40 \div 2^2$

81. $3 \cdot (18 + 3)^2 - 4 \cdot (8 - 3)^2$

82. $-\dfrac{7}{6}\left(\dfrac{1}{2}\right) + \dfrac{5}{9} \div \dfrac{7}{6}$

83. $\sqrt{5 \cdot 2 + 4^2 - 1}$

84. $\sqrt{6(5) + 17(3)} - 11.9$

85. $18 \div \sqrt{4 \cdot 5 + 2^4} \div |-3^4 + 13 \cdot 6|$

86. $\dfrac{7 \cdot 6 + 2^2}{31 - 2^3}$

87. $\dfrac{3(15) - 5}{(-2)^2 + 3}$

88. $\dfrac{2 - 7 \cdot 1}{3 \cdot 2 - 1} \div \dfrac{2 \cdot 7 + 1}{3 + 2 \cdot 1}$

89. $\dfrac{|6 - 11| + 4}{4^2 - 2(5)}$

90. $[5 + 3(10 - 7)^2]$

91. $5^2 - 3 \cdot [2 + 3 \cdot (1 + 5)]$

92. $|1 - 3|^3 \cdot \left[6 \cdot \dfrac{2}{3} + 4\left(1 + \dfrac{3}{4}\right)\right]$

93. $\dfrac{(3 + 2^3)^2 - 11}{7(4 - 5) - 3}$

94. $\dfrac{|-5^2 + (-3)^2| + 4 \cdot 5}{6 \div 2 - 3 \cdot 2^2}$

R.3 Algebraic Expressions

THINGS TO KNOW

Before working through this section, be sure you are familiar with the following concepts:

VIDEO   ANIMATION   INTERACTIVE

 You Try It   **1.** Perform Operations on Real Numbers (Section R.2, Objective 1)

 You Try It   **2.** Simplify Numeric Expressions Containing Exponents and Radicals (Section R.2, Objective 2)   ▶

 You Try It   **3.** Use Order of Operations to Evaluate Numeric Expressions (Section R.2, Objective 4)   ▶

OBJECTIVES

1 Evaluate Algebraic Expressions

2 Simplify Algebraic Expressions

3 Write Verbal Descriptions as Algebraic Expressions

··

OBJECTIVE 1   EVALUATE ALGEBRAIC EXPRESSIONS

A **variable** is a symbol (usually a letter) that represents a changeable value. For example, the letter x might represent 5 in one situation and 5000 in another. A **constant** has a value that never changes. For example, 3 is a constant because the value of 3 is always the same.

An **algebraic expression** is a **variable** or a combination of variables, **constants**, and **operations**. Examples of algebraic expressions include

$$x, \quad y, \quad 5z, \quad x^2, \quad \sqrt{y}, \quad w + 7, \quad x + y, \quad a^2 + 2ab + b^2, \quad \frac{3m - 4n}{2m + n}.$$

Algebraic expressions can be used to describe quantities in a general way. For example, if the average price for an iPhone app is $0.19 (*Source*: blog.flurry.com), then the algebraic expression $0.19x$ gives the average revenue earned for selling x apps.

To find the revenue from selling sixteen apps, substitute 16 in place of x and simplify the resulting **numeric expression**. This means to **evaluate** the algebraic expression and find the **value** of the expression $0.19x$ when $x = 16$:

$$0.19x = 0.19(16) = 3.04$$

So, the average revenue earned from selling 16 apps is $3.04.

The value of an algebraic expression changes for different values of the **variables**.

> ### Evaluate Algebraic Expressions
> To **evaluate an algebraic expression**, substitute the given values for the variables and simplify the resulting **numeric expression** using the **order of operations**.

Example 1 Evaluating Algebraic Expressions

Evaluate each algebraic expression for the given values of the variables.

a. $2x^3 - 7$ for $x = 3$ **b.** $a^2 + 3ab - 5b^2$ for $a = -2$ and $b = 4$.

Solutions

a. Begin with the original algebraic expression: $2x^3 - 7$

Substitute 3 for x: $= 2(3)^3 - 7$

Evaluate the exponent: $= 2(27) - 7$

Multiply: $= 54 - 7$

Subtract: $= 47$

My video summary ▶ **b.** Begin with the original algebraic expression: $a^2 + 3ab - 5b^2$

Substitute -2 for a and 4 for b: $= (-2)^2 + 3(-2)(4) - 5(4)^2$.

Try to finish **simplifying** on your own by following the **order of operations**. Check your **answer**, or watch this **video** for the complete solution.

You Try It   Work through this **You Try It** problem.

Work Exercises 1–20 in this eText or in the MyMathLab Study Plan.

OBJECTIVE 2   SIMPLIFY ALGEBRAIC EXPRESSIONS

The **terms** of an algebraic expression are the quantities being added. For example, in the expression $3x^2 + 5xy + 7$, the terms are $3x^2$, $5xy$, and 7. Because subtraction is defined by adding the opposite, subtracted quantities are "negative" terms. So, in the expression $9x - 4y$, the terms are $9x$ and $-4y$.

Two terms are **like terms** if they have **variable** factors that are exactly the same. Like terms can only differ by their numeric **factors**. For example, the terms $5x^2y$ and $4x^2y$ are like terms because they are alike except for the numeric factors 5 and 4. The numeric factor of a term is called the **coefficient** of the term.

When an algebraic expression contains like terms, it can be **simplified** by combining the like terms. To **combine like terms**, use the **distributive property** in reverse.

My video summary ▶ **Example 2 Combining Like Terms**

Simplify each algebraic expression by combining like terms.

a. $10x^2 + 2x^2$

b. $7xy + 8xy + xy$

c. $7y - 5y + 3y^2$

Solutions

Read the following, or watch this **video** for solutions to all three parts.

a. The terms $10x^2$ and $2x^2$ have the exact same variable factor x^2, so they are like terms.

$$\begin{aligned}
\text{Begin with the original expression:} \quad & 10x^2 + 2x^2 \\
\text{Reverse the distributive property:} \quad & = (10 + 2)x^2 \\
\text{Add 10 and 2:} \quad & = 12x^2
\end{aligned}$$

b. The three terms $7xy$, $8xy$, and xy are like terms.

$$\begin{aligned}
\text{Begin with the original expression:} \quad & 7xy + 8xy + xy \\
\text{Reverse the distributive property:} \quad & = (7 + 8 + 1)xy \\
\text{Add } 7 + 8 + 1: \quad & = 16xy
\end{aligned}$$

c. The terms $7y$ and $-5y$ are like terms, but not $3y^2$ because y and y^2 are not exactly the same. We can only combine the two like terms.

$$\begin{aligned}
\text{Begin with the original expression:} \quad & 7y - 5y + 3y^2 \\
\text{Reverse the distributive property:} \quad & = (7 - 5)y + 3y^2 \\
\text{Subtract } 7 - 5: \quad & = 2y + 3y^2
\end{aligned}$$

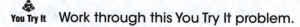 **You Try It** Work through this You Try It problem.

Work Exercises 21–24 in this eText or in the MyMathLab Study Plan.

Sometimes it is necessary to reorder the terms in an **algebraic expression** so that like terms are grouped together. Doing this is called **collecting like terms**. Collecting like terms is possible because of the **commutative and associative properties of addition**.

Example 3 Simplifying Algebraic Expressions

Simplify each algebraic expression.

a. $2x^2 + 9x + 4x^2 - 6x$ **b.** $8xy + 13 - 5xy + 3xy - 4$

Solutions

a. The terms $2x^2$ and $4x^2$ are like terms, as are $9x$ and $-6x$.

$$\begin{aligned}
\text{Begin with the original expression:} \quad & 2x^2 + 9x + 4x^2 - 6x \\
\text{Rearrange the terms to collect like terms:} \quad & = 2x^2 + 4x^2 + 9x - 6x \\
\text{Reverse the distributive property:} \quad & = (2 + 4)x^2 + (9 - 6)x \\
\text{Add and subtract:} \quad & = 6x^2 + 3x
\end{aligned}$$

My video summary ▶ **b.** $8xy$, $-5xy$, and $3xy$ are like terms, as are 13 and -4.

$$\begin{aligned}
\text{Begin with the original expression:} \quad & 8xy + 13 - 5xy + 3xy - 4 \\
\text{Rearrange the terms to collect like terms:} \quad & = 8xy - 5xy + 3xy + 13 - 4
\end{aligned}$$

Try to finish this problem on your own. Check your **answer**, or watch this **video** for a complete solution.

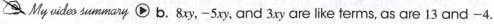 **You Try It** Work through this You Try It problem.

Work Exercises 25–29 in this eText or in the MyMathLab Study Plan.

It is often necessary to remove **grouping symbols** before **combining like terms.** This can be done by using the distributive property.

Example 4 Simplifying Algebraic Expressions

Simplify each algebraic expression.

a. $2(x - y) + 5x$ **b.** $7(2z + 4) - (5z - 7)$

Solutions

a.

Begin with the original expression:	$2(x - y) + 5x$
Distribute the 2:	$= 2x - 2y + 5x$
Rearrange the terms to collect like terms:	$= 2x + 5x - 2y$
Reverse the distributive property:	$= (2 + 5)x - 2y$
Add:	$= 7x - 2y$

My video summary ▶ **b.** Think of $-(5z - 7)$ as $-1(5z - 7)$.

Begin with the original expression:	$7(2z + 4) - (5z - 7)$
Rewrite $-(5z - 7)$ as $-1(5z - 7)$:	$= 7(2z + 4) - 1(5z - 7)$
Use the distributive property to remove the parentheses:	$= 14z + 28 - 5z + 7$

Try to finish this problem on your own. Check your **answer,** or watch this **video** for a complete solution.

You Try It Work through this You Try It problem.

Work Exercises 30–38 in this eText or in the MyMathLab Study Plan.

OBJECTIVE 3 WRITE VERBAL DESCRIPTIONS AS ALGEBRAIC EXPRESSIONS

In algebra, one key component of problem solving involves writing verbal descriptions as **algebraic expressions.** When translating the verbal descriptions, look for key word phrases that translate into **arithmetic operations.** Table 2 provides a list of phrases and their corresponding operations.

Table 2

Addition	Subtraction	Multiplication	Division
add	subtract	multiply	divide
plus	minus	times	divided by
sum	difference	product	quotient
increased by	decreased by	of	per
more than	less	double	ratio
total	less than	triple	
		twice	

My video summary ▶ **Example 5 Writing Verbal Descriptions as Algebraic Expression**

Write each word phrase as an algebraic expression.

a. Fourteen decreased by a number x

b. The sum of a number x and thirty

c. Eight more than twice a number x

d. Triple the sum of a number x and six

e. Three-fourths of a number x

f. The quotient of a number x and eight, increased by twice the number

Solutions

a. Looking at **Table 2**, the phrase "decreased by" indicates subtraction. What is being subtracted? Fourteen is decreased by x, so x is being subtracted from 14. "Fourteen decreased by a number x" is written algebraically as $14 - x$.

b. The word *sum* means addition. What is to be added? x and 30. So, "the sum of a number x and thirty" is written algebraically as $x + 30$.

c. "More than" means addition. What is being added? Twice x and 8. That is, $2x$ and 8. So, written algebraically, "eight more than twice a number x" is $2x + 8$.

d.–f. Try to finish these problems on your own. Check your **answers** to parts d–f, or watch this **video** for complete solutions to all six parts.

You Try It Work through this You Try It problem.

Work Exercises 39–46 in this eText or in the MyMathLab Study Plan.

R.3 Exercises

In Exercises 1–20, evaluate each algebraic expression for the given values of the variable.

1. $9w^2$ for $w = -5$

2. $8 - 5x$ for $x = -3$

3. $\dfrac{2}{3}x - 5$ for $x = -6$

4. $5x - 9y$ for $x = -3$ and $y = -7$

5. $36a + 60b$ for $a = -\dfrac{3}{4}$ and $b = \dfrac{2}{5}$

6. $\dfrac{3}{4}x - 9$ for $x = -\dfrac{5}{3}$

7. $\dfrac{2.5y + 3.8}{4}$ for $y = 1.4$

8. $-2(n + 4) + 9$ for $n = \dfrac{1}{2}$

9. $m^2 + 6mn + 5n^2$ for $m = 3$ and $n = -2$

10. $a^2 - b^2$ for $a = -3$ and $b = -5$

11. $a^3 - b^3$ for $a = 4$ and $b = -3$

12. $-z^2 - 6z$ for $z = -3$

13. $|8x^2 - 5y|$ for $x = -1$ and $y = 3$

14. $4|5 - x| + 1$ for $x = -8$

15. $\dfrac{2x + 6}{x - 5}$ for $x = 10$

16. $\dfrac{6b - 9b^2}{b^2 - 6}$ for $b = 3$

17. $5\sqrt{w} + 3z$ for $w = 100$ and $z = -2$

18. $\sqrt{b^2 - 4ac}$ for $a = 3, b = -5$, and $c = -2$

19. $mx + b$ for $m = \dfrac{1}{5}, x = 15$, and $b = -4$

20. $\dfrac{y_2 - y_1}{x_2 - x_1}$ for $x_1 = 4, x_2 = 6, y_1 = -2$, and $y_2 = 7$

In Exercises 21–38, simplify each algebraic expression.

21. $-8z + 3z$

22. $\dfrac{2}{3}z^2 - \dfrac{1}{4}z^2$

23. $10ab - ab + 4ab$

24. $17m - 5m + 13m^2$

25. $7a - 20 - 17a + 50$

26. $3k + 3k^2 + 7k + 7k^2$

27. $7x^2 - 8xy + y^2 - 3x^2 - 4y^2 + 6xy$

28. $\dfrac{3}{4}x - \dfrac{4}{3} + \dfrac{7}{2}x + \dfrac{5}{6}$

29. $-2c^2 + 0.7c^2 - 0.05c^2$

30. $6(x + 3) - 4$

31. $5(3t - 5) - 6t$

32. $-(4x - 6) + 9$

33. $-3(1 - 5x) + 4x - 5$

34. $2.5(y^2 + 2y - 3) + 1.4y^2 + 3.6$

35. $2(x^2 + 4) - (10 - x^2)$

36. $-5(3w - 9) + 2(2w + 7)$

37. $4(x^2 - 3x + 5) + 6(3x^2 + 5x - 2)$

38. $3(y^2 - 5) - 7(1 + 4y - 3y^2)$

In Exercises 39–46, write each word phrase as an algebraic expression.

39. The sum of a number x and ten

40. A number x decreased by seventeen

41. The product of five and a number x, increased by nine

42. The quotient of ten and a number x, increased by two

43. Twice the sum of a number x and four

44. Twelve more than five-eighths of a number x

45. The product of a number x and three, minus the product of nine and the number

46. The quotient of a number x and two, minus the product of four and the number

CHAPTER ONE
Equations and Inequalities in One Variable

CHAPTER ONE CONTENTS

1.1 Linear Equations in One Variable

THINGS TO KNOW

Before working through this section, be sure you are familiar with the following concepts:

VIDEO   ANIMATION   INTERACTIVE

You Try It 1. Evaluate Algebraic Expressions (Section R.3, Objective 1) ▶

You Try It 2. Simplify Algebraic Expressions (Section R.3, Objective 2) ▶

You Try It 3. Write Verbal Descriptions as Algebraic Expressions (Section R.3, Objective 3) ▶

OBJECTIVES

1 Determine If a Given Value Is a Solution to an Equation

2 Solve Linear Equations in One Variable

3 Identify Contradictions and Identities

4 Use Linear Equations to Solve Application Problems

OBJECTIVE 1   DETERMINE IF A GIVEN VALUE IS A SOLUTION TO AN EQUATION

An **equation** is a statement that two quantities are equal. When an equation contains one or more **variables**, it is called an **algebraic equation**. An algebraic equation indicates that two **algebraic expressions** are equal. Algebraic equations contain an equal sign ($=$), while algebraic expressions do not. For example, $3x + 2 = x + 10$ is an algebraic *equation*. It is made up of the two algebraic *expressions* $3x + 2$ and $x + 10$.

An equation is called an **equation in one variable** if it contains a single variable. Each of the following is an example of an equation in one variable:

$$\underbrace{4a - 1 = 3,}_{\text{One variable} \to a} \quad \underbrace{x^2 + 4x - 5 = 0,}_{\text{One variable} \to x} \text{ and } \underbrace{\frac{1}{3}y - \frac{3}{4} = \frac{1}{2}y}_{\text{One variable} \to y}$$

 TIP Even if the same variable appears multiple times, the equation is still considered to be "in one variable."

In this section, we focus on *linear* equations in one variable.

Definition Linear Equation in One Variable

A **linear equation in one variable** is an equation that can be written in the form $ax + b = c$, where a, b, and c are **real numbers** and $a \neq 0$.

Linear equations are also called **first-degree equations** because the **exponent** on the variable is **understood to be 1**. If an equation contains a variable raised to an exponent other than 1, then the equation is *nonlinear*. If an equation contains a variable under a **radical sign** or in the **denominator** of a fraction, then the equation is nonlinear also.

 TIP An equation does not have to be in the form $ax + b = c$ to be considered linear. Rather, the equation must be *able to be written in the form $ax + b = c$*.

View this **summary** to see other characteristics of nonlinear equations.

When solving a linear equation in one variable, we want to find all of its **solutions**. That is, find all values that, when substituted for the **variable**, make the equation true.

Example 1 Determining If a Given Value Is a Solution to an Equation

Determine if the given value is a solution to the equation.

a. $2x + 3 = 11$; 4

b. $3y + 8 = 5y - 4$; 2

c. $\frac{2}{3}w - \frac{1}{2} = \frac{1}{4}$; $\frac{3}{8}$

Solutions Substitute each given value for the variable and simplify. If the resulting statement is true, then the value is a solution to the equation.

a. Begin with the original equation: $2x + 3 = 11$

Substitute 4 for x: $2(4) + 3 \overset{?}{=} 11$

Simplify: $8 + 3 \overset{?}{=} 11$

$11 = 11$ True

The final statement is true, so 4 is a solution to the equation.

b. Begin with the original equation: $3y + 8 = 5y - 4$

Substitute 2 for y: $3(2) + 8 \stackrel{?}{=} 5(2) - 4$

Simplify: $6 + 8 \stackrel{?}{=} 10 - 4$

$$14 = 6 \qquad \text{False}$$

The final statement is false, so 2 is not a **solution** to the equation.

My video summary ▶ **c.** Begin with the original equation: $\dfrac{2}{3}w - \dfrac{1}{2} = \dfrac{1}{4}$

Substitute $\dfrac{3}{8}$ for w: $\dfrac{2}{3}\left(\dfrac{3}{8}\right) - \dfrac{1}{2} \stackrel{?}{=} \dfrac{1}{4}$

Finish the simplification. Then view the **answer**, or watch this **video** to check your work.

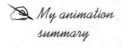 **Work through this You Try It problem.**

Work Exercises 1–8 in this eText or in the My MathLab Study Plan.

⚠ When substituting values for a variable, it is best to use parentheses around the substituted value to avoid mistakes in performing operations.

OBJECTIVE 2 SOLVE LINEAR EQUATIONS IN ONE VARIABLE

We now introduce some common terms to help us solve **linear equations in one variable**.

Once all **solutions** to an equation are found, they are listed in a **solution set**. Two or more equations with the same solution set are called **equivalent equations**. When solving a linear equation in one variable, look for simpler equivalent equations until one is found that ends with an **isolated variable** of the form:

$$variable = value$$

To find simpler equivalent equations, use the **properties of equality** to add, subtract, multiply, and/or divide both sides of an equation by the same quantity without changing its solution set.

⚠ Multiplying or dividing both sides of an equation by zero is not allowed. Do you see why?

My animation summary ✹ **Concept Animation** For a full explanation of the properties of equality, watch this animation.

Some equations will require both properties of equality in order to be solved.

Example 2 Solving Linear Equations in One Variable

Use the **properties of equality** to solve each equation.

a. $3x - 1 = 5$ **b.** $8 = \dfrac{1}{2}n + 3$

Solutions

a. Begin with the original equation: $\qquad 3x - 1 = 5$

Add 1 to both sides: $\quad 3x - 1 + 1 = 5 + 1 \leftarrow$ Addition property of equality

Simplify: $\qquad\qquad 3x = 6$

Divide both sides by 3: $\qquad \dfrac{3x}{3} = \dfrac{6}{3} \leftarrow$ Multiplication property of equality

Simplify: $\qquad\qquad x = 2$

The solution set is {2}. View the **check**.

📝 *My video summary* ▶ **b.** Try solving this equation on your own. Check your **answer**, or watch this **video** for the complete solution.

🔺 **You Try It** Work through this You Try It problem.

Work Exercises 9–20 in this eText or in the MyMathLab Study Plan.

Other equations will have **algebraic expressions** on both sides of the equal sign. When this occurs, we use the **properties of equality** to move all **variables** to one side.

📝 *My video summary* ▶ **Example 3 Solving Linear Equations with Variables on Both Sides**

Solve: $6x - 5 = 2x - 3$

Solution Work through the following, or watch this **video** for the complete solution.

Begin with the original equation: $\qquad 6x - 5 = 2x - 3$

Subtract $2x$ from both sides: $\quad 6x - 5 - 2x = 2x - 3 - 2x \leftarrow$ Addition property of equality

Simplify: $\qquad\qquad 4x - 5 = -3$

Add 5 to both sides: $\quad 4x - 5 + 5 = -3 + 5 \leftarrow$ Addition property of equality

Simplify: $\qquad\qquad 4x = 2$

Divide both sides by 4: $\qquad \dfrac{4x}{4} = \dfrac{2}{4} \leftarrow$ Multiplication property of equality

Simplify: $\qquad\qquad x = \dfrac{1}{2}$

Confirm that the solution set is $\left\{\dfrac{1}{2}\right\}$ by substituting $x = \dfrac{1}{2}$ into the original equation.

When finished, **check** your work.

🔺 **You Try It** Work through this You Try It problem.

Work Exercises 21–28 in this eText or in the MyMathLab Study Plan.

Sometimes, equations contain **non-simplified expressions**. When solving such equations, it is usually best to **simplify** each side first before using the **properties of equality** to isolate the variable.

My video summary ▶ **Example 4** Solving Linear Equations Containing
Non-Simplified Expressions

Solve: $5(x - 6) - 2x = 3 - (x + 1)$

Solution

Begin with the original equation:　$5(x - 6) - 2x = 3 - (x + 1)$

Use the distributive property:　$5x - 30 - 2x = 3 - x - 1 \leftarrow \boxed{5(x - 6) = 5 \cdot x - 5 \cdot 6}$

Combine like terms:　$3x - 30 = 2 - x$

Now the equation is simplified and looks like the equation in Example 3. Finish solving this equation. Check your **answer**, or watch this **video** for a fully worked solution.

TIP Though it is usually best to simplify each side of an equation before using the properties of equality to isolate the variable, it is not always necessary. Watch this
My video summary ▶ video for an explanation of why simplifying first would be a good idea.

You Try It Work through this **You Try It** problem.

Work Exercises 29–36 in this eText or in the MyMathLab Study Plan.

When an equation contains fractions, it is usually best to remove the fractions first. To do this, we multiply both sides of the equation by an appropriate common multiple of all the **denominators**, usually the **least common denominator (LCD)** of all the fractions.

My video summary ▶ **Example 5** Solving Linear Equations Containing Fractions

Solve: $\dfrac{x}{3} - \dfrac{5}{12} = \dfrac{5}{6}x - \dfrac{11}{12}$

Solution This equation contains fractions with the denominators 3, 6, and 12, so the LCD is 12. We multiply both sides of the equation by 12 to clear the fractions.

Begin with the original equation:　$\dfrac{x}{3} - \dfrac{5}{12} = \dfrac{5}{6}x - \dfrac{11}{12}$

Multiply both sides by 12:　$12\left(\dfrac{x}{3} - \dfrac{5}{12}\right) = 12\left(\dfrac{5}{6}x - \dfrac{11}{12}\right)$

Use the distributive property:　$12\left(\dfrac{x}{3}\right) - 12\left(\dfrac{5}{12}\right) = 12\left(\dfrac{5}{6}x\right) - 12\left(\dfrac{11}{12}\right)$

Clear the fractions:　$4x - 5 = 10x - 11$

Now the equation looks like the equation in Example 3. Finish solving this equation. Check your **answer**, or watch this **video** for a fully worked solution.

You Try It Work through this **You Try It** problem.

Work Exercises 37–41 in this eText or in the MyMathLab Study Plan.

✎ *My video summary* ▶ **Example 6** Solving Linear Equations Containing Fractions

Solve: $\frac{1}{3}(1 - x) - \frac{x + 1}{2} = -2$

Solution Try solving the equation on your own. Check your **answer**, or watch this **video** for a fully worked solution.

TIP **Note:** When solving an **equation** that contains fractions, it is usually best to **clear** the fractions first. However, clearing the fractions is not always necessary. View

✎ *My video summary* ▶ this **video** for an explanation.

🔺 **You Try It** Work through this You Try It problem.

Work Exercises 42–48 in this eText or in the MyMathLab Study Plan.

When an equation contains decimals, we remove the decimals by multiplying both sides of the equation by an appropriate power of 10, such as $10^1 = 10$, $10^2 = 100$, and $10^3 = 1000$.

✎ *My video summary* ▶ **Example 7** Solving Linear Equations Containing Decimals

Solve: $0.5n - 0.25 + 0.075n = 0.5 - 0.025n$

Solution The coefficients 0.075 and −0.025 each have three decimal places. This is the greatest number of decimal places. We can clear the decimals by multiplying both sides of the equation by $10^3 = 1000$.

Begin with the original equation:	$0.5n - 0.25 + 0.075n = 0.5 - 0.025n$
Multiply both sides by 1000:	$1000(0.5n - 0.25 + 0.075n) = 1000(0.5 - 0.025n)$
Distribute to clear the decimals:	$500n - 250 + 75n = 500 - 25n$
Combine like terms:	$575n - 250 = 500 - 25n$

Finish solving this equation. Check your **answer**, or watch this **video** for a fully worked solution.

🔺 **You Try It** Work through this You Try It problem.

Work Exercises 49–51 in this eText or in the MyMathLab Study Plan.

✎ *My video summary* ▶ **Example 8** Solving Linear Equations Containing Decimals

Solve: $0.1(y - 2) + 0.03(y - 4) = 0.02(10)$

Solution Try solving the equation on your own. Check your **answer**, or watch this **video** for a fully worked solution.

🔺 **You Try It** Work through this You Try It problem.

Work Exercises 52–54 in this eText or in the MyMathLab Study Plan.

In the previous examples, we used the following guidelines for solving **linear equations in one variable**. If a particular step does not apply to a given equation, simply skip it.

Guidelines for Solving Linear Equations in One Variable

Step 1 **Clear all fractions** from the equation by multiplying both sides by the **LCD**. Clear all decimals by multiplying both sides by the appropriate **power** of 10.

Step 2 Remove grouping symbols using the **distributive property**.

Step 3 **Simplify** each side of the equation by combining like terms.

Step 4 Use the **addition property of equality** to move all variable terms to one side of the equation and all constant terms to the other side.

Step 5 Use the **multiplication property of equality** to isolate the variable.

Step 6 Confirm that the result **satisfies** the original equation.

 If an equation contains fractions or decimals within grouping symbols, multiplying by the LCD or power of 10 may not clear the fractions or decimals within the grouping symbols. In such a situation, remove all grouping symbols and multiply both sides by the LCD for all remaining fractions or the appropriate power of 10 for all remaining decimals.

OBJECTIVE 3 IDENTIFY CONTRADICTIONS AND IDENTITIES

A **conditional equation** is an equation that is true for some values of the **variable** but not for others. So far, the equations we have solved have been conditional equations with one **solution**. However, not every **linear equation in one variable** has a single solution. There are two other possible cases: no solution and the set of all real numbers.

Consider the equation $x = x + 1$. No matter what value is **substituted** for x, the resulting value on the right side will always be one greater than the value on the left side. Therefore, the equation can never be true. Trying to solve the equation, we get

$$x = x + 1$$
$$x - x = x + 1 - x$$
$$0 = 1 \longleftarrow \boxed{\text{Variables are gone and statement is false.}}$$

No **variable terms** remain, and a false statement results. We call such an equation a **contradiction**, and it has *no solution*. Its **solution set** is the empty or **null set**, denoted by { } or $\varnothing$, respectively.

Now consider the equation $(x + 3) + (x - 8) = 2x - 5$. The expression on the left side of the equation **simplifies** to the expression on the right side. No matter what value is **substituted** for x, the resulting values on both the left and right sides will always be the same. Therefore, the equation is always true regardless of the value for the variable. Trying to solve this equation, we get

$$(x + 3) + (x - 8) = 2x - 5$$
$$x + 3 + x - 8 = 2x - 5$$
$$2x - 5 = 2x - 5$$
$$2x - 5 - 2x = 2x - 5 - 2x$$
$$-5 = -5 \longleftarrow \boxed{\text{Variables are gone and statement is true.}}$$

No **variable terms** remain, and a true statement results. We call such an equation an **identity**, and its **solution set** is the set of all real numbers, denoted by $\mathbb{R}$ or $\{x \mid x \text{ is a real number}\}$.

The next example shows how to identify **contradictions** and **identities** when solving equations.

Example 9 Identifying Contradictions and Identities

Determine if the equation is a **contradiction** or an **identity**. State the solution set.

a. $3x + 2(x - 4) = 5x + 7$ **b.** $3(x - 4) = x + 2(x - 6)$

Solutions

a. Begin with the original equation: $\quad 3x + 2(x - 4) = 5x + 7$

Use the **distributive property**: $\quad 3x + 2x - 8 = 5x + 7$

Combine like terms: $\quad 5x - 8 = 5x + 7$

Subtract $5x$ from both sides: $\quad 5x - 8 - 5x = 5x + 7 - 5x$

Simplify: $\quad -8 = 7 \quad$ False

All the **variable terms** drop out, leaving a false statement. Therefore, the equation is a contradiction and has no solution. Its solution set is { } or $\varnothing$.

My video summary ▶ **b.** Try to complete this problem on your own to verify that it is an identity. View this **popup** for the steps, or watch this **video** for the complete solution.

You Try It Work through this **You Try It** problem.

Work Exercises 55–60 in this eText or in the MyMathLab Study Plan.

OBJECTIVE 4 TRANSLATE SENTENCES INTO EQUATIONS

We have already seen how to translate *word phrases* into **algebraic expressions**. We can also translate *sentences* into **equations**. Like the **key words** that translate into **arithmetic operations**, there are key words that translate into an equal sign ($=$). See Table 1.

Table 1

Key Words That Translate to an Equal Sign			
is	was	will be	gives
yields	results in	equals	is equal to
is equivalent to	is the same as		

Consider the following sentence:

The **product** of 5 and a number is 45.

The key word "product" indicates multiplication, and the key word "is" indicates an equal sign. Letting x represent the unknown number, we can translate:

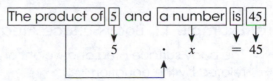

The equation is $5x = 45$.

Example 10 Translating Sentences into Equations

Translate each sentence into an **equation**. Use x to represent each unknown number.

a. Fifty-two less than a number results in -21.

b. Three-fourths of a number, increased by 8, gives the number.

c. The **difference** of 15 and a number is the same as the **sum** of the number and 1.

d. If the sum of a number and 4 is multiplied by 2, the result will be 2 less than the **product** of 4 and the number.

Solutions

a. The phrase "less than" indicates subtraction. "Fifty-two less than a number" translates into the **algebraic expression** $x - 52$. "Results in" indicates an equal sign.

$$\text{Fifty-two} \quad \text{less than} \quad \text{a number} \quad \text{results in} \quad -21.$$
$$x - 52 \qquad\qquad = \qquad -21$$

The equation is $x - 52 = -21$.

b. "Increased by" indicates addition. Three-fourths of x and 8 are the **terms** being added, which means $\frac{3}{4}x + 8$. The word "gives" means an equal sign, and "the number" means x.

$$\text{Three-fourths} \;\; \text{of} \;\; \text{a number}, \;\; \text{increased by 8}, \;\; \text{gives} \;\; \text{the number.}$$
$$\frac{3}{4} \qquad \cdot \quad x \qquad\qquad + \quad 8 \qquad = \qquad x$$

The equation is $\frac{3}{4}x + 8 = x$.

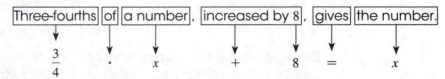 *My video summary* ▶ **c.–d.** Try to translate these sentences on your own, then check your **answers**. Watch this **video** for the complete solutions.

You Try It Work through this **You Try It** problem.

Work Exercises 61–68 in this eText or in the MyMathLab Study Plan.

OBJECTIVE 5   USE LINEAR EQUATIONS TO SOLVE APPLICATION PROBLEMS

We now look at how we can use linear equations to solve application problems. At this point you may wish to review the **Strategy for Solving Linear Equations in One Variable.**

My video summary ▶ **Example 11 Body Surface Area of Infants and Children**

The body surface area and weight of well proportioned infants and children are related by the equation

$$30S = W + 4$$

where S = the body surface area in square meters and W = weight in kilograms. Find the body surface area of a well-proportioned child that weighs 18 kg. (*Source: The Internet Journal of Anesthesiology*, Vol. 2, No. 2)

Solution Read the following, or watch this **video** for a detailed solution. We are given the child's weight, so begin by **substituting** 18 for W.

$$30S = (18) + 4$$

Simplifying on the right-hand side gives

$$30S = 22.$$

To solve for S, apply the **multiplication property of equality** and divide both sides of the equation by 30.

$$\frac{30S}{30} = \frac{22}{30}$$

$$S = \frac{22}{30} = \frac{11}{15} \approx 0.733$$

The child's body surface area is approximately 0.733 square meters.

🔺 **You Try It** Work through this You Try It problem.

Work Exercise 69 in this eText or in the MyMathLab Study Plan.

My video summary ▶ **Example 12 Red Meat vs. Poultry**

In the United States, the average pounds of red meat eaten, M, is related to the average pounds of poultry eaten, P, by the equation

$$100M = 14{,}000 - 42P.$$

Determine the average amount of poultry eaten if the average amount of red meat eaten is 100.1 pounds. (*Source:* U.S. Department of Agriculture)

Solution Substitute 100.1 for M in the equation, then solve the equation for P. Check your **answer**, or watch this **video** for the complete solution.

🔺 **You Try It** Work through this You Try It problem.

Work Exercises 70 and 71 in this eText or in the MyMathLab Study Plan.

A **mathematical model** uses the language of mathematics to describe a problem. Typically the model is an equation that describes a relationship within an application.

As you proceed through this course, you will learn how to model applied situations by using equations. We will use the following problem-solving strategy when translating, modeling, and solving applied problems involving **linear equations**.

Problem-Solving Strategy for Applications of Linear Equations

Step 1 Define the Problem. Read the problem carefully, or multiple times if necessary. Identify what you are trying to find and determine what information is available to help you find it.

Step 2 Assign Variables. Choose a variable to assign to an unknown quantity in the problem. For example, use p for price. If other unknown quantities exist, express them in terms of the selected variable.

Step 3 Translate into an Equation. Use the relationships among the known and unknown quantities to form an equation.

Step 4 Solve the Equation. Determine the value of the variable and use the result to find any other unknown quantities in the problem.

Step 5 Check the Reasonableness of Your Answer. Check to see if your answer makes sense within the context of the problem. If not, check your work for errors and try again.

Step 6 Answer the Question. Write a clear statement that answers the question(s) posed.

My video summary ▶ **Example 13 Cloud Storage**

Camille uses the cloud storage services Dropbox and Google Drive to store her photos in the cloud. The amount of storage she uses in Google Drive is 6 times the storage she uses in Dropbox. If she uses a total of 14 gigabytes of storage, how much storage does she use with each cloud service?

Solution Follow the problem-solving strategy, or watch this **video** to see the solution.

Step 1 We must find the amount of "storage", or number of gigabytes used with each cloud service. Camille uses 6 times as many gigabytes with Google Drive than she does with Dropbox, and her total storage usage is 14 gigabytes.

Step 2 Let d be the amount of storage she uses with Dropbox. Then $6d$ is the amount of storage she uses with Google Drive.

Step 3 Her total storage usage is 14 gigabytes. We add the two storage amounts to get 14.

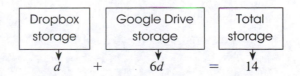

Dropbox storage		Google Drive storage		Total storage
d	$+$	$6d$	$=$	14

Step 4   Write the equation: $d + 6d = 14$

             Combine like terms: $7d = 14$

             Divide both sides by 7: $\dfrac{7d}{7} = \dfrac{14}{7}$

             Simplify: $d = 2$

If $d = 2$, then $6d = 6(2) = 12$.

Step 5 Because 12 is 6 times 2, and the sum of 12 and 2 is 14, these results make sense.

Step 6 Camile uses 2 gigabytes of storage with Dropbox and 12 gigabytes of storage with Google Drive.

You Try It **Work through this You Try It problem.**

Work Exercises 72–75 in this eText or in the MyMathLab Study Plan.

My video summary ▶ **Example 14 Movie Running Times**

Marvel Studios' *Iron Man 2* is 2 minutes shorter than the original *Iron Man*. *Iron Man 3* is 4 minutes longer than *Iron Man*. If the total running time for the three movies is 380 minutes, find the running time of each movie. (*Source*: imdb.com)

Solution

Step 1 We need to find the running time of each movie. The relationship between the quantities is that *Iron Man 2* is 2 minutes shorter than *Iron Man*, and *Iron Man 3* is 4 minutes longer than *Iron Man*. Also, the total running time of the three movies is 380 minutes.

Step 2 Let t represent the running time of Iron Man. Then $t - 2$ is the running time for Iron Man 2, and $t + 4$ is the running time for Iron Man 3.

Step 3 The sum of the three individual running times equals the total running time of 380 minutes.

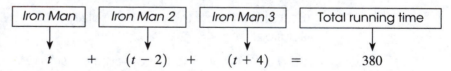

Iron Man	Iron Man 2	Iron Man 3	Total running time
↓	↓	↓	↓
t +	$(t - 2)$ +	$(t + 4)$ =	380

Complete the remaining steps of the **problem-solving strategy** to finish this problem on your own. View the **answer**, or watch this video for a complete solution.

You Try It **Work through this You Try It problem.**

Work Exercises 76 and 77 in this eText or in the MyMathLab Study Plan.

My video summary ▶ **Example 15 Fuel Stop**

Before filling up her car, Kenya sees the price of gasoline as $3.26 per gallon. When she finished filling the tank, Kenya owed $40.75. How many gallons of gas did she purchase?

Solution

Step 1 We need to find the amount of gas purchased. We know the price per gallon ($3.26) and the total amount spent ($40.75).

Step 2 Choose the variable g to represent the number of gallons of gas. This is the only unknown for the problem.

Step 3 We obtain the total cost by multiplying the price per gallon times the number of gallons. So,

Price per gallon	Number of gallons	Total cost
↓	↓	↓
3.26 ·	g =	40.75

Step 4 Try solving the equation and finishing the problem on your own. Check your answer, or watch this **video** for a fully worked solution.

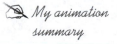
You Try It Work through this **You Try It** problem.

Work Exercises 78–83 in this eText or in the MyMathLab Study Plan.

My animation summary ✴ **Concept Animation** How would you use a variable to express **consecutive integers**? What about consecutive *even* or *odd* integers? Watch this **animation** to review this concept, then work through the following example.

Example 16 Solving a Consecutive Integer Problem

Three **consecutive even integers** add to 432. Find the three integers.

My video summary ▶ **Solution**

Step 1 We are looking for three consecutive even integers with a sum of 432.

Step 2 Let x be the first even integer. Then $x + 2$ is the 2nd consecutive even integer, and $x + 4$ is the third.

Step 3 The sum is 432, so we add the three integers and set the sum equal to 432.

1st even integer	2nd even integer	3rd even integer
↓	↓	↓
x +	$(x + 2)$ +	$(x + 4)$ = 432

Step 4 Write the equation: $x + (x + 2) + (x + 4) = 432$

Combine like terms: $3x + 6 = 432$

Finish solving this problem on your own. View the **answer**, or watch this **video** for the complete solution.

You Try It Work through this **You Try It** problem.

Work Exercises 84–87 in this eText or in the MyMathLab Study Plan.

1.1 Exercises

In Exercises 1–8, determine if the given value is a solution to the equation.

1. $3x - 8 = -23$; -5

2. $8y - 7 = 3(y - 1)$; -2

3. $z^2 - 7 = z + 5$; 4

4. $5(t + 1) - t = 6t - 7$; 6

5. $\dfrac{3}{5}n + \dfrac{17}{10} = \dfrac{3}{10} - \dfrac{1}{3}n$; $-\dfrac{3}{2}$

6. $0.12m + 0.25 = 0.07m$; 5

7. $2x - 1 = (-1 - x)^2$; -4

8. $5 - 2(y - 1) = 3y + (y - 5)$; 5

In Exercises 9–60, solve each equation and check your answer. Identify all identities and contradictions.

9. $c + 4 = 7$

10. $y - 2 = -5$

11. $\dfrac{x}{3} = 9$

12. $21 = -3x$

13. $3x + 32 = 71$

14. $-17 - 5w = -52$

15. $25 = 4a - 7$

16. $6y + 5 = 7$

17. $15 = 12 - \dfrac{1}{3}m$

18. $\dfrac{1}{4}z + 5 = 2$

19. $4 - 1.2x = 3$

20. $7.7 = 3x + 2.6$

21. $4x + 2 = 3x + 5$

22. $2z + 6 = 5z + 6$

23. $4.8 + 1.5x = 2.5x$

24. $\dfrac{1}{8} - \dfrac{5}{6}w = \dfrac{1}{6}w + \dfrac{9}{8}$

25. $9x + 7 = 3x - 11$

26. $5t - 3 = 7t + 11$

27. $11 - t = t - 9$

28. $3 - 8y = 12 - 2y$

29. $2x + 3 - 5x = 7 + x$

30. $9 + 2x + 8 = 2x + 12 - 5x$

31. $2x + 6 = 3(x + 2)$

32. $3x + 7(1 - x) = 15$

33. $6(a + 5) = 5(a + 6) - 4$

34. $3x - 2(x - 1) = 5x + 20 - 3x$

35. $2[3(x + 2) - 2] = x - 2$

36. $4(x + 1) - 7x = -(x - 7) + (2x + 3)$

37. $2y + \dfrac{1}{5} = \dfrac{4}{3}$

38. $2x + \dfrac{1}{2} = \dfrac{x}{3} + 3$

39. $\dfrac{n}{5} = \dfrac{n}{2} + \dfrac{3}{10}$

40. $\dfrac{x}{3} - \dfrac{5}{4} = \dfrac{x}{6} - \dfrac{1}{12}$

41. $\dfrac{a}{2} + \dfrac{2}{3} = \dfrac{3a}{4} + \dfrac{4}{5}$

42. $\dfrac{1}{3}(y + 1) = \dfrac{2}{3}y$

43. $\dfrac{1}{2}y - \dfrac{1}{3}(y - 1) = 5y$

44. $2w + \dfrac{7}{3} = \dfrac{w + 2}{3} - 2$

45. $\dfrac{w - 3}{4} + \dfrac{w + 5}{7} = \dfrac{5}{14}$

46. $\dfrac{x + 1}{2} + 3 = \dfrac{x + 4}{5}$

47. $\dfrac{2x + 2}{5} + 1 = \dfrac{5x + 1}{3} - 4$

48. $\dfrac{5x - 2}{3} - \dfrac{2x - 9}{4} = \dfrac{9}{2}$

49. $2.5x + 1.2 = 6.8$

50. $2.57y + 3.21 = 1.5y$

51. $0.3z + 5.4 = 0.1z + 7$

52. $2.3(5 + 2x) - 2.5 = 9$

53. $0.025p - 1 = 0.15(p - 5)$

54. $0.12x + 0.3(x - 4) = 0.01(2x - 3)$

55. $3(x + 2) - 2x + 4 = x + 5$

56. $2(3x - 5) = 6(x - 2)$

57. $5m + 2 - 7m = -2(m - 1)$

58. $-3(x + 1) + x = 2(x + 2) - 7 - 4x$

59. $8(m - 6) - 3m = 5(m - 9) - 3$

60. $7n + 1 - 9n = 3(n + 2) - 5n$

In Exercises 61–68, translate each sentence into an equation. Use x to represent each unknown number. Do not solve.

61. Thirteen more than a number is 38.

62. The sum of twice a number and 16 is equal to three times the number.

63. Subtracting 5 from the quotient of a number and 4 gives 31.

64. The product of 7 and the sum of a number and 4 results in the difference of 4 and the number.

65. The sum of a number and 2.7, divided by 6, yields the quotient of the number and 9.

66. If -7 is decreased by a number, the result will be three-fourths of the number.

67. The ratio of a number and 2 is equivalent to the difference of the number and 19.

68. Three times the sum of a number and 8, decreased by 10, is the same as five times the difference of the number and 1, increased by 14.

In Exercises 69–87, solve each application problem and check your answer.

69. **Child Height** The following equation shows the approximate linear relationship between the age (in years), a, of a child and the child's height, h (in inches):

$$h = 2.5433a + 28.465$$

If Payton, is 11.5 years old, approximately how tall might she be?
(*Source: The Merck Manual of Diagnosis and Therapy*, 15th ed)

70. **Chirp Rate** Temperature, T (in °F) and the *chirp rate of crickets*, C, (in chirps per minute) have an actual mathematical relationship as shown in the following equation:

$$T = 50 + \frac{C - 40}{4}$$

While sitting outside last night, I heard crickets chirping. If the temperature was about 52 degrees, how many chirps did I hear per minute?

71. **Sliders** The cost, C (in dollars), and the number of mini-burgers (or *sliders*), N, are related by the following equation:

$$40C = 27N + 2.$$

Find the number of sliders that can be purchased for $14.90.

72. **Cutting a Board** A 72-inch board is cut into two pieces so that one piece is four times as long as the other. Find the length of each piece.

73. **Inheritance** An inheritance of $35,000 will be divided between sisters Hannah and Taylor. If Hannah is to receive $3000 less than Taylor, how much does each sister receive?

74. **Skyscrapers** The Burj Khalifa (in Dubai, United Arab Emirates) is 320 meters taller than Taipei 101 (in Taipei, Taiwan). If the height of the Burj Khalifa is subtracted from twice the height of Taipei 101, the result is 188 meters. How tall is each skyscraper? (*Source*: Infoplease.com)

75. **Movie Classics** The running time for *Gone with the Wind* is 36 minutes more than twice the running time for *The Wizard of Oz*. If the running time for *Gone with the Wind* is subtracted from triple the running time for *The Wizard of Oz*, the result is 65 minutes. Find the running times for each movie. (*Source*: The Internet Movie Database, IMDb.com)

76. **Most Valuable Player** During his career, Michael Jordan was named the NBA's MVP one time less than Kareem Abdul-Jabbar. Bill Russell was MVP two more than half the times of Abdul-Jabbar. Together, the three men were named MVP 16 times. How many times were each of the men named MVP? (*Source*: NBA.com)

77. **Grand Slam Tennis** During his tennis career in singles play, Pete Sampras won 3 fewer Australian Open titles than U.S. Open titles and 2 more Wimbledon titles than U.S. Open titles. If he won 14 of these titles total, how many times did he win each one? (*Source*: petesampras.com)

78. **Parking Fees** It costs $8.00 to park in a parking garage for the first two hours. Then it costs $3.50 per hour for every hour afterwards. If Anna's parking fee is $22.00, how long has her car been parked?

79. **Using a Coupon** Judy purchased apples from a grocery store for $1.75 per pound. After using a coupon for $0.35 off the total price, her cost was $4.55. How many pounds of apples did Judy buy?

80. **Making Investments** Sabrina invested $20,000 in stocks and bonds. Her investment in bonds is $2000 more than half her investment in stocks. How much did Sabrina invest in stocks? How much did she invest in bonds?

81. **Grand Slam Titles** During her tennis career, Martina Navratilova won a total of 59 Grand Slam titles in three categories: women's singles, women's doubles, and mixed doubles. Out of this total, her number of wins in women's singles is eight more than her number of wins in mixed doubles. Her number of wins in women's doubles is one more than three times her number of wins in mixed doubles. How many Grand Slam titles did Martina win in each category?

82. **Buying a Jacket** Latoya purchased a leather jacket on sale for 20% off the original price. If the sale price of the jacket was $127.96, what was its original price?

83. **Bookstore Markup** A college bookstore marks up the price of textbooks by 35%. This results in the store selling textbooks for 35% more than the original cost to the store. If the bookstore charges $102.33 for your intermediate algebra textbook, what is the book's original cost to the store?

84. The sum of three consecutive integers is 258. Find the integers.

85. The sum of two consecutive odd integers is 116. Find the integers.

86. **An Open Book** The page numbers showing on an open book are consecutive integers with a sum of 317. Find the two page numbers.

87. **Office Numbers** The numbers on the doors of four adjacent offices are consecutive odd integers. If their sum is 880, find the four office numbers.

1.2 Linear Inequalities in One Variable

THINGS TO KNOW

Before working through this section, be sure you are familiar with the following concepts:

VIDEO   ANIMATION   INTERACTIVE

You Try It 1. Identify Sets (Section R.1, Objective 1)   ▶

You Try It 2. Use Order of Operations to Evaluate Numeric Expressions (Section R.2, Objective 4)   ▶

You Try It 3. Evaluate Algebraic Expressions (Section R.3, Objective 1)   ▶

You Try It 4. Simplify Algebraic Expressions (Section R.3, Objective 2)   ▶

OBJECTIVES

1 Determine If a Given Value Is a Solution to an Inequality

2 Graph the Solution Set of an Inequality on a Number Line

3 Use Interval Notation to Express the Solution Set of an Inequality

4 Solve Linear Inequalities in One Variable

5 Use Linear Inequalities to Solve Application Problems

OBJECTIVE 1 DETERMINE IF A GIVEN VALUE IS A SOLUTION TO AN INEQUALITY

While the equal sign is used in **equations** to indicate when two quantities are equal, **inequality symbols** are used in **inequalities** to indicate when two quantities are unequal. Examples of inequalities involving **algebraic expressions** include $x < 5$, $2x - 5 \geq 8$, $x^2 \leq 3x + 5$, and $|x + 6| - 9 > 3$.

As with equations, we can determine if a given value of a **variable** is a **solution** to an inequality by substituting the value for the variable and checking the resulting statement. If a false statement results, then the value *is not* a solution to the inequality. If a true statement results, then the value *is* a solution to the inequality.

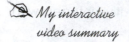 *My interactive video summary*

 Example 1 Determining if a Given Value Is a Solution to an Inequality

Determine if the given value is a **solution** to the inequality.

a. $3x + 4 < 8$; 2

b. $n^2 + 5n \geq 4$; -6

Solutions

a. This inequality is called a **strict** inequality because the left side is "strictly" less than the right side. Substitute 2 for x in the inequality. Since the resulting statement, $10 < 8$, is false, 2 is not a solution to the inequality.

b. This inequality is called a **non-strict** inequality because the left side is greater than *or* equal to the right side. There is a possibility of equality for some values of n. Substitute -6 for n in the inequality. Since the resulting statement, $6 \geq 4$, is true, -6 is a solution to the inequality.

To see all the steps for these solutions, watch this **interactive video**.

You Try It Work through this You Try It problem.

Work Exercises 1–6 in this eText or in the MyMathLab Study Plan.

OBJECTIVE 2 GRAPH THE SOLUTION SET OF AN INEQUALITY ON A NUMBER LINE

Although most **equations** have a finite number of solutions, most **inequalities** have an infinite number of solutions. For example, the inequality $x < 4$ has infinitely many values of x that are less than 4. The collection of all solutions to an inequality forms the **solution set** of the inequality. For an explanation of the difference between "finite" and "infinite," view this **popup**.

When a solution set contains an infinite number of values, we cannot list each solution. However, **set-builder notation** can be used to express the solution set. Also, the solution set can be graphed on a **number line**. When graphing a solution set, use an open circle (○) to indicate that a value is not included in the solution set and a closed circle (●) to indicate that a value is included in the solution set. For example, the graph of $\{x \,|\, x < 4\}$ is shown as follows:

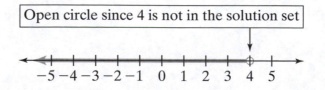

Open circle since 4 is not in the solution set

Example 2 Graphing the Solution Set of an Inequality on a Number Line

Graph each solution set on a number line.

a. $\{x \mid x \geq 0\}$ **b.** $\{x \mid 1 < x \leq 7\}$

c. $\{x \mid x < 3\}$ **d.** $\{x \mid 0 < x < 4\}$

e. $\{x \mid x \neq -2\}$ **f.** $\{x \mid -1 \leq x \leq 5\}$

g. $\{x \mid -3 \leq x < 2\}$ **h.** $\{x \mid x \text{ is any real number}\}$

Solutions

a. The **solution set** $\{x \mid x \geq 0\}$ reads as "the set of all values for x such that x is greater than or equal to 0." Because the **inequality** is **non-strict**, place a closed circle at 0 to show that 0 is a **solution**. Then shade the **number line** to the right, the direction that indicates "greater than," to show that all values greater than 0 are also solutions.

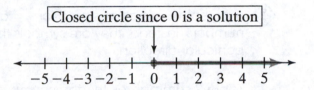

b. The solution set $\{x \mid 1 < x \leq 7\}$ reads as "the set of all values for x such that x is greater than 1 and less than or equal to 7." The inequality on the left is **strict**, so place an open circle at 1 to show that 1 is not a solution. Because the inequality on the right is non-strict, place a closed circle at 7 to show that 7 is a solution and then shade the number line between the two circles to indicate that all values between 1 and 7 are also solutions.

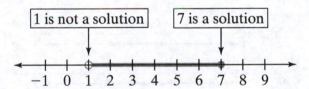

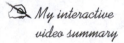 *My interactive video summary*

c.–h. View this **popup**, or watch this **interactive video**, to see the remaining solution sets and their corresponding graphs.

You Try It Work through this You Try It problem.

Work Exercises 7–14 in this eText or in the MyMathLab Study Plan.

OBJECTIVE 3 USE INTERVAL NOTATION TO EXPRESS THE SOLUTION SET OF AN INEQUALITY

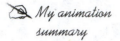

My animation summary

⊛ **Concept Animation** In addition to **set-builder notation**, the **solution set** of an inequality can also be expressed using **interval notation**. Watch this **animation** to review the concept.

First, each solution set has a **lower bound** and an **upper bound**, separated by a comma, which make up the **endpoints** of the interval. An interval will have the form:

lower bound, upper bound

Then we indicate whether or not these endpoints are included in the interval. A parenthesis shows that the endpoint *is not* included in the solution set. A square bracket shows that the endpoint *is* included in the solution set.

For example, the solution set in Example 2(b), $\{x \mid 1 < x \le 7\}$, is written as $(1, 7]$ in interval notation. A parenthesis is used on 1 because it *is not* included in the solution set, and a square bracket is used on 7 because it *is* included in the solution set.

⚠ A parenthesis is always used for $-\infty$ and ∞ because these symbols are not numbers. Instead, they are symbols that indicate the interval is **unbounded** in a particular direction.

Table 2 summarizes the three ways of expressing intervals used in this text: **set-builder notation**, **number line graph**, and **interval notation**. Typically interval notation and graphs will be used when expressing solution sets for inequalities.

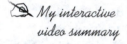

My interactive video summary

✍ Watch this **interactive video** to help you determine the lower and upper bounds of an interval or solution set.

Table 2

Graph	Interval Notation	Set-Builder Notation
○———○ a b	(a, b)	$\{x \mid a < x < b\}$
●———● a b	$[a, b]$	$\{x \mid a \le x \le b\}$
○———● a b ●———○ a b	$(a, b]$ $[a, b)$	$\{x \mid a < x \le b\}$ $\{x \mid a \le x < b\}$
○———→ a ←———○ b	(a, ∞) $(-\infty, b)$	$\{x \mid x > a\}$ $\{x \mid x < b\}$
●———→ a ←———● b	$[a, \infty)$ $(-\infty, b]$	$\{x \mid x \ge a\}$ $\{x \mid x \le b\}$

Example 3 Using Interval Notation to Express the Solution Set of an Inequality

Write each solution set using interval notation.

a. $\{x \mid x < 5\}$ **b.** $\{x \mid 2 \le x < 10\}$

c. $\{x \mid x \ge -3\}$ **d.** $\{x \mid -6 < x < 0\}$

e. $\{x \mid -1 \le x \le 5\}$ **f.** $\{x \mid x \text{ is any real number}\}$

Solutions

a. The solution set $\{x \mid x < 5\}$ has no lower bound and an upper bound of 5. The interval notation is $(-\infty, 5)$. Use a parenthesis on $-\infty$ to show that there is no lower bound. Use a parenthesis on 5, the upper bound, to show that 5 is not included in the solution set.

b. The solution set $\{x \mid 2 \le x < 10\}$ has a lower bound of 2 and an upper bound of 10. The interval notation is $[2, 10)$. Use a square bracket on 2, the lower bound, to show that 2 is included in the solution set and a parenthesis on the upper bound, 10, to show that 10 is not included in the solution set.

 **c.–f.** To see the remaining solution sets and their corresponding interval notation, view this **popup** or watch this **video** for the complete solutions to parts c–f.

You Try It Work through this **You Try It** problem.

Work Exercises 15–24 in this eText or in the MyMathLab Study Plan.

OBJECTIVE 4   SOLVE LINEAR INEQUALITIES IN ONE VARIABLE

The inequality $4x - 7 \ge 3$ is called a **linear inequality in one variable**.

Definition

A **linear inequality in one variable** is an inequality that can be written in the form $ax + b < c$, where a, b, and c are real numbers and $a \ne 0$.

Note: The inequality symbol $<$ can be replaced with $>$, $\le$, $\ge$, or $\ne$.

Why is it necessary to require $a \ne 0$ in our definition? View this **explanation**.

My interactive video summary   Watch this **interactive video** to learn how to identify linear and nonlinear inequalities.

Solving linear inequalities is similar to solving linear equations. The **properties of inequalities** are used to isolate the variable on either side of the inequality symbol.

Properties of Inequalities:

Let a, b, and c be real numbers.

1. **Addition Property of Inequality:** Adding or subtracting the same quantity from both sides of an inequality results in an equivalent inequality.

$$\text{If } a < b, \text{ then } a + c < b + c \text{ and } a - c < b - c.$$

---

2. **Multiplication Property of Inequality:** Multiplying or dividing both sides of an inequality by a positive number results in an equivalent inequality.

$$\text{If } a < b \text{ and } c > 0, \text{ then } ac < bc \text{ and } \frac{a}{c} < \frac{b}{c}.$$

Multiplying or dividing both sides of an inequality by a negative number, and switching the direction of the inequality, results in an equivalent inequality.

$$\text{If } a < b \text{ and } c < 0, \text{ then } ac > bc \text{ and } \frac{a}{c} > \frac{b}{c}.$$

Note: The inequality symbol $<$ can be replaced with $>$, $\leq$, $\geq$, or $\neq$.

Select any of the following to see an illustration of how to use the **inequality properties**.

Addition	Subtraction

Multiplication ($c > 0$)	Multiplication ($c < 0$)

Division ($c > 0$)	Division ($c < 0$)

Inequalities involving **non-simplified expressions**, fractions, or decimals are handled in the same way as equations. However, we need to be careful when multiplying or dividing both sides of an inequality by a negative number. Review the **guidelines for solving linear equations in one variable** before using a similar process when solving **linear inequalities in one variable**.

My video summary ▶ **Example 4 Solving a Linear Inequality in One Variable**

Solve the inequality $4x - 8 \geq 6x + 6$. Graph the **solution set** on a number line and write the solution set in **interval notation**.

Solution Our approach is similar to the one used when solving linear equations. We want to isolate the variable on one side of the inequality symbol. Work through the following, or watch this **video** for the complete solution.

Begin with the original inequality: $\qquad 4x - 8 \geq 6x + 6$

Subtract $6x$ from both sides: $\qquad 4x - 6x - 8 \geq 6x - 6x + 6$

Simplify: $\qquad -2x - 8 \geq 6$

Add 8 to both sides: $\qquad -2x - 8 + 8 \geq 6 + 8$

Simplify: $\qquad -2x \geq 14$

Divide both sides by -2, and switch the direction of the inequality: $\qquad \dfrac{-2x}{-2} \leq \dfrac{14}{-2}$

Direction is switched

Simplify: $\qquad x \leq -7$

The graph of the solution set appears on the following number line:

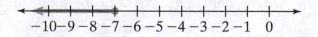

In interval notation, the solution set is $(-\infty, -7]$.

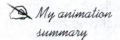
My animation summary

⊛ **Concept Animation** Remember to switch the direction of the inequality if you multiply or divide both sides of an inequality by a negative number. Watch this **animation** to see why this concept is important.

Check Since this **solution set** contains an **infinite** number of values, it is not possible to check every solution. Therefore, pick one **test value** from the proposed solution set as a "check." For example, let's choose $x = -10$ as our test value and substitute it for x in the original inequality to see if a true statement results. Since the resulting statement, $-48 \geq -54$, is true, -10 is a solution to the inequality. View the **steps** in this process.

⚠ Checking one solution, or even several solutions, does not guarantee that the solution set of the inequality is correct. It only determines whether or not the tested values are solutions. However, it is still a good idea to check one or two test values so that we are comfortable with the solution set.

You Try It Work through this You Try It problem.

Work Exercises 25–36 in this eText or in the MyMathLab Study Plan.

My video summary ▶ **Example 5 Solving a Linear Inequality**

Solve the inequality $2 - 5(x - 2) < 4(3 - 2x) + 7$. Write the solution set in set-builder notation.

Solution

Begin with the original inequality:	$2 - 5(x - 2) < 4(3 - 2x) + 7$
Use the distributive property:	$2 - 5x + 10 < 12 - 8x + 7$
Simplify:	$-5x + 12 < -8x + 19$
Add $8x$ to both sides:	$-5x + 8x + 12 < -8x + 8x + 19$
Simplify:	$3x + 12 < 19$

Finish solving the inequality on your own. Check your **answer**, or watch this **video** to see the fully worked solution.

You Try It Work through this You Try It problem.

Work Exercises 37–43 in this eText or in the MyMathLab Study Plan.

My video summary ▶ **Example 6 Solving a Linear Inequality Containing Fractions**

Solve the inequality $\frac{m}{2} - 5 + 2m > -\frac{m}{4} + \frac{1}{2}$ and write the solution set in **interval** notation.

Solution Check your **answer**, or watch this **video** to see the detailed solution.

You Try It Work through this You Try It problem.

Work Exercises 44–52 in this eText or in the MyMathLab **Study Plan**.

When solving inequalities involving fractions or decimals, some students like to clear these right away since errors are less likely when working with whole numbers. This, however, is not required. The same solution set will result whether or not fractions or decimals are cleared first.

 TIP As with equations, a linear inequality can have **no solution** or **all real numbers** as its solution. Remember that a contradiction has no solution, and an **identity** has the set of all real numbers as its solution set.

Example 7 Solving Special Cases of Linear Inequalities

Solve the following inequalities. Write each solution set in interval notation.

a. $3 + 4(x - 5) \leq 7x - 3(x + 8)$ **b.** $2(3 - x) - 7 > 4(x - 1) - 6x$

Solutions

a. Use the properties of inequalities to isolate the variable on one side of the inequality.

Begin with the original inequality:	$3 + 4(x - 5) \leq 7x - 3(x + 8)$
Use the distributive property:	$3 + 4x - 20 \leq 7x - 3x - 24$
Simplify:	$4x - 17 \leq 4x - 24$
Subtract $4x$ from both sides:	$4x - 4x - 17 \leq 4x - 4x - 24$
Simplify:	$-17 \leq -24$ False

Since the final statement is a **contradiction**, the inequality has no solution. The solution set is the empty set { } or **null set**, ∅.

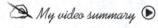 **b.** Watch the video to see the full solution for part b and confirm that the result is an **identity**.

 You Try It Work through this **You Try It** problem.

Work Exercises 53–56 in this eText or in the MyMathLab Study Plan.

An inequality statement can involve more than two parts. For example, a Google product search in April 2014 showed the range in price of an Xbox One bundle from $480 to $760. Letting X be the price of an Xbox One bundle, we can use the three-part inequality $480 \leq X \leq 760$ to show this relationship.

To solve a three-part inequality, use the **properties of inequalities** to isolate the variable between the two **inequality symbols**.

 When solving three-part inequalities, remember that what you do to one part of the inequality must be done to **all three parts** in order to obtain an equivalent inequality.

Example 8 Solving a Three-Part Inequality

Solve the inequality $-2 < \dfrac{3x - 5}{4} \leq 3$. Graph the solution set on a number line; write this solution set in interval notation.

My video summary ▶ **Solution** Use the properties of inequalities to isolate the variable between the inequality symbols. Watch this **video**, or work through the following for the complete solution.

Begin with the original inequality: $-2 < \dfrac{3x-5}{4} \leq 3$

Multiply all parts by 4 to clear the fraction: 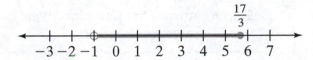 $4(-2) < 4 \cdot \dfrac{3x-5}{4} \leq 4(3)$

Simplify: $-8 < 3x - 5 \leq 12$

Add 5 to all parts: $-8 + 5 < 3x - 5 + 5 \leq 12 + 5$

Simplify: $-3 < 3x \leq 17$

Divide all parts by 3 to isolate the variable: $\dfrac{-3}{3} < \dfrac{3x}{3} \leq \dfrac{17}{3}$

Simplify: $-1 < x \leq \dfrac{17}{3}$

The graph of this solution set is:

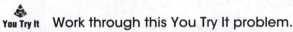

The solution set in interval notation is $\left(-1, \dfrac{17}{3} \right]$.

You Try It   Work through this You Try It problem.

Work Exercises 57–60 in this eText or in the My Math**Lab Study Plan.**

 My video summary ▶ **Example 9 Solving a Three-Part Inequality**

Solve the inequality $-1.4 < 5 - 3.2x < 3.4$ and write its solution set in interval notation.

Solution First, multiply all parts by 10 to clear the decimals and obtain the equivalent inequality $-14 < 50 - 32x < 34$. Solving this inequality results in the solution set $\left(\dfrac{1}{2}, 2 \right)$ in interval notation. Watch this **video** to see the solution worked out in detail.

You Try It   Work through this You Try It problem.

Work Exercises 61–64 in this eText or in the My Math**Lab Study Plan.**

⚠ When writing a solution set in **set-builder notation** or in **interval notation**, we want the values to increase as the interval is read from left to right.

At this point, you may wish to review the **strategy for solving application problems with linear equations**. In the next objective, we present a variation of this strategy using **mathematical models for linear inequalities**.

OBJECTIVE 5 USE LINEAR INEQUALITIES TO SOLVE APPLICATION PROBLEMS

Strategy for Solving Application Problems Involving Linear Inequalities

Step 1 Define the Problem. Read the problem carefully, or multiple times if necessary. Identify what you are trying to find and determine what information is available to help you find it.

Step 2 Assign Variables. Choose a variable to assign to an unknown quantity in the problem. For example, use p for price. If other unknown quantities exist, express them in terms of the selected variable.

Step 3 Translate into an Inequality. Use the relationships among the known and unknown quantities to form an inequality. It is helpful to consider different key words or phrases for each inequality symbol.

Step 4 Solve the Inequality. Determine the solution set of the inequality.

Step 5 Check the Reasonableness of Your Answer. Check to see if your results make sense within the context of the problem. If not, check your work for errors and try again.

Step 6 Answer the Question. Write a clear statement that answers the question(s) posed.

My video summary ▶ **Example 10 Wireless Plan Data Usage**

Suppose AT&T Wireless offers a monthly plan for a smartphone that includes 4 gigabytes (GB) of data for $110. Each additional gigabyte of data (or fraction thereof) costs $15. If Antoine subscribes to this plan, how many gigabytes of data can he use each month while keeping his total monthly cost to no more than $180 (before taxes)? (*Source*: att.com)

Solution Use the **problem-solving strategy** for applications of linear inequalities. Follow the worked solution below, or watch this **video**.

Step 1 We need to find the number of data gigabytes that Antoine can use each month. His monthly charge is $110, which includes 4 GB of data. Also, each additional gigabyte (or fraction thereof) costs $15. Antoine wants his total monthly cost to be no more than $180 before taxes.

Step 2 Choose the **variable** g to represent the unknown number of gigabytes that Antoine can use each month. The charge for any additional gigabytes used will be $15(g - 4)$.

Step 3 The sum of the monthly fee and the charge for additional gigabytes must not exceed $180. So, we have the following inequality:

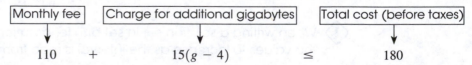

Monthly fee	Charge for additional gigabytes		Total cost (before taxes)
110	+	$15(g - 4)$ $\leq$	180

Step 4

$$110 + 15(g - 4) \leq 180$$

Distribute: $\quad 110 + 15g - 60 \leq 180$

Simplify: $\quad 15g + 50 \leq 180$

Subtract 50 from both sides: $\quad 15g + 50 - 50 \leq 180 - 50$

Simplify: $\quad 15g \leq 130$

Divide both sides by 15: $\quad \dfrac{15g}{15} \leq \dfrac{130}{15}$

Simplify: $\quad g \leq 8.\overline{6}$

Step 5 Because any fraction of a gigabyte is billed as a whole gigabyte, we must round our answer down to 8. Otherwise, the total cost would be slightly higher than $180.

Step 6 Antoine can use no more than 8 gigabytes of data per month if he wants to keep his monthly cost to no more than $180.

You Try It Work through this You Try It problem.

Work Exercises 65–67 in this eText or in the MyMathLab **Study Plan.**

My video summary ▶ **Example 11 Making a Profit**

An online retailer sells plush toys. She purchases the toys at the wholesale price of $2.75 each and sells them online for $7.25. If her fixed costs are $900, how many plush toys must she sell in order to make a **profit**? Solve the inequality $R > C$ with R as her **revenue** and C as her **cost**.

Solution

Step 1 We need to find the number of plush toys that must be sold in order to make a **profit**. We know that the price online is $7.25 each, and the wholesale cost to the retailer is $2.75 each. The fixed costs are $900.

Step 2 Let the variable x represent the number of plush toys that are sold. The retailer's **revenue**, R, is $7.25x$ because the number sold times the price online equals the revenue. The retailer's **cost**, C, is the sum of the fixed costs and the wholesale cost per toy. Therefore, C is $2.75x + 900$.

Step 3 To create a profit, the retailer's revenue must be greater than her cost. So, we have the following inequality:

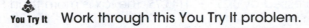

$$\underset{7.25x}{\boxed{\text{Revenue, } R}} \quad > \quad \underset{2.75x + 900}{\boxed{\text{Cost, } C}}$$

Step 4 The solution set is $\{x \mid x > 200\}$, or $(200, \infty)$ in interval notation. Watch this **video** to see the rest of the solution completed in detail. The online retailer needs to sell more than 200 plush toys to make a profit.

(eText Screens 1.2-1–1.2-30)

🔺 **You Try It**   Work through this You Try It problem.

Work Exercises 68–70 in this eText or in the MyMathLab Study Plan.

Now let's solve an application involving a three-part linear inequality in one variable.

📝 *My video summary* ▶ **Example 12 Taxable Income**

In 2014, a single person in the 25% tax bracket was required to pay a tax of $0.25x - 4143.75$, where x is the amount of taxable income. For this tax bracket, the tax due was more than $5081.25 but not more than $18,193.75. Determine the range of taxable income for this tax bracket.

Solution Watch this video, or work through the following solution using the problem-solving strategy for applications of linear inequalities.

We know that x is the amount of taxable income and that the tax in this bracket can be expressed by $0.25x - 4143.75$. The tax is more than $5081.25 but not more than $18,193.75.

Begin with the original inequality: $\qquad\qquad 5081.25 < 0.25x - 4143.75 \le 18,193.75$

Add 4143.75 to all 3 parts: $\quad 5081.25 + 4143.75 < 0.25x - 4143.75 + 4143.75$

$$\le 18,193.75 + 4143.75$$

Simplify: $\qquad\qquad 9225 < 0.25x \le 22,337.50$

Divide through by 0.25: $\qquad \dfrac{9225}{0.25} < \dfrac{0.25x}{0.25} \le \dfrac{22,337.50}{0.25}$

Simplify: $\qquad\qquad 36,900 < x \le 89,350$

Therefore, the 25% tax bracket covers taxable income that is more than $36,900 but not more than $89,350.

🔺 **You Try It**   Work through this You Try It problem.

Work Exercises 71 and 72 in this eText or in the MyMathLab Study Plan.

1.2 Exercises

In Exercises 1–6, determine if the given value is a solution to the inequality.

1. $4x - 7 \le 10$; 3

2. $10 - (3 - 5m) > 3m + 2$; -2

3. $3(5 - n) + 7 < 4 - 2(n + 3)$; 4

4. $y^2 + 4y \ge y + 10$; -5

5. $0.2(x + 0.5) + 3.25 \le 0.4x + 3.1$; 0.5

6. $\dfrac{2}{3}v + \dfrac{5}{4} > \dfrac{3}{4} - \dfrac{1}{6}v$; $\dfrac{3}{5}$

In Exercises 7–14, graph the solution set on a number line.

7. $\{x|x < 3\}$

8. $\{x|x > 0\}$

9. $\{x|x \geq -4\}$

10. $\{x|x \leq 5\}$

11. $\{x|-3 \leq x < 2\}$

12. $\{x|2 < x \leq 9\}$

13. $\{x|0 < x < 5\}$

14. $\{x|2 \leq x \leq 5\}$

In Exercises 15–24, write the solution set in interval notation.

15. $\{x|x \leq 1\}$

16. $\left\{x|x < -\dfrac{1}{3}\right\}$

17. $\{x|x > -5.4\}$

18. $\{x|x \geq -4\}$

19. $\{x|-6 \leq x \leq 11\}$

20. $\{y|-5 < y < -2\}$

21. $\left\{q|\dfrac{3}{2} \leq q < \dfrac{7}{3}\right\}$

22. $\{x|-7.2 < x \leq 5.9\}$

23. $\{y|0 < y\}$

24. $\{m|12 \geq m\}$

In Exercises 25–36, solve the inequality. Graph the solution set and write it in interval notation.

25. $x + 3 < 5$

26. $x - 1 \geq 4$

27. $5x > 10$

28. $\dfrac{y}{2} \leq -3$

29. $x + 3 \leq 2x + 1$

30. $12 + 3x \geq 2x - 3$

31. $4x + 15 > 9$

32. $7 - 3x \leq 1$

33. $2x + 1 > 5x + 7$

34. $15 - x < 3x - 1$

35. $3x - 2 < 8 + x$

36. $5x - 2 < 3x + 8$

In Exercises 37–64, solve the inequality. Write the solution set in interval notation.

37. $3(b + 1) - 7b > 5b + 3$

38. $4x - 5 < 3(x + 2) + 1$

39. $11 + 6(z + 1) \geq (5z - 6) - 8$

40. $2x - 1 \geq 5(x + 1) + 3$

41. $3(2 - x) + 8x < 12 + 2(x - 3)$

42. $5 + (x - 3) > 3 - (x - 5)$

43. $2 + 3(4 + x) < 2(7x + 3) - 3$

44. $2x + \dfrac{1}{2} > \dfrac{2}{3}x - 1$

45. $\dfrac{2 - 3w}{5} - \dfrac{w}{2} < -4$

46. $\dfrac{a}{8} - \dfrac{7}{4} \leq \dfrac{a}{4} - 1$

47. $\dfrac{7}{5}(b - 20) \geq \dfrac{3}{2}(4b - 9)$

48. $2.5x - 4 \geq 3.1 + 2.25x$

49. $0.2x + 1.2 > 0.4 - 0.2x$

50. $0.75 - (0.25m + 1) < 1.5 - 0.5m$

51. $2.9x + 1.2(2x + 3) \leq 3.9x - 4.1$

52. $\dfrac{3x}{4} - \dfrac{1}{2} \geq 0.25(2x + 1)$

53. $5n + 2 \leq 5(n - 3)$

54. $9(2x + 3) - 5x > 10x + 3(x + 5)$

55. $5x - (2x - 5) \geq 3x + 7$

56. $3(x - 1) + 2x < 2 + 5(x + 1)$

57. $-10 \leq \dfrac{3p + 1}{2} < 5$

58. $-6 < \dfrac{7 - 5x}{3} < -1$

59. $-4 \leq \dfrac{3(x - 1) + 4}{2} < 2$

60. $-1 \leq \dfrac{3 - 3x}{-3} \leq 3$

61. $-2 < 2x + 1 < 7$

62. $-4 < 1 - 3x < -1$

63. $1 < 5x + 6 \leq 16$

64. $6.4 \leq 3.5x - 2 < 14.1$

In Exercises 65–72, solve each application problem.

65. **Wireless Plan** Suppose Verizon Wireless offers a monthly plan for a smartphone that includes 5 gigabytes of data for $120. Each additional gigabyte (or fraction thereof) costs the user $10. If Logan subscribes to this plan, how many gigabytes of data can she use each month while keeping her total cost to no more than $165 (before taxes)?

66. **Checked Baggage** As of April 20, 2014, all checked baggage on American Airlines must not exceed 62 inches or an additional fee is charged. The size of baggage is determined by adding *length + width + height*. If a bag has a height of 14 inches and a width that is 12 inches less than the length, what lengths are acceptable for the bag to be checked free of charge? (*Source*: www.aa.com)

67. **Wedding Reception** Beverly is planning her wedding reception at the Missouri Botanical Gardens. The rental cost is $2000 plus $76 per person for catering. How many guests can she invite if her reception budget is $10,000? (*Source*: www.mobot.org)

68. **Retail Sales** An online retailer sells plush toys. He purchases the toys at the wholesale price of $3.15 each and sells them online for $14.75. If his fixed costs are $1060, how many plush toys must he sell in order to make a profit? That is, solve the inequality $R > C$, with R as his revenue and C as his cost.

69. **Truck Rental** Glider Truck Rentals will rent a 24-foot truck for a daily rate of $85 plus $0.10 per mile with 200 free miles. Y'All Rentals will rent a 24-foot truck for a daily rate of $40 plus $0.20 per mile. For what number of miles driven (in one day) will Glider Truck Rentals be cheaper?

70. **Flea Market** Nancy sells handcrafted bracelets at a flea market for $7. If her monthly fixed costs are $675 and each bracelet costs her $2.75 to make, how many bracelets must she sell in a month to make a profit?

71. **2014 Taxes** In 2014, a married couple filing jointly in the 28% tax bracket had to pay a tax of $0.28x - 12{,}753$, where x is the amount of taxable income. For this tax bracket, the tax due was more than \$28,925 but not more than \$50,765. Determine the range of taxable income for this tax bracket. (*Source*: www.irs.gov)

72. **Repair Costs** A mechanic's estimate for replacing a part on a car ranges from \$1300 to \$2260. If the part costs \$400 and the mechanic charges \$60 per hour for labor, what is the range of the mechanic's estimates for the time it will take to replace the part?

1.3 Compound Inequalities

THINGS TO KNOW

Before working through this section, be sure you are familiar with the following concepts:

VIDEO   ANIMATION   INTERACTIVE

 You Try It
1. Graph the Solution Set of an Inequality on a Number Line (Section 1.2, Objective 2)

 You Try It
2. Use Interval Notation to Express the Solution Set of an Inequality (Section 1.2, Objective 3)

OBJECTIVES

1 Find the Union and Intersection of Two Sets

2 Solve Compound Linear Inequalities in One Variable

....

OBJECTIVE 1   FIND THE UNION AND INTERSECTION OF TWO SETS

The words *and* and *or* are sometimes used when working with **sets of numbers.**

Intersection

For any two sets A and B, the **intersection** of A and B is given by $A \cap B$ and represents the **elements** that are in set A **and** in set B.

$$A \cap B = \{x \,|\, x \text{ is an element of } A \textbf{ and } \text{an element of } B\}$$

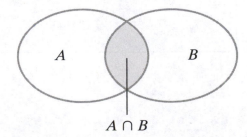

$A \cap B$

(eText Screens 1.3-1–1.3-16)

Notice that the intersection of two sets is the overlap of the two sets. Much like the intersection of two roads is the region common to both roads, the intersection of two sets is the set of elements that are common to both sets. If there is no overlap, the intersection is the **empty set**, and we write $A \cap B = \{\ \}$ or $A \cap B = \varnothing$.

My video summary ▶ **Example 1 Finding the Intersection of Two Sets**

Let $A = \{1, 3, 4, 5, 7, 10, 12\}$ and $B = \{2, 4, 6, 8, 10, 12\}$. Find $A \cap B$, the **intersection** of the two sets.

Solution Read the following, or watch this **video** for the solution.

The set $A \cap B$ is the set of elements that are in both A and B. Both sets contain the numbers 4, 10, and 12. So,

$$A \cap B = \{1, 3, 4, 5, 7, 10, 12\} \cap \{2, 4, 6, 8, 10, 12\} = \{4, 10, 12\}$$

You Try It Work through this You Try It problem.

Work Exercises 1 and 2 in this eText or in the MyMathLab Study Plan.

My video summary ▶ **Example 2 Finding the Intersection of Two Sets**

Let $A = \{x \mid x > -2\}$ and $B = \{x \mid x \le 5\}$. Find $A \cap B$, the intersection of the two sets.

Solution Check your answer, or watch the video for the solution.

You Try It Work through this You Try It problem.

Work Exercises 5–8 in this eText or in the MyMathLab Study Plan.

Union

For any two **sets** A and B, the **union** of A and B is given by $A \cup B$ and represents the **elements** that are in set A **or** in set B.

$$A \cup B = \{x \mid x \text{ is an element of } A \textbf{ or } \text{an element of } B\}$$

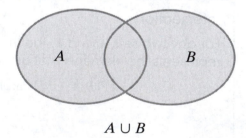

$A \cup B$

Notice that the union of two sets is the combination of the two sets. The union of two sets contains elements that are just in set A, just in set B, or in both A and B. Elements that appear in both sets A and B are only listed once when writing the union of the sets.

My video summary ▶ **Example 3   Finding the Union of Two Sets**

Let $A = \{1, 3, 4, 5, 7, 10, 12\}$ and $B = \{2, 4, 6, 8, 10, 12\}$. Find $A \cup B$, the union of the two sets.

Solution Read the following, or watch this video for the solution.

The union is the set of all unique values that are in either set A or in set B.

$$A \cup B = \{1, 2, 3, 4, 5, 6, 7, 8, 10, 12\}$$

Notice that while 4, 10, and 12 occurred in both sets, these numbers were listed only once in the union.

You Try It   Work through this You Try It problem.

Work Exercises 3 and 4 in this eText or in the MyMathLab Study Plan.

My video summary ▶ **Example 4 Finding the Union of Two Sets**

Let $A = \{x \mid x < -2\}$ and $B = \{x \mid x \geq 5\}$. Find $A \cup B$, the union of the two sets.

Solution Check your answer, or watch the video for the solution.

You Try It   Work through this You Try It problem.

Work Exercises 9–12 in this eText or in the MyMathLab Study Plan.

My video summary ▶ **Example 5   Finding the Intersection of Intervals**

Find the intersection of the following intervals and graph the set on a number line.

a. $[0, \infty) \cap (-\infty, 5]$ **b.** $\big((-\infty, -2) \cup (-2, \infty)\big) \cap [-4, \infty)$

Solutions Check your answer, or watch the video for the complete solutions to both parts.

You Try It   Work through this You Try It problem.

Work Exercises 13–16 in this eText or in the MyMathLab Study Plan.

OBJECTIVE 2   SOLVE COMPOUND LINEAR INEQUALITIES IN ONE VARIABLE

A **compound inequality** consists of two **inequalities** that are joined together using the words *and* or *or*. Figure 1 shows two examples of compound inequalities.

(a) $x + 2 < 5$ and $3x \geq -6$ **Figure 1**
(b) $x + 3 \leq 1$ or $2x - 5 > 7$ Examples of Compound Inequalities

The word *and* indicates **intersection**. A number is a **solution** to a compound inequality using *and* if it is a solution to *both* inequalities. For example, 2 is a solution to the compound inequality in Figure 1(a) because it is a solution to both $x + 2 < 5$ and $3x \geq -6$. View the **check**.

The word *or* indicates **union**. A number is a solution to a compound inequality using *or* if it is a solution to *either* inequality. For example, 8 is a solution to the compound inequality in Figure 1(b) because it is a solution to at least one of the inequalities. View the **check**.

My interactive video summary For more practice on checking solutions to compound inequalities, work through this **interactive video**.

The following general strategy can be used when solving compound inequalities.

Guidelines for Solving Compound Linear Inequalities

Step 1 Solve each inequality separately.

Step 2 Graph each **solution set** on a number line.

Step 3 For compound inequalities using *and*, the solution set is the **intersection** of the individual solution sets.

For compound inequalities using *or*, the solution set is the **union** of the individual solution sets.

My video summary ▶ **Example 6 Solving a Compound Linear Inequality Using *and***

Solve $3x - 5 < -2$ and $4x + 11 \geq 3$. Graph the solution set, and then write it in interval notation.

Solution Let's follow the guidelines for solving **compound linear inequalities**.

Step 1 Solve each inequality separately.

Original inequality:	$3x - 5 < -2$	and	Original inequality:		$4x + 11 \geq 3$
Add 5 to both sides:	$3x - 5 + 5 < -2 + 5$		Subtract 11 from both sides:		$4x + 11 - 11 \geq 3 - 11$
Simplify:	$3x < 3$		Simplify:		$4x \geq -8$
Divide both sides by 3:	$\dfrac{3x}{3} < \dfrac{3}{3}$		Divide both sides by 4:		$\dfrac{4x}{4} \geq \dfrac{-8}{4}$
Simplify:	$x < 1$		Simplify:		$x \geq -2$

Continue to follow the solution process as shown here, or watch this **video** to see the solution worked out in detail.

Step 2 Graph each solution set on a number line.

$\{x \mid x < 1\}$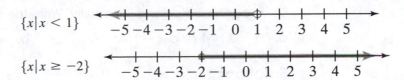

$\{x \mid x \geq -2\}$

Step 3 Since the **compound inequality** uses *and*, the **solution set** is the **intersection** of the two graphs. Look for all x-values that are common to both solution sets. The x-values common to both solution sets include all x-values that are greater than or equal to -2 *and* less than 1. That is, $\{x \mid -2 \leq x < 1\}$. The following graph shows this solution set:

The solution set of the compound inequality, written in **interval notation**, is $[-2, 1)$.

You Try It Work through this You Try It problem.

Work Exercises 17–20 in this eText or in the MyMathLab Study Plan.

My video summary ▶ **Example 7 Solving a Compound Linear Inequality Using *or***

Solve $9 - 4x < -7$ or $5x + 6 < 3(x + 2)$. Graph the solution set, and then write it in interval notation.

Solution

Step 1 Solve each inequality separately.

$$9 - 4x < -7 \qquad \text{or} \qquad 5x + 6 < 3(x + 2)$$
$$-4x < -16 \qquad\qquad\qquad 5x + 6 < 3x + 6$$
$$x > 4 \qquad\qquad\qquad\qquad 5x < 3x$$
$$5x - 3x < 3x - 3x$$
$$2x < 0$$
$$x < 0$$

> The inequality switched directions because we divided both sides by a negative number.

Steps 2–3 Graph each solution set on a number line. Since the **compound inequality** uses *or*, the solution set is the **union** of the two graphs. Look for all x values that appear in either of the two solution sets. Complete the solution; then check your **answer**. Or, watch the **video** for a detailed solution.

You Try It Work through this You Try It problem.

Work Exercises 21–24 in this eText or in the MyMathLab Study Plan.

Like other inequalities, a **compound inequality** can have no solution or the set of all real numbers as its solution set.

Example 8 Solving Special Case Compound Inequalities

Solve each compound inequality. Write the solution set in interval notation.

a. $2x - 3 \leq 1$ and $x - 7 \geq -3$ **b.** $10x + 7 > 2$ or $3x - 6 \leq 9$

Solutions

a. Start by solving each inequality separately.

$$2x - 3 \leq 1 \qquad \text{and} \qquad x - 7 \geq -3$$
$$2x \leq 4 \qquad\qquad\qquad\qquad x \geq 4$$
$$x \leq 2$$

Next, graph each solution set on a number line.

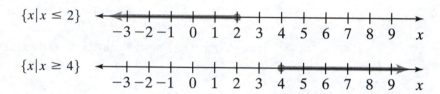

Since the compound inequality uses 'and,' the solution set contains values for x that are common to both solution sets. However, the graphs of the two solution sets do not intersect (overlap), so there are no x values in common. Thus, the compound inequality has no solution. Its solution set is the empty set $\{\,\}$, or **null set** $\varnothing$.

My video summary  ▶ **b.** The solution for this compound inequality is the set of all real numbers, $\mathbb{R}$, or in interval notation, $(-\infty, \infty)$. Try verifying this on your own, or watch this **video** for the complete solution to part (b).

🔺
You Try It Work through this You Try It problem.

Work Exercises 25–30 in this eText or in the MyMathLab Study Plan.

1.3 Exercises

In Exercises 1–4, find the indicated set given $A = \{-2, 0, 1, 5, 6, 9, 15\}$, $B = \{-6, -4, -2, 0, 2, 4, 6\}$, and $C = \{-1, 3, 7, 11, 20\}$.

1. $A \cap B$

2. $A \cap C$

3. $A \cup B$

4. $B \cup C$

In Exercises 5–8, find $A \cap B$.

5. $A = \{x | x < 9\}$; $B = \{x | x \geq 2\}$.

6. $A = \{x | x > 5\}$; $B = \{x | x \geq -3\}$

7. $A = \{x | x \leq -2\}$; $B = \{x | x \leq 0\}$

8. $A = \{x | x \leq 14\}$; $B = \{x | 0 < x < 18\}$

In Exercises 9–12, find $A \cup B$.

9. $A = \{x | x \leq -3\}$; $B = \{x | x \geq 5\}$.

10. $A = \{x | x \leq -8\}$; $B = \{x | x < 6.4\}$

11. $A = \{x | x < 10\}$; $B = \{x | x > -1\}$

12. $A = \{x | x \geq 0\}$; $B = \{x | -7 \leq x < 0\}$

In Exercises 13–16, find the intersection.

13. $(-8, 5] \cap (-12, 3)$

14. $(-\infty, 3) \cap [-2, \infty)$

15. $((-\infty, 0) \cup (0, \infty)) \cap [-3, \infty)$

16. $(7, \infty) \cap [4, \infty)$

In Exercises 17–30, solve the compound inequality. Write your answer in interval notation or state that there is no solution.

17. $2x - 3 \leq 5$ and $5x + 1 > 6$

18. $\dfrac{n - 1}{3} \geq 1$ and $\dfrac{4n - 2}{2} \leq 9$

19. $12v + 6 > 10v - 24$ and $4v + 1 > 8v - 7$

20. $1.3m - 2.4 > 7.2 - 1.1m$ and $3.9m - 4.2 \geq 2.4m - 1.2$

21. $4k + 2 \leq -10$ or $3k - 4 > 8$

22. $2x + 1 > 5$ or $5 - 3x < 11$

23. $12x - 9 \geq 2(8x - 5) + 1$ or $7 + 12x < 14x - 5$

24. $5x - 16 \leq -20 + 4x$ or $3 + 5x > 15x - 7$

25. $2y - 1 \leq 7$ or $3y - 5 > 1$

26. $3b - 5 \leq 1$ and $1 - b < -3$

27. $x + 1 < -5$ and $2x - 3 < -11$

28. $0.3x + 3 \leq x + 0.2$ or $4(2x - 5) > 3x + 5$

29. $0.25 + 1.75a \leq -2.75 + 0.25a$ or $-0.75a - 1.75 \geq 2a + 6.5$

30. $3 + 4z < 3(1 - 4z) + 6z$ and $7 - 12z \leq 4(6 - 2z) + 3$

1.4 Absolute Value Equations and Inequalities

THINGS TO KNOW

Before working through this section, be sure you are familiar with the following concepts:

| | | VIDEO | ANIMATION | INTERACTIVE |

 You Try It   1. Compute the Absolute Value of a Real Number (Section R.1, Objective 5)   ▶

 You Try It   2. Solve Linear Equations in One Variable (Section 1.1, Objective 2)   ▶

 You Try It   3. Solve Linear Inequalities in One Variable (Section 1.2, Objective 4)   ▶

 You Try It   4. Solve Compound Linear Inequalities in One Variable (Section 1.3, Objective 2)   ▶

OBJECTIVES

1 Solve Absolute Value Equations
2 Solve Absolute Value Inequalities

. . . .

OBJECTIVE 1   SOLVE ABSOLUTE VALUE EQUATIONS

Before reading further, you may want to review **absolute value**. Consider how to solve the equation $|x| = 5$. The solutions are the values of x that are 5 units away from zero on a number line. The two numbers that are 5 units away from zero on a number line are $x = -5$ and $x = 5$, as shown in Figure 2.

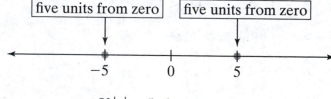

Figure 2
Solution to $|x| = 5$

If $|x| = 5$, then $x = -5$ or $x = 5$.
The solution set is $\{-5, 5\}$.

So how would we solve an equation such as $|x + 2| = 5$?

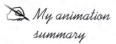 *My animation summary*   ✹ **Concept Animation** Watch this animation to see how the above concept can be applied to all **algebraic expressions**.

Absolute Value Equation Property

If u is an algebraic expression and c is a **real number** such that $c > 0$ then $|u| = c$ is equivalent to $u = -c$ or $u = c$.

Some special situations arise if $c = 0$ or $c < 0$. If $c = 0$, then $|u| = 0$ and $u = 0$. If $c < 0$, then $|u| = c$ has no solution. View an **explanation** of these results.

My video summary ▶ **Example 1 Solving an Absolute Value Equation**

Solve: $|m + 4| = 8$

Solution Watch this **video**, or work through the following.

Using the form $|u| = c$, where $u = m + 4$ and $c = 8$, look for values of m such that $m + 4$ is 8 units away from zero on a number line. The absolute value equation property shows that $|m + 4| = 8$ is equivalent to

$$m + 4 = -8 \quad \text{or} \quad m + 4 = 8$$

Now solve the two **equations** separately and combine the two **solution sets**.

$$m + 4 = -8 \quad \text{or} \quad m + 4 = 8$$
$$m = -12 \qquad\qquad m = 4$$

The solution set is $\{-12, 4\}$. View the **check**.

You Try It Work through this You Try It problem.

Work Exercises 1 and 2 in this eText or in the My MathLab Study Plan.

My video summary ▶ **Example 2 Solving an Absolute Value Equation**

Solve: $|1 - 3x| = 4$

Solution Check your **answer**, or watch this **video** for a detailed solution.

You Try It Work through this You Try It problem.

Work Exercises 3–6 in this eText or in the My MathLab Study Plan.

The following example illustrates an absolute value equation with a single solution.

My video summary ▶ **Example 3 Solving an Absolute Value Equation**

Solve: $|2x - 5| = 0$

Solution Check your **answer**, or watch this **video** for a detailed solution.

As with other equations, absolute value equations may also have no solution.

You Try It Work through this You Try It problem.

Work Exercises 7 and 8 in this eText or in the My MathLab Study Plan.

My video summary ▶ **Example 4 Solving an Absolute Value Equation**

Solve: $|3x + 7| = -4$

Solution Watch this video, or continue reading.

Using the form $|u| = c$, where $u = 3x + 7$ and $c = -4$, notice that the **absolute value** expression is equal to a negative number. However, the result of an absolute value can never be negative. This means that there is no solution, and the solution set is the empty set, { }.

You Try It Work through this You Try It problem.

Work Exercises 9 and 10 in this eText or in the MyMathLab Study Plan.

⚠ Care must be taken to not just blindly apply the absolute value equation property. Be sure to watch the video from Example 4 to see what happens if the absolute value equation property is mistakenly applied when $c < 0$.

When applying the absolute value equation property, remember to write the absolute value equation in the form $|u| = c$. If the equation is not in this form, then the absolute value expression must be isolated first before applying the property.

We now present a general strategy for solving **absolute value equations**.

Strategy for Solving Absolute Value Equations

Step 1 Isolate the absolute value expression on one side of the equation to obtain the form $|u| = c$.

Step 2 Apply the **absolute value equation property**.
   If $c > 0$, write $|u| = c$ as $u = -c$ or $u = c$.
   If $c = 0$, write $|u| = c$ as $u = c$.
   If $c < 0$, the equation has no solution.

Step 3 Solve any equations from Step 2 to find the **solution set**.

Step 4 Check your answer within the original absolute value equation to confirm the solution set.

My video summary ▶ **Example 5 Solving an Absolute Value Equation**

Solve: $2|w - 1| + 3 = 11$

Solution Follow the strategy for solving absolute value equations.

Start with the original equation: $\qquad 2|w - 1| + 3 = 11$

Subtract 3 from both sides: $\quad 2|w - 1| + 3 - 3 = 11 - 3$

Simplify: $\qquad\qquad\qquad 2|w - 1| = 8$

Divide both sides by 2: $\qquad \dfrac{2|w - 1|}{2} = \dfrac{8}{2}$

Simplify: $\qquad\qquad\qquad |w - 1| = 4$

Finish solving the equation. Check your **answer**, or watch the **video** to see the full solution.

You Try It Work through this You Try It problem.

Work Exercises 11–13 in this eText or in the MyMathLab Study Plan.

My video summary ▶ **Example 6** Solving an Absolute Value Equation

Solve: $-3|2 - m| + 8 = 2$

Solution Try solving this equation on your own. Check your **answer**, or watch the **video** to see the full solution.

You Try It Work through this You Try It problem.

Work Exercises 14–16 in this eText or in the MyMathLab Study Plan.

OBJECTIVE 2 SOLVE ABSOLUTE VALUE INEQUALITIES

Now that we know how to solve absolute value equations of the form $|u| = c$, let's turn our attention to solving absolute value inequalities of the forms $|u| < c$ or $|u| > c$.

My animation summary ✷ **Concept Animation** Watch this animation for an overview of this concept.

When solving $|x| = 5$, we looked for values of x that were 5 units away from zero on a number line. To solve the inequality $|x| < 5$, we look for values of x that are *less than* 5 units away from zero on a number line. Similarly, to solve the inequality $|x| > 5$, we look for values of x that are *greater than* 5 units away from zero on a number line. The solutions to these two inequalities are illustrated in **Figures 3** and **4**, respectively.

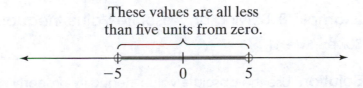

These values are all less than five units from zero.

Figure 3
Solution to $|x| < 5$

If $|x| < 5$, then $-5 < x < 5$.
The solution set is $\{x | -5 < x < 5\}$ in set-builder notation or $(-5, 5)$ in interval notation.

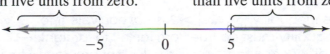

These values are more than five units from zero.   These values are more than five units from zero.

Figure 4
Solution to $|x| > 5$

If $|x| > 5$, then $x < -5$ or $x > 5$.
The solution set is $\{x | x < -5 \text{ or } x > 5\}$ in set-builder notation or $(-\infty, -5) \cup (5, \infty)$ in interval notation.

Absolute Value Inequality Property

If u is an algebraic expression and c is a real number such that $c > 0$, then

1. $|u| < c$ is equivalent to $-c < u < c$
   and
2. $|u| > c$ is equivalent to $u < -c$ or $u > c$

Similar forms exist for the non-strict cases $|u| \leq c$ and $|u| \geq c$.

As with absolute value equations, special cases arise when working with absolute value inequalities. Two special cases are $|u| < 0$ and $|u| \geq 0$. The **solution sets** of these two cases can be viewed here.

My video summary ▶ **Example 7 Solving an Absolute Value Inequality**

Solve: $|2m - 1| \leq 5$

Solution Using the form $|u| \leq c$, where $u = 2m - 1$ and $c = 5$, look for values of m such that $2m - 1$ is less than or equal to 5 units away from zero on a number line. Next use the **absolute value inequality property** to write $|2m - 1| \leq 5$ is equivalent to $-5 \leq 2m - 1 \leq 5$.

Now solve the resulting **three-part inequality**. Compare your **answer** and see a check, or watch this **video** to confirm your results.

You Try It Work through this You Try It problem.

Work Exercises 17–22 in this eText or in the MyMathLab Study Plan.

My video summary ▶ **Example 8 Solving an Absolute Value Inequality**

Solve: $|5x + 1| > 3$

Solution Use the absolute value inequality property to rewrite the statement as a compound inequality. Try solving the inequality on your own. Compare your **answer**, or watch this **video** to confirm your results.

You Try It Work through this You Try It problem.

Work Exercises 23–28 in this eText or in the MyMathLab Study Plan.

⚠ The inequality $|5x + 1| > 3$ is *not* equivalent to $-3 > 5x + 1 > 3$. A three-part inequality must be true from far left to far right: $-3 > 3$ is a false statement. Another common error in this type of problem is to write $5x + 1 > -3$ for the first inequality, instead of $5x + 1 < -3$. Think carefully about the meaning of the inequality before writing it.

As with absolute value equations, not every absolute value inequality will be written in a form that will allow us to use one of the **absolute value inequality properties** immediately. The following general strategy for solving absolute value inequalities can be used as a guide.

Strategy for Solving Absolute Value Inequalities

Step 1 Isolate the absolute value expression to obtain one of the following forms: $|u| < c$, $|u| > c$, $|u| \leq c$, or $|u| \geq c$.

Step 2 Apply the appropriate absolute value inequality property.

Step 3 Solve the resulting compound inequality.

My video summary ▶ **Example 9 Solving an Absolute Value Inequality**

Solve: $|4x - 3| + 2 \leq 7$

Solution Follow the strategy for solving absolute value inequalities, or watch this video for the complete solution.

Write the original inequality: $|4x - 3| + 2 \leq 7$

Subtract 2 from both sides: $|4x - 3| \leq 5$

Now we have the form $|u| \leq c$, where $u = 4x - 3$ and $c = 5$. We are looking for values of x such that $4x - 3$ is less than or equal to 5 units away from zero on a number line. Using the **absolute value inequality property**, $|4x - 3| \leq 5$ is equivalent to $-5 \leq 4x - 3 \leq 5$.

Find the solution set by solving the **three-part inequality**:

Write the three-part inequality: $-5 \leq 4x - 3 \leq 5$

Add 3 to all three parts: $-2 \leq 4x \leq 8$

Divide all three parts by 4 and simplify: $-\dfrac{1}{2} \leq x \leq 2$

The solution set is $\left\{ x \,\middle|\, -\dfrac{1}{2} \leq x \leq 2 \right\}$ in **set-builder notation** or $\left[-\dfrac{1}{2}, 2 \right]$ in **interval notation**. View the **check** for this solution set. The following shows the graph of the solution set on a number line:

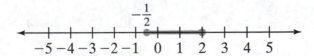

You Try It Work through this You Try It problem.

Work Exercises 29 and 30 in this eText or in the My MathLab Study Plan.

My interactive video summary ▶ **Example 10 Solving an Absolute Value Inequality**

Solve: $5|1 - 2x| - 3 > 12$

Solution Try solving this problem on your own. **Check** your solution set, or work through this **interactive video**.

You Try It Work through this You Try It problem.

Work Exercises 31–34 in this eText or in the My MathLab Study Plan.

1.4 Exercises

In Exercises 1–16, solve each absolute value equation.

1. $|x| = 3$

2. $|x - 1| = 5$

3. $|6x + 7| = 3$

4. $\left|\dfrac{3x + 1}{5}\right| = \dfrac{2}{3}$

5. $|8 - 3x| = 8$

6. $|-1.25x + 6.5| = 2$

7. $|7x + 3| = 0$

8. $\left|\dfrac{3x + 15}{7}\right| = 0$

9. $|1 - 6x| = -11$

10. $|x - 2| = -4$

11. $|3x + 4| - 6 = 11$

12. $|2x - 7| - 9 = -4$

13. $6|3 - 10x| = 24$

14. $3|-8x - 9| - 6 = 15$

15. $3|4x - 3| + 1 = 10$

16. $-2|3x + 2| + 7 = 5$

In Exercises 17–34, solve each absolute value inequality. Write your answer in interval notation.

17. $|9x| \leq 3$

18. $|x - 2| < 3$

19. $|3 - 4x| < 11$

20. $\left|\dfrac{2x - 5}{3}\right| \leq \dfrac{1}{5}$

21. $|3x - 7| \leq 0$

22. $|x - 1| < -2$

23. $|2x| > 6$

24. $|x + 4| > 5$

25. $\left|\dfrac{2x - 6}{7}\right| \geq \dfrac{3}{14}$

26. $|2 - 5x| \geq 3$

27. $|9x - 4| > 0$

28. $|3x + 1| \geq -5$

29. $|4x - 3| + 2 \leq 9$

30. $|1 - 2x| - 4 > 3$

31. $3|x + 5| - 1 \geq 8$

32. $-|3x + 2| + 5 > -6$

33. $-4|5x + 2| - 3 < 5$

34. $2|3 - 4x| + 1 < 4$

1.5 Formulas and Problem Solving

THINGS TO KNOW

Before working through this section, be sure you are familiar with the following concepts:

| | | VIDEO | ANIMATION | INTERACTIVE |

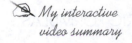 **You Try It** 1. Write Verbal Descriptions as Algebraic Expressions (Section R.3, Objective 3) ▶

You Try It 2. Solve Linear Equations in One Variable (Section 1.1, Objective 2) ▶ ✲

You Try It 3. Use Linear Equations to Solve Application Problems (Section 1.1, Objective 4) ▶

You Try It 4. Solve Linear Inequalities in One Variable (Section 1.2, Objective 4) ▶

You Try It 5. Use Linear Inequalities to Solve Application Problems (Section 1.2, Objective 5) ▶

OBJECTIVES

1 Solve a Formula for a Given Variable

2 Solve Application Problems Involving Geometric Formulas

3 Solve Application Problems Involving Money

4 Solve Application Problems Involving Uniform Motion

5 Solve Application Problems Involving Mixtures

6 Use Formulas to Solve Inequality Application Problems

....

OBJECTIVE 1   SOLVE A FORMULA FOR A GIVEN VARIABLE

A **formula** is an **equation** that describes the relationship between two or more **variables**. Typically formulas apply to physical or financial situations that relate quantities such as length, area, volume, time, speed, money, interest rates, and so on. For example, $P = 2l + 2w$ is the formula for finding the **perimeter** of a rectangle, and it relates three variables: perimeter P, length l, and width w. Knowing the values of any two of these variables, the formula can be used to find the third value.

My interactive video summary

Watch this **interactive video** to learn how to use a formula to find the value of a variable.

To **solve a formula for a given variable** means to **isolate the variable** on one side of the equation. For example the formula $P = 2l + 2w$ is solved for P.

When working with application problems, we may want to solve a formula for a different variable. For example if we need to find the length l of a rectangle when the perimeter P and width w are known, then we solve the formula $P = 2l + 2w$ for the variable l. To solve a formula for a given variable, we follow the steps for **solving an equation**.

Example 1 Solving a Formula for a Given Variable

Solve each formula for the given variable.

a. Area of a triangle: $A = \dfrac{1}{2}bh$ for b

b. Perimeter of a rectangle: $P = 2l + 2w$ for l

Solutions

a.

Begin with original formula: $A = \dfrac{1}{2}bh$

Multiply both sides by 2 to clear the fraction: $2(A) = 2\left(\dfrac{1}{2}bh\right)$

Simplify: $2A = bh$

Divide both sides by h: $\dfrac{2A}{h} = \dfrac{bh}{h}$

Simplify: $\dfrac{2A}{h} = b$

The area of a triangle formula, $b = \dfrac{2A}{h}$, is solved for the variable b.

My video summary ⊙ **b.** Begin with the original formula $P = 2l + 2w$. Subtract $2w$ from both sides of the equation and then divide both sides by 2. The result is $l = \dfrac{P - 2w}{2}$, which is the perimeter of a rectangle formula solved for the variable l. Watch this **video** to see a fully worked solution.

You Try It Work through this You Try It problem.

Work Exercises 1–10 in this eText or in the MyMathLab Study Plan.

⚠ To solve a formula for a given variable, the variable must be isolated on one side of the equation and must be the only variable of its type left in the equation.

Note: The relationship among the variables of a formula does not change when the formula is solved for a different variable. For example, the relationship among the perimeter, length, and width of a rectangle remains the same whether the formula is solved for l, w, or P. If we know any two of these variables, we can always find the third. See an **example** of why this is true.

OBJECTIVE 2 SOLVE APPLICATION PROBLEMS INVOLVING GEOMETRIC FORMULAS

For the rest of this section, we will use various types of formulas to solve application problems. In this objective, we use **formulas** from geometry. See this **review** of common formulas for area and perimeter or this **review** of common formulas for volume and surface area.

When using formulas to solve application problems, follow the **problem-solving strategy** for applications of linear equations.

🖊 *My video summary* ▶ **Example 2 Basketball Court Dimensions**

The length of a college basketball court (rectangle) is 6 feet less than twice its width. If the **perimeter** is 288 feet, then what are the dimensions of the court?

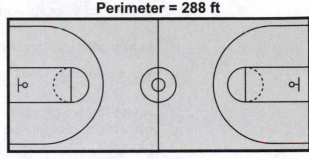

Perimeter = 288 ft

Solution

Step 1 We need to find both the length and the width of a college basketball court (rectangle). We know that the perimeter is 288 feet, and the length is 6 feet less than twice the width. The formula for the perimeter of a rectangle can be used to solve this problem.

Step 2 Letting w represent the width of the court, the length is $2w - 6$ (6 feet less than twice the width).

Step 3 The perimeter is 288 feet, so we have the **equation**:

Substitute $2w - 6$ for l ⟶ $2l + 2w = P$ ← Formula for perimeter of a rectangle

$$2(2w - 6) + 2w = 288$$

Step 4 Finish **solving** the equation to find the width of the court, then use the width to find the length. When finished, check your **answer**, or watch this **video** for a fully worked solution.

🔺
You Try It Work through this **You Try It** problem.

Work Exercises 11–16 in this eText or in the MyMathLab **Study Plan.**

🖊 *My video summary* ▶ **Example 3 Coffee Can Height**

A 13-ounce Maxwell House coffee can has a surface area of 186π cm^2. Find the height of the can if its radius is 5.0 cm.

Solution

Step 1 We need to find the height of the can. We know that the can is a circular cylinder, the radius is 5.0 cm, and the surface area is 186π cm^2. Refer to the **common formulas for volume and surface area** to find the formula for the surface area of a cylinder: $SA = 2\pi r^2 + 2\pi rh$.

Step 2 Values for all of the **variables** expect h are known. Let h represent the height of the can.

Step 3 Substituting for the variables gives

$$2\pi r^2 + 2\pi rh = SA$$

$$2\pi(5.0)^2 + 2\pi(5.0)h = 186\pi$$

Step 4   Finish **solving** the equation to find the height of the can. Check the result for reasonableness and write a clear answer. When finished, check your **answer**, or watch this **video** for a fully worked solution.

You Try It   Work through this You Try It problem.

Work Exercises 17–21 in this eText or in the MyMathLab Study Plan.

A well-known property of triangles is that the sum of the measures of the three **angles** equals 180°. This fact can be used as a **formula** to solve problems involving triangles.

My video summary ▶ **Example 4 A Triangular Field**

Ken owns a triangular-shaped plot of land. The **measure of the largest angle** is 15° more than twice the measure of the middle-sized angle. The measure of the smallest angle is two-thirds the measure of the middle-sized angle. Find the three angles of the triangular plot of land.

Solution   Work through the **Problem-Solving Strategy** to solve this problem on your own. Check your **answer**, or watch this **video** for a fully worked solution.

You Try It   Work through this You Try It problem.

Work Exercises 22–25 in this eText or in the MyMathLab Study Plan.

OBJECTIVE 3   SOLVE APPLICATION PROBLEMS INVOLVING MONEY

Some application problems involve working with money. For example, you may want to find the cost of a collection of items. Suppose you purchase 3 candy bars that cost $0.75 each. Your total cost for the 3 candy bars will be 3($0.75), or $2.25.

The following formula helps us to solve this problem:

$$\text{Total value} = (\text{value per item})(\text{number of items})$$

Note: We have used the term *total value* in the above formula instead of *total cost* because in an application, the "value" could be cost, profit, revenue, or earnings.

My video summary ▶ **Example 5 Candle Sale**

For a fundraiser, Anna's softball team sold candles in two sizes: large and small. The team earned $6.50 in profit for each large candle sold and $3.75 in profit for each small candle sold. If the team sold a total of 90 candles and earned $480.50 in profit, how many candles of each size were sold?

Solution

Step 1   We need to find the number of candles sold for each size. We know that a total of 90 candles were sold and that the team earned $480.50. We also know that the team earned a profit of $6.50 per large candle and $3.75 per small candle.

Step 2   Let x be the number of large candles sold. Then $90 - x$ is the number of small candles sold.

Step 3 The profit from selling x large candles is $6.50x$. The profit from selling $90 - x$ small candles is $3.75(90 - x)$. Adding these two amounts results in the total profit of $480.50. So, we have the **equation**

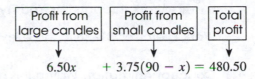

$$6.50x \quad + \quad 3.75(90 - x) = 480.50$$

Finish solving the equation to find the number of large candles sold, then use the result to find the number of small candles sold. Check your answer, or watch this **video** to see the complete solution.

You Try It Work through this You Try It problem.

Work Exercises 26 and 27 in this eText or in the MyMathLab Study Plan.

A common banking formula, $I = Prt$, is used for computing simple interest. In this formula, I is the *simple interest* earned or paid, P is the *principal*, r is the *interest rate* in decimal form, and t is the *time* that the money is invested or borrowed.

My video summary ▶ **Example 6 Investing Money**

Elizabeth inherited $25,000 from her grandparents with the promise that she invest the money and use it to attend college. She invests some of the money in a **certificate of deposit (CD)** that pays 5% simple interest and the rest in a **money market account** that pays 3.5% simple interest. If the total interest earned was $1130 after one year, how much did she invest in each account?

Solution Watch this video for a fully worked solution. Or, view this **popup** for the answer only.

You Try It Work through this You Try It problem.

Work Exercises 28–30 in this eText or in the MyMathLab Study Plan.

OBJECTIVE 4 SOLVE APPLICATION PROBLEMS INVOLVING UNIFORM MOTION

To solve **uniform motion** problems, we use the distance formula $d = rt$ or distance = rate × time. Using a table can often help to organize the information.

⚠ When working with the distance formula, the units for rate and time must be consistent. For example, if the unit for rate is miles per hour, then the unit for time must be hours.

My video summary ▶ **Example 7 Fishing Trip**

At 9:00 A.M., Rick left his house to go fishing. He rode his scooter at an average speed of 10 mph. At 9:15 A.M., his girlfriend, Deb, went to Rick's house and then followed his path on her bike at a rate of 15 mph. If Deb caught up with Rick at the same time that they both reached the fishing hole, how far is it from Rick's house to the fishing hole? At what time did Rick and Deb arrive at the fishing hole?

Solution Watch this **video** for a fully worked solution. View this **popup** for the answer only.

You Try It Work through this You Try It problem.

Work Exercises 31–36 in this eText or in the MyMathLab Study Plan.

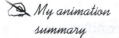
My animation summary

⚙ **Concept Animation** Watch this animation for an example in which distances are added.

OBJECTIVE 5 SOLVE APPLICATION PROBLEMS INVOLVING MIXTURES

Now let's look at mixture problems. Suppose you purchase a 50-pound bag of lawn fertilizer with a 12% nitrogen **concentration**. What does this **mean**?

We can find the amount of a particular component in a mixture by multiplying its concentration (in decimal form) by the amount of mixture:

$$\text{amount of component} = (\text{concentration})(\text{amount of mixture})$$

We can also write this in the following equivalent form:

$$\text{concentration} = \frac{\text{amount of component}}{\text{amount of mixture}}$$

The key to solving most mixture problems is recognizing that the total *amount* of a component does not change when two or more substances are mixed together. However, the *concentration* of that component might change.

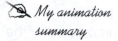
My animation summary

⚙ **Example 8 Bleach Concentration**

Suppose 2 gallons of a 10% bleach solution is mixed with 3 gallons of a 25% bleach solution. What is the **concentration** of bleach in the new 5-gallon mixture?

Solution Watch this **animation** to see a detailed solution to this problem. View the **answer**.

You Try It Work through this You Try It problem.

Work Exercises 37 and 38 in this eText or in the MyMathLab Study Plan.

⚠ The concentration of a mixture must always be in between the concentrations of the two mixed solutions. Do you see **why**?

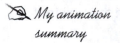
My animation summary

⚙ **Concept Animation** Watch this **animation** to better understand how to set up and solve mixture problems.

Example 9 Alcohol Concentration

How many milliliters of a 70% alcohol solution must be mixed with 30 mL of a 40% alcohol solution to result in a mixture that is 50% alcohol?

My video summary ▶ **Solution**

Step 1 Two solutions are being mixed together to obtain a third solution. We know that 30 mL of a 40% alcohol solution will be mixed with an unknown amount of a 70% alcohol solution to result in a 50% alcohol solution.

Step 2 Let x be the unknown amount of the 70% alcohol solution (in mL). Since this will be mixed with 30 mL of a 40% alcohol solution, the resulting amount of 50% alcohol solution is $30 + x$ (in mL). This is illustrated in **Figure 5**.

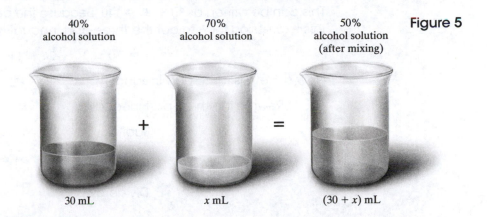

40% alcohol solution   70% alcohol solution   50% alcohol solution (after mixing)   **Figure 5**

30 mL x mL $(30 + x)$ mL

Step 3 To set up the **equation**, write the number of milliliters of pure alcohol in the 40% solution plus the number of milliliters of pure alcohol in the 70% solution and set this sum equal to the number of milliliters of pure alcohol in the 50% solution.

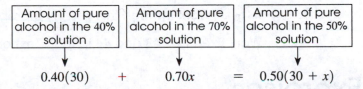

Amount of pure alcohol in the 40% solution	Amount of pure alcohol in the 70% solution	Amount of pure alcohol in the 50% solution

$$0.40(30) \quad + \quad 0.70x \quad = \quad 0.50(30 + x)$$

Step 4 Finish solving the equation and determine the solution to the problem. Check your **answer** or watch this **video** to see a detailed solution.

You Try It   **Work through this You Try It problem.**

Work Exercises 39–42 in this eText or in the MyMathLab Study Plan.

OBJECTIVE 6 USE FORMULAS TO SOLVE INEQUALITY APPLICATION PROBLEMS

Sometimes we may have to use formulas when solving applications of inequalities.

My video summary ▶ **Example 10 Perimeter of a Garden**

The perimeter of a rectangular garden must be at least 80 feet and no more than 140 feet. If the width of the garden is 8 feet, what is the range of values allowed for the length of the garden?

Solution Watch this **video**, or work through the following for the complete solution. We know that the width is 8 feet. Let l be the length of the rectangle, as shown in Figure 6. Therefore, the perimeter P of the rectangle is $P = 2(8) + 2l = 16 + 2l$.

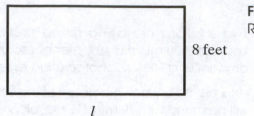

Figure 6
Rectangular Garden

8 feet

l

The perimeter must be greater than or equal to 80 and less than or equal to 140. This can be written as $80 \leq P \leq 140$. Because the perimeter is $16 + 2l$, we substitute this expression for P to get the **three-part inequality**

$$80 \leq 16 + 2l \leq 140$$

Now solve this **three-part inequality**.

Rewrite the three-part inequality:	$80 \leq 16 + 2l \leq 140$
Subtract 16:	$80 - 16 \leq 16 - 16 + 2l \leq 140 - 16$
Simplify:	$64 \leq 2l \leq 124$
Divide by 2:	$\dfrac{64}{2} \leq \dfrac{2l}{2} \leq \dfrac{124}{2}$
Simplify:	$32 \leq l \leq 62$

The length of the garden must be between 32 feet and 62 feet, inclusive. **Note:** The word *inclusive* means that the garden could be 32 feet or 62 feet as well as any value in between.

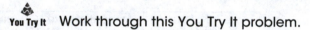

You Try It Work through this You Try It problem.

Work Exercises 43 and 44 in this eText or in the MyMathLab Study Plan.

1.5 Exercises

In Exercises 1–10, solve the formula for the given variable.

1. $A = lw$ for w

2. $I = Prt$ for P

3. $V = \dfrac{1}{3}\pi r^2 h$ for h

4. $Ax + By = C$ for y

5. $E = I(r + R)$ for r

6. $A = \dfrac{1}{2}h(a + b)$ for a

7. $A = 2\pi r^2 + 2\pi rh$ for h

8. $C = \dfrac{5}{9}(F - 32)$ for F

9. $z = \dfrac{x - \mu}{\sigma}$ for x

10. $A = 2lw + 2lh + 2wh$ for l

In Exercises 11–44, solve the application problem.

11. **Volleyball Court Dimensions** A volleyball court is twice as long as it is wide. If the perimeter of the rectangular court is 54 meters, determine the dimensions of the court.

12. **Solar Panel** A rectangular solar panel has a length that is 12 inches shorter than 3 times its width. If the perimeter of the panel is 136 inches, what are the dimensions of the panel?

13. **Lacrosse Field** The length of a lacrosse field is 10 yards less than twice its width, and the perimeter is 340 yards. The defensive area of the field is $\frac{7}{22}$ of the total field area. Find the defensive area of the lacrosse field.

14. **Patio Garden** One length of a rectangular garden lies along a patio wall. However, the rest of the garden is enclosed by 36 feet of fencing. If the length of the garden is twice its width, what is the area of the garden?

15. **Sail Height** The area of a triangular sail for a boat is 90 square feet. If the base of the sail is 10 feet long, find its height.

16. **Triangular Deck** A triangular-shaped deck has one side that is 5 feet longer than the shortest side and a third side that is 5 feet shorter than twice the length of the shorter side. If the perimeter of the deck is 60 feet, what are the lengths of the three sides?

17. **Height of a Cone** The volume of a cone is 84π cubic inches. If the radius of the cone is 6 inches, determine its height.

18. **Surface Area of a Box** A rectangular box is 36 cm wide and 15 cm high. If the surface area of the box is 3630 square centimeters, find the length of the box.

19. **Volume of a Brick** The width of a brick is half the length, which is 1 inch less than four times the height. If the sum of the three dimensions is 14.25 inches, find the volume of the brick.

20. **Building a Sidewalk** A concrete sidewalk 5 feet wide and 6 inches thick will be built around the square playground shown. How many cubic yards of concrete will be needed for the sidewalk?

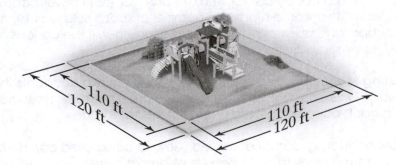

21. **Resurfacing a Driveway** A semicircular driveway (see figure) 10 feet wide will be resurfaced with asphalt. If the 2-inch thick asphalt costs $1.95 per square foot, find the cost of the driveway.

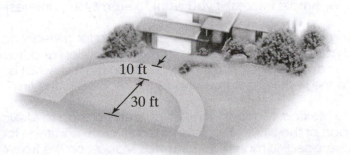

22. **Angles of a Triangle** The measure of the largest angle of a triangle is twice the measure of the smallest angle. The measure of the middle-sized angle is 20° more than the measure of the smallest angle. Find the measures of the three angles.

23. **Isosceles Triangles** An isosceles triangle has two equal angles. If the measure of the third angle is 20° less than the sum of the two equal angles, find the measures of the three angles.

24. **Triangular Sunshade** The smallest angle of a triangular sunshade measures 10 degrees less than the middle-sized angle. The largest angle measures 10 degrees more than twice the middle-sized angle. What are the angle measures?

25. **Support Cable** A platform on the back of an RV is anchored by two support cables. Each cable is attached to the front of the platform and the back wall of the RV. See the diagram. If angle B measures 30° less than twice the measure of angle A, what is the measure of angle B (the angle between the cable and the wall)?

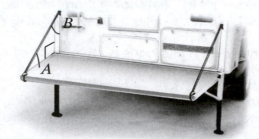

26. **Coin Collection** Jose has been collecting quarters and dimes. In his collection he has 33 coins (only quarters and dimes) worth $5.40. How many quarters and how many dimes does Jose have?

27. **Concession Stand** A basketball concession stand sells a soda for $1.25, a bag of popcorn for $0.75, and a candy bar for $1.00. At one game, 14 fewer candy bars were sold than bags of popcorn. That same night, the number of sodas sold was 10 more than the number of candy bars and bags of popcorn sold altogether. If the concession stand made $117 that night, how many of each item was sold?

28. **Investing Money** Juanita has $28,000 to invest in two accounts that pay simple interest annually. One account pays 4%, and the other pays 5%. How much would she have to invest in each account to earn a total interest of $1320?

29. **Borrowing Money** Jamal borrowed $4000 to buy a used car. He borrowed some of the money from a bank that charged 8.5% simple interest and the rest from a friend who charged 10% simple interest. If the total interest for a one-year loan was $362.50, how much did Jamal borrow at each rate?

30. **Simple Interest Principal** How much principal should be invested in a savings account paying 6% simple interest if you want to earn $450 in interest in 2 years.

31. **Catching Up** A carload of friends leaves for a long weekend road trip. Two hours later a second carload of friends leaves the same location and travels the same path. If the second car drives an average of 20 mph faster than the first, what is the average speed of each car if it takes the second car three hours to catch up to the first?

32. **Going to Mexico** Manuel traveled 31 hours nonstop to Mexico, a total of 2300 miles. He took a train part of the way, which averaged 80 mph, and then took a bus the remaining distance, which averaged 60 mph. How long was Manuel on the train?

33. **Hiking** Abbas and Muhammad are at opposite ends of a 16-mile hiking trail. If Abbas' average hiking speed on the trail is 1.5 mph whereas Muhammad's average hiking speed is 2.5 mph, how long will it be before they meet?

34. **River Transport** A grain barge travels on a river from point A to point B loading and unloading grain. The barge travels at a rate of 5 mph relative to the water. The river flows downstream at a rate of 1 mph. If the trip upstream takes two hours longer than the trip downstream, how far is it from point A to point B?

35. **Jogging Time** David leaves his house, jogging down the street at 6 mph. Once David has gone $\frac{1}{4}$ mile, Jacob comes out and follows him at a rate of 6.5 mph. How long will it take Jacob to catch David?

36. **Hiking Speed** Landon can climb a certain hill at a rate that is 2.5 mph slower than his rate coming down the hill. If it takes him 2 hours to climb the hill and 45 minutes to come down the hill, what is his rate coming down the hill?

37. **Mixing Alcohol** Suppose 8 pints of a 12% alcohol solution is mixed with 2 pints of a 60% alcohol solution. What is the concentration of alcohol in the new 10-pint mixture?

38. **Mixing Fertilizer** The granular fertilizer 12-12-12 is composed of 12% nitrogen, 12% phosphate, and 12% potassium. Similarly, the fertilizer 16-20-0 is composed of 16% nitrogen, 20% phosphate, and 0% potassium. If a gardener mixes a 10 pound bag of 12-12-12 with a 15 pound bag of 16-20-0, what are the concentrations of nitrogen, phosphate, and potassium in the mixture? Express the answers as percents.

39. **Mixing Orange Juice** How much of an 80% orange juice drink must be mixed with 20 gallons of a 20% orange juice drink to obtain a mixture that is 50% orange juice?

40. **Mixing Acid** How many liters (L) of pure water should be mixed with a 5-L solution of 80% acid to produce a mixture that is 70% water?

41. **Blending Coffee** A coffee shop owner blends a gourmet brand of coffee with a cheaper brand. The gourmet coffee usually sells for $9.00 per pound. The cheaper brand sells for $5.00 per pound. How much of each type should be mixed in order to have 30 pounds of coffee that is worth $6.50 per pound?

42. **Fuel Mixture** A two-cycle engine mechanic has 2 gallons of a fuel mixture that is 96% gasoline and 4% oil. How many gallons of gasoline should be added to form a fuel mixture that is 99% gasoline and 1% oil?

43. **Perimeter Inequality** The perimeter of a rectangular fence must be at least 100 feet and no more than 180 feet. If the width of the fence is 22 feet, what is the range of values allowed for the length of the fence?

44. **Perimeter Inequality** The perimeter of a rectangular fence must be at least 120 feet and no more than 180 feet. If the length of the fence must be twice the width, what is the range of values allowed for the width of the fence?

Graphs and Functions

CHAPTER TWO CONTENTS

2.1 The Rectangular Coordinate System and Graphing

THINGS TO KNOW

Before working through this section, be sure you are familiar with the following concepts:

VIDEO   ANIMATION   INTERACTIVE

You Try It   1. Plot Real Numbers on a Number Line (Section R.1, Objective 3)   ▶

You Try It   2. Determine If a Given Value Is a Solution to an Equation (Section 1.1, Objective 1)   ▶

You Try It   3. Solve Linear Equations in One Variable (Section 1.1, Objective 2)   ▶

OBJECTIVES

1 Plot Ordered Pairs in the Rectangular Coordinate System

2 Determine If an Ordered Pair Is a Solution to an Equation

3 Find Unknown Coordinates

4 Graph Equations by Plotting Points

5 Find x- and y-Intercepts

OBJECTIVE 1 PLOT ORDERED PAIRS IN THE RECTANGULAR COORDINATE SYSTEM

Graphs can be used to show relationships between two **variables**. Figure 1 shows the selling price of MasterCard, Inc., stock over time since it went public in May 2006. The graph shows how the value of the stock has changed over time. This information might be useful to someone who wants to invest in MasterCard, Inc.

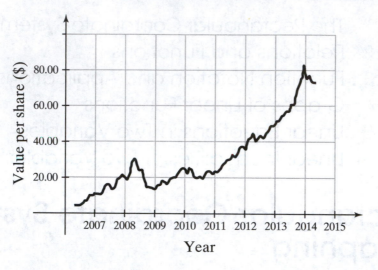

Figure 1
MasterCard, Inc., Stock Value
(*Data*: TD Ameritrade)

My interactive video summary

In algebra, graphs are used to show relationships defined by *equations in two variables*. An **equation in two variables** contains two distinct variables. Three examples of such equations follow:

$$p + q = 1, \quad y = x^2 + 3x - 4, \quad \text{and} \quad rt = 12$$

Watch this **interactive video** to practice identifying **equations in two variables**.

Such equations are graphed by using the **rectangular coordinate system**, also known as the *Cartesian coordinate system* in honor of its inventor **René Descartes**. As shown in Figure 2, the horizontal axis, or **x-axis**, and the vertical axis, or **y-axis**, intersect at the **origin** and divide the **coordinate plane** into four **quadrants**.

My animation summary

⊛ **Concept Animation** Watch this **animation** to review the rectangular coordinate system.

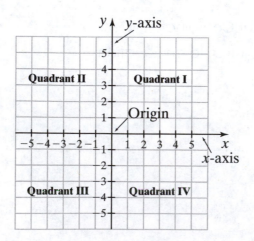

Figure 2
The Rectangular Coordinate System

Each position, or **point**, on the **coordinate plane** can be identified by an **ordered pair** of numbers in the form (x, y). For example in Figure 3, point A is identified by the ordered pair $(-3, 4)$. The first number -3 is called the **x-coordinate** or **abscissa** and indicates the point is located 3 units to the left of the **origin**. The second number 4 is called the **y-coordinate** or **ordinate** and indicates the point is located 4 units above the origin. Point B is identified by the ordered pair $(4, -3)$ because it lies 4 units to the right and 3 units below the origin.

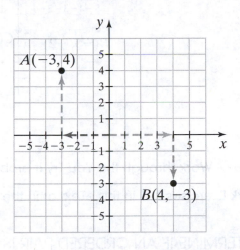

Figure 3

Compare the locations of points A and B in **Figure 3**. Both points include -3 and 4 as part of the **ordered pairs**, but the numbers, or **coordinates**, appear in reverse order. Also, point A is located in **Quadrant II**, whereas point B is located in Quadrant IV. So the *order* of the *pair* of coordinates is just as important as the coordinates themselves and helps to explain the meaning of the term *ordered pair*.

When a point lies in Quadrant II, its **x-coordinate** is negative, whereas its **y-coordinate** is positive $(x < 0, y > 0)$. When a point lies in Quadrant IV, its x-coordinate is positive, whereas its y-coordinate is negative $(x > 0, y < 0)$.

My animation summary

⊛ **Concept Animation** Watch this **animation** to see how the location of a point relates to the signs of the coordinates.

We **plot**, or **graph**, an ordered pair by placing a point (•) at it location on the coordinate plane.

My video summary ▶ **Example 1 Plotting Ordered Pairs**

Plot each ordered pair in the coordinate plane. In which quadrant or on which axis does each point lie?

a. $A(-4, 4)$ b. $B(-5, -2)$ c. $C(0, -2)$

d. $D\left(\dfrac{3}{2}, \dfrac{5}{2}\right)$ e. $E(3.5, -4.5)$ f. $F(2, 0)$

Solution The points are plotted in Figure 4. Point A lies in Quadrant II. Point B lies in Quadrant III. Point C lies on the y-axis. Point D lies in Quadrant I. Point E lies in Quadrant IV. Point F lies on the x-axis. Watch this **video** for a detailed solution.

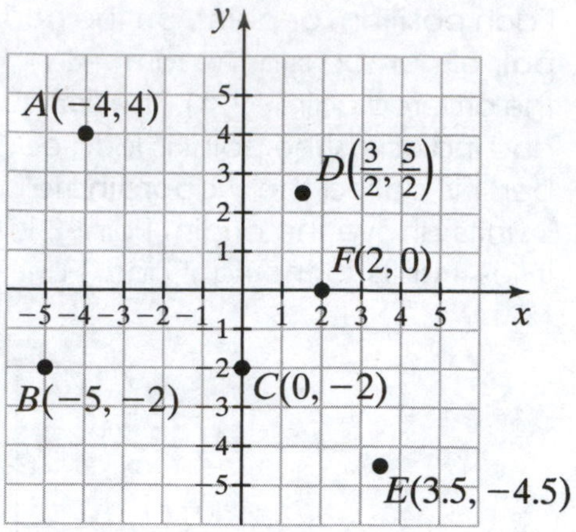

Figure 4

Figure 4 graph showing points: $A(-4, 4)$, $D\left(\frac{3}{2}, \frac{5}{2}\right)$, $F(2, 0)$, $B(-5, -2)$, $C(0, -2)$, $E(3.5, -4.5)$

You Try It Work through this You Try It problem.

Work Exercises 1–6 in this eText or in the MyMathLab Study Plan.

OBJECTIVE 2 DETERMINE IF AN ORDERED PAIR IS A SOLUTION TO AN EQUATION

In Section 1.1, we learned that a solution to an equation in one variable is a value that, when substituted for the variable, makes the equation true. We now expand on this idea and consider equations that contain two variables. For example, we may wish to find solutions to the equation $x + y = 4$.

Solutions to equations in two variables require *two* values—one value for each variable. For the equation $x + y = 4$, $x = 1$ and $y = 3$ form a solution because a true statement results if we substitute the values for the variables.

$$\text{Begin with the original equation:} \quad x + y = 4$$

$$\text{Substitute 1 for } x \text{ and 3 for } y: \quad 1 + 3 \overset{?}{=} 4$$

$$\text{Simplify:} \quad 4 = 4 \quad \text{True}$$

We can write this solution as the ordered pair $(1, 3)$, where the first number is the x-value and the second number is the y-value.

Solution to an Equation in Two Variables

A **solution to an equation in two variables** is an ordered pair of values that, when substituted for the variables, makes the equation true.

Example 2 Determining If an Ordered Pair Is a Solution to an Equation

Determine if each ordered pair is a solution to the equation $x + 2y = 8$.

a. $(-2, 5)$ **b.** $(2, 6)$ **c.** $\left(-11, \frac{3}{2}\right)$ **d.** $(0, 4)$

Solutions

a. Substitute −2 for x and 5 for y. Work through the simplification to determine if the ordered pair is a solution.

$$\text{Begin with the original equation:} \quad x + 2y = 8$$

$$\text{Substitute −2 for } x \text{ and 5 for } y: \quad (-2) + 2(5) \overset{?}{=} 8$$

$$\text{Simplify:} \quad -2 + 10 \overset{?}{=} 8$$

$$8 = 8 \quad \text{True}$$

The final statement is true, so $(-2, 5)$ is a solution to the equation.

b. Begin with the original equation: $\quad x + 2y = 8$

$$\text{Substitute 2 for } x \text{ and 6 for } y: \quad (2) + 2(6) \overset{?}{=} 8$$

$$\text{Simplify:} \quad 2 + 12 \overset{?}{=} 8$$

$$14 = 8 \quad \text{False}$$

The final statement is not true, so $(2, 6)$ is not a solution to the equation.

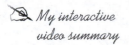 *My video summary* ⊙ **c.–d.** Substitute the values for the variables and work through the simplification to determine if each ordered pair is a solution. View the **answers**, or watch this **video** to see the details.

You Try It Work through this **You Try It** problem.

Work Exercises 7–9 in this eText or in the MyMathLab Study Plan.

OBJECTIVE 3 FIND UNKNOWN COORDINATES

The **graph of an equation in two variables** includes all points whose **coordinates** are **solutions** to the **equation**. To make such a graph, we can **plot** several points that **satisfy** the equation. Then we connect the points with a line or smooth curve.

In **Example 2**, we determined if a given **ordered pair** was a solution to an equation in two variables. But how do we find such ordered pair solutions if we only know one coordinate? We choose a convenient value for one **variable**, substitute it into the equation, and solve for the other variable. The resulting values for both variables together form an **ordered pair solution** to the equation.

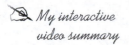 *My interactive video summary* ✑ **Example 3 Finding Unknown Coordinates**

Find the unknown coordinate so that each ordered pair satisfies $3x + 4y = 20$.

a. $(8, ?)$ **b.** $(?, 2)$ **c.** $\left(-\dfrac{2}{3}, ?\right)$

Solutions

a. To find the y-coordinate when the x-coordinate is 8, substitute 8 for x and solve for y.

$$\text{Begin with the original equation:} \quad 3x + 4y = 20$$

$$\text{Substitute 8 for } x: \quad 3(8) + 4y = 20$$

$$\text{Simplify:} \quad 24 + 4y = 20$$

Subtract 24 from both sides: $\qquad$ $4y = -4$

Divide both sides by 4: $\qquad$ $y = -1$

The ordered pair $(8, -1)$ is a solution to the equation $3x + 4y = 20$.

b. Begin with the original equation: $\qquad 3x + 4y = 20$

Substitute 2 for y: $\quad 3x + 4(2) = 20$

Simplify: $\qquad 3x + 8 = 20$

Finish solving for x to find the x-coordinate when the y-coordinate is 2. Check your **answer**, or watch this **interactive video** for a complete solution.

c. Try to find the y-coordinate on your own. View the **answer**, or watch this **interactive video** for a complete solution.

You Try It Work through this You Try It problem.

Work Exercises 10–12 in this eText or in the MyMathLab Study Plan.

OBJECTIVE 4 GRAPH EQUATIONS BY PLOTTING POINTS

Along with finding **solutions to equations in two variables**, we can identify other key points to help us graph equations. For example, an **endpoint** represents the end of the graph in one direction. Other key points include **maximum points** (high points on a graph) or **minimum points** (low points on a graph). Let's use the following strategy to graph equations by plotting points:

Strategy for Graphing Equations by Plotting Points

Step 1 Find several points that **satisfy** the equation. The exact number of points to find depends on the equation being graphed. Find enough to form a pattern. Try to locate key points such as endpoints and maximum or minimum points.

Step 2 Plot the points found in Step 1.

Step 3 Connect the points with a straight line or smooth curve, depending on the pattern formed.

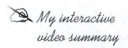 *My interactive video summary*

Example 4 Graphing Equations by Plotting Points

Graph each equation by plotting points.

a. $2x + y = 1$ $\qquad$ **b.** $y = x^2 - 4$ $\qquad$ **c.** $y = |x|$

Solutions Read through the following, or watch this **interactive video** for complete solutions.

a. Step 1 To find points that **satisfy** $2x + y = 1$, let $x = -2, -1, 0, 1$, and 2. Substitute each x-coordinate into the original equation and solve for y to find the y-coordinate. The results appear in Table 1.

Table 1

x	Original Equation: $2x + y = 1$	y	Point
-2	Substitute -2 for x: $2(-2) + y = 1$ Solve for y: $y = 5$	5	$(-2, 5)$
-1	Substitute -1 for x: $2(-1) + y = 1$ Solve for y: $y = 3$	3	$(-1, 3)$
0	Substitute 0 for x: $2(0) + y = 1$ Solve for y: $y = 1$	1	$(0, 1)$
1	Substitute 1 for x: $2(1) + y = 1$ Solve for y: $y = -1$	-1	$(1, -1)$
2	Substitute 2 for x: $2(2) + y = 1$ Solve for y: $y = -3$	-3	$(2, -3)$

Step 2 Plot the points found in Step 1. See Figure 5(a).

Step 3 The points form a straight pattern, so we connect them with a straight line. We draw arrows on each end to show that the line continues forever. See Figure 5(b).

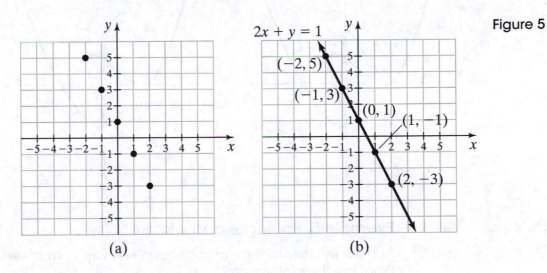

(a)   (b)

Figure 5

b. **Step 1** To find points that satisfy $y = x^2 - 4$, let $x = -3, -2, -1, 0, 1, 2,$ and 3. The results appear in Table 2.

Table 2

x	$y = x^2 - 4$	Point
-3	$y = (-3)^2 - 4 = 5$	$(-3, 5)$
-2	$y = (-2)^2 - 4 = 0$	$(-2, 0)$
-1	$y = (-1)^2 - 4 = -3$	$(-1, -3)$
0	$y = (0)^2 - 4 = -4$	$(0, -4)$
1	$y = (1)^2 - 4 = -3$	$(1, -3)$
2	$y = (-2)^2 - 4 = 0$	$(2, 0)$
3	$y = (3)^2 - 4 = 5$	$(3, 5)$

Step 2 Plot the points found in Step 1. **Check** your work.

Step 3 Do you see a pattern? Connect the points appropriately with a straight line or smooth curve. **Check** your graph.

c. The equation $y = |x|$ is called the **absolute value** equation. Try graphing the equation yourself by **plotting points**. Substitute some negative values for x as well as positive values. When done, **check** your graph.

You Try It Work through this You Try It problem.

Work Exercises 13–20 in this eText or in the MyMathLab Study Plan.

OBJECTIVE 5 FIND x- AND y-INTERCEPTS

The points where a graph crosses or touches the **axes** are called **intercepts**.

A **y-intercept** is the **y-coordinate** of a point where a graph crosses or touches the y-axis. An **x-intercept** is the **x-coordinate** of a point where a graph crosses or touches the x-axis. Figure 6 illustrates examples of x- and y-intercepts.

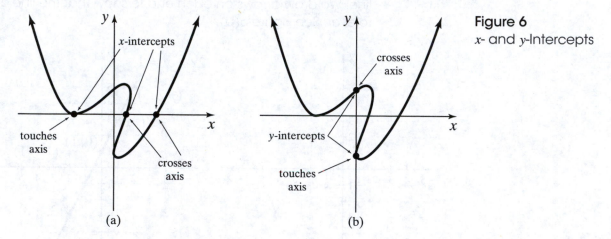

Figure 6
x- and y-Intercepts

(a)   (b)

My video summary ▶ **Example 5 Finding x- and y-Intercepts**

Find the **intercepts** of the graph shown in Figure 7. What are the **x-intercepts**? What are the **y-intercepts**?

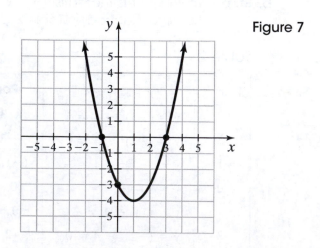

Figure 7

Solution Watch this video or read the following solution. The graph crosses the *y*-axis at $(0, -3)$ and crosses the *x*-axis at $(-1, 0)$ and $(3, 0)$, so the intercepts are $(0, -3)$, $(-1, 0)$, and $(3, 0)$. The *x*-intercepts are -1 and 3. The *y*-intercept is -3.

You Try It Work through this You Try It problem.

Work Exercises 21–26 in this eText or in the MyMathLab Study Plan.

TIP When the type of intercept is not specified, we must list the intercept as an ordered pair. If the type of intercept is specified, then we list only the coordinate as the intercept. For example, in **Figure 7**, one of the intercepts is $(3, 0)$. However, the *x*-intercepts are -1 and 3, and the *y*-intercept is -3.

2.1 Exercises

In Exercises 1–6, plot each ordered pair in the coordinate plane. State the quadrant or axis where each point lies.

1. $A(-1, 5)$

2. $B(0, 4)$

3. $C(1, 3)$

4. $D\left(\dfrac{7}{2}, -\dfrac{5}{2}\right)$

5. $E(-3, 0)$

6. $F(-5, -2)$

In Exercises 7–9, determine if the given ordered pairs are solutions to the equations.

7. $3x - 2y = 8$
   a. $(2, -1)$
   b. $(0, 4)$
   c. $\left(3, \dfrac{1}{2}\right)$
   d. $(-4, -10)$

8. $y = 2x^2 - x + 5$
   a. $(-1, 4)$
   b. $(-2, 11)$
   c. $(0, 5)$
   d. $\left(\dfrac{1}{2}, 5\right)$

9. $y = |x - 3|$
   a. $(2, 5)$
   b. $(-1, 4)$
   c. $(5, 2)$
   d. $(1, 2)$

In Exercises 10–12, find the unknown coordinate so that each ordered pair is a solution to the given equation.

10. $4x - 3y = 6$
    a. $(-3, ?)$
    b. $\left(?, -\dfrac{2}{3}\right)$

11. $y = 2x - 5$
    a. $\left(-\dfrac{1}{2}, ?\right)$
    b. $(?, 7)$

12. $y = \dfrac{2}{3}x + 2$
    a. $(-9, ?)$
    b. $(?, 5)$

In Exercises 13–20, graph each equation by plotting points.

13. $y = 2x + 1$

14. $y = \dfrac{1}{4}x - 2$

15. $3x - 2y = 8$

16. $3x + y = 3$

17. $y = 3x^2$

18. $y = x^2 - 2x - 3$

19. $y = |x + 3|$

20. $y = |x| - 2$

In Exercises 21–26, find the x- and y-intercepts of each graph.

21.

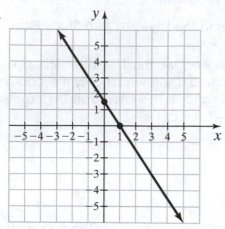

22.

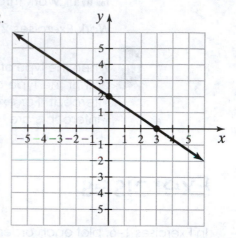

23.

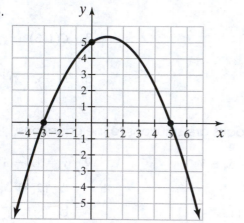

24.

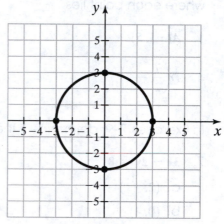

25.

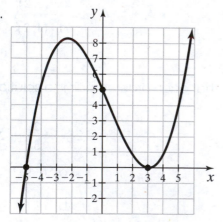

26.

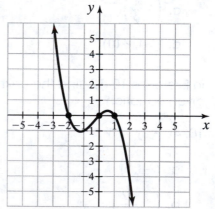

2.2 Relations and Functions

THINGS TO KNOW

Before working through this section, be sure you are familiar with the following concepts:

VIDEO   ANIMATION   INTERACTIVE

You Try It   1.   Identify Sets
(Section R.1, Objective 1)   ▶

You Try It   2.   Use Interval Notation to Express the
Solution Set of an Inequality
(Section 1.2, Objective 3)   ▶

You Try It   3.   Solve a Formula for a Given
Variable (Section 1.5, Objective 1)   ▶

OBJECTIVES

1   Identify Independent and Dependent Variables

2   Find the Domain and Range of a Relation

3   Determine If Relations Are Functions

4   Determine If Graphs Are Functions

5   Solve Application Problems Involving Relations and Functions

---

OBJECTIVE 1   IDENTIFY INDEPENDENT AND DEPENDENT VARIABLES

When applying math to everyday life, often we encounter situations where one quantity is related to another. For example, the **cost** to fill a gas tank is related to the number of gallons purchased.

When an **equation** is solved for a given variable, that variable is called the **dependent variable** because its value *depends on* the value(s) of the remaining variable(s). Any remaining variables are called **independent variables** because we are free to select their values.

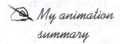 *My animation summary*   ✸ **Concept Animation** Watch this animation to explore this concept.

Note: When an equation involving x and y is not solved for either variable, like in the equation $3x + 4y = 12$, we will call x the independent variable and y the dependent variable.

 *My video summary* ▶ **Example 1** Identifying Independent and Dependent Variables

For each of the following equations, identify the **dependent variable** and the **independent variable(s)**.

a. $y = 3x + 5$ **b.** $w = ab + 3c^2$ **c.** $3x^2 + 9y = 12$

Solutions Watch this **video**, or continue reading for the solutions to all three parts.

a. Since the equation is solved for y, we identify y as the dependent variable. The remaining variable, x, is the independent variable.

b. Since the equation is solved for w, we identify w as the dependent variable. The remaining variables, a, b, and c, are independent variables.

c. Since the equation is not solved for either variable, we identify x as the independent variable and y as the dependent variable.

 You Try It Work through this You Try It problem.

Work Exercises 1–6 in this eText or in the MyMathLab Study Plan.

 TIP In Example 1, we see that it is possible to have more than one independent variable in an equation. However, we will limit our discussion mainly to situations involving one dependent and one independent variable.

OBJECTIVE 2 FIND THE DOMAIN AND RANGE OF A RELATION

Definition

A **relation** is a set of **ordered pairs**.

A relation shows the correspondence between a set of values for the **independent variable**, called the **domain** (or *input values*), and a set of values for the **dependent variable**, called the **range** (or *output values*).

Definitions

The **domain** is the set of all values for the independent variable. These are the first coordinates in the set of ordered pairs and are also known as input values.

The **range** is the set of all values for the dependent variable. These are the second coordinates in the set of ordered pairs and are also known as output values.

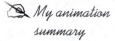 *My animation summary* ⊛ **Concept Animation** Watch this **animation** to review the concepts of relation, domain, and range.

My interactive video summary **Example 2 Finding the Domain and Range of a Relation**

Find the domain and range of each relation.

a. $\{(-5, 7), (3, 5), (6, 7), (12, -4)\}$

b.
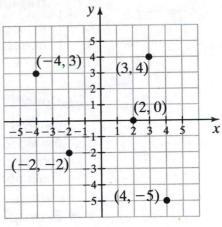

Solutions For parts a and b, identify the first coordinates of each ordered pair to find the domain and the second coordinates of each ordered pair to find the range. Check your answers, or watch this interactive video for detailed solutions.

You Try It Work through this You Try It problem.

Work Exercises 7 and 8 in this eText or in the MyMathLab Study Plan.

Because an equation in two variables defines a set of ordered pairs, such equations and their graphs are considered to be relations.

My interactive video summary **Example 3 Finding the Domain and Range of a Relation**

Find the domain and range of each relation.

a.
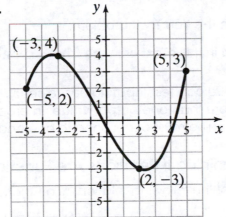

b.
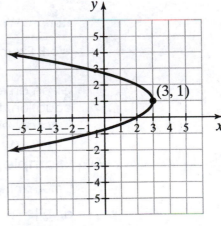

c. $y = |x - 1|$

Solutions Try to find the domain and range on your own. Check your answers, or watch this interactive video for detailed solutions.

You Try It Work through this You Try It problem.

Work Exercises 9 and 10 in this eText or in the MyMathLab Study Plan.

OBJECTIVE 3 DETERMINE IF RELATIONS ARE FUNCTIONS

 *My animation summary*

Definition

A **function** is a special type of **relation** in which each value in the **domain** corresponds to exactly one value in the **range**.

✺ **Concept Animation** Watch this animation to review this concept.

Given a set of ordered pairs, to determine whether the relation is a function, look at the *x*-coordinates. If no *x*-coordinate is repeated, then the relation is a function. If the same *x*-coordinate corresponds to two or more different *y*-coordinates, then the relation is not a function.

My video summary ▶ **Example 4 Determining If Relations Are Functions**

Determine if each of the following **relations** is a function.

a. $\{(-4, 5), (-2, 3), (0, 1), (3, -2), (7, -6)\}$

b. $\{(3, -2), (6, 5), (3, 8), (-1, -10)\}$

c. $\{(-6, 5), (-1, 5), (4, 5), (9, 5), (14, 5)\}$

Solutions

a. In the set of ordered pairs, no *x*-coordinate is repeated. This means that each value in the **domain** corresponds to exactly one value in the **range**, so the relation is a function.

b.–c. Try to work these parts on your own. View the **answers**, or watch this **video** for complete solutions to all three parts.

🔺 **You Try It** Work through this You Try It problem.

Work Exercises 11–14 in this eText or in the MyMathLab **Study Plan.**

Given an equation, we can test **input values** to see if, when substituted into the equation, there is more than one **output value**. In order for an equation to be a function, each input value must correspond to one and only one output value.

My video summary ▶ **Example 5 Determining If Relations Are Functions**

Determining if each of the following relations is a function.

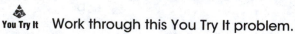

a. $|y| = x + 2$ b. $6x - 3y = 15$

Solutions To determine if the equation is a function, let's test for input values of *x*.

a. The absolute value bars around the dependent variable indicates that for some values of the independent variable, *x*, there will be two corresponding values for the dependent variable, *y*. For example, if $x = 3$, then $y = -5$ or $y = 5$ because $|-5| = 3 + 2$ and $|5| = 3 + 2$, which means the relation contains the points $(3, -5)$ and $(3, 5)$. Because one *x*-value (3) leads to more than one *y*-value (-5 and 5), the relation is not a function.

b. Try to work this problem on your own. View the **answer**, or watch this **video** for complete solutions to both parts.

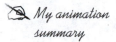 **You Try It** Work through this You Try It problem.

Work Exercises 15–20 in this eText or in the MyMathLab Study Plan.

OBJECTIVE 4 DETERMINE IF GRAPHS ARE FUNCTIONS

My animation summary If a **relation** appears as a graph, we can determine if the relation is a **function** by using the **vertical line test**.

> ### Vertical Line Test
>
> If a vertical line intersects (crosses or touches) the graph of a relation at more than one point, then the relation is not a function.
>
> If every vertical line intersects the graph of a relation at no more than one point, then the relation is a function.

Concept Animation Why does the vertical line test work? Watch this **animation** to find out.

My interactive video summary **Example 6 Determining If Graphs Are Functions**

Use the vertical line test to determine if each graph is a function.

a.
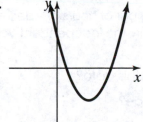

b.
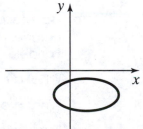

c.
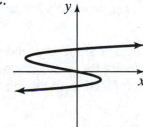

d.
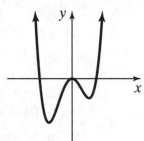

e.
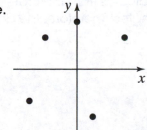

f.
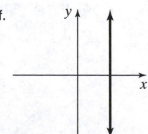

Solution Apply the vertical line test to each graph. Check your **answers**, or work through this **interactive video** for complete solutions.

You Try It Work through this **You Try It** problem.

Work Exercises 21–28 in this eText or in the MyMathLab Study Plan.

How many *x*- and *y*-intercepts can the graph of a function have? See the **answer**.

OBJECTIVE 5   SOLVE APPLICATION PROBLEMS INVOLVING RELATIONS AND FUNCTIONS

Mathematical models are used to describe many real-world situations involving relations and functions.

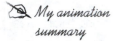 *My animation summary*

⊛ **Example 7 Video Entertainment and Sleep**

The data in the following table represent the average daily hours of sleep and average daily hours of video entertainment for six students at a local college.

Video	
Entertainment	**Sleep**
8	4
7	5
2	9

Video	
Entertainment	**Sleep**
5	7
4	8
7	6

a. If a researcher believes the number of hours of video entertainment affects the number of hours of sleep, identify the **independent variable** and the **dependent variable**.

b. What are the ordered pairs for this data?

c. What are the **domain and range**?

d. Is this **relation a function**? Explain.

Solutions Work through the following, or watch this **animation** for a complete solution.

a. Since the researcher believes hours of sleep are affected by, or depend on, the hours of video entertainment, the **independent variable** would be number of hours of video entertainment, and the **dependent variable** would be number of hours of sleep.

b. The corresponding **ordered pairs** are $\{(8, 4), (7, 5), (2, 9), (5, 7), (4, 8), (7, 6)\}$.

c. The **domain** is the set of *x*-coordinates from the ordered pairs in part (b), and the **range** is the set of *y*-coordinates. Therefore, the domain is $\{2, 4, 5, 7, 8\}$ and the range is $\{4, 5, 6, 7, 8, 9\}$.

d. This relation is not a **function** because one value from the domain, 7, corresponds to more than one value from the range, 5 and 6, as indicated in the ordered pairs (7, 5) and (7, 6).

 You Try It Work through this **You Try It** problem.

Work Exercise 29 in this eText or in the MyMathLab Study Plan.

Often when working with application problems, the **domain** must be restricted to use only those values that make sense within the context of the situation. This restricted domain is called the **feasible domain**. The feasible domain is the set of values for the **independent variable** that make sense, or are *feasible*, in the context of the application.

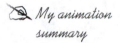 *My animation summary*

⊛ **Example 8 High-Speed Internet Access**

The percent of households, y, with broadband access in 2013 can be modeled by the equation $y = 0.46x + 50.64$, where x is the annual household income (in $1000s). (*Data*: pewinternet.org)

a. Identify the independent and dependent variables.

b. Use the **model** equation to estimate the percent of households in 2013 with broadband access (to the nearest whole percent) if the annual household income was $50,000. What point would this correspond to on the graph of the equation?

c. Is the relation a function? Explain.

d. Determine the feasible domain.

Solutions Try to work this problem on your own. View the **answer**, or watch this **animation** for the complete solution.

 You Try It Work through this **You Try It** problem.

Work Exercises 30–32 in this eText or in the MyMathLab Study Plan.

2.2 Exercises

In Exercises 1–6, identify the independent and dependent variables.

1. $y = 3x^2 - 7x + 5$

2. $k = 12v^2$

3. $A = \dfrac{1}{2}bh$

4. $d = 50t + 75$

5. $4x + 5y = 18$

6. $e = \dfrac{\sqrt{a^2 - b^2}}{a}$

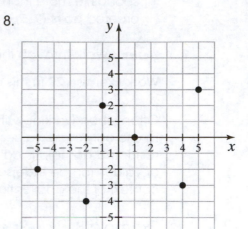

In Exercises 7–10, find the domain and range of each relation.

7. $\{(4, -2), (1, -1), (0, 0), (1, 1), (4, 2)\}$

8.

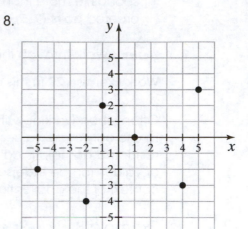

9.

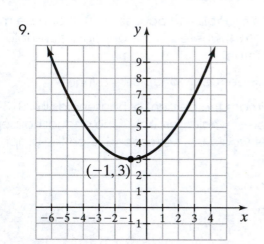

$(-1, 3)$

10.

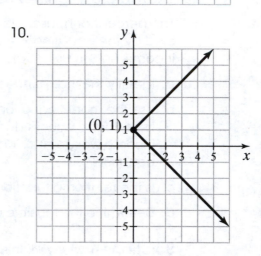

$(0, 1)$

In Exercises 11–18, determine if the relation is a function. Assume x is the independent variable.

11. $\{(2, 9), (3, 2), (7, 4), (8, 1), (10, -8)\}$

12. $\{(3, 4), (4, 5), (6, 3), (4, 1), (9, 12)\}$

13. $\{(3, 7), (5, 7), (8, 10), (11, 10), (15, 6)\}$

14. $\{(4, -5), (4, 0), (4, 3), (4, 7), (4, 11)\}$

15. $3x - y = 5$

16. $y = |x + 3| - 1$

17. $y = x^2 + 2x - 3$

18. $y^2 = x + 4$

19. $|y - 1| + x = 2$

20. $x^2 + y = 1$

In Exercises 21–28, use the vertical line test to determine if each graph is a function.

21.

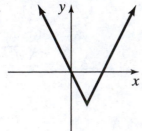

22.

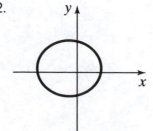

23.

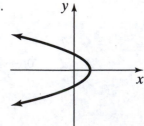

24.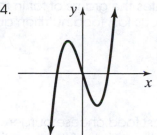

25.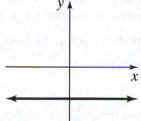

26.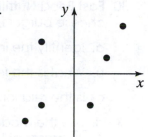

27.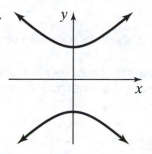

28.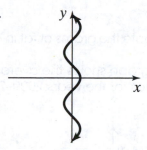

In Exercises 29–32, solve each application problem involving relations and functions.

29. **Population Data** The data in the following table represent the percent of the population with access to clean water and the percent undernourished for several South American countries in 2012. (*Data*: worldbank.org)

% Population with Access to Clean Water	% Undernourished
87	12
99	6
99	5
86	16
91	11

a. If a researcher believes the percent of the population with access to clean water affects the percent undernourished, identify the independent and dependent variables.

b. What are the corresponding ordered pairs for the data?

c. What are the domain and range?

d. Is this relation a function? Explain.

30. **Fast Food Nutrition Facts** The equation $y = 10.8x + 243$ relates the grams of fat in fast food cheese burgers, x, to the number of calories, y. (*Data*: Various fast food nutrition guides)

    a. Identify the independent and dependent variables.

    b. What is the feasible domain?

    c. Is the relation a function? Explain.

    d. Use the model to estimate the number of calories in a fast food cheese burger with 40 grams of fat. What point would this correspond to on the graph of the equation?

    e. Solve the equation for x.

    f. Use your model to estimate the grams of fat in a fast food cheese burger with 621 calories.

31. **Gas Prices** The following graph shows the average price per gallon of regular unleaded gasoline in the United States for the years 2009–2013. (*Source*: U.S. Energy Information Administration, 2014)

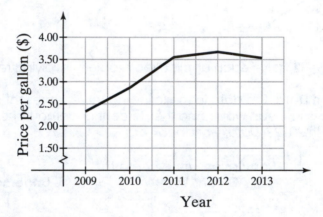

    a. Is this graph a function? Explain.

    b. Identify the independent and dependent variables.

    c. Use the graph to estimate the average price per gallon in 2012.

    d. If the data can be modeled by the equation $y = 0.314x - 0.298$, where x is the number of years since 2000 and y is the price per gallon (in dollars), in what year does the model predict that the price per gallon will reach $5?

32. **Part-Time Employment** The median usual weekly earnings of part-time workers can be modeled by the equation $y = -0.16x^2 + 15.40x - 71.89$, where x is the age of the worker. (*Data*: Bureau of Labor Statistics, 2013)

    a. Use the model to estimate the usual weekly earnings of a 20-year-old part-time worker.

    b. Use the model to estimate the usual weekly earnings of a 60-year-old part-time worker.

    c. What is the y-intercept of the model? Does the y-intercept make sense in the context of the problem?

    d. Sketch a graph of the equation for the restricted domain $16 \leq x \leq 75$ by plotting points.

2.3 Function Notation and Applications

THINGS TO KNOW

Before working through this section, be sure you are familiar with the following concepts:

VIDEO   ANIMATION   INTERACTIVE

You Try It
1. Use Order of Operations to Evaluate Numeric Expressions (Section R.2, Objective 4)

You Try It
2. Evaluate Algebraic Expressions (Section R.3, Objective 1)

You Try It
3. Solve a Formula for a Given Variable (Section 1.5, Objective 1)

OBJECTIVES

1 Express Equations of Functions Using Function Notation

2 Evaluate Functions

3 Graph Simple Functions by Plotting Points

4 Interpret Graphs of Functions

5 Solve Application Problems Involving Functions

....

OBJECTIVE 1   EXPRESS EQUATIONS OF FUNCTIONS USING FUNCTION NOTATION

Functions expressed as equations are often named using letters such as f, g, and h. The symbol $f(x)$ is read as "f of x" and is an example of **function notation**. We can use function notation in place of the **dependent variable** in the equation of a function. For example, the function $y = 2x + 3$ may be written as $f(x) = 2x + 3$.

The symbol $f(x)$ represents the value of the dependent variable (output) for a given value of the **independent variable** (input). For $y = f(x)$, we can interpret $f(x)$ as follows: f is the name of the function that relates the independent variable x to the dependent variable y.

⚠ Do not confuse function notation with multiplication. $f(x)$ does not mean $f \cdot x$.

Equivalent functions represent the same set of **ordered pairs**.

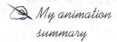
My animation summary

✹ **Concept Animation** Watch this **animation** to explore this concept.

Any equation of a function can be written in function notation.

Expressing Equations of Functions Using Function Notation

Step 1 Choose an appropriate name for the function.

Step 2 Solve the equation for the dependent variable.

Step 3 Replace the dependent variable with equivalent function notation.

 When letters are used to name functions, the case of the letters matters (upper or lowercase). For example, f and F are different symbols, so $f(x)$ and $F(x)$ represent two different functions.

Example 1 Expressing Equations of Functions Using Function Notation

Write each function using **function notation**. Let x be the independent variable and y be the dependent variable.

a. $y = 2x^2 - 4$ b. $y - \sqrt{x} = 0$ c. $3x + 2y = 6$

Solutions

a. Let's name this function f. The equation is already solved for y, so simply replace y with $f(x)$:

Begin with the original formula: $y = 2x^2 - 4$

Replace y with $f(x)$: $f(x) = 2x^2 - 4$

b. Let's name this function g. Solve for y, and then replace y with $g(x)$:

Begin with the original equation: $y - \sqrt{x} = 0$

Solve for y by adding $\sqrt{x}$ to both sides: $y = \sqrt{x}$

Replace y with $g(x)$: $g(x) = \sqrt{x}$

 *My video summary* 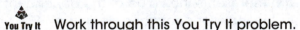 c. Name the function h. Try working through this process yourself. Check your answer, or watch this video to see a fully worked solution.

You Try It Work through this You Try It problem.

Work Exercises 1–10 in this eText or in the MyMathLab Study Plan.

OBJECTIVE 2 EVALUATE FUNCTIONS

For $y = f(x)$, the symbol $f(x)$ represents the value of the **dependent variable** y for a given value of the **independent variable** x. For this reason, we call $f(x)$ the **value of the function**. For example, $f(2)$ represents the value of the function f when $x = 2$. This process is called *evaluating the function*.

To **evaluate a function**, substitute the given value for the independent variable and simplify.

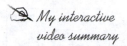 *My interactive video summary*

Example 2 Evaluating Functions

If $f(x) = 4x - 5$, $g(t) = 3t^2 - 2t + 1$, and $h(r) = \sqrt{r} - 9$, evaluate each of the following.

a. $f(3)$ **b.** $g(-1)$ **c.** $h(16)$ **d.** $f\left(\dfrac{1}{2}\right)$

Solutions Work through this **interactive video** for complete solutions, or read the following explanations.

a. The notation $f(3)$ represents the value of the function f when x is 3.

$$\text{Substitute 3 for } x \text{ in the function } f: \quad f(3) = 4(3) - 5$$
$$\text{Simplify:} \qquad = 12 - 5$$
$$= 7$$

So, $f(3) = 7$, meaning that the value of f is 7 when x is 3.

b. The notation $g(-1)$ represents the value of the function g when t is -1.

$$\text{Substitute } -1 \text{ for } t \text{ in the function } g: \quad g(-1) = 3(-1)^2 - 2(-1) + 1$$

Finish the problem by simplifying the right side. Check your **answer**, or watch this **interactive video** to see the fully worked solution.

c. Substitute 16 for r in function h and simplify. Try working this problem on your own. Check your **answer**, or watch this **interactive video** to see the fully worked solution.

d. Substitute $\dfrac{1}{2}$ for x in the function f: $\quad f\left(\dfrac{1}{2}\right) = 4\left(\dfrac{1}{2}\right) - 5$

Finish the problem on your own. Check your **answer**, or watch the **interactive video** to see the fully worked solution.

You Try It Work through this **You Try It** problem.

Work Exercises 11–18 in this eText or in the MyMathLab Study Plan.

OBJECTIVE 3 GRAPH SIMPLE FUNCTIONS BY PLOTTING POINTS

If $f(3) = 7$, then this means that an input of 3 into the function f results in an output of 7. Because this result corresponds to the ordered pair $(3, 7)$, we say that $(3, 7)$ **belongs** to the function f.

The **graph of a function** is the graph of all ordered pairs that belong to the function. Since $f(3) = 7$, the point $(3, 7)$ lies on the graph of f. This leads to the following **theorem**.

Theorem

The point (a, b) lies on the graph of a function f if and only if $f(a) = b$.

Let's graph simple functions by plotting ordered pairs that belong to the function.

Strategy for Graphing Simple Functions by Plotting Points

Step 1 Find several points that belong to the function. The exact number of points to find depends on the function being graphed. Find enough to form a pattern. Try to locate key points such as **endpoints** and **maximum and minimum points**.

Step 2 Plot the points found in Step 1.

Step 3 Connect the points with a straight line or smooth curve, depending on the pattern formed.

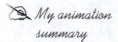 *My animation summary*

⊛ **Example 3 Graphing Simple Functions by Plotting Points**

Graph each function by plotting points.

a. $f(x) = 2x - 1$ **b.** $g(x) = x^2 + 2x - 3$ **c.** $h(x) = 2|x| - 1$

Solutions Read and work through the solutions to parts a–c, or watch this animation to see complete solutions.

a. Step 1 Note that the **domain** of f is the set of all real numbers. Evaluate $f(x)$ for $x = -2, -1, 0, 1$, and 2. See Table 3.

Table 3

x	$y = f(x)$	(x, y)
-2	$f(-2) = 2(-2) - 1 = -5$	$(-2, -5)$
-1	$f(-1) = 2(-1) - 1 = -3$	$(-1, -3)$
0	$f(0) = 2(0) - 1 = -1$	$(0, -1)$
1	$f(1) = 2(1) - 1 = 1$	$(1, 1)$
2	$f(2) = 2(2) - 1 = 3$	$(2, 3)$

Step 2 Plot the points found in Table 3.

Step 3 Do you see a pattern? Connect the points appropriately with a straight line or smooth curve. Check your **answer**, or work through this **animation** to see the complete solution.

b. Step 1 The **domain** of g is the set of all real numbers. Evaluate $g(x)$ for $x = -4, -3, -2, -1, 0, 1$, and 2 to create a table of **ordered pairs** that **belong** to the function. Check your **table**.

Step 2 Plot the points found in Step 1.

Step 3 Connect them with a line or smooth curve as appropriate. Check your **answer**, or work through this **animation** to see the fully worked solution.

c. Try to work this problem on your own. Check your **answer**, or work through this **animation** to see the fully worked solution.

You Try It Work through this You Try It problem.

Work Exercises 19–24 in this eText or in the 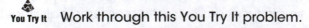 Study Plan.

OBJECTIVE 4   INTERPRET GRAPHS OF FUNCTIONS

Graphs of **functions** can be used when visualizing a variety of everyday situations. The next three examples show how graphs of functions can **model** common situations.

My video summary ▶ **Example 4 Straight-Line Depreciation**

Straight-line depreciation is an accounting method used for deducting losses in the value of equipment as it ages. Claiming a loss in **book value** allows the owner to avoid paying taxes on the amount of the loss. Figure 8 shows the straight-line depreciation of a machine as it ages.

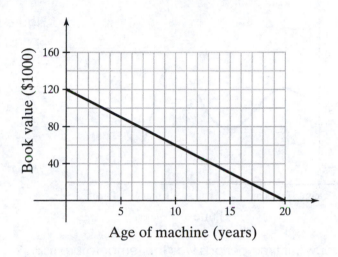

Figure 8
Straight-Line Depreciation

a. What was the book value of the machine when it was new?

b. How old will the machine be when it no longer has a book value?

c. What will the book value of the machine be when it is 5 years old?

d. At what age will the machine have lost half of its original book value?

e. By how much does the book value of the machine decrease each year?

Solutions Read through the following, or watch this **video** for complete solutions. View the **graph** as you work through each part.

a. The graph contains the point $(0, 120)$, which represents the **book value** of the machine when it was new. This book value was $120,000.

b. The point $(20, 0)$ shows that the machine will have a book value of $0 (no book value) after 20 years.

c. The point $(5, 90)$ represents the book value of the machine after 5 years. This book value will be $90,000.

d. From part (a), the original book value of the machine was $120,000. Half of this is $60,000. The point $(10, 60)$ indicates that the machine will have lost half of its original book value when it is 10 years old.

e. Since the machine's original value of $120,000 is lost evenly over a 20-year period, the book value decreases by $\frac{\$120,000}{20 \text{ years}} = \$6000$ each year.

You Try It Work through this You Try It problem.

Work Exercise 25 in this eText or in the MyMathLab Study Plan.

My video summary ▶ **Example 5 Spring Temperatures**

Figure 9 shows a graph of a function that describes the outside temperatures over one 24-hour period in spring.

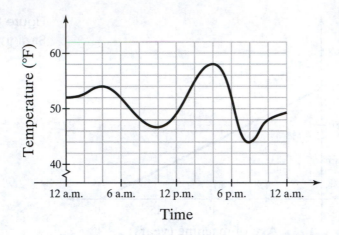

Figure 9
Spring Temperatures

a. Over what time periods was the temperature rising?

b. Over what time periods was the temperature falling?

c. What was the highest temperature for the day? At what time was it reached?

d. What was the lowest temperature for the day shown? At what time was it reached?

e. Over what time period did the temperature decrease most rapidly?

Solutions Try answering these questions on your own. View the **answers**, or watch this **video** for a complete solution.

You Try It Work through this You Try It problem.

Work Exercises 26–28 in this eText or in the MyMathLab Study Plan.

My video summary ▶ **Example 6 Flight Altitude**

A Boeing 757 jet took off and climbed steadily for 20 minutes until it reached an altitude of 18,000 feet. The jet maintained that altitude for 30 minutes. Then it climbed steadily for 10 minutes until it reached an altitude of 26,000 feet. The jet remained at 26,000 feet for 40 minutes. Then it descended steadily for 20 minutes until it reached an altitude of 20,000 feet, where it remained for 30 more minutes. During the final 20 minutes of the flight, the jet descended steadily until it landed at its destination airport. Draw a graph of the 757's altitude as a function of time.

Solution The 757's altitude is a **function** of time, so the **independent variable** *time* (in minutes) is represented by the horizontal axis and the **dependent variable** *altitude* is represented by the vertical axis. To **draw the graph**, identify key points. Plot these points and connect them. For example, the **ordered pair** (0, 0) represents an altitude of 0 feet (ground) at time 0 minutes (before takeoff). The ordered pair (20, 18000) represents an altitude of 18,000 feet at 20 minutes after takeoff. To show the initial climb of the airplane, we connect the points (0, 0) and (20, 18000) with a straight line segment. Continue this process. Check your **answer**, or watch this **video** to see the fully worked solution.

You Try It Work through this You Try It problem.

Work Exercises 29–32 in this eText or in the MyMathLab Study Plan.

OBJECTIVE 5   SOLVE APPLICATION PROBLEMS INVOLVING FUNCTIONS

Functions are used to **model** a variety of real-world applications in fields of study such as physics, biology, business, and economics. As you move through this text, you will learn methods for creating such models. In the next two examples, the models are provided.

My interactive video summary

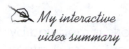 **Example 7 Falling Rock**

A rock is dropped from the top of a cliff. Its height, h, above the ground, in feet, at t seconds is given by the function $h(t) = -16t^2 + 900$. Use the model to answer the following questions.

a. Evaluate $h(0)$. What does this value represent?

b. Evaluate $h(2)$. What does this value represent? How far has the rock fallen at this time?

c. Evaluate $h(10)$. Is this possible? Explain.

d. Evaluate $h(7.5)$. Interpret this result.

e. Determine the **feasible domain** and **feasible range** within the context of the problem.

f. Graph the function.

Solutions Read through the following, or watch this **interactive video** for full solutions to all six parts.

a. $h(0) = -16(0)^2 + 900 = 900$

The height of the rock at $t = 0$ seconds (before the rock is dropped) is 900 feet above the foot of the cliff, so the height of the cliff is 900 feet.

b. $h(2) = -16(2)^2 + 900 = 836$

Two seconds after being dropped, the rock is 836 feet above the ground. Therefore, the rock has fallen $900 - 836 = 64$ feet.

c. $h(10) = -16(10)^2 + 900 = -700$

A negative height means that the rock has fallen below ground level. Our result shows that the rock would be 700 feet below the ground at the base of the cliff 10 seconds after being dropped. This result is not possible (assuming there is not a 700-foot-deep hole at the base of the cliff). Therefore, $t = 10$ seconds is outside the **feasible domain** of the function.

d. $h(7.5) = -16(7.5)^2 + 900 = 0$

After 7.5 seconds, the rock is 0 feet above the ground, or literally on the ground. Therefore, the rock hits the ground at the foot of the cliff 7.5 seconds after being dropped.

e. Since t represents time, its values are nonnegative. From part (d), we know that the rock hits the ground at $t = 7.5$ seconds. This means that function values for times after 7.5 seconds will not make sense. The **feasible domain** is $\{t \mid 0 \leq t \leq 7.5\}$, or $[0, 7.5]$. Height is non-negative. From part (a), we know that $h(0) = 900$, which is the largest possible value for h. Therefore, the **feasible range** is $\{h \mid 0 \leq h \leq 900\}$, or $[0, 900]$.

f. Try to graph this function on your own. Check your **answer**, or watch the **interactive video** for the fully worked solution.

You Try It Work through this You Try It problem.

Work Exercises 33 and 34 in this eText or in the MyMathLab Study Plan.

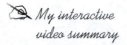 *My interactive video summary*

Example 8 Rent in Queens, New York

The average monthly rent, R, for apartments in Queens, New York, is **modeled by** the function $R(a) = 2.2a$, where a is the floor area of the apartment in square feet. Use the model to answer the following questions. (*Data*: zillow.com, 2014)

a. What is the average monthly rent for apartments in Queens, New York, with a floor area of 800 square feet?

b. What is the floor area of an apartment if its rent is $1430 per month?

c. Determine the **feasible domain** and the **feasible range** of the function.

d. Graph the function.

Solutions Read through the following solutions, or watch this **interactive video** for complete solutions.

a. $R(800) = 2.2(800) = 1760$

The average monthly rent for an 800-square-foot apartment in Queens is $1760.

b. In this case, we solve for a when $R(a) = 1430$.

Rewrite the original function: $R(a) = 2.2a$

Substitute 1430 for $R(a)$: $1430 = 2.2a$

Divide both sides by 2.2: $\dfrac{1430}{2.2} = \dfrac{2.2a}{2.2}$

Simplify: $650 = a$

So, the floor area of an apartment with a monthly rent of $1430 is 650 square feet.

    c. For the situation to make sense, the floor area of an apartment must not be negative. In a practical sense, it is not likely to find apartments with floor areas of 1 square foot or 100,000 square feet. However, there are no definite limits. Therefore, the **feasible domain** is $\{a \mid a \geq 0\}$ or $[0, \infty)$.

    Also, the problem does not make sense if the monthly rent is negative. As with floor area, you are not likely to find apartments with monthly rents of more than $1,000,000. However, there is no definite maximum limit. Therefore, the **feasible range** is $\{R \mid R \geq 0\}$ or $[0, \infty)$.

    d. Try to graph the function on your own. Check your **answer**, or watch this **interactive video** for the fully worked solution.

You Try It   Work through this You Try It problem.

Work Exercises 35 and 36 in this eText or in the MyMathLab **Study Plan.**

2.3 Exercises

In Exercises 1–10, write each function using function notation. Let x be the independent variable and y be the dependent variable. Use the letter f to name each function.

1. $y = |2x - 5|$

2. $y = 3x^2 + 2x - 5$

3. $y + \sqrt{x} = 1$

4. $2x + y = 3$

5. $3x + 4y = 12$

6. $-6x + 18y = 12$

7. $4x^2 - 2y = 10$

8. $3y + 6\sqrt{x - 5} = 0$

9. $\dfrac{3y - 7}{2} = 3x^2 + 1$

10. $\dfrac{5y - 8}{3} = \dfrac{10x^2 + 4}{6}$

In Exercises 11–18, evaluate each function.

11. $f(x) = 3x - 5$; $f(6)$

12. $h(x) = 2x^2 + 5x - 17$; $h(-4)$

13. $F(z) = 2|z - 3| - 5$; $F(0)$

14. $T(t) = \dfrac{5}{6}t + \dfrac{1}{3}$; $T(8)$

15. $r(x) = 3 + \sqrt{x - 5}$; $r(9)$

16. $c(x) = \sqrt{25 - x^2}$; $c(3)$

17. $\Phi(p) = (p - 1)p^3$; $\Phi(3)$

18. $R(x) = 8x^2 - 2x + 1$; $R\left(-\dfrac{1}{2}\right)$

In Exercises 19–24, graph each function by plotting points.

19. $f(x) = 2x - 3$

20. $g(x) = -\dfrac{1}{2}x + 4$

21. $h(x) = x^2 - 4x + 3$

22. $F(x) = |x + 1| - 3$

23. $G(x) = -2|x| + 3$

24. $H(x) = \sqrt{x + 4}$

In Exercises 25–28, solve each application problem.

25. **Straight-Line Depreciation** A rancher bought a new truck and will use the truck's depreciation in book value as a tax write-off. The graph of the function shown illustrates the rancher's plan.

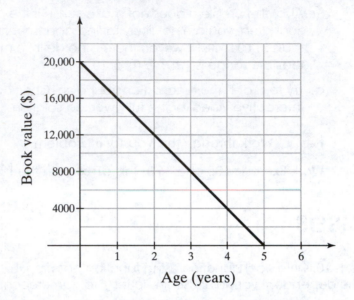

a. What did the rancher pay for the truck?

b. How old will the truck be when its book value is $0?

c. What are the feasible domain and the feasible range of this function?

d. What will the book value be when the truck is 4 years old?

e. How much money (in book value depreciation) can the rancher write off on taxes each year?

26. **Car Rental** The graph of the function shown illustrates the cost of a weekend rental for a full-size car.

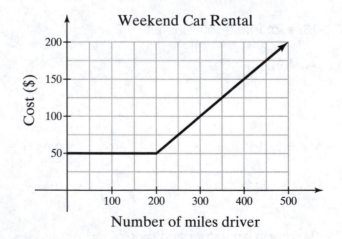

a. Under this plan, what is the cost for the first 200 miles?

b. After the first 200 miles, what is the cost per mile?

c. What is the cost for a 400-mile trip?

d. What are the feasible domain and the feasible range of this function?

e. A second rental option offers unlimited miles for a cost of $175. Describe the number of miles that must be driven to make this second option a better deal.

27. **Dow Jones Industrial Average** The graph of the function shown illustrates the 2013 Dow Jones Industrial Average (DJIA). Each tick mark on the horizontal axis represents the first opening day of the market for that given month.

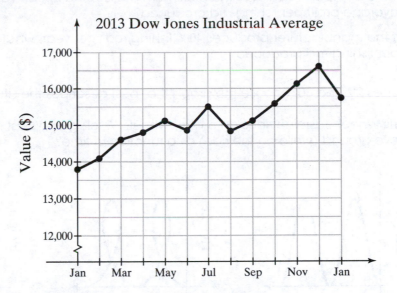

2013 Dow Jones Industrial Average

a. Over which month(s) did the DJIA increase?

b. Over which month(s) did the DJIA decrease?

c. Over which month did the DJIA decrease most rapidly?

d. Approximate the DJIA's value at the beginning of 2013.

e. Approximate its value at the end of 2013.

f. Approximate the average gain per month over 2013.

28. **Fishing Rod Production** A fishing rod manufacturer developed a model for the profit expected when a given number of fishing rods is produced. The graph of the profit function is shown.

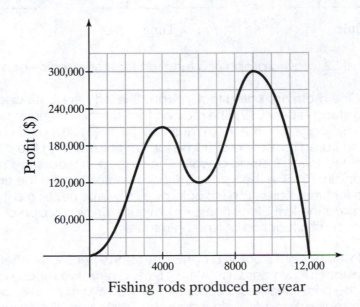

Fishing rods produced per year

a. If the manufacturer produces 7000 fishing rods per year, what is the expected profit?

b. For which interval(s) of fishing rods produced will the expected profit increase?

c. For which interval(s) of fishing rods produced will the expected profit decrease?

2.3 Function Notation and Applications   **2-31**

d. What is the maximum profit that can be expected in a year? How many fishing rods should it produce to achieve this profit?

e. If the company achieves the maximum profit found in part (d), what would be the average profit per fishing rod produced?

f. If the manufacturer produces 4000 fishing rods per year, what would be the average profit per fishing rod produced?

In Exercises 29 and 30, choose the graph that best describes the situation.

29. **Ferris Wheel** Jasmine rides on a Ferris wheel. Which graph best represents Jasmine's height above ground (while on the ride) as a function of time?

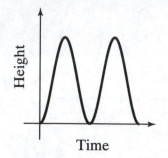 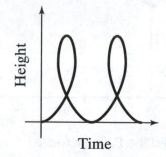 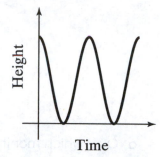

30. **Falling Ball** While playing in his tree house, Tyrone drops a ball out of its window. Which graph best represents the height of the ball above ground as a function of time?

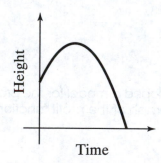 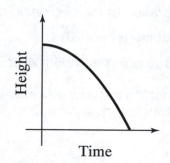 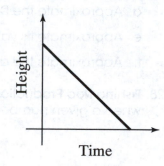

In Exercises 31 and 32, draw a graph of a function that represents each situation.

31. **Exercise** (1) One morning, Karen took a walk. She left her home and walked 4 blocks in 4 minutes at a steady speed. (2) She realized that she did not have her cell phone. Being on call for work, she jogged home in 2 minutes to get it. (3) Just as she arrived home, her cell phone rang, so Karen remained at home for 3 minutes during the call. (4) Now pressed for time, she decided to jog instead of walk. She jogged 10 blocks in 5 minutes when her cell phone rang again. (5) She then stopped for 4 minutes to take the call. (6) Now being even more pressed for time, Karen ran 9 blocks back towards her home in 3 minutes. (7) To cool down, she walked the last block home in 1 minute. Draw a graph of Karen's distance from home (in blocks) as a function of time (in minutes).

32. **Stock Price** (1) Bill bought stock in a company for $10 per share. After his purchase, the price remained constant for two days. (2) At that time, a negative report was released about the company, so the price declined steadily for three days until it lost a total of $6 per share. (3) The price remained steady for four days. (4) Then, a positive report was released about the company, so the price rose sharply for two days, gaining $8 per share. (5) The price continued gaining gradually for four more days, ultimately rising $1 more per share. (6) At that point, the price began to decline slowly for six days, losing $2 per share. Then, Bill sold the stock. Draw a graph of the stock's value (in dollars per share) as a function of the time (in days).

In Exercises 33–36, solve each application problem.

33. **Willis Tower** If an object is dropped from the roof of the Willis Tower (formerly known as the Sears Tower), its height, h, in meters, at t seconds would be given approximately by the function $h(t) = -4.9t^2 + 442.225$.

   a. Evaluate $h(0)$. What does this value represent?

   b. Evaluate $h(5)$. What does this value represent?

   c. Evaluate $h(10)$. Is this possible? Explain.

   d. Evaluate $h(9.5)$. What does this value represent?

   e. Determine the feasible domain and the feasible range of the function in the context of this problem.

   f. Graph the function by plotting points.

34. **Area of a Circle** The area A of a circle with circumference C is given by the function $A(C) = \dfrac{C^2}{4\pi}$.

   a. Evaluate $A(4\pi)$. What does this value represent?

   b. Evaluate $A(12)$. What does this value represent?

   c. Determine the area of a circle with a circumference of 16π (in inches).

   d. Determine the feasible domain and the feasible range of the function in the context of this problem.

   e. Graph the function by plotting points. (Hint: To make graphing easier, substitute multiples of π for C.)

35. **Car Payments** Car companies occasionally offer qualified buyers an incentive of 0% financing and no money down. Ethan just purchased a new automobile under such an agreement. The amount of money, A, that he will still owe after n monthly payments is given by the function, $A(n) = 34{,}515 - 575.25n$.

   a. Evaluate $A(0)$. What does this value represent?

   b. Evaluate $A(24)$. What does this value represent?

   c. Evaluate $A(72)$. Is this possible? Explain.

   d. After 3 years of making payments, how much will Ethan still owe on the car?

   e. How long will it take Ethan to pay off the car?

   f. Determine the feasible domain and the feasible range of the function in the context of the problem.

   g. Graph the function by plotting points.

36. **Gas Mileage** The amount of gas, G (in gallons), that remains in the tank of a hybrid car after it has been driven m miles, starting with a full tank, is given by the function, $G(m) = \dfrac{571.2 - m}{48}$.

   a. Evaluate $G(100)$. What does this value represent?

   b. Evaluate $G(200)$. What does this value represent?

   c. How much gas does a full tank hold?

   d. How much gas will remain in the tank after it has been driven 400 miles, starting with a full tank?

   e. After filling up, how far can the car travel before it runs out of gas?

   f. Determine the feasible domain and the feasible range of the function in the context of this problem.

   g. Graph the function by plotting points.

2.4 Graphs of Linear Functions

THINGS TO KNOW

Before working through this section, be sure you are familiar with the following concepts:

VIDEO   ANIMATION   INTERACTIVE

 You Try It   1.   Graph Equations by Plotting Points (Section 2.1, Objective 4)

 You Try It   2.   Find x- and y-intercepts (Section 2.1, Objective 5)

 You Try It   3.   Graph Simple Functions by Plotting Points (Section 2.3, Objective 3)

OBJECTIVES

1   Graph Linear Functions by Plotting Points

2   Graph Linear Functions by Using Intercepts

3   Graph Vertical and Horizontal Lines

OBJECTIVE 1   GRAPH LINEAR FUNCTIONS BY PLOTTING POINTS

In Example 4 of **Section 2.1**, we graphed the equation $2x + y = 1$, which is an example of a *linear equation in two variables.* Review the **graph**.

> **Definition**   Linear Equation in Two Variables (Standard Form)
>
> A **linear equation in two variables** is an equation that can be written in the standard form $Ax + By = C$, where A, B, and C are real numbers, and A and B are not both equal to 0.

Looking at the graph of $2x + y = 1$, notice that the graph is a **function** because it passes the **vertical line test**. To write the equation in **function notation**, first solve for y to get $y = -2x + 1$. Then replace y with $f(x)$: $f(x) = -2x + 1$. This is an example of a *linear function.*

> **Definition**   Linear Function
>
> A **linear function** is a function of the form $f(x) = ax + b$, where a and b are real numbers.

If we evaluate a linear function for any **real number** x, the result is a real number. This means that the **domain** of every linear function is the set of all real numbers, $\mathbb{R}$, or $(-\infty, \infty)$ in interval notation.

A line can be completely determined by two points. So, to graph a linear function in two variables, we find two **ordered pair solutions**, plot the corresponding points, and connect the points with a line. In practice, we will generally find a third solution to serve as a **check**. If the corresponding third point also lies on the **graph**, we can feel comfortable with our solution.

My video summary ▶ **Example 1** Graphing Linear Functions by Plotting Points

Graph $3x - y = 2$ by plotting points.

Solution Work through the following, or watch this **video** for a complete solution.

First, determine some **ordered pair solutions** to the equation. Select a value for one of the variables and then solve for the remaining variable.

Let $x = 1$:

Original equation:	$3x - y = 2$
Substitute 1 for x:	$3(1) - y = 2$
Simplify:	$3 - y = 2$
Subtract 3 from both sides:	$-y = -1$
Divide both sides by -1:	$y = 1$

Let $y = -2$:

Original equation:	$3x - y = 2$
Substitute -2 for y:	$3x - (-2) = 2$
Simplify:	$3x + 2 = 2$
Subtract 2 from both sides:	$3x = 0$
Divide both sides by 3:	$x = 0$

When $x = 1$, we get $y = 1$. When $y = -2$, we get $x = 0$. So, the ordered pairs $(1, 1)$ and $(0, -2)$ belong to the function and are points on the graph.

The two points are enough to sketch the graph, but we want to find a third point as a **check**.

Let $x = -1$:

Original equation:	$3x - y = 2$
Substitute -1 for x:	$3(-1) - y = 2$
Simplify:	$-3 - y = 2$
Add 3 to both sides:	$-y = 5$
Divide both sides by -1:	$y = -5$

The ordered pairs $(-1, -5)$, $(1, 1)$, and $(0, -2)$ are all solutions to the equation $3x - y = 2$ and points on the graph of this function.

We plot the three points in Figure 10(a). Connecting the points with a line, the complete graph of $3x - y = 2$ is given in Figure 10(b).

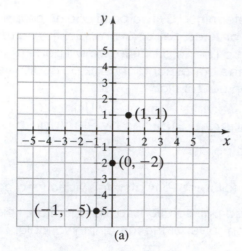

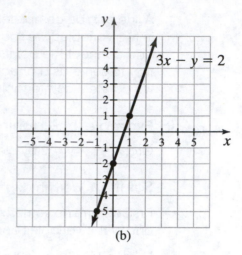

Figure 10

(a)   (b)

🔺 **You Try It**   Work through this You Try It problem.

Work Exercises 1–3 in this eText or in the MyMathLab Study Plan.

✎ *My video summary* ▶ **Example 2 Graphing Linear Functions by Plotting Points**

Graph $f(x) = -\dfrac{3}{5}x + 2$ by plotting points.

Solution Create a table of values by selecting three distinct values for x and evaluating the function for each value. Using multiples of 5 can simplify the computations since the coefficient of x has a denominator of 5. Plot the resulting points and connect them with a straight line. View the **graph**, or watch this video to see a detailed solution.

🔺 **You Try It**   Work through this You Try It problem.

Work Exercises 4–12 in this eText or in the MyMathLab Study Plan.

OBJECTIVE 2 GRAPH LINEAR FUNCTIONS BY USING INTERCEPTS

✎ *My animation summary*   ✳ **Concept Animation** Watch this animation to review the concept of intercepts.

Let's learn how to find intercepts algebraically so that we can use them to graph linear functions.

Because every point on the x-axis has a y-coordinate of 0, to find *x*-intercepts of an equation, we can let $y = 0$ and solve for x. Because every point on the y-axis has an x-coordinate of 0, to find *y*-intercepts, we can let $x = 0$ and solve for y.

Finding x- and y-Intercepts of a Graph Given an Equation

- To find an x-intercept, let $y = 0$ (or $f(x) = 0$) and solve for x.
- To find a y-intercept, let $x = 0$ and solve for y (or evaluate $f(0)$).

Intercepts are often easy to find and **plot** because one of the **coordinates** is 0. This makes them useful points to find when graphing equations or **functions**. Since the graph of a **linear function** can be drawn using only two points, we can use the *x*- and *y*-intercepts along with a third point to check.

My video summary ▶ **Example 3 Graphing Linear Functions by Using Intercepts**

Graph $2x - 5y = 8$ by using **intercepts**.

Solution Watch this **video**, or continue reading for a complete solution.

x-intercept:

Let $y = 0$: $2x - 5(0) = 8$

Simplify: $2x = 8$

Divide both sides by 2: $x = 4$

The *x*-intercept is 4, so the corresponding point is $(4, 0)$.

y-intercept:

Let $x = 0$: $2(0) - 5y = 8$

Simplify: $-5y = 8$

Divide both sides by -5: $y = -\dfrac{8}{5}$

The *y*-intercept is $-\dfrac{8}{5}$, so the corresponding point is $\left(0, -\dfrac{8}{5}\right)$.

Check point:

Let $x = -1$: $2(-1) - 5y = 8$

Simplify: $-2 - 5y = 8$

Add 2 to both sides: $-5y = 10$

Divide both sides by -5: $y = -2$

The corresponding **check point** is $(-1, -2)$.

Plot the three points and connect them with a straight line. The resulting graph is shown in Figure 11.

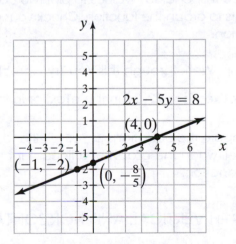

Figure 11

You Try It Work through this You Try It problem.

Work Exercises 13–17 in this eText or in the MyMathLab Study Plan.

My video summary ▶ **Example 4 Graphing Linear Functions by Using Intercepts**

Graph $f(x) = -\dfrac{2}{3}x + 2$ by using intercepts.

Solution Find the *x*-intercept by letting $f(x) = 0$ and solving for *x*. Then find the *y*-intercept by letting $x = 0$ and evaluating $f(0)$. As a check, evaluate the function for a third value, such as $x = -3$, and find the corresponding point. Plot the intercepts and the **check point**. Then connect the points with a straight line. Check your **graph**, or watch this **video** for a detailed solution.

You Try It Work through this You Try It problem.

Work Exercises 18–21 in this eText or in the MyMathLab Study Plan.

The **domain** of a **linear function** is all real numbers, so every linear function contains at least one intercept, the *y*-intercept, because $f(0)$ exists. When the graph of a linear function has only one intercept $(0, 0)$, we need to find at least one more point that is not an intercept in order to graph the function.

My video summary ▶ **Example 5 Graphing Linear Functions by Using Intercepts**

Graph $6x = 4y$ by using **intercepts**.

Solution Find the *x*-intercept by letting $y = 0$ and solving for *x*. Find the *y*-intercept by letting $x = 0$ and solving for *y*.

Let $y = 0$:	$6x = 4(0)$	Let $x = 0$:	$6(0) = 4y$
Simplify:	$6x = 0$	Simplify:	$0 = 4y$
Divide by 6:	$x = 0$	Divide by 4:	$0 = y$
	The *x*-intercept is 0.		The *y*-intercept is 0.

The graph has only one intercept: $(0, 0)$. Select two more values for *x* and find the corresponding *y*-values to plot two additional points on the graph. Use the points to graph the function. Check your **graph**, or watch this **video** for a detailed solution.

You Try It Work through this You Try It problem.

Work Exercises 22–24 in this eText or in the MyMathLab Study Plan.

TIP Compare the equations $2x - 5y = 8$ and $6x = 4y$ from Examples 3 and 5. How can we tell when the graph of a linear function will go through the origin? View this explanation.

OBJECTIVE 3 GRAPH VERTICAL AND HORIZONTAL LINES

When graphed, the special linear equations $x = a$ and $y = b$ are **vertical lines** and **horizontal lines**, respectively.

My animation summary ✦ **Concept Animation** Watch this animation to explore these concepts.

My video summary ⊙ **Example 6** Graphing Horizontal Lines

Graph $y = 2$.

Solution Watch this **video** or continue reading for a complete solution.

The equation $y = 2$ can be written as $0x + y = 2$ to show that it is a **linear equation in two variables**. For any **real number** x in this equation, y will always equal 2. Therefore, every point on the graph of the equation will have a **y-coordinate** of 2. If we choose x-values of -3, 0, and 4, the corresponding ordered pairs are $(-3, 2)$, $(0, 2)$, and $(4, 2)$. The resulting graph, shown in Figure 12, is a horizontal line with **y-intercept** 2 and no **x-intercept**.

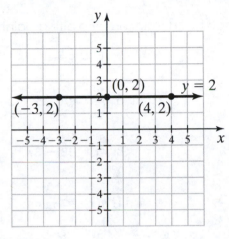

Figure 12

👆
You Try It Work through this You Try It problem.

Work Exercises 25–27 in this eText or in the My MathLab **Study Plan.**

In all of the examples so far, the graphs of **linear functions** have had one or two intercepts. When does the graph have an **infinite** number of intercepts? View this **explanation.**

My video summary ⊙ **Example 7** Graphing Vertical Lines

Graph $x = -4$.

Solution The equation $x = -4$ can be written as $x + 0y = -4$ to show that it is a linear equation in two variables. For any **real number** y in this equation, x will always equal -4. Therefore, every point on the graph of the equation will have an **x-coordinate** of -4. Choose three values for y and find the corresponding points. Plot the points and connect them with a line. Check your **graph,** or watch this **video** to see a detailed solution.

👆
You Try It Work through this You Try It problem.

Work Exercises 28–30 in this eText or in the My MathLab **Study Plan.**

Are horizontal lines functions? What about vertical lines? Sketch a graph for each type of line and apply the **vertical line test** to find out. View this **explanation** to check.

> **Graphs of Vertical and Horizontal Lines**
>
> The graph of $y = b$ is a **horizontal line** through the point $(0, b)$. The constant b is the **y-intercept** of the graph. Horizontal lines are **functions** and can be written in **function notation** as $f(x) = b$.
>
> The graph of $x = a$ is a **vertical line** through the point $(a, 0)$. The constant a is the **x-intercept** of the graph. Vertical lines are not functions.

2.4 Exercises

In Exercises 1–12, graph each function by plotting points.

1. $y = 2x - 4$

2. $3x - 2y = -16$

3. $0.2x + 0.1y = 2$

4. $f(x) = -4x + 1$

5. $f(x) = 0.25x + 2$

6. $g(x) = \dfrac{3x - 1}{4}$

7. $g(x) = -\dfrac{2}{5}x - 3$

8. $4x = 12 - 3y$

9. $2x + y = 6$

10. $h(x) = \dfrac{3}{4}x - 4$

11. $h(x) = -\dfrac{3}{2}x$

12. $2x + 5y = 3$

In Exercises 13–24, graph each function by using intercepts.

13. $x - 3y = -9$

14. $3x + y = 6$

15. $y = 3.2x + 1.5$

16. $-x + 2y = 5$

17. $1.2x - 2.4y = 4.8$

18. $f(x) = x - 5$

19. $g(x) = -4x - 3$

20. $f(x) = -\dfrac{2}{5}x + 1$

21. $g(x) = \dfrac{1}{3}x + \dfrac{1}{4}$

22. $y = 3x$

23. $5y = -3x$

24. $f(x) = \dfrac{1}{2}x$

In Exercises 25–30, graph each equation.

25. $y = 5$

26. $f(x) = 3$

27. $6y + 12 = 0$

28. $x = 3$

29. $x = -2$

30. $2x - 5 = 0$

2.5 Linear Equations in Two Variables

THINGS TO KNOW

Before working through this section, be sure you are familiar with the following concepts:

VIDEO ANIMATION INTERACTIVE

You Try It

1. Solve a Formula for a Given Variable (Section 1.5, Objective 1)

You Try It
2. Find x- and y-Intercepts
(Section 2.1, Objective 5)

You Try It
3. Graph Linear Functions by Plotting
Points (Section 2.4, Objective 1)

OBJECTIVES

1 Find the Slope of a Line

2 Find the Slopes of Horizontal and Vertical Lines

3 Find the Slope and y-Intercept of a Line from an Equation

4 Graph a Line Using the Slope and a Point

5 Determine the Relationship between Two Lines

6 Write the Equation of a Line from Given Information

7 Write Equations of Parallel and Perpendicular Lines

8 Use Linear Models to Solve Application Problems; Direct Variation

---

OBJECTIVE 1 FIND THE SLOPE OF A LINE

Although the graph of every linear equation is a straight line, there can be many differences between the graphs. A key feature of a line is its **slant** or **steepness**. We measure the slant or steepness of a line using *slope*.

My animation summary

⊛ **Concept Animation** Watch this animation to explore this concept.

> **Definition** Slope
>
> The **slope** of a line is the **ratio** of the vertical change in y, or **rise**, to the horizontal change in x, or **run**.
>
> $$\text{slope} = \frac{\text{vertical change}}{\text{horizontal change}} = \frac{\text{change in } y}{\text{change in } x} = \frac{\text{rise}}{\text{run}}$$

Slope is traditionally identified with the letter m. The following **formula** for finding slope uses **subscript notation** to identify two general points on the line, (x_1, y_1) and (x_2, y_2). Read x_1 as "x sub one" and y_2 as "y sub two."

> **Slope Formula**
>
> Given two points, (x_1, y_1) and (x_2, y_2), on the graph of a line, the **slope m** of the line containing the two points is given by the formula
>
> $$m = \frac{\text{Change in } y}{\text{Change in } x} = \frac{\text{Rise}}{\text{Run}} = \frac{y_2 - y_1}{x_2 - x_1},$$
>
> where $x_1 \neq x_2$.

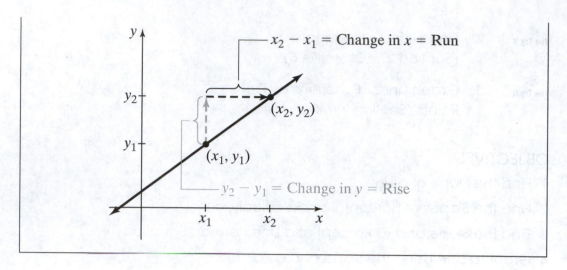

If the graph of a line is rising or **increasing**, then the **slope** is positive. If the graph is falling or **decreasing**, then the slope is negative. If the graph is staying the same, then it is **constant**, and the slope is zero.

My video summary ▶ **Example 1 Finding the Slope of a Line Given Two Points**

Find the slope of the line containing the points $(-2, 1)$ and $(3, 5)$.

Solution Continue reading, or watch this **video** for a complete solution.

Let $(x_1, y_1) = (-2, 1)$ and $(x_2, y_2) = (3, 5)$. Now substitute the **coordinates** into the slope formula:

$$m = \frac{\text{rise}}{\text{run}} = \frac{\text{change in } y}{\text{change in } x} = \frac{y_2 - y_1}{x_2 - x_1}$$

$$= \frac{5 - 1}{3 - (-2)} = \frac{5 - 1}{3 + 2} = \frac{4}{5}$$

The slope of the line is $\frac{4}{5}$. This means that for every horizontal change, or **run**, of

5 units, there is a corresponding vertical change, or **rise**, of 4. Since the slope is positive, the graph of the line **increases** from left to right. The slope of this line is shown in Figure 13.

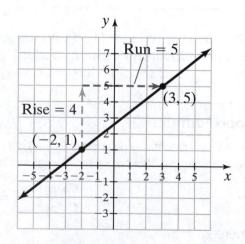

Figure 13

You Try It Work through this You Try It problem.

Work Exercises 1 and 2 in this eText or in the MyMathLab Study Plan.

TIP When the **slope** is a fraction, we consider the fraction as rise over run. How do we interpret the slope $m = 3$? View this **explanation**.

My video summary ▶ **Example 2 Finding the Slope of a Line Given Two Points**

Find the slope of the line containing the points $(-4, 5)$ and $(3, -4)$.

Solution Substitute the values for the coordinates into the **slope formula** to find the slope on your own. Check your **answer**, or watch this **video** for a detailed solution.

 You Try It Work through this You Try It problem.

Work Exercises 3–10 in this eText or in the MyMathLab Study Plan.

Would the result from Example 2 be different if we switched the order of the two points? View this **explanation**.

OBJECTIVE 2 FIND THE SLOPES OF HORIZONTAL AND VERTICAL LINES

Let's look at the **slopes** of lines that do not slant upward or downward from left to right.

My video summary ▶ **Example 3 Finding the Slope of a Line Given Two Points**

Find the slope of the line containing the given points. Simplify if possible.

a. $(-3, 2)$ and $(1, 2)$ **b.** $(4, 2)$ and $(4, -5)$

Solutions Read through the following, or watch this **video** for detailed solutions to both parts.

a. Use the **slope formula**. Let $(x_1, y_1) = (-3, 2)$ and $(x_2, y_2) = (1, 2)$.

Write the slope formula: $\quad m = \dfrac{y_2 - y_1}{x_2 - x_1}$

Substitute $x_1 = -3, y_1 = 2, x_2 = 1$, and $y_2 = 2$: $\quad = \dfrac{2 - 2}{1 - (-3)}$

Simplify: $\quad = \dfrac{0}{4}$

$\quad = 0$

The slope of the line is 0. Figure 14 shows that the line through these points is horizontal. It does not slant upward or downward.

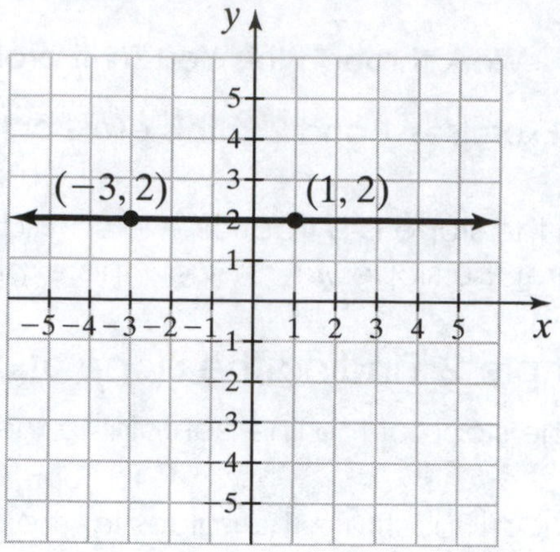

Figure 14
Slope 0

b. Let $(x_1, y_1) = (4, 2)$ and $(x_2, y_2) = (4, -5)$.

Write the slope formula: $\quad m = \dfrac{y_2 - y_1}{x_2 - x_1}$

Substitute $x_1 = 4$, $y_1 = 2$, $x_2 = 4$, and $y_2 = -5$: $\quad = \dfrac{-5 - 2}{4 - 4}$

Simplify: $\quad = \dfrac{-7}{0}$

Division by 0 is undefined, so this line has an *undefined slope*. Figure 15 shows that the graph of the line through these points is vertical. It does not slant upward nor downward.

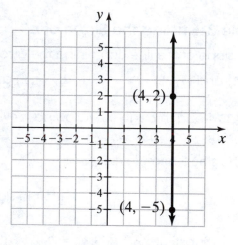

Figure 15
Undefined Slope

![You Try It] **You Try It** Work through this You Try It problem.

Work Exercises 11–14 in this eText or in the MyMathLab Study Plan.

Horizontal lines have no vertical change. This makes the **numerator** in the slope formula 0, which makes the slope of any horizontal line 0. Similarly, **vertical lines** have no horizontal change. This makes the **denominator** of the slope formula 0, which makes the slope of any vertical line undefined.

Slopes of Horizontal and Vertical Lines

All horizontal lines (which have equations of the form $y = b$) have **slope 0**.

All vertical lines (which have equations of the form $x = a$) have **undefined slope**.

 Avoid using the term "no slope." Technically, "no slope" means undefined slope (vertical line), but it can easily be confused with zero slope (horizontal line). Therefore, it is better to clearly state "zero slope" or "undefined slope" and avoid "no slope."

Figure 16 summarizes the relationship between the **slope** and the graph of a linear equation.

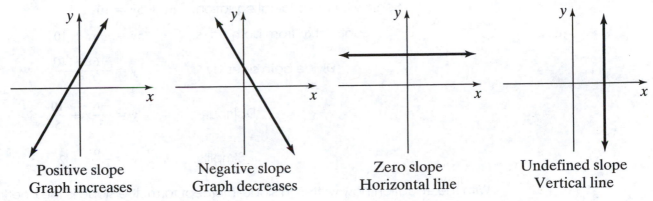

Positive slope
Graph increases

Negative slope
Graph decreases

Zero slope
Horizontal line

Undefined slope
Vertical line

Figure 16
Types of Lines

OBJECTIVE 3   FIND THE SLOPE AND y-INTERCEPT OF A LINE FROM AN EQUATION

In **Example 4** of Section 2.4, we used intercepts to plot the graph of the linear function $f(x) = -\frac{2}{3}x + 2$ or $y = -\frac{2}{3}x + 2$. The x-intercept is 3, and the y-intercept is 2, so the points $(3, 0)$ and $(0, 2)$ are located on the graph of the line. Using these two points, we can find the **slope** of the line, $m = \dfrac{2 - 0}{0 - 3} = \dfrac{2}{-3} = -\dfrac{2}{3}$. Notice the slope of this linear function is equal to the **coefficient** of x in the function and the y-intercept is the constant term. This result can be generalized as follows:

Slope-Intercept Form

A linear equation in two variables of the form

$$y = mx + b \text{ or } f(x) = mx + b$$

is written in **slope-intercept form**, where m is the slope of the line and b is the y-intercept.

✎ *My video summary* ▶ **Concept Video** Watch this **video** to see how the slope-intercept form is derived from the slope formula.

⚠ Be careful not to assume that the coefficient of x is always the slope. This is true when the equation is in slope-intercept form, but it is not true in general.

▶ **Example 4** Finding the Slope and y-Intercept of a Line from an Equation

Find the slope and y-intercept of the line $6x + 5y = 10$.

✎ *My video summary* **Solution** Read through the following, or watch this **video** for a detailed solution.

To find the slope and y-intercept of the given line, solve the equation for y and write it in **slope-intercept form**.

$$\text{Begin with the original equation:} \quad 6x + 5y = 10$$

$$\text{Subtract } 6x \text{ from both sides:} \quad 5y = -6x + 10$$

$$\text{Divide both sides by 5:} \quad y = \frac{-6x + 10}{5}$$

$$\text{Distribute:} \quad y = \frac{-6}{5}x + \frac{10}{5}$$

$$\text{Simplify:} \quad y = -\frac{6}{5}x + 2$$

With the equation now written in slope-intercept form, the slope is the **coefficient** of x, $m = -\dfrac{6}{5}$, and the **constant term**, $b = 2$, is the y-intercept.

🔺 **You Try It** Work through this You Try It problem.

Work Exercise 15 in this eText or in the MyMathLab Study Plan.

✎ *My video summary* ▶ **Example 5** Finding the Slope and y-Intercept of a Line from an Equation

Find the slope and y-intercept of the line $4x - 2y = -14$.

Solution Try to work this problem on your own. View the **answer**, or watch this **video** for a detailed solution.

🔺 **You Try It** Work through this You Try It problem.

Work Exercises 16–22 in this eText or in the MyMathLab Study Plan.

Not every linear equation in two variables can be written in **slope-intercept-form**. Think about what lines cannot be written in this form. Then view this **example** of such a line.

OBJECTIVE 4 GRAPH A LINE USING THE SLOPE AND A POINT

My animation summary

Section 2.4 taught us that only two points are required to find the graph of a line. Given one point and the **slope**, additional points on the graph can be found.

⊛ **Concept Animation** Watch the animation to explore this concept.

⊙ **Example 6** Graphing a Line Using the Slope and a Point

Graph the line that has slope $m = \dfrac{3}{2}$ and passes through the point $(1, -2)$.

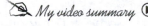
My video summary

Solution Read below, or watch this **video** for a complete solution.

Note that the slope is positive, so the line will slant upward from left to right. The slope is $m = \dfrac{3}{2}$, so rise = 3 and run = 2. Plot the given point $(1, -2)$. From this point, move up (rise) 3 units and move right (run) 2 units. This brings us to a second point on the line, $(3, 1)$. Draw the line between $(1, -2)$ and $(3, 1)$, as shown in Figure 17.

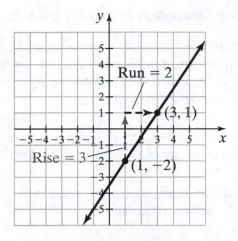

Figure 17

You Try It Work through this You Try It problem.

Work Exercises 23 and 24 in this eText or in the MyMathLab Study Plan.

TIP If desired, additional points on the line can be plotted using the slope. View this **popup** to see how to find and plot more points for Example 6.

My video summary ⊙ **Example 7** Graphing a Line Using the Slope and a Point

Graph the line that has slope $m = -3$ and passes through the point $(2, -1)$.

Solution The slope is negative, so the line will slant downward from left to right. The slope is $m = -3 = \dfrac{-3}{1} = \dfrac{3}{-1}$. We can use either rise = -3 and run = 1, or rise = 3 and run = -1. Try to graph this line on your own. View the **graph**, or watch this **video** for a detailed solution.

You Try It Work through this You Try It problem.

Work Exercises 25–28 in this eText or in the MyMathLab Study Plan.

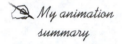 *My video summary* ▶ **Example 8 Graphing a Linear Equation Using Slope-Intercept Form**

Graph the equation $2x + 3y = 9$ using the **slope** and *y*-intercept.

Solution Start by writing the equation in **slope-intercept form**, $y = -\frac{2}{3}x + 3$. The slope is $m = -\frac{2}{3}$ and the *y*-intercept is $b = 3$, so the point $(0, 3)$ is on the graph of the equation. Use this point and the slope to sketch the graph of the equation. Compare your graph to the **answer**, or watch this **video** for a detailed solution.

You Try It Work through this You Try It problem.

Work Exercises 29–34 in this eText or in the MyMathLab Study Plan.

OBJECTIVE 5 DETERMINE THE RELATIONSHIP BETWEEN TWO LINES

Slopes and *y*-intercepts can be used to determine the relationship between two lines.

 My animation summary ✹ **Concept Animation** Watch this **animation** to explore these relationships.

Parallel lines have the same slope but different *y*-intercepts. **Coinciding lines** have the same slope *and* the same *y*-intercept. **Perpendicular lines** have slopes that are **opposite reciprocals**, so the product of their slopes is -1. **Intersecting lines** have different slopes.

⚠ Perpendicular lines do intersect. We consider them separately from **only intersecting lines** because they intersect in a unique way (right angles).

Table 4 summarizes how the slope and *y*-intercept allow us to determine the relationship between two lines.

Table 4 Relationship between Two Lines

	Parallel Lines	Coinciding Lines	Intersecting and Perpendicular Lines	Only Intersecting Lines (not Perpendicular)
Slopes are	Same	Same	*opposite-reciprocals* $\left(m_1 \cdot m_2 = -1 \text{ or } m_1 = -\frac{1}{m_2}\right)$	Different (but $m_1 \cdot m_2 \neq -1$)
y-Intercepts are	Different	Same	Same or Different	Same or Different

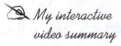 *My interactive video summary* 🐾 **Example 9 Determining the Relationship between Two Lines**

For each pair of lines, determine if the lines are **parallel**, **perpendicular**, **coinciding**, or **only intersecting**.

a. $3y = -4x + 10$
   $16x + 12y = 27$

b. $y = \frac{3}{2}x - 7$
   $6x - 4y = 28$

c. $2x - y = 3$
   $3x + 6y = 5$

d. $5x + 2y = 8$
   $x - 3y = 7$

Solutions

a. Begin by writing each equation in **slope-intercept form**.

Original equation: $3y = -4x + 10$

Divide by 3: $y = -\dfrac{4}{3}x + \dfrac{10}{3}$

Original equation: $16x + 12y = 27$

Subtract 16x: $12y = -16x + 27$

Divide by 12: $y = -\dfrac{4}{3}x + \dfrac{9}{4}$

The slope-intercept forms of the equations show that the two lines have the same slope, $-\dfrac{4}{3}$, so the lines are either **parallel** or **coinciding**. The lines have different y-intercepts, $\dfrac{10}{3}$ and $\dfrac{9}{4}$. Therefore, these lines are parallel because they have the same slope but different y-intercepts.

b.–d. Try to work these problems on your own. Check your **answers**, or watch this **interactive video** for detailed solutions to all four parts.

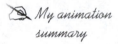 **You Try It**   Work through this You Try It problem.

Work Exercises 35–40 in this eText or in the MyMathLab Study Plan.

OBJECTIVE 6   WRITE THE EQUATION OF A LINE FROM GIVEN INFORMATION

My animation summary

How to find the equation of a line depends on what information is known. For example, if the **slope** of a line and a point on the line are known, the equation can be found by using the **point-slope form**.

> **Point-Slope Form**
>
> Given the slope m of a line and a point (x_1, y_1) on the line, the **point-slope form** of the equation of the line is given by
>
> $$y - y_1 = m(x - x_1).$$

✺ **Concept Animation** Watch this animation to see how the point-slope form is derived.

To find the equation of a line, the slope and a point on the line are required. With this information, use the point-slope form to write the equation of the line. Typically, at least one point on the line will be known, but the slope may have to be found by using the **slope formula** and a second point, or by using other given information (such as that the line is **parallel** to some given line).

My video summary ▶ **Example 10 Writing the Equation of a Line Given the Slope and the y-Intercept**

Write the equation of the line with slope $m = 6$ and passing through the point $(0, 4)$.

Solution Watch this **video**, or continue reading for a complete solution.

The slope $m = 6$ and a point $(0, 4)$ are known. Notice that the x-coordinate of the point is 0 so the y-coordinate, 4, is the y-intercept, b. Using the **slope-intercept**

(eText Screens 2.5-1–2.5-38)

form, $y = mx + b$, substitute in the values of the slope and y-intercept to obtain the equation $y = 6x + 4$.

You Try It Work through this You Try It problem.

Work Exercises 41–44 in this eText or in the MyMathLab Study Plan.

My video summary ▶ **Example 11 Writing the Equation of a Line Given the Slope and a Point**

Find the equation of the line with slope $m = \dfrac{2}{5}$ and passing through the point $(5, 8)$. Write the answer in **slope-intercept** form.

Solution Watch this **video**, or continue reading for a complete solution.

Given the slope $m = \dfrac{2}{5}$ and a point on the graph, $(x_1, y_1) = (5, 8)$, let's use the **point-slope form**, $y - y_1 = m(x - x_1)$. Substitute the value of the **slope** and the **coordinates** of the point to obtain the equation.

$$\text{Point-slope form:} \quad y - 8 = \frac{2}{5}(x - 5)$$

$$\text{Distribute:} \quad y - 8 = \frac{2}{5}x - 2$$

$$\text{Add 8 to both sides:} \quad y = \frac{2}{5}x + 6$$

The equation in slope-intercept form is $y = \dfrac{2}{5}x + 6$.

You Try It Work through this You Try It problem.

Work Exercises 45–48 in this eText or in the MyMathLab Study Plan.

Given the slope m and a point (x_1, y_1), the **slope-intercept form** can be used as an alternative to the **point-slope form** to find the equation of the line.

My video summary ▶ To see an alternative solution of Example 11 solved using the slope-intercept form, view this **popup** or watch this **video**.

My interactive video summary **Example 12 Writing the Equation of a Line Given Two Points**

Write the equation of the line passing through the points $(3, -8)$ and $(5, -2)$.

Solution Two points on the line are given but not the slope. Compute the **slope** by using the **slope formula**: $m = \dfrac{-2 - (-8)}{5 - 3} = \dfrac{-2 + 8}{5 - 3} = \dfrac{6}{2} = 3$. Now that the slope is known, use it together with *either* of the two given points to write the equation of the line. Choose one of the points and find the equation of the line. Check your **answer**, or watch this **interactive video** for a detailed solution.

(eText Screens 2.5-1–2.5-38)

 You Try It Work through this You Try It problem.

Work Exercises 49–52 in this eText or in the MyMathLab Study Plan.

TIP When more than one point on a line is known, it does not matter which point is used in the point-slope form or the slope-intercept form to find the equation of the line.

My video summary ▶ **Example 13 Writing the Equation of a Vertical or Horizontal Line**

Write the equation of a **vertical line** passing through the point (4, 10), and then write the equation of a **horizontal line** passing through the same point.

Solution Watch this video, or read through the following for a complete solution.

Vertical lines have the form $x = a$, which means that every point on the line has the same x-coordinate, a. The line passes through (4, 10), which has x-coordinate 4. Therefore, $a = 4$. The equation of the vertical line through (4, 10) is $x = 4$.

Horizontal lines have the form $y = b$, which means that every point on the line has the same y-coordinate, b. The line passes through (4, 10), which has y-coordinate 10. Therefore, $b = 10$. The equation of the horizontal line through (4, 10) is $y = 10$.

 You Try It Work through this You Try It problem.

Work Exercises 53–56 in this eText or in the MyMathLab Study Plan.

OBJECTIVE 7 WRITE EQUATIONS OF PARALLEL AND PERPENDICULAR LINES

At this point, you may want to review **parallel lines** and **perpendicular lines**. When writing the equation of a parallel line or perpendicular line, the key ideas to remember are that parallel lines have the same **slope** and perpendicular lines have **opposite-reciprocal** slopes.

My interactive video summary 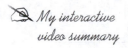 **Example 14 Writing the Equation of a Perpendicular Line**

Write the equation of the line that passes through the point (−3, 1) and is perpendicular to $7x − 3y = 2$.

Solution A point on the graph of the line is known but not the slope. However, the line is perpendicular to $7x − 3y = 2$, which provides information about the slope. Perpendicular lines have opposite-reciprocal slopes, so determine the slope of the given line first.

$$\text{Original equation:} \quad 7x - 3y = 2$$
$$\text{Subtract } 7x \text{ from each side:} \quad -3y = -7x + 2$$
$$\text{Divide both sides by } -3: \quad y = \frac{7}{3}x - \frac{2}{3}$$

The slope of the given line is $\frac{7}{3}$. The **opposite reciprocal** is $-\frac{3}{7}$, so the slope of the desired line is $-\frac{3}{7}$. Now that the slope and a point on the graph are known, follow the process for writing the equation of a line as in the previous examples. View the **answer**, or watch this **interactive video** for a detailed solution.

You Try It Work through this You Try It problem.

Work Exercises 57–58 in this eText or in the MyMathLab Study Plan.

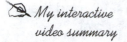
My interactive video summary

Example 15 Writing the Equation of a Parallel Line

Write the equation of the line that passes through the point $(3, -2)$ and is **parallel** to $y = -3x + 5$.

Solution Try to find this equation on your own. Check your **answer**, or watch this **interactive video** for the detailed solution.

You Try It Work through this You Try It problem.

Work Exercises 59–60 in this eText or in the MyMathLab Study Plan.

Table 5 summarizes the different forms for equations of lines.

Table 5 Equations of Lines

$m = \dfrac{y_2 - y_1}{x_2 - x_1}$	**Slope** Rate of change
$y - y_1 = m(x - x_1)$	**Point-Slope Form** Slope is m, and (x_1, y_1) is a point on the line.
$y = mx + b$	**Slope-Intercept Form** Slope is m, and y-intercept is b.
$Ax + By = C$	**Standard Form** A, B, and C are real numbers, with A and B not both zero and $A \geq 0$.
$y = b$	**Horizontal Line** Slope is zero, and y-intercept is b.
$x = a$	**Vertical Line** Slope is undefined, and x-intercept is a.
Perpendicular Lines	Opposite-Reciprocal Slopes $(m_1 \cdot m_2 = -1)$
Parallel Lines	Same slope, different y-intercept

OBJECTIVE 8 USE LINEAR MODELS TO SOLVE APPLICATION PROBLEMS; DIRECT VARIATION

Graphs that display data as a set of points are called **scatter plots**. For example, Figure 18 shows a scatter plot of the percent of Internet users in the years 2010–2014 who use Chrome as their primary browser (*Data*: statcounter.com).

Figure 18

Years after 2010

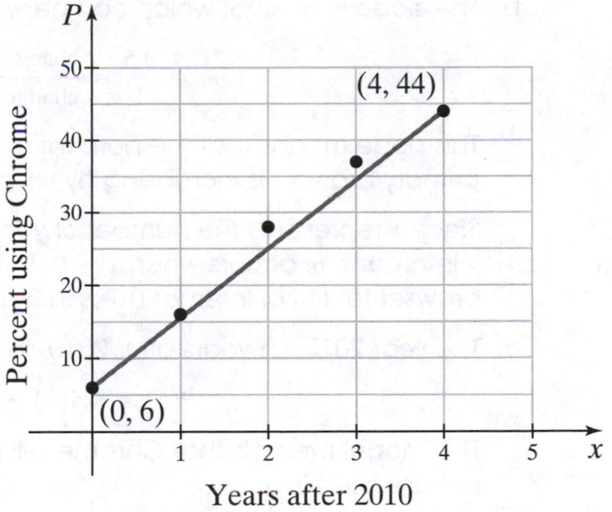

Scatter plots can be used to show overall trends in data. When a trend is identified, analysts summarize the data by fitting an appropriate **model** to the data. A linear model $y = mx + b$ can help to summarize data that fall on or near a line such as the data in Figure 18. Once the model is fit, statements can be made about data trends and the model can be used to make predictions.

Example 16 Primary Web Browser Chrome

a. Use the points labeled in **Figure 18** to write a **function** of the form $P(x) = mx + b$ that models the percent, P, of Internet users who use Chrome as their primary browser x years after 2010.

b. Interpret the **slope** and the y-intercept in the context of the problem.

c. Use the model to predict the percent of Internet users using Chrome as their primary browser in 2015.

Solution Read through the following, or watch this **video** for a complete solution.

a. Looking at the scatter plot, note that the line passes through the points $(0, 6)$ and $(4, 44)$. Using these points, find the slope:

$$m = \frac{44 - 6}{4 - 0} = \frac{38}{4} = \frac{19}{2} \text{ or } 9.5$$

With this slope and the point $(4, 44)$, use the **point-slope form** of the equation of a line to obtain

Point-slope form: $\quad y - y_1 = m(x - x_1)$

Substitute in the slope and point: $\quad y - 44 = 9.5(x - 4)$

Distribute: $\quad y - 44 = 9.5x - 38$

Add 44 to both sides: $\quad y = 9.5x + 6$

Write in function form: $\quad P(x) = 9.5x + 6$

Notice that the y-intercept is 6, which could have been recognized from the given point $(0, 6)$.

b. The **slope** is $m = 9.5$, which can be written as

$$m = \frac{9.5}{1} = \frac{\text{change in } y}{\text{change in } x} = \frac{\text{change in percent}}{\text{change in years}}$$

This slope means that the percent of Internet users who use Chrome as their primary browser is increasing by 9.5 percentage points each year.

Since x represents the number of years after 2010, $x = 0$ represents 2010. The *y*-intercept, 6, occurs when $x = 0$. This means that Chrome was the primary browser for 6% of Internet users in 2010.

c. The year 2015 is 5 years after 2010, so $x = 5$. Substitute 5 for x in $P(x) = 9.5x + 6$:

$$P(5) = 9.5(5) + 6 = 47.5 + 6 = 53.5$$

The **model** predicts that Chrome will be the primary browser for 53.5% of Internet users in 2015.

You Try It Work through this You Try It problem.

Work Exercises 61 and 62 in this eText or in the MyMathLab Study Plan.

Another application of linear models is called **direct variation**. Direct variation means that one variable is a constant multiple of another.

Direct Variation

A linear model of the form

$$y = kx,$$

means that y **varies directly** with x, or y is *proportional to* x. The constant k is called the **constant of variation** or the *proportionality constant*.

We can use the **strategy** for solving application problems involving linear equations in the following examples.

My video summary ▶ **Example 17 Water Pressure on a Scuba Diver**

During scuba diving, water pressure varies directly with the depth of the diver. A diver 40 feet below the surface will face 17.8 pounds per square inch (psi) of pressure from the water. How much water pressure will a diver face if she is 100 feet below the surface?

Solution Watch this video, or continue reading for a complete solution.

Step 1 We want to find the water pressure on the diver at 100 feet below the surface if the pressure is 17.8 psi at 40 feet below the surface. We know that pressure **varies directly** with the depth of the diver.

Step 2 Let p = water pressure (psi) and d = depth (feet).

Step 3 Since water pressure p varies directly with depth d, write the relation as

$$p = kd,$$

where k is the **constant of variation**. To determine k, use the fact that the water pressure is 17.8 psi at a depth of 40 feet.

$$p = kd$$

$$17.8 = k(40)$$

$$\frac{17.8}{40} = k$$

$$0.445 = k$$

So, the constant of variation is 0.445. The equation is $p = 0.445d$.

Step 4 To determine the water pressure at a depth of 100 feet, substitute 100 for d and solve for p.

$$p = 0.445d$$

$$p = 0.445(100)$$

$$p = 44.5$$

Step 5 The constant of variation (0.445) is slightly less than $\frac{1}{2}$, so the numeric value for water pressure (44.5) should be slightly less than half the numeric value for depth (100). Since 44.5 is slightly less than half of 100, the result seems reasonable.

Step 6 The diver will experience 44.5 psi of water pressure if she is 100 feet below the surface.

You Try It   Work through this You Try It problem.

Work Exercise 63 in this eText or in the MyMathLab Study Plan.

My video summary ▶ **Example 18 Simple Interest Earned**

Simple interest after 1 year varies directly with the amount of **principal**. If $3,200 earns $240 in **interest**, then how much interest will $4,800 earn after one year? (**Note:** In this problem, the **constant of variation** represents the annual interest rate in decimal form.)

Solution Try to solve this problem on your own. Check your **answer**, or watch this **video** for a detailed solution.

You Try It   Work through this You Try It problem.

Work Exercise 64 in this eText or in the MyMathLab Study Plan.

How is direct variation related to the slope-intercept form of a line? See this popup for the details.

2.5 Exercises

In Exercises 1–14, find the slope of the line containing the given points.

1. $(-5, 2)$ and $(2, -6)$

2. $(0, -7)$ and $(-4, -9)$

3. $(-1, 4)$ and $(3, 7)$

4. $(1, -2)$ and $(3, 2)$

5. $(-3, 5)$ and $(2, -1)$

6. $(-1, 7)$ and $(5, 3)$

7. $(5, 0)$ and $(-1, -3)$

8. $(2, 13)$ and $(8, 5)$

9.

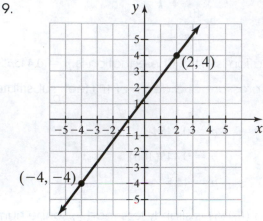

10.

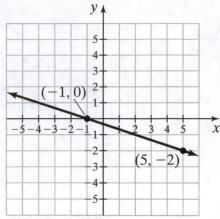

11. $(4, -2)$ and $(-1, -2)$

12. $(5, -3)$ and $(5, 4)$

13. $\left(-\dfrac{3}{2}, -\dfrac{1}{4}\right)$ and $\left(\dfrac{5}{3}, -\dfrac{1}{4}\right)$

14. $(4, 9)$ and $(4, 5)$

In Exercises 15–22, find the slope and y-intercept for each equation.

15. $4x - y = 12$

16. $3x + y = 9$

17. $6x + 5y = -4$

18. $2x - 5y = 3$

19. $5x + 7y = 12$

20. $3x - 8y = 0$

21. $y = 10$

22. $x = -2$

In Exercises 23–28, graph the line given the slope and a point on the line.

23. Slope $= \dfrac{1}{2}$; $(-2, -2)$

24. Slope $= -\dfrac{4}{3}$; $(3, 8)$

25. $m = 2$; $(0, 4)$

26. $m = -\dfrac{3}{4}$; $(0, 0)$

27. $m = 0$; $(4, -3)$

28. Undefined slope; $(-2, 5)$

In Exercises 29–34, find the slope and y-intercept and use them to graph the linear function.

29. $f(x) = \dfrac{1}{3}x + 3$

30. $f(x) = -\dfrac{3}{5}x + 4$

31. $y = 3x + 2$

32. $y = -\dfrac{1}{2}x$

33. $5x - 3y = -12$

34. $2x + 3y = 4$

In Exercises 35–40, determine if the two lines are parallel, perpendicular, coinciding, or only intersecting.

35. $-4x - 2y = -9$
    $5x - 10y = 7$

36. $y = 3x - 2$
    $9x - 3y = 1$

37. $y = -2$
    $3y = 12$

38. $x = -1$
    $y = -1$

39. $4x - 3y = 7$
    $3x - 4y = -6$

40. $2x + 3y = 3$
    $y = -\dfrac{2}{3}x + 1$

In Exercises 41–60, write the equation of the line from the given information. Write the line in slope-intercept form and standard form.

41. Slope $= 1$; passes through $(0, -2)$

42. Slope $= -\dfrac{1}{6}$; y-intercept $= \dfrac{1}{2}$

43. Slope $= \dfrac{7}{9}$; passes through $\left(0, -\dfrac{8}{7}\right)$

44. Slope $= 0$; y-intercept $= 5$

45. Slope $= 3$; passes through $(-5, 3)$

46. Slope $= -3$; passes through $(-2, 3)$

47. Slope $= -\dfrac{4}{3}$; passes through $(3, 8)$

48. Slope $= -\dfrac{2}{3}$; passes through the origin

49. Passing through $(1, 2)$ and $(-2, 5)$

50. Passing through $(-5, 7)$ and $(3, -5)$

51. Passing through $\left(-\dfrac{1}{2}, 1\right)$ and $\left(-3, \dfrac{2}{3}\right)$

52. Passing through $(-3, 4)$ and $(2, 4)$

53. Write the equation of the horizontal line passing through the point $(5, -2)$.

54. Write the equation of the vertical line passing through the point $(5, -2)$.

55. Slope $= 0$; passes through $(4, 1)$.

56. Undefined slope; passes through $(-5, 9)$

57. Write the equation of the line perpendicular to $y = \dfrac{1}{4}x - 2$ that passes through the point $(-2, 3)$.

58. Write the equation of the line perpendicular to $3x - 5y = 1$ that passes through the point $(1, -4)$.

59. Write the equation of the line parallel to $y = \dfrac{1}{4}x - 2$ that passes through the point $(-2, 3)$.

60. Write the equation of the line parallel to $3x - 5y = 1$ that passes through the point $(1, -4)$.

In Exercises 61–64, use linear models to solve application problems.

61. **College Tuition and Fees** The following scatter plot shows the average annual tuition and fees for public 4-year colleges and universities in the years 2009–2013. (*Data*: College Board)

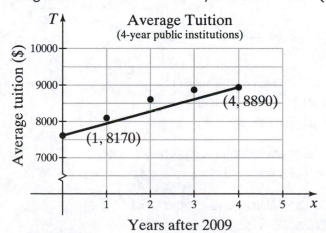

a. Use the two points labeled in the scatter plot to write a function of the form $T(x) = mx + b$ that models the average annual tuition and fees T at 4-year public colleges and universities x years after 2009.

b. Interpret the slope and the y-intercept in the context of the problem.

c. Use the model to predict the average annual tuition and fees at 4-year colleges and universities in 2018.

62. **Daily Newspapers** The following scatter plot shows the number of daily U.S. newspapers in the years 1980–2010. (*Data*: Statista.com)

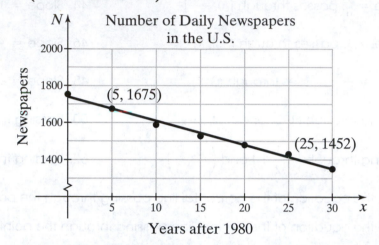

a. Use the two points labeled in the scatter plot to write a function of the form $N(x) = mx + b$ that models the number N of daily U.S. newspapers x years after 1980.

b. Interpret the slope and the y-intercept in the context of the problem.

c. Use the model to predict the number of daily U.S. newspapers in 2020.

63. **Hooke's Law** The amount that a spring is stretched when a load is applied varies directly with the amount of the load. If a load of 50 pounds stretches a spring by 2 inches, how much will the spring stretch if a 180-pound load is applied?

64. **Calories from Fat** The number of calories from fat in peanut butter varies directly with the number of ounces. If a 3-ounce serving of peanut butter contains 435 calories from fat, how many calories from fat are there in a 7-ounce serving?

2.6 Linear Inequalities in Two Variables

THINGS TO KNOW

Before working through this section, be sure you are familiar with the following concepts:

| | | VIDEO | ANIMATION | INTERACTIVE |

You Try It

1. Determine If a Given Value Is a Solution to an Inequality (Section 1.2, Objective 1)

You Try It

2. Solve Linear Inequalities in One Variable (Section 1.2, Objective 4)

You Try It

3. Graph a Line Using the Slope and a Point (Section 2.5, Objective 4)

OBJECTIVES

1 Determine If an Ordered Pair Is a Solution to a Linear Inequality in Two Variables

2 Graph a Linear Inequality in Two Variables

3 Solve Applications Involving Linear Inequalities in Two Variables

OBJECTIVE 1 **DETERMINE IF AN ORDERED PAIR IS A SOLUTION TO A LINEAR INEQUALITY IN TWO VARIABLES**

In Section 1.2, we solved **linear inequalities in one variable**. The **solution set** for such an inequality is the set of all values that make the inequality true. Typically, the solution set of a linear inequality in one variable is graphed on a **number line**.

This section looks at *linear inequalities in two variables*. A linear inequality in two variables looks much like a **linear equation in two variables** except that an **inequality symbol** replaces the equal sign.

Definition Linear Inequality in Two Variables

A **linear inequality in two variables** is an inequality that can be written in the form $Ax + By < C$, where A, B, and C are real numbers, and A and B are not both equal to zero.

Note: The inequality symbol "$<$" can be replaced with $>$, $\leq$, or $\geq$.

An ordered pair is a **solution to a linear inequality in two variables** if, when substituted for the **variables**, it makes the inequality true.

My video summary ▶ **Example 1 Determining If an Ordered Pair Is a Solution to a Linear Inequality in Two Variables**

Determine if the given ordered pair is a solution to the inequality $2x - 3y < 6$.

a. $(-1,-2)$ **b.** $(4,-1)$ **c.** $(6,2)$

Solutions We substitute the x- and y-coordinates for the **variables** and simplify. If the resulting statement is true, then the ordered pair is a solution to the inequality.

a. Begin with the original inequality: $\qquad 2x - 3y < 6$

Substitute -1 for x and -2 for y: $\quad 2(-1) - 3(-2) \overset{?}{<} 6$

Simplify: $\qquad\qquad -2 + 6 \overset{?}{<} 6$

$\qquad\qquad\qquad 4 < 6 \quad$ True

The final statement is true, so $(-1,-2)$ is a solution to the inequality.

b.–c. Try to determine if these ordered pairs are solutions on your own. View the answers, or watch this video for detailed solutions to all three parts.

You Try It Work through this You Try It problem.

Work Exercises 1–6 in this eText or in the MyMathLab Study Plan.

OBJECTIVE 2   GRAPH A LINEAR INEQUALITY IN TWO VARIABLES

Now let's look for all solutions to a linear inequality in two variables.

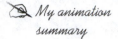 *My animation summary*   ⊛ **Concept Animation** Watch this animation to explore this concept.

First, focus on the related **linear equation.** For example, to find solutions to $x + y > 3$ or $x + y < 3$, look at the graph of the equation $x + y = 3$. See Figure 19. Note that the line divides the **coordinate plane** into two **half-planes,** an *upper half-plane* shaded blue and a *lower half-plane* shaded pink. The graph of the equation $x + y = 3$ acts as a **boundary line** that separates the solutions of the two inequalities $x + y > 3$ and $x + y < 3$.

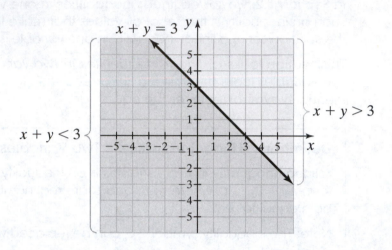

Figure 19

The preceding information leads to a set of **steps for graphing linear inequalities in two variables.**

Steps for Graphing Linear Inequalities in Two Variables

Step 1 Find the boundary line for the inequality by replacing the inequality symbol with an equal sign and graphing the resulting equation. If the inequality is **strict,** graph the boundary using a dashed line. If the inequality is **non-strict,** graph the boundary using a solid line.

Step 2 Choose a **test point** that does not belong to the boundary line and determine if it is a solution to the inequality.

Step 3 If the test point is a solution to the inequality, then shade the half-plane that contains the test point. If the test point is not a solution to the inequality, then shade the half-plane that does not contain the test point. The shaded area represents the set of all **ordered pair solutions** to the inequality.

 TIP When graphing a **linear inequality in two variables,** think of a dashed line as being similar to an open circle when graphing a **linear inequality in one variable** on a **number line.** Likewise, think of a solid line as being similar to a solid circle.

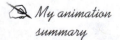 *My animation summary*   ⊛ **Concept Animation** To see why, watch this animation.

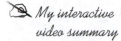 *My interactive video summary*

Example 2 Graphing a Linear Inequality in Two Variables

Graph each inequality.

a. $x - 2y \geq 4$ b. $3y < 2x$ c. $x < -2$

Solutions

a. Follow the **three-step process**, or watch this **interactive video** for a detailed solution.

Step 1 The **boundary line** is $x - 2y = 4$. Graph the line as a solid line because the inequality is **non-strict**. See **Figure 20(a)**.

Step 2 Choose the **test point** $(0, 0)$, and check to see if it **satisfies** the inequality.

Begin with the original inequality: $x - 2y \geq 4$

Substitute 0 for x and for y: $0 - 2(0) \overset{?}{\geq} 4$

Simplify: $0 \overset{?}{\geq} 4$ False

The point $(0, 0)$ is not a solution to the inequality.

Step 3 Because the test point is not a solution to the inequality, shade the **half-plane** that does not contain $(0, 0)$. See **Figure 20(b)**. The shaded region, including the boundary line, represents all ordered pair solutions to the inequality $x - 2y \geq 4$.

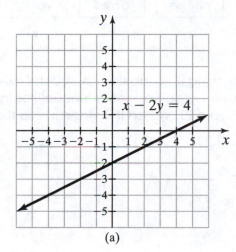

(a)

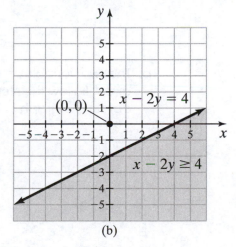

(b)

Figure 20

b. The **boundary line** is $3y = 2x$. Because the inequality is **strict**, graph the boundary line using a **dashed line**. Choose a **test point** and complete the graph. See the **answer**, or watch this **interactive video** for a detailed solution.

c. Try to graph the inequality $x < -2$ on your own. Note that the boundary line is the **vertical line** $x = -2$. See the **answer**, or watch this **interactive video** for a detailed solution.

 You Try It   Work through this **You Try It** problem.

Work Exercises 7–20 in this eText or in the MyMathLab Study Plan.

⚠ A test point cannot belong to the boundary line. Do you see why?

OBJECTIVE 3 SOLVE APPLICATIONS INVOLVING LINEAR INEQUALITIES IN TWO VARIABLES

Sometimes real-life applications can be **modeled** by **linear inequalities in two variables.**

My video summary ▶ **Example 3 A Piggy Bank**

A piggy bank contains only nickels and dimes with a total value of less than $9. Let n = the number of nickels and d = the number of dimes.

a. Write an inequality describing the possible numbers of coins in the bank.

b. Graph the inequality. Because n and d must be **whole numbers**, restrict the graph to **Quadrant I.**

c. Could the piggy bank contain 90 nickels and 60 dimes?

Solution Watch this **video**, or continue reading for a complete solution.

a. The value of n nickels, in cents, is $5n$, and the value of d dimes is $10d$, so the total value of the coins in the bank is $5n + 10d$. The total value is less than $9, or 900¢, so the inequality is:

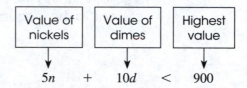

$$5n \quad + \quad 10d \quad < \quad 900$$

b. The two variables are n and d, so label the horizontal axis n and the vertical axis d. Follow the **steps for graphing linear inequalities in two variables** to obtain the graph in Figure 21. Watch this **video** to see the details.

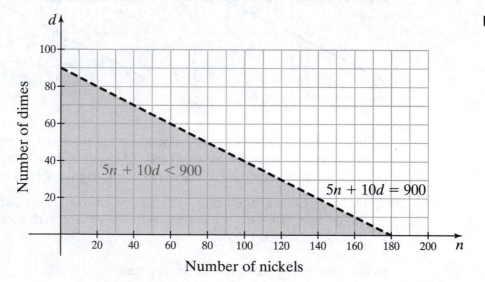

Figure 21

The number of nickels and dimes must be **whole numbers**, so only those points in the shaded region that have whole number coordinates are acceptable solutions to this application.

c. Substitute $(n, d) = (90, 60)$ into the inequality to see if a **true** statement results.

Write the inequality: $5n + 10d < 900$

Substitute $n = 90$ and $d = 60$: $5(90) + 10(60) \overset{?}{<} 900$

Simplify: $1050 < 900$ False

The piggy bank cannot contain 90 nickels and 60 dimes.

 You Try It Work through this You Try It problem.

Work Exercises 21–24 in this eText or in the MyMathLab Study Plan.

2.6 Exercises

In Exercises 1–6, determine if each ordered pair is a solution to the given inequality.

1. $x - 4y < 8$
  a. $(2, -3)$
  b. $(4, -1)$
  c. $(-6, -2)$

2. $5x + 2y \geq 20$
  a. $(5, -1)$
  b. $(3, 2)$
  c. $(2, 5)$

3. $2x + 5y > 10$
  a. $(-5, 4)$
  b. $(2, 3)$
  c. $\left(\frac{5}{2}, \frac{3}{5}\right)$

4. $6x - 5y \leq 30$
  a. $(0, 0)$
  b. $(2, -4)$
  c. $(2.5, -1.5)$

5. $y < \frac{4}{3}x - 2$
  a. $(-6, -5)$
  b. $(3, 0)$
  c. $\left(\frac{3}{2}, \frac{1}{2}\right)$

6. $3x - 5y \geq 45$
  a. $(7, -5)$
  b. $(-2, -10)$
  c. $(7.5, -4.5)$

In Exercises 7–20, graph each inequality.

7. $2x - y \geq -3$

8. $3x + y \leq 6$

9. $3x + 2y > 6$

10. $5x - 3y > 15$

11. $-x + 3y < -9$

12. $x + y \leq 0$

13. $x + 2y < 0$

14. $y \geq -2x$

15. $y \le \frac{5}{2}x - 1$

16. $4y < -3x$

17. $\frac{1}{2}x + \frac{2}{3}y > \frac{5}{6}$

18. $0.6x - 1.8y \le 2.4$

19. $x \ge -1$

20. $y < 3$

In Exercises 21–24, solve each application by using an inequality.

21. **Growing Crops** In Illinois the annual cost to grow corn is about $450 per acre, and the annual cost to grow soybeans is about $300 per acre (not counting the cost of land). Suppose a small farmer has at most $90,000 to spend on growing corn and soybeans this year. Let c = acres of corn and s = acres of soybeans. (*Source: Farmdoc*, University of Illinois)

a. Write an inequality that describes the possible acres of corn and soybeans this farmer can afford to grow.

b. Graph the inequality. Because c and s must be whole numbers, restrict the graph to Quadrant I.

c. Can this farmer afford to grow 125 acres of corn and 125 acres of soybeans?

22. **Transporting Crops** An agriculture company needs to ship its crops. Corn weighs about 50 pounds per bushel, while soybeans weigh about 60 pounds per bushel. A river barge can safely carry up to 3,000,000 pounds of cargo. Let c = bushels of corn and s = bushels of soybeans. (*Source:* powderandbulk.com)

a. Write an inequality that describes the bushels of corn and soybeans that can be shipped together on this barge.

b. Graph the inequality. Because c and s must be non-negative, restrict the graph to Quadrant I.

c. Can the company ship 25,000 bushels of corn and 25,000 bushels of soybeans on this barge?

23. **Coin Collection** A coin collection consists of only dimes and quarters with a value of more than $12. Let d = the number of dimes and q = the number of quarters.

a. Write an inequality that describes the possible numbers of dimes and quarters in the collection.

b. Graph the inequality. Because d and q must be whole numbers, restrict the graph to Quadrant I.

c. Could this collection consist of 35 dimes and 45 quarters?

24. **Car Rental** While on a business trip, Bruce pays $22.50 per day to rent a car, plus $0.45 per mile driven. Bruce's company will reimburse him at most $270 for the entire rental. Let d = number of days and m = number of miles.

a. Write an inequality that describes Bruce's car rental limitations.

b. Graph the inequality. Because d and m must be non-negative, restrict the graph to Quadrant I.

c. If Bruce keeps the car for 4 days and drives 500 miles, will he be within his budget?

Systems of Linear Equations and Inequalities

CHAPTER THREE CONTENTS

3.1 Systems of Linear Equations in Two Variables

THINGS TO KNOW

Before working through this section, be sure you are familiar with the following concepts:

		VIDEO	ANIMATION	INTERACTIVE
You Try It	1. Determine If an Ordered Pair Is a Solution to an Equation (Section 2.1, Objective 2)	▶		
You Try It	2. Graph a Line Using the Slope and a Point (Section 2.5, Objective 4)	▶	✳	
You Try It	3. Determine the Relationship between Two Lines (Section 2.5, Objective 5)		✳	🖰

OBJECTIVES

1 Determine If an Ordered Pair Is a Solution to a System of Linear Equations in Two Variables

2 Solve Systems of Linear Equations in Two Variables by Graphing

3 Solve Systems of Linear Equations in Two Variables by Substitution

4 Solve Systems of Linear Equations in Two Variables by Elimination

5 Solve Inconsistent and Dependent Systems

6 Use Systems of Linear Equations in Two Variables to Solve Application Problems

OBJECTIVE 1 DETERMINE IF AN ORDERED PAIR IS A SOLUTION TO A SYSTEM OF LINEAR EQUATIONS IN TWO VARIABLES

Figure 1 shows that the approximate global revenue for digital music is projected to increase steadily through 2017, while global revenue for physical music (CDs and vinyl) is projected to decrease.

The *x*-coordinate of the point where the two lines **intersect** gives the year when revenues are expected to be equal for both music types, and the *y*-coordinate of the intersection point gives the expected revenue when the values are equal for both music types.

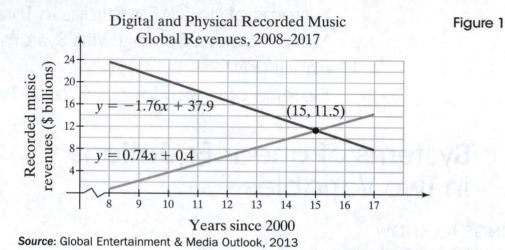

Digital and Physical Recorded Music Global Revenues, 2008–2017

Figure 1

Source: Global Entertainment & Media Outlook, 2013

Figure 1 illustrates the graph of a **system of linear equations in two variables**.

Definition System of Linear Equations in Two Variables

A **system of linear equations in two variables** is a collection of two or more linear equations in two variables considered together.

The system graphed in Figure 1 is given by the equations $\begin{cases} y = -1.76x + 37.9 \\ y = 0.74x + 0.4 \end{cases}$.

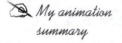

My animation summary

⊛ **Concept Animation** Watch this animation for an introduction to systems of linear equations in two variables.

The **intersection point** in Figure 1, (15, 11.5), lies on the graph of both equations, so it is a solution to both equations. Substituting 15 for *x* and 11.5 for *y* in either equation results in a true statement (view the **checks**). The ordered pair (15, 11.5) is an example of a **solution to a system of linear equations**.

Definition Solution to a System of Linear Equations in Two Variables

A **solution to a system of linear equations** in two variables is an ordered pair that, when substituted for the variables, makes all equations in the system true.

To determine if an ordered pair is a solution to a system, we check to see if the ordered pair makes *all* equations true when substituted for the variables. It is not enough to check only one equation in the system.

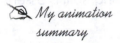
My interactive video summary

✎ **Example 1 Determining If an Ordered Pair Is a Solution to a System of Linear Equations in Two Variables**

Determine if each ordered pair is a solution to the following system:

$$\begin{cases} 2x + 3y = 12 \\ x + 2y = 7 \end{cases}$$

a. $(-3, 6)$ **b.** $(3, 2)$

Solutions Check each ordered pair to see if it makes both equations true. If both equations are true, then the ordered pair is a **solution to the system**. If either equation is false, then the ordered pair is not a solution to the system.

a.

	First Equation	Second Equation
Begin with the original equations:	$2x + 3y = 12$	$x + 2y = 7$
Substitute -3 for x and 6 for y in each equation:	$2(-3) + 3(6) \overset{?}{=} 12$	$(-3) + 2(6) \overset{?}{=} 7$

Finish simplifying both equations to see if $(-3, 6)$ is a solution to the system. Check your **answer**, or watch this **interactive video** for the full solution to both parts.

b. **Substitute** 3 for x and 2 for y in each equation. Simplify to see if the resulting equations are true. Check your **answer**, or watch this **interactive video** for a full solution.

You Try It Work through this You Try It problem.

Work Exercises 1–4 in this eText or in the MyMathLab Study Plan.

OBJECTIVE 2 SOLVE SYSTEMS OF LINEAR EQUATIONS IN TWO VARIABLES BY GRAPHING

In this section, we look at three methods for solving systems of linear equations in two variables: *graphing, substitution,* and *elimination*.

To **solve a system of linear equations in two variables by graphing**, graph each line and find the **intersection point**, if any.

My animation summary

⊛ **Concept Animation** Watch this animation to see why.

When two **linear equations** are graphed, there are three possible outcomes:

1. The two lines intersect at one point. See **Figure 2(a)**. The system has one **solution**.

2. The two lines are **parallel** and do not intersect at all. See **Figure 2(b)**. The system has no solution.

3. The two lines **coincide** and have an **infinite** number of intersection points. See **Figure 2(c)**. The system has an infinite number of solutions.

A system with at least one solution is **consistent**. A system without a solution is **inconsistent**. So the intersecting and coinciding lines reveal consistent systems, while the parallel lines indicate an inconsistent system.

When the equations in a system are coinciding lines, the system is **dependent**. When the equations in a system are different lines, the system is **independent**.

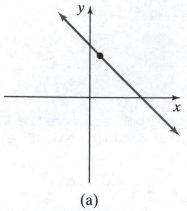

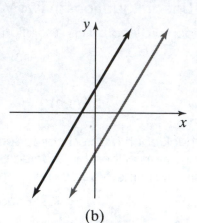

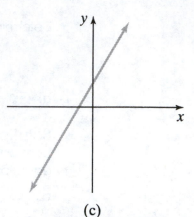

(a)
Intersecting Lines
One solution
Consistent, Independent

(b)
Parallel Lines
No solution
Inconsistent, Independent

(c)
Coinciding Lines
Infinitely many solutions
Consistent, Dependent

Figure 2

 *My animation summary* ⊛ **Concept Animation** Watch this animation to see how the slope and *y*-intercept can be used to determine the number of solutions to a system of linear equations in two variables.

To solve **systems of linear equations** by graphing, use the following three-step process.

Solving Systems of Linear Equations in Two Variables by Graphing

Step 1 Graph the two equations on the same set of **axes**.

Step 2 If the lines intersect, then find the **coordinates** of the **intersection point**. The ordered pair is the **solution** to the **system**.

Step 3 Check the ordered-pair solution in both of the original equations.

Note: If the lines are **parallel**, then the system has no solution. If the lines **coincide**, then the system has infinitely many solutions.

⚠ It can be difficult to identify the exact intersection point by graphing, so it is essential to check the solution.

 My video summary ▷ **Example 2 Solving Systems of Linear Equations in Two Variables by Graphing**

Solve the following system by graphing:

$$\begin{cases} 2x + y = -4 \\ x + 3y = 3 \end{cases}$$

Solution Follow the **three-step process.**

Step 1 Graph each line, as shown in Figure 3.

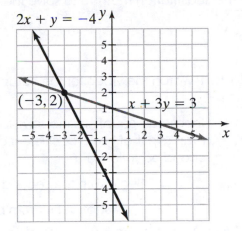

$2x + y = -4$

$(-3, 2)$

$x + 3y = 3$

Figure 3
Graph of System
$$\begin{cases} 2x + y = -4 \\ x + 3y = 3 \end{cases}$$

Step 2 The intersection point is $(-3, 2)$.

Step 3 Check $(-3, 2)$ in both equations to see if it is the solution to the system.

Watch this **video** to see the fully worked solution.

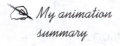 **Work through this You Try It problem.**

Work Exercises 5–15 in this eText or in the MyMathLab Study Plan.

OBJECTIVE 3 SOLVE SYSTEMS OF LINEAR EQUATIONS IN TWO VARIABLES BY SUBSTITUTION

My animation summary

⊛ **Concept Animation** The **substitution method** involves solving one of the equations for one **variable**, substituting the resulting **expression** into the other equation, and then **solving** for the remaining variable. Watch this **animation** to see how this works.

The substitution method can be summarized in four steps.

Solving Systems of Linear Equations in Two Variables by Substitution

Step 1 Choose an equation and solve for one variable in terms of the other variable.

Step 2 Substitute the expression from Step 1 into the other equation.

Step 3 Solve the equation in one variable from Step 2.

Step 4 Substitute the solution from Step 3 into one of the original equations to find the value of the other variable.

⚠ It is good practice to check the ordered-pair solution by substituting it into the original equation that was not used in Step 4.

My video summary ▶ **Example 3 Solving Systems of Linear Equations in Two Variables by Substitution**

Use the substitution method to solve the following system:

$$\begin{cases} 5x + 2y = 8 \\ x + 6y = 10 \end{cases}$$

Solution Watch the video for the complete solution, or follow the process as shown.

Step 1   Choose an equation and solve for one of the **variables**. If possible, choose a variable that has a **coefficient** of 1. In this example, it is easiest to solve the second equation for x:

Begin with the second equation: $x + 6y = 10$

Subtract $6y$ from both sides: $x = -6y + 10$

Step 2 Substitute $-6y + 10$ for x in the first equation.

Rewrite the first equation: $5x + 2y = 8$

Substitute $-6y + 10$ for x: $5(-6y + 10) + 2y = 8$

Step 3 Solve for y.

Equation from Step 2: $5(-6y + 10) + 2y = 8$

Distribute: $-30y + 50 + 2y = 8$

Simplify: $-28y + 50 = 8$

Subtract 50 from both sides: $-28y = -42$

Divide both sides by -28 and simplify: $y = \dfrac{-42}{-28} = \dfrac{3}{2}$

Step 4 Find x by substituting $y = \dfrac{3}{2}$ into one of the original equations.

Begin with the original second equation: $x + 6y = 10$

Substitute $\dfrac{3}{2}$ for y: $x + 6\left(\dfrac{3}{2}\right) = 10$

Simplify: $x + 9 = 10$

Subtract 9 from both sides: $x = 1$

The solution to this system is the ordered pair $\left(1, \dfrac{3}{2}\right)$. Even though we used the second original equation in Step 4, the first original equation is used to check the final answer. See the **check**, or watch this **video** for a fully worked solution.

Figure 4 shows the graph of the system in Example 3. Notice that the solution $\left(1, \dfrac{3}{2}\right)$ is the **intersection point** of the two lines in the system.

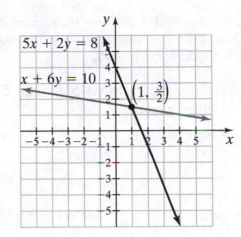

Figure 4
Graph of System
$$\begin{cases} 5x + 2y = 8 \\ x + 6y = 10 \end{cases}$$

You Try It Work through this You Try It problem.

Work Exercises 16–24 in this eText or in the MyMathLab Study Plan.

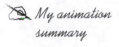 *My interactive video summary*

Example 4 Solving Systems of Linear Equations by Substitution

Use the substitution method to solve the following system:

$$\begin{cases} 6x - 3y = -33 \\ 2x + 4y = 4 \end{cases}$$

Solution In this system, we do not see any variables with a **coefficient** of 1 or −1. We can select either equation and solve for either variable. Notice that if we divide both sides of the first equation by 3, we get

$$\frac{6x}{3} - \frac{3y}{3} = \frac{-33}{3} \rightarrow 2x - y = -11.$$

Dividing both sides of the second equation by 2 gives

$$\frac{2x}{2} + \frac{4y}{2} = \frac{4}{2} \rightarrow x + 2y = 2.$$

Try solving this system on your own. View the **answer**, or watch this **interactive video** for a fully worked solution.

You Try It Work through this You Try It problem.

Work Exercises 25–30 in this eText or in the MyMathLab Study Plan.

OBJECTIVE 4 SOLVE SYSTEMS OF LINEAR EQUATIONS IN TWO VARIABLES BY ELIMINATION

My animation summary

Concept Animation The **elimination method** involves adding the **two equations** together in a way that will *eliminate* one of the **variables**. Watch this **animation** to see the logic on which this method is based.

This logic means that if two true equations are added, then the result will be a third true equation. Because equations are added, the elimination method is also known as the **addition method**.

Example 5 Solving Systems of Linear Equations in Two Variables by Elimination

Solve the following **system**:

$$\begin{cases} x - y = 6 \\ x + y = 12 \end{cases}$$

Solution Watch this video to see a complete solution, or read the following.

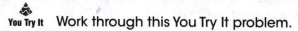 *My video summary* ▶ Based on the logic for the elimination method, we can add two true equations together to result in a third true equation.

$$x - y = 6$$
$$x + y = 12$$
$$2x + 0 = 18$$

Notice that y is eliminated. Solving the resulting equation for x gives the x-**coordinate** of the **ordered-pair solution**.

Rewrite the equation: $2x + 0 = 18$

Simplify: $2x = 18$

Divide both sides by 2: $x = 9$

The value of x is 9. By substituting 9 for x in one of the original equations, we can find the value of y.

Begin with the second original equation: $x + y = 12$

Substitute 9 for x: $9 + y = 12$

Subtract 9 from both sides: $y = 3$

The ordered-pair solution is $(9, 3)$. Check $(9, 3)$ in both of the original equations. View the **check**.

You Try It Work through this You Try It problem.

Work Exercises 31–34 in this eText or in the MyMathLab Study Plan.

In Example 5, the y-variable was eliminated easily when we added the two equations together. Unfortunately, a variable will not always be eliminated so conveniently. Consider the system from **Example 3**. Adding the two original equations does not eliminate a variable.

$$5x + 2y = 8$$
$$x + 6y = 10$$
$$6x + 8y = 18$$

In this case, we must write **equivalent equations** in the system before adding so that a variable can be eliminated.

To eliminate a variable by adding, the **coefficients** of the variable must differ only in **sign**. We can make this happen by multiplying one or both of the equations by a suitable nonzero **constant**.

📧 *My video summary* ▶ **Example 6** Solving Systems of Linear Equations in Two Variables by Elimination

Use **elimination** to solve the following **system** from Example 3:

$$\begin{cases} 5x + 2y = 8 \\ x + 6y = 10 \end{cases}$$

Solution Work through the following, or watch this **video** for the complete solution. If we multiply both sides of the second equation by -5, this will change the **coefficient** of the x-variable in the second equation to the **opposite** of the coefficient of the x-variable in the first equation. When the result is added to the first equation, the x-variable will be eliminated.

$$5x + 2y = 8 \xrightarrow{\text{No change needed}} 5x + 2y = 8 \xrightarrow{\text{No change needed}} 5x + 2y = 8$$

$$x + 6y = 10 \xrightarrow{\text{Multiply by } -5} -5(x + 6y) = -5(10) \xrightarrow{\text{Simplify}} -5x - 30y = -50$$

Add the new equations: $-28y = -42$

Solve for y: $y = \dfrac{-42}{-28} = \dfrac{3}{2}$

The value of y is $\dfrac{3}{2}$. We find the value of x as we did in **Step 4** of Example 3. So, the value of x is 1, and the solution to the system is the ordered pair $\left(1, \dfrac{3}{2}\right)$.

🔺 **You Try It** Work through this **You Try It** problem.

Work Exercises 35–38 in this eText or in the MyMathLab Study Plan.

The elimination method can be summarized in five steps.

Solving Systems of Linear Equations in Two Variables by Elimination

Step 1 Choose a variable to eliminate.

Step 2 Multiply one or both **equations** by an appropriate nonzero **constant** so that the sum of the **coefficients** of one of the variables is zero.

Step 3 Add the two equations from Step 2 together to obtain an equation in one variable.

Step 4 Solve the equation in one variable from Step 3.

Step 5 Substitute the value found in Step 4 into one of the original equations to solve for the other variable.

⚠ It is good practice to check the ordered-pair solution by **substituting** it into the original equation that was not used in Step 5.

My video summary ▶ **Example 7 Solving Systems of Linear Equations in Two Variables by Elimination**

Use the **elimination method** to solve the following system:

$$\begin{cases} 5x - 2y = -12 \\ 3x + 4y = -2 \end{cases}$$

Solution Watch this video, or continue reading, for the fully worked solution.

Step 1 Choose which variable to eliminate. We eliminate the y-variable because the **coefficients** of the y-variables will be **opposites** if we multiply the first equation by 2.

Step 2 Multiply both sides of the first equation by 2.

$5x - 2y = -12 \xrightarrow{\text{Multiply by 2}} 2(5x - 2y) = 2(-12) \xrightarrow{\text{Simplify}} 10x - 4y = -24$

$3x + 4y = -2 \xrightarrow{\text{No change}} 3x + 4y = -2 \xrightarrow{\text{No change}} 3x + 4y = -2$

Step 3 Add the two new equations together.

$$\begin{aligned} 10x - 4y &= -24 \\ 3x + 4y &= -2 \\ \hline 13x \quad\quad &= -26 \end{aligned}$$

Step 4 Solve the resulting equation for x.

Rewrite the resulting equation from Step 3: $13x = -26$

Divide both sides by 13: $x = -2$

Step 5 Find y by substituting $x = -2$ into one of the original equations.

Begin with the original first equation: $5x - 2y = -12$

Substitute -2 for x: $5(-2) - 2y = -12$

Simplify: $-10 - 2y = -12$

Add 10 to both sides: $-2y = -2$

Divide both sides by -2: $y = 1$

The **solution** to this system is the ordered pair $(-2, 1)$.

Since the original first equation was used in Step 5, use the original second equation to check the final answer. See the **check**.

To solve some systems, we must multiply both equations by nonzero constants.

My video summary ⊙ **Example 8** Solving Systems of Linear Equations in Two Variables by Elimination

Use the elimination method to solve the following system:

$$\begin{cases} 5x + 4y = 10 \\ -4x + 3y = 23 \end{cases}$$

Solution Notice that, if we multiply the first equation by 4 and the second equation by 5, we can eliminate the x-variable. Try this and finish solving this system on your own. Check your **answer**, or watch this **video** for a fully worked solution.

You Try It Work through this You Try It problem.

Work Exercises 39–42 in this eText or in the MyMathLab Study Plan.

If the equations in a system have **coefficients** involving **fractions**, then it may be helpful to **clear the fractions** before proceeding. Likewise, it may be helpful to **clear decimal** coefficients.

My video summary ⊙ **Example 9** Solving Systems of Linear Equations in Two Variables Involving Fractions

Use the **elimination** method to solve the following system:

$$\begin{cases} \dfrac{2}{9}x - \dfrac{5}{6}y = -\dfrac{8}{3} \\ \dfrac{3}{4}x - \dfrac{1}{3}y = \dfrac{11}{12} \end{cases}$$

Solution Try solving this system on your own. Check your **answer**, or watch this **video** for a fully worked solution.

You Try It Work through this You Try It problem.

Work Exercises 43–46 in this eText or in the MyMathLab Study Plan.

OBJECTIVE 5 SOLVE INCONSISTENT AND DEPENDENT SYSTEMS

Each **system** in Examples 2 through 9 had only one **solution**. We now look at systems without a solution and systems with an infinite number of solutions. Recall from **Objective 1** that a system without a solution is **inconsistent**, and a system with an infinite number of solutions is **dependent**.

My interactive video summary

Example 10 Solving an Inconsistent System

Solve the following system:

$$\begin{cases} x - 2y = 11 \\ -2x + 4y = 8 \end{cases}$$

Solution We can solve this system by **graphing**, **substitution**, or **elimination**. Work through this solution using the elimination method, or watch this **interactive video** for the complete solution by graphing or substitution.

Step 1 We choose to eliminate the x-variable because it has a **coefficient** of 1.

Step 2 Multiply both sides of the first equation by 2. This makes the coefficients of the x-terms in the two equations 2 and -2, respectively.

$$x - 2y = 11 \xrightarrow{\text{Multiply by 2}} 2x - 4y = 22$$
$$-2x + 4y = 8 \xrightarrow{\text{No change}} -2x + 4y = 8$$

Step 3 Add the two equations from Step 2 together.

$$\begin{array}{r} 2x - 4y = 22 \\ \underline{-2x + 4y = 8} \\ 0 = 30 \quad \text{False} \end{array}$$

Since both **variables** are eliminated, we are left with the **contradiction** $0 = 30$. When this happens, the system has no solution and is an **inconsistent system**. We use the null symbol, $\varnothing$, or empty set, $\{\ \}$, to show that a system has no solution.

You Try It Work through this **You Try It** problem.

Work Exercises 47 and 48 in this eText or in the MyMathLab Study Plan.

My interactive video summary

Example 11 Solving a Dependent System

Solve the following system:

$$\begin{cases} -3x + 6y = 9 \\ x - 2y = -3 \end{cases}$$

Solution We can solve this system by **graphing**, **substitution**, or **elimination**. Work through this solution using the substitution method, or watch this **interactive video** for the complete solution by graphing or elimination.

Step 1 Solve the second equation for x.

Begin with the second equation: $\quad x - 2y = -3$

Add $2y$ to both sides: $\quad x = 2y - 3$

Step 2 Substitute $2y - 3$ for x in the first equation.

Rewrite the first equation: $\quad -3x + 6y = 9$

Substitute $2y - 3$ for x: $\quad -3(2y - 3) + 6y = 9$

Step 3 Solve for y.

$$\text{Distribute: } -6y + 9 + 6y = 9$$

$$\text{Combine like terms: } \qquad 9 = 9 \quad \text{True}$$

The equation simplifies to the **identity** $9 = 9$, which is true for any value of y. The system has an infinite number of solutions and is a **dependent system**. Any ordered pair that is a solution to one of the equations is a solution to both equations.

There are two common ways to write the solution to a dependent system: **set-builder notation** and **ordered-pair notation**.

With *set-builder notation*, we indicate the solutions to a **dependent system** by using one of the equations from the system (or any **equivalent equation**). With *ordered-pair notation*, we indicate the solutions to a dependent system by expressing one coordinate in terms of the other in an ordered pair.

My animation summary ✵ **Concept Animation** When writing the solution to a system of equations, slope-intercept form can be useful. Watch this **animation** to see how.

The solution to the system in Example 11 can be written as $\left\{ (x, y) \,\middle|\, y = \dfrac{1}{2}x + \dfrac{3}{2} \right\}$ in set-builder notation. We read this as "the set of all **ordered pairs** (x, y), such that $y = \dfrac{1}{2}x + \dfrac{3}{2}$. Also, we can write the solution as $\left(x, \dfrac{1}{2}x + \dfrac{3}{2} \right)$ in ordered pair notation.

Selecting a value for x will allow us to determine a solution to the system. For example, if $x = 3$, then $y = \dfrac{1}{2}(3) + \dfrac{3}{2} = 3$ and $(3, 3)$ is a solution to the system. If $x = -5$, then $y = \dfrac{1}{2}(-5) + \dfrac{3}{2} = -1$ and $(-5, -1)$ is a solution. And so on.

⚠ It is incorrect to say that the solution to a dependent system is "the set of all real numbers" or "the set of all ordered pairs." Although dependent systems have an infinite number of ordered-pair solutions, the solution set of the system consists only of those ordered pairs, which lie on the coinciding lines.

 You Try It Work through this **You Try It** problem.

Work Exercises 49–55 in this eText or in the MyMathLab **Study Plan.**

OBJECTIVE 6 USE SYSTEMS OF LINEAR EQUATIONS IN TWO VARIABLES TO SOLVE APPLICATION PROBLEMS

Sometimes, we can use a single **equation** in one **variable** to solve application problems. However, it is often easier to use two variables and create a **system of linear equations**. The following steps are used to solve problems for applications of linear systems of equations in two variables.

Problem-Solving Strategy for Applications Using Systems of Linear Equations

Step 1 Define the Problem. Read the problem carefully; multiple times if necessary. Identify what you need to find and determine what information is available to help you find it.

Step 2 Assign Variables. Choose variables that describe each unknown quantity.

Step 3 Translate into a System of Equations. Use the relationships among the known and unknown quantities to form a system of equations.

Step 4 Solve the System. Use graphing, substitution, or elimination to solve the system.

Step 5 Check the Reasonableness of Your Answers. Check to see if your answers make sense within the context of the problem. If not, check your work for errors and try again.

Step 6 Answer the Question. Write a clear statement that answers the question(s) posed.

My video summary ▶ **Example 12 Smartphone Users**

The number of smartphone users in the United States in 2017 is projected to be 36 million more than four times the number in the United Kingdom. If the combined number of users is projected to be 253 million, how many users are projected for the U.S.? How many for the U.K.?

Solution Follow the **problem-solving strategy** using systems of equations shown below, or watch this **video** to see the solution worked out.

Step 1 We are looking for the projected number of smartphone users in the U.S. and the U.K. in 2017. The number in the U.S. is 36 million more than four times the number in the U.K. and the combined number is 253 million.

Step 2 Let s = the number of smartphone users in the U.S. (in millions), and k = the number of smartphone users in the U.K. (in millions).

Step 3 The combined users is 253 million, so we can write

U.S. users	U.K. users	Total users
↓	↓	↓
s +	k =	253

We also know that the number in the U.S. is 36 million more than four times the number in the U.K., so we can write $s = 4k + 36$.

The two equations together form the system

$$\begin{cases} s + k = 253 \\ s = 4k + 36 \end{cases}$$

Step 4 Since the second equation is already solved for s, the substitution method seems most appropriate. Substitute $4k + 36$ for s in the first equation and solve for k.

$$\text{Begin with the original first equation:} \qquad s + k = 253$$

$$\text{Substitute } 4k + 36 \text{ for } s: \quad (4k + 36) + k = 253$$

$$\text{Simplify:} \qquad 5k + 36 = 253$$

$$\text{Subtract 36 from both sides:} \qquad 5k = 217$$

$$\text{Divide both sides by 5:} \qquad k = 43.4$$

Substituting this result into the second equation gives

$$s = 4k + 36 = 4(43.4) + 36 = 209.6.$$

Step 5 Since the sum of the two values is $43.4 + 209.6 = 253$, the results seem reasonable.

Step 6 The projected number of smartphone users in 2017 is 209.6 million in the U.S. and 43.4 million in the U.K.

You Try It Work through this You Try It problem.

Work Exercises 56–61 in this eText or in the MyMathLab Study Plan.

When using **systems of linear equations** to solve real-world problems, the solutions do not always work out to nice, exact answers. Sometimes we need to approximate or round the answer.

It is best to wait to round your answer until the end of the problem-solving process, if possible. Otherwise, if a rounded answer is used in later calculations, then those calculations will have errors due to rounding.

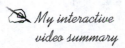
My interactive video summary

Example 13 Media Usage

The average number of minutes per day spent by U.S. adults on mobile devices can be **modeled** by $y = 41.3x + 10$, where x is the number of years after 2010. The average number of minutes spent per day watching television can be modeled by $y = -2.5x + 279$. Use these models to find the year when the times are equal, and describe what this means in the context of the problem.

Solution The two models together form the following system:

$$\begin{cases} y = 41.3x + 10 \\ y = -2.5x + 279 \end{cases}$$

Follow the **problem-solving strategy** using systems of equations. Solve this system by **substitution** or **elimination**. Remember not to round until the end of the problem, if possible. See the answer, or watch this **interactive video** for a fully worked solution.

You Try It Work through this You Try It problem.

Work Exercises 62–65 in this eText or in the MyMathLab Study Plan.

3.1 Exercises

In Exercises 1–4, determine if each ordered pair is a solution to the given system.

1. $\begin{cases} 3x - 2y = -2 \\ 4x + 3y = 37 \end{cases}$ a. $(4, 7)$ b. $(-4, -5)$

2. $\begin{cases} 3x + 18y = 15 \\ -5x - 30y = -25 \end{cases}$ a. $(12.8, -1.3)$ b. $\left(\dfrac{1}{2}, \dfrac{3}{4}\right)$

3. $\begin{cases} 0.3x + 0.4y = 2.3 \\ 0.5x - 0.2y = -0.5 \end{cases}$ a. $(-3, 8)$ b. $(1, 5)$

4. $\begin{cases} \dfrac{1}{9}x - \dfrac{1}{4}y = -\dfrac{2}{3} \\ \dfrac{1}{3}x - \dfrac{1}{2}y = -1 \end{cases}$ a. $(3, 4)$ b. $(12, 8)$

In Exercises 5–15, solve each system by graphing.

5. $\begin{cases} y = 2x \\ y = -3x - 5 \end{cases}$

6. $\begin{cases} y = \dfrac{3}{2}x - 4 \\ y = 2x - 6 \end{cases}$

7. $\begin{cases} 3x - y = -5 \\ 4x + y = -2 \end{cases}$

8. $\begin{cases} 4x + 2y = -8 \\ -7x + y = 5 \end{cases}$

9. $\begin{cases} 3x - 4y = 0 \\ 9x - 4y = 24 \end{cases}$

10. $\begin{cases} 2x - 6y = -6 \\ 2x + 3y = 12 \end{cases}$

11. $\begin{cases} 2x - 3y = -9 \\ x + 0.5y = -2.5 \end{cases}$

12. $\begin{cases} x + 0.5y = 6 \\ 1.8x + 1.5y = 12 \end{cases}$

13. $\begin{cases} 9x - 2y - 8 = 0 \\ y = 2x + 1 \end{cases}$

14. $\begin{cases} 4x - \dfrac{5}{3}y = -15 \\ -5y = x + 20 \end{cases}$

15. $\begin{cases} 7x + \dfrac{7}{2}y = 28 \\ \dfrac{3}{2}x + \dfrac{5}{4}y = 5 \end{cases}$

In Exercises 16–30, solve each system by substitution.

16. $\begin{cases} y = 3x - 2 \\ 3x + 5y = 44 \end{cases}$

17. $\begin{cases} 5x - 2y = -1 \\ y = 3x \end{cases}$

18. $\begin{cases} y = 2x - 5 \\ y = 5x - 2 \end{cases}$

19. $\begin{cases} x = 4 - y \\ 3x - 2y = -3 \end{cases}$

20. $\begin{cases} x + 4y = 5 \\ 2x - y = -8 \end{cases}$

21. $\begin{cases} 2x - y = 3 \\ -8x + 3y = -8 \end{cases}$

22. $\begin{cases} 5x - 2y = 0 \\ 3x + y = 0 \end{cases}$

23. $\begin{cases} 2x + \dfrac{3}{2}y = 1 \\ x + 3y = 7 \end{cases}$

24. $\begin{cases} 1.5x + y = 1.1 \\ 2.4x - 3y = 8.4 \end{cases}$

25. $\begin{cases} 5x + 4y = -6 \\ 8x - 4y = -72 \end{cases}$

26. $\begin{cases} 2x + 4y = 6 \\ 5x + 2y = -1 \end{cases}$

27. $\begin{cases} 3x + 6y = \dfrac{19}{2} \\ \dfrac{8}{5}x + 3y = 5 \end{cases}$

28. $\begin{cases} 1.7x + 3.2y = 0.22 \\ 2.1x - 2.1y = 6.93 \end{cases}$

29. $\begin{cases} \dfrac{1}{2}x + \dfrac{5}{3}y = \dfrac{2}{3} \\[2mm] 3x - 2y = \dfrac{8}{5} \end{cases}$

30. $\begin{cases} -\dfrac{2}{3}x + \dfrac{1}{2}y = -\dfrac{1}{3} \\[2mm] -\dfrac{1}{2}x + \dfrac{3}{4}y = -\dfrac{1}{2} \end{cases}$

In Exercises 31–46, solve each system by elimination.

31. $\begin{cases} x + y = 8 \\ x - y = 4 \end{cases}$

32. $\begin{cases} x + 2y = 10 \\ x - 2y = -6 \end{cases}$

33. $\begin{cases} -4x + 3y = -4 \\ 4x + 5y = -28 \end{cases}$

34. $\begin{cases} -3x + 4y = -15 \\ y = \dfrac{1}{2}x - \dfrac{5}{2} \end{cases}$

35. $\begin{cases} x + 3y = 4 \\ 2x - y = -6 \end{cases}$

36. $\begin{cases} 5x - y = 7 \\ 3x + 4y = 18 \end{cases}$

37. $\begin{cases} 8x - 3y = -17 \\ y = -2x + 1 \end{cases}$

38. $\begin{cases} 10x - 3y = 45 \\ 2x - 5y = -13 \end{cases}$

39. $\begin{cases} 8x - 6y = -6 \\ 5x - 7y = -33 \end{cases}$

40. $\begin{cases} 8x - 9y = 12 \\ 4x + 6y = -1 \end{cases}$

41. $\begin{cases} 2x + 3y = 13 \\ 5x + 2y = -6 \end{cases}$

42. $\begin{cases} 4x - 5y = 51 \\ 3x + 7y = -37 \end{cases}$

43. $\begin{cases} \dfrac{3}{2}x + \dfrac{5}{6}y = \dfrac{2}{3} \\[2mm] \dfrac{5}{8}x - \dfrac{1}{6}y = \dfrac{19}{24} \end{cases}$

44. $\begin{cases} 0.6x + 0.4y = 21 \\ 0.3x - 0.6y = -1.5 \end{cases}$

45. $\begin{cases} \dfrac{1}{4}x - y = \dfrac{9}{8} \\[2mm] x + \dfrac{3}{5}y = -\dfrac{1}{10} \end{cases}$

46. $\begin{cases} 0.3x + 0.2y = 1 \\ 0.1x - 0.3y = 1.8 \end{cases}$

In Exercises 47–55, use the substitution or elimination method to solve each system. If the system is dependent, write the solution in ordered-pair notation or set-builder notation.

47. $\begin{cases} x + y = 3 \\ -2x - 2y = 1 \end{cases}$

48. $\begin{cases} 8x - y = -13 \\ y = 8x \end{cases}$

49. $\begin{cases} 2x - y = -6 \\ 8x - 4y = -24 \end{cases}$

50. $\begin{cases} \dfrac{3}{4}x + \dfrac{1}{2}y = \dfrac{5}{8} \\[2mm] 0.75x + 0.5y = 0.625 \end{cases}$

51. $\begin{cases} 3x - 12y = 6 \\ -2x + 8y = -4 \end{cases}$

52. $\begin{cases} 8x + 4y = 5 \\ 2x + y = -4 \end{cases}$

53. $\begin{cases} y = \dfrac{5}{7}x - 9 \\[2mm] -15x + 21y = -189 \end{cases}$

54. $\begin{cases} 3.4x - 5.1y = -1.7 \\ -2.4x + 3.6y = 1.2 \end{cases}$

55. $\begin{cases} \dfrac{2}{3}x + \dfrac{7}{3}y = \dfrac{13}{3} \\[2mm] \dfrac{4}{5}x + \dfrac{14}{5}y = \dfrac{18}{5} \end{cases}$

In Exercises 56–65, use a system of linear equations in two variables to solve each application problem.

56. **Basketball Score** During a game of one-on-one, Rodney and Justin scored 47 points altogether. If Justin lost the game by 5 points, how many points did each player score?

57. **Two Brothers** Clayton is 2 years younger than his brother, Josh. If the sum of their ages is 26, how old is each boy?

58. **Number Problem** The sum of two numbers is 121, while the difference of the two numbers is 15. What are the two numbers?

59. **Coin Collection** Susan has a collection of 50 nickels and dimes. If the number of nickels is four times the number of dimes, how many nickels and how many dimes does she have?

60. **Heights of Landmarks** The height of the St. Louis Arch is 19 feet more than twice the height of the Statue of Liberty (from the base of its pedestal). If the difference in their heights is 324.5 feet, find the heights of the two landmarks. (*Sources*: gatewayarch.com and statueofliberty.org)

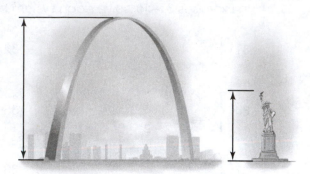

61. **River Lengths** The Missouri River is 200 miles longer than the Mississippi River. If the sum of their lengths is 4880 miles, what is the length of each river? (*Source*: U.S. Geological Survey)

62. **Electronic Devices** Global shipments of tablets in 2017 is projected to be 183 million less than three times the number of laptops shipped. If the combined number of units shipped is projected to be 603.4 million, how many of each type are expected to ship?

63. **Earning Power** In 2013, the average starting salary in the U.S. for a social science major was $6700 more than half the average starting salary for an engineering major. If the combined salaries equal $100,600, what was the average starting salary for each major?

64. **Gift of Love** Based on the results of a survey, the average amount of money, in dollars, spent on Valentine's Day gifts for family members can be modeled by $y = 2x + 35.52$, where x is the number of years after 2010. The average amount of money spent on Valentine's Day gifts for friends can be modeled by $y = 2.8x + 23.31$. In what year will the average amount spent on family and friends be the same, and what is that amount? Describe your answer in terms of the context of the problem.

65. **Non-Housing Debt** The total amount of student loan debt in the U.S. (in trillions of dollars) can be modeled by $y = 0.09x - 0.07$, where x is the number of years after 2000. The total amount of credit card debt in the U.S. (in trillions of dollars) can be modeled by $y = -0.03x + 1.05$. In what year was total student loan debt equal to total credit card debt, and what was that amount? Describe your answer in terms of the context of the problem.

3.2 Systems of Linear Equations in Three Variables

THINGS TO KNOW

Before working through this section, be sure you are familiar with the following concepts:

VIDEO   ANIMATION   INTERACTIVE

You Try It

1. Determine If an Ordered Pair Is a Solution To a System of Linear Equations in Two Variables (Section 3.1, Objective 1)

2. Solve Systems of Linear Equations in Two
   Variables by Elimination
   (Section 3.1, Objective 4)

3. Solve Inconsistent and Dependent Systems
   (Section 3.1, Objective 5)

4. Use Systems of Linear Equations in Two
   Variables to Solve Application Problems
   (Section 3.1, Objective 6)

OBJECTIVES

1 Determine If an Ordered Triple Is a Solution to a System of Linear Equations in Three Variables

2 Solve Systems of Linear Equations in Three Variables

3 Solve Inconsistent and Dependent Systems in Three Variables

4 Use Systems of Linear Equations in Three Variables to Solve Application Problems

---

OBJECTIVE 1   DETERMINE IF AN ORDERED TRIPLE IS A SOLUTION TO A SYSTEM OF LINEAR EQUATIONS IN THREE VARIABLES

Now that we know how to solve **systems of linear equations in two variables**, we can apply this knowledge to **systems of linear equations in three variables**. An equation such as $4x + 2y - 7z = 12$ is called a **linear equation in three variables** because there are three **variables** and each variable term is **linear**.

> **Definition**   Linear Equation in Three Variables
>
> A **linear equation in three variables** is an equation that can be written in the form $Ax + By + Cz = D$, where A, B, C, and D are **real numbers**, and A, B, and C are not all equal to 0.

A solution to a linear equation in three variables is an **ordered triple** (x, y, z) that makes the equation true. If we consider a collection of linear equations in three variables together, we have a **system of linear equations in three variables**.

> **Definition**   System of Linear Equations in Three Variables
>
> A **system of linear equations in three variables** is a collection of **linear equations in three variables** considered together. A **solution to a system** of linear equations in three variables is an **ordered triple** that satisfies all equations in the system.

 *My animation summary*

⊛ **Concept Animation** Watch this animation to see how a three variable system relates to a two variable system.

Example 1 Determining If an Ordered Triple Is a Solution to a System

Determine if each ordered triple is a **solution** to the given system:

$$\begin{cases} 3x + y - 2z = 4 \\ 2x - 2y + 3z = 9 \\ x + y - z = 5 \end{cases}$$

a. $(3, 9, 7)$ **b.** $(2, -4, -1)$

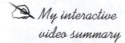 *My interactive video summary*

✎ **Solutions**

a. To determine if the **ordered triple** $(3, 9, 7)$ is a **solution** to the system, substitute the values for the variables into each equation to see if a true statement results. To be a solution to the system, the ordered triple must be a solution to each equation in the system.

	First Equation	Second Equation	Third Equation
Original equation:	$3x + y - 2z = 4$	$2x - 2y + 3z = 9$	$x + y - z = 5$
Substitute:	$3(3) + (9) - 2(7) \overset{?}{=} 4$	$2(3) - 2(9) + 3(7) \overset{?}{=} 9$	$(3) + (9) - (7) \overset{?}{=} 5$
Simplify:	$9 + 9 - 14 \overset{?}{=} 4$	$6 - 18 + 21 \overset{?}{=} 9$	$3 + 9 - 7 \overset{?}{=} 5$
	$4 = 4$ True	$9 = 9$ True	$5 = 5$ True

Since the ordered triple makes each of the three equations true, $(3, 9, 7)$ is a solution to the system.

b. Check the ordered triple $(2, -4, -1)$ on your own. Confirm your **answer**, or watch this **interactive video** for a detailed solution to both parts.

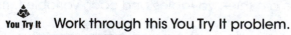
You Try It Work through this You Try It problem.

Work Exercises 1 and 2 in this eText or in the MyMathLab Study Plan.

OBJECTIVE 2   SOLVE SYSTEMS OF LINEAR EQUATIONS IN THREE VARIABLES

Graphically, we can visualize solutions to systems of linear equations in two variables as intersection points of lines. These systems could have one solution (intersecting lines), no solution (parallel lines), or infinitely many solutions (coinciding lines). These same situations apply when solving **systems of linear equations in three variables**, but graphically, we view the solutions as intersection points of **planes**.

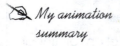 *My animation summary*

⊛ **Concept Animation** Watch this **animation** for an overview of the possibilities for systems of linear equations in three variables. **Figure 5** summaries the possibilities.

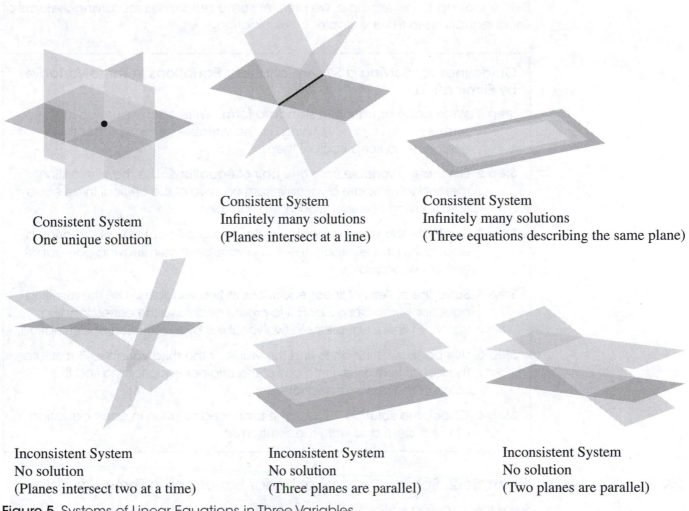

Consistent System
One unique solution

Consistent System
Infinitely many solutions
(Planes intersect at a line)

Consistent System
Infinitely many solutions
(Three equations describing the same plane)

Inconsistent System
No solution
(Planes intersect two at a time)

Inconsistent System
No solution
(Three planes are parallel)

Inconsistent System
No solution
(Two planes are parallel)

Figure 5 Systems of Linear Equations in Three Variables

To solve **systems of linear equations in three variables** by graphing would require us to graph **planes** in three dimensions, which is not a practical task. Instead, we will focus on solving systems of linear equations in three variables by applying the **elimination method** used for solving systems of linear equations in two variables.

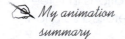 *My animation summary* ⊛ **Concept Animation** Watch this animation to see how a process called **back substitution** was used to solve systems of linear equations in two variables.

A similar process is used to solve a **system of linear equations in three variables**. The goal is to reduce the system of three equations in three variables down to a system of two equations in two variables. At that point, we can reduce the two-equation system to a single equation in one variable and easily solve that equation. Using **back substitution**, we can find the values of the other two variables. See Figure 6.

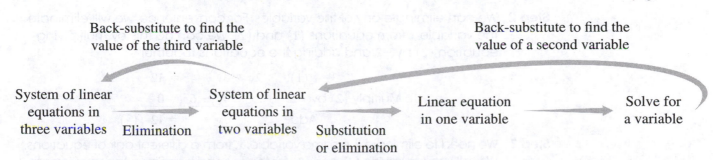

Back-substitute to find the value of the third variable

Back-substitute to find the value of a second variable

System of linear equations in three variables → Elimination → System of linear equations in two variables → Substitution or elimination → Linear equation in one variable → Solve for a variable

Figure 6 Solving a System of Linear Equations in Three Variables

Before looking at an example, we present some **guidelines** for solving systems of linear equations in three variables by elimination.

Guidelines for Solving a System of Linear Equations in Three Variables by Elimination

Step 1 **Write each equation in standard form.** Write each equation in the form $Ax + By + Cz = D$, lining up the variable terms. Number the equations to keep track of them.

Step 2 **Eliminate a variable from one pair of equations.** Use the elimination method to eliminate a variable from any two of the original three equations, leaving one equation in two variables.

Step 3 **Eliminate the same variable again.** Use a different pair of the original equations and eliminate the same variable again, leaving one equation in two variables.

Step 4 **Solve the system of linear equations in two variables.** Use the resulting equations from Steps 2 and 3 to create and solve the corresponding system of linear equations in two variables by **substitution** or **elimination**.

Step 5 **Use** back substitution **to find the value of the third variable.** Substitute the results from Step 4 into any of the original equations to find the value of the remaining variable.

Step 6 **Check the solution.** Check the proposed solution in each equation of the system and write the solution set.

📝 *My video summary* ▶ **Example 2 Solving Systems of Linear Equations in Three Variables**

Solve the following system:

$$\begin{cases} 2x + 3y + 4z = 12 \\ x - 2y + 3z = 0 \\ -x + y - 2z = -1 \end{cases}$$

Solution Follow the guidelines for solving systems of linear equations in three variables. Work through the following, or watch this **video** for a detailed solution.

Step 1 The equations are already in **standard form** and all the variables are lined up. Rewrite the system and number each equation.

$$\begin{cases} 2x + 3y + 4z = 12 & (1) \\ x - 2y + 3z = 0 & (2) \\ -x + y - 2z = -1 & (3) \end{cases}$$

Step 2 We can eliminate any of the variables. For convenience, we will eliminate the variable x from equations (1) and (2). We can do this by multiplying equation (2) by -2 and adding the equations together.

$$\begin{array}{rl} (1): & 2x + 3y + 4z = 12 \\ \text{Multiply (2) by } -2: & \underline{-2x + 4y - 6z = 0} \\ \text{Add:} & 7y - 2z = 12 \quad (4) \end{array}$$

Step 3 We need to eliminate the same variable, x, from a different pair of equations. We will use equations (2) and (3) for the second pair. Since the **coefficients**

of x in these equations have the same **absolute value**, but opposite signs, we can eliminate the variable by simply adding the equations.

$$(2): \quad x - 2y + 3z = 0$$
$$(3): \quad \underline{-x + y - 2z = -1}$$
$$\text{Add:} \qquad -y + z = -1 \quad (5)$$

Step 4 Combining equations (4) and (5), form a system of linear equations in two variables.

$$\begin{cases} 7y - 2z = 12 & (4) \\ -y + z = -1 & (5) \end{cases}$$

To solve this system, use the **elimination method** by multiplying equation (5) by 7 and adding the result to equation (4) to eliminate the variable y.

$$(4): \qquad 7y - 2z = 12$$
$$\text{Multiply (5) by 7:} \quad \underline{-7y + 7z = -7}$$
$$\text{Add:} \qquad\qquad 5z = 5$$
$$\text{Divide by 5:} \qquad\qquad z = 1$$

Since $z = 1$, back-substitute this into (5) to solve for y.

$$(5): \qquad -y + z = -1$$
$$\text{Substitute 1 for } z: \quad -y + (1) = -1$$
$$\text{Simplify:} \qquad -y + 1 = -1$$
$$\text{Subtract 1:} \qquad\qquad -y = -2$$
$$\text{Divide by } -1: \qquad\qquad y = 2$$

When **back-substituting**, we can use either of the equations, but often one equation is preferred over another. Can you see why we chose to use equation (5) instead of equation (4)? View this **explanation**.

Step 5 Substitute 2 for y and 1 for z in any of the original equations, (1), (2), or (3), and solve for x. Back-substitute using equation (2).

$$(2): \qquad x - 2y + 3z = 0$$
$$\text{Substitute 2 for } y \text{ and 1 for } z: \quad x - 2(2) + 3(1) = 0$$
$$\text{Multiply:} \qquad x - 4 + 3 = 0$$
$$\text{Simplify:} \qquad x - 1 = 0$$
$$\text{Add 1:} \qquad\qquad x = 1$$

The solution to the system is the **ordered triple** $(1, 2, 1)$.

Step 6 View the check.

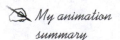 **You Try It** Work through this You Try It problem.

Work Exercises 3–5 in this eText or in the MyMathLab Study Plan.

My animation summary

⊛ **Concept Animation** Step 2 in the previous example involved choosing one of the three variables in the system to eliminate. How do you know which variable to eliminate? Watch this **animation** to find out.

Now let's solve a system when one or more equations has missing terms.

My video summary ▶ **Example 3 Solving Systems of Linear Equations in Three Variables with Missing Terms**

Solve the following system:

$$\begin{cases} 2x + y = 13 \\ 3x - 2y + z = 8 \\ x + 2y - 3z = 5 \end{cases}$$

Solution Follow the guidelines for solving a system of linear equations in three variables.

Step 1 First, we rewrite and number each equation, lining up the variables. Notice that the first equation has a variable term missing, so we put a gap in its place.

$$\begin{cases} 2x + y = 13 \quad (1) \\ 3x - 2y + z = 8 \quad (2) \\ x + 2y - 3z = 5 \quad (3) \end{cases}$$

Step 2 We can eliminate any of the variables, but notice that one equation already has z eliminated. By selecting z as the variable to eliminate, we can move directly to Step 3.

Step 3 Looking at equations (2) and (3), it is tempting to add these equations to eliminate y. However, in Step 2, since we selected z as the variable to eliminate, we need to eliminate z again in this step.

Continue working through the solution on your own. Check your **answer**, or watch this **video** for a detailed solution.

You Try It Work through this **You Try It** problem.

Work Exercises 6–10 in this eText or in the MyMathLab Study Plan.

What if the system of equations includes fractions as coefficients?

My video summary ▶ **Example 4 Solving Systems of Linear Equations in Three Variables Involving Fractions**

Solve the following system:

$$\begin{cases} \dfrac{1}{2}x + y + \dfrac{2}{3}z = 2 \\[2mm] \dfrac{3}{4}x + \dfrac{5}{2}y - 2z = -7 \\[2mm] x + 4y + 2z = 4 \end{cases}$$

Solution If an equation in the system contains fractions, then it is often helpful to **clear the fractions** first. After doing this, follow the **guidelines** for solving a system of linear equations in three variables. Check your **answer**, or watch this **video** for a detailed solution.

You Try It Work through this **You Try It** problem.

Work Exercises 11–14 in this eText or in the MyMathLab Study Plan.

OBJECTIVE 3   SOLVE INCONSISTENT AND DEPENDENT SYSTEMS IN THREE VARIABLES

Systems of linear equations in three variables may be **inconsistent** or may include **dependent equations**. If, in our solution process, we find a **contradiction**, then the system is inconsistent and has no solution. In the two-variable case, this occurred if we had **parallel lines** because the lines had no points in common.

For the three-variable case, the system will be inconsistent if all three **planes** have no points in common. This would happen if all the planes were parallel, but would also happen if two or none of the planes are parallel, as shown in Figure 7.

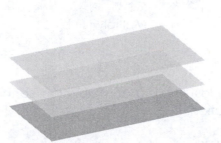

Inconsistent System
No solution
(Planes intersect two at a time)

Inconsistent System
No solution
(Three planes are parallel)

Inconsistent System
No solution
(Two planes are parallel)

Figure 7

When solving **systems of linear equations in two variables**, encountering an **identity** meant the system was **dependent** and had an infinite number of **ordered-pair solutions**. This happened when the lines were **coinciding**.

Identifying dependent systems in three variables takes more work because we must consider all possible pairings of the equations in the system. Obtaining an identity with one pairing is not sufficient to say the system is dependent. This idea is illustrated in the following example.

My video summary ▶ **Example 5 Solving Systems of Linear Equations in Three Variables with No Solution**

Solve the following system:

$$\begin{cases} x - y + 2z = 5 \\ 3x - 3y + 6z = 15 \\ -2x + 2y - 4z = 7 \end{cases}$$

Solution Use the guidelines for solving a system of linear equations in three variables. Work through the following, or watch this **video** for the complete solution.

Step 1 The equations are already in standard form with all the variables lined up. Rewrite the system and number each equation.

$$\begin{cases} x - y + 2z = 5 & (1) \\ 3x - 3y + 6z = 15 & (2) \\ -2x + 2y - 4z = 7 & (3) \end{cases}$$

Step 2 For convenience, eliminate the variable x from equations (1) and (2). To do this, multiply equation (1) by -3 and add the equations together.

$$\text{Multiply (1) by } -3: \quad -3x + 3y - 6z = -15$$
$$(2): \quad \underline{3x - 3y + 6z = 15}$$
$$\text{Add:} \qquad\qquad\qquad\quad 0 = 0 \qquad \text{True}$$

The last line is an **identity**. In the two-variable case, we would stop and say that the system had an infinite number of solutions. However, in systems of three variables, this is not necessarily the case.

Step 3 Continue the process and eliminate the variable x from the pairing of equations (1) and (3). To do this, multiply equation (1) by 2 and add the equations.

$$\text{Multiply (1) by } 2: \quad 2x - 2y + 4z = 10$$
$$(3): \quad \underline{-2x + 2y - 4z = 7}$$
$$\text{Add:} \qquad\qquad\qquad\quad 0 = 17 \quad \text{False}$$

The last line is a **contradiction**, so the system is **inconsistent** and has no solution.

You Try It Work through this You Try It problem.

Work Exercises 15 and 16 in this eText or in the MyMathLab Study Plan.

My video summary ▶ **Concept Video** As we have just seen, finding an **identity** in the three-variable case does not necessarily mean the system is **dependent** and has an infinite number of solutions. Watch this **video** for a more detailed explanation.

Example 6 Solving Systems of Dependent Linear Equations in Three Variables

Solve the following system:

$$\begin{cases} x - y + 2z = 5 \\ 3x - 3y + 6z = 15 \\ -2x + 2y - 4z = -10 \end{cases}$$

My video summary ▶ **Solution** Work through the following, or watch this **video** for the complete solution.

$$\begin{cases} x - y + 2z = 5 & (1) \\ 3x - 3y + 6z = 15 & (2) \\ -2x + 2y - 4z = -10 & (3) \end{cases}$$

For this system, notice that multiplying equation (1) by 3 gives us equation (2) and multiplying equation (1) by -2 gives us equation (3). Therefore, all three equations describe the same **plane** and the system is dependent.

Recall from the previous section that we can write the solutions to a dependent system in two ways: **set-builder notation** and **ordered-pair notation** (or in the three-variable case, **ordered-triple notation**). The solution set for this system can be written as $\{(x, y, z) \mid x - y + 2z = 5\}$ in set-builder notation or $(5 + y - 2z, y, z)$ in ordered-triple notation. To see some variations, select one of the following:

Set-Builder Notation   Ordered-Triple Notation

You Try It Work through this You Try It problem.

Work Exercises 17 and 18 in this eText or in the MyMathLab Study Plan.

Which notation should be used to express the solution of a dependent system? View this suggestion.

Example 7 Solving Systems of Dependent Linear Equations in Three Variables

Solve the following system:

$$\begin{cases} 4x + y - 5z = -1 & (1) \\ -2x - y + z = 3 & (2) \\ x + y + z = -4 & (3) \end{cases}$$

Solution Work through the following, or watch this **video** for the complete solution.

Eliminating y using equations (1) and (2) results in the equation $2x - 4z = 2$. Eliminating y again using equations (1) and (3) results in the equation $3x - 6z = 3$. Combining these two equations gives a two-variable system:

$$\begin{cases} 2x - 4z = 2 & (4) \\ 3x - 6z = 3 & (5) \end{cases}$$

If we divide equation (4) by 2 and divide equation (5) by 3, then we see that the two equations in the system are equivalent. Therefore, this two-variable system has an infinite number of solutions.

Solving either equation (4) or (5) for x gives the relationship $x = 2z + 1$. When we substitute this result into any of the three original equations and solve for y, we have $y = -3z - 5$. Since two of the variables are written in terms of the third, the **ordered-triple solution** is $(2z + 1, -3z - 5, z)$.

 You Try It Work through this You Try It problem.

Work Exercises 19 and 20 in this eText or in the MyMathLab Study Plan.

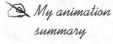 *My animation summary*

 Concept Animation Having an infinite number of ordered-triple solutions does not mean that all ordered triples are solutions to a system of linear equations in three variables. Watch this animation for an explanation.

OBJECTIVE 4 USE SYSTEMS OF LINEAR EQUATIONS IN THREE VARIABLES TO SOLVE APPLICATION PROBLEMS

Building on the **problem-solving strategy** using systems of equations in two variables, we now look at applications involving systems of linear equations in three variables.

My video summary ▶ **Example 8 Real-Time Strategy Game**

While playing a real-time strategy game, Joel created military units to defend his town: warriors, skirmishers, and archers. Warriors require 20 units of food and 50 units of gold. Skirmishers require 25 units of food and 35 units of wood. Archers require 32 units of wood and 32 units of gold. If Joel used 506 units of gold, 606 units of wood, and 350 units of food to create the units, how many of each type of military unit did he create?

Solution

Step 1 **Define the Problem.** We want to find the number of each type of military unit created. There are three types of units: warriors, skirmishers, and archers. Each unit requires a certain amount of gold, wood, and food. We know how much of each resource is needed for each unit, and we know the total amount of each resource that is used.

This information can be summarized in the following table.

	Each Warrior	Each Skirmisher	Each Archer	Total
Units of Gold	50	0	32	506
Units of Wood	0	35	32	606
Units of Food	20	25	0	350

Step 2 **Assign Variables.** Let W, S, and A represent the number of warrior, skirmisher, and archer units, respectively.

Step 3 **Translate into a System of Equations.** We need to translate the given information into three equations to form a system. When writing a system of equations, totals are a good place to start. We know the total units of gold, wood, and food used, and we know how much of each resource is required by the individual units. Therefore, we can write one equation based on total gold, a second equation based on total wood, and a third equation based on total food. A total of 506 units of gold were used. Since warriors require 50 units of gold and archers require 32 units of gold, we write the following equation.

$$\overbrace{50W}^{\text{Gold for warriors}} + \overbrace{32A}^{\text{Gold for archers}} = \overbrace{506}^{\text{Total gold}}$$

Note that there is no variable term involving S because skirmishers do not require gold.

A total of 606 units of wood were used. Since skirmishers require 35 units of wood and archers require 32 units of wood, the second equation is:

$$\overbrace{35S}^{\text{Wood for skirmishers}} + \overbrace{32A}^{\text{Wood for archers}} = \overbrace{606}^{\text{Total wood}}$$

Again, notice that there is a variable term missing because warriors do not require wood.

A total of 350 units of food were used. Since warriors require 20 units of food and skirmishers require 25 units of food, the third equation is:

$$\overbrace{20W}^{\text{Food for warriors}} + \overbrace{25S}^{\text{Food for skirmishers}} = \overbrace{350}^{\text{Total food}}$$

Step 4 Solve the System. Using the three equations, we form the system

$$\begin{cases} 50W \quad\;\;\; + 32A = 506 \\ \qquad 35S + 32A = 606 \\ 20W + 25S \qquad\;\; = 350 \end{cases}$$

Now solve this system using the **elimination method**. Remember to check the reasonableness of your answer. Check your **answer**, or watch this **video** for a detailed solution.

 You Try It Work through this You Try It problem.

Work Exercises 21 and 22 in this eText or in the MyMathLab Study Plan.

My video summary ▶ **Example 9 Buy Clothes Online**

Wendy ordered 30 T-shirts online for her three children. Small T-shirts cost $4 each, medium T-shirts cost $5 each, and large T-shirts are $6 each. She spent $40 more for the large T-shirts than for the small T-shirts. Wendy's total bill was $154. How many T-shirts of each size did she buy?

Solution Use the problem-solving strategy. Begin by defining some variables. Let S, M, and L represent the number of small, medium, and large T-shirts, respectively. With the given information, we can write and solve a system of three linear equations.

One equation uses the total number of shirts purchased; a second equation uses the total amount spent, and a third equation uses the fact that Wendy spent $40 more on large T-shirts than on small T-shirts. Finish the problem on your own. Check your **answer**, or watch this **video** for a detailed solution.

 You Try It Work through this You Try It problem.

Work Exercises 23–29 in this eText or in the MyMathLab Study Plan.

3.2 Exercises

In Exercises 1 and 2, two ordered triples are given. Determine if each ordered triple is a solution to the given system.

1. $(-1, 1, -2), (1, -1, 2)$

$$\begin{cases} x + y + z = -2 \\ -x - 3y - 2z = 2 \\ 2x - 2y + 5z = -14 \end{cases}$$

2. $(2, -1, 4), (-2, 1, -4)$

$$\begin{cases} \dfrac{1}{2}x + 3y - z = 6 \\ -x + y + \dfrac{1}{4}z = 2 \\ x - 4y + z = -10 \end{cases}$$

In Exercises 3–14, solve each system of linear equations.

3. $\begin{cases} x + y + z = 4 \\ 2x - y - 2z = -10 \\ -x - y + 3z = 8 \end{cases}$

4. $\begin{cases} x - 2y + z = 6 \\ 2x + y - 3z = -3 \\ x - 3y + 3z = 10 \end{cases}$

5. $\begin{cases} x - 2y + 2z = 2 \\ 3x + 2y - 2z = -1 \\ x - y - 2z = 0 \end{cases}$

6. $\begin{cases} -4x + 5y + 9z = -9 \\ x - 2y + z = 0 \\ 2y - 8z = 8 \end{cases}$

7. $\begin{cases} 2x + 2y + z = 9 \\ x \qquad + z = 4 \\ 4y - 3z = 17 \end{cases}$

8. $\begin{cases} x - y = 7 \\ y - z = 2 \\ x + z = 1 \end{cases}$

3.2 Systems of Linear Equations in Three Variables **3-29**

9. $\begin{cases} 4x + 6y - 6z = -12 \\ -2x - 3y + 6z = 9 \\ 6x - 2z = -2 \end{cases}$

10. $\begin{cases} x - 5y + 3z = -17 \\ x -   z = 0 \\ 4x + y = -11 \end{cases}$

11. $\begin{cases} x - \dfrac{1}{2}y + \dfrac{1}{2}z = -3 \\ x + \phantom{\dfrac{1}{2}}y - \phantom{\dfrac{1}{2}}z = 0 \\ -3x - 3y + 4z = 1 \end{cases}$

12. $\begin{cases} \dfrac{1}{3}x - \dfrac{2}{3}y + z = 0 \\ \dfrac{1}{2}x - \dfrac{3}{4}y + z = -\dfrac{1}{2} \\ -2x - y + z = 7 \end{cases}$

13. $\begin{cases} x + y + 10z = 3 \\ \dfrac{1}{2}x - y + z = -\dfrac{5}{6} \\ -2x + 3y - 5z = \dfrac{7}{3} \end{cases}$

14. $\begin{cases} x + \dfrac{1}{3}y - z = 6 \\ y + \dfrac{1}{2}z = 1 \\ \dfrac{1}{2}x + y - \dfrac{1}{4}z = \dfrac{7}{2} \end{cases}$

In Exercises 15–20, determine if the system has no solution or infinitely many solutions. If the system has infinitely many solutions, describe the solution with the equation of a plane or an ordered triple in terms of one variable.

15. $\begin{cases} x - 4y + 2z = 7 \\ \dfrac{1}{2}x - 2y + z = 1 \\ -3x + y - 4z = 9 \end{cases}$

16. $\begin{cases} 4x - y + z = 8 \\ x + y + 3z = 2 \\ 3x - 2y - 2z = 5 \end{cases}$

17. $\begin{cases} 2x + 6y - 4z = 8 \\ -x - 3y + 2z = -4 \\ x + 3y - 2z = 4 \end{cases}$

18. $\begin{cases} 2x - y + z = -6 \\ x - \dfrac{1}{2}y + \dfrac{1}{2}z = -3 \\ 4x - 2y + 2z = -12 \end{cases}$

19. $\begin{cases} x + 2y - z = 11 \\ x + 3y - 2z = 14 \\ 3x + 7y - 4z = 36 \end{cases}$

20. $\begin{cases} x - y + z = 5 \\ 2x + 3y - 3z = -5 \\ 3x + 2y - 2z = 0 \end{cases}$

In Exercises 21–29, solve using a system of linear equations in three variables.

21. **Real-Time Strategy Game** While playing a real-time strategy game, Arvin created military units for a battle: long swordsmen, spearmen, and crossbowmen. Long swordsmen require 60 units of food and 20 units of gold. Spearmen require 35 units of food and 25 units of wood. Crossbowmen require 25 units of wood and 45 units of gold. If Arvin used 1975 units of gold, 1375 units of wood, and 1900 units of food to create the units, how many of each type of military unit did he create?

22. **Concession Stand** The concession stand at a school basketball tournament sells hot dogs, hamburgers, and chicken sandwiches. During one game, the stand sold 16 hot dogs, 14 hamburgers, and 8 chicken sandwiches for a total of $89.00. During a second game, the stand sold 10 hot dogs, 13 hamburgers, and 5 chicken sandwiches for a total of $66.25. During a third game, the stand sold 4 hot dogs, 7 hamburgers, and 7 chicken sandwiches for a total of $49.75. Determine the price of each product.

23. **Ordering Pizza** Ben ordered 35 pizzas for an office party. He ordered three types: cheese, supreme, and pepperoni. Cheese pizza costs $9 each, pepperoni pizza costs $12 each, and supreme pizza costs $15 each. He spent exactly twice as much on the pepperoni pizzas as he did on the cheese pizzas. If Ben spent $420, how many pizzas of each type did he buy?

24. **Theater Tickets** On opening night of a production of *Annie*, 577 tickets were sold for a total of $79,527. Adult tickets cost $158 each, children's tickets cost $63 each, and senior citizen tickets cost $121 each. If the total number of adult and children tickets sold exceeded twice the number of senior citizen tickets sold by 211 tickets, then how many tickets of each type were sold?

25. **NCAA Basketball** Shabazz Napier was the leading scorer of the 2014 NCAA Basketball champions, the Connecticut Huskies. Napier scored a total of 720 points during the 2014 season. He made 81 more one-point free throws than two-point field goals, and his number of two-point field goals was 48 less than twice his number of three-point field goals. How many free throws, two-point field goals, and three-point field goals did Napier make during the 2014 season? (*Source*: espn.com)

26. **Instagram Users** The number of Instagram users, y (in millions), in 2014 can be modeled by the equation $y = ax^2 + bx + c$, where x represents the age of the user. Using the ordered-pair solutions $(20, 7)$, $(30, 9)$, and $(50, 1)$, create a system of linear equations in three variables for a, b, and c. Do this by substituting each ordered-pair solution into the model, creating an equation in three variables. Solve the resulting system to find the coefficients of the model. Then use the model to predict the number of Instagram users who were 24 years old in 2014. (*Source*: www.emarketer.com)

27. **Rental Properties** Chris decides to take advantage of the poor housing market and invest in rental properties. She purchases some townhouses that rent for $820 per month, some condos that rent for $750 per month, and some houses that rent for $1025 per month. She owns 28 units with the number of houses being four more than the number of condos. If all her units are rented, she brings in $24,995 each month in rent. How many of each type of unit does she own?

28. **Loose Change** Brigitte cleans out the loose change from the cup holder in her car (pennies, nickels, and dimes). She finds three more nickels than dimes, and five more pennies than twice the number of nickels. If the total value of the coins is $2.30, how many of each coin did she have in her cup holder?

29. **College costs** Last year, Blake spent $17,850 on his education. He spent $1400 more on room and board than on tuition, and the total he spent on tuition and course materials was $150 less than what he spent on room and board. How much did he spend on tuition, course materials, and room and board?

3.3 More Problem Solving with Systems of Linear Equations

THINGS TO KNOW

Before working through this section, be sure you are familiar with the following concepts:

VIDEO   ANIMATION   INTERACTIVE

You Try It
1. Solve a Formula for a Given Variable (Section 1.5, Objective 1)   ▶

You Try It
2. Solve Application Problems Involving Geometric Formulas (Section 1.5, Objective 2)   ▶

You Try It
3. Solve Application Problems Involving Uniform Motion (Section 1.5, Objective 4)   ▶   ✺

4. Solve Application Problems Involving Mixtures (Section 1.5, **Objective 5**)

 5. Use Systems of Linear Equations in Two Variables to Solve Application Problems (Section 3.1, **Objective 6**)

 6. Use Systems of Linear Equations in Three Variables to Solve Application Problems (Section 3.2, **Objective 4**) ▶

OBJECTIVES

1 Use Systems of Linear Equations to Solve Uniform Motion Problems

2 Use Systems of Linear Equations to Solve Geometry Problems

3 Use Systems of Linear Equations to Solve Mixture Problems

4 Use Systems of Linear Equations to Solve Simple Interest Problems

---

OBJECTIVE 1 **USE SYSTEMS OF LINEAR EQUATIONS TO SOLVE UNIFORM MOTION PROBLEMS**

When the **problem-solving strategy** for applications of linear equations was first introduced, we saw situations with more than one unknown quantity. At that time, we defined one **variable** and expressed all other unknowns in terms of that one variable. In this chapter, we have used more than one variable when solving applications involving **systems of linear equations**. Now, let's apply the **problem-solving strategy** for systems of linear equations to earlier types of application problems where we used a single equation and one variable. We start with **uniform motion** problems.

My video summary ▶ **Example 1 Geocaching**

Geocaching is a worldwide hide-and-seek treasure hunt in which players (called *geocachers*) hide a cache (collection of items or trinkets such as coins or buttons) and share its GPS location online. Other players use a GPS device to help them locate the cache.

Andrea receives an e-mail about a new geocache with a Travel Bug (a trackable item) that is within walking distance. She leaves to find the cache at the same time that another geocacher, Colin, leaves to locate the cache. They both arrive at the cache 45 minutes later. If the combined distance traveled is 4.5 miles and Andrea walks 0.4 mph faster than Colin, find their average speeds.

Solution We apply our problem-solving strategy using systems of linear equations.

Step 1 **Define the Problem.** We want to find the average speed of both Andrea and Colin. The total distance they walk is 4.5 miles and Andrea walks 0.4 mph faster than Colin. We also know they each walk for 45 minutes (0.75 hours).

(eText Screens 3.3-1–3.3-34)

Step 2 **Assign Variables.** We know the time traveled and the distance traveled but not the average speeds. We can define two **variables** to represent the average speeds.

$$x = \text{Andrea's average speed}$$

$$y = \text{Colin's average speed}$$

Step 3 **Translate into a System of Equations.** This is a **uniform motion** problem, so we need the distance formula. The distance traveled by each person is given by the product of the person's speed and time traveled. The total distance is 4.5 miles, so we write the following equation:

$$\underbrace{0.75x}_{\substack{\text{Distance Andrea}\\\text{traveled}}} + \underbrace{0.75y}_{\substack{\text{Distance Colin}\\\text{traveled}}} = 4.5$$

We also know that Andrea walks 0.4 mph faster than Colin. This gives the following second equation:

$$\underbrace{x}_{\substack{\text{Andrea's}\\\text{speed}}} = \underbrace{y}_{\substack{\text{Colin's}\\\text{speed}}} + 0.4$$

Writing the two equations together gives the system:

$$\begin{cases} 0.75x + 0.75y = 4.5 \\ x = y + 0.4 \end{cases}$$

Note that in the first equation we used 0.75 (hours) instead of 45 (minutes). Why?

Step 4 **Solve the System.** Solve the system by **graphing**, **substitution**, or **elimination** on your own. The second equation is already solved for a variable, so substitution may be preferable. Then complete Steps 5 and 6. Check your answer, or watch this **video** for a detailed solution.

You Try It Work through this **You Try It** problem.

Work Exercises 1 and 2 in this eText or in the MyMathLab Study Plan.

My video summary ▶ **Example 2 High Adventure Trek**

Dimitri and Ricardo joined a high adventure trek in the Joshua Tree backcountry. One day the participants decided to hike to a site where they would go rock climbing. Ricardo's group left for the site at 8:30 A.M. and hiked at a rate of 3 mph. Dimitri's group had to finish packing gear and did not leave for the site until 25 minutes later, following the same path. If Dimitri's group, hiking at a rate of 4 mph, reached the rock climbing site at the same time as Ricardo's group, how far was the site from their base camp? At what time did the two groups arrive at the rock climbing site?

Solution Apply the problem-solving strategy using systems of linear equations.

Step 1 **Define the Problem.** We must find the distance between the camp and the rock climbing site, and the time when the groups arrived at the site. We know that Ricardo's group left at 8:30 A.M. traveling at 3 mph, while Dimitri's group left 25 minutes later traveling at 4 mph. Since they traveled the same path, the distance is the same for both groups.

Step 2 **Assign Variables.** We know the rate of travel, so we can find the distance traveled if we know the travel time. (Note that 1 hour = 60 minutes.) Let's define two variables to represent the travel times for Ricardo's group and Dimitri's group.

$$x = \text{hiking time for Ricardo's group}$$

$$y = \text{hiking time for Dimitri's group}$$

Step 3 **Translate into a System of Equations.** This is a uniform motion problem, so use the **distance formula** $d = r \cdot t$. The distance traveled by each group was the same, so we get the following equation:

$$\underbrace{3x}_{\substack{\text{Distance for} \\ \text{Ricardo's group}}} = \underbrace{4y}_{\substack{\text{Distance for} \\ \text{Dimitri's group}}}$$

Ricardo's group hiked 25 minutes longer than Dimitri's group. Since 25 minutes $= \dfrac{25}{60} = \dfrac{5}{12}$ hours, we get the following second equation:

$$\underbrace{x}_{\substack{\text{Time for} \\ \text{Ricardo's} \\ \text{group}}} = \underbrace{y}_{\substack{\text{Time for} \\ \text{Dimitri's} \\ \text{group}}} + \dfrac{5}{12}$$

Writing the two equations together gives the following system:

$$\begin{cases} 3x = 4y \\ x = y + \dfrac{5}{12} \end{cases}$$

Step 4 **Solve the System.** Solve the system by **graphing, substitution,** or **elimination.** Since the variables are travel times, multiply one of the times by the corresponding rate to find the distance traveled. To get the arrival time, add the travel time to the appropriate starting time. Complete Steps 5 and 6 on your own. See the **answer,** or watch this **video** for a detailed solution.

You Try It Work through this You Try It problem.

Work Exercises 3 and 4 in this eText or in the MyMathLab Study Plan.

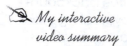
My interactive video summary

Example 3 Finding Different Times

Shawn is training for the Dirty Duo running-and-bicycling race. During a three-hour training session, his total distance cycling and running was 33 miles. If he cycled at a rate of 18 miles per hour and ran at a rate of 6 miles per hour, how much time did he spend doing each activity?

Solution Apply the problem-solving strategy for applications using systems.

Step 1 We want to find the time spent cycling and the time spent running. The total distance is 33 miles and the total time is 3 hours. We also know that Shawn can cycle 18 miles in one hour and run 6 miles in one hour.

Step 2   We know the time traveled and the distance traveled but not the individual times. Define two variables to represent the times.

$$x = \text{time spent cycling}$$

$$y = \text{time spent running}$$

Step 3   The distance traveled for each activity is given by the product of the person's speed and time traveled. Because the total distance is 33 miles, we can write the following equation:

$$\underbrace{18x}_{\text{Distance cycling}} + \underbrace{6y}_{\text{Distance running}} = \underbrace{33}_{\text{Total distance}}$$

And, because the total time was three hours, we can write the following second equation:

$$\underbrace{x}_{\text{Time cycling}} + \underbrace{y}_{\text{Time running}} = \underbrace{3}_{\text{Total time}}$$

Writing the two equations together gives the system:

$$\begin{cases} 18x + 6y = 33 \\ x + y = 3 \end{cases}$$

Step 4   Solve this system by **substitution** or **elimination**, then complete Steps 5 and 6 on your own. View the **answer**, or watch this **interactive video** for a detailed solution.

You Try It   Work through this You Try It problem.

Work Exercises 5 and 6 in this eText or in the My Math Lab **Study Plan.**

Some uniform motion problems involve one motion that works with or against another. Walking up the down escalator or paddling upstream are examples of two motions working against each other. Kicking a football with the wind or walking in the direction of motion on a moving sidewalk are examples of two motions working together. For these problems, remember that when motions work together, the **rates** are added, but when motions work against each other, the rates are subtracted.

My interactive video summary   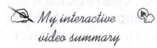 **Example 4 Finding Different Rates**

A jet plane travels 1950 miles in 3.9 hours going with the wind. On the return trip, the plane must fly into the wind and the travel time increases to 5 hours. Find the speed of the jet plane in still air and the speed of the wind. Assume the wind speed is the same for both trips.

Solution   Apply the **problem-solving strategy** for applications using systems.

Step 1   We want to find the speed of the plane in still air and the speed of the wind. We know that the time of travel is 3.9 hours with the wind and 5 hours against (or into) the wind. We also know that the plane travels 1950 miles each way.

Step 2   We know the time traveled and the distance traveled but not the speeds. Define two variables to represent the speeds.

$$r = \text{speed of the plane in still air}$$

$$w = \text{wind speed}$$

Step 3 The distance traveled by the plane is given by the product of the plane's speed and time traveled. When flying with the wind, the plane's speed is found by adding the speed of the plane in still air to the wind speed. This leads to the following equation:

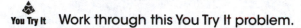

$$\underbrace{1950}_{\substack{\text{Distance}\\ \text{plane}\\ \text{traveled}}} = \underbrace{(r + w)}_{\substack{\text{Plane}\\ \text{speed}\\ \text{with the}\\ \text{wind}}} \cdot \underbrace{(3.9)}_{\substack{\text{Time of}\\ \text{travel}\\ \text{with the}\\ \text{wind}}}$$

When flying into (or against) the wind, the plane's speed is found by subtracting the speed of the wind from the speed of the plane in still air. The distance traveled is the same, so we get the following second equation:

$$\underbrace{1950}_{\substack{\text{Distance}\\ \text{plane}\\ \text{traveled}\\ \text{on return}\\ \text{trip}}} = \underbrace{(r - w)}_{\substack{\text{Plane}\\ \text{speed}\\ \text{against}\\ \text{the wind}}} \cdot \underbrace{(5)}_{\substack{\text{Time of}\\ \text{travel}\\ \text{against}\\ \text{the wind}}}$$

Writing the two equations together gives the system:

$$\begin{cases} 1950 = (r + w)(3.9) \\ 1950 = (r - w)(5) \end{cases}$$

Step 4 Solve the system by **substitution** or **elimination**, then complete Steps 5 and 6 on your own. View the **answer**, or watch this **interactive video** for a detailed solution.

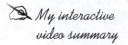 **You Try It** Work through this **You Try It** problem.

Work Exercises **7** and **8** in this eText or in the My**Math**Lab **Study Plan.**

OBJECTIVE 2 USE SYSTEMS OF LINEAR EQUATIONS TO SOLVE GEOMETRY PROBLEMS

Geometry formulas, such as those for **area** and **perimeter**, often involve more than one **variable**. In Chapter 1, we expressed all unknown quantities in terms of one variable for such problems. Now we can use **systems of linear equations in more than one variable** to solve these same types of problems.

My interactive video summary 📖 **Example 5 A Calculator Display Panel**

The display panel of a graphing calculator has the shape of a rectangle with a **perimeter of 264 millimeters.** If the length of the display panel is 18 millimeters longer than the width, find its dimensions.

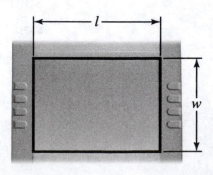

Solution

Step 1 We need to find the length and width of the display panel. We know that the perimeter is 264 mm and the length is 18 mm longer than the width.

Step 2   Let l = length and w = width.

Step 3   The perimeter of a rectangle is given by the formula $P = 2l + 2w$, so we have $2l + 2w = 264$.

Also, the length is 18 mm longer than the width, so a second equation is $l = w + 18$.

The two equations together give the following system:

$$\begin{cases} 2l + 2w = 264 \\ l = w + 18 \end{cases}$$

Step 4   Finish solving this system of equations on your own. See the **answer**, or watch this **interactive video** for a fully worked solution.

You Try It   Work through this **You Try It** problem.

Work Exercises 9–13 in this eText or in the MyMathLab **Study Plan.**

Geometry problems sometimes involve *complementary* and *supplementary* angles. The measures of two **complementary angles** add to 90°. The measures of two **supplementary angles** add to 180°. Complementary angles form a **right angle**, and supplementary angles form a **straight angle**.

My interactive video summary   **Example 6   Complementary Angles**

Find the measures of two **complementary angles** if the measure of the smaller angle is 15 degrees more than half the measure of the larger angle.

Solution   See the **answer**, or watch this **interactive video** for a detailed solution.

You Try It   Work through this **You Try It** problem.

Work Exercises 14 and 15 in this eText or in the MyMathLab **Study Plan.**

My interactive video summary   **Example 7   Supplementary Angles**

Find the measures of two **supplementary angles** if the measure of the larger angle is 20 degrees less than three times the measure of the smaller angle.

Solution   Follow the **problem-solving strategy** for applications using systems and try to solve this problem on your own. View the **answer**, or watch this **interactive video** for a detailed solution.

You Try It   Work through this **You Try It** problem.

Work Exercises 16–18 in this eText or in the MyMathLab **Study Plan.**

My video summary   **Example 8   Interior Angle Measures**

Find the measures of the three **interior angles**, marked by x, y, and z in the following triangle. Remember that two angles forming a straight line are supplementary angles, and the measures of the three interior angles of a triangle must add to 180°.

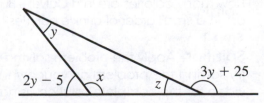

Solution

Step 1 Define the Problem. We want to find the measures of the three interior angles in the given triangle. We know that the measures of the three interior angles add to 180° and that two angles forming a straight line are supplementary angles, which means that the sum of their measures is also 180°.

Step 2 Assign Variables. We want to find the measures of the three interior angles. In the given triangle, these angles are already defined as the variables x, y, and z.

Step 3 Translate into a System of Equations. With three variables, we need a system of three linear equations. The sum of the measures of the three interior angles is 180°, so our first equation is

$$x + y + z = 180.$$

There are two sets of **supplementary angles**, which we can use to write the remaining two equations. Since supplementary angles have measures that add to 180°, we get:

$$x + (2y - 5) = 180$$
$$z + (3y + 25) = 180$$

The three equations form the following system:

$$\begin{cases} x + y + z = 180 \\ x + (2y - 5) = 180 \\ z + (3y + 25) = 180 \end{cases}$$

Writing each equation in standard form, $Ax + By + Cz = D$, gives

$$\begin{cases} x + y + z = 180 \\ x + 2y = 185 \\ 3y + z = 155 \end{cases}$$

Step 4 Solve the System. Solve the system by **elimination**. Complete Steps 5 and 6 on your own. See the **answer**, or watch this **video** for a detailed solution.

You Try It Work through this **You Try It** problem.

Work Exercises 19 and 20 in this eText or in the My**Math**Lab Study Plan.

OBJECTIVE 3   USE SYSTEMS OF LINEAR EQUATIONS TO SOLVE MIXTURE PROBLEMS

Let's explore using systems of linear equations to solve **mixture problems**.

My interactive video summary   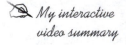 **Example 9 Counting Calories**

Logan and Payton went to Culver's for lunch. Logan ate two Butterburgers with cheese and a small order of crinkle cut fries for a total of 1042 calories. Payton ate one Butterburger with cheese and two small orders of fries for a total of 881 calories. How many calories are in a Culver's Butterburger with cheese? How many calories are in a small order of crinkle cut fries?

Solution Apply the **problem-solving strategy** for applications using systems and try solving this problem on your own. View the **answer**, or watch this **interactive video** for the complete solution using either **substitution** or **elimination**.

You Try It Work through this You Try It problem.

Work Exercises 21–23 in this eText or in the MyMathLab Study Plan.

My interactive video summary

Example 10 Mailing Packages

A shipping company delivered 160 packages one day. The cost of regular delivery is $6.50, and the cost for express delivery is $17.50. Total shipping revenue for the day was $1513. How many of each kind of delivery were made?

Solution Apply the problem-solving strategy for systems of linear equations.

Step 1 We want to find the number of regular deliveries and the number of express deliveries. We know that the total number of deliveries was 160 with a total revenue of $1513. We also know that regular deliveries cost $6.50 and express deliveries cost $17.50.

Step 2 We know the total number of deliveries and total revenue, but not the number of each delivery type. Define two variables to represent the number of each type of delivery.

$$R = \text{number of regular deliveries in one day}$$

$$E = \text{number of express deliveries in one day}$$

Step 3 For mixture problems, *totals* are good places to start when trying to write equations in a system. The first sentence tells us about the *total* number of deliveries.

A shipping company delivered 160 packages one day.

The total number of deliveries is the sum of the numbers for each delivery type. Therefore, the first equation is

$$\underbrace{R}_{\substack{\text{Regular}\\\text{deliveries}}} + \underbrace{E}_{\substack{\text{Express}\\\text{deliveries}}} = \underbrace{160}_{\substack{\text{Total}\\\text{deliveries}}}$$

The second and third sentences tell us about the *total* revenue.

The cost of regular delivery is $6.50, and the cost for express delivery is $17.50. Total shipping revenue for the day was $1513.

Revenue is found by multiplying the number of packages delivered by the charge per package. The total revenue is the sum of the revenue from each delivery type. This is our second equation.

$$\underbrace{6.50R}_{\substack{\text{Regular delivery}\\\text{revenue}}} + \underbrace{17.50E}_{\substack{\text{Express delivery}\\\text{revenue}}} = \underbrace{1513}_{\text{Total revenue}}$$

Writing the two equations together gives the system:

$$\begin{cases} R + E = 160 \\ 6.50R + 17.50E = 1513 \end{cases}$$

Step 4 Solve this system by **substitution** or **elimination**, then complete Steps 5 and 6 on your own. View the **answer**, or watch this **interactive video** for a detailed solution.

(eText Screens 3.3-1–3.3-34)

You Try It Work through this You Try It problem.

Work Exercises 24–27 in this eText or in the MyMathLab Study Plan.

My interactive video summary

Example 11 Mixing Solutions

A chemist needs 8 liters of a 50% alcohol solution but only has a 30% solution and an 80% solution available. How many liters of each solution should be mixed to form the needed solution?

Solution Apply the **problem-solving strategy** for applications using systems.

Step 1 We want to find the number of liters of a 30% alcohol solution and the number of liters of an 80% alcohol solution that must be mixed to form 8 liters of a 50% alcohol solution.

Step 2 We know the total number of liters needed for the mixture, but not the number of liters for each type of solution. Define two variables to represent the number of liters of each type.

$$x = \text{liters of 30\% alcohol solution}$$
$$y = \text{liters of 80\% alcohol solution}$$

Step 3 We know the total number of liters, so the first equation deals with the number of liters.

$$x + y = 8$$

The second equation deals with the percents, but let's consider the amount of alcohol contained in each solution. The total amount of alcohol in the mixture will equal the sum of the alcohol from each solution. To find the number of liters of alcohol from each solution, multiply each percent by the number of liters of that solution type. In words,

$$\underbrace{(\text{percent})(\text{liters of 30\%})}_{\text{Alcohol in 30\% solution}} + \underbrace{(\text{percent})(\text{liters of 80\%})}_{\text{Alcohol in 80\% solution}} = \underbrace{(\text{percent})(\text{total liters of 50\%})}_{\text{Alcohol in 50\% mixture}}$$

Use the given percents and the defined variables to find the second equation:

$$0.3(x) + 0.8(y) = 0.5(8)$$

or

$$0.3x + 0.8y = 4$$

So the two equations in the system are

$$\begin{cases} x + y = 8 \\ 0.3x + 0.8y = 4 \end{cases}$$

Step 4 Solve this system by **substitution** or **elimination**, then complete Steps 5 and 6 on your own. View the **answer**, or watch this **interactive video** for a detailed solution.

You Try It Work through this You Try It problem.

Work Exercises 28–32 in this eText or in the MyMathLab Study Plan.

OBJECTIVE 4 USE SYSTEMS OF LINEAR EQUATIONS TO SOLVE SIMPLE INTEREST PROBLEMS

Simple interest problems sometimes involve the use of linear systems in three variables.

My video summary ▶ **Example 12 Investing Money**

Tremayne had $40,000 to invest and decided to split up the money into three different accounts that pay 3%, 5%, and 6% in annual simple interest. The combined amount invested in the 5% account and the 6% account was three times the amount invested in the 3% account. If the total interest for one year was $1980, how much did Tremayne invest in each account?

Solution

Step 1 Define the Problem. We want to find the amount invested in each of the three accounts. The accounts pay 3%, 5%, and 6% in annual simple interest. Tremayne had $40,000 to invest, and total interest for one year was $1980. The combined amount in the 5% and 6% accounts was three times the amount in the 3% account.

Step 2 Assign Variables. There are three unknowns: the amount invested in each account. Let's define these three **variables**:

$$x = \text{amount invested at 3\%}$$
$$y = \text{amount invested at 5\%}$$
$$z = \text{amount invested at 6\%}$$

Step 3 Translate into a System of Equations. There are three **variables**, so there must be three **equations** in our system. As with other systems, totals are a good place to start. We know the total amount invested. Also, we know that the sum of the three amounts invested must equal the total amount invested, so the first equation is

x	$+$	y	$+$	z	$=$	40,000
Amt invested at 3%		Amt invested at 5%		Amt invested at 6%		Total amt invested

We also know the total interest after one year. The total interest must equal the sum of the interest from each investment. The second equation is

$0.03x$	$+$	$0.05y$	$+$	$0.06z$	$=$	1980
Interest from 3% investment		Interest from 5% investment		Interest from 6% investment		Total interest

To find the third equation, recall that the combined amounts in the 5% and 6% accounts was three times the amount in the 3% account. The third equation is

y	$+$	z	$=$	$3x$
Amt invested at 5%		Amt invested at 6%		Three times the amt invested at 3%

Writing the three equations together gives the following system:

$$\begin{cases} x + y + z = 40{,}000 \\ 0.03x + 0.05y + 0.06z = 1980 \\ y + z = 3x \end{cases}$$

3.3 More Problem Solving with Systems of Linear Equations **3-41**

Writing each equation in standard form, $Ax + By + Cz = D$, gives

$$\begin{cases} x + \quad y + \quad z = 40{,}000 \\ 0.03x + 0.05y + 0.06z = 1980 \\ 3x - \quad y - \quad z = 0 \end{cases}$$

Step 4 Solve the System. Solve the system by **elimination**, then complete Steps 5 and 6 on your own. See the **answer**, or watch this **video** for a detailed solution.

You Try It Work through this You Try It problem.

Work Exercises 33 and 34 in this eText or in the MyMathLab Study Plan.

3.3 Exercises

In Exercises 1–34, solve each application problem using the problem-solving strategy for systems of linear equations.

1. **Traveling Speed** Yessica leaves her house on her bicycle to locate a geocache at the same time that Zoe leaves her house heading for the same cache. Zoe lives further away than Yessica so she takes her car and travels 16 mph faster than Yessica. Their combined distance traveled is 17 miles. If both women arrive at the geocache site after 30 minutes, find their average speeds.

2. **Jogging Speed** Starting at the same point, Angie and Payton jog in opposite directions. Angie can jog 2.5 miles per hour faster than Payton. After 1.5 hours, they are 16 miles apart. How fast can each person jog?

3. **Amazing Chase** Rachel and Lindsay are in third place in the Amazing Race. Due to a time penalty, they must start the next leg 90 minutes after the second-place team. If Rachel and Lindsay travel at an average rate of 13.2 mph and the second-place team travels at an average rate of 11 mph, how long will it take the girls to catch up to the second-place team?

4. **Aerobic Exercise** As part of her training routine for basketball, Shaylle alternates between cycling and running for exercise. She cycles at a rate of 16 mph and runs at a rate of 6 mph. If she spends 8.5 hours exercising and covers a total of 96 miles, how much time did she spend on each exercise?

5. **Aquathon** Nikola competed in an aquathon (swimming and running) competition. He swam at a rate of 3 km/hr and ran at a rate of 10 km/hr for a total distance traveled of 11.5 km. If he completed the race in 1.5 hours, how long did he take to complete each part of the race?

6. **Traveling Apart** Two friends, Thaddeus and Ian, start at the same location and drive in opposite directions, but leave at different times. When they are 620 miles apart, their combined travel time is 10 hours. If Thaddeus drives at a rate of 60 miles per hour and Ian drives at a rate of 65 miles per hour, how long had each been driving?

7. **Puddle Jumper** A small plane flies roundtrip between Wichita, KS, and Columbus, OH. The trip is 784 miles each way. The trip with the wind (the wind going in the same direction) takes 3.5 hours, while the trip against the wind takes 4 hours. What is the speed of the plane in still air? What is the speed of the wind?

8. **Canoe Trip** Nicholas paddles his canoe downstream from the Lodge to Big Bend in 4 hours and then back upstream to the Lodge in 6 hours. If the distance from the Lodge to Big Bend is 18 miles, find Nicholas' speed in still water and the speed of the current.

9. **TV Dimensions** The flat screen of an LED television is 49 centimeters longer than it is wide. If the perimeter of the screen is 398 centimeters, find its dimensions.

10. **Tennis Court** A doubles tennis court has a perimeter of 228 feet. If 6 times the length of the court equals 13 times the width, what are its dimensions?

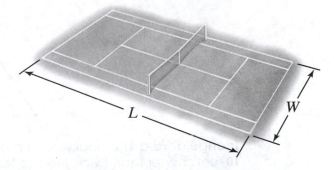

11. **Isosceles Triangle** The perimeter of an isosceles triangle is 80 centimeters. If the non-equal side is 10 centimeters shorter than each of the two equal sides, find the lengths of the three sides.

12. **Soccer Field Dimensions** The St. Louis Athletica women's soccer team began their inaugural season at Korte Stadium in Edwardsville, IL. The perimeter of the soccer field is 390 yards. If the length is 30 yards less than twice the width, what are the dimensions of the field? (*Source*: SIU-Edwardsville)

13. **Isosceles Triangle** An isosceles triangle has an angle that measures 20 degrees more than twice the two common angles. What are the three angle measures?

14. **Complementary Angles** The measure of the larger of two complementary angles is 2° more than three times the measure of the smaller angle. What are the angle measures?

15. **Complementary Angles** Two angles are complementary. The measure of the larger angle is 35 degrees less than four times the measure of smaller angle. Find the measure of each angle.

16. **Supplementary Angles** Two angles are supplementary. The measure of the larger angle is 8 degrees more than three times the measure of the smaller angle. Find the measure of each angle.

17. **Supplementary Angles** A 2009 intersection study for the City of Marysville, OH, found that Milford Avenue and Maple Street intersected at an angle much smaller than design standards. Find the measures of the angle of intersection and its supplement if the supplement measures 30 degrees more than nine times the angle of intersection. (*Source*: www.marysvilleohio.org/DocumentCenter/Home/View/81)

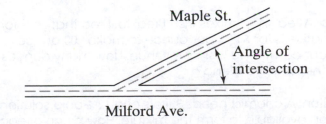

18. **Right Triangle** Recall that a right triangle contains a 90-degree angle. Also recall that the measures of the three angles in any triangle add to 180 degrees. In the figure shown, the measure of angle y is 6 degrees larger than twice the measure of angle x. Find the measures of angles x and y.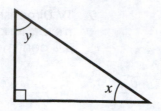

19. **Interior Angles** Find the measures of the three interior angles in the following triangle.

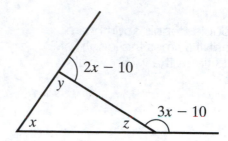

20. **Triangular Yard** The backyard of a house is in the shape of a triangle. The longest side is 10 feet shorter than twice the shortest side. The third side is 10 feet less than the longest side. If the perimeter of the yard is 167 feet, what are the lengths of the sides?

21. **Counting Carbs** Two glasses of milk and 3 snack bars have a total of 61 carbs, and 1 glass of milk and 5 snack bars have a total of 76 carbs. How many carbs are in one glass of milk? How many carbs are in one snack bar?

22. **Pastry Purchase** Bill and Mary Ann went to the Viola bakery. Bill bought 5 danish and 7 donuts for $13.30. Mary Ann bought 6 of each for $13.08. What is the price of each type of pastry?

23. **Counting Calories** At In-N-Out Burger, 6 cheeseburgers w/onion and 3 orders of french fries contain 4065 calories. 3 cheeseburgers w/onion and 2 orders of french fries contain 2230 calories. How many calories are in a cheeseburger w/onion? How many calories are in one order of french fries?

24. **Soda Mix** Elodie is buying cans of soda for the school picnic. Her choices are Coke at $0.50 per can and Dr. Pepper at $0.55 per can. She wants to provide a mixture of choices but has a limited budget. If she buys 271 cans and spends $141, how many cans of each flavor did she buy?

25. **Pancake Breakfast** A youth club held a pancake breakfast to raise money for a trip. Tickets were $3 for children and $5.50 for adults. If 279 tickets were sold and the group took in $1322, find the number of adults and the number of children that attended the breakfast.

26. **Trail Mix** Chloe purchased some dried mango and dried kiwi to make some trail mix. The dried mango cost $3.86 per pound and the dried kiwi cost $4.60 per pound. If she spent $17.29 for 4 pounds of dried fruit, how many pounds of each fruit did she have?

27. **Mixing Tea** A tea shop mixes Irish Breakfast tea that sells for $2.50 per ounce with buttered rum tea that sells for $1.75 per ounce to make 100 ounces of a new tea blend that sells for $2.10 per ounce with no loss of revenue. How many ounces of each type should be mixed to create the new tea blend?

28. **Acid Solution** A chemist needs 3 liters of a 12% acid solution. He has a 10% solution and a 20% solution available to form the mixture. How much of each should be used to form the 12% solution?

29. **Gold Jewelry** A jeweler wants to make gold chains that are 60% gold. She has a supply of 18-carat gold (75% gold) and 12-carat gold (50% gold) that can be melted to form the desired alloy. How much 18-carat gold and how much 12-carat gold should be used to form 350 ounces of the 60% alloy?

30. **Fruit Juice** To be considered "nectar" a drink must contain 40% fruit juice. Find how much pure (100%) fruit juice and how much fruit drink (10% juice) must be mixed to make 9 gallons of "nectar."

31. **Mixing Massage Oils** Kristen wants to make 4.5 ounces of a 25% jasmine essential oil. She only has a 15% jasmine mixture and a 30% jasmine mixture available. How much of each mixture should be combined to create the desired oil?

32. **Blending Wines** A winery blends two wines to make a new 12% alcohol product. The base wine is 9% alcohol and the fortifier wine is 14% alcohol. If the winery wants to make 250 gallons of the new blend, how much of the base and fortifier wines should be mixed?

33. **Investing Money** Amanda invested a total of $2200 into three separate accounts that pay 6%, 8%, and 9% annual simple interest. Amanda has three times as much money invested in the 9% account as she does in the 6% account. If the total interest for the year is $178, how much money did Amanda invest in each account?

34. **Investments** Scott invested a total of $3000 at two separate banks. One bank pays simple interest at 3% per year, while the other bank pays simple interest at 4% per year. If Scott earned $110.50 in interest during a single year, how much did he deposit at each bank?

3.4 Systems of Linear Inequalities in Two Variables

THINGS TO KNOW

Before working through this section, be sure you are familiar with the following concepts:

| | VIDEO | ANIMATION | INTERACTIVE |

You Try It 1. Determine If an Ordered Pair Is a Solution to a Linear Inequality in Two Variables (Section 2.6, Objective 1) ▶

You Try It 2. Graph a Linear Inequality in Two Variables (Section 2.6, Objective 2) ✳ ☞

You Try It 3. Solve Applications Involving Linear Inequalities in Two Variables (Section 2.6, Objective 3) ▶

OBJECTIVES

1 Determine If an Ordered Pair Is a Solution to a System of Linear Inequalities in Two Variables

2 Graph Systems of Linear Inequalities

3 Solve Applications Involving Systems of Linear Inequalities

OBJECTIVE 1 DETERMINE IF AN ORDERED PAIR IS A SOLUTION TO A SYSTEM OF LINEAR INEQUALITIES IN TWO VARIABLES

In Section 2.6, we learned about linear inequalities in two variables. In this section, we explore systems of linear inequalities in two variables.

> **Definition** System of Linear Inequalities in Two Variables
>
> A system of linear inequalities in two variables is a collection of two or more linear inequalities in two variables considered together.

The following three examples all represent systems of linear inequalities in two variables.

$$\begin{cases} 2x - 3y < 6 \\ 3x + 2y > 12 \end{cases} \qquad \begin{cases} q \ge -\dfrac{5}{2}p + 6 \\ q \le 3 \end{cases} \qquad \begin{cases} y < -x + 5 \\ x \ge 0 \\ y \ge 0 \end{cases}$$

Notice that a system of linear inequalities looks like a system of linear equations, except inequality symbols replace the equal signs.

> **Definition** Solution to a System of Linear Inequalities in Two Variables
>
> A solution to a system of linear inequalities in two variables is an ordered pair that, when substituted for the variables, makes all inequalities in the system true.

My video summary ▶ **Example 1** Determining If an Ordered Pair Is a Solution to a System of Linear Inequalities in Two Variables

Determine if each ordered pair is a solution to the following system of inequalities.

$$\begin{cases} 2x + y \ge -3 \\ x - 4y \le 12 \end{cases}$$

a. $(4, 2)$ **b.** $(2, -5)$ **c.** $(0, -3)$

Solutions Substitute the x- and y-coordinates for the variables in each inequality and simplify. If the resulting statements are both true, then the ordered pair is a solution to the system.

a.

	First inequality	Second inequality
Write the original inequalities:	$2x + y \ge -3$	$x - 4y \le 12$
Substitute 4 for x and 2 for y:	$2(4) + 2 \overset{?}{\ge} -3$	$4 - 4(2) \overset{?}{\le} 12$
Multiply:	$8 + 2 \overset{?}{\ge} -3$	$4 - 8 \overset{?}{\le} 12$
Add or subtract:	$10 \ge -3$ True	$-4 \le 12$ True

The ordered pair makes both inequalities true, so $(4, 2)$ is a solution to the system of inequalities.

b.–c. Try working through these parts on your own, then check your **answers**. Watch this **video** for detailed solutions to all three parts.

 You Try It Work through this You Try It problem.

Work Exercises 1–6 in this eText or in the MyMathLab Study Plan.

OBJECTIVE 2 GRAPH SYSTEMS OF LINEAR INEQUALITIES

The **graph of a system of linear inequalities in two variables** is the intersection of the graphs of each inequality in the system. This graph represents the set of all solutions to the system of inequalities.

To graph a system of linear inequalities, graph each inequality in the system and find the region they all have in common, if any. This common region is called the **solution region**.

 *My animation summary*

✽ **Concept Animation** Watch this **animation** for an overview of graphing systems of linear inequalities, using the system $\begin{cases} 2x - 3y \le 9 \\ 2x - y > -1 \end{cases}$.

Steps for Graphing Systems of Linear Inequalities

Step 1 Use the steps for graphing linear inequalities in two variables to graph each linear inequality in the system on the same **coordinate plane**.

Step 2 Determine the region where the shaded areas overlap, if any. This region represents the set of all solutions to the system of inequalities.

⚠ Remember to use a dashed **boundary line** when the inequality is **strict** and a solid boundary line when the inequality is **non-strict**.

Note: For the remainder of this section, we will not find or label the exact **intersection point** of boundary lines. If desired, however, these points could be found by using the **substitution** or **elimination** methods for solving **systems of linear equations.**

My video summary ▶ **Example 2 Graphing Systems of Linear Inequalities in Two Variables**

Graph the system of linear inequalities in two variables.

$$\begin{cases} x + y > 2 \\ 2x - y \le 6 \end{cases}$$

Solution Follow the steps for graphing systems of linear inequalities in two variables, or watch this **video** for a detailed solution.

Step 1 **Graph $x + y > 2$.** The inequality $x + y > 2$ is **strict**, so graph the boundary line $x + y = 2$ using a **dashed** line. Use the **test point** $(0, 0)$. Do you see **why**? Since $0 + 0 > 2$ is false, shade the **half-plane** that does not contain $(0, 0)$. See **Figure 8(a)**.

Graph $2x - y \le 6$. The inequality $2x - y \le 6$ is **non-strict**, so graph the boundary line $2x - y = 6$ using a **solid line**. The test point $(0, 0)$ satisfies the inequality $(2 \cdot 0 - 0 \leq 6$ is true), so shade the half-plane that contains $(0, 0)$. See Figure 8(b).

Step 2 Determine the region shared by both inequalities in the system. This is the purple-shaded region in Figure 8(c). Any point in this region is a **solution to the system of inequalities.**

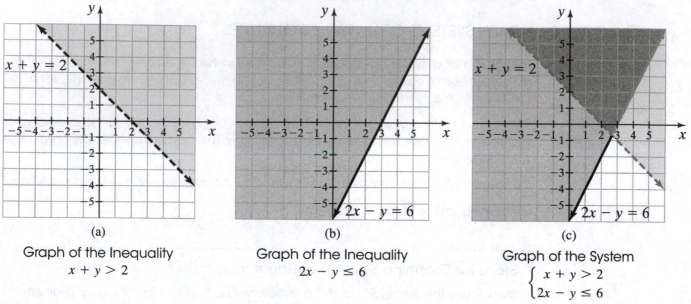

(a)
Graph of the Inequality
$x + y > 2$

(b)
Graph of the Inequality
$2x - y \leq 6$

(c)
Graph of the System
$$\begin{cases} x + y > 2 \\ 2x - y \leq 6 \end{cases}$$

Figure 8

Only the ordered pairs in the purple-shaded region in Figure 8(c), including the purple-shaded portion of the solid line $2x - y = 6$, are solutions to the system. Note that none of the points on the boundary line $x + y = 2$ are solutions to the system, including the intersection point of the boundary lines, $\left(\dfrac{8}{3}, -\dfrac{2}{3}\right)$.

You Try It Work through this You Try It problem.

Work Exercises 7–10 in this eText or in the MyMathLab Study Plan.

My video summary ▶ **Example 3 Graphing Systems of Linear Inequalities**

Graph the system of linear inequalities.

$$\begin{cases} x + y < 4 \\ x - 2y < -2 \end{cases}$$

Solution Both inequalities in this system are strict, so the **boundary lines** $x + y = 4$ and $x - 2y = -2$ are dashed lines. Graph the inequalities on the same **coordinate plane** and find the region they have in common. View the graph, or watch this **video** for a detailed solution.

The points on the two dashed boundary lines $x + y = 4$ and $x - 2y = -2$, including the intersection point, are not solutions to this system.

You Try It Work through this You Try It problem.

Work Exercises 11–15 in this eText or in the MyMathLab Study Plan.

A system of linear inequalities typically has an infinite number of solutions, but a system of inequalities with no solution, called an **inconsistent system of inequalities**, is also possible.

My video summary ▶ **Example 4 Identifying Inconsistent Systems of Linear Inequalities**

Graph the system of linear inequalities in two variables.

$$\begin{cases} x - 3y > 6 \\ 2x - 6y < -9 \end{cases}$$

Solution Follow the steps for graphing systems of linear inequalities in two variables, or watch this video for a more detailed solution.

Step 1 **Graph $x - 3y > 6$.** The inequality $x - 3y > 6$ is strict, so graph the boundary line $x - 3y = 6$ using a dashed line. The test point $(0, 0)$ does not satisfy the inequality, so shade the **half-plane** that does not contain $(0, 0)$. See **Figure 9(a)**.

Graph $2x - 6y < -9$. The inequality $2x - 6y < -9$ is strict, so graph the boundary line $2x - 6y = -9$ using a dashed line. The test point $(0, 0)$ does not satisfy the inequality, so shade the half-plane that does not contain $(0, 0)$. See **Figure 9(b)**.

Step 2 The boundary lines $x - 3y = 6$ and $2x - 6y = -9$ are parallel, and the shaded regions are on opposite sides of these parallel lines. Therefore, no region is shared by both inequalities in the system. See **Figure 9(c)**. There are no ordered pairs that satisfy both inequalities, so there is no solution to the system. We indicate this with the null symbol $\varnothing$ or empty set $\{\,\}$.

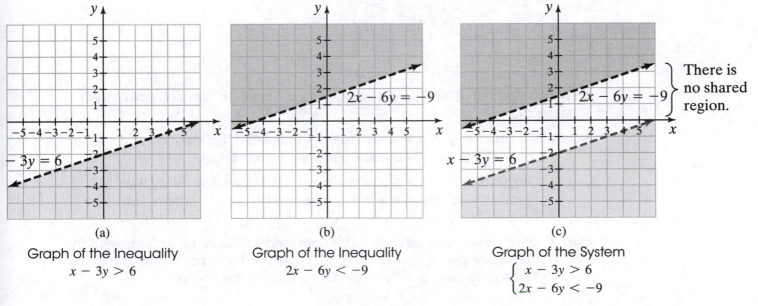

(a)
Graph of the Inequality
$x - 3y > 6$

(b)
Graph of the Inequality
$2x - 6y < -9$

(c)
Graph of the System
$$\begin{cases} x - 3y > 6 \\ 2x - 6y < -9 \end{cases}$$

There is no shared region.

Figure 9

You Try It Work through this You Try It problem.

Work Exercises 16–18 in this eText or in the MyMathLab Study Plan.

In **Example 4**, the **system** contained two linear inequalities whose **boundary lines** were **parallel**. Since the shaded regions for the two inequalities did not overlap, the system had no solution.

In **Example 5**, we consider three systems of linear inequalities in two variables with parallel boundary lines similar to those in **Example 4**. However, the results are very different.

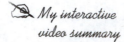
My interactive video summary

Example 5 Graphing Systems of Linear Inequalities with Parallel Boundary Lines

Graph each system of linear inequalities in two variables.

a. $\begin{cases} x - 3y < 6 \\ 2x - 6y < -9 \end{cases}$ **b.** $\begin{cases} x - 3y < 6 \\ 2x - 6y > -9 \end{cases}$ **c.** $\begin{cases} x - 3y > 6 \\ 2x - 6y > -9 \end{cases}$

Solutions The three systems in this example are very similar to each other and to the system in **Example 4**. The graphs of all three systems have the **boundary lines** $x - 3y = 6$ and $2x - 6y = -9$. Since all of the inequalities are strict, the boundary lines will be dashed lines. See Figure 10.

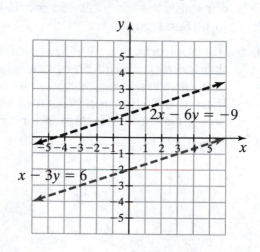

Figure 10

a. Watch part (a) of this **interactive video** for a detailed solution, or follow the steps shown.

Step 1 Graph $x - 3y < 6$. The test point $(0, 0)$ satisfies the inequality, so shade the half-plane that contains $(0, 0)$. See **Figure 11(a)**.

Graph $2x - 6y < -9$. The test point $(0, 0)$ does not satisfy the inequality, so shade the half-plane that does not contain $(0, 0)$. See **Figure 11(b)**.

Step 2 Determine the region shared by both inequalities in the system. This is the purple-shaded region in **Figure 11(c)**. Any ordered pair that is a solution to the inequality $2x - 6y < -9$ is a solution to the system.

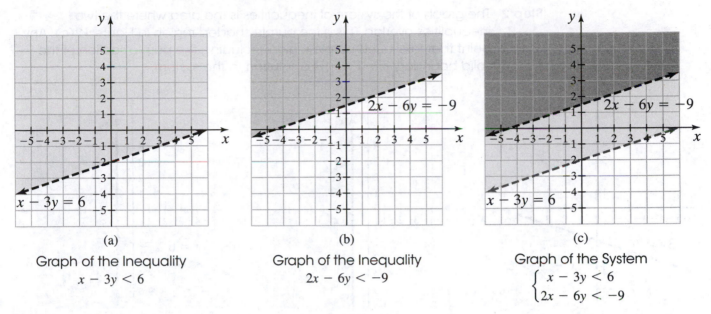

(a)

Graph of the Inequality
$$x - 3y < 6$$

(b)

Graph of the Inequality
$$2x - 6y < -9$$

(c)

Graph of the System
$$\begin{cases} x - 3y < 6 \\ 2x - 6y < -9 \end{cases}$$

Figure 11

b. Try graphing this **system of inequalities** on your own. View the **answer**, or watch part (b) of this **interactive video** for a detailed solution.

c. Graph this system of inequalities on your own. View the **answer**, or watch part (c) of this **interactive video** for a detailed solution.

You Try It   Work through this You Try It problem.

Work Exercises 19–21 in this eText or in the MyMathLab Study Plan.

Some systems of linear inequalities involve **horizontal** or **vertical** boundary lines.

My video summary ▶ **Example 6 Graphing Systems of Linear Inequalities**

Graph the system of linear inequalities.

$$\begin{cases} x - 3y > 6 \\ x \geq 1 \end{cases}$$

Solution Watch this video for a detailed solution, or follow the **steps for graphing systems of linear inequalities**.

Step 1 Graph $x - 3y > 6$. The inequality is **strict**, so use a dashed line to graph its **boundary line** $x - 3y = 6$. The **test point** $(0, 0)$ does not **satisfy the** inequality because $0 - 3(0) > 6$ is false, so shade the **half-plane** that does not contain $(0, 0)$. See **Figure 12(a)**.

Graph $x \geq 1$. The inequality is **non-strict**, so use a solid line to graph its **vertical boundary line** $x = 1$. The test point $(0, 0)$ does not satisfy the inequality because $0 \geq 1$ is false, so shade the half-plane that does not contain $(0, 0)$. See **Figure 12(b)**.

Step 2 The graph of the system of inequalities is the area where the two inequalities overlap. This is the purple shaded region in **Figure 12(c)**. Any point that falls in this purple region, including the purple section of the solid boundary line $x = 1$, is a **solution** to the system.

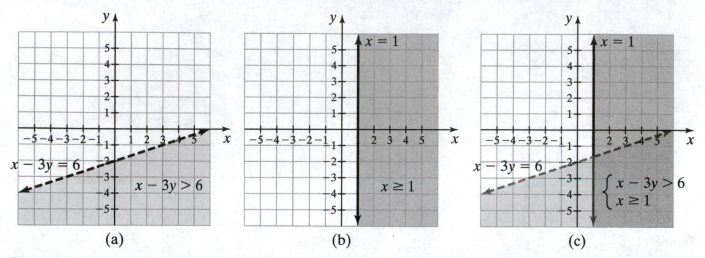

(a)   (b)   (c)

Figure 12

⚠ The points on the dashed boundary line $x - 3y = 6$, including the intersection point with the solid boundary line $x = 1$, are not solutions to this system.

You Try It Work through this You Try It problem.

Work Exercises 22–27 in this eText or in the MyMathLab **Study Plan.**

Sometimes a system contains more than two inequalities.

My video summary ⏵ **Example 7** Graphing Systems with More Than Two Linear Inequalities

Graph the system of linear inequalities in two variables.

$$\begin{cases} 4x > y \\ x - 3y < 9 \\ x + y < 4 \end{cases}$$

Solution Each inequality in the system is strict, so all of the boundary lines are dashed lines. Graph all three inequalities on the same coordinate plane and find the region shared by all three inequalities, if any. View the **answer**, or watch this **video** to see a detailed solution.

You Try It Work through this You Try It problem.

Work Exercises 28–33 in this eText or in the MyMathLab **Study Plan.**

OBJECTIVE 3 SOLVE APPLICATIONS INVOLVING SYSTEMS OF LINEAR INEQUALITIES

Let's look at a real-life application that is **modeled** by a **system of linear inequalities.**

My video summary ▶ **Example 8** Planning a Barbeque

Savannah is planning a barbeque for her family and friends. She will spend $150 or less to buy hamburger patties that cost $3 per pound and boneless chicken breast that cost $5 per pound. To limit waste, she will purchase at most 40 pounds of meat all together. Also, the amount of hamburger and chicken purchased must be **non-negative.** A system of linear inequalities that models this situation is

$$\begin{cases} 3h + 5c \le 150 \\ h + c \le 40 \\ h \ge 0 \\ c \ge 0 \end{cases}$$

where h = pounds of hamburger patties and c = pounds of chicken breast.

a. Graph the system of linear inequalities.

b. Can Savannah purchase 20 pounds of hamburger patties and 15 pounds of chicken breast for the barbeque?

c. Can Savannah purchase 10 pounds of hamburger patties and 30 pounds of chicken breast for the barbeque?

Solutions Work through the following, or watch this **video** for a detailed solution.

a. Label the horizontal axis h and the vertical axis c. The inequalities $h \ge 0$ and $c \ge 0$ restrict the graph to **Quadrant I.** Follow the **steps for graphing systems of linear inequalities** to obtain the graph in Figure 13.

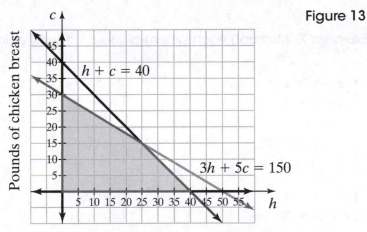

Figure 13

Pounds of chicken breast

$h + c = 40$

$3h + 5c = 150$

Pounds of hamburger patties

b. Yes. The ordered pair $(h, c) = (20, 15)$ lies within the shaded region in Figure 13, so Savannah can purchase 20 pounds of hamburger patties and 15 pounds of chicken.

c. No. The ordered pair $(h, c) = (10, 30)$ lies outside of the shaded region in Figure 13, so Savannah cannot purchase 10 pounds of hamburger patties and 30 pounds of chicken.

🔺 **You Try It** Work through this You Try It problem.

Work Exercises 34–37 in this eText or in the MyMathLab Study Plan.

3.4 Exercises

In Exercises 1–6, determine if each ordered pair is a solution to the given system of linear inequalities in two variables.

1. $\begin{cases} x - 2y < 6 \\ 3x + y > 2 \end{cases}$

   a. $(1, 0)$

   b. $(4, -1)$

   c. $(3, -2)$

2. $\begin{cases} 3x - 2y \geq 6 \\ 4x + 3y \leq 9 \end{cases}$

   a. $(5, -3)$

   b. $(0, 0)$

   c. $(2, -2)$

3. $\begin{cases} x - y > -8 \\ 5x - 2y \leq 10 \\ 4x + 3y > 12 \end{cases}$

   a. $(2, 4)$

   b. $(1, 1)$

   c. $(-1, 6)$

4. $\begin{cases} 2x + 3y < 6 \\ 3x - 4y \leq -4 \\ 3x - y \geq -6 \end{cases}$

   a. $(1, 3)$

   b. $(-2, 0)$

   c. $(3, 2)$

5. $\begin{cases} x + 2y \leq 6 \\ 4x + y \leq 10 \\ x \geq 0 \\ y \geq 0 \end{cases}$

   a. $(1, 6)$

   b. $(3, 4)$

   c. $(2, 2)$

6. $\begin{cases} x + y < 8 \\ x + y < 3 \\ x < 5 \\ y < 7 \end{cases}$

   a. $(5, -3)$

   b. $(0, 0)$

   c. $(2, -2)$

In Exercises 7–33, graph each system of linear inequalities in two variables.

7. $\begin{cases} x + y \geq 4 \\ 2x - y \leq 5 \end{cases}$

8. $\begin{cases} y < -2x \\ y > -3x \end{cases}$

9. $\begin{cases} 2x - y < 6 \\ y \geq \frac{3}{2}x + 3 \end{cases}$

10. $\begin{cases} 3x - 2y \leq -2 \\ x + 2y > 6 \end{cases}$

11. $\begin{cases} y > 2x - 1 \\ y \leq x + 2 \end{cases}$

12. $\begin{cases} x + y \geq 2 \\ x - y \leq 4 \end{cases}$

13. $\begin{cases} 3x + y < 2 \\ x + 2y < 8 \end{cases}$

14. $\begin{cases} y < 2x \\ y > -3x \end{cases}$

15. $\begin{cases} 3x - 2y \leq -2 \\ x + 2y \geq 6 \end{cases}$

16. $\begin{cases} y < x - 2 \\ y \geq x + 4 \end{cases}$

17. $\begin{cases} 2x + y \geq 4 \\ y \leq -2x - 1 \end{cases}$

18. $\begin{cases} 3x - 4y > 12 \\ 6x - 8y < -32 \end{cases}$

19. $\begin{cases} 6x - 8y < 4 \\ -3x + 4y \leq 12 \end{cases}$

20. $\begin{cases} 2x - y > 7 \\ y < 2x + 5 \end{cases}$

21. $\begin{cases} 2x - 3y \geq 12 \\ -x + 1.5y < 6 \end{cases}$

22. $\begin{cases} 4x - y > 2 \\ y \ge -1 \end{cases}$

23. $\begin{cases} x \ge -1 \\ y \le 3 \end{cases}$

24. $\begin{cases} \dfrac{1}{2}x - \dfrac{1}{3}y \le 1 \\ x \ge -1 \end{cases}$

25. $\begin{cases} x - 3y \le 6 \\ x < 3 \end{cases}$

26. $\begin{cases} x > 3 \\ x < -2 \end{cases}$

27. $\begin{cases} y \ge 2 \\ y \le 5 \end{cases}$

28. $\begin{cases} 3x + 2y \le 4 \\ 3x - 2y \ge -16 \\ x - 2y \le 4 \end{cases}$

29. $\begin{cases} 2x - y > -3 \\ 5x + y > 0 \\ x > 1 \end{cases}$

30. $\begin{cases} x + y \le 5 \\ x \ge 0 \\ y \ge 0 \end{cases}$

31. $\begin{cases} x + y > -2 \\ x - y < 2 \\ y < 2 \end{cases}$

32. $\begin{cases} x + y < 6 \\ 3x - y > -2 \\ x \ge 0 \\ y \ge 0 \end{cases}$

33. $\begin{cases} 2x + y \le 4 \\ x + 3y \le 6 \\ x \ge 0 \\ y \ge 0 \end{cases}$

In Exercises 34–37, solve each application by using a system of linear inequalities in two variables.

34. **Investing** Judy has $50,000 to invest. Her financial advisor recommends that she place at least $30,000 in mutual funds and no more than $20,000 in bonds. A system of linear inequalities that models this situation is

$$\begin{cases} m + b \le 50{,}000 \\ m \ge 30{,}000 \\ b \le 20{,}000 \\ b \ge 0 \end{cases}$$

where m = the dollar amount invested in mutual funds and b = the dollar amount invested in bonds.

a. Graph the system of linear inequalities.

b. Can Judy invest $20,000 in mutual funds and $20,000 in bonds?

c. Can Judy invest $38,000 in mutual funds and $12,000 in bonds?

35. **Getting Cash** For an upcoming shopping trip, Elaine will withdraw cash from her checking account which contains $100. She will want the cash in only five-dollar and ten-dollar bills, and she will carry no more than 12 bills all together. A system of linear inequalities that models this situation is

$$\begin{cases} 5f + 10t \le 100 \\ f + t \le 12 \\ f \ge 0 \\ t \ge 0 \end{cases}$$

where f = the number of $5 bills and t = the number of $10 bills Elaine receives.

a. Graph the system of linear inequalities.

b. Can Elaine receive 4 five-dollar bills and 8 ten-dollar bills?

c. Can Elaine receive 7 five-dollar bills and 3 ten-dollar bills?

36. **Growing Crops** In Illinois, the annual cost to grow corn is about $450 per acre, and the annual cost to grow soybeans is about $300 per acre (not counting the cost of land). A small farmer has at most 150 acres for growing corn and soybeans. His budget for growing these crops is at most $60,000. A system of linear inequalities that models this situation is

$$\begin{cases} c + s \le 150 \\ 450c + 300s \le 60{,}000 \\ c \ge 0 \\ s \ge 0 \end{cases}$$

where c = acres of corn and s = acres of soybeans. (*Source: Farmdoc*, University of Illinois)

a. Graph the system of linear inequalities.

b. Can this farmer grow 75 acres of corn and 50 acres of soybeans?

c. Can this farmer grow 120 acres of corn and 30 acres of soybeans?

37. **Dimensions of a Garden** Austin is planning to build a rectangular garden. Its perimeter will be at most 36 feet, and its length will be no more than twice its width. A system of linear inequalities that models this situation is

$$\begin{cases} 2l + 2w \le 36 \\ l \le 2w \\ l \ge w \end{cases}$$

where l = the length and w = the width of the garden.

a. Graph the system of linear inequalities.

b. Can Austin's garden be 15 feet long and 10 feet wide?

c. Can Austin's garden be 10 feet long and 8 feet wide?

CHAPTER FOUR

Polynomial Expressions and Functions

CHAPTER FOUR CONTENTS

4.1 Rules for Exponents

THINGS TO KNOW

Before working through this section, be sure you are familiar with the following concepts:

| | | VIDEO | ANIMATION | INTERACTIVE |

You Try It

1. Perform Operations on Real Numbers (Section R.2, Objective 1) ▶

You Try It

2. Simplify Numeric Expressions Containing Exponents and Radicals (Section R.2, Objective 2) ▶

OBJECTIVES

1 Simplify Exponential Expressions Using the Product Rule

2 Simplify Exponential Expressions Using the Quotient Rule

3 Use the Zero-Power Rule

4 Use the Negative-Power Rule

5 Use the Power-to-Power Rule

6 Use the Product-to-Power and Quotient-to-Power Rules

7 Simplify Exponential Expressions Using a Combination of Rules

8 Use Rules for Exponents with Scientific Notation

....

OBJECTIVE 1   SIMPLIFY EXPONENTIAL EXPRESSIONS USING THE PRODUCT RULE

An **exponential expression** is a constant or algebraic expression that is raised to a **power**. The constant or algebraic expression makes up the **base**, and the power is the **exponent** on the base.

In the following expressions, the exponent 5 means there are five **factors** of the base x, and the exponent 3 means there are three factors of the base $2y$.

$$\underbrace{x \cdot x \cdot x \cdot x \cdot x}_{5 \text{ factors of } x} = x^5 \xleftarrow{} \text{Exponent} \qquad \underbrace{(2y) \cdot (2y) \cdot (2y)}_{3 \text{ factors of } 2y} = (2y)^3 \xleftarrow{} \text{Exponent}$$

Base   Base

 When an algebraic expression such as $2y$ is the base of an exponential expression, parentheses (or other **grouping symbols**) must be used to show that the entire expression $2y$ is raised to the third power, not just the variable y.

$$(2y) \cdot (2y) \cdot (2y) = (2y)^3 \neq 2y^3 = 2 \cdot y \cdot y \cdot y$$

If **exponential expressions** with the same **base** are multiplied, then the **product** can be simplified by adding the **exponents**. For example,

$$x^3 \cdot x^4 = x^{3+4} = x^7.$$

 *My animation summary*   ✪ **Concept Animation** Watch this **animation** to explore this concept.

> ### The Product Rule for Exponents
>
> When multiplying exponential expressions with the same base, add the exponents and keep the common base.
>
> $$a^m \cdot a^n = a^{m+n}$$

TIP When a factor has no written exponent, it is understood to be 1. For example, $x = x^1$.

 My video summary ▶ **Example 1 Using the Product Rule for Exponents**

Use the **product rule** to simplify each expression.

a. $5^4 \cdot 5^6$ **b.** $x^5 \cdot x^7$ **c.** $y^3 \cdot y$ **d.** $b^3 \cdot b^5 \cdot b^4$

Solutions

a. Since the two exponential expressions 5^4 and 5^6 have the same base 5, add the exponents:

| Add the exponents. |

$$5^4 \cdot 5^6 = 5^{4+6} = 5^{10}$$

| Keep the common base. |

b. $x^5 \cdot x^7 = x^{5+7} = x^{12}$

c.–d. Try to simplify these expressions on your own. View the **answers**, or watch this **video** for detailed solutions to all four parts.

You Try It Work through this **You Try It** problem.

Work Exercises 1–6 in this eText or in the MyMathLab Study Plan.

⚠ When using the product rule for exponents, do not multiply the bases. Instead, keep the common base and add the exponents. For example,

Incorrect	**Correct**
$2^8 \cdot 2^6 = 4^{14}$	$2^8 \cdot 2^6 = 2^{14}$

When the exponential expressions being multiplied involve more than one base, first use the **commutative** and **associative** properties of multiplication to group **like bases** together and then apply the **product rule**.

✎ *My video summary* ▶ **Example 2 Using the Product Rule for Exponents**

Simplify using the product rule.

a. $(4x^2)(7x^3)$ **b.** $(m^4n^2)(m^3n^6)$ **c.** $(-3a^5b^3)(-8a^2b)$

Solutions

a. Begin with the original expression: $(4x^2)(7x^3)$

Rearrange factors to group like bases: $= 4 \cdot 7 \cdot x^2 \cdot x^3$

Multiply constants; apply product rule: $= 28 \cdot x^{2+3}$

Simplify: $= 28x^5$

b. Begin with the original expression: $(m^4n^2)(m^3n^6)$

Rearrange factors to group like bases: $= m^4 \cdot m^3 \cdot n^2 \cdot n^6$

Apply the product rule: $= m^{4+3} \cdot n^{2+6}$

Simplify: $= m^7n^8$

c. Try to simplify this expression on your own. View the **answer**, or watch this **video** for detailed solutions to all three parts.

 You Try It Work through this **You Try It** problem.

Work Exercises 7–12 in this eText or in the MyMathLab Study Plan.

OBJECTIVE 2 SIMPLIFY EXPONENTIAL EXPRESSIONS USING THE QUOTIENT RULE

If exponential expressions with the same nonzero **base** are divided, then the **quotient** can be found by subtracting the exponents. For example, if $x \neq 0$, then

$$\frac{x^5}{x^2} = x^{5-2} = x^3.$$

✎ *My animation summary* ✵ **Concept Animation** Watch this animation to explore this concept.

> **The Quotient Rule for Exponents**
>
> When dividing exponential expressions with the same nonzero base, subtract the **denominator** exponent from the **numerator** exponent and keep the common base.
>
> $$\frac{a^m}{a^n} = a^{m-n} \quad (a \neq 0)$$

My video summary ⏵ **Example 3 Using the Quotient Rule for Exponents**

Use the quotient rule to simplify each expression.

a. $\dfrac{t^9}{t^5}$ b. $\dfrac{7^5}{7^3}$ c. $\dfrac{y^{24}}{y^{15}}$ d. $\dfrac{(-4)^{14}}{(-4)^{11}}$

Solutions

a. The two exponential expressions t^9 and t^5 have the same base t, so we subtract the exponents:

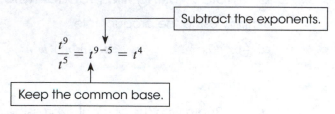

Subtract the exponents.

$$\dfrac{t^9}{t^5} = t^{9-5} = t^4$$

Keep the common base.

b. $\dfrac{7^5}{7^3} = 7^{5-3} = 7^2 = 49$

c.–d. Try to simplify these expressions on your own. View the **answers**, or watch this **video** for detailed solutions to all four parts of this example.

You Try It Work through this You Try It problem.

Work Exercises 13–16 in this eText or in the MyMathLab **Study Plan.**

When the **exponential expressions** being divided involve more than one **base**, group **like bases** into individual **quotients**. Then apply the **quotient rule** to each quotient and simplify.

My video summary ⏵ **Example 4 Using the Quotient Rule for Exponents**

Simplify using the quotient rule.

a. $\dfrac{15x^6}{3x^2}$ b. $\dfrac{a^4 b^9 c^5}{a^2 b^3 c}$ c. $\dfrac{4m^6 n^7}{12m^5 n^2}$

Solutions

a.   Begin with the original expression: $\dfrac{15x^6}{3x^2}$

   Group like bases into individual quotients: $= \dfrac{15}{3} \cdot \dfrac{x^6}{x^2}$

   Divide the constants; apply the quotient rule: $= 5 \cdot x^{6-2}$

   Simplify: $= 5x^4$

b. $\dfrac{a^4 b^9 c^5}{a^2 b^3 c} = \dfrac{a^4}{a^2} \cdot \dfrac{b^9}{b^3} \cdot \dfrac{c^5}{c^1} = a^{4-2} \cdot b^{9-3} \cdot c^{5-1} = a^2 b^6 c^4$

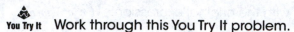

Understood exponent of 1

c. Try to simplify this expression on your own. View the **answer**, or watch this **video** for detailed solutions to all three parts of this example.

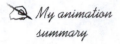 **You Try It** Work through this You Try It problem.

Work Exercises 17–22 in this eText or in the MyMathLab Study Plan.

OBJECTIVE 3 USE THE ZERO-POWER RULE

If an exponential expression has a nonzero **base** and an exponent of 0, the result is 1. For example, if $x \neq 0$, then $x^0 = 1$.

My animation summary ⊛ **Concept Animation** Watch this animation to explore this concept.

> **The Zero-Power Rule**
>
> A nonzero base raised to the 0 power equals 1.
>
> $$a^0 = 1 \quad (a \neq 0)$$

My video summary ▶ **Example 5 Using the Zero-Power Rule**

Evaluate. Assume that all variable bases are nonzero.

a. 10^0 b. -10^0 c. $5x^0$ d. $(5x)^0$

e. $(x + 3)^0$ f. $a^0 + b^0$ g. $(x^2y^5z)^0$

Solutions

a. The base 10 is nonzero, so $10^0 = 1$.

b. The base is 10, not -10, so we multiply the exponential expression 10^0 by -1. Using the **order of operations**, we get

$$-10^0 = -1(10)^0 = -1 \cdot 1 = -1.$$

c. We multiply the exponential expression x^0 by 5 and assume that the variable base x is nonzero. Applying the zero-power rule,

$$5x^0 = 5 \cdot x^0 = 5 \cdot 1 = 5.$$

d.–g. Try to work the remaining parts on your own. View the **answers**, or watch this **video** for detailed solutions to all seven parts.

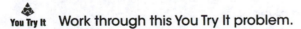 **You Try It** Work through this You Try It problem.

Work Exercises 23–28 in this eText or in the MyMathLab Study Plan.

OBJECTIVE 4 USE THE NEGATIVE-POWER RULE

So far, we have encountered only exponential expressions with non-negative exponents. But what if a base is raised to a negative exponent, such as x^{-2} or $\dfrac{1}{x^{-3}}$?

My animation summary

⊛ **Concept Animation** Watch this animation to explore this concept.

> **The Negative-Power Rule**
>
> To remove a negative exponent, switch the location of the base (**numerator** or **denominator**) and change to a positive exponent.
>
> $$a^{-n} = \frac{1}{a^n} \quad \text{and} \quad \frac{1}{a^{-n}} = a^n \quad (a \neq 0)$$

⚠ A negative exponent does not mean the result will be negative. For example, $2^{-1} = \frac{1}{2}$, which is positive.

My interactive video summary

✎ **Example 6 Using the Negative-Power Rule**

Simplify. Use only positive exponents in your answers.

a. $(3)^{-2}$ **b.** m^{-5} **c.** $\dfrac{1}{5x^{-4}}$ **d.** $(7n)^{-1}$ **e.** $8a^{-1}$

f. -2^{-4} **g.** $4a^{-3}b^2$ **h.** $\dfrac{x^{-2}y}{z^{-5}}$ **i.** $2^{-1} - 3^{-1}$

Solutions In each case, use the **negative-power rule** to rewrite the expression with all positive exponents. Then simplify if necessary.

a. $(3)^{-2} = \dfrac{1}{3^2} = \dfrac{1}{9}$

b. $m^{-5} = \dfrac{1}{m^5}$

c. The **negative exponent** in the denominator is only on the **variable** x. Rewriting, we get $\dfrac{1}{5x^{-4}} = \dfrac{1}{5} \cdot \dfrac{1}{x^{-4}}$. In the fraction on the right, move the base x to the numerator, and make the exponent positive.

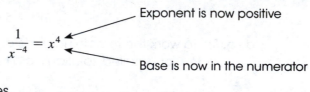

$$\frac{1}{x^{-4}} = x^4$$

Exponent is now positive

Base is now in the numerator

Simplifying gives

$$\frac{1}{5x^{-4}} = \frac{1}{5} \cdot \frac{1}{x^{-4}} = \frac{1}{5} \cdot x^4 = \frac{x^4}{5}.$$

d.–i. Try to work the remaining parts on your own. View the **answers**, or watch this interactive video for detailed solutions to all nine parts.

▲ **You Try It** Work through this **You Try It** problem.

Work Exercises 29–44 in this eText or in the MyMathLab **Study Plan**.

OBJECTIVE 5   USE THE POWER-TO-POWER RULE

If an **exponential expression** itself is raised to a **power**, then the entire expression can be simplified by multiplying the **exponents**.

 *My animation summary*

⊛ **Concept Animation** Watch this animation to explore this concept.

The Power-to-Power Rule

When an exponential expression is raised to a power, multiply the exponents.

$$(a^m)^n = a^{m \cdot n}$$

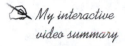 *My interactive video summary*

✍ **Example 7** Using the Power-to-Power Rule

Simplify. Use only positive exponents in your answers.

a. $(a^3)^7$ **b.** $(x^{-2})^4$ **c.** $(m^{-3})^{-2}$

Solutions Watch this interactive video, or continue reading for complete solutions

a. Because the exponential expression a^3 is being raised to the seventh power, apply the **power-to-power rule** and multiply the exponents.

$$(a^3)^7 = a^{3 \cdot 7} = a^{21}$$

b. Applying the power-to-power rule gives $(x^{-2})^4 = x^{-2 \cdot 4} = x^{-8}$. Because the resulting exponent is **negative**, apply the **negative-power rule** to obtain $\frac{1}{x^8}$.

c. Apply the power-to-power rule.

$$(m^{-3})^{-2} = m^{(-3)(-2)} = m^6$$

You Try It   Work through this You Try It problem.

Work Exercises 45–50 in this eText or in the MyMathLab Study Plan.

OBJECTIVE 6   USE THE PRODUCT-TO-POWER AND QUOTIENT-TO-POWER RULES

PRODUCT-TO-POWER RULE
If the **base** of an exponential expression is a **product**, then the expression can be simplified by raising each factor of the base to the common **power**.

 *My animation summary*

⊛ **Concept Animation** Watch this animation to explore this concept.

The Product-to-Power Rule

When raising a product to a power, raise each factor of the base to the common exponent.

$$(ab)^n = a^n b^n$$

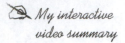
My interactive
video summary

Example 8 Using the Product-to-Power Rule

Simplify. Use only positive exponents in your answers.

a. $(2x)^3$ **b.** $(c^2d)^5$ **c.** $(-2a^3b^4)^4$ **d.** $(x^{-3}y^2)^{-2}$

Solutions

a. The base is the product $2x$. Apply the **product-to-power rule** and simplify.

$$(2x)^3 \;\; = \;\; 2^3x^3 \;\; = 8x^3$$

$\underbrace{}_{\text{Product-to-power}}$ $\underbrace{}_{\text{Simplify}}$

b. The base is the product c^2d. Apply the **product-to-power rule**, raising each of the factors, c^2 and d, to the common exponent 5. To simplify, apply the **power-to-power rule.**

$$(c^2d)^5 \;\; = \;\; (c^2)^5d^5 \;\; = c^{10}d^5$$

$\underbrace{}_{\text{Product-to-power}}$ $\underbrace{}_{\text{Power-to-power}}$

c.–d. Try to simplify these expressions on your own. View the **answers**, or watch this **interactive video** for detailed solutions to all four parts.

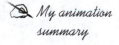
You Try It Work through this You Try It problem.

Work Exercises 51–58 in this eText or in the MyMathLab Study Plan.

QUOTIENT-TO-POWER RULE

If the base of an exponential expression is a **quotient**, then the expression can be simplified by raising both the **numerator** and the **denominator** of the base to the **common power.**

My animation
summary

Concept Animation Watch this animation to explore this concept.

> ### The Quotient-to-Power Rule
>
> When raising a quotient to a power, raise both the numerator and denominator to the common exponent.
>
> $$\left(\frac{a}{b}\right)^n = \frac{a^n}{b^n} \quad (b \neq 0)$$

For a quotient raised to a negative power, a variation of the quotient-to-power rule is possible:

$$\left(\frac{a}{b}\right)^{-n} = \left(\frac{b}{a}\right)^n = \frac{b^n}{a^n} \quad (a \neq 0, b \neq 0)$$

View this **explanation.**

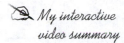
My interactive video summary

Example 9 Using the Quotient-to-Power Rule

Simplify. Use only positive exponents in your answers.

a. $\left(\dfrac{y^3}{z^2}\right)^5$ b. $\left(\dfrac{2x^2y}{3z}\right)^3$ c. $\left(\dfrac{-mn^4}{p^2}\right)^3$ d. $\left(\dfrac{x^4}{y^3}\right)^{-2}$ e. $\left(\dfrac{5a^3}{b^{-4}}\right)^3$

Solutions

a. The base is the quotient $\dfrac{y^3}{z^2}$.

Apply the quotient-to-power rule: $\left(\dfrac{y^3}{z^2}\right)^5 = \dfrac{(y^3)^5}{(z^2)^5}$

Apply the power-to-power rule: $\qquad\qquad = \dfrac{y^{15}}{z^{10}}$

b. The base is the quotient $\dfrac{2x^2y}{3z}$.

Apply the quotient-to-power rule: $\left(\dfrac{2x^2y}{3z}\right)^3 = \dfrac{(2x^2y)^3}{(3z)^3}$

Apply the product-to-power rule: $\qquad\qquad = \dfrac{2^3(x^2)^3y^3}{3^3z^3}$

Apply the power-to-power rule: $\qquad\qquad = \dfrac{2^3x^6y^3}{3^3z^3}$

Simplify: $\qquad\qquad = \dfrac{8x^6y^3}{27z^3}$

c.–e. Try to simplify these expressions on your own. View the **answers**, or watch this **interactive video** for detailed solutions to all five points.

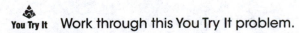
You Try It Work through this You Try It problem.

Work Exercises 59–66 in this eText or in the **MyMathLab** Study Plan.

OBJECTIVE 7 SIMPLIFY EXPONENTIAL EXPRESSIONS USING A COMBINATION OF RULES

In some of the earlier examples, more than one rule for exponents was used to simplify the **exponential expressions**. To be considered **simplified** an exponential expression must meet the following conditions:

- No parentheses or other **grouping symbols** are present.

- No zero or **negative exponents** are present.

- No **powers** are raised to powers.

- Each **base** occurs only once.

The following table summarizes the rules for exponents.

Rules for Exponents		
Product Rule	$a^m \cdot a^n = a^{m+n}$	
Quotient Rule	$\dfrac{a^m}{a^n} = a^{m-n}$	$(a \neq 0)$
Zero-Power Rule	$a^0 = 1$	$(a \neq 0)$
Negative-Power Rule	$a^{-n} = \dfrac{1}{a^n}$ or $\dfrac{1}{a^{-n}} = a^n$	$(a \neq 0)$
Power-to-Power Rule	$(a^m)^n = a^{m \cdot n}$	
Product-to-Power Rule	$(ab)^n = a^n b^n$	
Quotient-to-Power Rule	$\left(\dfrac{a}{b}\right)^n = \dfrac{a^n}{b^n}$	$(b \neq 0)$

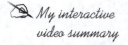 *My interactive video summary*

 Example 10 Simplifying Exponential Expressions Using a Combination of Rules

Simplify using the **rules for exponents.** Use only positive exponents in your answers.

a. $\dfrac{x^2 y^0 z^{-3}}{x^4 z^{-2}}$

b. $\dfrac{24x^5 y^{-4}}{-3x^{-3} y^6}$

c. $\left(\dfrac{a^3 b^{-4}}{2a^4 b^3}\right)^3$

d. $\left(\dfrac{8m^6 n^3}{4m^{-3} n^7}\right)^{-4}$

e. $\left(-3m^3 n^{-8}\right)\left(-2m^{-5} n^3\right)^4$

f. $\left(\dfrac{3a^2 b}{c^{-3}}\right)^2 \left(\dfrac{-4a^{-5} b^3}{c^5}\right)^3$

Solutions Try to simplify these expressions on your own. View the **answers**, or watch this **interactive video** for detailed solutions to all six parts.

 You Try It Work through this You Try It problem.

Work Exercises 67–78 in this eText or in the MyMathLab Study Plan.

TIP There is often more than one way to simplify an exponential expression. You do not need to always work problems using the exact same steps as in the examples. Just make sure that your solutions meet the conditions for simplified exponential expressions.

OBJECTIVE 8 USE RULES FOR EXPONENTS WITH SCIENTIFIC NOTATION

Scientists frequently work with very large or very small numbers. For example, the distance from the Sun to the center of the Milky Way is about 19,200,000,000,000 miles, and the mass of a neutron is about 0.00000000000000000000001675 grams.

Calculating with such numbers in **standard form** can be difficult. To make the numbers more manageable, we can use **scientific notation**.

Scientific Notation

A number is written in **scientific notation** if it has the form

$$a \times 10^n,$$

where a is a **real number**, such that $1 \le |a| < 10$, and n is an **integer**.

In scientific notation, the distance from the sun to the center of the Milky Way is 1.92×10^{13} miles, and the mass of a neutron is 1.675×10^{-24} grams.

 TIP For very large numbers written in scientific notation, the **exponent** on the 10 is positive. For very small numbers the exponent on the 10 is negative.

The following procedure can be used to convert a **real number** from **standard form** to **scientific notation**.

Converting from Standard Form to Scientific Notation

1. Write the real number factor a by moving the **decimal point** so that $|a|$ is greater than or equal to 1 but less than 10. To do this, place the decimal point to the right of the first nonzero **digit**.

2. Multiply the number by 10^n, where $|n|$ is the number of places that the decimal point moves. If the decimal point moves to the left, then $n > 0$ (n is positive). If the decimal point moves to the right, then $n < 0$ (n is negative). Remove any zeros lying to the right of the last nonzero digit or to the left of the first nonzero digit.

My video summary ▶ **Example 11 Converting from Standard Form to Scientific Notation**

Write each number in scientific notation.

a. 56,800,000,000,000,000

b. 0.0000000467

c. 0.00009012

d. 200,000,000

Solutions

a. Move the **decimal point** to the right of the 5, the first nonzero **digit**.

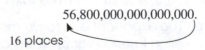
56,800,000,000,000,000.

16 places

The decimal point has moved sixteen places to the left, so the exponent on the 10 will be 16. Remove all of the zeros behind the last nonzero digit 8.

$$56,800,000,000,000,000 = 5.68 \times 10^{16}$$

b. Move the decimal point to the right of the 4, the first nonzero digit.

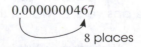
0.0000000467

8 places

The decimal point has moved eight places to the right, so the exponent on the 10 will be −8. Remove all of the zeros in front of the 4.

$$0.0000000467 = 4.67 \times 10^{-8}$$

c.-d. Try to convert these numbers to scientific notation on your own. View the answers, or watch this **video** for detailed solutions to all four parts.

 You Try It Work through this You Try It problem.

Work Exercises 79–86 in this eText or in the MyMathLab Study Plan.

Use the following procedure to convert from scientific notation to standard form.

Converting from Scientific Notation to Standard Form

1. Remove the exponential factor, 10^n.

2. Move the decimal point $|n|$ places, inserting zero placeholders as needed. If $n > 0$ (n is positive), move the decimal point to the right. If $n < 0$ (n is negative), move the decimal point to the left.

TIP Remember, if the **exponent** on 10 is positive, then the **absolute value of the number** is larger than 1. If the exponent on 10 is negative, then the absolute value of the number is smaller than 1.

My video summary ▶ **Example 12 Converting from Scientific Notation to Standard Form**

Write each number in standard form.

a. 4.98×10^{-5} **b.** 9.4×10^{7}

c. -3.015×10^{9} **d.** 1.203×10^{-4}

Solutions

a. The exponent is −5, so we move the decimal point five places to the left, filling in zero placeholders as needed.

$$00004.98$$

5 places

So, $4.98 \times 10^{-5} = 0.0000498$.

b. The exponent is 7, so we move the decimal point seven places to the right, filling in zero placeholders as needed.

$$9.4000000$$

7 places

So, $9.4 \times 10^{7} = 94,000,000$.

c.-d. Try to convert these numbers to standard form on your own. View the answers, or watch this **video** for detailed solutions to all four parts.

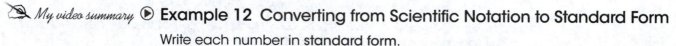 **You Try It** Work through this You Try It problem.

Work Exercises 87–94 in this eText or in the MyMathLab Study Plan.

Rules for Exponents can be used when multiplying or dividing numbers written in scientific notation.

My interactive video summary

Example 13 Using Rules for Exponents with Scientific Notation

Perform the indicated operations. Write your results in scientific notation.

a. $(2 \times 10^6)(4 \times 10^{-3})$ b. $(7.2 \times 10^{-4})(9.5 \times 10^9)$ c. $\dfrac{8.4 \times 10^7}{2.1 \times 10^3}$

d. $\dfrac{1.8 \times 10^5}{4.5 \times 10^{-3}}$ e. $(520{,}000{,}000)(0.00004)$ f. $\dfrac{0.000000056}{0.0000000004}$

Hint: In parts (e) and (f), write each number in scientific notation before completing the operation.

Solutions

a. Begin with original expression: $(2 \times 10^6)(4 \times 10^{-3})$

Regroup factors: $(2 \times 4) \times (10^6 \times 10^{-3})$

Multiply constant factors: $8 \times (10^6 \times 10^{-3})$

Apply the product rule: $8 \times (10^{6+(-3)})$

Simplify: 8×10^3

b. Begin with original expression: $(7.2 \times 10^{-4})(9.5 \times 10^9)$

Regroup factors: $(7.2 \times 9.5) \times (10^{-4} \times 10^9)$

Multiply constant factors: $68.4 \times (10^{-4} \times 10^9)$

Write 68.4 in scientific notation: $6.84 \times 10^1 \times (10^{-4} \times 10^9)$

Apply the product rule: $6.84 \times 10^{1+(-4)+9}$

Simplify: 6.84×10^6

c. Begin with original expression: $\dfrac{8.4 \times 10^7}{2.1 \times 10^3}$

Regroup factors: $\left(\dfrac{8.4}{2.1}\right) \times \left(\dfrac{10^7}{10^3}\right)$

Divide constant factors: $4 \times \left(\dfrac{10^7}{10^3}\right)$

Apply the quotient rule: $4 \times 10^{7-3}$

Simplify: 4×10^4

d.–f. Try to work these problems on your own. Check your **answers**, or watch this interactive video for detailed solutions.

You Try It Work through this You Try It problem.

Work Exercises 95–100 in this eText or in the MyMathLab Study Plan.

4.1 Exercises

In Exercises 1–12, simplify using the product rule.

1. $7^4 \cdot 7^2$ 2. $(-3)^3 \cdot (-3)^2$ 3. $y^3 \cdot y^8$ 4. $x^8 \cdot x^5$

5. $w^5 \cdot w$ 6. $t^6 \cdot t^2 \cdot t^5$ 7. $(6a^3)(5a^7)$ 8. $(x^2 y^7)(x^4 y^3)$

9. $(4a^3 b^2)(5a^4 b^7)$ 10. $(-2m^9 n)(-9m^4 n^5)$ 11. $(-5x^4 y)(3x^2 z^3)$ 12. $(-6p^2)(5p^4)(2p^3)$

In Exercises 13–22, simplify using the quotient rule.

13. $\dfrac{b^{12}}{b^8}$ 14. $\dfrac{5^9}{5^7}$ 15. $\dfrac{(-10)^7}{(-10)^4}$ 16. $\dfrac{y^{17}}{y^{16}}$

17. $\dfrac{24q^{15}}{6q^3}$ 18. $\dfrac{60a^4 b^{13}}{12ab^9}$ 19. $\dfrac{x^7 y^9 z^8}{x^5 y^3 z^7}$ 20. $\dfrac{10m^{21} n^9}{15m^{13} n^2}$

21. $\dfrac{2x^5 y^6}{-8x^3 y^2}$ 22. $\dfrac{15m^8 n^3}{5m^7}$

In Exercises 23–28, simplify using the zero-power rule.

23. 4^0 24. $(-8)^0$ 25. -12^0

26. $6m^0$ 27. $(3x)^0$ 28. $3p^0 + 2q^0$

In Exercises 29–44, write each expression using only positive exponents. Simplify if possible.

29. y^{-3} 30. 5^{-3} 31. $6w^{-5}$ 32. $-3t^{-8}$

33. $(-8)^{-2}$ 34. -10^{-4} 35. $2^{-3} + 4^{-1}$ 36. $3^{-1} \cdot 5^{-2}$

37. $\dfrac{1}{z^{-5}}$ 38. $\dfrac{1}{4^{-3}}$ 39. $\dfrac{1}{(-3)^{-4}}$ 40. $\dfrac{5}{7x^{-4}}$

41. $\dfrac{-12}{t^{-3}}$ 42. $\dfrac{a^{-8}}{b^{-15}}$ 43. $\dfrac{9^{-2}}{2^{-5}}$ 44. $\dfrac{x^{-2} y^7}{z^{-5}}$

In Exercises 45–50, simplify using the power-to-power rule. Use only positive exponents in your answers.

45. $(x^3)^8$ 46. $(q^{-5})^3$ 47. $(y^{-7})^{-2}$

48. $(y^9)^{-4}$ 49. $[(-2)^4]^3$ 50. $[(m^4)^5]^3$

In Exercises 51–58, simplify using the product-to-power rule. Use only positive exponents in your answers.

51. $(3x)^5$ 52. $(pq)^6$ 53. $(-4m^3)^4$ 54. $(a^3 b)^7$

55. $(4a^2 b^4)^3$ 56. $(-7r^3 s^9)^2$ 57. $(2x^3 y^2 z)^4$ 58. $(2x^{-3} y^4)^{-3}$

In Exercises 59–66, simplify using the quotient-to-power rule. Use only positive exponents in your answers.

59. $\left(\dfrac{t}{4}\right)^3$

60. $\left(\dfrac{u^5}{2}\right)^7$

61. $\left(\dfrac{m^4}{n^7}\right)^5$

62. $\left(\dfrac{2a^2}{6b^5}\right)^3$

63. $\left(\dfrac{-4x^2}{y^5}\right)^3$

64. $\left(\dfrac{x}{2}\right)^{-5}$

65. $\left(\dfrac{p^3}{q^9}\right)^{-4}$

66. $\left(\dfrac{2a^3}{b^{-2}}\right)^5$

In Exercises 67–78, simplify using the rules for exponents. Use only positive exponents in your answers.

67. $(6x^{-2}y^5)^{-3}$

68. $\dfrac{a^3b^{-4}c^0}{a^5c^3}$

69. $\dfrac{-15x^{-6}y^7}{5x^{-2}y^3}$

70. $\dfrac{(5ac)^{-4}}{(3bc)^{-2}}$

71. $\left(\dfrac{p^2}{q^7}\right)^{-8}$

72. $\left(\dfrac{4m^{-2}n^{-1}}{m^6n^5}\right)^3$

73. $\left(\dfrac{6x^{-4}y^6}{42x^2y^{-11}}\right)^{-2}$

74. $(2x^4)^{-3}(-3y^{-7})$

75. $\dfrac{(3a^5b^{-6})^4}{9a^{-7}b^{-1}}$

76. $\left(\dfrac{x^2}{y}\right)^3\left(\dfrac{x^3}{y^2}\right)^2$

77. $(7a^4b^{-6})(-4m^7n^{-5})^3$

78. $\left(\dfrac{-2a^4b^{-3}}{c^4}\right)^2\left(\dfrac{9a^{-1}b^8}{c^{-2}}\right)^{-2}$

In Exercises 79–86, write each number in scientific notation.

79. 89,000,000,000

80. 1,025,000,000,000,000

81. 0.000000406

82. −0.000005

83. **Microns** A micron is a metric unit for measuring very small lengths or distances. There are 25,400 microns in 1 inch. Write 25,400 in scientific notation.

84. **X-Rays** An X-ray has a wavelength of 0.000000724 millimeters. Write 0.000000724 in scientific notation.

85. **Response Time** A computer's RAM response time is 0.000000005 second. Write 0.000000005 in scientific notation.

86. **National Debt** On June 18, 2014, the U.S. national debt was approximately 17,500,000,000,000. Write 17,500,000,000,000 in scientific notation. (*Data*: usdebtclock.org)

In Exercises 87–94, write each number in standard form.

87. 2.97×10^{-6}

88. 9.63×10^{13}

89. 8×10^{-10}

90. -6.045×10^8

91. **Light-Year** One light-year is the distance light can travel in one year, which is approximately 5.878×10^{12} miles. Write 5.878×10^{12} in standard form.

92. **Chickenpox** A chickenpox virus measures 1.6×10^{-4} millimeters in diameter. Write 1.6×10^{-4} in standard form.

93. **Dust Mites** A dust mite weighs 5.3×10^{-5} grams. Write 5.3×10^{-5} in standard form.

94. **Google Search** In 2013, an average of 1.49×10^{11} searches per month were conducted using the Google search engine. Write 1.49×10^{11} in standard form. (*Data*: factshunt.com)

In Exercises 95–100, perform the indicated operations. Write your results in scientific notation.

95. $(3.2 \times 10^4)(2.5 \times 10^7)$
96. $(8.3 \times 10^{-4})(5.6 \times 10^8)$
97. $\dfrac{7.5 \times 10^{11}}{1.5 \times 10^6}$

98. $\dfrac{1.74 \times 10^{10}}{2.4 \times 10^{-4}}$
99. $(71{,}000{,}000)(0.000006)$
100. $\dfrac{0.00000189}{0.000009}$

4.2 Introduction to Polynomial Functions

THINGS TO KNOW

Before working through this section, be sure you are familiar with the following concepts:

| | | VIDEO | ANIMATION | INTERACTIVE |

You Try It
1. Simplify Algebraic Expressions
(Section R.3, Objective 2)

You Try It
2. Evaluate Functions
(Section 2.3, Objective 2)

OBJECTIVES

1 Find the Degree and Coefficient of a Monomial
2 Find the Degree and Leading Coefficient of a Polynomial
3 Evaluate a Polynomial Function for a Given Value
4 Add Polynomials
5 Subtract Polynomials
6 Add and Subtract Polynomial Functions

OBJECTIVE 1 FIND THE DEGREE AND COEFFICIENT OF A MONOMIAL

A **term** is a **constant**, a **variable**, or the **product** of a constant and one or more variables raised to **powers**. If a term contains a single constant **factor** (possibly a simplified fraction) and if none of the variable factors can be combined using the **rules for exponents**, then it is a **simplified term**. Examples of simplified terms include:

$$4x^3, \ -\frac{2}{5}x^2y^3, \ x^2, \ 35x^{-3}, \ -3yz^4, \ 9y^{1/2}, \ \frac{8}{x}, \ \text{and } 10.$$

The terms $7x^3x^2$ and $\dfrac{4x^4}{6}$ are not simplified. Do you see **why**?

Some simplified terms are also *monomials*.

> **Definition   Monomial**
>
> A **monomial** is a simplified term in which all variables are raised to non-negative integer powers and no variables appear in any denominator.

My animation summary   ⊛ **Concept Animation** Watch this animation to explore this concept.

Every monomial has both a *degree* and a *coefficient*.

> **Definition   Degree of a Monomial**
>
> The **degree of a monomial** is the sum of the **exponents** on the variables.

For example, the degree of $4x^3$ is 3, and the degree of $-\dfrac{2}{5}x^2y^3$ is $2 + 3 = 5$.

> **Definition   Coefficient of a Monomial**
>
> The **coefficient of a monomial** is the **constant factor**. It is sometimes called the **numerical coefficient**.

For example, the coefficient of $4x^3$ is 4, and the coefficient $-\dfrac{2}{5}x^2y^3$ is $-\dfrac{2}{5}$.

My video summary ⊙ **Example 1 Finding the Degree and Coefficient of a Monomial**

Find the **degree** and **coefficient** of each monomial.

a. $4x^5y^2$

b. x^3

c. $\dfrac{3}{5}x$

d. $-x^4y^2z^5$

e. 8.5

f. $3.5xyz$

Solutions

a. The degree of $4x^5y^2$ is the **sum** of the **exponents** on the **variables**: $5 + 2 = 7$. The coefficient is the **constant factor** 4.

b. Because the exponent is 3, the degree is 3. Because $x^3 = 1x^3$, the coefficient is 1.

c. The exponent of x is **understood** to be 1, so the degree is 1. The coefficient of $\dfrac{3}{5}x$ is $\dfrac{3}{5}$.

d.–f. Try answering these questions on your own. Check your **answers**, or watch this **video** for detailed solutions to all six parts.

👣 **You Try It** Work through this You Try It problem.

Work Exercises 1–8 in this eText or in the MyMathLab Study Plan.

 TIP The degree of any nonzero constant term is 0.

OBJECTIVE 2 FIND THE DEGREE AND LEADING COEFFICIENT OF A POLYNOMIAL

Let's look at special kinds of **algebraic expressions** called *polynomials*.

Definition Polynomial

A **polynomial** is a monomial or a finite sum of monomials.

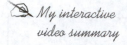 *My interactive video summary*

Watch this interactive video to determine if an algebraic expression is a polynomial.

The monomials that make up a polynomial are called the **terms of the polynomial**. A polynomial is a **simplified polynomial** if all of its terms are simplified and none of its terms are like terms.

Definition Degree of a Polynomial

The **degree of a polynomial** is the largest **degree** of its terms.

My video summary ▶ **Example 2 Finding the Degree of a Polynomial**

Find the **degree** of each polynomial.

a. $7x - 5x^2y + 9x^3y^2$

b. $16x^4 - 8x^3 + 4x^2 - 2x + 1$

Solutions Watch this video, or continue reading for solutions to both parts.

a. The degree of $7x$ is 1, the degree of $-5x^2y$ is $2 + 1 = 3$, and the degree of $9x^3y^2$ is $3 + 2 = 5$. The largest degree of the monomials is 5, so the degree of the polynomial is 5.

b. The degree of each term is shown below:

Term	$16x^4$	$-8x^3$	$4x^2$	$-2x$	1
Degree	4	3	2	1	0

The largest degree of the monomials is 4, so the degree of the polynomial is 4.

You Try It Work through this You Try It problem.

Work Exercises 9–12 in this eText or in the MyMathLab Study Plan.

The polynomial $16x^4 - 8x^3 + 4x^2 - 2x + 1$ is an example of a **polynomial in one variable**. It is written in **descending order** by the degree of its terms. This is called the **standard form** for polynomials in one variable. When a polynomial is in standard form we can find its *leading coefficient*.

> **Definition** Leading Coefficient of a Polynomial in One Variable
>
> When a polynomial in one variable is written in standard form, the **coefficient** of the first term (the term with the highest degree) is called the **leading coefficient**.

My interactive video summary

 Example 3 Finding the Degree and Leading Coefficient of a Polynomial in One Variable

Write each **polynomial** in standard form. Then find its **degree** and **leading coefficient**.

a. $13 - 8x^2 + 9x^4 + 7x^3 - \dfrac{1}{2}x$

b. $\dfrac{1}{2}p + \dfrac{2}{3}p^2 - \dfrac{5}{8}p^3$

Solutions

a. To write the polynomial in standard form, arrange the terms in order from the largest degree, 4, to the smallest degree, 0: $9x^4 + 7x^3 - 8x^2 - \dfrac{1}{2}x + 13$. The degree of the polynomial is 4, the largest degree. The leading coefficient is 9, the coefficient of the first term.

b. Try working this problem on your own. View the **answers**, or watch this **interactive video** for detailed solutions to both parts.

You Try It Work through this **You Try It** problem.

Work Exercises 13–18 in this eText or in the MyMathLab Study Plan.

OBJECTIVE 3 EVALUATE A POLYNOMIAL FUNCTION FOR A GIVEN VALUE

A function is a **polynomial function** if it is defined by a **polynomial**. For example, $P(x) = 2x^2 - 5x + 3$ is a **polynomial function** because $2x^2 - 5x + 3$ is a polynomial.

To **evaluate** a polynomial function, substitute the given value for the **variable** and simplify.

My interactive video summary

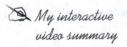 **Example 4 Evaluating Polynomial Functions**

If $P(x) = 4x^3 - 2x^2 + 8x + 7$, evaluate each of the following.

a. $P(4)$

b. $P(-2)$

c. $P\left(-\dfrac{1}{2}\right)$

Solutions

a.

Substitute 4 for x in P:	$P(4) = 4(4)^3 - 2(4)^2 + 8(4) + 7$
Simplify the exponents:	$= 4(64) - 2(16) + 8(4) + 7$
Simplify the multiplication:	$= 256 - 32 + 32 + 7$
Add and subtract:	$= 263$

The value of P is 263 when x is 4.

b.–c. Try to work these problems on your own. View the **answers**, or watch this **interactive video** for complete solutions to all three parts.

 You Try It Work through this You Try It problem.

Work Exercises 19–24 in this eText or in the MyMathLab Study Plan.

OBJECTIVE 4   ADD POLYNOMIALS

In **Section R.3**, we learned how to simplify **algebraic expressions**. This process often involves **combining like terms**.

> **Adding Polynomials**
>
> To add **polynomials**, remove all **grouping symbols** and **combine like terms**.

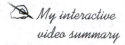
My interactive video summary

Example 5 Adding Polynomials

Add.

a. $(4x^2 - 3x + 6) + (9x^2 + 3x + 1)$

b. $(2y^3 + y^2 - 7) + (4y^3 - 3y^2 + 2y) + (6y^2 + 2)$

c. $(5p^2q + 7pq + 4pq^2) + (p^2q - 8pq - 3pq^2)$

Solutions Follow the procedure for **adding polynomials**.

a. Write the original expression: $(4x^2 - 3x + 6) + (9x^2 + 3x + 1)$

Remove grouping symbols: $= 4x^2 - 3x + 6 + 9x^2 + 3x + 1$

Rearrange terms: $= 4x^2 + 9x^2 - 3x + 3x + 6 + 1$

Combine like terms: $= 13x^2 + 7$

b. Write the original expression: $(2y^3 + y^2 - 7) + (4y^3 - 3y^2 + 2y) + (6y^2 + 2)$

Remove grouping symbols: $= 2y^3 + y^2 - 7 + 4y^3 - 3y^2 + 2y + 6y^2 + 2$

Rearrange terms: $= 2y^3 + 4y^3 + y^2 - 3y^2 + 6y^2 + 2y - 7 + 2$

Finish this problem on your own. Check your **answer**, or watch this **interactive video** to see the complete solution.

c. Try adding these polynomials on your own. Check your **answer**, or watch this **interactive video** to see the complete solution.

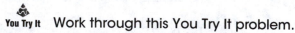
You Try It Work through this You Try It problem.

Work Exercises 25–34 in this eText or in the MyMathLab Study Plan.

TIP Sometimes it is helpful to **add polynomials vertically**. To do this, line up like terms in columns and then **combine like terms**. Example 5a is repeated below to show this method:

$$
\begin{array}{r}
4x^2 - 3x + 6 \\
+\ 9x^2 + 3x + 1 \\
\hline
13x^2 \qquad + 7
\end{array}
$$

Repeat Examples 5b and c by adding vertically. View their **solutions**.

OBJECTIVE 5   SUBTRACT POLYNOMIALS

Two real numbers are **opposites** if their **sum** is zero. In the same way, two **polynomials** are **opposite polynomials** if they add to zero. For example, the polynomials $2x^2 - 8x - 7$ and $-2x^2 + 8x + 7$ are opposites because

$$(2x^2 - 8x - 7) + (-2x^2 + 8x + 7) = 0$$

A negative sign can also be used to indicate opposite polynomials. For example, $-(2x^2 - 8x - 7)$ represents the opposite of $2x^2 - 8x - 7$. So,

$$-(2x^2 - 8x - 7) = -2x^2 + 8x + 7.$$

Note that $2x^2 - 8x - 7$ and $-2x^2 + 8x + 7$ look the same except each pair of **like terms** have opposite signs. Therefore, to find the **opposite of a polynomial**, change the sign of each term.

TIP Finding the opposite of a polynomial can be thought of as **distributing** the negative sign through the polynomial.

$$-(2x^2 - 8x - 7) = -2x^2 - (-8x) - (-7) = -2x^2 + 8x + 7$$

Finding the opposite is important for subtracting polynomials.

Subtracting Polynomials

To subtract a polynomial, add its opposite polynomial.

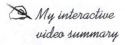 *My interactive video summary*

 Example 6 Subtracting Polynomials

Subtract.

a. $(10x^2 - 7x + 9) - (8x^2 - 4x + 1)$

b. $(2a^2 + 5ab - 9b^2) - (-4a^2 + 8ab - 5b^2)$

Solutions Follow the procedure for **subtracting polynomials**.

a.　　　　Begin with the original expression:　$(10x^2 - 7x + 9) - (8x^2 - 4x + 1)$

Change to add the opposite polynomial:　$= (10x^2 - 7x + 9) + (-8x^2 + 4x - 1)$

Remove grouping symbols:　$= 10x^2 - 7x + 9 - 8x^2 + 4x - 1$

Rearrange terms:　$= 10x^2 - 8x^2 - 7x + 4x + 9 - 1$

Combine like terms:　$= 2x^2 - 3x + 8$

b.　　　　Begin with the original expression:　$(2a^2 + 5ab - 9b^2) - (-4a^2 + 8ab - 5b^2)$

Change to add the opposite polynomial:　$= (2a^2 + 5ab - 9b^2) + (+4a^2 - 8ab + 5b^2)$

Try finishing this problem on your own. Check your **answer**, or watch this interactive video for detailed solutions to both parts.

You Try It　Work through this You Try It problem.

Work Exercises 35–44 in this eText or in the MyMathLab Study Plan.

TIP As with addition, polynomials may be subtracted vertically. **Example 6a** is repeated below to show this method:

$$
\begin{array}{r}
10x^2 - 7x + 9 \\
- (8x^2 - 4x + 1)
\end{array}
\quad
\xrightarrow[\text{polynomial}]{\substack{\text{Add the}\\\text{opposite}}}
\quad
\begin{array}{r}
10x^2 - 7x + 9 \\
+ \ -8x^2 + 4x - 1 \\
\hline
2x^2 - 3x + 8
\end{array}
$$

Repeat **Example 6b** on your own by subtracting vertically. See the **solution**.

OBJECTIVE 6 ADD AND SUBTRACT POLYNOMIAL FUNCTIONS

The procedures used to add and subtract polynomials can be used to add and subtract polynomial functions.

Definition The Sum and Difference of Functions

Let f and g represent two functions.

1. The **sum of f and g** is $(f + g)(x) = f(x) + g(x)$.

2. The **difference of f and g** is $(f - g)(x) = f(x) - g(x)$.

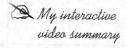 *My interactive video summary*

Example 7 Adding and Subtracting Polynomial Functions

For $P(x) = x^4 - 9x^2 + 7$ and $Q(x) = 3x^4 - 4x^2 + 2x - 10$, find each of the following.

a. $(P + Q)(x)$ b. $(P - Q)(x)$

Solutions

a. Rewrite as a sum of 2 functions: $(P + Q)(x) = P(x) + Q(x)$

Substitute for the function notation: $= (x^4 - 9x^2 + 7) + (3x^4 - 4x^2 + 2x - 10)$

Remove grouping symbols: $= x^4 - 9x^2 + 7 + 3x^4 - 4x^2 + 2x - 10$

Rearrange terms: $= x^4 + 3x^4 - 9x^2 - 4x^2 + 2x + 7 - 10$

Combine like terms: $= 4x^4 - 13x^2 + 2x - 3$

b. Rewrite as a difference of 2 functions: $(P - Q)(x) = P(x) - Q(x)$

Substitute for the function notation: $= (x^4 - 9x^2 + 7) - (3x^4 - 4x^2 + 2x - 10)$

Add the opposite polynomial: $= (x^4 - 9x^2 + 7) + (-3x^4 + 4x^2 - 2x + 10)$

Try to finish this problem on your own. Check your **answer**, or watch this interactive video to see the detailed solution.

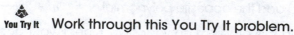 **You Try It** Work through this **You Try It** problem.

Work Exercises 45–50 in this eText or in the MyMathLab Study Plan.

4.2 Exercises

In Exercises 1–8, find the degree and coefficient of each monomial.

1. $6x^4$

2. $25a^5b^3$

3. $\dfrac{3}{8}x$

4. x^4y^3

5. -13

6. $-xy$

7. $4x^3y$

8. $-2.8x^3y^6z^7$

In Exercises 9–12, find the degree of each polynomial.

9. $15x^3y^4 + 13x^2y^5 - 11x^7y$

10. $27p^3 - 54p^2q + 36pq^2 - 8q^3$

11. $\dfrac{9}{10}x^2y^2z + \dfrac{8}{11}x^3y^3 - \dfrac{7}{12}xy^2z$

12. $17 - 10x$

In Exercises 13–18, write each polynomial in standard form. Then find its degree and leading coefficient.

13. $8 - 5m + m^5 - 9m^3 - \dfrac{2}{3}m^7$

14. $5 + 2y - 9y^2 + 6y^3$

15. $5x^2 - 9x + 4$

16. $4w - w^5 + 0.7w^2 - 10.4$

17. $9 - 3x$

18. $4.1x^3 - 6.7x + 3.8 + x^2$

In Exercises 19–24, evaluate each polynomial function for the given value.

19. $P(x) = 18x^2 - 6x + 5;\quad P\left(\dfrac{1}{3}\right)$

20. $Q(t) = t^5 + 3t^4 - 5t^3 - 8t^2 + t - 15;\quad Q(-2)$

21. $R(p) = 0.4p^3 - 2.6p^2 - 6.8p + 52;\quad R(10)$

22. $S(x) = 16.4x^4 - 10.6x^2 + 20.1;\quad S(2.5)$

23. $T(t) = \dfrac{5}{6}t + \dfrac{1}{3};\quad T(8)$

24. $R(x) = x^2 + \dfrac{2}{5}x - 2;\quad R\left(\dfrac{3}{2}\right)$

In Exercises 25–34, add.

25. $(14x^3 - 5x^2 - 9x - 7) + (-6x^3 + 2x^2 + 4x - 8)$

26. $(9x^2y - 6xy) + (8x^2y - xy)$

27. $(4w^2 - 7w + 5) + (3w^3 + 5w - 8)$

28. $(5y^2 - 3y) + (2y^2 - y)$

29. $(2a^2 - 11a + 9) + (8a^2 - 13) + (2a - 3)$

30. $(0.4w^2 - 5.6w + 3.1) + (1.3w^2 + 2.1w - 4.0)$

31. $(a^3 + b^3) + (2a^3 - 3a^2b + 9b^3)$

32. $(5x^3 + xy - 7) + (-2x^3 - 4xy + 3)$

33. $(0.4m^3n^2 - 6.8mn) + (1.2m^2n^3 + 2.4mn)$

34. $\left(\frac{2}{5}x^2 + \frac{5}{6}xy - \frac{3}{8}y^2\right) + \left(-\frac{4}{5}x^2 + \frac{1}{9}xy - \frac{1}{4}y^2\right)$

In Exercises 35–44, subtract.

35. $(10y^3 - 3y + 8) - (-2y^3 - 5y^2 + 5y - 13)$

36. $(17p^2q + 21pq^2) - (10pq^2 - 13p^2q)$

37. $(m^3 - 2m + 8) - (-3 + m^2 - 6m^3)$

38. $(2t^2 + 13t - 17) - (-3t^2 + 11t - 13)$

39. $(7.1x^2 - 3.2x + 5.8) - (4.8x^2 + 2.9)$

40. $(2m^2 + n) - (4m^2 - 16mn - 5n)$

41. $(x - y) - (4x + 2xy - 3y)$

42. $(3a^4b^2 + 7a^2b^2 - 5ab) - (2a^4b^2 - 8a^2b^2 + ab)$

43. $\left(\frac{9}{10}a - \frac{1}{4}b\right) - \left(\frac{3}{10}a - \frac{5}{8}b\right)$

44. $(m^2 - 9m + 1) - (3m^2 - 7) - (-2m + 3)$

In Exercises 45–50, let $P(x) = x^3 - 5x^2 + 2x - 1$, $Q(x) = x^3 - 9x^2 - 12$, and $R(x) = 9x^2 + 4x - 8$. Add or subtract the polynomial functions, as indicated.

45. $(P + Q)(x)$ 46. $(Q - P)(x)$ 47. $(Q + R)(x)$

48. $(P - R)(x)$ 49. $(Q - R)(x)$ 50. $(P + R - Q)(x)$

4.3 Multiplying Polynomials

THINGS TO KNOW

Before working through this section, be sure you are familiar with the following concepts:

			VIDEO	ANIMATION	INTERACTIVE

 You Try It   1.   Simplify Algebraic Expressions (Section R.3, Objective 2)

 You Try It   2.   Simplify Exponential Expressions Using the Product Rule (Section 4.1, Objective 1)

 You Try It   3.   Use the Power-to-Power Rule (Section 4.1, Objective 5)

 You Try It   4.   Use the Product-to-Power and Quotient-to-Power Rules (Section 4.1, Objective 6)

OBJECTIVES

1   Multiply Monomials

2   Multiply a Polynomial by a Monomial

3   Multiply Two Binomials

4   Multiply Two Binomials Using Special Product Rules

5   Multiply Two or More Polynomials

6   Multiply Polynomial Functions

....

OBJECTIVE 1   MULTIPLY MONOMIALS

To multiply **monomials** use the **commutative** and **associative** properties to group **factors** with **like bases** and then apply the **product rule for exponents**. This is similar to the way **exponential expressions** were simplified in **Section 4.1**.

> ### Multiplying Monomials
>
> Rearrange the factors to group the **coefficients** and to group like bases. Then, multiply the coefficients and apply the product rule for exponents.

My video summary ▶ **Example 1 Multiplying Monomials**

Multiply.

a. $(5x^4)(9x^3)$

b. $(-8p^3q^2r^4)\left(-\frac{1}{2}pq^6\right)$

Solutions Work through the following or watch this **video** for complete solutions.

a.

Begin with the original expression:	$(5x^4)(9x^3)$
Rearrange the factors to group coefficients and like bases:	$= (5 \cdot 9)(x^4 x^3)$
Multiply coefficients; apply the product rule:	$= 45x^{4+3}$
Simplify:	$= 45x^7$

b. $\left(-8p^3q^2r^4\right)\left(-\dfrac{1}{2}pq^6\right) = \left[(-8)\left(-\dfrac{1}{2}\right)\right](p^3 p)(q^2 q^6)r^4$

$$= 4p^{3+1}q^{2+6}r^4$$

$$= 4p^4 q^8 r^4$$

You Try It Work through this You Try It problem.

Work Exercises 1–6 in this eText or in the MyMathLab Study Plan.

OBJECTIVE 2 MULTIPLY A POLYNOMIAL BY A MONOMIAL

To multiply a **monomial** and a **polynomial** with more than one **term**, use the distributive property.

Multiplying Polynomials by Monomials

To multiply a polynomial by a monomial, use the distributive property to multiply each term of the polynomial by the monomial. Then simplify using the method for **multiplying monomials**.

If m is a monomial and $p_1, p_2, p_3, \ldots, p_n$ are the terms of a polynomial, then

$$m \cdot (p_1 + p_2 + p_3 + \ldots + p_n) = m \cdot p_1 + m \cdot p_2 + m \cdot p_3 + \ldots + m \cdot p_n$$

Example 2 Multiplying a Polynomial by a Monomial

Multiply.

a. $2x^3(3x^2 - 7x + 9)$ **b.** $-0.7xy^2(6x^3y - 5x^2y^2)$

Solutions

a.

Begin with the original expression:	$2x^3(3x^2 - 7x + 9)$
Distribute the monomial $2x^3$:	$= 2x^3 \cdot 3x^2 + 2x^3 \cdot (-7x) + 2x^3 \cdot 9$
Rearrange the factors:	$= 2 \cdot 3 \cdot x^3 \cdot x^2 + 2 \cdot (-7) \cdot x^3 \cdot x + 2 \cdot 9 \cdot x^3$
Multiply coefficients; apply the product rule:	$= 6x^{3+2} - 14x^{3+1} + 18x^3$
Add the exponents:	$= 6x^5 - 14x^4 + 18x^3$

My video summary ▶ **b.** Begin with the original expression: $-0.7xy^2\,(6x^3y - 5x^2y^2)$

Distribute the monomial $-0.7xy^2$: $= (-0.7xy^2) \cdot 6x^3y + (-0.7xy^2) \cdot (-5x^2y^2)$

Try to finish this problem on your own. View the **answer**, or watch this **video** to see a complete solution to part b.

You Try It Work through this You Try It problem.

Work Exercises 7–14 in this eText or in the MyMathLab Study Plan.

OBJECTIVE 3   MULTIPLY TWO BINOMIALS

A simplified **polynomial** with two terms is called a **binomial**. Two binomials can be multiplied by using the **distributive property** twice.

My animation summary ✷ **Concept Animation** Watch this **animation**, or continue reading to explore this concept.

Distribute the first binomial to each **term** in the second binomial. Then distribute each term from the second binomial through the first binomial (from the back).

$$(a + b)(c + d) = (a + b)c + (a + b)d = ac + bc + ad + bd$$

The end result is that each term in the first binomial gets multiplied by each term in the second binomial.

$$(a + b)(c + d) = ac + ad + bc + bd$$

Multiplying Two Binomials

To multiply two binomials, multiply each term of the first binomial by each term of the second binomial. To simplify, **combine like terms**, if any.

My video summary ▶ **Example 3 Multiplying Two Binomials**

Multiply $(2x + 5)(3x + 4)$.

Solution Watch this **video**, or read below for a complete solution.

Multiply each term in the first **binomial** by each term in the second binomial.

Begin with the original expression: $(2x + 5)(3x + 4)$

Multiply $2x$ by $3x$ and 4; multiply 5 by $3x$ and 4: $= 2x \cdot 3x + 2x \cdot 4 + 5 \cdot 3x + 5 \cdot 4$

Simplify: $= 6x^2 + 8x + 15x + 20$

Combine like terms: $= 6x^2 + 23x + 20$

You Try It Work through this You Try It problem.

Work Exercises 15–18 in this eText or in the MyMathLab Study Plan.

TIP As with addition and subtraction of **polynomials**, the multiplication of binomials can be organized vertically. View a **vertical solution** of Example 3.

The process used in Example 3 follows a specific order for multiplying **binomials**. The acronym **FOIL** summarizes the process.

 *My animation summary*

⊛ **Concept Animation** Watch this animation to explore this concept.

The FOIL Method

FOIL reminds us to multiply the two **First** terms, the two **Outside** terms, the two **Inside** terms, and the two **Last** terms.

⚠ After using FOIL, be sure to simplify by combining any like terms.

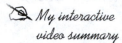 *My interactive video summary*

✍ **Example 4 Multiplying Two Binomials**

Multiply using the FOIL method.

a. $(x - 4)(2x + 3)$　　**b.** $\left(\frac{1}{2}x - 6\right)(3x - 4)$　　**c.** $(5x + 7)(4x + 3)$

Solutions Apply the FOIL method.

a. The two first terms are x and $2x$. The two outside terms are x and 3. The two inside terms are -4 and $2x$. The two last terms are -4 and 3.

$$(x - 4)(2x + 3) = x \cdot 2x + x \cdot 3 + (-4) \cdot 2x + (-4) \cdot 3$$
$$= 2x^2 + 3x + (-8x) + (-12)$$
$$= 2x^2 + 3x - 8x - 12$$
$$= 2x^2 - 5x - 12 \quad \leftarrow \boxed{3x - 8x = -5x}$$

b. $\left(\frac{1}{2}x - 6\right)(3x - 4) = \frac{1}{2}x \cdot 3x + \frac{1}{2}x \cdot (-4) + (-6) \cdot 3x + (-6) \cdot (-4)$

Try finishing this problem on your own. View the **answer**, or watch this **interactive video** to see the complete solution to all three parts.

c. Try working this problem on your own. View the **answer**, or watch this **interactive video** to see the complete solution to all three parts.

 You Try It Work through this You Try It problem.

Work Exercises 19–21 in this eText or in the MyMathLab Study Plan.

My video summary ▶ **Example 5 Multiplying Two Binomials Using the FOIL Method**

Multiply using **FOIL**.

$$(4m^2 - 5n)(3m^2 - 9n)$$

Solution Read the following, or watch this **video** for a complete solution.

$$(4m^2 - 5n)(3m^2 - 9n) = \overbrace{4m^2 \cdot 3m^2}^{F} + \overbrace{4m^2 \cdot (-9n)}^{O} + \overbrace{(-5n) \cdot 3m^2}^{I} + \overbrace{(-5n) \cdot (-9n)}^{L}$$

$$= 12m^4 - 36m^2 n - 15m^2 n + 45n^2$$

$$= 12m^4 - 51m^2 n + 45n^2$$

⚠ The FOIL method can be used only when multiplying two binomials.

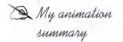 **You Try It** Work through this You Try It problem.

Work Exercises 22–30 in this eText or in the MyMathLab Study Plan.

OBJECTIVE 4 MULTIPLY TWO BINOMIALS USING SPECIAL PRODUCT RULES

Recall that a binomial is a simplified **polynomial** with two terms. Likewise, a **trinomial** is a simplified polynomial with three terms.

A **binomial sum** is a binomial in which the two terms are added. A **binomial difference** is a binomial in which one term is subtracted from the other. When a binomial sum and a binomial difference are made from the same two terms, they are called **conjugates** of each other. For example, $5x + 3y$ and $5x - 3y$ are conjugates.

My animation summary ✷ **Concept Animation** Watch this animation to derive rules for special products involving binomial sums and differences.

The results $A^2 + 2AB + B^2$ and $A^2 - 2AB + B^2$ from **squaring** binomials are called **perfect square trinomials**, and the result $A^2 - B^2$ from multiplying the sum and difference of two terms (or **conjugates**) is a called a **difference of two squares**.

Special Product Rules for Binomials

Square of a binomial sum	$(A + B)^2 = A^2 + 2AB + B^2$	Perfect square trinomial
Square of a binomial difference	$(A - B)^2 = A^2 - 2AB + B^2$	Perfect square trinomial
Product of the sum and difference of two terms (product of conjugates)	$(A + B)(A - B) = A^2 - B^2$	Difference of two squares

---

Recognizing the special cases for binomials will allow their products to be found more quickly. Let's practice using the **rules** in the next three examples.

My video summary ▶ **Example 6 Finding the Square of a Binomial Sum**

Multiply $(4x + 5)^2$.

Solution Watch this video, or continue reading for a complete solution.

$(4x + 5)^2$ is the **square of a binomial sum**. Use the rule for $(A + B)^2$ with $A = 4x$ and $B = 5$.

Write the square of a binomial sum rule: $(A + B)^2 = A^2 + 2AB + B^2$

Substitute $4x$ for A and 5 for B: $(4x + 5)^2 = (4x)^2 + 2(4x)(5) + (5)^2$

Simplify: $= 16x^2 + 40x + 25$

You Try It Work through this You Try It problem.

Work Exercises 31–34 in this eText or in the MyMathLab Study Plan.

My video summary ▶ **Example 7 Finding the Square of a Binomial Difference**

Multiply $(9p^3 - 11q)^2$.

Solution Watch this video, or continue reading for a complete solution.

$(9p^3 - 11q)^2$ is the **square of a binomial difference**. Use the rule for $(A - B)^2$ with $A = 9p^3$ and $B = 11q$.

Write the square of a binomial
difference rule: $(A + B)^2 = A^2 - 2AB + B^2$

Substitute $4x$ for A and 7 for B: $(9p^3 - 11q)^2 = (9p^3)^2 - 2(9p^3)(11q) + (11q)^2$

Simplify: $= 81p^6 - 198p^3q + 121q^2$

You Try It Work through this You Try It problem.

Work Exercises 35–37 in this eText or in the MyMathLab Study Plan.

Example 8 Multiplying the Sum and Difference of Two Terms (Product of Conjugates)

Multiply.

a. $(2x + 7)(2x - 7)$ **b.** $(10a^2 + b^3)(10a^2 - b^3)$

Solutions

a. Use the rule for multiplying the sum and difference of two terms (product of conjugates). $(A + B)(A - B)$ with $A = 2x$ and $B = 7$.

Write the product of conjugates rule: $(A + B)(A - B) = A^2 - B^2$

Substitute $2x$ for A and 7 for B: $(2x + 7)(2x - 7) = (2x)^2 - (7)^2$

Simplify: $= 4x^2 - 49$

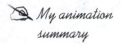 *My video summary* ▶ **b.** Try working this problem on your own. View the **answer**, or watch this **video** for a complete solution.

     You Try It   Work through this You Try It problem.

Work Exercises 38–42 in this eText or in the MyMathLab Study Plan.

OBJECTIVE 5   MULTIPLY TWO OR MORE POLYNOMIALS

We can apply what we have learned so far about multiplying **binomials** to multiply any two **polynomials**.

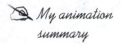 *My animation summary*   ✤ **Concept Animation** Watch this animation to review this concept.

> **Multiplying Two or More Polynomials**
>
> To multiply two polynomials, multiply each term of the first polynomial by each term of the second polynomial. **Combine like terms.** If there are more than two polynomials, multiply two polynomials at a time following this procedure.

Example 9 Multiplying Two Polynomials

Multiply.

a. $(3x + 5)(2x^2 + 7x + 9)$ **b.** $(y^2 + 2y - 9)(2y^2 - 4y + 7)$

Solutions

a. Multiply each term in the binomial by each term in the **trinomial**.

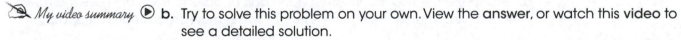

$$\text{Begin with the original expression:} \quad (3x + 5)(2x^2 + 7x + 9)$$

Multiply $3x$ by $2x^2$, $7x$, and 9;
multiply 5 by $2x^2$, $7x$, and 9: $\quad = 3x \cdot 2x^2 + 3x \cdot 7x + 3x \cdot 9 + 5 \cdot 2x^2 + 5 \cdot 7x + 5 \cdot 9$

$$\text{Simplify:} \quad = 6x^3 + 21x^2 + 27x + 10x^2 + 35x + 45$$

$$\text{Combine like terms:} \quad = 6x^3 + 31x^2 + 62x + 45$$

View an **alternate solution** of this problem done by multiplying vertically.

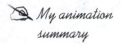 *My video summary* ▶ **b.** Try to solve this problem on your own. View the **answer**, or watch this **video** to see a detailed solution.

     You Try It   Work through this You Try It problem.

Work Exercises 43–48 in this eText or in the MyMathLab Study Plan.

Two **polynomials** can also be multiplied by using a **rectangular array**. This approach is sometimes called the **box method**.

 *My animation summary*   ✤ **Concept Animation** Watch this animation to explore this approach.

To multiply three or more polynomials, multiply two polynomials at a time.

My interactive video summary

Example 10 Multiplying Three or More Polynomials

Multiply.

a. $2x^2(5x + 7)(x - 8)$ **b.** $(x + 5)^3$

Solutions

a. Because multiplication is associative, we can group the multiplication in the order we wish. For this problem, we choose to multiply the two binomials and then distribute the monomial through the resulting product.

Begin with the original expression: $2x^2(5x + 7)(x - 8)$

Multiply $5x + 7$ and $x - 8$: $= 2x^2(5x^2 - 33x - 56)$

Distribute $2x^2$: $= 10x^4 - 66x^3 - 112x^2$

Watch this interactive video to see the solution worked out in detail.

b. $(x + 5)^3$ is an exponential expression with the base $x + 5$ and the exponent 3, so multiply three factors of $x + 5$.

Begin with the original expression: $(x + 5)^3$

Group two factors to multiply: $= (x + 5)^2(x + 5)$

Multiply out $(x + 5)^2$: $= (x^2 + 10x + 25)(x + 5)$

Try to finish this problem on your own. View the **answer**, or watch this interactive video for the complete solution.

You Try It Work through this You Try It problem.

Work Exercises 49–58 in this eText or in the MyMathLab Study Plan.

In some cases, special product rules for binomials can be used to multiply two trinomials.

My interactive video summary

Example 11 Using Special Product Rules to Multiply Trinomials

Use special product rules to multiply.

a. $(x + 8 + 3y)(x + 8 - 3y)$ **b.** $(2x + 3y + 5)^2$

Solutions

a. Group the first two terms within each pair of parentheses, and use the rule for multiplying the **sum and difference of two terms** (product of **conjugates**).

$$(x + 8 + 3y)(x + 8 - 3y) = [(x + 8) + 3y][(x + 8) - 3y]$$

Write the problem in the form $(A + B)(A - B)$ with $A = x + 8$ and $B = 3y$.

Write the special product rule: $(A + B)(A - B) = A^2 - B^2$

Substitute $(x + 8)$ for A and $3y$ for B: $[(x + 8) + 3y][(x + 8) - 3y]$

$$= (x + 8)^2 - (3y)^2$$

Simplify the resulting expression to finish the problem. Check your answer, or watch this interactive video for a complete solution.

b. Group the first two terms, and use the rule for the **square of a binomial sum.**

$$(2x + 3y + 5)^2 = [(2x + 3y) + 5]^2$$

Write the problem in the form $(A + B)^2$ with $A = 2x + 3y$ and $B = 5$. Try to finish this problem on your own. View the **answer**, or watch this **interactive video** to see a complete solution.

You Try It Work through this You Try It problem.

Work Exercises 59–60 in this eText or in the MyMathLab Study Plan.

OBJECTIVE 6 MULTIPLY POLYNOMIAL FUNCTIONS

After learning how to **multiply polynomials**, we now can multiply **polynomial functions**.

> **Definition** The Product of Functions
>
> If f and g represent functions, then the **product of f and g** is $(f \cdot g)(x) = f(x) \cdot g(x)$.

My video summary ▶ **Example 12 Multiplying Polynomial Functions**

For $P(x) = 5x - 2$ and $Q(x) = 4x - 9$, find $(P \cdot Q)(x)$.

Solution Read through the following, or watch this **video** for a complete solution.

Define the original problem: $(P \cdot Q)(x) = P(x) \cdot Q(x)$

Substitute for the function notation: $= (5x - 2)(4x - 9)$

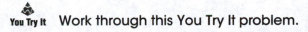

Use the FOIL method: $= \overbrace{5x \cdot 4x}^{F} + \overbrace{5x \cdot (-9)}^{O} + \overbrace{(-2) \cdot 4x}^{I} + \overbrace{(-2) \cdot (-9)}^{L}$

Simplify the multiplication: $= 20x^2 - 45x - 8x + 18$

Combine like terms: $= 20x^2 - 53x + 18$

You Try It Work through this You Try It problem.

Work Exercises 61–64 in this eText or in the MyMathLab Study Plan.

4.3 Exercises

In Exercises 1–6, multiply the monomials.

1. $(5x^3)(6x^4)$

2. $(-3y^5)(-2y^9)$

3. $(3.2w^3)(2.5w^2)$

4. $(3x^2y^3)(5x^4y^2)$

5. $\left(-\dfrac{3}{4}a^5b^3\right)\left(\dfrac{2}{5}a^2b^6\right)$

6. $(8x^3yz^5)(3xy^4z^2)$

In Exercises 7–14, multiply the monomial and the polynomial.

7. $-8x(4x - 9)$

8. $7x^3(5x^2 - 9)$

9. $-\dfrac{1}{2}x^2(2x^2 + 4x + 6)$

10. $2x^2(3x^5 - 2x^3 + 7x - 4)$

11. $3x^2y(x + y)$

12. $-4y^3(4xy + 15y^2)$

13. $3.4a^3b(2a^2b^3 + 7ab)$

14. $-2x^3y^2z(6xy - 12x^2yz^3 + 5y^2z^4)$

In Exercises 15–18, multiply using the distributive property twice.

15. $(x + 2)(x + 5)$

16. $(w - 3)(2w - 5)$

17. $\left(y + \dfrac{3}{5}\right)\left(y - \dfrac{4}{3}\right)$

18. $(2x^2 - 3)(3x + 2)$

In Exercises 19–30, multiply the binomials.

19. $(x + 10)(x + 3)$

20. $(y + 9)(y - 4)$

21. $(3x + 2)(2x + 5)$

22. $(6x - 5y)(2x + 7y)$

23. $(3m^2 - 2n^2)(m^2 - 4n^2)$

24. $(a - 4)(a^2 - 5)$

25. $\left(x + \dfrac{1}{3}\right)\left(x - \dfrac{1}{2}\right)$

26. $(7z + 6)(8z - 3)$

27. $(2x^2 - 3)(3x - 2)$

28. $(x + 7y)(3x + 8y)$

29. $(0.2a^2 + 3b)(3a - 0.4b^2)$

30. $\left(\dfrac{1}{3}x - 3y\right)\left(\dfrac{2}{3}x + 2y\right)$

In Exercises 31–60, multiply using the special product rules for binomials.

31. $(y + 5)^2$

32. $(4z + 5)^2$

33. $(10x^2 + 11y)^2$

34. $\left(\dfrac{2}{5}xy + 3\right)^2$

35. $(x - 8)^2$

36. $(11y - 4)^2$

37. $(7p - 2q)^2$

38. $(3a^2 - b^2)(3a^2 + b^2)$

39. $(x - 10)(x + 10)$

40. $(7x + 9)(7x - 9)$

41. $(3m^2 + 2n)(3m^2 - 2n)$

42. $(x^2y^3 - 4xy^2)(x^2y^3 + 4xy^2)$

43. $(x + 4)(x^2 - 3x + 5)$

44. $(3m + 2)(9m^2 - 11m + 1)$

45. $(4a + b)(3a^2 - 5ab + 2b^2)$

46. $(x^2 - 3x + 7)(x^2 + x - 5)$

47. $(2x - y)(x^2 + 3xy - y^2)$

48. $(x + 5y)(x^2 - 5xy + 25y^2)$

49. $3x^2(4x + 9)(x - 5)$

50. $(2y + 3)(2y - 3)(4y^2 + 9)$

51. $(x + 1)(x + 2)(x + 3)$

52. $(x - 3)(x^2 + 3x + 9)$

53. $(x + 4)^3$

54. $(x - 3)^4$

55. $-2x(x - y)(3x^2 + y)$

56. $(2x)(-3x)(x^2 + x + 1)$

57. $(x + 4)(2x - 1)(3x - 2)$

58. $(5x)(3x - 2)(2x + 3)$

59. $(3x + 2y - 5)(3x + 2y + 5)$

60. $(x + y + 7)^2$

In Exercises 61–64, find $(P \cdot Q)(x)$.

61. $P(x) = 2x - 5;\quad Q(x) = x + 8$

62. $P(x) = 10x + 4;\quad Q(x) = -3x + 2$

63. $P(x) = 0.3x - 1.6;\quad Q(x) = 1.4x + 5.4$

64. $P(x) = x - 4;\quad Q(x) = x^2 + 3x - 5$

4.4 Dividing Polynomials

THINGS TO KNOW

Before working through this section, be sure you are familiar with the following concepts:

| | VIDEO | ANIMATION | INTERACTIVE |

 1. Perform Operations on Real Numbers (Section R.2, **Objective 1**)

 2. Multiply Two or More Polynomials (Section 4.3, **Objective 5**)

OBJECTIVES

1 Divide Monomials

2 Divide a Polynomial by a Monomial

3 Divide Polynomials Using Long Division

4 Divide Polynomials Using Synthetic Division

5 Divide Polynomial Functions

6 Use the Remainder and Factor Theorems

....

OBJECTIVE 1 DIVIDE MONOMIALS

In **Section 4.1**, we grouped **exponential expressions** with **like bases** into individual quotients and then applied the **quotient rule** to simplify. Dividing monomials is a similar process.

> **Dividing Monomials**
>
> Group the **coefficients** and group like bases into individual quotients. Then divide the coefficients and apply the quotient rule for exponents.

My video summary ⊙ **Example 1 Dividing Monomials**

Divide.

a. $\dfrac{32x^7}{4x^3}$

b. $\dfrac{9y^4}{45y^4}$

c. $\dfrac{60y}{5y^4}$

Solutions Watch this **video**, or read the following for complete solutions to all three parts.

a. Begin with the original expression: $\dfrac{32x^7}{4x^3}$

Group like bases into individual quotients: $= \dfrac{32}{4} \cdot \dfrac{x^7}{x^3}$

Divide the coefficients; apply the quotient rule: $= 8 \cdot x^{7-3}$

Simplify: $= 8x^4$

b. Begin with the original expression: $\dfrac{9y^4}{45y^4}$

Group like bases into individual quotients: $= \dfrac{9}{45} \cdot \dfrac{y^4}{y^4}$

Divide the **coefficients**; apply the quotient rule: $= \dfrac{1}{5} \cdot y^0$ ← $\boxed{4-4}$

Apply the **zero-power** rule: $= \dfrac{1}{5}$ ← $\boxed{\frac{1}{5} \cdot y^0 = \frac{1}{5} \cdot 1 = \frac{1}{5}}$

c. Begin with the original expression: $\dfrac{60y}{5y^4}$

Group like bases into individual quotients: $= \dfrac{60}{5} \cdot \dfrac{y^1}{y^4}$ ← $\boxed{y = y^1}$

Divide the **coefficients**; apply the quotient rule: $= 12 \cdot y^{-3}$ ← $\boxed{1-4}$

Apply the **negative-power** rule: $= \dfrac{12}{y^3}$

 You Try It Work through this You Try It problem.

Work Exercises 1–4 in this eText or in the MyMathLab Study Plan.

OBJECTIVE 2 DIVIDE A POLYNOMIAL BY A MONOMIAL

From Section 4.3, we learned how to multiply a **polynomial** by a **monomial**. Using the **distributive property**, we multiplied each term of the polynomial by the monomial and then simplified the expression. Dividing a polynomial by a monomial is a similar process. We divide each term of the polynomial by the monomial.

Dividing Polynomials by Monomials

To divide a polynomial by a monomial, divide each term of the polynomial by the monomial and simplify using the method for **dividing monomials**.

TIP Division answers can be checked by multiplying. For example, $\dfrac{10}{2} = 5$ checks because $2 \cdot 5 = 10$.

In general, if $\dfrac{\text{dividend}}{\text{divisor}} = \text{quotient}$, then $(\text{divisor})(\text{quotient}) = \text{dividend}$.

Example 2 Dividing a Polynomial by a Monomial

Divide.

a. $\dfrac{12x^3 - 28x^2}{4x^2}$

b. $(9m^5 - 15m^4 + 18m^3) \div 3m^3$

Solutions

a. Begin with the original expression: $\dfrac{12x^3 - 28x^2}{4x^2}$

Divide each term of the **polynomial** by the **monomial**: $= \dfrac{12x^3}{4x^2} - \dfrac{28x^2}{4x^2}$

Use the method for **dividing monomials**: $= \dfrac{12}{4}x^{3-2} - \dfrac{28}{4}x^{2-2}$

Simplify: $= 3x^1 - 7x^0$

$= 3x - 7$

Check: If $\dfrac{12x^3 - 28x^2}{4x^2} = 3x - 7$, then $4x^2(3x - 7)$ must equal $12x^3 - 28x^2$. We multiply to check.

$$4x^2(3x - 7) = 4x^2 \cdot 3x - 4x^2 \cdot 7 = 12x^3 - 28x^2$$

So, our quotient $3x - 7$ checks.

 ⓑ **b.** Try to work this problem on your own. View the **answer**, or watch this **video** for a complete solution to part b.

You Try It Work through this You Try It problem.

Work Exercises 5–10 in this eText or in the MyMathLab Study Plan.

Example 3 Dividing a Polynomial by a Monomial

My video summary ⓑ Divide $\dfrac{9a^3b^4 - 12a^2b^5 + 6ab^7}{6a^2b^3}$.

Solution

Begin with the original expression: $\dfrac{9a^3b^4 - 12a^2b^5 + 6ab^7}{6a^2b^3}$

Divide each term of the polynomial by the monomial: $= \dfrac{9a^3b^4}{6a^2b^3} - \dfrac{12a^2b^5}{6a^2b^3} + \dfrac{6ab^7}{6a^2b^3}$

Try to finish this problem on your own. View the **answer**, or watch this **video** for the complete solution.

You Try It Work through this You Try It problem.

Work Exercises 11–14 in this eText or in the MyMathLab Study Plan.

OBJECTIVE 3 DIVIDE POLYNOMIALS USING LONG DIVISION

When dividing a **polynomial** by a polynomial, we use **long division**. **Polynomial long division** follows the same approach as long division for **real numbers**. View this **example** to review using long division to divide real numbers.

Long division is a four-step process: **divide**, **multiply**, **subtract**, and **drop**.

Before performing polynomial long division, write the **dividend** and **divisor** in descending order. If any **powers** are missing, then insert them with a **coefficient** of 0 as a placeholder.

Process for Polynomial Long Division

1. **Divide** the first term of the dividend by the first term of the divisor.

2. **Multiply** the result of the division by the divisor. Write this result under the dividend (line up like terms).

3. **Subtract** the result of the multiplication from the dividend.

4. **Drop** down the next term from the original dividend to form a reduced polynomial.

Repeat these four steps using the reduced polynomial. Continue repeating until the **remainder** can no longer be divided. This occurs when the **degree** of the remainder is less than the degree of the divisor

---

The final result can be written in the form

$$\frac{\text{dividend}}{\text{divisor}} = \text{quotient} + \frac{\text{remainder}}{\text{divisor}}.$$

We check this result by noting that

$$\text{dividend} = (\text{divisor})(\text{quotient}) + \text{remainder}.$$

The following examples demonstrate **polynomial long division**.

Example 4 Dividing Polynomials Using Long Division

 ▶ Divide $(8x^2 - 6x - 38) \div (2x - 5)$.

Solution Watch this **video**, or continue reading for a complete solution.

First, note that the **dividend** is $8x^2 - 6x - 38$ and the **divisor** is $2x - 5$. Next, apply the **four-step process** until the **remainder** can no longer be divided.

Divide the first **term** of the dividend by the first term of the divisor. So, we divide $8x^2$ by $2x$. Because $\dfrac{8x^2}{2x} = 4x$, $4x$ is the first term in the quotient.

$$\begin{array}{r} 4x \\ 2x - 5 \overline{)8x^2 - 6x - 38} \end{array}$$

Multiply the result of the division, $4x$, by the divisor, $2x - 5$, to get $4x(2x - 5) = 8x^2 - 20x$. Write this result under the dividend, making sure to line up **like terms**.

$$\begin{array}{r} 4x \\ 2x - 5 \overline{)8x^2 - 6x - 38} \\ 8x^2 - 20x \end{array}$$

Subtract the result of the multiplication, $8x^2 - 20x$, from the first two terms of the **dividend**, $8x^2 - 6x$. To do the subtraction, change the sign of each term in $(8x^2 - 20x)$ and add.

$$
\begin{array}{r}
4x \\
2x - 5\overline{)8x^2 - 6x - 38} \\
\underline{-(8x^2 - 20x)}
\end{array}
\qquad \rightarrow \qquad
\begin{array}{r}
4x \\
2x - 5\overline{)8x^2 - 6x - 38} \\
\text{Add} \rightarrow \underline{-8x^2 + 20x} \\
14x
\end{array}
$$

Drop down the next term, -38, and repeat the process.

$$
\begin{array}{r}
4x \\
2x - 5\overline{)8x^2 - 6x - 38} \\
\underline{-(8x^2 - 20x)} \\
14x - 38
\end{array}
$$

Divide $14x$ by $2x$. Because $\dfrac{14x}{2x} = 7$, the next term in the quotient is 7.

$$
\begin{array}{r}
4x + 7 \\
2x - 5\overline{)8x^2 - 6x - 38} \\
\underline{-(8x^2 - 20x)} \\
14x - 38
\end{array}
$$

Multiply 7 by $2x - 5$ to get $7(2x - 5) = 14x - 35$.

$$
\begin{array}{r}
4x + 7 \\
2x - 5\overline{)8x^2 - 6x - 38} \\
\underline{-(8x^2 - 20x)} \\
14x - 38 \\
14x - 35
\end{array}
$$

Subtract $14x - 35$ from $14x - 38$. Do this by changing the sign of each term in $(14x - 35)$ and adding.

$$
\begin{array}{r}
4x + 7 \\
2x - 5\overline{)8x^2 - 6x - 38} \\
\underline{-(8x^2 - 20x)} \\
14x - 38 \\
\underline{-(14x - 35)}
\end{array}
\qquad \rightarrow \qquad
\begin{array}{r}
4x + 7 \\
2x - 5\overline{)8x^2 - 6x - 38} \\
\underline{-(8x^2 - 20x)} \\
14x - 38 \\
\text{Add} \rightarrow \underline{-14x + 35} \\
-3 \leftarrow \text{Remainder}
\end{array}
$$

Since the **degree** of the remainder is less than the degree of the **divisor**, our division is complete. The quotient is $4x + 7$, and the remainder is -3. So,

$$\frac{8x^2 - 6x - 38}{2x - 5} = 4x + 7 - \frac{3}{2x - 5}$$

Check: We check our result by showing that
$$\text{dividend} = (\text{divisor})(\text{quotient}) + \text{remainder}.$$

$$\underbrace{(2x - 5)(4x + 7)}_{\text{Divisor} \quad \text{Quotient}} + \underbrace{(-3)}_{\text{Remainder}} = 8x^2 + 14x - 20x - 35 - 3 = \underbrace{8x^2 - 6x - 38}_{\text{Dividend}}$$

Our result checks, so the division is correct.

👤 **You Try It**   Work through this You Try It problem.

Work Exercises 15–18 in this eText or in the MyMathLab Study Plan.

🖋 *My video summary* ▶ **Example 5 Dividing Polynomials Using Long Division**

Divide $\dfrac{x^2 + 26x - 6x^3 - 12}{2x - 3}$.

Solution First, rewrite the dividend in descending order: $-6x^3 + x^2 + 26x - 12$. The divisor is $2x - 3$. Apply the long division process of divide, multiply, subtract, and drop until the remainder can no longer be divided.

Divide the first term of the dividend, $-6x^3$, by the first term of the divisor, $2x$. Because $\dfrac{-6x^3}{2x} = -3x^2$, $-3x^2$ is the first term in the quotient.

$$
\begin{array}{r}
-3x^2 \\
2x - 3 \overline{)\,-6x^3 + x^2 + 26x - 12\,}
\end{array}
$$

Multiply the result of the division, $-3x^2$, by the divisor, $2x - 3$, to get $-3x^2(2x - 3) = -6x^3 + 9x^2$. Write this result under the dividend, making sure to line up like terms.

$$
\begin{array}{r}
-3x^2 \\
2x - 3 \overline{)\,-6x^3 + x^2 + 26x - 12\,} \\
-6x^3 + 9x^2
\end{array}
$$

Subtract the result of the multiplication, $-6x^3 + 9x^2$, from the first two terms of the dividend. To do the subtraction, change the sign of each term in $(-6x^3 + 9x^2)$ and add.

$$
\begin{array}{r}
-3x^2 \\
2x - 3 \overline{)\,-6x^3 + x^2 + 26x - 12\,} \\
-(-6x^3 + 9x^2)
\end{array}
\quad \rightarrow \quad
\begin{array}{r}
-3x^2 \\
2x - 3 \overline{)\,-6x^3 + x^2 + 26x - 12\,} \\
\text{Add} \rightarrow \quad +6x^3 - 9x^2 \\
\hline
-8x
\end{array}
$$

Drop down the next term, $26x$, and repeat the process.

$$
\begin{array}{r}
-3x^2 \\
2x - 3 \overline{)\,-6x^3 + x^2 + 26x - 12\,} \\
+6x^3 - 9x^2 \\
\hline
-8x^2 + 26x
\end{array}
$$

Try finishing the problem on your own. See the **answer**, or watch this **video** for the complete solution.

Check: Show that dividend = (divisor)(quotient) + remainder. View the **check**.

👤 **You Try It**   Work through this You Try It problem.

Work Exercises 19–22 in this eText or in the MyMathLab Study Plan.

My video summary ▶ **Example 6 Dividing Polynomials Using Long Division**

Divide $3x^4 + 5x^3 - 6x - 15$ by $x^2 - 2$.

Solution The dividend is $3x^4 + 5x^3 - 6x - 15$ and the divisor is $x^2 - 2$. Before applying the **four-step process**, write both **polynomials** in **standard form**, inserting **coefficients** of 0 for any missing **terms**.

$$x^2 + 0x - 2 \overline{)3x^4 + 5x^3 + 0x^2 - 6x - 15}$$

Try finishing this problem on your own. View the **answer**, or watch this **video** for the complete solution.

You Try It Work through this **You Try It** problem.

Work Exercises 23–28 in this eText or in the MyMathLab Study Plan.

OBJECTIVE 4   DIVIDE POLYNOMIALS USING SYNTHETIC DIVISION

When dividing a **polynomial** by a **binomial** of the form $x - c$, a shortcut method called **synthetic division** can be used instead of **long division**. Synthetic division is less repetitive than long division. Using zeros as placeholders for missing **powers** and lining up **like terms**, we only need to work with the **coefficients** because the position of the coefficient determines the power of the term.

My animation summary ✷ **Concept Animation** Watch this animation to see how synthetic division works as compared to polynomial long division.

Example 7 Dividing Polynomials Using Synthetic Division

Divide $2x^4 + 9x^3 - 12x + 1$ by $x + 5$ using synthetic division.

My video summary ▶ **Solution** Read through the following, or watch this **video** for a detailed solution.

Step 1 Write both polynomials in **standard form**. Both polynomials are already in standard form, so we move on to Step 2.

Step 2 Rewrite the divisor as a **binomial** of the form $x - c$. Because $x + 5 = x - (-5)$, $c = -5$.

Step 3 Write down c and the coefficients of the dividend. If there are powers "missing," insert a 0 coefficient as a placeholder for the missing term(s).

$$\begin{array}{c} c \\ \searrow \\ -5 \overline{)\ 2\ \ 9\ \ 0\ \ -12\ \ 1} \longleftarrow \text{Coefficients of dividend} \\ \uparrow \\ \text{Placeholder} \end{array}$$

Step 4 Leave some space under the coefficients, draw a horizontal line, and drop down the **leading coefficient** under the line.

$$\begin{array}{lll} -5\overline{)\ 2\ \ 9\ \ 0\ \ -12\ \ 1} & \text{Row 1} \\ \ \ \ \ \downarrow \ \ \underline{} & \text{Row 2} \\ \ \ \ \ 2 & \text{Row 3} \end{array}$$

Step 5 Multiply c by the entry in Row 3 and put the result in the next position to the right in Row 2. Add the values from Rows 1 and 2 to get the next entry in Row 3.

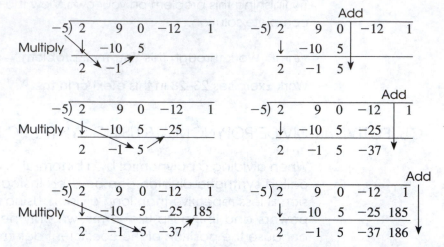

Step 6 Continue with Step 5 until Row 3 is complete.

The **quotient** and **remainder** are determined by looking at the completed Row 3 of the **synthetic division**. Because the **divisor** is a **polynomial** of **degree** 1, the quotient will have a degree that is one less than the **dividend**. Since the dividend has a degree of 4, the quotient will have a degree of 3. The last entry in the last row is the remainder. The remaining entries in that row are the coefficients of the quotient when written in **standard form**.

$$
\begin{array}{r|rrrrr}
-5) & 2 & 9 & 0 & -12 & 1 \\
& \downarrow & -10 & 5 & -25 & 185 \\
\hline
& 2 & -1 & 5 & -37 & 186
\end{array}
$$

$$\underbrace{}_{\text{Coefficients of quotient}} \quad \underbrace{}_{\text{Remainder}}$$

Here, the quotient is $2x^3 - 1x^2 + 5x - 37$, and the remainder is 186.

$$\frac{2x^4 + 9x^3 - 12x + 1}{x + 5} = 2x^3 - x^2 + 5x - 37 + \frac{186}{x + 5}$$

You Try It Work through this **You Try It** problem.

Work Exercises 29 and 30 in this eText or in the MyMathLab Study Plan.

My video summary ▶ **Example 8 Dividing Polynomials Using Synthetic Division**

Divide $2x^4 - 3x^2 + 5x - 30$ by $x - 2$ using synthetic division.

Solution The divisor is a binomial of the form $x - c$ where $c = 2$. Use the synthetic division process as follows:

$$
\begin{array}{r|rrrrr}
2) & 2 & 0 & -3 & 5 & -30 \\
& \downarrow & & & & \\
\hline
& 2 & & & &
\end{array}
$$
← Products

← Sums

Complete this process on your own. Check your **answer**, or watch this **video** for a detailed solution.

You Try It Work through this You Try It problem.

Work Exercises 31–36 in this eText or in the MyMathLab Study Plan.

My video summary ▶ **Example 9 Dividing Polynomials Using Synthetic Division**

Divide $4x^3 - 8x^2 + 7x - 4$ by $x - \dfrac{1}{2}$ using synthetic division.

Solution The divisor is a binomial of the form $x - c$ with $c = \dfrac{1}{2}$. Try to complete the synthetic division on your own. View the **answer**, or watch this **video** for the full solution.

You Try It Work through this You Try It problem.

Work Exercises 37 and 38 in this eText or in the MyMathLab Study Plan.

⚠️ Synthetic division can only be used when dividing a **polynomial** by a binomial of the form $x - c$.

OBJECTIVE 5   DIVIDE POLYNOMIAL FUNCTIONS

After learning how to divide polynomials, we now can divide **polynomial functions**.

Definition   **The Quotient of Functions**

Let f and g represent two functions such that the **quotient of f and g** is

$\left(\dfrac{f}{g}\right)(x) = \dfrac{f(x)}{g(x)}$, as long as $g(x) \neq 0$.

My video summary ▶ **Example 10 Dividing Polynomial Functions**

For $P(x) = 15x^3 + 41x^2 + 4x + 3$ and $Q(x) = 5x + 2$, find $\left(\dfrac{P}{Q}\right)(x)$. State any values that cannot be included in the domain of $\left(\dfrac{P}{Q}\right)(x)$. (Note that $Q(x)$ cannot be 0.)

Solution

Begin with the original problem: $\left(\dfrac{P}{Q}\right)(x) = \dfrac{P(x)}{Q(x)}$

Substitute for the function notation: $= \dfrac{15x^3 + 41x^2 + 4x + 3}{5x + 2}$

Try to complete this problem using **polynomial long division**. To find the values that cannot be included in the domain, set $Q(x)$ equal to 0 and solve. View the **answer**, or watch this **video** for a detailed solution.

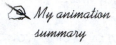 **You Try It** Work through this You Try It problem.

Work Exercises 39–42 in this eText or in the MyMathLab Study Plan.

OBJECTIVE 6 USE THE REMAINDER AND FACTOR THEOREMS

THE REMAINDER THEOREM

Let $f(x) = 2x^4 + 9x^3 - 12x + 1$. Then $f(-5) = 2(-5)^4 + 9(-5)^3 - 12(-5) + 1 = 186$.

In **Example 7**, we used synthetic division to divide the polynomial $2x^4 + 9x^3 - 12x + 1$ by the binomial $x + 5$, with $c = -5$, and we found a remainder of 186.

These results demonstrate the **Remainder Theorem**.

The Remainder Theorem

If a polynomial $f(x)$ is divided by $x - c$, then the remainder is $f(c)$.

My animation summary ⊛ **Concept Animation** Watch this animation to see why the remainder theorem is true.

Example 11 Using the Remainder Theorem

Use the remainder theorem to find the remainder when $f(x)$ is divided by $x - c$,

a. $f(x) = 5x^4 - 8x^2 + 3x - 1$; $x - 2$

b. $f(x) = 3x^3 + 5x^2 - 5x - 6$; $x + 2$

My video summary ▶ **Solutions** Continue reading, or watch this **video** for complete solutions to both parts.

a. The remainder theorem says that the remainder must be $f(2)$ because $c = 2$. Since $f(2) = 53$, the remainder is 53. See the **work**.

b. To find the remainder when $f(x) = 3x^3 + 5x^2 - 5x - 6$ is divided by $x + 2 = x - (-2)$, evaluate $f(-2)$ because $c = -2$. Since $f(-2) = 0$, the remainder is 0. See the **work**.

 **You Try It** Work through this You Try It problem.

Work Exercises 43–46 in this eText or in the MyMathLab Study Plan.

THE FACTOR THEOREM

When $f(x) = 3x^3 + 5x^2 - 5x - 6$ is divided by $x + 2$ in Example 11, the remainder is 0. The remainder of 0 means that $x + 2$ divides $f(x)$ evenly, and the binomial $x + 2$ is a factor of the polynomial function $f(x)$. Related to the remainder theorem, the **Factor Theorem** follows:

The Factor Theorem

For a polynomial function $f(x)$,

1. If $f(c) = 0$, then $x - c$ is a factor of $f(x)$.

2. If $x - c$ is a factor of $f(x)$, then $f(c) = 0$.

(eText Screens 4.4-1–4.4-31)

My video summary ▶ **Example 12 Using the Factor Theorem**

Use the factor theorem to determine if $x - c$ is a factor of $f(x)$. If so, use **synthetic division** to write $f(x)$ in the form $f(x) = (x - c) \cdot q(x)$.

a. $f(x) = 2x^3 - 5x^2 + 7x - 4; \quad x - 1$

b. $f(x) = x^4 - 3x^3 + 15x + 45; \quad x + 3$

Solutions

a. The divisor is $x - 1$, so $c = 1$. Evaluate $f(1)$.

$$f(1) = 2(1)^3 - 5(1)^2 + 7(1) - 4 = 2 - 5 + 7 - 4 = 0$$

Since $f(1) = 0$, the **factor theorem** states that $x - 1$ is a factor of $f(x)$. Use synthetic division to find the **quotient** when $f(x)$ is divided by $x - 1$, and then write $f(x)$ in the form $f(x) = (x - 1) \cdot q(x)$.

Check your **answer**, or watch this **video** for a complete solution to both parts.

b. Read the following, or watch this **video** for the complete solution. The divisor is $x + 3 = x - (-3)$, so we have $c = -3$. Evaluate $f(-3)$.

$$f(-3) = (-3)^4 - 3(-3)^3 + 15(-3) + 45 = 81 + 81 - 45 + 45 = 162$$

Since $f(-3) \neq 0$, $x + 3$ is not a factor of $f(x)$.

You Try It Work through this You Try It problem.

Work Exercises 47–50 in this eText or in the MyMathLab Study Plan.

4.4 Exercises

In Exercises 1–4, divide the monomials.

1. $\dfrac{54x^{11}}{9x^3}$
2. $\dfrac{-108z^{36}}{27z^{12}}$
3. $\dfrac{5m^3}{30m^3}$
4. $\dfrac{72t^4}{8t^9}$

In Exercises 5–14, divide the polynomial by the monomial.

5. $\dfrac{14x^4 - 35x^3}{7x^3}$

6. $\dfrac{12m^4 - 16m^3 + 32m^2}{4m}$

7. $(48y^7 - 24y^5 + 6y^2) \div 6y^2$

8. $\dfrac{-10p^5 + 75p^4}{-5p^4}$

9. $\dfrac{20x^7 + 12x^5 - 8x^4 + 32x^2}{4x^2}$

10. $\dfrac{8y^5 - 3y^4 + 2y^2}{-4y^3}$

11. $\dfrac{18a^3 - 63a^2 - 27a}{9a^2}$

12. $\dfrac{8w^3 - 14w + 6}{-2w^2}$

13. $\dfrac{8x^4y^2 + x^3y^3 - 2x^2y^4}{2x^2y^2}$

14. $\dfrac{a^5b^3 - 2a^4b + 3a^2b^2}{3a^3b^2}$

4.4 Dividing Polynomials 4-45

In Exercises 15–28, divide the polynomials using long division.

15. $\dfrac{x^2 + 18x + 72}{x + 6}$

16. $\dfrac{6t^2 - t - 43}{2t + 5}$

17. $\dfrac{x^3 - 2x^2 - 8x + 21}{x + 3}$

18. $\dfrac{6a^3 - 5a^2 - 11a + 13}{3a - 1}$

19. $\dfrac{2x^2 + x^3 - 9x - 4}{x + 5}$

20. $\dfrac{23x^2 - 13x^3 + 6x^4 - 57x + 30}{3x - 5}$

21. $\dfrac{20 - 11x - 3x^2}{4 - 3x}$

22. $\dfrac{30 + 5x - 6x^2}{5 - 2x}$

23. $\dfrac{3x^4 - 11x^3 + 9x^2 + 22x - 33}{x^2 - 2}$

24. $\dfrac{2x^5 + 3x^4 + 5x^3 + 6x^2 + 2x}{x^2 + 2}$

25. $\dfrac{4m^3 + 5m - 7}{m + 2}$

26. $\dfrac{9x^4 - 2x^2 + 2x + 9}{x^2 + 5}$

27. $\dfrac{6x^6 - x^4 + 6x^3 - x^2 + 2x}{2x^3 - x + 2}$

28. $\dfrac{x^4 - 2x^3 - 13x^2 + 24x - 36}{x - 6}$

In Exercises 29–38, divide using synthetic division.

29. $(4x^3 + 5x - 7) \div (x + 2)$

30. $(3x^3 + 11x^2 + 2x + 31) \div (x + 4)$

31. $x^2 - x - 17$ divided by $x - 5$

32. $\dfrac{7x^4 - 12x^3 - 8x^2 + 7x + 2}{x - 2}$

33. $\dfrac{x^3 - 12x + 9}{x - 3}$

34. $x^2 + \dfrac{16}{3}x - 4$ divided by $x + 6$

35. $\dfrac{x^4 + 5x^2 - 41}{x - 2}$

36. $\dfrac{x^4 + 15x^2 - 16}{x + 1}$

37. $\dfrac{15x^3 + 14x^2 - 43x + 14}{x - \dfrac{2}{5}}$

38. $\dfrac{3x^3 - 20x^2 - x - 7}{x + \dfrac{1}{3}}$

In Exercises 39–42, find $\left(\dfrac{P}{Q}\right)(x)$ and state any values that cannot be included in the domain of $\left(\dfrac{P}{Q}\right)(x)$.

39. $P(x) = 3x^2 - 7x + 5; \quad Q(x) = x - 2$

40. $P(x) = 6x^3 + 11x^2 - 9x - 2; \quad Q(x) = 3x + 1$

41. $P(x) = 6x^3 + 11x^2 - 9x - 4; \quad Q(x) = 3x + 1$

42. $P(x) = 5x^3 - 37x^2 - 23x + 56; \quad Q(x) = x - 8$

In Exercises 43–46, use the remainder theorem to find the remainder if $f(x)$ is divided by $x - c$.

43. $f(x) = 3x^2 - 7x + 12;\quad x - 1$

44. $f(x) = 3x^2 - x - 14;\quad x + 2$

45. $f(x) = x^4 + 5x^3 - 5x + 29;\quad x + 3$

46. $f(x) = 20x^3 - 13x^2 + 14x - 3;\quad x - \dfrac{1}{4}$

In Exercises 47–50, use the factor theorem to determine if $x - c$ is a factor of $f(x)$. If so, use synthetic division to write $f(x)$ in the form $f(x) = (x - c) \cdot q(x)$.

47. $f(x) = x^2 - 11x + 24;\quad x - 8$

48. $f(x) = x^3 + 8;\quad x + 2$

49. $f(x) = 2x^3 - 5x^2 + 7x - 4;\quad x + 1$

50. $f(x) = 4x^3 - 5x^2 - 8x + 4;\quad x - 2$

CHAPTER FIVE

Factoring

CHAPTER FIVE CONTENTS

5.1 Greatest Common Factor and Factoring by Grouping

THINGS TO KNOW

Before working through this section, be sure you are familiar with the following concepts:

			VIDEO	ANIMATION	INTERACTIVE
You Try It	1.	Identify and Use the Properties of Real Numbers (Section R.2, Objective 3)	▶		
You Try It	2.	Multiply a Polynomial by a Monomial (Section 4.3, Objective 2)	▶		
You Try It	3.	Multiply Two Binomials (Section 4.3, Objective 3)	▶	✹	☞
You Try It	4.	Divide a Polynomial by a Monomial (Section 4.4, Objective 2)	▶		

OBJECTIVES

1 Find the Greatest Common Factor of a Group of Monomials

2 Factor Out the Greatest Common Factor from a Polynomial

3 Factor by Grouping

OBJECTIVE 1   FIND THE GREATEST COMMON FACTOR OF A GROUP OF MONOMIALS

Factoring a polynomial is the reverse process of **multiplying polynomials**

 *My animation summary*   ✦ **Concept Animation** Watch this animation to see the connection between factoring and multiplying polynomials.

When a polynomial is written as an **equivalent expression** that is a **product** of polynomials, we say that the polynomial has been **factored** or written in **factored form**.

When factoring a polynomial, first we find the *greatest common factor* or GCF of its terms. In this eText, we always **factor over the integers**, which means that all **coefficients** in both the original polynomial and its factors will be **integers**.

The **greatest common factor (GCF) of a group of integers** is the largest **integer** that divides evenly into each integer in the group. For example, the GCF of 12 and 18 is 6 because 6 is the largest integer that divides both 12 and 18 evenly ($12 \div 6 = 2$ and $18 \div 6 = 3$).

My animation summary   ✦ **Concept Animation** Watch this animation to review how to find the GCF of a group of integers.

The **greatest common factor (GCF) of a group of monomials** is the **monomial** with the largest coefficient and highest **degree** that divides each monomial evenly.

Consider the monomials x^3y^5 and x^4y^2, each of which has a coefficient of 1.

$$x^3y^5 = \underbrace{x \cdot x \cdot x} \cdot \underbrace{y \cdot y} \cdot y \cdot y \cdot y \qquad x^4y^2 = \underbrace{x \cdot x \cdot x} \cdot x \cdot \underbrace{y \cdot y}$$

Common Factors

The monomials have three factors of the base x in common, and two factors of the base y. So, the greatest common *variable factor* of the expressions is $x \cdot x \cdot x \cdot y \cdot y = x^3y^2$. Notice that in this GCF, each variable base has the lowest **power** that occurs on that variable in any of the monomials. For x, this is x^3. For y, this is y^2. This is true in general.

> ### Common Variable Factors for a GCF
>
> If a variable is a **common factor** of a group of monomials, then the lowest power of that variable in the group will be a factor of the GCF.

My video summary   ⊙ **Example 1 Finding the GCF of a Group of Exponential Expressions**

Find the GCF of each group of exponential expressions.

a. x^4 and x^7 **b.** a^8b^3c and a^2b^4 **c.** w^6z^2, w^3z^5, and w^5z^4

Solutions Watch this **video**, or continue reading for the solutions.

a. The variable x is a common factor of both expressions. The lowest power is 4, so the GCF is x^4.

b. The variables a and b are common factors of both expressions. The lowest power of a is 2 and the lowest power of b is 3, so the GCF is a^2b^3.

c. The variables w and z are common factors of all three expressions. The lowest power of w is 3 and the lowest power of z is 2, so the GCF is $w^3 z^2$.

You Try It Work through this You Try It problem.

Work Exercises 1–4 in this eText or in the MyMathLab Study Plan.

We can use the following process to find the **GCF** of a group of monomials.

Finding the GCF of a Group of Monomials

Step 1 Find the GCF of the **coefficients**.

Step 2 Find the lowest power for each **common variable factor**.

Step 3 The GCF is the **monomial** with the coefficient from Step 1 and the variable factor(s) from Step 2.

Example 2 Finding the GCF of a Group of Monomials

Find the GCF of each group of monomials.

a. $14x^6$ and $21x^8$

b. $6a^2$, $10ab$, and $14b^2$

c. $40x^5y^6$, $-48x^9y$, and $24x^2y^4$

d. $14m^3n^2$, $6m^5n$, and $9m^4$

My video summary ▶ **Solutions**

a. We follow the steps for finding the GCF of a group of monomials.

Step 1 The prime factorizations of the coefficients 14 and 21 are

$$14 = 2 \cdot 7 \text{ and } 21 = 3 \cdot 7.$$

So, the GCF for the coefficients is 7.

Step 2 The variable x is a **common factor** of both expressions, and the lowest power is 6. So, x^6 is a factor of the GCF.

Step 3 Combining Steps 1 and 2, the GCF of the group of monomials is $7x^6$.

b. **Step 1** The prime factorizations of the coefficients 6, 10, and 14 are

$$6 = 2 \cdot 3, 10 = 2 \cdot 5, \text{ and } 14 = 2 \cdot 7.$$

The GCF for the coefficients is 2.

Step 2 There is no common variable factor for all three monomials. (**Note:** $6a^2$ does not have a factor of b, and $14b^2$ does not have a factor of a.) So, no variables are included in the GCF.

Step 3 The GCF for the group of monomials is 2.

c.–d. Try finding these GCFs on your own, then check your **answers**. Watch this video for complete solutions to parts c and d.

You Try It Work through this You Try It problem.

Work Exercises 5–10 in this eText or in the MyMathLab Study Plan.

OBJECTIVE 2 FACTOR OUT THE GREATEST COMMON FACTOR
FROM A POLYNOMIAL

The **greatest common factor (GCF) of a polynomial** is the expression with the largest **coefficient** and highest **degree** that divides each **term** of the **polynomial** evenly.

Consider the **binomial** $6x^2 + 10x$. The **GCF** of the two **monomials** $6x^2$ and $10x$ is $2x$, so $2x$ is the GCF of $6x^2 + 10x$. Once we know the GCF, we can **factor** it out of the binomial. We write each term of the binomial as a product that includes the **factor** $2x$.

$$6x^2 + 10x = 2x \cdot 3x + 2x \cdot 5$$

Reversing the **distributive property**, we factor $2x$ out of each term.

$$2x \cdot 3x + 2x \cdot 5 = 2x(3x + 5)$$

So, $6x^2 + 10x = 2x(3x + 5)$.

Notice that the final result is the GCF times a **binomial** factor. This is expected because the original expression is a binomial. When the GCF is factored from a polynomial, the **polynomial factor** of the resulting product will have the same number of terms as the original polynomial.

Generalizing our process for this example gives the following steps for factoring out the GCF from a polynomial.

Steps for Factoring Out the Greatest Common Factor from a Polynomial

Step 1 Find the GCF of all terms in the polynomial.

Step 2 Write each term as a product that includes the GCF.

Step 3 Use the **distributive property** in reverse to factor out the GCF.

Step 4 Check the answer. The terms of the **polynomial factor** should have no more **common factors**. Multiplying the answer back out should give the original polynomial.

My video summary ▶ **Example 3 Factoring Out the Greatest Common Factor from a Polynomial**

Factor out the GCF.

a. $24m^5 + 32m^3 + 8m^2$ **b.** $18x^3y^5 - 36x^2y^7 + 27xy^9$

Solutions Follow the four-step process.

a. Step 1 The polynomial has three terms: $24m^5$, $32m^3$, and $8m^2$. The largest **integer** that divides evenly into the three **coefficients** 24, 32, and 8 is 8. The variable m appears in all three terms and its smallest **power** is m^2. So, the GCF is $8m^2$.

        Step 2 Write each term as the product of $8m^2$ and another factor. Since the third term $8m^2$ is the GCF, write it multiplied by 1.

$$24m^5 + 32m^3 + 8m^2 = 8m^2 \cdot 3m^3 + 8m^2 \cdot 4m + 8m^2 \cdot 1$$

Step 3 Factor out $8m^2$ using the **distributive property** in reverse.

$$8m^2 \cdot 3m^3 + 8m^2 \cdot 4m + 8m^2 \cdot 1 = 8m^2(3m^3 + 4m + 1)$$

Step 4 The terms of the polynomial factor $3m^3 + 4m + 1$ have no **common** factors, so our **GCF** should be correct. Multiply the final answer back out.

$$8m^2(3m^3 + 4m + 1) = 8m^2 \cdot 3m^3 + 8m^2 \cdot 4m + 8m^2 \cdot 1$$
$$= 24m^5 + 32m^3 + 8m^2$$

This is the original polynomial, so our answer checks. So, $24m^5 + 32m^3 + 8m^2 = 8m^2(3m^3 + 4m + 1)$.

b. Try factoring out the GCF on your own. Check your **answer**, or watch this **video** for a complete solution.

My animation summary

⚙ ⚠ **Concept Animation** When checking answers after factoring out the GCF, multiplying is not enough. We must also make sure that the terms of the polynomial factor have no more common factors. Watch this **animation** to explore this concept in more detail.

You Try It Work through this **You Try It** problem.

Work Exercises 11-18 in this eText or in the MyMathLab Study Plan.

When the **leading coefficient** of a polynomial is negative, we may wish to **factor** out the negative sign with the GCF.

My video summary ▶ **Example 4 Factoring Out a Negative Sign with the GCF**

Factor out the negative sign with the GCF: $-25x^4 + 15x$

Solution Watch this video, or follow the **four-step process** below.

Step 1 The largest integer that divides into -25 and 15 is 5. The variable x appears in both terms with its smallest **power** as x. Since the leading coefficient is negative, we include a negative sign as part of the GCF. So, $-5x$ is the GCF.

Step 2 Write each term as the product of $-5x$ and another factor. Since the second term, $15x$, has a positive coefficient, both factors must be negative.

$$-25x^4 + 15x = (-5x) \cdot 5x^3 + (-5x) \cdot (-3)$$

Step 3 Factor out $-5x$ using the **distributive property** in reverse.

$$(-5x) \cdot 5x^3 + (-5x) \cdot (-3) = -5x(5x^3 - 3)$$

So, $-25x^4 + 15x = -5x(5x^3 - 3)$. See the **check**.

You Try It Work through this **You Try It** problem.

Work Exercises 19–22 in this eText or in the MyMathLab Study Plan.

In each example so far, the GCF has been a **monomial**. In the next example, we factor out a **binomial** as the GCF.

Example 5 Factoring Out a Binomial as the GCF

Factor out the common binomial factor as the GCF.

a. $5x(2x + 3) + 6y(2x + 3)$ **b.** $7x(x + y) - (x + y)$

Solutions

a. Treating $5x(2x + 3)$ and $6y(2x + 3)$ as **terms**, the only **common factor** is the binomial factor $2x + 3$. So, the GCF is $2x + 3$. Each term is already written as a product of $2x + 3$, so factor it out as follows:

$$5x(2x + 3) + 6y(2x + 3) = (2x + 3)(5x + 6y)$$

So, $5x(2x + 3) + 6y(2x + 3) = (2x + 3)(5x + 6y)$.

My video summary ▶ **b.** Try factoring out the common binomial factor as the GCF on your own. Check your answer, or watch this video for a complete solution.

You Try It Work through this You Try It problem.

Work Exercises 23–26 in this eText or in the MyMathLab Study Plan.

OBJECTIVE 3 FACTOR BY GROUPING

Suppose the **polynomial** $5x(2x + 3) + 6y(2x + 3)$ from **Example 5a** had been simplified so that its terms did not contain the common **binomial factor** $2x + 3$. Can we **factor** the simplified polynomial $10x^2 + 15x + 12xy + 18y$? The answer is yes. To do this, we use a method called **factoring by grouping**.

My animation summary ✸ **Concept Animation** Watch this animation to see how to factor the simplified polynomial by grouping.

Factoring by grouping involves grouping terms together and then looking for the GCF of each group separately. With four terms, we often try grouping the first two terms together and the last two terms together. However, terms can be rearranged to create better groupings if needed.

We summarize how to *factor by grouping* with the following steps:

Factoring a Polynomial by Grouping

Step 1 Group terms with a common factor. It may be necessary to rearrange the terms.

Step 2 For each group, factor out the greatest common factor.

Step 3 Factor out the common polynomial factor, if there is one.

Step 4 Check your answer by multiplying out the factors.

My video summary ▶ **Example 6 Factoring by Grouping**

    a. $x^3 - 3x^2 + 5x - 15$ **b.** $2h^5 - 5h^3 - 4h^2 + 10$

Solutions Watch this video, or work through the solutions below.

a. The four terms have no **common factor**, so try **factoring by grouping**. The first two terms have a common factor of x^2, and the last two terms have a common factor of 5.

Begin with the original polynomial expression:	$x^3 - 3x^2 + 5x - 15$
Group the first two terms and the last two terms:	$= (x^3 - 3x^2) + (5x - 15)$
Factor out the GCF from each group:	$= x^2(x - 3) + 5(x - 3)$
Factor out the common binomial factor:	$= (x - 3)(x^2 + 5)$

This answer **checks**, so $x^3 - 3x^2 + 5x - 15 = (x - 3)(x^2 + 5)$.

b. The four terms have no **common factor**, so try **factoring by grouping**. The first two terms have a common factor of h^3, and the last two terms have a common factor of -2.

Begin with the original polynomial expression:	$2h^5 - 5h^3 - 4h^2 + 10$
Group the first two terms and the last two terms:	$= (2h^5 - 5h^3) + (-4h^2 + 10)$
Factor out the GCF from each group:	$= h^3(2h^2 - 5) - 2(2h^2 - 5)$
Factor out the common binomial factor:	$= (2h^2 - 5)(h^3 - 2)$

This answer **checks**, so $2h^5 - 5h^3 - 4h^2 + 10 = (2h^2 - 5)(h^3 - 2)$.

⚠ If minus signs are involved when factoring by grouping, pay close attention when grouping terms. For example, in part b it is incorrect to write the grouping as $(2h^5 - 5h^3) - (4h^2 + 10)$ because $-(4h^2 + 10) = -4h^2 - 10$, not $-4h^2 + 10$.

You Try It Work through this You Try It problem.

Work Exercises 27–30 in this eText or in the MyMathLab Study Plan.

⚠ If no arrangement of terms leads to a common polynomial factor for Step 3, then the polynomial cannot be factored by grouping.

My interactive video summary 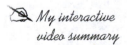 **Example 7 Factoring by Grouping**

Factor by grouping.

    a. $2x^2 - 6x + xy - 3y$ **b.** $5xy + 6 + 5x + 6y$

    c. $3m^2 + 3m - 2mn - 2n$ **d.** $4w^3 - 14w^2 - 10w + 35$

Solutions

a. Begin with the original polynomial expression: $2x^2 - 6x + xy - 3y$

      Group the first two terms and last two terms: $= (2x^2 - 6x) + (xy - 3y)$

Factor out the GCF from each group: $= 2x(x - 3) + y(x - 3)$

Factor out the common binomial factor: $= (x - 3)(2x + y)$

This answer **checks**, so $2x^2 - 6x + xy - 3y = (x - 3)(2x + y)$.

b. For $5xy + 6 + 5x + 6y$, the first two terms and the last two terms have no common factors other than 1. Rearrange the terms to group terms with common factors. Notice that $5xy$ and $5x$ have a common factor of $5x$ and that $6y$ and 6 have a common factor of 6.

Begin with the original polynomial expression: $5xy + 6 + 5x + 6y$

Rearrange to group terms with common factors: $= (5xy + 5x) + (6y + 6)$

Factor out the GCF from each group: $= 5x(y + 1) + 6(y + 1)$

Factor out the common binomial factor: $= (y + 1)(5x + 6)$

This answer **checks**, so $5xy + 6 + 5x + 6y = (y + 1)(5x + 6)$.

c.–d. Try factoring these polynomials on your own. View the **answers**, or watch this **interactive video** for complete solutions to all four parts.

You Try It   Work through this You Try It problem.

Work Exercises 31–40 in this eText or in the MyMathLab Study Plan.

5.1 Exercises

In Exercises 1–10, find the GCF of each group of monomials.

1. x^6 and x^2

2. w^4, w^7, and w^{10}

3. a^3b^9 and a^5b

4. x^4y^7, x^8y^6, and $x^{12}y^5$

5. $12x^8$ and $30x^5$

6. $32y^2$ and $27y^5$

7. $10p^2$, $25pq$, and $15q^2$

8. $8x^3$, $16x^5$, and $4x^7$

9. $54m^3n^4$, $-36mn^5$, and $72m^8n^6$

10. $24a^2b^4$, $15ab^4$, and $25b^5$

In Exercises 11–18, factor out the GCF from each polynomial.

11. $20x + 8$

12. $5y^6 + 10x^2y^8$

13. $12x^2y + 4xy$

14. $28m^6n^2 + 32m^4n^5$

15. $4x^4 - 20x^3 + 12x^2$

16. $9a^3b^5 - 6a^2b^4 + 3a^2b^2$

17. $16x^2y^3z^6 - 12x^4y^4z^4 + 20x^3y^2z^5$

18. $24w^6y^4 - 40w^2y^2 + 16w^2y - 48w^3y^4$

In Exercises 19–22, factor out the negative sign with the GCF.

19. $-8x^3 + 48x^2$

20. $-14x^5y^6 - 21x^2y^8$

21. $-4x^2 - 16x + 18$

22. $-2x^3 + 10x^2 - 14x$

In Exercises 23–26, factor out the common binomial factor as the GCF.

23. $11(x + 7) + 3a(x + 7)$

24. $7x(z + 5) + (z + 5)$

25. $9y(x^2 + 11) - 7(x^2 + 11)$

26. $(x + 5y)(x + 7) + (3x - 2y)(x + 7)$

In Exercises 27–40, factor by grouping.

27. $x^3 - 8x^2 + 3x - 24$

28. $x^3 - x^2 - 7x + 7$

29. $4m^5 + 12m^3 - 3m^2 - 9$

30. $m^4 + m^3 + 2m + 2$

31. $xy + 5x + 3y + 15$

32. $ab + 7a - 2b - 14$

33. $9xy + 12x + 6y + 8$

34. $3xy - 2x - 12y + 8$

35. $5x^2 + 4xy + 15x + 12y$

36. $3x^2 + 3xy - 2x - 2y$

37. $x^3 + 9x^2 + 4x + 36$

38. $6x^5 + 20x^3 - 15x^2 - 50$

39. $x^3 + 8x^2 + x + 8$

40. $a^2 + 4b + ab + 4a$

5.2 Factoring Trinomials

THINGS TO KNOW

Before working through this section, be sure you are familiar with the following concepts:

VIDEO   ANIMATION   INTERACTIVE

 You Try It

1. Find the Degree and Leading Coefficient of a Polynomial (Section 4.2, Objective 2)

 You Try It

2. Multiply Two Binomials (Section 4.3, Objective 3)

 You Try It

3. Factor Out the Greatest Common Factor from a Polynomial (Section 5.1, Objective 2)

 You Try It

4. Factor by Grouping (Section 5.1, Objective 3)

OBJECTIVES

1 Factor Trinomials of the Form $x^2 + bx + c$

2 Factor Trinomials of the Form $ax^2 + bx + c$ Using Trial and Error

3 Factor Trinomials of the Form $ax^2 + bx + c$ Using the *ac* Method

4 Factor Trinomials Using Substitution

OBJECTIVE 1   FACTOR TRINOMIALS OF THE FORM $x^2 + bx + c$

In **Section 4.3**, we saw that the **product of two binomials** is often a trinomial. Since factoring is the reverse of multiplication, trinomials will often factor into the product of two binomials.

Let's factor trinomials with a **leading coefficient** of 1 and a **degree** of 2. These trinomials have the form $x^2 + bx + c$.

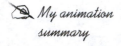 *My animation summary*

⊛ **Concept Animation** Watch this **animation** for an introduction to factoring trinomials.

The relationship between the **terms** of the trinomial and the terms of the **binomial factors** leads us to the following steps for factoring trinomials of the form $x^2 + bx + c$. Remember that we are **factoring over integers**.

Factoring Trinomials of the Form $x^2 + bx + c$

Step 1 Find two integers, n_1 and n_2, whose product is the **constant** term c and whose sum is the **coefficient** b. So, $n_1 \cdot n_2 = c$ and $n_1 + n_2 = b$.

Step 2 Write the trinomial in the **factored form** $(x + n_1)(x + n_2)$.

Step 3 Check the answer by multiplying out the factored form.

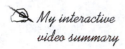 *My interactive video summary*

✎ **Example 1 Factoring Trinomials of the Form $x^2 + bx + c$**

Factor.

a. $x^2 + 16x + 28$ **b.** $x^2 + 11x + 28$ **c.** $x^2 + 29x + 28$

Solutions Follow the three-step process.

a. Step 1 We need to find two **integers** whose product is 28 and whose sum is 16. Since 28 and 16 are both positive, we begin by listing the pairs of **positive factors** for 28:

Positive Factors of 28	Sum of Factors
1, 28	$1 + 28 = 29$
2, 14	$2 + 14 = 16$ ← This is the pair.
4, 7	$4 + 7 = 11$

Step 2 Since $2 \cdot 14 = 28$ and $2 + 14 = 16$, we have:

$$x^2 + 16x + 28 = (x + 2)(x + 14)$$

Step 3 Check by multiplying.

$$\text{Check: } (x + 2)(x + 14) = \overset{F}{\overbrace{x^2}} + \overset{O}{\overbrace{14x}} + \overset{I}{\overbrace{2x}} + \overset{L}{\overbrace{28}}$$

$$= x^2 + 16x + 28 \leftarrow \text{Original trinomial}$$

Our result checks, so $x^2 + 16x + 28 = (x + 2)(x + 14)$.

Note: Since multiplication is **commutative**, the order in which the factors are listed in the product does not matter. For example, we can also write this answer as $x^2 + 16x + 28 = (x + 14)(x + 2)$.

b. Notice that $x^2 + 11x + 28$ has the same first and last **terms** as the **trinomial** in part a. The difference is the middle term. Looking at the pairs of **factors** of 28 listed in the solution to part a, the factors 4 and 7 have the required **sum** of 11. Use this information to finish factoring the trinomial on your own. Check your **answer**, or watch this **interactive video** for the complete solution.

c. Try factoring this trinomial on your own. Check your **answer**, or watch this **interactive video** for the complete solution.

You Try It Work through this You Try It problem.

Work Exercises 1–3 in this eText or in the MyMathLab Study Plan.

Recall that a **prime number** is a **whole number** greater than 1 whose only whole number factors are 1 and itself. For example, the first ten prime numbers are 2, 3, 5, 7, 11, 13, 17, 19, 23, and 29. As with numbers, there are also *prime polynomials*.

Definition   Prime Polynomial

A polynomial is a **prime polynomial** if its only factors over the integers are 1 and itself.

For example, the **binomial** $2x + 5$ is a prime polynomial because, over the integers, it cannot be written as the product of any two factors other than 1 and itself. The binomial $2x + 6$ is not prime because it can be factored into $2(x + 3)$.

⚠ When deciding if a polynomial is **prime**, consider only **factors** over the **integers**. For example, even though $2x + 5 = 2(x + 2.5)$, the binomial $2x + 5$ is prime. Because 2.5 is not an integer, $2(x + 2.5)$ is not factored over the integers.

My video summary ▶ **Example 2 Recognizing a Prime Trinomial of the Form** $x^2 + bx + c$

Factor $x^2 + 8x + 14$.

Solution Watch this video, or follow the **three-step process**.

Step 1 We need to find two integers whose **product** is 14 and whose **sum** is 8. Because 14 and 8 are both positive, we list the pairs of positive factors for 14:

Positive Factors of 14	Sum of Factors
1, 14	$1 + 14 = 15$
2, 7	$2 + 7 = 9$

Step 2 Since neither of the two pairs of factors has a sum of 8, we cannot factor the trinomial into the form $(x + n_1)(x + n_2)$, and the trinomial is prime.

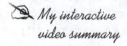 **You Try It** Work through this You Try It problem.

Work Exercise 4 in this eText or in the MyMathLab Study Plan.

In our first two examples involving $x^2 + bx + c$, the coefficients b and c were positive integers, but these coefficients can also be negative. When this happens, we must consider the sign of the factors of c as we try to find the pair of factors whose sum is b.

My interactive video summary

Example 3 Factoring Trinomials of the Form $x^2 + bx + c$

Factor.

a. $x^2 - 10x + 24$ **b.** $x^2 - 2x - 24$ **c.** $x^2 + 5x - 24$

Solutions Follow the three-step process.

a. Step 1 We need to find two **integers** whose **product** is 24 and whose **sum** is -10. The only way that the product can be positive and the sum can be negative is if both integers are negative. Begin by listing the pairs of negative factors for 24:

Negative Factors of 24	Sum of Factors
$-1, -24$	$(-1) + (-24) = -25$
$-2, -12$	$(-2) + (-12) = -14$
$-3, -8$	$(-3) + (-8) = -11$
$-4, -6$	$(-4) + (-6) = -10 \leftarrow$ This is the pair.

Step 2 Since $(-4)(-6) = 24$ and $(-4) + (-6) = -10$, we write:

$$x^2 - 10x + 24 = (x - 4)(x - 6)$$

Step 3 Check the answer by multiplying. See the **check**.

b. We need to find two **integers** whose **product** is -24 and whose **sum** is -2. For the product to be negative, one of the integers must be positive and the other must be negative. For the sum to be negative, the integer with the larger absolute value must be negative. We list such pairs of **factors** for -24:

Factors of -24	Sum of Factors
$1, -24$	$1 + (-24) = -23$
$2, -12$	$2 + (-12) = -10$
$3, -8$	$3 + (-8) = -5$
$4, -6$	$4 + (-6) = -2$

Now finish factoring the trinomial on your own. Check your **answer**, or watch this **interactive video** for the complete solution.

c. Try factoring this trinomial on your own. Check your **answer**, or watch this **interactive video** for the complete solution.

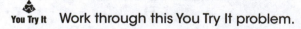 **You Try It** Work through this You Try It problem.

Work Exercises 5–18 in this eText or in the MyMathLab Study Plan.

A polynomial is **factored completely** if it is written as the product of all **prime polynomials**. For example, we factored each **trinomial** in Example 3 as a product of prime **binomials**, so each trinomial was factored completely.

In **Section 5.1**, we learned how to factor out the **greatest common factor** from a polynomial. If a trinomial has a **common factor**, we factor it out first.

Example 4 Factoring Trinomials with a Common Factor

Factor completely.

a. $5y^2 - 25y + 20$ **b.** $3m^3 - 12m^2 - 96m$

Solutions

a. The three **terms** have a GCF of 5. First, factor out the 5, and then factor the remaining trinomial.

$$\text{Begin with the original trinomial:} \quad 5y^2 - 25y + 20$$

$$\text{Write each term as a product of the GCF 5:} \quad = 5 \cdot y^2 - 5 \cdot 5y + 5 \cdot 4$$

$$\text{Factor out the 5:} \quad = 5(y^2 - 5y + 4)$$

The trinomial $y^2 - 5y + 4$ is of the form $x^2 + bx + c$, so we use the **three-step process**. We need to find two **integers** whose product is 4 and whose sum is -5. Since the product is positive and the sum is negative, both integers must be negative.

The pairs of **negative factors** for 4 are:

Negative Factors of 4	Sum of Factors
$-1, -4$	$(-1) + (-4) = -5 \leftarrow$ This is the pair.
$-2, -2$	$(-2) + (-2) = -4$

Since $(-1)(-4) = 4$ and $(-1) + (-4) = -5$, we have:

$$5(y^2 - 5y + 4) = 5(y - 1)(y - 4)$$

Therefore, $5y^2 - 25y + 20 = 5(y - 1)(y - 4)$.

Check the answer by multiplying. See the **check**.

My video summary ▶ **b.** Try factoring this **trinomial** on your own. Remember to factor out the **GCF** first. Check your **answer**, or watch this **video** for the complete solution.

You Try It **Work through this You Try It problem.**

Work Exercises 19–24 in this eText or in the MyMathLab Study Plan.

A trinomial will sometimes have a **leading coefficient** of -1, such as $-x^2 - 8x + 9$. Since it is much easier to factor a trinomial when the leading coefficient is 1, we often begin by factoring out -1.

My video summary ▶ ### Example 5 Factoring a Trinomial with a Leading Coefficient of -1

Factor $-x^2 - 8x + 9$.

Solution Since the leading coefficient is -1, we factor it out.

$$-x^2 - 8x + 9 = -(x^2 + 8x - 9)$$

The trinomial $x^2 + 8x - 9$ is of the form $x^2 + bx + c$. Try to finish factoring this trinomial on your own. Check your **answer**, or watch this **video** for a complete solution.

You Try It Work through this You Try It problem.

Work Exercises 25–28 in this eText or in the MyMathLab Study Plan.

We can factor a **trinomial** in two **variables** using the same approach as that for factoring a trinomial in one variable. For example, we factor a trinomial of the form $x^2 + bxy + cy^2$ just like we would factor $x^2 + bx + c$. The difference is that the second **term** of each **binomial factor** must contain the y variable. So, we find two integers, n_1 and n_2, whose product is the coefficient c and whose sum is the coefficient b. Then we write:

$$x^2 + bxy + cy^2 = (x + n_1 y)(x + n_2 y)$$

My video summary ▶ **Example 6 Factoring a Trinomial of the Form $x^2 + bxy + cy^2$**

Factor.

a. $x^2 + 10xy + 24y^2$ **b.** $m^2 + 22mn - 48n^2$

Solutions

a. We need to find two integers whose product is 24 and whose sum is 10. Both 24 and 10 are positive, so we consider the pairs of **positive factors** for 24.

Positive Factors of 24	Sum of Factors
1, 24	$1 + 24 = 25$
2, 12	$2 + 12 = 14$
3, 8	$3 + 8 = 11$
4, 6	$4 + 6 = 10$ ← This is the pair.

Because $4 \cdot 6 = 24$ and $4 + 6 = 10$, we write

$$x^2 + 10xy + 24y^2 = (x + 4y)(x + 6y).$$

Check the answer by multiplying. View the **check**.

b. Try factoring this **trinomial** on your own. View the **answer**, or watch this **video** for a detailed solution.

You Try It Work through this You Try It problem.

Work Exercise 29–32 in this eText or in the MyMathLab Study Plan.

OBJECTIVE 2 FACTOR TRINOMIALS OF THE FORM $ax^2 + bx + c$ USING TRIAL AND ERROR

If the **leading coefficient** of a trinomial is not 1, we must consider the coefficient as we look for the **binomial factors**.

My animation summary ✲ **Concept Animation** Work through this **animation** to again see how factoring is related to polynomial multiplication.

If a trinomial of the form $ax^2 + bx + c$ can be factored, it will factor to the form $(m_1 x + n_1)(m_2 x + n_2)$. We can find the integers $m_1, m_2, n_1,$ and n_2 by using the following *trial-and-error strategy*.

Trial-and-Error Strategy for Factoring Trinomials of the Form $ax^2 + bx + c$

Step 1 Find all pairs of factors for the **leading coefficient** a.

Step 2 Find all pairs of factors for the **constant term** c.

Step 3 By trial and error, check different combinations of factors from Step 1 and factors from Step 2 in the form $(\square x + \square)(\square x + \square)$ until the correct middle term bx is found by adding the "outside" and "inside" products. If no such combination of factors exists, the trinomial is **prime**.

Step 4 Check your answer by multiplying out the **factored form**.

My video summary ▶ **Example 7 Factoring Trinomials of the Form $ax^2 + bx + c$ Using Trial and Error**

Factor $3x^2 + 7x + 2$.

Solution Watch this video, or follow the trial-and-error strategy below. Because all of the **coefficients** are positive, we need to check only combinations of positive factors.

Step 1 The leading coefficient a is 3. The only pair of positive factors is $3 \cdot 1$.

Step 2 The **constant term** c is 2. The only pair of positive factors is $2 \cdot 1$.

Step 3 Using the form $(\square x + \square)(\square x + \square)$, try different combinations of the possible factors until one is found that gives the middle term $7x$.

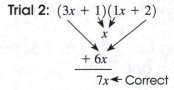

Trial 1: $(3x + 2)(1x + 1)$
$2x$
$+ 3x$
$5x$ ← Not $7x$

The "outside" product is $3x \cdot 1 = 3x$ and the "inside" product is $2 \cdot 1x = 2x$, giving a sum of $3x + 2x = 5x$. This is not the correct combination.

Trial 2: $(3x + 1)(1x + 2)$
x
$+ 6x$
$7x$ ← Correct

The "outside" product is $3x \cdot 2 = 6x$ and the "inside" product is $1 \cdot 1x = x$, giving a sum of $6x + x = 7x$. This is the correct combination.

Step 4 Check the answer by multiplying out $(3x + 1)(x + 2)$. View the **check**.

You Try It Work through this You Try It problem.

Work Exercises 33 and 34 in this eText or in the MyMathLab Study Plan.

⚠ Remember, because multiplication is **commutative**, the order in which the **binomial factors** are listed does not matter. So, we could also write the answer to Example 7 as $3x^2 + 7x + 2 = (x + 2)(3x + 1)$.

In Example 7, there were only two possible combinations of factors to check. Often, there are many more combinations to consider.

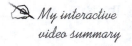 *My video summary* ▶ **Example 8 Factoring Trinomials of the Form $ax^2 + bx + c$ Using Trial and Error**

Factor $2x^2 + 13x + 15$.

Solution Watch this video, or follow the trial-and-error strategy.

Since all of the coefficients are positive, we only need to check combinations of positive factors.

Step 1 The leading coefficient a is 2. The only pair of positive factors is $2 \cdot 1$.

Step 2 The constant term c is 15. The pairs of positive factors are $1 \cdot 15$ and $3 \cdot 5$.

Step 3 Using the form $(\Box x + \Box)(\Box x + \Box)$, try different combinations of the possible factors until one is found that gives the middle term $13x$.

> **Trial 1:** $(2x + 15)(1x + 1)$. The "outside" product is $2x \cdot 1 = 2x$ and the "inside" product is $15 \cdot 1x = 15x$, giving a sum of $2x + 15x = 17x$. This is not the correct combination.

> **Trial 2:** $(2x + 1)(1x + 15)$. The "outside" product is $2x \cdot 15 = 30x$ and the "inside" product is $1 \cdot 1x = x$, giving a sum of $30x + x = 31x$. This is not the correct combination.

> **Trial 3:** $(2x + 5)(1x + 3)$. The "outside" product is $2x \cdot 3 = 6x$ and the "inside" product is $5 \cdot 1x = 5x$, giving a sum of $6x + 5x = 11x$. Again, this is not the correct combination.

> **Trial 4:** $(2x + 3)(1x + 5)$. The "outside" product is $2x \cdot 5 = 10x$ and the "inside" product is $3 \cdot 1x = 3x$, giving a sum of $10x + 3x = 13x$. This is the correct combination.

Step 4 Check the answer by multiplying. View the check.

You Try It Work through this You Try It problem.

Work Exercises 35 and 36 in this eText or in the MyMathLab Study Plan.

Now let's try factoring trinomials with negative coefficients.

My interactive video summary **Example 9 Factoring Trinomials of the Form $ax^2 + bx + c$ Using Trial and Error**

Factor.

a. $10x^2 + 11x - 6$ **b.** $8x^2 - 34x + 21$

Solutions

a. We follow the trial-and-error strategy.

> **Step 1** The leading coefficient a is 10. For convenience, we will only consider the positive factors of 10, which are $10 \cdot 1$ and $5 \cdot 2$.

> **Step 2** The constant term c is -6. In this case, we must consider the negative sign. The pairs of factors are $1 \cdot (-6)$, $-1 \cdot 6$, $2 \cdot (-3)$, and $-2 \cdot 3$.

Step 3 Use the form $(\square x + \square)(\square x + \square)$ with different combinations of the factors from Step 2 until one is found that gives a middle term of $11x$. Try to complete this problem on your own. Check your **answer**, or watch this **interactive video** for the complete solution.

b. Try factoring this trinomial on your own. Check your **answer**, or watch this **interactive video** for the complete solution.

You Try It Work through this **You Try It** problem.

Work Exercises 37–40 in this eText or in the MyMathLab Study Plan.

My animation summary

⊛ **Concept Animation** The number of possible combinations in the trial-and-error method can sometimes be reduced. Watch this **animation** to see how.

A trinomial is **prime** if no combination of binomial factors results in the correct middle term.

Example 10 Recognizing a Prime Trinomial of the Form $ax^2 + bx + c$

Factor $6x^2 - 2x + 5$.

My video summary ▶ **Solution** Watch this video, or follow the trial-and-error strategy below.

Step 1 The leading coefficient a is 6. The positive factors of 6 are $6 \cdot 1$ and $3 \cdot 2$.

Step 2 The constant term c is 5. Notice that the middle term has a negative coefficient, so we must consider the negative sign when choosing factors. Since the middle term comes from adding the outside and inside products of the **binomial factors**, we only need to consider the pairs of two negative factors for 5, which are $-1 \cdot (-5)$.

Step 3 Using the form $(\square x + \square)(\square x + \square)$, try different combinations of the factors until one is found that gives the middle term $-2x$.

Trial 1: $(6x - 1)(1x - 5)$. The "outside" product is $6x \cdot (-5) = -30x$ and the "inside" product is $-1 \cdot 1x = -x$, giving a sum of $-30x + (-x) = -31x$. This is not the correct combination.

Trial 2: $(6x - 5)(1x - 1)$. The "outside" product is $6x \cdot (-1) = -6x$ and the "inside" product is $-5 \cdot 1x = -5x$, giving a sum of $-6x + (-5x) = -11x$. This is not the correct combination.

Trial 3: $(3x - 1)(2x - 5)$. The "outside" product is $3x \cdot (-5) = -15x$ and the "inside" product is $-1 \cdot 2x = -2x$, giving a sum of $-15x + (-2x) = -17x$. This is not the correct combination.

Trial 4: $(3x - 5)(2x - 1)$. The "outside" product is $3x \cdot (-1) = -3x$ and the "inside" product is $-5 \cdot 2x = -10x$, giving a sum of $-3x + (-10x) = -13x$. This is not the correct combination.

We have checked all the possible combinations of **binomial factors**, and none result in the middle term $-2x$. This means that $6x^2 - 2x + 5$ is a **prime trinomial**.

You Try It Work through this **You Try It** problem.

Work Exercises 41–46 in this eText or in the MyMathLab Study Plan.

We factor a **trinomial** of the form $ax^2 + bxy + cy^2$ just like we would factor $ax^2 + bx + c$. The difference is that the second **term** of each binomial factor must contain the y variable. So, we use **trial and error** to check different combinations of factors for a and factors for c in the form $(\Box x + \Box y)(\Box x + \Box y)$ until the correct middle term bxy is found.

Example 11 Factoring a Trinomial of the Form $ax^2 + bxy + cy^2$

Factor.

a. $6x^2 + 17xy - 3y^2$ **b.** $2m^2 + 11mn + 12n^2$

Solutions

a. The pairs of **positive factors** of the leading coefficient 6 are $6 \cdot 1$ and $3 \cdot 2$. The coefficient of the last term is -3, and its factors are $1 \cdot (-3)$ and $-1 \cdot 3$. Using the form $(\Box x + \Box y)(\Box x + \Box y)$, try different combinations of the factors until finding one that gives the middle term $17xy$.

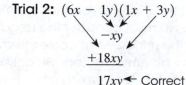

Trial 1: $(6x + 1y)(1x - 3y)$

xy

$+(-18xy)$

$-17xy$ ← Wrong sign

The "outside" product is $6x \cdot (-3y) = -18x$ and the "inside" product is $1y \cdot 1x = xy$, giving a sum of $-18xy + xy = -17xy$. This is not the correct combination. Only the sign is wrong, so switch the signs in the binomial factors.

Trial 2: $(6x - 1y)(1x + 3y)$

$-xy$

$+18xy$

$17xy$ ← Correct

The "outside" product is $6x \cdot 3y = 18xy$ and the "inside" product is $-1y \cdot 1x = -xy$, giving a sum of $18xy - xy = 17xy$. This is the correct combination.

Because we have found the correct factored form, we do not need to test any more combinations. Check the answer by multiplying out $(6x - y)(x + 3y)$. View the **check**.

The answer does check, so $6x^2 + 17xy - 3y^2 = (6x - y)(x + 3y)$.

 *My video summary* ▶ **b.** Try factoring this trinomial on your own, then compare your **answer**. Work through this **video** for a complete solution.

You Try It Work through this **You Try It** problem.

Work Exercises 47–50 in this eText or in the MyMathLab Study Plan.

⚠ When factoring a trinomial in two variables, make sure that the middle term contains both variables. For example, $6x^2 + 17x - 3y^2$ cannot be factored because the middle term does not contain the variable y.

OBJECTIVE 3 FACTOR TRINOMIALS OF THE FORM $ax^2 + bx + c$ USING THE ac METHOD

The trial-and-error method for factoring trinomials can be time consuming. A second method, known as the ac **method**, can also be used when factoring trinomials of the form $ax^2 + bx + c$.

My animation summary   ⊛ **Concept Animation** Watch this animation to see how the ac method works.

The ac Method for Factoring Trinomials of the Form $ax^2 + bx + c$

Step 1 Multiply $a \cdot c$.

Step 2 Find two integers, n_1 and n_2, whose product is ac and whose sum is b. So, $n_1 \cdot n_2 = ac$ and $n_1 + n_2 = b$. If no such pair of integers exists, the trinomial is **prime**.

Step 3 Rewrite the middle term as the sum of two terms using the integers found in Step 2. So, $ax^2 + bx + c = ax^2 + n_1x + n_2x + c$.

Step 4 Factor by grouping.

Step 5 Check your answer by multiplying out the **factored form**.

The ac method is also known as the **grouping method** or the **expansion method**. It is favored by some because its approach is more systematic, but it can still be time consuming. We suggest that you practice using both methods and then choose the one that you like best.

My video summary ▶ **Example 12 Factoring Trinomials of the Form $ax^2 + bx + c$ Using the ac Method**

Factor $3x^2 + 14x + 8$ using the ac method.

Solution For $3x^2 + 14x + 8$, we have $a = 3$, $b = 14$, and $c = 8$.

Step 1 $a \cdot c = 3 \cdot 8 = 24$.

Step 2 We must find two integers whose product is $ac = 24$ and whose sum is $b = 14$. Because the product and sum are both positive, we consider only the positive factors of 24.

Positive Factors of 24	Sum of Factors
1, 24	$1 + 24 = 25$
2, 12	$2 + 12 = 14$ ← This is the pair.
3, 8	$3 + 8 = 11$
4, 6	$4 + 6 = 10$

From the list above, 2 and 12 are the integers we need.

Step 3 $3x^2 + 14x + 8 = 3x^2 + 2x + 12x + 8$

Step 4 Begin with the new polynomial from Step 3: $3x^2 + 2x + 12x + 8$

Group the first two terms and last two terms: $= (3x^2 + 2x) + (12x + 8)$

Factor out the GCF from each group: $= x(3x + 2) + 4(3x + 2)$

Factor out the common binomial factor: $= (3x + 2)(x + 4)$

Step 5 Check the answer by multiplying out $(3x + 2)(x + 4)$.

View the **check**, or watch this **video** for a complete solution.

The answer checks, so $3x^2 + 14x + 8 = (3x + 2)(x + 4)$.

You Try It Work through this **You Try It** problem.

Work Exercises 51 and 52 in this eText or in the MyMathLab Study Plan.

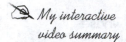 *My interactive video summary*

Example 13 Factoring Trinomials of the Form $ax^2 + bx + c$ Using the ac Method

Factor each trinomial using the ac method. If the trinomial is **prime**, state this as your answer.

a. $2x^2 + 9x - 18$ **b.** $6x^2 - 23x + 20$ **c.** $5x^2 + x + 6$

Solutions

a. For $2x^2 + 9x - 18$, we have $a = 2$, $b = 9$, and $c = -18$.

Step 1 $a \cdot c = 2(-18) = -36$.

Step 2 We need to find two **integers** whose product is $ac = -36$ and whose sum is $b = 9$. For the product to be negative, one of the integers must be positive and the other must be negative. For the sum to be positive, the integer with the larger **absolute value** must be positive. View this **popup** to see a list of such pairs of **factors** for -36.

Step 3 From the list in Step 2, -3 and 12 are the integers we need.
$2x^2 + 9x - 18 = 2x^2 - 3x + 12x - 18$.

Step 4 Begin with the new polynomial from step 3: $2x^2 - 3x + 12x - 18$

Group the first two terms and last two terms: $= (2x^2 - 3x) + (12x - 18)$

Factor out the GCF from each group: $= x(2x - 3) + 6(2x - 3)$

Factor out the common **binomial** factor: $= (2x - 3)(x + 6)$

Step 5 Check the answer by multiplying. View the **check**.

b. For $6x^2 - 23x + 20$, we have $a = 6$, $b = -23$, and $c = 20$. Follow the ac method.

Step 1 $a \cdot c = 6 \cdot 20 = 120$

Step 2 We need to find two **integers** whose product is $ac = 120$ and whose sum is $b = -23$. Since the product is positive and the sum is negative, both integers must be negative. View this **popup** to see a list of the pairs of **negative factors** for 120.

From the list, -8 and -15 are the integers we need. Try to complete this problem on your own. Check your **answer**, or watch this **interactive video** for the complete solution.

c. Try factoring $5x^2 + x + 6$ on your own. Check your **answer**, or watch this **inter-active video** for the complete solution.

You Try It Work through this You Try It problem.

Work Exercises 53–60 in this eText or in the MyMathLab Study Plan.

In **Example 11**, we used trial and error to factor a trinomial of the form $ax^2 + bxy + cy^2$. We can also use the *ac* method to factor such trinomials.

My video summary ▶ **Example 14 Factoring a Trinomial of the Form $ax^2 + bxy + cy^2$ Using the *ac* Method**

Factor $6m^2 - 7mn - 10n^2$ using the *ac* method.

Solution For this trinomial, $a = 6$, $b = -7$, and $c = -10$.

Step 1 $a \cdot c = 6(-10) = -60$.

Step 2 We need to find two integers whose **product** is $ac = -60$ and whose **sum** is $b = -7$. These integers are 5 and -12. Do you see **why**?

Step 3 $6m^2 - 7mn - 10n^2 = 6m^2 + 5mn - 12mn - 10n^2$

Step 4 Factor by grouping to complete this problem on your own. Remember to check your answer by multiplying out your final result. Compare your answer, or watch this video for a complete solution.

You Try It Work through this You Try It problem.

Work Exercise 61–64 in this eText or in the MyMathLab Study Plan.

When a trinomial has a common factor, factor it out first.

My interactive video summary 🖑 **Example 15 Factoring Trinomials with a Common Factor**

Factor completely: $24t^5 - 52t^4 - 20t^3$

Solution The terms in this trinomial have a GCF of $4t^3$. First, factor out $4t^3$, and then factor the remaining trinomial.

Begin with the original trinomial: $24t^5 - 52t^4 - 20t^3$

Write each term as a product of the GCF $4t^3$: $= 4t^3 \cdot 6t^2 - 4t^3 \cdot 13t - 4t^3 \cdot 5$

Factor out the $4t^3$: $= 4t^3(6t^2 - 13t - 5)$

Try to finish the problem on your own by factoring $6t^2 - 13t - 5$. Using either the *ac* method or trial and error. Check your answer, or watch this **interactive video** for the complete solution.

You Try It Work through this You Try It problem.

Work Exercises 65–70 in this eText or in the MyMathLab Study Plan.

Consider the trinomial $-2x^2 + 9x + 35$. Notice that the three terms do not have a common factor. However, when a trinomial has a negative number for a **leading coefficient**, we typically begin by factoring out a -1 like we did in **Example 5**. We do this because it is much easier to factor a trinomial when the leading coefficient is positive.

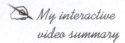 *My interactive video summary*

 Example 16 Factoring Trinomials with a Common Factor

Factor completely: $-2x^2 + 9x + 35$

Solution Since the leading coefficient is negative, factor out a -1.

$$-2x^2 + 9x + 35 = -1(2x^2 - 9x - 35) = -(2x^2 - 9x - 35)$$

Try to finish the problem on your own by factoring $2x^2 - 9x - 35$. You can use either the *ac* method or trial and error. Check your answer, or watch this **interactive video** for the complete solution.

You Try It Work through this You Try It problem.

Work Exercises 71–74 in this eText or in the MyMathLab Study Plan.

OBJECTIVE 4   FACTOR TRINOMIALS USING SUBSTITUTION

In **Objective 3**, we factored **second-degree trinomials** of the form $ax^2 + bx + c$, also known as **quadratic trinomials**. Using **substitution**, we can sometimes factor more complicated trinomials, such as those where the **degree** of the leading **term** is twice the degree of the middle term. Trinomials like this are called **quadratic in form** because the **variable** part of the first term can be written as the **square** of the variable factor in the middle term.

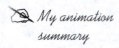 *My animation summary*

Concept Animation Watch this animation to explore this concept further.

Use the following steps to factor a trinomial that is **quadratic in form**:

Using Substitution to Factor Trinomials That Are Quadratic in Form

Step 1 Write the **variable** part of the leading **term** as the **square** of the variable factor of the middle term.

Step 2 Substitute u for the variable part from Step 1.

Step 3 Factor the resulting trinomial, if possible.

Step 4 Substitute the variable factor from Step 1 for u in the **factored** form. Simplify this result if necessary.

Step 5 Check your answer by multiplying out the factored form.

Example 17 Factoring Trinomials Using Substitution

Use substitution to factor each trinomial.

a. $x^4 - 11x^2 - 26$ b. $2(x + y)^2 + 13(x + y) + 21$

Solutions

a.   Begin with the original trinomial: $x^4 - 11x^2 - 26$

Rewrite the first term in terms of x^2: $= (x^2)^2 - 11(x^2) - 26$

Substitute u for x^2: $= u^2 - 11u - 26$

Factor: $= (u - 13)(u + 2)$

Substitute x^2 for u: $= (x^2 - 13)(x^2 + 2)$

See the check.

My video summary ▶ **b.** Substitute u for $x + y$ to result in the **quadratic trinomial** $2u^2 + 13u + 21$. Try to finish the factoring process on your own. Check your **answer**, or watch this **video** for the complete solution.

You Try It Work through this You Try It problem.

Work Exercises 75–78 in this eText or in the MyMathLab Study Plan.

5.2 Exercises

In Exercises 1–18, factor each trinomial, or state that the trinomial is prime.

1. $x^2 + 8x + 15$

2. $x^2 + 15x + 50$

3. $z^2 + 16z + 48$

4. $b^2 + 12b + 18$

5. $y^2 + 3y - 18$

6. $w^2 - 18w + 56$

7. $z^2 - 4z - 32$

8. $m^2 + 3m - 54$

9. $p^2 - 24p + 44$

10. $x^2 + 12x + 36$

11. $y^2 + 10y - 36$

12. $z^2 - 10z + 25$

13. $n^2 + 38n + 240$

14. $w^2 - 12w - 108$

15. $p^2 + 8p - 15$

16. $x^2 + 10x - 144$

17. $y^2 - 42y + 392$

18. $m^2 + 16m + 54$

In Exercises 19–32, factor completely.

19. $8x^2 + 8x - 96$

20. $9x^2 + 45x + 14$

21. $2y^4 + 34y^3 + 120y^2$

22. $9y^2z + 18yz - 315z$

23. $m^5 - 17m^4 + 60m^3$

24. $5m^3 - 15m^2 + 20m$

25. $-z^2 - 4z + 96$

26. $35 + 2w - w^2$

27. $-h^2 - 10h + 96$

28. $-5x^4 + 90x^3 - 360x^2$

29. $x^2 - 20xy + 36y^2$

30. $p^2 - 18pq - 40q^2$

31. $6x^3y^2 - 72x^2y^3 + 192xy^4$

32. $25w^3x^2 - 75w^3x - 450w^3$

In Exercises 33–50, use trial and error to factor each trinomial, or state that the trinomial is prime.

33. $5x^2 + 16x + 3$

34. $6y^2 + 7y + 2$

35. $21w^2 + 20w + 4$

36. $6x^2 + 31x + 35$

37. $5y^2 - 3y - 8$

38. $8w^2 + 14w - 15$

39. $6p^2 - 19p + 3$

40. $2q^2 + 15q + 7$

41. $4z^2 - 23z - 18$

42. $25t^2 - 60t + 36$

43. $18z^2 - 21z - 4$

44. $9x^2 + 15x + 8$

45. $22y^2 + 13y - 30$

46. $2w^2 - 21w + 27$

47. $12m^2 - 17mn + 6n^2$

48. $9x^2 - 6xy - 8y^2$

49. $36m^2 + 84mn + 49n^2$

50. $20a^2 + 17ab + 3b^2$

In Exercises 51–64, use the *ac* method to factor each trinomial, or state that the trinomial is prime.

51. $9x^2 + 18x + 5$

52. $13y^2 + 27y + 2$

53. $12m^2 - 47m + 40$

54. $4x^2 + 16x + 15$

55. $8x^2 - 6x - 9$

56. $10y^2 - 23y + 12$

57. $22m^2 + 81m + 14$

58. $5t^2 + 8t - 21$

59. $15a^2 + 16a - 7$

60. $8n^2 - 51n - 56$

61. $12p^2 - 52pq - 9q^2$

62. $3m^2 - 23mn + 40n^2$

63. $7m^2 + 32mn + 9n^2$

64. $9a^2 + 15ab - 14b^2$

In Exercises 65–74, factor completely.

65. $24x^2 - 4x - 60$

66. $18y^5 - 93y^4 + 84y^3$

67. $5x^4 + 22x^3 - 15x^2$

68. $18x^3 + 108x^2 + 120x$

69. $36x^4y^2 + 60x^3y^3 - 75x^2y^4$

70. $90m^5n^3 + 306m^5n^2 + 108m^5n$

71. $-20m^2 - 7m + 6$

72. $12 - x - 20x^2$

73. $-16wx^2 - 68wx - 42w$

74. $-24ab^2 + 294ab - 72a^2b$

In Exercises 75–78, use substitution to factor each trinomial.

75. $x^6 - x^3 - 20$

76. $6x^8 - 5x^4 - 25$

77. $5(x + y)^2 - 38(x + y) - 120$

78. $9(m - n)^2 + 78(m - n) - 27$

5.3 Special-Case Factoring; A General Factoring Strategy

THINGS TO KNOW

Before working through this section, be sure you are familiar with the following concepts:

		VIDEO	ANIMATION	INTERACTIVE
You Try It	1. Multiply Two Binomials Using Special Product Rules (Section 4.3, Objective 4)	▶	✳	
You Try It	2. Find the Greatest Common Factor of a Group of Monomials (Section 5.1, Objective 1)	▶	✳	
You Try It	3. Factor Out the Greatest Common Factor from a Polynomial (Section 5.1, Objective 2)	▶		
You Try It	4. Factor by Grouping (Section 5.1, Objective 3)			👆
You Try It	5. Factor Trinomials of the Form $x^2 + bx + c$ (Section 5.2, Objective 1)	▶	✳	👆
You Try It	6. Factor Trinomials of the Form $ax^2 + bx + c$ Using Trial and Error (Section 5.2, Objective 2)	▶	✳	👆
You Try It	7. Factor Trinomials of the Form $ax^2 + bx + c$ Using the ac Method (Section 5.2, Objective 3)	▶	✳	👆
You Try It	8. Factor Trinomials Using Substitution (Section 5.2, Objective 4)	▶		

OBJECTIVES

1 Factor the Difference of Two Squares

2 Factor Perfect Square Trinomials

3 Factor the Sum or Difference of Two Cubes

4 Factor Polynomials Completely

OBJECTIVE 1 FACTOR THE DIFFERENCE OF TWO SQUARES

In this section, we focus on the *form*. For example, the **standard form** for a **linear equation in two variables** is $Ax + By = C$, where A and B are not both zero. The equations $y = -3x$ and $4 + 5y = 8x$ may look different and have different **coefficients**, but they are both linear equations in two variables because they can be written in the same standard form, $Ax + By = C$.

Equation	Standard Form	Coefficients
$y = -3x$	$3x + y = 0$	$A = 3, B = 1, C = 0$
$4 + 5y = 8x$	$8x - 5y = 4$	$A = 8, B = -5, C = 4$

Focusing on *form* is particularly helpful with **factoring**. In **Section 4.3**, we used **special product rules** to multiply two **binomials**. In that section, recognizing how two binomial factors fit a special form helped us to quickly find the product.

Recall that the product of a binomial and its **conjugate** (the sum and difference of two terms) is the difference of the **squares** of the **terms** in the binomial. So, $(A + B)(A - B) = A^2 - B^2$. We can use this result to form a rule for factoring the difference of two squares.

Factoring the Difference of Two Squares

If A and B are real **numbers**, **variables**, or **algebraic expressions**, then the difference of their squares can be factored into the product of the sum and difference of the two quantities.

$$A^2 - B^2 = (A + B)(A - B)$$

Note: Because multiplication is **commutative**, we could also write

$$A^2 - B^2 = (A - B)(A + B).$$

 Remember to focus on *form*. The quantities A and B could be numbers, variables, or algebraic expressions.

Example 1 Factoring the Difference of Two Squares

Factor each expression completely.

a. $x^2 - 9$ 　　　　　　　　　　**b.** $16 - y^2$

Solutions

a. Start by rewriting the expression as the difference of two squares.

$$x^2 - 9 = x^2 - (3)^2$$

Here, we have $A = x$ (a **variable**) and $B = 3$ (a **real number**). Applying the difference of two squares rule, we get

$$x^2 - 9 = \underbrace{x^2 - 3^2}_{A^2 - B^2} = \underbrace{(x + 3)}_{(A + B)}\underbrace{(x - 3)}_{(A - B)}.$$

My video summary ▶ **b.** Try factoring this expression on your own. Check your **answer**, or watch this **video** for the complete solution.

🔺 **You Try It** Work through this **You Try It** problem.

Work Exercises 1–4 in this eText or in the MyMathLab Study Plan.

To use the **difference of two squares rule**, both terms must be **perfect squares**.

My animation summary ✹ **Concept Animation** Watch this animation to review the concept of a perfect square.

Example 2 Factoring the Difference of Two Squares

Factor each expression completely.

a. $z^2 - \dfrac{25}{16}$ **b.** $36x^2 - 25$ **c.** $4 - 49n^6$

Solutions

a. Start by rewriting the expression as the difference of two squares.

$$z^2 - \frac{25}{16} = z^2 - \left(\frac{5}{4}\right)^2$$

Here, we have $A = z$ (a variable) and $B = \dfrac{5}{4}$ (a real number).

Applying the **difference of two squares rule** gives

$$z^2 - \frac{25}{16} = \underbrace{z^2 - \left(\frac{5}{4}\right)^2}_{A^2 - B^2} = \underbrace{\left(z + \frac{5}{4}\right)}_{(A + B)}\underbrace{\left(z - \frac{5}{4}\right)}_{(A - B)}.$$

b. Start by rewriting the expression as the difference of two squares.

$$36x^2 - 25 = 6^2x^2 - 5^2 = (6x)^2 - 5^2$$

Here, we have $A = 6x$ (an **algebraic expression**) and $B = 5$ (a real number). Apply the difference of two squares rule, then view the **answer** to see if you are correct.

My video summary ▶ **c.** Try factoring this expression on your own using the **difference of two squares rule**. View the **answers**, or watch this **video** for the detailed solution.

🔺 **You Try It** Work through this **You Try It** problem.

Work Exercises 5–8 in this eText or in the MyMathLab Study Plan.

Example 3 Factoring the Difference of Two Squares

Factor each expression completely.

a. $81m^2 - 49n^2$ **b.** $(x - 4)^2 - 25$

My video summary ▶ **Solutions** Rewrite each expression as the difference of two perfect squares.

a. $81m^2 - 49n^2 = (9m)^2 - (7n)^2$

b. $(x - 4)^2 - 25 = (x - 4)^2 - 5^2$

Factor each **difference of two squares**, simplifying each **factor** if necessary. Check your answers, or watch this **video** for a detailed solution to both parts.

You Try It Work through this You Try It problem.

Work Exercises 9–12 in this eText or in the MyMathLab **Study Plan.**

Example 4 Factoring the Difference of Two Squares with a Greatest Common Factor

Factor each expression completely.

a. $3x^2 - 75$

b. $36x^3 - 64x$

Solutions

a. Start by factoring out the greatest common factor, 3.

$$3x^2 - 75 = 3 \cdot x^2 - 3 \cdot 25$$

$$= 3(x^2 - 25)$$

Next, rewrite the expression in parentheses as the **difference of two squares**.

$$3(x^2 - 25) = 3(x^2 - 5^2)$$

Here, we have $A = x$ and $B = 5$. Applying the **difference of two squares rule**, gives

$$3(\underbrace{x^2 - 5^2}_{A^2 - B^2}) = \underbrace{3(x + 5)}_{(A + B)}\underbrace{(x - 5)}_{(A - B)}.$$

So, $3x^2 - 75 = 3(x + 5)(x - 5)$.

My video summary ▶ b. Try factoring this expression on your own. Begin by looking for a **greatest common factor**, then use the difference of two squares rule. View the **answer**, or watch this **video** for a detailed solution to part b.

You Try It Work through this You Try It problem.

Work Exercises 13–18 in this eText or in the MyMathLab **Study Plan.**

⚠ The *sum* of two **perfect squares** cannot be factored using real numbers other than to factor out the greatest common factor.

When factoring, we want to make sure we always factor *completely*. This means we should always check each factor to see if it can be factored further.

🖊 *My video summary* ▶ **Example 5 Factoring the Difference of Two Squares More than Once**

Factor completely: $16x^4 - 81$

Solution Work through the following, or watch this **video** solution.

Start by rewriting the expression as the **difference of two squares**.

$$16x^4 - 81 = (4x^2)^2 - (9)^2$$

Here, we have $A = 4x^2$ and $B = 9$. Applying the **difference of two squares rule** gives

$$16x^4 - 81 = \underbrace{(4x^2)^2 - (9)^2}_{A^2 - B^2} = \underbrace{(4x^2 + 9)}_{(A + B)}\underbrace{(4x^2 - 9)}_{(A - B)}.$$

Notice that the factor $(4x^2 - 9)$ can be written as $(2x)^2 - 3^2$, so it too is the **difference of two squares** and can be factored further. The factor $(4x^2 + 9)$ is the sum of two squares, so it cannot be factored further.

$$4x^2 - 9 = \underbrace{(2x)^2 - (3)^2}_{A^2 - B^2} = \underbrace{(2x + 3)}_{(A + B)}\underbrace{(2x - 3)}_{(A - B)}$$

Thus,

$$16x^4 - 81 = (4x^2 + 9)\overbrace{(2x + 3)(2x - 3)}^{4x^2 - 9}.$$

The polynomial is now factored completely.

🔺 **You Try It** Work through this **You Try It** problem.

Work Exercises 19–24 in this eText or in the MyMathLab Study Plan.

OBJECTIVE 2   FACTOR PERFECT SQUARE TRINOMIALS

In **Section 4.3**, we had two **special product rules** for squaring the sum or difference of two terms. The results of these products were called **perfect square trinomials**. We can use these special product rules to factor perfect square trinomials by reversing the process.

Factoring Perfect Square Trinomials

If A and B are real numbers, variables, or algebraic expressions, then

$$A^2 - 2AB + B^2 = (A - B)(A - B) = (A - B)^2 \text{ and}$$
$$A^2 + 2AB + B^2 = (A + B)(A + B) = (A + B)^2.$$

Again, we focus on the form of the **trinomial**. To be a perfect square trinomial, the first and last **terms** must be **perfect squares**, and the middle term must be twice the product of the two quantities being squared or the opposite of the product. To **factor** a perfect square trinomial, first identify the quantities being squared in the first and last terms and then apply the appropriate rule.

Example 6 Factoring Perfect Square Trinomials

Factor each expression completely.

a. $x^2 + 6x + 9$

b. $25y^2 - 60y + 36$

Solutions

a. The first term is a **perfect square**, $x^2 = (x)^2$, and the last term is also a perfect square, $9 = 3^2$. Because $6x = 2(x)(3)$, we have a **perfect square trinomial** with a positive middle term.

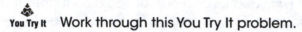

$$x^2 + 6x + 9 = \underbrace{(x)^2 + 2(x)(3) + 3^2}_{A^2 + 2 \cdot A \cdot B + B^2} = \underbrace{(x + 3)^2}_{(A + B)^2}$$

 *My video summary* ▶ b. Try to do this factoring on your own. When finished, view the **answer**, or watch this **video** for a detailed solution to part b.

You Try It Work through this **You Try It** problem.

Work Exercises 25–29 in this eText or in the MyMathLab Study Plan.

Perfect square trinomials may contain several variables as shown in Example 7a.

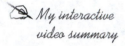 *My interactive video summary* 🖑 ## Example 7 Factoring Perfect Square Trinomials

Factor each expression completely.

a. $16x^2 + 24xy + 9y^2$

b. $m^4 - 12m^2 + 36$

Solutions

a. Write the first and last terms as **perfect squares** to determine A and B. Now complete the factorization on your own. View the **answer**, or watch this **interactive video** for a detailed solution.

b. Using rules for exponents, write $m^4 = (m^2)^2$. Now complete the factorization on your own. View the **answer**, or watch this **interactive video** for a detailed solution.

You Try It Work through this **You Try It** problem.

Work Exercises 30–36 in this eText or in the MyMathLab Study Plan.

OBJECTIVE 3   FACTOR THE SUM OR DIFFERENCE OF TWO CUBES

Like the difference of two **squares**, the difference of two **cubes** is the result of a special product rule. However, unlike the sum of two squares, the sum of two cubes *can* be factored. These two new factor rules are given next.

Factoring the Sum and Difference of Two Cubes

If A and B are real numbers, variables, or algebraic expressions, then the sum or difference of their cubes can be factored as follows:

$$A^3 + B^3 = (A + B)(A^2 - AB + B^2)$$
$$A^3 - B^3 = (A - B)(A^2 + AB + B^2)$$

In order to use these rules, both terms in the expression must be **perfect cubes**. An integer is a perfect cube if it is the cube of another **integer**. View this **popup** to review some **perfect cube** integers. A **monomial** with a perfect cube coefficient and variables raised to powers that are multiples of 3 is also considered a perfect cube (Example: $27x^3$). An algebraic expression is a perfect cube if it is raised to a power that is a multiple of 3 (Example: $(2m - 1)^3$).

To factor the sum or difference of two cubes, first identify the quantities being cubed, and then apply the appropriate rule.

Example 8 Factoring the Sum or Difference of Two Cubes

Factor each expression completely.

a. $x^3 + 64$ **b.** $z^3 - 8$

 *My video summary* ▶ **Solutions**

a. The first term is a **perfect cube**, $x^3 = (x)^3$, and the second term is also a perfect cube, $64 = 4^3$. We have the **sum of two cubes**, so apply the rule for $A^3 + B^3$.

$$A^3 + B^3 = (A + B)(A^2 - AB + B^2)$$
$$x^3 + 64 = (x)^3 + (4)^3 = (x + 4)(x^2 - 4x + 4^2)$$
$$= (x + 4)(x^2 - 4x + 16)$$

b. Try writing this expression as the **difference of two cubes**, and factoring on your own. Check your **answer**, or watch this **video** for the complete solution.

You Try It Work through this You Try It problem.

Work Exercises 37 and 38 in this eText or in the MyMathLab Study Plan.

Example 9 Factoring the Sum or Difference of Two Cubes

Factor each expression completely.

a. $125y^3 + 1$ **b.** $64y^3 - 216z^3$ **c.** $8x^3y^2 + y^5$

Solutions

a. Because $125y^3 = (5y)^3$ and $1 = 1^3$, we have the **sum of two cubes**. Apply the rule for $A^3 + B^3$.

$$A^3 + B^3 = (A + B)(A^2 - AB + B^2)$$
$$125y^3 + 1 = (5y)^3 + (1)^3 = (5y + 1)((5y)^2 - (5y)(1) + (1)^2)$$
$$= (5y + 1)(25y^2 - 5y + 1)$$

b. First, factor out the **GCF**, 8. Write the remaining two **terms** as **perfect cubes** to determine the quantities $A = 2y$ and $B = 3z$. Try to complete the factorization on your own. When finished, view the rest of the **solution**.

My video summary ▶ **c.** First, factor out the GCF. Using **rules for exponents**, write $y^5 = y^2 \cdot y^3$. Now complete this factorization on your own. Check your **answer**, or watch this **video** for a detailed solution.

🔺
You Try It **Work through this You Try It problem.**

Work Exercises 39–48 in this eText or in the MyMathLab **Study Plan.**

OBJECTIVE 4 FACTOR POLYNOMIALS COMPLETELY

Now we know several techniques and rules for factoring **polynomials**. We can use this knowledge to create a general strategy for **factoring polynomials completely**. A polynomial is factored completely if all its polynomial **factors**, other than **monomials**, are **prime**.

General Strategy for Factoring Polynomials Completely

Step 1 If necessary, factor out the **greatest common factor**. If the **leading coefficient** is negative, factor out a **common factor** with a negative coefficient.

Step 2 Select a strategy based on the number of **terms**.

    a. If there are two terms, try to use one of the following special factor rules:

        Difference of two squares: $A^2 - B^2 = (A + B)(A - B)$

        Sum of two cubes: $A^3 + B^3 = (A + B)(A^2 - AB + B^2)$

        Difference of two cubes: $A^3 - B^3 = (A - B)(A^2 + AB + B^2)$

    b. If there are three terms, see if the **trinomial** is a **perfect square trinomial**. If so, factor using one of these special factor rules:

$$A^2 + 2AB + B^2 = (A + B)^2$$
$$A^2 - 2AB + B^2 = (A - B)^2$$

    If the trinomial is not a perfect square, factor using **trial and error** or the ac **method**.

    c. If there are four or more terms, try **factoring by grouping**.

Step 3 Check to see if any factors can be factored further. Check each factor, other than **monomial factors**, to make sure they are **prime**.

Step 4 Check your answer by multiplying. Multiply out your result to see if it equals the original expression.

My interactive video summary

Example 10 Factoring a Polynomial Completely

Factor each expression completely.

a. $w^2 - w - 20$

b. $4y^4 - 32y$

c. $x^2 - 14x + 49$

d. $3z^3 - 15z^2 - 42z$

Solutions

a. Follow the general factoring strategy.

Step 1 Looking at each term, notice there is no common factor.

Step 2 The expression has three terms. The first and last terms are not perfect squares, so this is not one of the special forms.

The **leading coefficient** of the trinomial factor is 1, so we need two factors whose product is the **constant**, -20, and whose sum is the **middle coefficient** -1.

The required factors are 4 and -5. View this **popup** to see why. Factor the trinomial as $w^2 - w - 20 = (w + 4)(w - 5)$.

Step 3 All the factors are **prime**, so the factorization is complete.

Step 4 Check by multiplying.

b. Follow the **general factoring strategy**.

Step 1 Looking at each **term**, we can factor out a **common factor** of $4y$.

$$4y^4 - 32y = 4y \cdot y^3 - 4y \cdot 8 = 4y(y^3 - 8)$$

Step 2 The expression in parentheses, $y^3 - 8$, has two terms that are both cubes $(8 = 2^3)$, so this is one of the special forms. Factor this expression using the **difference of two cubes** rule with $A = y$ and $B = 2$.

$$y^3 - 8 = y^3 - 2^3$$
$$= (y - 2)(y^2 + 2 \cdot y + 2^2)$$
$$= (y - 2)(y^2 + 2y + 4)$$

Thus, $4y^4 - 32y = 4y(y - 2)(y^2 + 2y + 4)$.

Step 3 All the factors, other than the **monomial factor**, are prime, so the factorization is complete.

Step 4 Check by multiplying.

c.–d. Follow the general factoring strategy to factor these trinomials on your own. Once finished, view the **answers**. If you need help, watch this **interactive video** for complete solutions to all four parts.

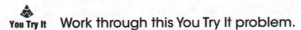
You Try It Work through this You Try It problem.

Work Exercises 49–56 in this eText or in the MyMathLab Study Plan.

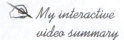
My interactive video summary

 Example 11 Factoring a Polynomial Completely

Factor each expression completely.

a. $2x^3 - 5x^2 - 8x + 20$　　　b. $3a^2 - 10a - 8$　　　c. $3z^2 + z - 1$

Solutions

a. Use the general factoring strategy. Other than 1 or -1, there is no **common factor**. There are four **terms**, so consider **factoring by grouping**. Try grouping the first two terms and the last two terms.

$$2x^3 - 5x^2 - 8x + 20$$

From the first two terms, factor out x^2. From the last two terms, factor out -4.

$$2x^3 - 5x^2 - 8x + 20 = (2x^3 - 5x^2) + (-8x + 20)$$
$$= x^2(2x - 5) - 4(2x - 5)$$

Try to complete this factorization on your own. View the **answer**, or watch this **interactive video** for the full solution.

b.–c. Use the general factoring strategy. Other than 1 or -1, there is no **common factor**. There are three **terms** in both cases, but neither is a **perfect square trinomial**. Try factoring on your own using **trial and error** or the ac **method**. View the **answers**, or watch this **interactive video** for complete solutions to all three parts.

You Try It　Work through this You Try It problem.

Work Exercises 57–64 in this eText or in the MyMathLab Study Plan.

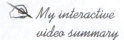
My interactive video summary

 Example 12 Factoring a Polynomial Completely

Factor each expression completely.

a. $(m - 2)^2 - (n + 4)^2$　　　　　b. $-45p^2 + 150pq - 125q^2$

c. $27x^5y^5 + 64x^2y^5$　　　　　d. $8z^3x - 27y + 8z^3y - 27x$

e. $3x^2 - 7x - 6$　　　　　f. $4a^2 - 12ab + 9b^2 - c^2$

Solutions Try to complete each factorization on your own. Check your **answers**, or watch this **interactive video** to see fully worked solutions.

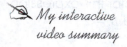
You Try It　Work through this You Try It problem.

Work Exercises 65–78 in this eText or in the MyMathLab Study Plan.

5.3 Exercises

In Exercises 1–78, factor each expression completely. If a polynomial is prime, then state this as your answer.

1. $y^2 - 16$

2. $49 - z^2$

3. $x^2 - y^2$

4. $-p^2 + q^2$

5. $x^{10} - 4$

6. $x^2 - \dfrac{49}{9}$

7. $9y^2 - 1$

8. $16 - 25m^2$

9. $4x^2 - 49y^2$

10. $x^2 y^2 - 25$

11. $4x^2 y^4 - \dfrac{9}{4} z^8$

12. $(m + 2)^2 - 81$

13. $12x^2 - 27$

14. $50 - 8y^2$

15. $20x^2 + 45$

16. $8a^2 - 18b^2$

17. $3x^2 y - 12y$

18. $a^3 b - ab^3$

19. $x^4 - 1$

20. $81x^4 - 16$

21. $x^4 - y^4$

22. $m^8 - 16n^4$

23. $5x^8 - 5$

24. $y^{16} - z^{16}$

25. $x^2 - 2x + 1$

26. $m^2 + 16m + 64$

27. $4x^2 + 28x + 49$

28. $9h^2 - 30h + 25$

29. $x^2 + 4x + 16$

30. $x^2 + 2xy + y^2$

31. $m^2 - 14mn + 49n^2$

32. $4a^2 - 20ab + 25b^2$

33. $25xy^2 + 70xy + 49x$

34. $4x^2 y^2 + 20xy + 25$

35. $z^6 + 10z^3 + 25$

36. $16a^4 - 72a^2 b^2 + 81b^4$

37. $y^3 - 125$

38. $z^3 + 27$

39. $8x^3 - 1$

40. $5t^3 + 40$

41. $8z^3 + 125$

42. $64p^3 + 121q^3$

43. $27x^3 - y^3$

44. $3z^5 - 24z^2$

45. $x^3 y^3 - 27$

46. $a^3 b^3 + 8c^3$

47. $2a^4 b^2 - 16ab^5$

48. $x^9 - 1$

49. $3x^2 - 12$

50. $2y^3 - 10y^2 - 28y$

51. $a^3 + 8b^3$

52. $y^2 + 3y + 3$

53. $x^2 + x - 6$

54. $w^2 + 12w + 20$

55. $5z^2 - 30z + 45$

56. $-4m^4 + 4m$

57. $4x^2 - 20x + 25$

58. $2x^3 + 5x^2 - 8x - 20$

59. $5x^3 + 14x^2 - 3x$

60. $6x^2 - x - 7$

61. $3w^2 - 5w + 4$

62. $9t^2 - 15t - 36$

63. $9x^2 + 12x + 4$

64. $3x^4 + 3x^3 - x^2 - x$

65. $-6p^2 + 9pq + 6q^2$

66. $5y^2 z^3 - 20z$

67. $9y + 18y^2$

68. $6xz^2 - 14xz - 40x$

69. $25m^2 + 40mn + 16n^2$

70. $6x^2 - 5xy - 6y^2$

71. $(x + 1)^2 - (y - 1)^2$

72. $-6b^2 + 3ab + a - 2b$

73. $x^2 - x - 20$

74. $1 - a^4$

75. $48y^3 + 6y^2 - 9y$

76. $r^3(t^3 - 8) + 8(t^3 - 8)$

77. $p^2 - 10p + 9$

78. $y^5 - y^3 + y^2 - 1$

5.4 Polynomial Equations and Models

THINGS TO KNOW

Before working through this section, be sure you are familiar with the following concepts:

| | VIDEO | ANIMATION | INTERACTIVE |

 You Try It 1. Solve Linear Equations in One Variable (Section 1.1, Objective 2)

 You Try It 2. Use Linear Equations to Solve Application Problems (Section 1.1, Objective 4)

 You Try It 3. Express Equations of Functions Using Function Notation (Section 2.3, Objective 1)

 You Try It 4. Factor Polynomials Completely (Section 5.3, Objective 4)

OBJECTIVES

1 Solve Polynomial Equations by Factoring

2 Find the Zeros of a Polynomial Function

3 Use Polynomial Equations and Models to Solve Application Problems

....

OBJECTIVE 1 SOLVE POLYNOMIAL EQUATIONS BY FACTORING

A **polynomial equation** results when we set two **polynomials** equal to each other. Some examples of polynomial equations are

$$2x - 7 = 4, \quad 3x^2 + 5x = x - 2, \quad \text{and} \quad 2x^3 + 7 = 3x^2 - x.$$

A polynomial equation is in **standard form** if one side equals zero and the other side is a simplified polynomial written in **descending order**.

Definition Polynomial Equation

A **polynomial equation** in standard form is written as

$$P(x) = 0$$

where $P(x)$ is a simplified polynomial in descending order.

The standard forms of the above polynomial equations are

$$2x - 11 = 0, \quad 3x^2 + 4x + 2 = 0, \quad \text{and} \quad 2x^3 - 3x^2 + x + 7 = 0.$$

The **degree of a polynomial equation** in standard form is the same as the highest degree of any of its terms. Notice that a polynomial of degree one, as in our first example, is a linear equation. We learned how to solve these types of equations in Section 1.1. To solve polynomial equations of degree 2 or higher, we can use the factoring techniques discussed in this chapter. To do so, we rely on the **zero product property.**

Zero Product Property

If A and B are real numbers or algebraic expressions and $A \cdot B = 0$, then $A = 0$ or $B = 0$.

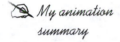 *My animation summary*

⊛ **Concept Animation** Watch this animation to learn more about the zero product property and how it can be used to solve a polynomial equation.

Example 1 Using the Zero Product Property to Solve a Polynomial Equation

Solve each equation.

a. $(x + 3)(x - 7) = 0$ **b.** $x(2x - 5) = 0$

Solutions

a. We have a product of two factors that is equal to zero. Using the **zero product property**, set each factor equal to zero and solve the resulting equations.

$$x + 3 = 0 \quad \text{or} \quad x - 7 = 0$$
$$x = -3 \qquad\qquad x = 7$$

To check that these values are **solutions**, substitute them back into the original equation to see if a true statement results.

Check $x = -3$: $(x + 3)(x - 7) = 0$
$$(-3 + 3)(-3 - 7) \overset{?}{=} 0$$
$$(0)(-10) \overset{?}{=} 0$$
$$0 = 0 \quad \text{True}$$

Check $x = 7$: $(x + 3)(x - 7) = 0$
$$(7 + 3)(7 - 7) \overset{?}{=} 0$$
$$(10)(0) \overset{?}{=} 0$$
$$0 = 0 \quad \text{True}$$

Both values check, so the **solution set** is $\{-3, 7\}$.

✍ *My video summary* ▶ **b.** Try applying the zero product property on your own to solve this equation. Check your **answer**, or watch this **video** for the complete solution.

🔺 **You Try It** Work through this **You Try It** problem.

Work Exercises **1–6** in this eText or in the MyMathLab Study Plan.

Remember that the zero product property only works when the product equals zero.

Example 2 Solving Polynomial Equations by Factoring

Solve each equation by factoring.

a. $z^2 + 4z - 12 = 0$ b. $-4x^2 + 28x - 40 = 0$ c. $z^3 + z^2 = z + 1$

Solutions Follow the four-step process for solving polynomial equations by factoring.

a. **Step 1** The equation is already in **standard form**, so move on to Step 2.

Step 2 Using the **general strategy for factoring**, we need two numbers whose product is the constant -12 and whose sum is the middle coefficient 4. Because the product is negative, the two numbers have opposite signs. Because the sum is positive, the number with the larger **absolute value** is positive.

Factor 1	Factor 2	Sum
-1	12	11
-2	6	4 ← This is the desired sum.
-3	4	1

The required numbers are -2 and 6, which can be used to factor the polynomial.

$$z^2 + 4z - 12 = 0$$
$$(z - 2)(z + 6) = 0$$

Step 3 Set each factor equal to zero and solve the equations.

$$z - 2 = 0 \quad \text{or} \quad z + 6 = 0$$
$$z = 2 \qquad\qquad z = -6$$

Step 4 Check these values on your own to confirm that the **solution set** is $\{-6, 2\}$.

 b. The equation is in standard form. We factor the expression on the left by first factoring out the **greatest common factor**, -4.

$$-4x^2 + 28x - 40 = 0$$
$$-4(x^2 - 7x + 10) = 0$$

Now factor $x^2 - 7x + 10$. Set each variable factor equal to 0 and solve the resulting equations to form the solution set. View the **answer**, or watch this **video** for the complete solution.

 c. Move the terms on the right-hand side to the left so the right side is 0.

Factor the left side by **grouping**, set each **variable factor** equal to zero, and solve the resulting equations on your own. Check your **answer**, or watch this **video** for a detailed solution.

You Try It Work through this You Try It problem.

Work Exercises 7–22 in this eText or in the MyMathLab Study Plan.

Note in Example 2b that the factor -4 is a constant and will never equal 0. When solving equations by factoring, we only set factors that contain a **variable** equal to 0.

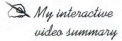 *My interactive video summary*

 Example 3 Solving Polynomial Equations by Factoring

Solve each equation by factoring.

a. $(x + 2)(x - 5) = 18$ **b.** $(x + 3)(3x - 5) = 5(x + 1) - 10$

Solutions

We follow the **four-step process** for solving polynomial equations by factoring.

a. Begin by simplifying both sides; then write the equation in **standard form**.

Begin with the original equation: $(x + 2)(x - 5) = 18$

Expand the left side: $x^2 - 5x + 2x - 10 = 18$

Simplify: $x^2 - 3x - 10 = 18$

Subtract 18 from both sides: $x^2 - 3x - 28 = 0$

Solve the equation on your own. Check your **answer**, or watch this **interactive video** for a detailed solution.

b. Begin by simplifying both sides; then write the equation in **standard form**. Solve the equation on your own. Check your **answer**, or watch this **interactive video** for a detailed solution.

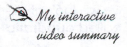 **You Try It**   Work through this You Try It problem.

Work Exercises 23–32 in this eText or in the MyMathLab Study Plan.

My interactive video summary

Example 4 Solving Polynomial Equations by Factoring

Solve each equation by factoring.

a. $\dfrac{3}{2}k^2 = 10 - \dfrac{1}{4}k$ **b.** $0.02x^2 + 0.07x = 0.3$

Solutions

a. Since we want to **factor over integers** only, first **clear the fractions** by multiplying both sides of the equation by the **LCD**. In this case, the LCD is 4. Multiplying both sides of the equation by 4 gives

$$4\left(\frac{3}{2}k^2\right) = 4\left(10 - \frac{1}{4}k\right)$$

$$6k^2 = 40 - k.$$

Continue solving the equation using the **four-step process**. Check your **answer**, or watch this **interactive video** for a detailed solution.

b. Begin by **clearing the decimals**; then continue solving the equation using the **four-step process**. Check your **answer**, or watch this **interactive video** for a detailed solution.

You Try It Work through this You Try It problem.

Work Exercises 33–36 in this eText or in the MyMathLab Study Plan.

My interactive video summary **Example 5 Solving Polynomial Equations by Factoring**

Solve the equations by factoring.

a. $4x^2 = 64x$ b. $4x - x^2 = 2x - 15$ c. $-h^3 + h^2 = -6h$

Solutions

a. Although there is a **common factor** of x on both sides of the equation, do not divide both sides of the equation by x. See why. Continue following the **four-step process** to solve the equation. Check your **answer**, or watch this **interactive video** for a detailed solution.

b.–c. In these two cases, the **leading coefficient** is negative. Either multiply both sides of the equation by -1 to change all the signs, or write the **polynomial** on the other side of the equation. Try to solve these equations on your own. Check your **answers**, or watch this **interactive video** for detailed solutions.

You Try It Work through this You Try It problem.

Work Exercises 37–46 in this eText or in the MyMathLab Study Plan.

We have seen how to **solve linear equations** in Section 1.1 and have used **factoring** to solve **polynomial equations** of higher degrees. Graphs and tables can also be used to solve polynomial equations. In Chapter 8, we will look at different methods for solving polynomial equations of degree 2, called **quadratic equations**.

If $P(x)$ is prime, does this mean the polynomial equation $P(x) = 0$ has no solutions? View this **popup** to find out.

My video summary **Example 6 Solving Polynomial Equations by Factoring**

Solve by factoring: $(2x - 9)(3x^2 - 16x - 12) = 0$

Solution

Begin with the original equation: $(2x - 9)(3x^2 - 16x - 12) = 0$

Factor the trinomial: $(2x - 9)(x - 6)(3x + 2) = 0$

Set each **variable factor** equal to 0 and solve the resulting equations and confirm that the **solution set** is $\left\{ -\dfrac{2}{3}, \dfrac{9}{2}, 6 \right\}$.

Watch this **video** for the complete solution.

You Try It Work through this You Try It problem.

Work Exercises 47–52 in this eText or in the MyMathLab Study Plan.

Note that in Example 6, we did not begin by multiplying the factors on the left together as we did in **Example 3a**. This was because the right side of the equation was already 0 as required by the **zero product property**.

OBJECTIVE 2 FIND THE ZEROS OF A POLYNOMIAL FUNCTION

A **zero**, or **root**, of a function is a value for the **independent variable** that makes the value of the function equal to zero.

> **Definition** Zero of a Function
>
> If c is a real number such that $f(c) = 0$, then c is called a **zero**, or **root**, of the function f.

When studying function notation we found that $y = f(x)$ is the value of the **dependent variable** y for a given value of the independent variable x. And we found **x-intercepts** by letting $y = 0$, or $f(x) = 0$, and solving for x. Therefore, the zeros of a function are the same as the x-intercepts of the graph of the function.

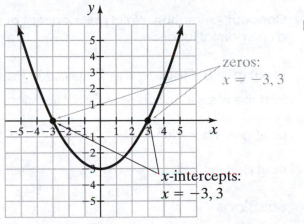

zeros:
$x = -3, 3$

x-intercepts:
$x = -3, 3$

Figure 1

Example 7 Finding the Zeros of a Polynomial Function

Find the zeros for each polynomial function using its graph.

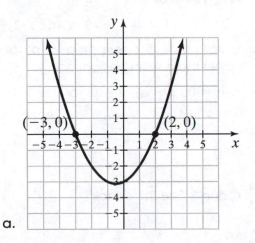

a.

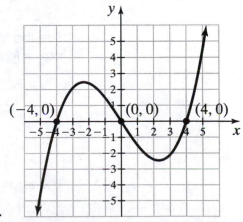

b.

 *My video summary* ▶ **Solutions** Watch this **video**, or continue reading for the full solutions.

    a. The zeros of a polynomial function are the same as the **x-intercepts** of its graph. Using the graph, the *x*-intercepts are −3 and 2. Therefore, the zeros of the function are −3 and 2.

    b. Since the *x*-intercepts are −4, 0, and 4, the zeros of the function are −4, 0, and 4.

     You Try It Work through this You Try It problem.

Work Exercises 53 and 54 in this eText or in the MyMathLab **Study Plan.**

When solving polynomial equations in standard form, we are actually looking for the zeros of the **polynomial function** $P(x)$.

Zeros of a Polynomial Function

Let $P(x)$ be a simplified polynomial function. If c is a **real number** such that $P(c) = 0$, then c is called a **zero**, or **root**, of the polynomial function. The zeros of a polynomial function are the **solutions** to the equation $P(x) = 0$.

 *My animation summary* ✹ **Concept Animation** Watch this concept animation to see how to find the zeros of a polynomial function.

Example 8 Finding the Zeros of a Polynomial Function

Find the zeros for each polynomial function.

    a. $g(x) = 12x^3 - 8x^2 - 32x$ **b.** $h(x) = \dfrac{5}{6}x^2 - 3x - \dfrac{4}{3}$

    c. $f(x) = 3x^2 - 24x + 48$ **d.** $p(x) = x^3 + 2x^2 - 9x - 18$

Solutions

    a. First, set the function equal to zero and solve the resulting equation.

$$\text{Begin with } g(x): \quad g(x) = 12x^3 - 8x^2 - 32x$$

$$\text{Set } g(x) \text{ equal to 0:} \quad 0 = 12x^3 - 8x^2 - 32x$$

$$\text{Factor:} \quad 0 = 4x(x - 2)(3x + 4)$$

See the factoring **details**. Using the **zero product property**, set each **factor** equal to zero and solve the resulting equations.

$$4x = 0 \quad \text{or} \quad x - 2 = 0 \quad \text{or} \quad 3x + 4 = 0$$

$$x = 0 \qquad\qquad x = 2 \qquad\qquad 3x = -4$$

$$x = -\frac{4}{3}$$

The zeros are $-\dfrac{4}{3}$, 0, and 2. See the **check**.

b. First, set the function equal to zero and solve the resulting equation.

$$\text{Begin with } h(x): \quad h(x) = \frac{5}{6}x^2 - 3x - \frac{4}{3}$$

$$\text{Set } h(x) \text{ equal to } 0: \quad 0 = \frac{5}{6}x^2 - 3x - \frac{4}{3}$$

Since the equation involves fractions, **clear the fractions** by multiplying both sides of the equation by the **LCD** 6.

$$\text{Multiply by the LCD:} \quad 0 = 5x^2 - 18x - 8$$

$$\text{Factor:} \quad 0 = (x - 4)(5x + 2)$$

See the factoring **details**. Using the **zero-product property**, set each **factor** equal to zero and solve the resulting equations.

$$x - 4 = 0 \quad \text{or} \quad 5x + 2 = 0$$
$$x = 4 \qquad\qquad 5x = -2$$
$$x = -\frac{2}{5}$$

The zeros are $-\frac{2}{5}$ and 4. See the **check**.

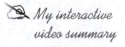 *My interactive video summary*

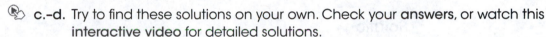 **c.–d.** Try to find these solutions on your own. Check your **answers**, or watch this interactive video for detailed solutions.

You Try It Work through this You Try It problem.

Work Exercises 55–64 in this eText or in the MyMathLab Study Plan.

Using **factoring** to find the **zeros** of a **polynomial function** should not be too surprising.

From the **factor theorem**, we can say that $(x - c)$ is a **factor** of a polynomial function $P(x)$ if $P(c) = 0$. Therefore, if c is a zero of a polynomial function, then $x - c$ is a factor of the polynomial function.

The following is a summary of the connection between zeros, **x-intercepts**, and factors.

Zeros, x-Intercepts, and Factors

Let $P(x)$ be a simplified polynomial function and c be a **real number** such that $P(c) = 0$. The following statements are all equivalent:

1. c is a zero of the function.

2. c is an x-intercept of the graph of the function.

3. $x - c$ is a factor of the function.

OBJECTIVE 3   USE POLYNOMIAL EQUATIONS AND MODELS TO SOLVE APPLICATION PROBLEMS

Some real-world situations can be **modeled** by **polynomial equations**. To solve these types of applications, we follow the same **problem-solving strategy** used for **linear equations**.

⚠️ Recall that we need to pay attention to the **feasible domain** when working with applications. Based on the problem's context, not every **solution** to an equation will be a solution to the problem. We will need to discard any solutions that do not make sense.

 My video summary ▶ **Example 9 Falling Object**

The Burj Khalifa, the world's tallest building at 2717 feet, has an observation deck on the 124th floor. An object is thrown upward with an initial velocity of 12 feet per second off the edge of the observation deck. The height of the object h, in feet, after t seconds is given by the function

$$h(t) = -16t^2 + 12t + 1480.$$

How long will it take for the object to hit the ground?

Solution

Step 1 Given the initial velocity and a function that describes the height of the object after some time, we must find the time it takes for the object to hit the ground.

Work through the following, or watch this **video** for the solution.

Step 2 We know that h is the height of the object in feet and t is the time in seconds after the object is thrown. Height is given as a function of time through the relation $h(t) = -16t^2 + 12t + 1480$.

Step 3 When the object hits the ground, it has a height of 0. Therefore, solve the equation $h(t) = 0$. Substituting the expression for $h(t)$, write $-16t^2 + 12t + 1480 = 0$.

Step 4 Begin with the original equation: $-16t^2 + 12t + 1480 = 0$

Factor out the GCF: $-4(4t^2 - 3t - 370) = 0$

Factor the trinomial: $-4(t - 10)(4t + 37) = 0$

Set each **variable factor** equal to zero and solve the resulting equations.

$$t - 10 = 0 \quad \text{or} \quad 4t + 37 = 0$$

$$t = 10 \qquad t = -\frac{37}{4}$$

Step 5 Since the time for the object to fall cannot be negative, we discard the negative solution. The only reasonable solution is $t = 10$ seconds.

Step 6 The object will hit the ground 10 seconds after it is thrown.

🔺 **You Try It** Work through this You Try It problem.

Work Exercises 65–72 in this eText or in the MyMathLab Study Plan.

We have already explored **common formulas** for geometric figures such as triangles. **Right triangles** are triangles with a 90° angle, or **right angle**. The **hypotenuse** of a right triangle is the side opposite the right angle and is the longest of the three sides. The other two sides are called the **legs** of the triangle.

Figure 2

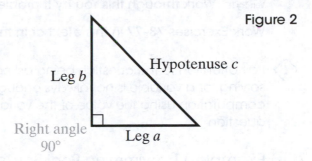

Hypotenuse c

Leg b

Right angle
90°

Leg a

Pythagorean Theorem

For right triangles, the sum of the **squares** of the lengths of the legs of the triangle equals the square of the length of the hypotenuse.

$$a^2 + b^2 = c^2$$

My video summary ▶ **Example 10 Zip Line Adventure**

Zip line rides are popular activities at many vacation destinations. A zip line consists of a pulley mounted on a cable and set at an incline. The zip line uses gravity to propel a rider from one end to the other. For one such ride, the length of the zip line is 30 feet shorter than seven times the rise of the line. The run of the line is 30 feet longer than six times the rise of the line. See Figure 3. Find the length of the zip line.

Figure 3

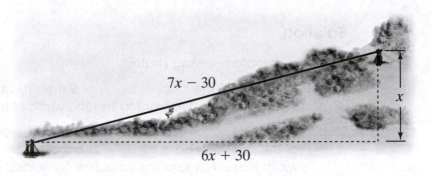

$7x - 30$

x

$6x + 30$

Solution

Use the **problem-solving strategy**. Figure 3 indicates we can use a right triangle to describe this situation. One **leg** is the **rise** of the line, and the other leg is the **run**. The length of the zip line is the **hypotenuse**. The **Pythagorean theorem** can be used to find the length of the line.

Begin with the Pythagorean theorem: $\qquad\qquad\qquad\qquad a^2 + b^2 = c^2$

Substitute expressions for the lengths: $\qquad\qquad (6x + 30)^2 + (x)^2 = (7x - 30)^2$

Expand using special products: $\quad 36x^2 + 360x + 900 + x^2 = 49x^2 - 420x + 900$

Simplify the left side: $\qquad\qquad 37x^2 + 360x + 900 = 49x^2 - 420x + 900$

Finish solving this problem on your own. Be careful! In this problem, the value for the **variable** does not directly answer the question. Check your **answer**, or watch this **video** for a detailed solution.

 You Try It Work through this You Try It problem.

Work Exercises 73–77 in this eText or in the MyMathLab Study Plan.

⚠️ Pay attention to the question being asked. When solving application problems, solving for a variable is not always enough. Sometimes, as in Example 10, some computation using the value of the variable will be needed to answer the question.

✎ *My video summary* ▶ **Example 11 Swimming Pool Border**

A swimming pool is 20 feet wide and 30 feet long. A sidewalk border around the pool has uniform width and an **area** that is equal to the area of the pool. See Figure 4. Find the width of the border.

Figure 4

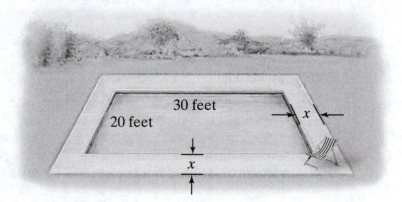

Solution

Follow the **problem-solving strategy.**

Step 1 The area of the pool is $20(30) = 600$ sq ft, which is also the area of the sidewalk border. We want to find the width of the border.

Step 2 From the figure, the width of the sidewalk border is x. The combined width of the pool and border is $20 + 2x$ because we are adding one border width to each side of the pool. The combined length is $30 + 2x$ for the same reason. The area of the pool and the area of the border are both 600 ft^2, so the combined area is 1200 ft^2.

Step 3 Use the formula for the **area of a rectangle** to write an equation for the total combined area.

Begin with the rectangle area formula: $A = lw$

Substitute: $1200 = (20 + 2x)(30 + 2x)$

Finish the problem on your own by solving the equation to find the width of the sidewalk border, x. View the **answer**, or watch this **video** for a detailed solution.

 You Try It Work through this You Try It problem.

Work Exercises 78–82 in this eText or in the MyMathLab Study Plan.

5.4 Exercises

In Exercises 1–6, solve each equation using the zero product property.

1. $(x + 1)(x - 2) = 0$

2. $w(w + 6) = 0$

3. $(2x + 1)(x + 3) = 0$

4. $5(2x - 1)(4x + 9) = 0$

5. $(2r - 3)(r + 1)(3r + 5) = 0$

6. $x(4x + 1)(x - 2) = 0$

In Exercises 7–52, use factoring techniques to solve each equation.

7. $x^2 + 3x + 2 = 0$

8. $x^2 - 5x + 6 = 0$

9. $2q^2 - 5q - 3 = 0$

10. $y^2 + 3y - 10 = 0$

11. $x^3 + 3x^2 + 2x = 0$

12. $2x^2 + 10x + 12 = 0$

13. $z^2 + 9z + 14 = 0$

14. $5x^2 + 10x = 0$

15. $y^2 - 3y = 28$

16. $4x^2 + 1 = 4x$

17. $x^2 + 3 = 12$

18. $2w^2 + 14 = 11w$

19. $2x^2 = 50$

20. $2x^2 + 5x + 3 = 1$

21. $3q^2 + 5q - 2 = 2q - 2$

22. $6m^3 + 5m^2 - 6m = 6m^2 - m$

23. $(x + 1)(x - 2) = 4$

24. $(x - 1)(3x + 4) = -4$

25. $9x(x + 3) = 3x - 16$

26. $7x(x + 3) = 2x + 6$

27. $x(3x + 1) = 2(x + 5)$

28. $(x + 3)(x - 2) = x + 3$

29. $(x - 4)(x + 6) = 7x$

30. $x(x^2 + 2) + 8 = 6(x^2 + x) - 4x^2$

31. $(2x + 1)(x - 4) = (x - 3)(x + 8)$

32. $(x + 10)(x - 1) = 2(x - 5) - 12$

33. $\dfrac{x^2}{4} - \dfrac{5x}{6} - \dfrac{2}{3} = 0$

34. $0.4x^2 + 0.3x = 1$

35. $0.06h^2 + 0.05h = 0.5$

36. $\dfrac{m^2}{10} = \dfrac{m}{5} + \dfrac{3}{2}$

37. $x^2 = 7x$

38. $25k^2 = 9$

39. $-5y^2 = 40y$

40. $3x - 2x^2 = 3x^2 - 2$

41. $x^3 + x^2 = 2x$

42. $6y^3 + 7y^2 = 8y^2 + 5y$

43. $x^2(7x + 3) - 4(7x + 3) = 0$

44. $z^3 + 4 = z^2 + 4z$

45. $12x^3 - 75x = 25 - 4x^2$

46. $8h^3 + 41 = 14$

47. $(x - 2)(x^2 - 4x + 3) = 0$

48. $(2x + 1)(3x^2 + 8x - 3) = 0$

49. $(x^2 - 3x - 10)(3x + 5) = 0$

50. $(x^2 - 9)(x^2 - 25) = 0$

51. $(x^2 - 1)(x^2 - 4x + 4) = 0$

52. $x^2 + 4 = 5x^2$

In Exercises 53 and 54, find the zeros for the function in each graph.

53.

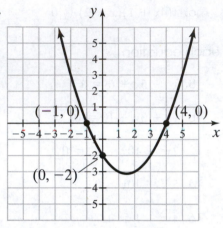

54.

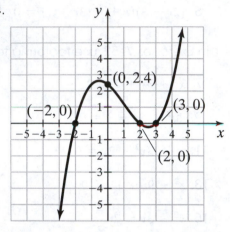

In Exercises 55–64, find the zeros of each polynomial function.

55. $p(x) = 9(x + 2)(x - 3)$

56. $p(x) = x(x - 1)(x - 2)$

57. $p(x) = x^2 - 9$

58. $p(x) = x^2 + 2x - 15$

59. $p(x) = 3x^2 + 6x + 3$

60. $p(x) = 2x^3 + x^2 - 18x - 9$

61. $p(x) = x^3 - 7x^2 + 12x$

62. $p(x) = 4x^2 + 3x - 7$

63. $p(x) = x^4 - 13x^2 + 36$

64. $p(x) = (x^2 + 2x + 1)(x - 5)$

In Exercises 65–82, solve each application problem.

65. **Observation Deck** The Canton Tower has the world's highest outdoor observation deck at 1601 feet. An object is thrown upward with an initial velocity of 20 feet per second off the edge of the observation deck. The height h of the object in feet after t seconds is given by the function

$$h(t) = -16t^2 + 20t + 1601.$$

How long will it take for the object to be 737 feet above the ground?

66. **Falling Object** The Infinity Room at Wisconsin's House on the Rock is an observation deck that extends 200 feet out above a scenic valley. The deck sits 150 feet above the valley floor. If an object is dropped from the observation deck, its height h in feet, after t seconds, is given by

$$h = -16t^2 + 150.$$

How long will it take for the object to be 6 feet above the valley floor?

67. **Projectile Motion** A cannonball is fired across a field. Its height above the ground after t seconds is given by $-16t^2 + 40t$. How long does it take for the ball to reach its maximum height of 25 feet?

68. **Projectile Motion** Cierra stands on the edge of her dorm roof and throws a ball into the air. The ball's height above the ground after t seconds is given by

$$h = -16t^2 + 32t + 48.$$

When will the ball hit the ground?

69. **Humans vs. Zombies** During one semester at Truman State University, 300 students participated in a one-week Human vs. Zombies game. The number of zombies, z, after d days of playing can be modeled by the function

$$z(d) = -d^2 + 60d - 75.$$

How many days did it take for the number of zombies to reach 200?

70. **Population** A councilwoman conducts a study and finds the population of her city can be estimated by the equation $p = x^2 - 5x + 1$, where p is thousands of people and x is the number of years after 2000. In what year did the city have a population of 7000?

71. **Wrench Cost** Eduardo notices that the number of oil filter wrenches n that he sells at his hardware store each week is related to the price p (in dollars) by the model,

$$n = -p^2 - 2p + 263.$$

What should Eduardo charge for the oil filter wrench if he wants to sell 200 per week?

72. **Rental Cost** The manager of a 50-unit apartment complex is trying to decide what rent to charge. He knows that all units will be rented if he charges $600 per month, but for every increase of $25 over the $600, there will be a unit vacant. Find the number of occupied units if his total monthly rental income is $33,600 and the rent per unit is less than $1000.

73. **Signal Tower Support** A 120-ft signal tower sits on the roof of a tall building. For support, a guy wire is attached to the top of the tower and anchored to the roof. The length of the wire is 30 feet more than twice the distance between the anchor and the base of the tower. How long is the wire?

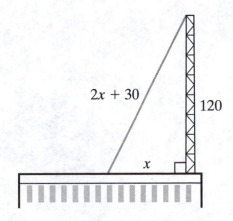

74. **Dining Canopy** A dining canopy has corner posts that are anchored by support lines. The length of each support line is 5 feet longer than the height of the post, and the support line is anchored at a distance that is 5 feet less than twice the height of the post. Determine the height of the corner post.

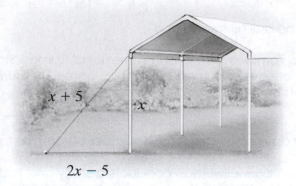

75. **Modern Architecture** An art studio purchased a triangular plot of land in downtown San Diego and constructed a new building whose base is in the shape of a right triangle. The two legs have lengths of $x - 2$ feet and $x + 5$ feet. The hypotenuse is $2x - 27$ feet. Find the lengths of the three sides.

76. **Desktop Organizer** A dorm room desktop organizer is roughly in the shape of a right triangle. The height and base are the solutions to the equation $x^2 + 200 = 28x + 8$. Find the hypotenuse, h.

77. **Bike Ride** At 8 AM, Todd rides his bike due north from campus at 10 mph. One hour later, Chad rides his bike due east from the same point, also at 10 mph. When will they be 50 miles apart?

78. **Garden Path** Aliesha has a rectangular plot of land where she can plant a garden. She wants to have a path around the garden with a uniform width of x. If the plot of land measures 30 feet by 45 feet and Aliesha has a usable garden area of 700 square feet, what is the width of the path?

79. **Picture Frame** A portrait has dimensions 20 inches by 30 inches and is surrounded by a frame of uniform width. If the total area (portrait plus frame) is 1064 square inches, find the width of the frame.

80. **Rectangular Table Top** A rectangular table top has a perimeter of 38 inches and an area of 84 square inches. Find its dimensions.

81. **Triangular Sail** The height of a triangular sail is 4 yards less than twice the length of its base. If the area of the sail is 48 square yards, find the dimensions of the sail.

82. **Banner Size** The length of a rectangular banner is 5 feet longer than its width. If the area is 36 square feet, find the dimensions.

CHAPTER SIX
Rational Expressions, Equations, and Functions

6.1 Introduction to Rational Expressions and Functions

THINGS TO KNOW

Before working through this section, be sure you are familiar with the following concepts:

		VIDEO	ANIMATION	INTERACTIVE

 You Try It 1. Find the Domain and Range of a Relation (Section 2.2, Objective 2)

 You Try It 2. Evaluate Functions (Section 2.3, Objective 2)

 You Try It 3. Factor Polynomials Completely (Section 5.3, Objective 4)

OBJECTIVES

1 Find the Domain of a Rational Function
2 Evaluate Rational Functions
3 Simplify Rational Expressions

....

OBJECTIVE 1 FIND THE DOMAIN OF A RATIONAL FUNCTION

Recall that a number is a **rational number** if it can be written as the **quotient** $\frac{p}{q}$

of two integers p and q, as long as $q \neq 0$. A *rational expression* has a similar definition.

Definition Rational Expression

A rational expression is an **expression** that can be written as the quotient $\dfrac{P}{Q}$ of two **polynomials** P and Q, as long as $Q \neq 0$.

Some examples of rational expressions are

$$\frac{5 - x}{3}, \quad \frac{x + 2}{x - 8}, \quad \frac{x^2 - x - 12}{x^2 - 3x - 4}, \quad \text{and} \quad \frac{x + 3y}{x - 3y}.$$

The first three expressions are *rational expressions in one variable, x.* The last expression is a *rational expression in two variables, x and y.*

 In the definition of a rational expression, the statement $Q \neq 0$ means that no value of the variable(s) can cause the polynomial Q in the **denominator** to equal zero.

Definition Rational Function

A **rational function** is a function defined by a **rational expression**.

For example the rational function $f(x) = \dfrac{x + 2}{x - 8}$ is defined by the rational expression $\dfrac{x + 2}{x - 8}$. The **domain of a rational function** is the set of all **real numbers** except for those values that cause the **denominator** to equal zero. The domain of $f(x) = \dfrac{x + 2}{x - 8}$, using **set-builder notation**, is $\{x \mid x \neq 8\}$. Using **interval notation**, this domain is $(-\infty, 8) \cup (8, \infty)$.

TIP Typically, set-builder notation is less messy when stating the domain of rational functions, so primarily, we will use set-builder notation in this section.

Finding the Domain of a Rational Function

To find the domain of a rational function, set the denominator equal to zero. Then solve the resulting **equation** and exclude the **solutions** from the domain. If the equation has no real solution, then the domain is the set of all real numbers.

 My video summary ▶ **Example 1 Finding the Domain of a Rational Function**

Find the **domain** of $f(x) = \dfrac{2x + 9}{5x - 4}$.

Solution Read the following, or watch this **video** for a complete solution.

Follow the **process** for finding the domain of a **rational function**.

$$\text{Set the denominator equal to 0:} \quad 5x - 4 = 0$$

$$\text{Add 4 to both sides:} \quad 5x = 4$$

$$\text{Divide both sides by 5:} \quad x = \frac{4}{5}$$

Exclude the value $\frac{4}{5}$ from the domain, so the domain of f is $\left\{ x \mid x \neq \frac{4}{5} \right\}$.

View the **answer** in interval notation.

You Try It Work through this You Try It problem.

Work Exercises 1–4 in this eText or in the MyMathLab Study Plan.

My video summary ▶ **Example 2** Finding the Domain of a Rational Function

Find the domain of $g(x) = \dfrac{x^2 + 2x - 15}{x^2 + 5x - 24}$.

Solution Follow the same process.

$$\text{Set the polynomial in the denominator equal to 0:} \quad x^2 + 5x - 24 = 0$$

Finish finding the domain by solving this equation and excluding the solutions from the domain. Check your **answer**, or watch this **video** for a complete solution.

You Try It Work through this You Try It problem.

Work Exercises 5–8 in this eText or in the MyMathLab Study Plan.

My video summary ▶ **Example 3** Finding the Domain of a Rational Function

Find the domain of each rational function.

a. $R(x) = \dfrac{5x - 8}{7}$

b. $h(x) = \dfrac{2x - 1}{x^2 + 4}$

Solutions Try to find each domain on your own. Check your **answers**, or watch this **video** for detailed solutions.

You Try It Work through this You Try It problem.

Work Exercises 9 and 10 in this eText or in the MyMathLab Study Plan.

OBJECTIVE 2 **EVALUATE RATIONAL FUNCTIONS**

We evaluate **rational functions** just like we evaluated other functions, as described in **Section 2.3**. We substitute the given value for the **variable** and simplify.

My video summary ▶ **Example 4** Evaluating a Rational Function

If $R(x) = \dfrac{5x^2 - 9}{7x + 3}$, evaluate each of the following.

a. $R(1)$ **b.** $R(-3)$

Solutions

a. Substitute 1 for x in the function R: $R(1) = \dfrac{5(1)^2 - 9}{7(1) + 3}$

        Simplify the exponent: $= \dfrac{5(1) - 9}{7(1) + 3}$

       Simplify the multiplication: $= \dfrac{5 - 9}{7 + 3}$

            Add and subtract: $= \dfrac{-4}{10}$

         Simplify the fraction: $= -\dfrac{2}{5}$

So, $R(1) = -\dfrac{2}{5}$. The value of R is $-\dfrac{2}{5}$ when x is 1.

b. Substitute -3 for x in the function R: $R(-3) = \dfrac{5(-3)^2 - 9}{7(-3) + 3}$

Try to finish simplifying on your own. Check your **answer**, or watch this **video** to see the complete solution.

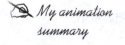 **You Try It** Work through this **You Try It** problem.

Work Exercises 11–16 in this eText or in the MyMathLab **Study Plan.**

        OBJECTIVE 3   SIMPLIFY RATIONAL EXPRESSIONS

A fraction is written in **simplest form**, or **lowest terms**, if the numerator and denominator have no common factors other than 1.

My animation summary   ✱ **Concept Animation** Watch this animation for a review of how to simplify fractions.

Simplifying rational expressions is done in the same way as simplifying fractions. **Divide out** the common factors.

Simplification Principle for Rational Expressions

If P, Q, and R are polynomials, then $\dfrac{P \cdot R}{Q \cdot R} = \dfrac{P \cdot \overset{1}{\cancel{R}}}{Q \cdot \underset{1}{\cancel{R}}} = \dfrac{P \cdot 1}{Q \cdot 1} = \dfrac{P}{Q}$ for

$Q \neq 0$ and $R \neq 0$.

The following steps are used to simplify rational expressions.

Simplifying Rational Expressions

Step 1 Factor the numerator and denominator completely.

Step 2 Divide out each common factor of the numerator and denominator.

Step 3 Separately in the numerator and denominator, multiply any factors that were not divided out to obtain the simplified rational expression. If all **factors** in the numerator divide out, the numerator will be 1.

TIP In most cases, we will not multiply out the polynomial factors in a simplified rational expression. Instead, we will leave them in **factored form**.

 My video summary ⊙ **Example 5 Simplifying a Rational Expression**

Simplify $\dfrac{2x^2 - 6x}{7x - 21}$.

Solution

Read through the following, or watch this **video** for a complete solution.

Follow the **three-step process**.

Factor the numerator and denominator: $\dfrac{2x^2 - 6x}{7x - 21} = \dfrac{2x(x - 3)}{7(x - 3)}$

Divide out the common factor $x - 3$: $= \dfrac{2x(x - \overset{1}{\cancel{3}})}{7(x - \underset{1}{\cancel{3}})}$

Write the simplified rational expression: $= \dfrac{2x}{7}$ for $x \neq 3$

You Try It Work through this **You Try It** problem.

Work Exercises 17–19 in this eText or in the MyMathLab **Study Plan.**

⚠ The original expression $\dfrac{2x^2 - 6x}{7x - 21}$ and the simplified expression $\dfrac{2x}{7}$ are only equal when $x \neq 3$. This is why we write "for $x \neq 3$." Do you see why? Typically, we will not list such restricted values when simplifying rational expressions.

 My video summary ⊙ **Example 6 Simplifying a Rational Expression**

Simplify $\dfrac{5x}{x^2 + 5x}$.

Solution Read through the following, or watch this **video** for a complete solution.

Factor the denominator: $\dfrac{5x}{x^2 + 5x} = \dfrac{5x}{x(x+5)}$

Divide out the common factor x: $= \dfrac{5\overset{1}{\cancel{x}}}{\underset{1}{\cancel{x}}(x+5)}$

Write the simplified rational expression: $= \dfrac{5}{x+5}$

 You Try It Work through this You Try It problem.

Work Exercises 20–22 in this eText or in the MyMathLab **Study Plan.**

⚠ Only **common factors** of an expression can be divided out. It is incorrect to divide out **terms.** In Example 6, $5x$ is a term of the original denominator, and 5 is a term of the simplified denominator. Neither can be divided out.

Incorrect   **Incorrect**

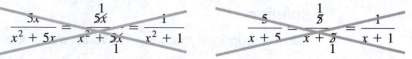

TIP When simplifying a **rational expression**, replacing each divided-out **factor** with a 1 can cause unnecessary clutter. Therefore, it is fine to leave out the 1's. However, if every factor in the numerator divides out, be sure to include a 1 in the numerator of the simplified rational expression.

My video summary ▶ **Example 7 Simplifying a Rational Expression**

Simplify $\dfrac{y^2 + 2y - 24}{y^2 + 4y - 32}$.

Solution Follow the three-step process.

Factor the numerator and denominator: $\dfrac{y^2 + 2y - 24}{y^2 + 4y - 32} = \dfrac{(y-4)(y+6)}{(y-4)(y+8)}$

Try to finish simplifying this rational expression on your own. View the **answer,** or watch this **video** for a complete solution.

 You Try It Work through this You Try It problem.

Work Exercises 23–28 in this eText or in the MyMathLab **Study Plan.**

My video summary ▶ **Example 8** Simplifying a Rational Expression

Simplify $\dfrac{2m^2 + m - 15}{2m^3 - 5m^2 - 18m + 45}$.

Solution Follow the three-step process, and try to simplify this **rational expression** on your own. View the **answer**, or watch this **video** for a complete solution.

You Try It Work through this You Try It problem.

Work Exercises 29–32 in this eText or in the MyMathLab Study Plan.

My video summary ▶ **Example 9** Simplifying a Rational Expression in Two Variables

Simplify $\dfrac{x^2 - xy - 12y^2}{2x^2 + 7xy + 3y^2}$.

Solution

Factor the numerator and denominator: $\dfrac{x^2 - xy - 12y^2}{2x^2 + 7xy + 3y^2} = \dfrac{(x - 4y)(x + 3y)}{(2x + y)(x + 3y)}$

Try to finish this problem on your own. View the **answer**, or watch this **video** for a complete solution.

You Try It Work through this You Try It problem.

Work Exercises 33 and 34 in this eText or in the MyMathLab Study Plan.

The **commutative property of addition** states that a sum is not affected by the order of the **terms**. So, $a + b = b + a$. Sometimes when factoring the numerator and denominator of a rational expression, the terms of **common factors** may be arranged in a different order. We can still **divide out** these common factors.

My video summary ▶ **Example 10** Simplifying a Rational Expression

Simplify $\dfrac{w^2 - y^2}{2xy + 2xw}$.

Solution Read through the following, or watch this **video** for a complete solution.

Factor the numerator and denominator: $\dfrac{w^2 - y^2}{2xy + 2xw} = \dfrac{(w + y)(w - y)}{2x(y + w)}$

Divide out the common factors $w + y = y + w$: $= \dfrac{\cancel{(w + y)}(w - y)}{2x\cancel{(y + w)}}$

Write the simplified rational expression: $= \dfrac{w - y}{2x}$

You Try It Work through this You Try It problem.

Work Exercises 35 and 36 in this eText or in the MyMathLab Study Plan.

If the numerator and denominator of a **rational expression** have factors that are opposite polynomials, we can simplify by first factoring −1 from one of them.

📖 *My video summary* ▶ **Example 11 Simplifying a Rational Expression Involving Opposites**

Simplify $\dfrac{3x - 10}{10 - 3x}$.

Solution Read the following, or watch this **video** for a complete solution. Recognize that $3x - 10$ and $10 - 3x$ are opposite polynomials, and factor −1 from $10 - 3x$:

$$10 - 3x = -1(-10 + 3x) = -1(3x - 10)$$

This gives

$$\frac{3x - 10}{10 - 3x} = \frac{3x - 10}{-1(3x - 10)} = \frac{3x - 10}{-1(3x - 10)} = \frac{1}{-1} = -1.$$

🔺 **You Try It** Work through this You Try It problem.

Work Exercises 37 and 38 in this eText or in the MyMathLab Study Plan.

Example 11 shows us that opposite polynomials in the numerator and denominator of a rational expression will **divide out**, leaving a factor of −1.

📖 *My video summary* ▶ **Example 12 Simplifying a Rational Expression Involving Opposites**

Simplify $\dfrac{2x^2 - 27x + 70}{49 - 4x^2}$.

Solution

Factor the numerator and denominator: $\dfrac{2x^2 - 27x + 70}{49 - 4x^2} = \dfrac{(2x - 7)(x - 10)}{(7 - 2x)(7 + 2x)}$

Note that $2x - 7$ and $7 - 2x$ are opposite polynomials. Finish simplifying the rational expression on your own. View the **answer**, or watch this **video** for a complete solution.

🔺 **You Try It** Work through this You Try It problem.

Work Exercises 39–42 in this eText or in the MyMathLab Study Plan.

📖 *My video summary* ▶ **Example 13 Simplifying a Rational Expression**

Simplify $\dfrac{w^3 + 64}{w^2 - 4w + 16}$.

Solution Try to simplify this rational expression on your own. View the **answer**, or watch this **video** for a complete solution.

🔺 **You Try It** Work through this You Try It problem.

Work Exercises 43–50 in this eText or in the MyMathLab Study Plan.

6.1 Exercises

In Exercises 1–10, find the domain of each rational function. Write your answer in set-builder notation.

1. $f(x) = \dfrac{10}{x + 2}$

2. $g(x) = \dfrac{x - 4}{x - 6}$

3. $h(x) = \dfrac{3x - 7}{5x + 2}$

4. $s(t) = \dfrac{3t - 7}{2t}$

5. $f(x) = \dfrac{x + 9}{(x + 2)(x - 5)}$

6. $g(x) = \dfrac{8x}{x^2 + 3x - 88}$

7. $R(x) = \dfrac{x + 4}{3x^2 + 10x - 8}$

8. $h(x) = \dfrac{3x - 1}{x^3 - 4x^2 - 12x}$

9. $f(x) = \dfrac{7x - 2}{5}$

10. $R(x) = \dfrac{x + 4}{2x^2 + 1}$

In Exercises 11–16, evaluate the given rational function as indicated. If the function value is undefined, state this as your answer.

11. If $h(x) = \dfrac{x + 5}{x - 1}$, find $h(2)$, $h(3)$, and $h(-3)$.

12. If $F(p) = \dfrac{2p + 1}{3p - 1}$, find $F(0)$, $F(-1)$, and $F(4)$.

13. If $R(x) = \dfrac{x + 2}{x^2 - 4}$, find $R(1)$, $R(0)$, and $R(-2)$.

14. If $Q(t) = \dfrac{t^3 - 1}{t^2 - 1}$, find $Q(-2)$, $Q(2)$, and $Q(4)$.

15. If $f(x) = \dfrac{x^2 - 5x - 14}{5 - x}$, find $f(-2)$, $f(3)$, and $f(5)$.

16. If $g(t) = \dfrac{2t^3 - 7}{t^2 + 5}$, find $g(-1)$, $g(0)$, and $g(2)$.

In Exercises 17–50, simplify each rational expression. If the expression is already in simplest form, state this in your answer.

17. $\dfrac{3m - 9}{5m - 15}$

18. $\dfrac{5x^2 - 10x}{13x - 26}$

19. $\dfrac{4n + 24}{9n^3 + 54n^2}$

20. $\dfrac{6y}{y^2 + 6y}$

21. $\dfrac{3n^2}{3n^2 - 6n}$

22. $\dfrac{x^2 - x}{7x}$

23. $\dfrac{z^2 - 2z - 15}{z^2 + 10z + 21}$

24. $\dfrac{q^2 - q - 42}{q^2 - 9q + 14}$

25. $\dfrac{m^2 + 4m - 117}{m^2 + 18m + 80}$

26. $\dfrac{t^2 + 5t - 24}{2t^2 - 11t + 15}$

27. $\dfrac{6w^2 + 5w - 4}{3w^2 + 13w + 12}$

28. $\dfrac{3z^2 - 2z - 8}{3z^2 - 14z - 24}$

29. $\dfrac{2x^2 + 11x + 12}{2x^3 + 3x^2 - 32x - 48}$

30. $\dfrac{12a^3 + 4a^2 + 9a + 3}{3a^2 - 11a - 4}$

31. $\dfrac{4x^2 - 7x - 2}{8x^3 + 2x^2 + 4x + 1}$

32. $\dfrac{m^3 - 4m^2 + 6m - 24}{m^2 - 8m + 16}$

33. $\dfrac{x^2 - 7xy - 8y^2}{x^2 - 3xy - 4y^2}$

34. $\dfrac{4p^2 + 5pq - 6q^2}{3p^2 + 5pq - 2q^2}$

35. $\dfrac{12 + 5x}{5x^2 + 22x + 24}$

36. $\dfrac{a^2 - 49b^2}{56bc + 8ac}$

37. $\dfrac{2m - 9}{9 - 2m}$

38. $\dfrac{3x + 8}{8 - 3x}$

39. $\dfrac{w^2 - 6w - 16}{64 - w^2}$

40. $\dfrac{2 - z}{z^2 + z - 6}$

41. $\dfrac{7k - 14}{28 - 14k}$

42. $\dfrac{x^2 - 4}{-2 - x}$

43. $\dfrac{36x^2 - 42x + 49}{216x^3 + 343}$

44. $\dfrac{x^3 - 125}{3x - 15}$

45. $\dfrac{a^3 - 8}{a^2 - 4}$

46. $\dfrac{w^3 + 6w^2 - 27w}{2w^4 - 18w^2}$

47. $\dfrac{x^2 - 9}{9 + x}$

48. $\dfrac{x^2 + 3xy - 4y^2}{2x^2 + 5xy - 12y^2}$

49. $\dfrac{x^2 + x - 6}{x + 3}$

50. $\dfrac{5x + 6}{3 + 5x}$

6.2 Multiplying and Dividing Rational Expressions

THINGS TO KNOW

Before working through this section, be sure you are familiar with the following concepts:

		VIDEO	ANIMATION	INTERACTIVE-

You Try It 1. Simplify Exponential Expressions Using the Quotient Rule (Section 4.1, Objective 2)   ▶

You Try It 2. Factor Polynomials Completely (Section 5.3, Objective 4)

You Try It 3. Simplify Rational Expressions (Section 6.1, Objective 3)   ▶

OBJECTIVES

1 Multiply Rational Expressions

2 Divide Rational Expressions

OBJECTIVE 1   MULTIPLY RATIONAL EXPRESSIONS

Recall that to multiply **rational numbers** (fractions), we can multiply straight across the **numerators**, multiply straight across the **denominators**, and **simplify**. An often easier approach is to first write the numerator and the denominator as products of **prime factors**, and then divide out any common factors before multiplying.

✎ My animation summary   ⊛ **Concept Animation** Watch this **animation** to review this concept.

Follow the same approach when multiplying **rational expressions**.

If $\dfrac{P}{Q}$ and $\dfrac{R}{S}$ are rational expressions, then

$$\frac{P}{Q} \cdot \frac{R}{S} = \frac{PR}{QS}.$$

Multiplying Rational Expressions

Step 1 Factor each numerator and denominator completely into prime factors.

Step 2 Divide out common factors.

Step 3 Multiply remaining factors in the numerators and multiply remaining factors in the denominators.

TIP Keep the result in **factored form** just as we did when simplifying rational expressions.

✎ My video summary ▶ **Example 1 Multiplying Rational Expressions**

Multiply $\dfrac{5x^2}{2y} \cdot \dfrac{6y^2}{25x^3}.$

Solution Read through the following, or watch this **video** for a complete solution. Follow the **three-step process** for multiplying **rational expressions**.

Begin with the original expression: $\dfrac{5x^2}{2y} \cdot \dfrac{6y^2}{25x^3}$

Factor numerators and denominators: $= \dfrac{5 \cdot x \cdot x}{2 \cdot y} \cdot \dfrac{2 \cdot 3 \cdot y \cdot y}{5 \cdot 5 \cdot x \cdot x \cdot x}$

Divide out common factors: $= \dfrac{5 \cdot x \cdot x}{2 \cdot y} \cdot \dfrac{2 \cdot 3 \cdot y \cdot y}{5 \cdot 5 \cdot x \cdot x \cdot x}$

Multiply remaining factors: $= \dfrac{3 \cdot y}{5 \cdot x} = \dfrac{3y}{5x}$

You Try It Work through this **You Try It** problem.

Work Exercises 1–4 in this eText or in the **My**MathLab Study Plan.

(eText Screens 6.2-1–6.2-21)

🖎 *My video summary* ⊙ Note that Example 1 also can be worked by using the **rules for exponents**. Watch this **video** to see an alternate solution.

Example 2 Multiplying Rational Expressions

Multiply $\dfrac{3x - 6}{2x} \cdot \dfrac{8}{5x - 10}$.

🖎 *My video summary* ⊙ **Solution** Read through the following, or watch this **video** for a complete solution. Follow the **three-step process** for multiplying **rational expressions**.

$$\text{Begin with the original expression:} \quad \frac{3x - 6}{2x} \cdot \frac{8}{5x - 10}$$

$$\text{Factor numerators and denominators:} \quad = \frac{3(x - 2)}{2 \cdot x} \cdot \frac{2 \cdot 4}{5(x - 2)}$$

$$\text{Divide out common factors:} \quad = \frac{3\cancel{(x-2)}}{2 \cdot x} \cdot \frac{2 \cdot 4}{5\cancel{(x-2)}}$$

$$\text{Multiply remaining factors:} \quad = \frac{3 \cdot 4}{x \cdot 5} = \frac{12}{5x}$$

You Try It Work through this You Try It problem.

Work Exercises 5–10 in this eText or in the MyMathLab Study Plan.

🖎 *My video summary* ⊙ ### Example 3 Multiplying Rational Expressions

Multiply $\dfrac{x^2 - 4}{x^2 + 2x - 35} \cdot \dfrac{x^2 - 25}{x + 2}$.

Solution Follow the **three-step process** for multiplying rational expressions.

$$\text{Begin with the original expression:} \quad \frac{x^2 - 4}{x^2 + 2x - 35} \cdot \frac{x^2 - 25}{x + 2}$$

$$\text{Factor numerators and denominators:} \quad = \frac{(x + 2)(x - 2)}{(x + 7)(x - 5)} \cdot \frac{(x + 5)(x - 5)}{(x + 2)}$$

Try to finish this problem on your own by dividing out **common factors**, multiplying remaining factors in the numerator, and multiplying remaining factors in the denominator. View the **answer**, or watch this **video** for a complete solution.

You Try It Work through this You Try It problem.

Work Exercises 11–16 in this eText or in the MyMathLab Study Plan.

🖎 *My video summary* ⊙ ### Example 4 Multiplying Rational Expressions

Multiply $\dfrac{2x^2 + 3x - 2}{3x^2 - 2x - 1} \cdot \dfrac{3x^2 + 4x + 1}{2x^2 + x - 1}$.

Solution Work through this **interactive video** to complete the solution by factoring the expressions and **dividing out** common factors.

You Try It Work through this You Try It problem.

Work Exercises 17–24 in this eText or in the MyMathLab Study Plan.

Common factors are not limited to constants or **binomial factors**. It is possible to have other common **polynomial factors**, as shown in the next example.

Example 5 Multiplying Rational Expressions

Multiply $\dfrac{3x^2 + 9x + 27}{x - 1} \cdot \dfrac{x + 3}{x^3 - 27}$.

Solution Read the following, or watch this **video** for a complete solution. Follow the **three-step process** for multiplying rational expressions.

Begin with the original expression: $\dfrac{3x^2 + 9x + 27}{x - 1} \cdot \dfrac{x + 3}{x^3 - 27}$

Factor numerators and denominators: $= \dfrac{3(x^2 + 3x + 9)}{(x - 1)} \cdot \dfrac{(x + 3)}{(x - 3)(x^2 + 3x + 9)}$

Notice there is a common **trinomial factor** in a numerator and denominator. This trinomial factor happens to be prime. So, these factors **divide out** just like any other common factor.

Divide out common factors: $\dfrac{3\cancel{(x^2 + 3x + 9)}}{(x - 1)} \cdot \dfrac{(x + 3)}{(x - 3)\cancel{(x^2 + 3x + 9)}}$

Multiply remaining factors: $= \dfrac{3(x + 3)}{(x - 1)(x - 3)}$

You Try It Work through this You Try It problem.

Work Exercises 25–28 in this eText or in the MyMathLab Study Plan.

Sometimes, factors in the numerator and denominator are opposites (or **additive inverses**). Recall that the quotient of opposites equals −1. For example,

$$\frac{4x - 7}{7 - 4x} = -1.$$

 My video summary ▶ **Example 6 Multiplying Rational Expressions**

Multiply $\dfrac{3x^2 + 10x - 8}{2x - 3x^2} \cdot \dfrac{4x + 1}{x + 4}$.

Solution Try to multiply these expressions on your own. View the **answer**, or watch this **video** for a complete solution.

You Try It Work through this You Try It problem.

Work Exercises 29–32 in this eText or in the MyMathLab Study Plan.

TIP Usually, we want the factors to have positive **leading coefficients**. If a leading coefficient is negative, we can factor out a **negative constant**.

As we saw when factoring in **Section 5.2**, rational expressions may contain more than one **variable**.

My video summary ▶ **Example 7 Multiplying Rational Expressions**

Multiply $\dfrac{x^2 + xy}{3x + y} \cdot \dfrac{3x^2 + 7xy + 2y^2}{x^2 - y^2}$.

Solution For polynomials of the form $ax^2 + bxy + cy^2$, remember that the second term in each binomial factor must contain the variable y.

Begin with the original expression: $\dfrac{x^2 + xy}{3x + y} \cdot \dfrac{3x^2 + 7xy + 2y^2}{x^2 - y^2}$

Factor the numerators and denominators: $= \dfrac{x(x + y)}{3x + y} \cdot \dfrac{(3x + y)(x + 2y)}{(x + y)(x - y)}$

Try to finish this problem on your own. View the **answer**, or watch this **video** for a complete solution.

 You Try It Work through this **You Try It** problem.

Work Exercises 33–40 in this eText or in the MyMathLab Study Plan.

OBJECTIVE 2 DIVIDE RATIONAL EXPRESSIONS

Recall that to divide by a rational number, we multiply by its **reciprocal**.

My animation summary ✸ **Concept Animation** Watch this **animation** to review this concept.

Follow the same approach when dividing **rational expressions**.

Dividing Rational Expressions

Step 1 Change the division to multiplication and replace the **divisor** by its reciprocal.

Step 2 Multiply the expressions.

If $\dfrac{P}{Q}$ and $\dfrac{R}{S}$ are rational expressions, then

$$\frac{P}{Q} \div \frac{R}{S} = \frac{P}{Q} \cdot \frac{S}{R} = \frac{PS}{QR}.$$

My video summary ▶ **Example 8 Dividing Rational Expressions**

Divide $\dfrac{6x^5}{9y^3} \div \dfrac{5x^4}{3y^2}$.

Solution Read the following, or watch this **video** for a complete solution.

Begin with the original expression: $\dfrac{6x^5}{9y^3} \div \dfrac{5x^4}{3y^2}$

Change to multiplication by the reciprocal: $= \dfrac{6x^5}{9y^3} \cdot \dfrac{3y^2}{5x^4}$ ← $\dfrac{3y^2}{5x^4}$ is the reciprocal of $\dfrac{5x^4}{3y^2}$

Factor: $= \dfrac{2 \cdot 3 \cdot x \cdot x \cdot x \cdot x \cdot x}{3 \cdot 3 \cdot y \cdot y \cdot y} \cdot \dfrac{3 \cdot y \cdot y}{5 \cdot x \cdot x \cdot x \cdot x}$

Divide out common factors: $= \dfrac{2 \cdot \cancel{3} \cdot \cancel{x} \cdot \cancel{x} \cdot \cancel{x} \cdot \cancel{x} \cdot x}{\cancel{3} \cdot \cancel{3} \cdot \cancel{y} \cdot \cancel{y} \cdot y} \cdot \dfrac{\cancel{3} \cdot \cancel{y} \cdot \cancel{y}}{5 \cdot \cancel{x} \cdot \cancel{x} \cdot \cancel{x} \cdot \cancel{x}}$

Multiply remaining factors: $= \dfrac{2 \cdot x}{y \cdot 5} = \dfrac{2x}{5y}$

You Try It Work through this You Try It problem.

Work Exercises 41 and 42 in this eText or in the MyMathLab Study Plan.

My video summary ▶ **Example 9 Dividing Rational Expressions**

Divide $\dfrac{9y^2 - 81}{4y^2} \div \dfrac{y + 3}{8}$.

Solution Begin with the original expression: $\dfrac{9y^2 - 81}{4y^2} \div \dfrac{y + 3}{8}$

Change to multiplication by the reciprocal: $= \dfrac{9y^2 - 81}{4y^2} \cdot \dfrac{8}{y + 3}$

Factor out the GCF: $= \dfrac{9(y^2 - 9)}{4y^2} \cdot \dfrac{8}{y + 3}$

Factor difference of two squares: $= \dfrac{9(y - 3)(y + 3)}{4y^2} \cdot \dfrac{8}{y + 3}$

Try to finish this problem on your own. View the **answer**, or watch this **video** for a complete solution.

You Try It Work through this You Try It problem.

Work Exercises 43–48 in this eText or in the MyMathLab Study Plan.

My video summary ▶ **Example 10 Dividing Rational Expressions**

Divide $\dfrac{2x^2 + 21x + 40}{3x^2 + 23x - 8} \div \dfrac{4x^2 + 16x + 15}{x + 2}$.

Solution Begin with the original expression: $\dfrac{2x^2 + 21x + 40}{3x^2 + 23x - 8} \div \dfrac{4x^2 + 16x + 15}{x + 2}$

Change to multiplication by the reciprocal: $= \dfrac{2x^2 + 21x + 40}{3x^2 + 23x - 8} \cdot \dfrac{x + 2}{4x^2 + 16x + 15}$

Factor the numerators and denominators and then multiply. View the **answer**, or watch this **video** for a complete solution.

You Try It Work through this You Try It problem.

Work Exercises 49–52 in this eText or in the MyMathLab Study Plan.

My video summary ▶ **Example 11 Dividing Rational Expressions**

Divide $\dfrac{x^3 - 8}{2x^2 - x - 6} \div \dfrac{x^2 + 2x + 4}{6x^2 + 11x + 3}$.

Solution Work through this **interactive video** to complete the solution, or view the **answer**.

You Try It Work through this You Try It problem.

Work Exercises 53–56 in this eText or in the MyMathLab Study Plan.

As with multiplication, when dividing rational expressions, it is possible to have more than one variable as shown in the next example.

My video summary ▶ **Example 12 Dividing Rational Expressions**

Divide $\dfrac{x^3 - 8y^3}{3x + y} \div \dfrac{4x - 8y}{6x^2 + 17xy + 5y^2}$.

Solution   Begin with the original expression: $\dfrac{x^3 - 8y^3}{3x + y} \div \dfrac{4x - 8y}{6x^2 + 17xy + 5y^2}$

Change to multiplication: $= \dfrac{x^3 - 8y^3}{3x + y} \cdot \dfrac{6x^2 + 17xy + 5y^2}{4x - 8y}$

Factor: $= \dfrac{(x - 2y)(x^2 + 2xy + 4y^2)}{3x + y} \cdot \dfrac{(3x + y)(2x + 5y)}{4(x - 2y)}$

Try to finish this problem on your own. View the **answer**, or watch the **video** for a complete solution.

You Try It Work through this You Try It problem.

Work Exercises 57–60 in this eText or in the MyMathLab Study Plan.

If multiplying or dividing more than two rational expressions, we first change any divisions to multiplication using the appropriate reciprocal. Then we follow the three-step process for multiplying rational expressions.

✎ *My video summary* ▶ **Example 13 Multiplying and Dividing Rational Expressions**

Perform the indicated operations.

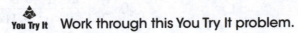

$$\frac{x^2 + 2x - 15}{x^2 + 2x - 8} \cdot \frac{x^2 + 3x + 2}{x^2 + 4x - 21} \div \frac{x + 2}{x^2 + 9x + 14}$$

Solution Try to work this problem on your own. Remember to first change all divisions to multiplication using the reciprocal of the divisor. View the **answer**, or watch the **video** to see a complete solution.

You Try It Work through this You Try It problem.

Work Exercises 61–70 in this eText or in the MyMathLab **Study Plan.**

6.2 Exercises

In Exercises 1–40, multiply the rational expressions.

1. $\dfrac{6a^3}{2b^2} \cdot \dfrac{5b^3}{3a^4}$

2. $\dfrac{14x^2y}{3y^2x} \cdot \dfrac{9x^3y^2}{7x^2y^3}$

3. $\dfrac{2xyz}{5y^2z^2} \cdot \dfrac{15xz^2}{6x^2y}$

4. $\dfrac{-3x^2z}{2y^3z^2} \cdot \dfrac{4xy^2}{z^3}$

5. $\dfrac{5x}{3x + 6} \cdot \dfrac{x + 2}{7}$

6. $\dfrac{3m + 9}{10m} \cdot \dfrac{5m^2}{2m + 6}$

7. $\dfrac{6x^2}{2x - 2} \cdot \dfrac{x^2 - x}{5x^3}$

8. $\dfrac{x - 3}{x + 5} \cdot \dfrac{4x + 20}{9x - 27}$

9. $\dfrac{2x + 2}{4x} \cdot \dfrac{5x - 35}{8x + 16}$

10. $\dfrac{3x + 9}{2x^2 - 6x} \cdot \dfrac{x^2 - 3x}{5x + 15}$

11. $\dfrac{x + 1}{x^2 - 4} \cdot \dfrac{x - 2}{x^2 - 1}$

12. $\dfrac{x - 6}{x^2 + 3x - 10} \cdot \dfrac{2x - 4}{x^2 - 36}$

13. $\dfrac{3r^2 + 12r}{r - 1} \cdot \dfrac{r^2 - 5r + 4}{r + 2}$

14. $\dfrac{x^2 - 3x}{x^2 + x} \cdot \dfrac{x^2 + 3x + 2}{x^2 - 2x - 3}$

15. $\dfrac{x^2 + 4x + 3}{x^2 + 5x + 6} \cdot \dfrac{x + 2}{x - 1}$

16. $\dfrac{x^2 + 12x + 36}{x + 1} \cdot \dfrac{4x + 3}{x^2 + 2x - 24}$

17. $\dfrac{x^2 - 1}{3x^2 + 11x + 6} \cdot \dfrac{3x^2 + 7x - 6}{x^2 - 2x + 1}$

18. $\dfrac{x^2 + 3x}{x^2 - 4x - 5} \cdot \dfrac{x^2 - 25}{2x^2 + x}$

19. $\dfrac{3b^2 + 4b - 4}{b^2 + b - 2} \cdot \dfrac{b^2 - 1}{3b^2 + b - 2}$

20. $\dfrac{4x^2 + 20x - 24}{6x^2 + 7x - 3} \cdot \dfrac{4x^2 + 20x + 21}{9x^2 + 51x - 18}$

21. $\dfrac{3x - 21}{x + 5} \cdot \dfrac{x^2 - 2x - 15}{4x^2 + 12x}$

22. $\dfrac{6x^2 - x - 1}{x^2 - 7x - 18} \cdot \dfrac{3x^2 + 10x + 8}{3x^2 + 10x + 3}$

23. $\dfrac{x^2 + 4x + 4}{3x^2 + 3x - 18} \cdot \dfrac{5x^2 - 45}{x^2 - 3x - 10}$

24. $\dfrac{6x^2 - 33x - 120}{x^2 + 9x + 14} \cdot \dfrac{x^2 + 16x + 63}{30x^2 - 222x - 144}$

25. $\dfrac{x^3 - 8}{x^2 + 4x + 3} \cdot \dfrac{x^2 + 3x}{x^3 + 2x^2 + 4x}$

26. $\dfrac{x + 10}{2x^2 - 4x + 8} \cdot \dfrac{3x^3 + 24}{x - 5}$

27. $\dfrac{5x^2 - 5x + 5}{10x^2 + 20x} \cdot \dfrac{x^2 - 2x}{x^3 + 1}$

28. $\dfrac{2n^3 + 2n^2 + n + 1}{n^2 + 5n + 6} \cdot \dfrac{n^2 - 2n}{n^3 + 1}$

29. $\dfrac{4 - x^2}{x^2 - x - 6} \cdot \dfrac{5x^2 - 14x - 3}{2x^2 - 3x - 2}$

30. $\dfrac{7 - 28x}{2x - 7} \cdot \dfrac{x^2 + 7x + 12}{4x^2 + 11x - 3}$

31. $\dfrac{p^2 - p}{3p - 2p^2} \cdot \dfrac{2p^2 - p - 3}{1 - p^2}$

32. $\dfrac{3 - 8x - 3x^2}{9 - x^2} \cdot \dfrac{x^2 - x - 6}{9x^2 - 1}$

33. $\dfrac{x^2 + 2xy - 3y^2}{3x^2 + 2xy - y^2} \cdot \dfrac{x^2 + 3xy + 2y^2}{x^2 + xy - 2y^2}$

34. $\dfrac{a^3 + a^2b + a + b}{2a^3 + 2a} \cdot \dfrac{18a^2}{6a^2 - 6b^2}$

35. $\dfrac{x^2 - y^2}{2x^2 + xy - y^2} \cdot \dfrac{2xy - y^2}{x^3 - y^3}$

36. $\dfrac{2x^2 + xy - y^2}{2x^2 - 3xy + y^2} \cdot \dfrac{x^2 - y^2}{x^2 + 2xy + y^2}$

37. $\dfrac{2xy + y}{3y^2z} \cdot \dfrac{9yz^2}{2xz + z}$

38. $\dfrac{12 - 2x - 2x^2}{2y^2} \cdot \dfrac{x + 3y}{-x^2 + xy - y^2}$

39. $\dfrac{x^3 + y^3}{2y^2} \cdot \dfrac{x + 3y}{-x^2 + xy - y^2}$

40. $\dfrac{20x^3 + 30x^2 - 10x}{x^2 - 2x - 8} \cdot \dfrac{x^2 - 5x + 4}{2x^2 + 3x - 1}$

In Exercises 41–60, divide the rational expressions.

41. $\dfrac{4y^3}{3x^2} \div \dfrac{8y^2}{15x^3}$

42. $\dfrac{6x^4y^2}{5z^3} \div \dfrac{3y^6}{10x^2z}$

43. $\dfrac{2x^2}{3x^2 - 3} \div \dfrac{14}{x + 1}$

44. $\dfrac{3m}{4m - 12} \div \dfrac{m^2 + m}{6 - 2m}$

45. $\dfrac{x - 2}{3x - 9} \div \dfrac{x^2 - 2x}{x - 3}$

46. $\dfrac{5x}{6} \div \dfrac{10x + 15}{6x + 18}$

47. $\dfrac{a^2 - 16}{a + 3} \div \dfrac{2a + 8}{a - 3}$

48. $\dfrac{6y^2}{3y^2 + 5y + 2} \div \dfrac{10y^2 - 6y}{2y^2 - 3y - 5}$

49. $\dfrac{x^2 + x - 6}{x + 5} \div \dfrac{x - 2}{x^2 + 4x - 5}$

50. $\dfrac{x^2 - 12x + 36}{x^2 - x - 42} \div \dfrac{x^2 - 36}{4}$

51. $\dfrac{2x^2 + x - 3}{x^2 + 3x + 2} \div \dfrac{x^2 + 4x - 5}{3x^2 + 5x + 2}$

52. $\dfrac{x^2 - 2x - 3}{2x^2 + x - 1} \div \dfrac{3x^2 - 7x - 6}{6x^2 + x - 2}$

53. $\dfrac{x^2 - 4}{x^3 - 27} \div \dfrac{x^3 - 8}{x^2 - 9}$

54. $\dfrac{5x - x^2}{x^3 - 125} \div \dfrac{x}{x^2 + 4x + 8}$

55. $\dfrac{3h^2 - 3h - 60}{7h^2 - 37h + 10} \div \dfrac{12h^2 + 20h + 3}{2h^2 - h - 6}$

56. $\dfrac{10x^2 + 3x - 1}{3x + 2} \div \dfrac{2x + 1}{9x^2 + 3x - 2}$

57. $\dfrac{x^2 - y^2}{4x + 2y} \div \dfrac{x^2 - 3xy - 4y^2}{2x^2 - 7xy - 4y^2}$

58. $\dfrac{x^3 - y^3}{x^2 - xy + y^2} \div \dfrac{x^2 - y^2}{x^3 + y^3}$

59. $\dfrac{27x^3 - y^3}{4x} \div \dfrac{27x^2 + 9xy + 3y^2}{4x^2 + 16xy}$

60. $\dfrac{16 + 6x - x^2}{3x^2 - 192} \div \dfrac{x^2 + 10x + 16}{x^2 + 16x + 64}$

In Exercises 61–70, perform the indicated operations.

61. $\dfrac{5xy}{x^3} \cdot \dfrac{xy^3}{8y} \div \dfrac{2y^3}{3x^4y}$

62. $\dfrac{7}{x^2} \div \dfrac{9x^2y}{x^3} \cdot \dfrac{4x}{x^8}$

63. $\dfrac{x^2 - 2x - 3}{x - 1} \cdot \dfrac{x + 4}{x^2 + 2x - 15} \cdot \dfrac{x^2 + 4x - 5}{x^2 - x - 2}$

64. $\dfrac{x^2 + 6x + 8}{x^2 + 6x + 5} \div \dfrac{x^2 - 5x - 14}{x^2 + 3x - 10} \cdot \dfrac{x^2 - 6x - 7}{x^2 + x - 6}$

65. $\dfrac{x^2 + 5x + 4}{x - 3} \cdot \dfrac{x^2 - 4x + 3}{x^2 + 6x + 5} \div \dfrac{x^2 - 3x + 2}{x + 5}$

66. $\dfrac{x^2 + x - 2}{x^2 - 5x + 6} \div \dfrac{x^2 + 4x + 3}{x^2 + 2x - 15} \div \dfrac{(x + 2)^2}{x^2 + 3x - 10}$

67. $\dfrac{12x^2 + 8x + 1}{y^2 - 2y - 3} \cdot \dfrac{y^2 - 7y + 12}{6x^2 + 7x + 1} \div \dfrac{4x^2 - 4x - 3}{3x^2 - 2x - 5}$

68. $\dfrac{4a^2 - 64}{2a^2 - 4a} \div \dfrac{a^3 + 4a^2}{5a^2 - 10a} \cdot \dfrac{6a^3 + 4a^2}{4a^2 - 16a}$

69. $\dfrac{x + 2}{7x + 2} \cdot \dfrac{3x + 8}{x^2 + x - 2} \cdot \dfrac{7x^2 - 5x - 2}{5x + 10}$

70. $\dfrac{2x - 12}{4x} \div \dfrac{x^2 - 36}{10x} \div \dfrac{x + 1}{x + 6}$

6.3 Adding and Subtracting Rational Expressions

THINGS TO KNOW

Before working through this section, be sure you are familiar with the following concepts:

VIDEO   ANIMATION   INTERACTIVE

 You Try It
1. Add Polynomials
   (Section 4.2, Objective 4)

You Try It
2. Subtract Polynomials
   (Section 4.2, Objective 5)

You Try It
3. Factor Polynomials Completely
   (Section 5.3, Objective 4)

You Try It
4. Simplify Rational Expressions
   (Section 6.1, Objective 3)

OBJECTIVES

1 Add and Subtract Rational Expressions with Common Denominators

2 Find the Least Common Denominator of Rational Expressions

3 Write Equivalent Rational Expressions

4 Add and Subtract Rational Expressions with Unlike Denominators

....

OBJECTIVE 1 ADD AND SUBTRACT RATIONAL EXPRESSIONS WITH COMMON DENOMINATORS

Recall that when adding or subtracting fractions we need a **common denominator**. Then we add or subtract the **numerators** and keep the common denominator. We follow this general approach:

$$\frac{p}{q} + \frac{r}{q} = \frac{p+r}{q} \quad \text{or} \quad \frac{p}{q} - \frac{r}{q} = \frac{p-r}{q}.$$

For example,

$$\frac{3}{7} + \frac{12}{7} = \frac{3+12}{7} = \frac{15}{7} \quad \text{or} \quad \frac{10}{3} - \frac{8}{3} = \frac{10-8}{3} = \frac{2}{3}.$$

The same is true when working with **rational expressions**. Once the expressions have a common denominator, add or subtract the numerators and keep the common denominator.

Adding and Subtracting Rational Expressions with Common Denominators

If $\dfrac{P}{Q}$ and $\dfrac{R}{Q}$ are rational expressions, then

$$\frac{P}{Q} + \frac{R}{Q} = \frac{P+R}{Q} \quad \text{and} \quad \frac{P}{Q} - \frac{R}{Q} = \frac{P-R}{Q}.$$

My video summary ⊙ **Example 1 Adding and Subtracting Rational Expressions with Common Denominators**

Add or subtract.

a. $\dfrac{4z}{3} + \dfrac{5z}{3}$

b. $\dfrac{3r}{7s^2} - \dfrac{2r}{7s^2}$

Solutions Read through the following, or watch this **video** for complete solutions to both parts.

a. The **rational expressions** have a **common denominator**, so write the sum of the **numerators** over the common denominator and **simplify**.

Begin with the original expression: $\dfrac{4z}{3} + \dfrac{5z}{3}$

Add the numerators: $= \dfrac{4z + 5z}{3}$

$$\text{Combine like terms:} \quad = \frac{9z}{3}$$

$$\text{Divide out the common factor of 3:} \quad = \frac{{}^{3}9z}{3_{1}}$$

$$= 3z$$

b. The rational expressions have a **common denominator**, so write the difference of the **numerators** over the common denominator and **simplify**.

$$\text{Begin with the original expression:} \quad \frac{3r}{7s^2} - \frac{2r}{7s^2}$$

$$\text{Subtract the numerators:} \quad = \frac{3r - 2r}{7s^2}$$

$$\text{Simplify:} \quad = \frac{r}{7s^2}$$

You Try It Work through this You Try It problem.

Work Exercises 1–6 in this eText or in the MyMathLab Study Plan.

⚠ When subtracting rational expressions, it is a good idea to use **grouping symbols** around each numerator that involves more than one term. This will help remind us to subtract each term in the second numerator. An illustration of this point is seen in the next example.

My video summary ▶ **Example 2 Adding and Subtracting Rational Expressions with Common Denominators**

Add or subtract.

a. $\dfrac{2x}{x+3} + \dfrac{4x-1}{x+3}$ 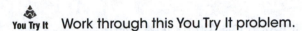 b. $\dfrac{8x-15}{x-7} - \dfrac{5x+6}{x-7}$

Solutions Watch this video, or read the following for complete solutions to both parts.

a. The rational expressions have **common denominators**, so write the sum of the **numerators** over the common denominator and **simplify**.

$$\text{Begin with the original expression:} \quad \frac{2x}{x+3} + \frac{4x-1}{x+3}$$

$$\text{Add the numerators:} \quad = \frac{2x + (4x-1)}{x+3}$$

$$\text{Combine like terms:} \quad = \frac{6x-1}{x+3}$$

b. The rational expressions have **common denominators**, so write the difference of the **numerators** over the common denominator and **simplify**.

Begin with the original expression: $\dfrac{8x - 15}{x - 7} - \dfrac{5x + 6}{x - 7}$

Subtract the numerators: $= \dfrac{(8x - 15) - (5x + 6)}{x - 7}$

Use the distributive property: $= \dfrac{8x - 15 - 5x - 6}{x - 7}$

Combine like terms: $= \dfrac{3x - 21}{x - 7}$

Factor the numerator: $= \dfrac{3(x - 7)}{x - 7}$

Divide out the common factor $x - 7$: $= \dfrac{3(\cancel{x - 7})}{\cancel{x - 7}}$

$= 3$

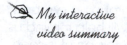 **You Try It** Work through this You Try It problem.

Work Exercises 7–12 in this eText or in the MyMathLab Study Plan.

My interactive video summary 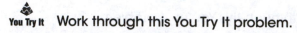 **Example 3 Adding and Subtracting Rational Expressions with Common Denominators**

Add or subtract.

a. $\dfrac{4}{x^2 + 2x - 8} + \dfrac{x}{x^2 + 2x - 8}$

b. $\dfrac{x}{x + 2} - \dfrac{x - 3}{x + 2}$

c. $\dfrac{x^2 - 2}{x - 5} - \dfrac{4x + 3}{x - 5}$

Solutions Try to perform the operations on your own. Remember to check your result for any **common factors**. View the answers, or watch this **interactive video** for complete solutions to all three parts.

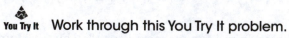 **You Try It** Work through this You Try It problem.

Work Exercises 13–18 in this eText or in the MyMathLab Study Plan.

OBJECTIVE 2 **FIND THE LEAST COMMON DENOMINATOR OF RATIONAL EXPRESSIONS**

We have seen that adding or subtracting **rational expressions** requires a **common denominator**. If the rational expressions have unlike denominators, then a common denominator must be found. Typically, the **least common denominator** (LCD) is used, which is the smallest multiple of all the **denominators** under consideration.

My animation summary ⊛ **Concept Animation** Watch this animation to review the process of finding the LCD of rational numbers (fractions).

A similar process is used to find the LCD of rational expressions.

> **Finding the Least Common Denominator (LCD) of Rational Expressions**
>
> **Step 1** Factor each denominator completely.
>
> **Step 2** List each unique factor from any denominator.
>
> **Step 3** The least common denominator is the product of the unique factors, each raised to a **power** equivalent to the largest number of times that the factor occurs in any denominator.

⚠ Don't forget numerical factors when forming the LCD.

My video summary ▶ **Example 4 Finding the Least Common Denominator of Rational Expressions**

Find the **LCD** of the rational expressions.

a. $\dfrac{7}{10x^3}, \dfrac{3}{5x^2}$ 　　　　b. $\dfrac{x+2}{3x}, \dfrac{x-1}{2x^2+6x}$

Solutions Watch this video, or continue reading for complete solutions to both parts.

a. Follow the three-step process.

　Step 1 $10x^3 = 2 \cdot 5 \cdot x \cdot x \cdot x$

　　　　　$5x^2 = 5 \cdot x \cdot x$

　Step 2 The unique factors are 2, 5, and x.

　Step 3 The factors 2 and 5 each occur at most once, and the factor x occurs at most three times.

$$\text{LCD} = 2 \cdot 5 \cdot x \cdot x \cdot x = 10x^3$$

b. **Step 1** $3x = 3 \cdot x$

　　　　　$2x^2 + 6x = 2 \cdot x(x+3)$

　Step 2 The unique factors are 3, x, 2, and $(x+3)$. Note that both x and $x+3$ are listed because x is not a factor of $x+3$.

　Step 3 The factors 3, x, 2, and $(x+3)$ each occur at most once.

$$\text{LCD} = 3 \cdot 2 \cdot x \cdot (x+3) = 6x(x+3)$$

You Try It Work through this **You Try It** problem.

Work Exercises 19–22 in this eText or in the MyMathLab Study Plan.

Example 5 Finding the Least Common Denominator of Rational Expressions

Find the **LCD** of the rational expressions.

a. $\dfrac{z^2}{6-z}, \dfrac{9}{2z-12}$ 　　　　b. $\dfrac{y+2}{y^2+2y-3}, \dfrac{2y}{y^2+5y+6}$

Solutions

a. Follow the **three-step process.**

Step 1 $6 - z = -1 \cdot (z - 6)$
$2z - 12 = 2 \cdot (z - 6)$

Step 2 The unique factors are -1, $(z - 6)$, and 2. Note that we factored out a -1 in the first denominator. Do this when the **leading coefficient** is negative to avoid including **opposite factors** in the LCD. In this case, include -1 and $z - 6$ instead of $z - 6$ and $6 - z$.

Step 3 The factors -1, $(z - 6)$, and 2 each occur at most once.

$$LCD = -1 \cdot (z - 6) \cdot 2 = -2(z - 6)$$

My video summary ▶ **b. Step 1** $y^2 + 2y - 3 = (y + 3)(y - 1)$

$y^2 + 5y + 6 = (y + 3)(y + 2)$

Try to finish this problem on your own. View the **answer**, or watch this **video** for a complete solution to part b.

You Try It Work through this You Try It problem.

Work Exercises 23–28 in this eText or in the MyMathLab Study Plan.

My interactive video summary ✎ **Example 6 Finding the Least Common Denominator of Rational Expressions**

Find the LCD of the rational expressions.

a. $\dfrac{4x}{10x^2 - 7x - 12}$, $\dfrac{2x - 3}{5x^2 - 11x - 12}$

b. $\dfrac{10 - x}{6x^2 + 5x + 1}$, $\dfrac{-4}{9x^2 + 6x + 1}$, $\dfrac{x^2 - 7x}{10x^2 - x - 3}$

Solutions Try to find the LCDs on your own. View the **answers**, or watch this **interactive video** for complete solutions to both parts.

You Try It Work through this You Try It problem.

Work Exercises 29–36 in this eText or in the MyMathLab Study Plan.

OBJECTIVE 3 WRITE EQUIVALENT RATIONAL EXPRESSIONS

To add or subtract rational expressions with unlike denominators, first write each expression as an **equivalent rational expression** using the LCD. To do this, multiply the original expression by 1, but in such a way as to obtain the desired LCD. First, ask the question,

"What should we multiply the original denominator by to get the LCD?"

Then multiply both the **numerator** and the denominator by the answer.

Writing Equivalent Rational Expressions

To write an equivalent rational expression with a desired denominator:

Step 1 Factor the given denominator and the desired denominator.

Step 2 Determine the missing factor(s) from the given denominator that must be multiplied to get the desired denominator.

Step 3 Multiply the numerator and denominator of the given rational expression by the missing factor(s) from Step 2.

TIP The order of the factors in either the numerator or the denominator does not matter. For example,

$$\frac{(x+1)(x-4)}{(x+3)(2x-1)} = \frac{(x-4)(x+1)}{(2x-1)(x+3)}.$$

 My video summary ▶ **Example 7 Writing Equivalent Rational Expressions**

Write each rational expression as an equivalent rational expression with the desired denominator.

a. $\dfrac{3}{2x} = \dfrac{}{10x^3}$

b. $\dfrac{x+2}{3x+15} = \dfrac{}{3(x-1)(x+5)}$

Solutions Read the following, or watch this **video** for complete solutions to both parts. Follow the **three-step process**.

a. Step 1 $2x = 2 \cdot x$
$10x^3 = 2 \cdot 5 \cdot x \cdot x \cdot x$

Step 2 The given denominator is missing $5 \cdot x \cdot x$. What do we multiply $2x$ by to get $10x^3$? The answer is $5 \cdot x \cdot x = 5x^2$.

Step 3 $\dfrac{3}{2x} = \dfrac{3}{2x} \cdot \underbrace{\dfrac{5x^2}{5x^2}}_{1} = \dfrac{15x^2}{10x^3}$

So, $\dfrac{3}{2x} = \dfrac{15x^2}{10x^3}$.

b. Step 1 $3x + 15 = 3 \cdot (x+5)$
$3(x-1)(x+5)$ is already factored.

Step 2 The given denominator is missing $(x-1)$. What do we multiply $3x + 15$ by to get $3(x-1)(x+5)$? The answer is $(x-1)$.

Step 3 $\dfrac{x+2}{3x+15} = \dfrac{x+2}{3(x+5)} \cdot \underbrace{\dfrac{(x-1)}{(x-1)}}_{1} = \dfrac{(x+2)(x-1)}{3(x+5)(x-1)}$ or $\underbrace{\dfrac{(x+2)(x-1)}{3(x-1)(x+5)}}_{\substack{\text{Commutative}\\ \text{property of}\\ \text{multiplication}}}$

So, $\dfrac{x+2}{3x+15} = \dfrac{(x+2)(x-1)}{3(x-1)(x+5)}$ or $\dfrac{x^2+x-2}{3(x-1)(x+5)}$.

You Try It   Work through this You Try It problem.

Work Exercises 37–42 in this eText or in the MyMathLab Study Plan.

Example 8 Writing Equivalent Rational Expressions

Write each rational expression as an equivalent rational expression with the desired denominator.

a. $\dfrac{-7}{1 - 4y} = \dfrac{}{8y^2 - 2y}$
b. $\dfrac{5z}{z^2 + z - 6} = \dfrac{}{(z - 4)(z - 2)(z + 3)}$

Solutions

a. **Step 1** $1 - 4y = -1 \cdot (4y - 1)$
$8y^2 - 2y = 2 \cdot y \cdot (4y - 1)$

Step 2 What do we multiply $1 - 4y$ by to get $8y^2 - 2y$? The answer is $-2 \cdot y = -2y$. Note that -2 is used because $-1 \cdot -2 = 2$.

Step 3 $\dfrac{-7}{1 - 4y} = \dfrac{-7}{-1(4y - 1)} \cdot \underbrace{\dfrac{-2y}{-2y}}_{1} = \dfrac{14y}{2y(4y - 1)}$

My video summary ▶ b. Try to work this problem on your own. View the **answer**, or watch this **video** for a complete solution.

You Try It   Work through this You Try It problem.

Work Exercises 43–46 in this eText or in the MyMathLab Study Plan.

OBJECTIVE 4   ADD AND SUBTRACT RATIONAL EXPRESSIONS WITH UNLIKE DENOMINATORS

To add or subtract rational expressions with unlike denominators, first, we write each fraction as an **equivalent expression** using the LCD. Then we can add or subtract the numerators, keeping the common denominator.

Adding and Subtracting Rational Expressions with Unlike Denominators

Step 1 Find the LCD for all expressions being added or subtracted.

Step 2 Write equivalent expressions for each term using the LCD as the denominator.

Step 3 Add/subtract the numerators, but keep the denominator the same (the LCD).

Step 4 Simplify if possible.

 We focus on the LCD because it is the easiest **common denominator** to use. Any common denominator will allow us to add or subtract rational expressions.

Example 9 Adding and Subtracting Rational Expressions

Perform the indicated operations and simplify.

a. $\dfrac{7}{6x} + \dfrac{3}{2x^3}$

b. $\dfrac{3x}{x-3} - \dfrac{x-2}{x+3}$

Solutions Follow the four-step process.

a. **Step 1**

$$6x = 2 \cdot 3 \cdot x$$

$$2x^3 = 2 \cdot x \cdot x \cdot x$$

The **unique factors** are 2, 3, and x. Include each factor the largest number of times that it occurs in any **denominator**.

$$\text{LCD} = 2 \cdot 3 \cdot x \cdot x \cdot x = 6x^3$$

Step 2 In the first expression, multiplying the denominator by x^2 gives the LCD. Multiply both the numerator and denominator by x^2 to obtain an **equivalent expression**. In the second expression, multiplying the denominator by 3 gives the LCD. Multiply both the numerator and denominator by 3 to obtain an equivalent expression.

The overall expression becomes

$$\frac{7}{6x} + \frac{3}{2x^3} = \frac{7}{6x} \cdot \frac{x^2}{x^2} + \frac{3}{2x^3} \cdot \frac{3}{3}$$

$$= \frac{7x^2}{6x^3} + \frac{9}{6x^3}.$$

Step 3 Now that there is a **common denominator**, add:

$$\frac{7x^2}{6x^3} + \frac{9}{6x^3} = \frac{7x^2 + 9}{6x^3}$$

Step 4 The expression cannot be simplified further, so

$$\frac{7}{6x} + \frac{3}{2x^3} = \frac{7x^2 + 9}{6x^3}.$$

My video summary ▶ b. **Step 1** The denominators cannot be factored further.

The **unique factors** are $x - 3$ and $x + 3$. Include each factor one time.

$$\text{LCD} = (x-3)(x+3)$$

Step 2 In the first expression, obtain an **equivalent expression** by multiplying both the **numerator** and denominator by $x + 3$. In the second expression, obtain an equivalent expression by multiplying both the numerator and denominator by $x - 3$.

$$\frac{3x}{x-3} - \frac{x-2}{x+3} = \frac{3x}{x-3} \cdot \frac{x+3}{x+3} - \frac{x-2}{x+3} \cdot \frac{x-3}{x-3}$$

$$= \frac{3x(x+3)}{(x-3)(x+3)} - \frac{(x-2)(x-3)}{(x-3)(x+3)}.$$

Carry out the subtraction on your own and **simplify** if necessary. View the answer, or watch this **video** for a complete solution.

You Try It Work through this You Try It problem.

Work Exercises 47–52 in this eText or in the MyMathLab Study Plan.

Example 10 Adding and Subtracting Rational Expressions

Perform the indicated operations and simplify.

a. $\dfrac{z + 2}{3z} - \dfrac{5}{3z + 12}$

b. $\dfrac{5}{4m - 12} + \dfrac{3}{2m}$

Solutions

a. Step 1

$$3z = 3 \cdot z$$

$$3z + 12 = 3(z + 4)$$

The **unique factors** are 3, z, and $(z + 4)$. Include each factor the largest number of times that it occurs in any **denominator**.

$$\text{LCD} = 3 \cdot z \cdot (z + 4) = 3z(z + 4).$$

Step 2 In the first expression, obtain an **equivalent expression** by multiplying both the numerator and denominator by $z + 4$. In the second expression, obtain an equivalent expression by multiplying both the numerator and denominator by z. Then

$$\frac{z + 2}{3z} - \frac{5}{3z + 12} = \frac{z + 2}{3z} \cdot \frac{z + 4}{z + 4} - \frac{5}{3(z + 4)} \cdot \frac{z}{z}$$

$$= \frac{z^2 + 6z + 8}{3z(z + 4)} - \frac{5z}{3z(z + 4)}$$

Step 3 Now that there is a **common denominator**, subtract:

$$\text{Subtract:} \quad = \frac{z^2 + 6z + 8 - 5z}{3z(z + 4)}$$

$$\text{Simplify:} \quad = \frac{z^2 + z + 8}{3z(z + 4)}$$

Step 4 The expression cannot be simplified further, so

$$\frac{z + 2}{3z} - \frac{5}{3z + 12} = \frac{z^2 + z + 8}{3z(z + 4)}.$$

My video summary ▶ **b.** Write equivalent expressions using the LCD and carry out the addition on your own. **Simplify** if necessary, then view the **answer**, or watch this **video** for a complete solution.

You Try It Work through this You Try It problem.

Work Exercises 53–56 in this eText or in the MyMathLab Study Plan.

Example 11 Adding and Subtracting Rational Expressions

Perform the indicated operations and simplify.

a. $2 + \dfrac{4}{x - 5}$

b. $\dfrac{x^2 - 2}{x^2 + 6x + 8} - \dfrac{x - 3}{x + 4}$

Solutions Follow the four-step process.

a. **Step 1** Note that the first term is a rational expression since it can be written

as $2 = \dfrac{2}{1}$. The **LCD** is $x - 5$.

Step 2 In the first expression, obtain an **equivalent expression** by multiplying both the numerator and denominator by $x - 5$.

$$2 + \frac{4}{x-5} = \frac{2}{1} + \frac{4}{x-5}$$

$$= \frac{2}{1} \cdot \frac{x-5}{x-5} + \frac{4}{x-5}$$

$$= \frac{2(x-5)}{x-5} + \frac{4}{x-5}.$$

Step 3 Now that there is a **common denominator**, add:

$$\text{Add:} \quad = \frac{2(x-5)+4}{x-5}$$

$$\text{Distribute:} \quad = \frac{2x-10+4}{x-5}$$

$$\text{Simplify:} \quad = \frac{2x-6}{x-5}$$

$$= \frac{2(x-3)}{x-5}$$

Step 4 The expression cannot be simplified further, so

$$2 + \frac{4}{x-5} = \frac{2(x-3)}{x-5}.$$

My video summary ▶ b. **Step 1** Factor the denominators and determine the LCD. The second denominator is **prime**, but the first denominator can be factored as $x^2 + 6x + 8 = (x+4)(x+2)$.

The LCD is $(x+4)(x+2)$.

Step 2 The first expression already contains the LCD. In the second expression, obtain an **equivalent expression** by multiplying both the **numerator** and **denominator** by $x + 2$.

Doing this gives

$$\frac{x^2-2}{x^2+6x+8} - \frac{x-3}{x+4} = \frac{x^2-2}{(x+4)(x+2)} - \frac{x-3}{x+4} \cdot \frac{x+2}{x+2}$$

$$= \frac{x^2-2}{(x+4)(x+2)} - \frac{(x-3)(x+2)}{(x+4)(x+2)}.$$

Try to carry out the subtraction on your own and **simplify** if necessary. View the answer, or watch this **video** for a complete solution.

You Try It Work through this **You Try It** problem.

Work Exercises 57–62 in this eText or in the MyMathLab Study Plan.

Example 12 Adding and Subtracting Rational Expressions with Unlike Denominators

Perform the indicated operations and simplify.

a. $\dfrac{x+7}{x^2-9} + \dfrac{3}{x+3}$

b. $\dfrac{x+1}{2x^2+5x-3} - \dfrac{x}{2x^2+3x-2}$

Solutions Perform the operations on your own and simplify. View the **answers**, or watch this **interactive video** to see complete solutions.

You Try It Work through this You Try It problem.

Work Exercises 63–72 in this eText or in the MyMathLab Study Plan.

If adding or subtracting rational expressions whose denominators are **opposites**, find a **common denominator** by multiplying the numerator and denominator of either rational expression (but not both) by -1.

▶ **Example 13 Adding and Subtracting Rational Expressions with Unlike Denominators**

Perform the indicated operation and simplify.

$$\dfrac{2y}{y-5} + \dfrac{y-1}{5-y}$$

Solution Read the following, or watch this **video** for a complete solution.

The denominators are opposites: $\dfrac{2y}{y-5} + \dfrac{y-1}{5-y}$

Multiply the numerator and denominator of the second rational expression by -1: $= \dfrac{2y}{y-5} + \dfrac{(-1)}{(-1)}\cdot\dfrac{y-1}{5-y}$

Distribute: $= \dfrac{2y}{y-5} + \dfrac{-y+1}{-5+y}$

Rewrite the denominator in second expression: $= \dfrac{2y}{y-5} + \dfrac{-y+1}{y-5}$

Add: $= \dfrac{2y-y+1}{y-5} = \dfrac{y+1}{y-5}$

You Try It Work through this You Try It problem.

Work Exercises 73 and 74 in this eText or in the MyMathLab Study Plan.

▶ **Example 14 Adding and Subtracting Rational Expressions with Unlike Denominators**

Perform the indicated operations and simplify.

$$\dfrac{x+1}{x^2-6x+9} + \dfrac{3}{x-3} - \dfrac{6}{x^2-9}$$

Solution Following the four-step process, start by finding the **LCD** for all the terms.

$x^2 - 6x + 9 = (x - 3)(x - 3)$; $x - 3$ is prime; $x^2 - 9 = (x + 3)(x - 3)$

The LCD is $(x - 3)(x - 3)(x + 3)$.

Next, rewrite each term as an equivalent expression using the LCD.

$$\frac{x + 1}{x^2 - 6x + 9} + \frac{3}{x - 3} - \frac{6}{x^2 - 9}$$

$$= \frac{x + 1}{(x - 3)(x - 3)} \cdot \frac{x + 3}{x + 3} + \frac{3}{x - 3} \cdot \frac{(x - 3)(x + 3)}{(x - 3)(x + 3)} - \frac{6}{(x + 3)(x - 3)} \cdot \frac{x - 3}{x - 3}$$

$$= \frac{(x + 1)(x + 3)}{(x - 3)(x - 3)(x + 3)} + \frac{3(x - 3)(x + 3)}{(x - 3)(x - 3)(x + 3)} - \frac{6(x - 3)}{(x - 3)(x - 3)(x + 3)}$$

Now combine the three numerators and keep the **common denominator**.

$$= \frac{(x + 1)(x + 3) + 3(x - 3)(x + 3) - 6(x - 3)}{(x - 3)(x - 3)(x + 3)}$$

Try to finish simplifying the expression on your own. View the **answer**, or watch this video for a complete solution.

You Try It Work through this You Try It problem.

Work Exercises 75–80 in this eText or in the MyMathLab Study Plan.

My video summary ▶ **Example 15** Adding and Subtracting Rational Expressions with Unlike Denominators

Perform the indicated operation and simplify.

$$\frac{4x + 2y}{x^2 + 4xy - 5y^2} - \frac{3}{x + 5y}$$

Solution Try to work this problem on your own. View the **answer**, or watch this video for a complete solution.

You Try It Work through this You Try It problem.

Work Exercises 81 and 82 in this eText or in the MyMathLab Study Plan.

6.3 Exercises

In Exercises 1–18, add or subtract as indicated.

1. $\dfrac{4x}{7} + \dfrac{2x}{7}$

2. $\dfrac{8}{7y^3} + \dfrac{6}{7y^3}$

3. $\dfrac{a^2 + 1}{5c^2} + \dfrac{b^2 + 2}{5c^2}$

4. $\dfrac{5w}{3ab^2} - \dfrac{2w}{3ab^2}$

5. $\dfrac{5x + 1}{9a^3} - \dfrac{2x + 1}{9a^3}$

6. $\dfrac{2x - 3}{3x} - \dfrac{2x + 9}{3x}$

7. $\dfrac{2x}{8y^3 + 2} - \dfrac{x}{8y^3 + 2}$

8. $\dfrac{x + 1}{2x + 3} + \dfrac{2x + 1}{2x + 3}$

9. $\dfrac{x^2 - 8}{x - 3} + \dfrac{2x - 7}{x - 3}$

10. $\dfrac{5x + 1}{3x + 2} + \dfrac{x + 3}{3x + 2}$

11. $\dfrac{7x + 3}{5x - 1} - \dfrac{2x + 5}{5x - 1}$

12. $\dfrac{x^2}{2x + 1} + \dfrac{x^2 + x}{2x + 1}$

13. $\dfrac{x^2 + 9}{x^2 - 9} + \dfrac{6x}{x^2 - 9}$

14. $\dfrac{2}{x^2 + 4x + 3} + \dfrac{2x}{x^2 + 4x + 3}$

15. $\dfrac{x}{x^2 - 4} - \dfrac{2}{x^2 - 4}$

16. $\dfrac{x^2 + x + 6}{x^2 - x - 6} - \dfrac{3x + 9}{x^2 - x - 6}$

17. $\dfrac{2x}{x^2 - 3x - 28} - \dfrac{14}{x^2 - 3x - 28}$

18. $\dfrac{n^2}{n^2 - 64} + \dfrac{64 - 16n}{n^2 - 64}$

In Exercises 19–36, find the LCD for the given rational expressions.

19. $\dfrac{5}{6x^5}, \dfrac{4}{9x^2}$

20. $\dfrac{6}{5a^3b^4}, \dfrac{7}{15a^2b^8}$

21. $\dfrac{3}{7a^3c^3}, \dfrac{5}{14b^5c^2}$

22. $\dfrac{x + 1}{4x^2}, \dfrac{x - 2}{3x^2 + 9x}$

23. $\dfrac{3x - 5}{6x^2 - 3x}, \dfrac{2x + 1}{12x^2 + 4x}$

24. $\dfrac{5}{3x^3 + 9x^2 + 6x}, \dfrac{7x}{2x^3 - 2x}$

25. $\dfrac{x}{3 - x}, \dfrac{2x + 1}{x^2 - x - 6}$

26. $\dfrac{3x + 1}{x^2 - 4x - 5}, \dfrac{x + 7}{2 + x - x^2}$

27. $\dfrac{2x}{3 + 2x - x^2}, \dfrac{5x - 1}{9 - x^2}$

28. $\dfrac{y - 5}{y^2 - y - 2}, \dfrac{y + 5}{y^2 + y - 6}$

29. $\dfrac{2y - 1}{y^2 - 9}, \dfrac{1 - 5y}{y^2 + 6y + 9}$

30. $\dfrac{2x}{4 - x^2}, \dfrac{1 - x}{6 + x - x^2}$

31. $\dfrac{7 - x}{2x^2 - 5x - 3}, \dfrac{2x + 3}{3x^2 - 11x + 6}$

32. $\dfrac{3x^2 + 2x + 1}{10x^2 - 15x}, \dfrac{2x^2 - 7x}{6x^2 - 13x + 6}$

33. $\dfrac{5x + 1}{6x^3 + 26x^2 - 20x}, \dfrac{5x - 1}{6x^4 + 33x^3 + 15x^2}$

34. $\dfrac{7}{x^2 - 16}, \dfrac{x - 1}{3x}, \dfrac{2x}{20 - 5x}$

35. $\dfrac{7 + x}{4 - x^2}, \dfrac{6 + x}{x^2 + x - 2}, \dfrac{5 + x}{x^2 - x - 2}$

36. $\dfrac{x + 1}{2x^2 + x - 3}, \dfrac{x + 2}{3x^2 + 13x - 10}, \dfrac{x + 3}{6x^2 + 5x - 6}$

In Exercises 37–46, write each rational expression as an equivalent rational expression with the desired denominator.

37. $\dfrac{2}{3x} = \dfrac{}{15x^4}$

38. $\dfrac{3}{5x^2y^3} = \dfrac{}{15x^4y^5}$

39. $\dfrac{2b}{3a^2c^3} = \dfrac{}{21a^5b^3c^3}$

40. $\dfrac{x+1}{2x+6} = \dfrac{}{2(x-1)(x+3)}$

41. $\dfrac{7}{x^2-x-6} = \dfrac{}{(x+2)(x-2)(x+3)(x-3)}$

42. $\dfrac{3x-5}{2x+3} = \dfrac{}{(2x+3)(x+2)}$

43. $\dfrac{5}{2-x}, \dfrac{}{(x^2-4)(x^2+x)}$

44. $\dfrac{x+1}{9x-3x^2} = \dfrac{}{3x(x-1)(x-2)(x-3)}$

45. $\dfrac{2x+1}{x^2+x-2} = \dfrac{}{(x-1)(x+1)(x+2)}$

46. $\dfrac{3x-5}{2x^2-4x-6} = \dfrac{}{10(x-3)(x-2)(x+1)}$

In Exercises 47–82, add or subtract. Simplify if possible.

47. $\dfrac{x+5}{5x^2} + \dfrac{3}{10x}$

48. $\dfrac{2}{3x} + \dfrac{3}{2x}$

49. $\dfrac{5x}{x+2} + \dfrac{7}{2x+1}$

50. $\dfrac{x+2}{x-2} - \dfrac{x-2}{x+2}$

51. $\dfrac{2x+1}{x} - \dfrac{2x}{x+1}$

52. $\dfrac{8}{x-1} - \dfrac{6}{x+8}$

53. $\dfrac{x+1}{2x} + \dfrac{3x}{5x-10}$

54. $\dfrac{2m}{3m+3} + \dfrac{m+1}{2m}$

55. $\dfrac{x+2}{2x} - \dfrac{3}{2x+10}$

56. $\dfrac{3}{2x+2} - \dfrac{2}{3x+3}$

57. $3 + \dfrac{x}{x-2}$

58. $\dfrac{3x}{x+1} + 2$

59. $\dfrac{2(x+11)}{2x^2-x-10} + \dfrac{x+4}{x+2}$

60. $\dfrac{3x-4}{x+3} - \dfrac{6-24x}{x^2-9}$

61. $\dfrac{x^2-3}{x^2-5x+6} - \dfrac{x-3}{x-2}$

62. $\dfrac{x}{2} - \dfrac{2}{x}$

63. $\dfrac{x-1}{x^2-4} + \dfrac{5}{x-2}$

64. $\dfrac{4}{x^2-4} + \dfrac{2x}{x+2}$

65. $\dfrac{1-12x}{2x^2-5x-3} + \dfrac{x+2}{x-3}$

66. $\dfrac{3x-4}{x+3} - \dfrac{6-24x}{x^2-9}$

67. $\dfrac{2x}{x^2-x-2} - \dfrac{x}{x^2-1}$

68. $\dfrac{2x-1}{x+1} + \dfrac{3(3x+1)}{x^2-1}$

69. $\dfrac{8}{x-2} - \dfrac{3x+26}{x^2-4}$

70. $\dfrac{x-8}{x^2+12x+27} + \dfrac{x-9}{x^2-9}$

71. $\dfrac{6x}{x^2+4x-12} - \dfrac{x}{x^2-36}$

72. $\dfrac{y-8}{2y^2+9y+7} - \dfrac{y+9}{2y^2-3y-35}$

73. $\dfrac{3x}{x-1} + \dfrac{2x+3}{1-x}$

74. $\dfrac{x^2}{x-3} + \dfrac{9}{3-x}$

75. $\dfrac{1}{x^2+x} + \dfrac{1}{x^2-x} + \dfrac{1}{x^2-1}$

76. $\dfrac{1}{x} + \dfrac{1}{x+1} - \dfrac{1}{x-1}$

77. $\dfrac{x-3}{x^2+12x+36} + \dfrac{1}{x+6} - \dfrac{2x+3}{2x^2+3x-54}$

78. $\dfrac{1}{x+1} + \dfrac{2}{x+2} - \dfrac{3}{x+3}$

79. $\dfrac{1}{x+2} + \dfrac{1}{x-2} - \dfrac{4}{x^2-4}$

80. $\dfrac{5}{y} - \dfrac{5}{y-3} + \dfrac{16}{(y-3)^2}$

81. $\dfrac{9}{x-y} - \dfrac{x+3y}{x^2-y^2}$

82. $\dfrac{6x}{2x^2-6xy+4y^2} - \dfrac{3y}{x^2-3xy+2y^2}$

6.4 Complex Rational Expressions

THINGS TO KNOW

Before working through this section, be sure you are familiar with the following concepts:

| | | VIDEO | ANIMATION | INTERACTIVE |

 You Try It 1. Use the Negative-Power Rule (Section 4.1, Objective 4)

 You Try It 2. Simplify Rational Expressions (Section 6.1, Objective 3)   ▶

 You Try It 3. Multiply Rational Expressions (Section 6.2, Objective 1)   ▶

You Try It 4. Divide Rational Expressions (Section 6.2, Objective 2)   ▶

You Try It 5. Find the Least Common Denominator of Rational Expressions (Section 6.3, Objective 2)   ▶

You Try It 6. Add and Subtract Rational Expressions with Unlike Denominators (Section 6.3, Objective 4)   ▶

OBJECTIVES

1 Simplify Complex Rational Expressions by First Simplifying the Numerator and Denominator

2 Simplify Complex Rational Expressions by Multiplying by a Common Denominator

....

OBJECTIVE 1   SIMPLIFY COMPLEX RATIONAL EXPRESSIONS BY FIRST SIMPLIFYING THE NUMERATOR AND DENOMINATOR

Sometimes the **numerator** and/or **denominator** of a **rational expression** will contain one or more rational expressions. In this section, we learn to simplify such *complex rational expressions.*

Definition   Complex Rational Expression

A **complex rational expression**, or **complex fraction**, is a rational expression in which the numerator and/or denominator contain(s) rational expressions.

Some examples of **complex rational expressions** are

$$\frac{\dfrac{2}{9x}}{\dfrac{5}{6xy}}, \quad \frac{\dfrac{x+2}{x-4}}{\dfrac{x^2-4}{x+1}}, \quad \frac{\dfrac{1}{y}+\dfrac{y+1}{y-1}}{\dfrac{1}{y}-\dfrac{y-1}{y+1}}, \quad \frac{\dfrac{1}{x}+\dfrac{1}{y}}{z}, \quad \text{and} \quad \frac{z^{-2}+z^{-1}}{z^{-1}-z^{-2}}.$$

The rational expressions within the numerator and denominator are called **minor rational expressions** or **minor fractions**. The numerator and denominator of the complex rational expression are separated by the **main fraction bar**.

$$\text{Minor rational expressions} \quad \begin{cases} \nearrow \dfrac{x+2}{x-4} \\ \searrow \dfrac{x^2-4}{x+1} \end{cases}$$

← Numerator of the complex rational expression
← Main fraction bar
← Denominator of the complex rational expression

To **simplify a complex rational expression** means to rewrite it in the form $\dfrac{P}{Q}, Q \neq 0$,

where P and Q are **polynomials** with no **common factors**. In this section, two methods are shown for simplifying complex rational expressions. These are called *Method I* and *Method II*. Method I results from recognizing that the **main fraction bar** is a division symbol. So, the **numerator** is divided by the **denominator**.

My video summary ▶ **Example 1 Simplifying a Complex Rational Expression**

Simplify $\dfrac{\dfrac{2}{9x}}{\dfrac{5}{6xy}}$.

Solution Read the following, or watch this video for a complete solution. Divide the minor rational expression from the **numerator** by the minor rational expression from the **denominator**. To do this, multiply the numerator by the **reciprocal** of the denominator.

Begin with the original expression:
$$\frac{\dfrac{2}{9x}}{\dfrac{5}{6xy}} = \frac{2}{9x} \div \frac{5}{6xy}$$

Change to multiplication by the reciprocal:
$$= \frac{2}{9x} \cdot \frac{6xy}{5}$$

Factor:
$$= \frac{2}{3\cdot3\cdot x} \cdot \frac{2\cdot3\cdot x\cdot y}{5}$$

Divide out common factors:
$$= \frac{2}{\cancel{3}\cdot3\cdot \cancel{x}} \cdot \frac{2\cdot\cancel{3}\cdot\cancel{x}\cdot y}{5}$$

Multiply the remaining factors:
$$= \frac{4y}{15}$$

You Try It Work through this You Try It problem.

Work Exercises 1–3 in this eText or in the MyMathLab Study Plan.

In addition to acting as a division symbol, the main fraction bar of a complex fraction also serves as a grouping symbol. If the numerator and denominator are not each written as single rational expressions, then they must be simplified as such before dividing.

Method I for Simplifying Complex Rational Expressions

Step 1 Simplify the expression in the numerator into a single rational expression.

Step 2 Simplify the expression in the denominator into a single rational expression.

Step 3 Divide the expression in the numerator by the expression in the denominator. To do this, multiply the minor rational expression in the numerator by the reciprocal of the minor rational expression in the denominator. Simplify if possible.

Example 2 Simplifying a Complex Rational Expression Using Method I

Use Method I to simplify each complex rational expression.

a. $\dfrac{\dfrac{1}{3} - \dfrac{1}{x}}{\dfrac{1}{9} - \dfrac{1}{x^2}}$

b. $\dfrac{4 - \dfrac{5}{x-1}}{\dfrac{6}{x-1} - 7}$

Solutions

a. **Step 1** The LCD for the minor fractions $\dfrac{1}{3}$ and $-\dfrac{1}{x}$ in the numerator is $3x$.

$$\frac{1}{3} - \frac{1}{x} = \frac{1}{3} \cdot \frac{x}{x} - \frac{1}{x} \cdot \frac{3}{3} = \frac{x}{3x} - \frac{3}{3x} = \frac{x-3}{3x}$$

Step 2 The LCD for the minor fractions $\dfrac{1}{9}$ and $-\dfrac{1}{x^2}$ in the denominator is $9x^2$.

$$\frac{1}{9} - \frac{1}{x^2} = \frac{1}{9} \cdot \frac{x^2}{x^2} - \frac{1}{x^2} \cdot \frac{9}{9} = \frac{x^2}{9x^2} - \frac{9}{9x^2} = \frac{x^2-9}{9x^2}$$

Step 3 Substitute the simplified rational expressions from Steps 1 and 2 for the numerator and denominator of the complex fraction, and then divide.

$$\frac{\dfrac{1}{3} - \dfrac{1}{x}}{\dfrac{1}{9} - \dfrac{1}{x^2}} = \frac{\dfrac{x-3}{3x}}{\dfrac{x^2-9}{9x^2}} = \frac{x-3}{3x} \div \frac{x^2-9}{9x^2}$$

Change to multiplication by the reciprocal: $= \dfrac{x-3}{3x} \cdot \dfrac{9x^2}{x^2-9}$

Factor: $= \dfrac{x-3}{3x} \cdot \dfrac{3 \cdot 3 \cdot x \cdot x}{(x+3)(x-3)}$

Divide out common factors: $= \dfrac{\cancel{x-3}}{3x} \cdot \dfrac{3 \cdot 3 \cdot x \cdot x}{(x+3)\cancel{(x-3)}}$

Multiply remaining factors: $= \dfrac{3x}{x+3}$

My video summary ▶ **b.** **Step 1** The LCD for the numerator is $x-1$.

$$4 - \frac{5}{x-1} = 4 \cdot \frac{x-1}{x-1} - \frac{5}{x-1} = \frac{4(x-1)-5}{x-1} = \frac{4x-4-5}{x-1} = \frac{4x-9}{x-1}$$

Step 2 The LCD for the denominator is $x-1$.

$$\frac{6}{x-1} - 7 = \frac{6}{x-1} - 7 \cdot \frac{x-1}{x-1} = \frac{6-7(x-1)}{x-1} = \frac{6-7x+7}{x-1} = \frac{13-7x}{x-1}$$

Step 3 Try to finish simplifying the complex rational expression on your own. View the **answer**, or watch this **video** for a complete solution to part b.

You Try It Work through this **You Try It** problem.

Work Exercises 4–6 in this eText or in the MyMathLab Study Plan.

OBJECTIVE 2 SIMPLIFY COMPLEX RATIONAL EXPRESSIONS BY MULTIPLYING BY A COMMON DENOMINATOR

In **Section 6.3**, we saw that multiplying the **numerator** and **denominator** of a rational expression by the same nonzero expression resulted in an equivalent expression. For example,

$$\frac{3}{4} = \frac{3}{4} \cdot \frac{7}{7} = \frac{21}{28} \text{ because } \frac{7}{7} = 1.$$

The second method for simplifying complex rational expressions is based on this same concept. Let's take another look at the complex fraction from **Example 1**.

My video summary ▶ **Example 3** Simplifying a Complex Rational Expression

Simplify $\dfrac{\dfrac{2}{9x}}{\dfrac{5}{6xy}}$.

Solution Read the following, or watch this **video** for a complete solution. Multiply the numerator and denominator of the complex rational expression by $18xy$, the **LCD** of $\dfrac{2}{9x}$ and $\dfrac{5}{6xy}$.

Begin with the original expression: $\dfrac{\dfrac{2}{9x}}{\dfrac{5}{6xy}}$

Multiply the numerator and denominator by $18xy$: $= \dfrac{\dfrac{2}{9x} \cdot 18xy}{\dfrac{5}{6xy} \cdot 18xy}$

Divide out common factors: $= \dfrac{\dfrac{2}{\cancel{9x}} \cdot \overset{2}{\cancel{18xy}}}{\dfrac{5}{\cancel{6xy}} \cdot \overset{3}{\cancel{18xy}}}$

Rewrite the remaining factors: $= \dfrac{2 \cdot 2 \cdot y}{5 \cdot 3}$

Multiply: $= \dfrac{4y}{15}$

You Try It Work through this You Try It problem.

Work Exercises 7–9 in this eText or in the MyMathLab Study Plan.

In **Example 3**, we multiplied the **numerator** and **denominator** of the **complex rational expression** by the **LCD** of its **minor rational expressions**. Multiplying by the LCD is the foundation of Method II for simplifying complex rational expressions.

Method II for Simplifying Complex Rational Expressions

Step 1 Determine the LCD of all the minor rational expressions within the complex rational expression.

Step 2 Multiply the numerator and denominator of the complex rational expression by the LCD from Step 1.

Step 3 Simplify.

For comparison, let's revisit the complex rational expressions from **Example 2**.

Example 4 Simplifying a Complex Rational Expression by Using Method II

Use **Method II** to simplify each complex rational expression.

a. $\dfrac{\dfrac{1}{3} - \dfrac{1}{x}}{\dfrac{1}{9} - \dfrac{1}{x^2}}$

b. $\dfrac{4 - \dfrac{5}{x-1}}{\dfrac{6}{x-1} - 7}$

Solutions

a. Step 1 The denominators of all the minor fractions are 3, x, 9, and x^2, so the LCD is $9x^2$.

Step 2 Multiply the **numerator** and denominator by $9x^2$.

$$\frac{\dfrac{1}{3} - \dfrac{1}{x}}{\dfrac{1}{9} - \dfrac{1}{x^2}} \cdot \frac{9x^2}{9x^2}$$

Step 3 Use the distributive property:

$$= \frac{\dfrac{1}{3} \cdot 9x^2 - \dfrac{1}{x} \cdot 9x^2}{\dfrac{1}{9} \cdot 9x^2 - \dfrac{1}{x^2} \cdot 9x^2}$$

Divide out common factors:

$$= \frac{\dfrac{1}{\cancel{3}} \cdot \cancel{9}^{3}x^2 - \dfrac{1}{\cancel{x}} \cdot 9x^{\cancel{2}x}}{\dfrac{1}{\cancel{9}} \cdot \cancel{9}x^2 - \dfrac{1}{\cancel{x^2}} \cdot 9x^2}$$

Simplify:

$$= \frac{3x^2 - 9x}{x^2 - 9}$$

Factor each polynomial:

$$= \frac{3x(x - 3)}{(x + 3)(x - 3)}$$

Divide out common factors:

$$= \frac{3x\cancel{(x - 3)}}{(x + 3)\cancel{(x - 3)}}$$

Simplify:

$$= \frac{3x}{x + 3}$$

📝 *My video summary* ▶ **b. Step 1** The LCD for all the minor fractions is $x - 1$.

Step 2 Multiply the **numerator** and denominator by $x - 1$.

$$\frac{4 - \dfrac{5}{x - 1}}{\dfrac{6}{x - 1} - 7} \cdot \frac{x - 1}{x - 1}$$

Step 3 Try to finish simplifying the **complex rational expression** on your own. View the **answer**, or watch this **video** for a complete solution.

You Try It Work through this You Try It problem.

Work Exercises **10–12** in this eText or in the MyMathLab Study Plan.

As you have seen in Examples 1–4, **Method I** and **Method II** both give the same simplification. Have you developed a preference for one method over the other? Whether you have or not, you should practice using both methods for a while to discover when one method might be better suited than the other.

My interactive
video summary

Example 5 Simplifying a Complex Rational Expression

Simplify the complex rational expression using Method I or Method II.

$$\frac{\dfrac{5}{n-2} - \dfrac{3}{n}}{\dfrac{6}{n^2-2n} + \dfrac{2}{n}}$$

Solution Try simplifying on your own. View the answer, or watch this interactive video for a complete solution.

You Try It Work through this You Try It problem.

Work Exercises 13–24 in this eText or in the MyMathLab Study Plan.

Complex rational expressions can be written using negative exponents.

For example, the expression $\dfrac{3^{-1} - x^{-1}}{3^{-2} - x^{-2}}$ is equivalent to the complex fraction $\dfrac{\dfrac{1}{3} - \dfrac{1}{x}}{\dfrac{1}{9} - \dfrac{1}{x^2}}$

from **Example 2.**

When a complex rational expression contains negative exponents, rewrite it as an equivalent expression with positive exponents by using the **negative power rule.** Then simplify it by using Method I or Method II.

My interactive
video summary

Example 6 Simplifying a Complex Rational Expression Containing Negative Exponents

Simplify the complex rational expression.

$$\frac{1 - 9y^{-1} + 14y^{-2}}{1 + 3y^{-1} - 10y^{-2}}$$

Solution Use the negative power rule to rewrite the expression with positive exponents.

$$\frac{1 - 9y^{-1} + 14y^{-2}}{1 + 3y^{-1} - 10y^{-2}} = \frac{1 - \dfrac{9}{y} + \dfrac{14}{y^2}}{1 + \dfrac{3}{y} - \dfrac{10}{y^2}}$$

Try to finish simplifying the **complex rational expression** on your own using Method I or Method II. View the answer, or watch this **interactive video** for a complete solution.

You Try It Work through this You Try It problem.

Work Exercises 25–30 in this eText or in the MyMathLab Study Plan.

6.4 Exercises

In Exercises 1–6, use Method I to simplify each complex rational expression.

1. $\dfrac{\dfrac{14}{5x}}{\dfrac{7}{15x}}$

2. $\dfrac{\dfrac{x+7}{2}}{\dfrac{3x-1}{10}}$

3. $\dfrac{\dfrac{5x+5}{x^2-9}}{\dfrac{x^2-1}{x+3}}$

4. $\dfrac{\dfrac{5}{6}+\dfrac{3}{4}}{\dfrac{8}{3}-\dfrac{5}{9}}$

5. $\dfrac{\dfrac{3}{x}+2}{\dfrac{9}{x^2}-4}$

6. $\dfrac{\dfrac{9}{x-4}+\dfrac{6}{x-5}}{\dfrac{5x-23}{x^2-9x+20}}$

In Exercises 7–12, use Method II to simplify each complex rational expression.

7. $\dfrac{\dfrac{x+2}{15x}}{\dfrac{2x-1}{10x}}$

8. $\dfrac{\dfrac{8x}{x+y}}{\dfrac{2x^3}{y}}$

9. $\dfrac{\dfrac{3}{x^2-x-2}}{\dfrac{x+4}{x-2}}$

10. $\dfrac{2-\dfrac{a}{b}}{\dfrac{a^2}{b^2}-4}$

11. $\dfrac{3-\dfrac{7}{x-2}}{\dfrac{8}{x-2}-5}$

12. $\dfrac{\dfrac{x+5}{x}-\dfrac{8}{x-1}}{\dfrac{x+1}{x}+\dfrac{x+1}{x-1}}$

In Exercises 13–26, simplify each complex rational expression using Method I or Method II.

13. $\dfrac{7+\dfrac{1}{x}}{7-\dfrac{1}{x}}$

14. $\dfrac{\dfrac{1}{x}+3}{\dfrac{1}{x^2}-9}$

15. $\dfrac{\dfrac{x+2}{x-6}-\dfrac{x+12}{x+5}}{x+82}$

16. $\dfrac{\dfrac{6}{x}+\dfrac{5}{y}}{\dfrac{5}{x}-\dfrac{6}{y}}$

17. $\dfrac{\dfrac{1}{16}-\dfrac{1}{x^2}}{\dfrac{1}{4}+\dfrac{1}{x}}$

18. $\dfrac{\dfrac{8}{x+4}-\dfrac{2}{x+7}}{\dfrac{x+8}{x+4}}$

19. $\dfrac{\dfrac{2}{x+13}+\dfrac{1}{x-5}}{2-\dfrac{x+25}{x+13}}$

20. $\dfrac{\dfrac{x-5}{x+5}+\dfrac{x-5}{x-9}}{1+\dfrac{x+5}{x-9}}$

21. $\dfrac{1+\dfrac{4}{x}-\dfrac{5}{x^2}}{1-\dfrac{2}{x}-\dfrac{35}{x^2}}$

22. $\dfrac{\dfrac{x-4}{x^2-25}}{1+\dfrac{1}{x-5}}$

23. $\dfrac{\dfrac{x-5}{x-3}+\dfrac{x-5}{x+3}}{\dfrac{x-3}{x+3}+\dfrac{x+3}{x-3}}$

23. $\dfrac{\dfrac{x+3}{x-3}-\dfrac{x-3}{x+3}}{\dfrac{x-3}{x+3}+\dfrac{x+3}{x-3}}$

24. $\dfrac{\dfrac{3}{a^2}-\dfrac{1}{ab}-\dfrac{2}{b^2}}{\dfrac{2}{a^2}-\dfrac{5}{ab}+\dfrac{3}{b^2}}$

25. $\dfrac{6x^{-1}+6y^{-1}}{xy^{-1}-x^{-1}y}$

26. $\dfrac{7x^{-1}-3y^{-1}}{49x^{-2}-9y^{-2}}$

27. $\dfrac{7^{-1}+n^{-1}}{7^{-2}-n^{-2}}$

28. $\dfrac{\dfrac{x-6}{x+3}}{\dfrac{9}{x+3}-\dfrac{2}{x-4}}$

29. $\dfrac{\dfrac{x^2+15x+50}{2x^2+3x+1}}{\dfrac{x^2+5x-50}{2x^2+5x+2}}$

30. $\dfrac{\dfrac{x-5}{x-3}-\dfrac{3}{x-7}}{\dfrac{x^2+7x-44}{x^2-10x+21}}$

6.5 Rational Equations and Models

THINGS TO KNOW

Before working through this section, be sure you are familiar with the following concepts:

	VIDEO	ANIMATION	INTERACTIVE

 You Try It   1. Solve Linear Equations in One Variable
   (Section 1.1, Objective 2)

 You Try It   2. Factor Polynomials Completely
   (Section 5.3, Objective 4)

 You Try It   3. Solve Polynomial Equations by Factoring
   (Section 5.4, Objective 1)

 You Try It   4. Find the Zeros of a Polynomial Function
   (Section 5.4, Objective 2)

 You Try It   5. Use Polynomial Equations and Models to
   Solve Application Problems
   (Section 5.4, Objective 3)

 You Try It   6. Find the Domain of a Rational Function
   (Section 6.1, Objective 1)

 You Try It   7. Evaluate Rational Functions
   (Section 6.1, Objective 2)

 You Try It   8. Add and Subtract Rational Expressions
   with Unlike Denominators
   (Section 6.3, Objective 4)

OBJECTIVES

1 Identify Rational Equations
2 Solve Rational Equations
3 Identify and Solve Proportions
4 Find the Zeros of a Rational Function
5 Use Rational Equations to Solve Application Problems

OBJECTIVE 1   IDENTIFY RATIONAL EQUATIONS

We have previously defined linear equations and polynomial equations. Now, let's define a *rational equation*.

Definition   Rational Equation

A **rational equation** is a statement in which two **rational expressions** are set equal to each other.

Some examples of rational equations are

$$\underbrace{\frac{2x}{x-1}}_{\substack{\text{Rational} \\ \text{expression}}} \underbrace{=}_{\substack{\text{Equal} \\ \text{sign}}} \underbrace{\frac{2x+3}{x}}_{\substack{\text{Rational} \\ \text{expression}}} \quad \text{and} \quad \underbrace{\frac{1}{x-2}+\frac{1}{x+2}}_{\substack{\text{Rational} \\ \text{expression}}} \underbrace{=}_{\substack{\text{Equal} \\ \text{sign}}} \underbrace{\frac{4}{x^2-4}}_{\substack{\text{Rational} \\ \text{expression}}}.$$

My video summary ▶ **Example 1 Identifying Rational Equations**

Determine if each statement is a rational equation. If not, state why.

a. $\dfrac{x-4}{x}+\dfrac{4}{x+5}=\dfrac{6}{x}$

b. $\dfrac{5}{y}+\dfrac{7}{y+2}$

c. $\dfrac{\sqrt{k+1}}{k+3}=\dfrac{k-5}{k+4}$

d. $5n^{-1}=3n^{-2}$

Solutions Read the following, or watch this **video** for complete solutions to all four parts.

a. Both $\dfrac{x-4}{x}+\dfrac{4}{x+5}$ and $\dfrac{6}{x}$ are rational expressions. They are set equal to each other, so the statement is a **rational equation**.

b. This statement $\dfrac{5}{y}+\dfrac{7}{y+2}$ does not contain an equal sign, so it is not a rational equation. Instead, it is a rational *expression*.

c. This statement is not a rational equation because $\dfrac{\sqrt{k+1}}{k+3}$ is not a rational expression. Note that the numerator is not a **polynomial** because there is a **variable** under a radical.

d. Note that $5n^{-1}=\dfrac{5}{n}$ and $3n^{-2}=\dfrac{3}{n^2}$, so they are both rational expressions. They are set equal to each other, so this statement is a rational equation.

You Try It Work through this **You Try It** problem.

Work Exercises 1–4 in this eText or in the MyMathLab Study Plan.

OBJECTIVE 2 SOLVE RATIONAL EQUATIONS

In **Section 1.1**, we learned to **clear fractions** from **equations** by multiplying both sides of the equation by the **LCD**. We then solved the resulting equation. This idea is the key to solving **rational equations**.

Example 2 Solving Rational Equations

Solve.

a. $\dfrac{1}{2}x + \dfrac{2}{3} = \dfrac{3}{4}$

b. $\dfrac{1}{x} + \dfrac{1}{2} = \dfrac{1}{3}$

Solutions

a. The LCD of the fractions $\dfrac{1}{2}$, $\dfrac{2}{3}$, and $\dfrac{3}{4}$ is 12.

Begin with the original equation:
$$\frac{1}{2}x + \frac{2}{3} = \frac{3}{4}$$

Multiply both sides by the LCD:
$$12\left(\frac{1}{2}x + \frac{2}{3}\right) = 12\left(\frac{3}{4}\right)$$

Distribute:
$$12\left(\frac{1}{2}x\right) + 12\left(\frac{2}{3}\right) = 12\left(\frac{3}{4}\right)$$

Simplify:
$$6x + 8 = 9$$

Subtract 8 from both sides:
$$6x = 1$$

Divide both sides by 6:
$$x = \frac{1}{6}$$

View this **popup** to see that this answer checks. The **solution set** is $\left\{\dfrac{1}{6}\right\}$.

My video summary ▶ b. The **LCD** of $\dfrac{1}{x}$, $\dfrac{1}{2}$, and $\dfrac{1}{3}$ is $6x$.

Begin with the original equation:
$$\frac{1}{x} + \frac{1}{2} = \frac{1}{3}$$

Multiply both sides by the LCD:
$$6x\left(\frac{1}{x} + \frac{1}{2}\right) = 6x\left(\frac{1}{3}\right)$$

Distribute:
$$6x \cdot \frac{1}{x} + 6x \cdot \frac{1}{2} = 6x \cdot \frac{1}{3}$$

Divide out common factors:
$$6\cancel{x} \cdot \frac{1}{\cancel{x}} + \overset{3}{\cancel{6}}x \cdot \frac{1}{\cancel{2}_{1}} = \overset{2}{\cancel{6}}x \cdot \frac{1}{\cancel{3}_{1}}$$

Simplify:
$$6 + 3x = 2x$$

Finish solving this equation on your own. Remember to check your result. View the **answer**, or watch this **video** for a complete solution to part b.

You Try It Work through this **You Try It** problem.

Work Exercises 5–8 in this eText or in the MyMathLab Study Plan.

Any value that causes the **denominator** of a **rational expression** to equal zero is a **restricted value**. A restricted value cannot be a **solution** of a **rational equation**. However, sometimes when we multiply both sides of a rational equation by the **LCD**, a solution of the resulting **linear** or **polynomial equation** will be a restricted value of the original rational equation. Such "solutions" are called **extraneous solutions**, and they must be excluded from the **solution set**.

My animation summary   ✱ **Concept Animation** Watch this animation to explore the concept of extraneous solutions.

The following steps can be used to solve rational equations.

Solving Rational Equations

Step 1 List all restricted values.

Step 2 Determine the LCD of all denominators in the equation.

Step 3 Multiply both sides of the equation by the LCD.

Step 4 Solve the resulting polynomial equation.

Step 5 Discard any restricted values and check the remaining **solutions** in the original equation.

My video summary ▶ **Example 3 Solving Rational Equations**

Solve $\dfrac{2}{x} - \dfrac{x-3}{2x} = 3$.

Solution Read the following, or watch this **video** for a complete solution. Follow the five-step process.

Step 1 To find the restricted values, look for values that make any denominator equal to zero.

$$\frac{2}{x} - \frac{x-3}{2x} = \frac{3}{1}$$

Examining the variable factors in the denominators, we see that the only restricted value is 0.

Step 2 There are unique factors of 2 and x in the denominators, each occurring at most once. The LCD is $2x$.

Step 3 Multiply both sides of the equation by the LCD.

Multiply both sides by the LCD: $2x\left(\dfrac{2}{x} - \dfrac{x-3}{2x}\right) = 2x(3)$

Distribute: $2x \cdot \dfrac{2}{x} - 2x \cdot \dfrac{x-3}{2x} = 2x \cdot 3$

Divide out common factors: $2\cancel{x} \cdot \dfrac{2}{\cancel{x}} - \cancel{2x} \cdot \dfrac{x-3}{\cancel{2x}} = 2x \cdot 3$

Multiply: $4 - (x-3) = 6x$

Step 4 Solve the resulting equation.

Distribute: $4 - x + 3 = 6x$

Simplify: $7 - x = 6x$

Add x to both sides: $7 = 7x$

Divide both sides by 7: $1 = x$

Step 5 The potential solution $x = 1$ is not a **restricted value**, so do not discard it. View this **popup** to see the check of $x = 1$ in the original equation. The solution set is $\{1\}$.

You Try It Work through this **You Try It** problem.

Work Exercises 9 and 10 in this eText or in the MyMathLab Study Plan.

My video summary ⊙ **Example 4 Solving Rational Equations**

Solve $\dfrac{4}{5} - \dfrac{3}{x-3} = \dfrac{1}{x}$.

Solution Follow the five-step process.

Step 1 To find the **restricted values**, look for values that make any **denominator** equal to zero. Set each denominator with a **variable** equal to zero and solve.

$$x - 3 = 0 \quad \text{or} \quad x = 0$$
$$x = 3$$

The restricted values are 0 and 3.

Step 2 There are unique factors of 5, x, and $x - 3$ in the denominators, each occurring at most once. The LCD is $5x(x - 3)$.

Step 3 Multiply both sides of the equation by the LCD.

Multiply both sides by the LCD: $5x(x - 3)\left(\dfrac{4}{5} - \dfrac{3}{x-3}\right) = 5x(x - 3)\left(\dfrac{1}{x}\right)$

Distribute: $5x(x-3)\cdot\dfrac{4}{5} - 5x(x-3)\cdot\dfrac{3}{x-3} = 5x(x-3)\cdot\dfrac{1}{x}$

Divide out common factors: $5x(x-3)\cdot\dfrac{4}{\cancel{5}} - 5x\cancel{(x-3)}\cdot\dfrac{3}{\cancel{x-3}} = 5\cancel{x}(x-3)\cdot\dfrac{1}{\cancel{x}}$

Multiply: $4x(x - 3) - 15x = 5(x - 3)$

Try to finish solving the equation on your own. View the **answer**, or watch this **video** for a complete solution.

You Try It Work through this **You Try It** problem.

Work Exercises 11 and 12 in this eText or in the MyMathLab Study Plan.

My video summary ⊙ **Example 5 Solving Rational Equations**

Solve $\dfrac{m}{m+2} + \dfrac{5}{m-2} = \dfrac{20}{m^2 - 4}$.

Solution Follow the five-step process.

Step 1 To find the restricted values, look for values that make any denominator equal to zero. First, factor the denominators.

$$\frac{m}{m + 2} + \frac{5}{m - 2} = \frac{20}{(m - 2)(m + 2)}$$

Now set each **variable factor** equal to zero and solve.

$$m + 2 = 0 \quad \text{or} \quad m - 2 = 0$$
$$m = -2 \qquad\qquad m = 2$$

The restricted values are -2 and 2.

Step 2 There are unique factors of $m - 2$ and $m + 2$ in the denominators, each occurring at most once. The **LCD** is $(m - 2)(m + 2)$.

Step 3 Multiply both sides of the equation by the **LCD** and **divide out** common factors.

$$(m - 2)(m + 2)\left(\frac{m}{m + 2} + \frac{5}{m - 2}\right) = (m - 2)(m + 2) \cdot \frac{20}{(m - 2)(m + 2)}$$

$$(m - 2)(m + 2) \cdot \frac{m}{m + 2} + (m - 2)(m + 2) \cdot \frac{5}{m - 2} = (m - 2)(m + 2) \cdot \frac{20}{(m - 2)(m + 2)}$$

$$(m - 2)\cancel{(m + 2)} \cdot \frac{m}{\cancel{m + 2}} + \cancel{(m - 2)}(m + 2) \cdot \frac{5}{\cancel{m - 2}} = \cancel{(m - 2)}\cancel{(m + 2)} \cdot \frac{20}{\cancel{(m - 2)}\cancel{(m + 2)}}$$

$$m(m - 2) + 5(m + 2) = 20$$

Try to finish solving the equation on your own, making sure to check for **restricted values**. View the **answer**, or watch this **video** for a complete solution.

You Try It Work through this You Try It problem.

Work Exercises 13 and 14 in this eText or in the MyMathLab Study Plan.

My video summary ▶ **Example 6 Solving Rational Equations**

Solve $\dfrac{2}{x - 3} - \dfrac{4}{x^2 - 2x - 3} = \dfrac{1}{x + 1}$.

Solution Watch this video, or continue reading for a complete solution. Following the five-step process, start by finding the **restricted values**. Factoring the denominators gives

$$\frac{2}{x - 3} - \frac{4}{(x - 3)(x + 1)} = \frac{1}{x + 1}.$$

Setting each **variable factor** in the denominators equal to zero and solving leads to the restricted values -1 and 3.

The **LCD** is $(x - 3)(x + 1)$. Multiplying both sides of the equation by the LCD gives the **polynomial equation**

$$2(x + 1) - 4 = x - 3.$$

Solving the polynomial equation leads to $x = -1$. However, $x = -1$ is a restricted value, so it must be discarded. Since no other possible **solution** exists, the equation has no solution. The **solution set** is { } or ∅.

 You Try It Work through this You Try It problem.

Work Exercises 15–18 in this eText or in the MyMathLab Study Plan.

TIP If an **identity** occurs when solving a rational equation, the solution will be the set of all real numbers except the restricted values, if any.

OBJECTIVE 3 IDENTIFY AND SOLVE PROPORTIONS

A **ratio** is the **quotient** of two numbers or **algebraic expressions**. A **rational expression** is the quotient of two **polynomials**, so rational expressions are ratios.

A **proportion** is a statement that two ratios are equal, such as $\dfrac{P}{Q} = \dfrac{R}{S}$.

When a rational equation is a proportion, it can be solved by first **cross multiplying** to get $P \cdot S = Q \cdot R$ and then solving the resulting **polynomial equation**.

Cross Multiplying

$$\frac{P}{Q} \diagup\!\!\!\!\diagdown \frac{R}{S}$$

$$P \cdot S = Q \cdot R$$

Note that any "solution" that is a **restricted value** in the original equation must be discarded. Proportions can be used to solve many types of applications, as we will see later in this section.

Example 7 Solving Proportions

Solve.

a. $\dfrac{8}{x + 3} = \dfrac{5}{x}$

b. $\dfrac{x}{6} = \dfrac{2}{x - 1}$

Solutions

a. This rational equation is a proportion.

First, identify the restricted values by setting each denominator equal to zero and solving.

$$x + 3 = 0 \quad \text{or} \quad x = 0$$
$$x = -3$$

There are two restricted values, -3 and 0.

Next, **cross multiply** and solve the resulting **polynomial equation**.

Begin with the original equation: $\dfrac{8}{x+3} = \dfrac{5}{x}$

Cross multiply: $8x = 5(x+3)$

Distribute: $8x = 5x + 15$

Subtract $5x$ from both sides: $3x = 15$

Divide both sides by 3: $x = 5$

Because 5 is not a restricted value, the solution set is $\{5\}$. View this **popup** to see that this result checks.

My video summary ▶ **b.** This rational equation is a proportion.

The denominator $x - 1$ is the only one that contains a variable, so the only restricted value is 1.

Begin with the original equation: $\dfrac{x}{6} = \dfrac{2}{x-1}$

Cross multiply: $x(x-1) = 6(2)$

Try to solve this resulting equation on your own. Be sure to check for **extraneous solutions**. View the **answer**, or watch this video for a complete solution.

 You Try It Work through this You Try It problem.

Work Exercises 19–22 in this eText or in the MyMathLab Study Plan.

⚠ Only proportions can be solved by cross multiplying. If a rational equation is not a proportion, do not try to cross multiply.

OBJECTIVE 4   FIND THE ZEROS OF A RATIONAL FUNCTION

In Section 5.4, we found the zeros of a polynomial function by setting the function equal to zero and solving the resulting equation. We follow the same procedure for finding the zeros of a rational function, but we must consider the restricted values.

Let $R(x)$ be a rational function. The zeros of R are the **solutions** to the equation $R(x) = 0$.

Because a rational function is defined by a rational **expression**, solving $R(x) = 0$ is equivalent to solving $\dfrac{P(x)}{Q(x)} = 0$. The quotient will equal zero if the numerator is 0 and the denominator is not 0.

This leads us to the following strategy for finding zeros of a rational function.

Finding the Zeros of a Rational Function

To find the **zeros**, or **roots**, of a rational function of the form $\dfrac{P(x)}{Q(x)}$, do the following:

Step 1 Find the restricted values by setting the denominator equal to zero and solving the resulting equation, $Q(x) = 0$.

(continued)

Step 2 Set the numerator equal to zero and solve the resulting equation, $P(x) = 0$.

Step 3 Any solution from Step 2 that is not a restricted value is a zero of the function.

Example 8 Finding the Zeros of a Rational Function

Find the zeros for each rational function.

a. $R(x) = \dfrac{3x^2 - 5x - 2}{x^2 - 4}$

b. $R(x) = \dfrac{3x^2 + 2x - 21}{6x^2 + 7x - 20}$

Solutions

a. First, find the **restricted values** of the function. Set the denominator equal to zero and solve the resulting equation.

Set denominator equal to 0: $\qquad\qquad x^2 - 4 = 0$

Factor: $\qquad\qquad (x + 2)(x - 2) = 0$

Set each factor equal to 0 and solve: $\quad x + 2 = 0 \quad$ or $\quad x - 2 = 0$

$$x = -2 \qquad\qquad x = 2$$

The restricted values are −2 and 2.

Next, set the numerator equal to zero and solve.

Set numerator equal to 0: $\qquad\qquad 3x^2 - 5x - 2 = 0$

Factor: $\qquad\qquad (x - 2)(3x + 1) = 0$

Set each factor equal to 0 and solve: $\quad x - 2 = 0 \quad$ or $\quad 3x + 1 = 0$

$$x = 2 \qquad\qquad 3x = -1$$

$$x = -\frac{1}{3}$$

The solution $x = 2$ is a **restricted value**, so it must be discarded. The remaining solution, $x = -\dfrac{1}{3}$, is not a restricted value, so it is a **zero** of the function (the only one).

 *My video summary* ⊙ b. Try to determine the **zeros** of this function on your own. View the **answer**, or watch this **video** for a complete solution.

❧ **You Try It** Work through this **You Try It** problem.

Work Exercises **23–26** in this eText or in the MyMathLab **Study Plan.**

OBJECTIVE 5 USE RATIONAL EQUATIONS TO SOLVE APPLICATION PROBLEMS

Rational equations appear frequently in applications. We follow the usual problem-solving strategy when solving application problems involving rational equations.

My video summary ▶ **Example 9 Road Trip**

Payton's car holds 13 gallons of gas and can travel 351 miles on a full tank. Currently, she has 4 gallons of gas in her tank. Assuming the same gas mileage, how many gallons of gas will she need to purchase on a 540-mile road trip to Chicago?

Solution Watch this video, or read the following for a complete solution.

Follow the problem-solving strategy.

Step 1 We want to find the number of gallons of gas Payton needs to purchase for a 540-mile road trip. We know she can travel 351 miles on 13 gallons of gas, and we know she currently has 4 gallons of gas in the car.

Step 2 Let $x =$ the number of gallons purchased.

Step 3 We know the car's gas mileage remains constant, so we can solve the problem using a **proportion**.

$$\underset{\dfrac{\text{Miles}}{\text{Gallons}}}{\overset{\text{Full tank}}{\downarrow}} = \underset{\dfrac{\text{Miles}}{\text{Gallons}}}{\overset{\text{Road trip}}{\downarrow}}$$

$$\frac{351}{13} = \frac{540}{4 + x}$$

Step 4 Cross-multiply and solve the resulting equation.

Cross multiply: $\dfrac{351}{13} \nearrow\hspace{-1em}\searrow \dfrac{540}{4 + x}$

$$351(4 + x) = 13(540)$$

Distribute: $1404 + 351x = 7020$

Subtract 1404 from both sides: $351x = 5616$

Divide both sides by 351: $x = 16$

Step 5 Note that 540 is about 1.5 times 351 and the gas mileage remains constant. We expect Payton will need about 1.5 tanks, or $1.5(13) = 19.5$ gallons for the trip. Her tank had 4 gallons already, so she would need about 15.5 more gallons. Our result is reasonable.

Step 6 Payton will need to purchase 16 gallons of gas to complete the road trip.

You Try It Work through this You Try It problem.

Work Exercises 27–30 in this eText or in the MyMathLab Study Plan.

My video summary ▶ **Example 10 Resistance in Parallel Circuits**

In electronics, the total resistance R of a circuit containing two resistors in parallel is given by the formula $\dfrac{1}{R} = \dfrac{1}{R_1} + \dfrac{1}{R_2}$, where R_1 and R_2 are the two individual resistances. If the total resistance is 10 ohms and one resistor has twice the resistance of the other, find the resistance of each circuit.

Solution Watch this video, or read the following for a complete solution.

Step 1 We want to find the individual resistances for a circuit containing two resistors in parallel. We know that the total resistance is 10 ohms and one resistor has twice the resistance of the other.

Step 2 We are given the resistance of one resistor in terms of the other. If we let x = the resistance of the first resistor, then $2x$ = the resistance of the second.

Step 3 Using the given formula, we substitute in the total resistance and expressions for the individual resistances. This gives the equation

$$\frac{1}{10} = \frac{1}{x} + \frac{1}{2x}.$$

Step 4 The LCD is $10x$. Note that the only restricted value is $x = 0$.

Begin with the original equation: $\quad \frac{1}{10} = \frac{1}{x} + \frac{1}{2x}$

Multiply both sides by the LCD: $\quad 10x\left(\frac{1}{10}\right) = 10x\left(\frac{1}{x} + \frac{1}{2x}\right)$

Distribute: $\quad 10x \cdot \frac{1}{10} = 10x \cdot \frac{1}{x} + 10x \cdot \frac{1}{2x}$

Simplify: $\quad \cancel{10}x \cdot \frac{1}{\cancel{10}} = 10\cancel{x} \cdot \frac{1}{\cancel{x}} + \overset{5}{\cancel{10}}\cancel{x} \cdot \frac{1}{2\cancel{x}}$

$$x \cdot 1 = 10 \cdot 1 + 5 \cdot 1$$

$$x = 10 + 5$$

$$x = 15$$

Step 5 Check the answer using the equation $\frac{1}{10} = \frac{1}{x} + \frac{1}{2x}$.

$$\frac{1}{15} + \frac{1}{2(15)} = \frac{1}{15} + \frac{1}{30} = \frac{2}{30} + \frac{1}{30} = \frac{3}{30} = \frac{1}{10}$$

The total resistance is 10 ohms, so this result is reasonable and it checks.

Step 6 The resistances are 15 ohms and $2(15) = 30$ ohms.

You Try It Work through this You Try It problem.

Work Exercises 31 and 32 in this eText or in the MyMathLab Study Plan.

We have studied **uniform motion problems** and used the **formula** $d = rt$.

Solving this formula for t gives $t = \frac{d}{r}$, or time $= \frac{\text{distance}}{\text{rate}}$. This **rational equation** can be used to solve problems.

My video summary ▶ **Example 11 Boat Speed**

Emalie can travel 16 miles upriver in the same amount of time it takes her to travel 24 miles downriver. If the speed of the current is 4 mph, how fast can her boat travel in still water?

Solution Follow the problem-solving strategy.

Step 1 We want to find the speed of Emalie's boat in still water. We know she can travel 16 miles upriver in the same amount of time she can travel 24 miles downriver, and we know the speed of the current is 4 mph.

Step 2 Let r = the speed of Emalie's boat in still water. Then her speed upriver is $r - 4$ (going against the current) and her speed downriver is $r + 4$ (going with the current).

Step 3 The amount of time traveled is the same in either direction.

$$\text{time}_{\text{upriver}} = \text{time}_{\text{downriver}}$$

Solving the **distance formula** for t gives us $t = \dfrac{d}{r}$, or $\text{time} = \dfrac{\text{distance}}{\text{rate}}$. So, we rewrite our equation as follows:

$$\left(\frac{\text{distance}}{\text{rate}}\right)_{\text{upriver}} = \left(\frac{\text{distance}}{\text{rate}}\right)_{\text{downriver}}$$

Substituting the given distances and our expressions for rates from Step 2, we get the equation

$$\frac{16}{r - 4} = \frac{24}{r + 4}.$$

Try to solve this equation to finish this problem on your own. View the **answer**, or watch this **video** for a complete solution.

You Try It Work through this You Try It problem.

Work Exercises 33 and 34 in this eText or in the MyMathLab Study Plan.

My video summary ▶ **Example 12 Train Speed**

Fatima rode an express train 223.6 miles from Boston to New York City and then rode a passenger train 218.4 miles from New York City to Washington, DC. If the express train travels 30 miles per hour faster than the passenger train and her total trip took 6.5 hours, what was the average speed of the express train?

Solution Follow the problem-solving strategy.

Step 1 We want to find the average speed of the express train. We know the express train traveled 30 miles per hour faster than the passenger train. We also know that the distances traveled on the express and passenger trains were 223.6 and 218.4 miles, respectively. Finally, we know the time of the trip was 6.5 hours.

Step 2 Let r = the average speed of the express train. Then $r - 30$ = the average speed of the passenger train.

Step 3 The total time of Fatima's trip must be the **sum** of her time on the express train and her time on the passenger train.

$$\text{time}_{\text{express}} + \text{time}_{\text{passenger}} = \text{time}_{\text{total}}$$

Solving the **distance formula** for t gives us $t = \dfrac{d}{r}$, or time $= \dfrac{\text{distance}}{\text{rate}}$. So, we rewrite our equation as follows:

$$\left(\frac{\text{distance}}{\text{rate}}\right)_{\text{express}} + \left(\frac{\text{distance}}{\text{rate}}\right)_{\text{passenger}} = \text{time}_{\text{total}}$$

Substituting the known distances, the expressions for train speeds from Step 2, and the total time, we get the equation

$$\frac{223.6}{r} + \frac{218.4}{r - 30} = 6.5.$$

Step 4 The LCD is $r(r - 30)$. Multiply both sides of the equation by the LCD to obtain a **polynomial equation**.

Write the original equation:

$$\frac{223.6}{r} + \frac{218.4}{r - 30} = 6.5$$

Multiply by the LCD:

$$r(r - 30)\left(\frac{223.6}{r} + \frac{218.4}{r - 30}\right) = r(r - 30)(6.5)$$

Distribute:

$$r(r - 30)\cdot\frac{223.6}{r} + r(r - 30)\cdot\frac{218.4}{r - 30} = r(r - 30)(6.5)$$

Divide out common factors:

$$\not{r}(r - 30)\cdot\frac{223.6}{\not{r}} + r\not{(r - 30)}\cdot\frac{218.4}{\not{r - 30}} = r(r - 30)(6.5)$$

Multiply:

$$223.6(r - 30) + 218.4r = r(r - 30)(6.5)$$

Finish solving the equation on your own to answer the question. View the **answer**, or watch this **video** for a complete solution.

You Try It Work through this You Try It problem.

Work Exercises 35 and 36 in this eText or in the MyMathLab Study Plan.

For work problems involving multiple workers, such as people, copiers, pumps, etc., it is often helpful to consider **rate of work**.

Rate of Work

The **rate of work** is the number of jobs that can be completed in a given unit of time. If one job can be completed in t units of time, then the rate of work is given by $\dfrac{1}{t}$.

When dealing with two workers, use the following formula:

$$\underbrace{\frac{1}{t_1}}_{\substack{\text{Work rate of}\\\text{1st worker}}} + \underbrace{\frac{1}{t_2}}_{\substack{\text{Work rate of}\\\text{2nd worker}}} = \underbrace{\frac{1}{t}}_{\substack{\text{Work rate}\\\text{together}}}$$

Here, t_1 and t_2 are the individual times to complete one job, and t is the time to complete the job when working together.

✎ *My video summary* ▶ ## Example 13 Painting a Room

Avril can paint a room in 4 hours if she works alone. Anisa can paint the same room in 2 hours if she works alone. How long will it take the two women to paint the room if they work together?

Solution Watch this **video**, or read the following for a complete solution.

Follow the **problem-solving strategy**.

Step 1 We want to find how long it will take Avril and Anisa to paint the room if they work together. Working alone, Avril can paint the room in 4 hours, while Anisa can paint the room in 2 hours.

Step 2 Let x = the time to paint the room when working together.

Step 3 Avril can paint the room in 4 hours, so her **rate of work** is $\frac{1}{4}$ room per hour.

Anisa can paint the room in 2 hours, so her rate of work is $\frac{1}{2}$ room per hour.

Together, they can paint the room in x hours, so their combined rate of work is $\frac{1}{x}$ room per hour.

The **sum** of the two individual work rates results in the combined work rate:

$$\underbrace{\frac{1}{4}}_{\text{Avril's rate}} + \underbrace{\frac{1}{2}}_{\text{Anisha's rate}} = \underbrace{\frac{1}{x}}_{\text{Combined rate}}.$$

Step 4 The LCD is $4x$.

Write the original equation: $\quad \frac{1}{4} + \frac{1}{2} = \frac{1}{x}$

Multiply both sides by the LCD: $\quad 4x\left(\frac{1}{4} + \frac{1}{2}\right) = 4x\left(\frac{1}{x}\right)$

Distribute: $\quad 4x \cdot \frac{1}{4} + 4x \cdot \frac{1}{2} = 4x \cdot \frac{1}{x}$

Divide out common factors: $\quad \cancel{4}x \cdot \frac{1}{\cancel{4}} + \overset{2}{\cancel{4}}x \cdot \frac{1}{\cancel{2}} = 4\cancel{x} \cdot \frac{1}{\cancel{x}}$

Simplify: $\quad x + 2x = 4$

Combine like terms: $\quad 3x = 4$

Divide both sides by 3: $\quad x = \frac{4}{3}$

Step 5 The result $\frac{4}{3}$ hours is less than the 2 hours needed for the faster worker alone, so the result is reasonable.

Step 6 Together, Avril and Anisa can paint the room in $\frac{4}{3}$ hours, or 1 hour and 20 minutes.

You Try It Work through this You Try It problem.

Work Exercises 37 and 38 in this eText or in the MyMathLab Study Plan.

My video summary ▶ **Example 14 Emptying a Pool**

A small pump takes 8 more hours than a larger pump to empty a pool. Together, the pumps can empty the pool in 3 hours. How long will it take the larger pump to empty the pool if it works alone?

Solution Follow the problem-solving strategy.

Step 1 We want to find the time it will take the larger pump to empty the pool by itself. We know that the two pumps take 3 hours to empty the pool and the smaller pump takes 8 more hours than the larger pump to empty the pool by itself.

Step 2 Let $x =$ the time for the larger pump to empty the pool. Then $x + 8 =$ the time for the smaller pump to empty the pool.

Step 3 Since we are combining **rates of work**, we get

$$\underbrace{\frac{1}{t_1}}_{\text{Larger pump rate}} + \underbrace{\frac{1}{t_2}}_{\text{Smaller pump rate}} = \underbrace{\frac{1}{t}}_{\text{Combined rate}}.$$

Substituting in the given combined time, and expressions for individual times, we get

$$\underbrace{\frac{1}{x}}_{\text{Larger pump rate}} + \underbrace{\frac{1}{x+8}}_{\text{Smaller pump rate}} = \underbrace{\frac{1}{3}}_{\text{Combined rate}}.$$

Step 4 The LCD is $3x(x + 8)$. Multiply both sides of the equation by the LCD and solve the resulting **polynomial equation** to finish this problem on your own. View the **answer**, or watch this **video** for a complete solution.

You Try It Work through this You Try It problem.

Work Exercises 39 and 40 in this eText or in the MyMathLab Study Plan.

My animation summary ✪ **Example 15 Filling a Pond**

A garden hose can fill a pond in 2 hours whereas an outlet pipe can drain the pond in 10 hours. If the outlet pipe is accidentally left open, how long would it take to fill the pond?

Solution **Concept Animation** Watch this animation, or continue reading for a complete solution.

We want to find the time required to fill the pond if the hose is running and the outlet pipe is open. Let t = the time required to fill the pond.

We are combining **rates of work**, but the rates are not working together. Rates that work towards completion of a task are positive actions, which means positive values. Rates that work against the completion of the task are negative actions, which means negative values.

$$\underbrace{\frac{1}{t_1}}_{\text{Garden hose rate}} + \underbrace{\frac{-1}{t_2}}_{\text{Outlet pipe rate}} = \underbrace{\frac{1}{t}}_{\text{Combined rate}}$$

Substituting in the given individual times, we get

$$\underbrace{\frac{1}{2}}_{\text{Garden hose rate}} + \underbrace{\frac{-1}{10}}_{\text{Outlet pipe rate}} = \underbrace{\frac{1}{t}}_{\text{Combined rate}}$$

or

$$\underbrace{\frac{1}{2}}_{\text{Garden hose rate}} - \underbrace{\frac{1}{10}}_{\text{Outlet pipe rate}} = \underbrace{\frac{1}{t}}_{\text{Combined rate}}.$$

Because the outlet pipe is working against the garden hose, we subtract the rates. This is similar to a boat traveling upstream or downstream. When going downstream (with the current), we add rates. However, when going upstream (against the current), we subtract rates because the current is working against the boat.

The **LCD** is $10t$. Multiply both sides of the equation by the LCD and solve the resulting equation for t.

Original equation:	$\dfrac{1}{2} - \dfrac{1}{10} = \dfrac{1}{t}$
Multiply by the LCD:	$10t\left(\dfrac{1}{2} - \dfrac{1}{10}\right) = 10t\left(\dfrac{1}{t}\right)$
Distribute:	$5t - t = 10$
Simplify:	$4t = 10$
Divide both sides by 4:	$t = \dfrac{10}{4} = 2.5$

It will take 2.5 hours to fill the pond with the outlet pipe open.

You Try It Work through this **You Try It** problem.

Work Exercises 41 and 42 in this eText or in the MyMathLab Study Plan.

6.5 Exercises

In Exercises 1–4, determine if the statement is a rational equation. If not, state why.

1. $\dfrac{x - 3}{x^2 + 1} = \dfrac{\sqrt{x^2 - 4x + 1}}{x - 5}$

2. $4y^{-2} = 15 - 17y^{-1}$

3. $\dfrac{x^2 + 3x - 4}{2x + 5} = \dfrac{x - 2}{3x + 1}$

4. $\dfrac{x^3 - 4x^2 + 5x}{2x - 9}$

In Exercises 5–22, solve the rational equation. If there is no solution, state this as your answer.

5. $\dfrac{1}{5}x - \dfrac{5}{6} = \dfrac{2}{3}$

6. $\dfrac{n - 7}{8} = \dfrac{5n}{3} - \dfrac{7}{12}$

7. $\dfrac{1}{y} + \dfrac{1}{6} = \dfrac{1}{4}$

8. $\dfrac{5}{m} - \dfrac{3}{10} = \dfrac{3}{2m} - \dfrac{9}{4}$

9. $\dfrac{4}{3x} - \dfrac{x + 1}{x} = \dfrac{2}{5}$

10. $\dfrac{x - 1}{x} = \dfrac{3}{4} - \dfrac{3}{2x}$

11. $\dfrac{3}{z} - \dfrac{9}{z - 5} = \dfrac{1}{4}$

12. $\dfrac{5}{3p} + \dfrac{2p}{p - 5} = -\dfrac{4}{3}$

13. $\dfrac{b}{b - 5} + \dfrac{2}{b + 5} = \dfrac{50}{b^2 - 25}$

14. $\dfrac{3}{3m + 2} - \dfrac{2}{m - 6} = \dfrac{1}{3m^2 - 16m - 12}$

15. $\dfrac{1}{n + 4} + \dfrac{3}{n + 5} = \dfrac{1}{n^2 + 9n + 20}$

16. $\dfrac{7x + 2}{x^2 - 4} - \dfrac{3x + 1}{x + 2} = \dfrac{-3x^2 + 12x + 4}{x^2 - 4}$

17. $\dfrac{2w + 3}{4w^2 - 9} + 4 = \dfrac{16w^2 + 2w - 33}{4w^2 - 9}$

18. $\dfrac{6}{y^2 - 25} = \dfrac{3}{y^2 - 5y}$

19. $\dfrac{3}{t + 2} = \dfrac{6}{t + 16}$

20. $\dfrac{k}{7} = \dfrac{5}{k - 2}$

21. $\dfrac{x + 2}{x - 5} = \dfrac{x - 6}{x + 1}$

22. $\dfrac{x}{3x + 2} = \dfrac{x + 5}{3x - 1}$

In Exercises 23–26, find the zeros of the rational function.

23. $R(x) = \dfrac{3x + 4}{3x^2 + 13x + 4}$

24. $R(x) = \dfrac{x^2 - 5x - 24}{x + 3}$

25. $R(x) = \dfrac{2x^2 + 5x - 3}{x^2 - 9}$

26. $R(x) = \dfrac{6x^2 - 5x - 6}{4x^2 - 4x - 15}$

In Exercises 27–42, solve the application problems.

27. **Time to Complete an Order** Olivia works at a craft store and can make 2 stone crafts in 45 minutes. If she has been working for 30 minutes on an order for 7 stone crafts, how much longer does she need to work to complete the order?

28. **Commissions** Nadia, a personal trainer, sells 4 training packages and earns $700 in commissions. How much would she make in commissions if she sold 13 training packages?

29. **Planting Corn** To plant a 250-acre field, a farmer used 3500 pounds of corn seed. If the farmer still has 500 pounds of corn seed on hand, how many more pounds of corn seed will he need to buy in order to plant an additional 180-acre field?

30. **Similar Triangles** For two similar triangles, such as those shown in the figure below, the ratios of the lengths of corresponding sides are equal.

For example, $\dfrac{a}{A} = \dfrac{b}{B}$.

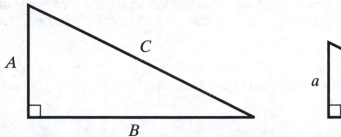

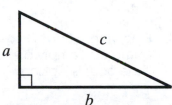

A right triangle with a hypotenuse measuring 10 inches and one leg measuring 4 inches has its hypotenuse increased to 12.5 inches. If the larger triangle is to be similar to the smaller triangle, by how much must the 4-inch leg be increased?

31. **Resistors in Parallel** If the total resistance of two circuits in parallel is 9 ohms and one circuit has three times the resistance of the other, find the resistance of each circuit. Use $\dfrac{1}{R} = \dfrac{1}{R_1} + \dfrac{1}{R_2}$, where R_1 and R_2 are the two individual resistances.

32. **Focal Length** The focal length f of a lens is given by the equation $\dfrac{1}{f} = \dfrac{1}{x_o} + \dfrac{1}{x_i}$, where x_o is the distance from the object to the lens, and x_i is the distance from the lens to the image of the object. If the focal length of a lens is 12 cm and an object is three times as far from the lens as its image, then how far is the object from the lens?

33. **Current Speed** On a float trip, Kim traveled 30 miles downstream in the same amount of time that it would take her to row 12 miles upstream. If her speed in still water is 7 mph, find the speed of the current.

34. **Plane Speed** A plane can fly 500 miles against the wind in the same amount of time that it can fly 725 miles with the wind. If the wind speed is 45 mph, find the speed of the plane in still air.

35. **Walking Speed** For exercise, Mae jogs 10 miles then walks an additional 2 miles to cool down. If her jogging speed is 4.5 miles per hour faster than her walking speed and her total exercise time is 3 hours, what is her walking speed?

36. **Moving Sidewalk** Choi can walk 108 m in 72 seconds. Standing on a moving sidewalk at Thailand's Suvarnabhumi International Airport, she can travel 108 m in 40 seconds. How long will it take her to travel 108 m if she walks on the moving sidewalk? Round your answer to the nearest tenth.

37. **Making Copies** Beverly can copy final exams in 30 minutes using a new copy machine. Using an old copy machine, the same job takes 45 minutes. If both copy machines are used, how long will it take to copy the final exams?

38. **Mowing the Lawn** By himself, Chris can mow his lawn in 60 minutes. If his daughter Claudia helps, they can mow the lawn together in 40 minutes. How long would it take Claudia to mow the lawn by herself?

39. **Staining a Deck** It takes Shawn 2 more hours to stain a deck than Michelle. Together it takes them 2.4 hours to complete the work. How long would it take Shawn to stain the deck by himself?

40. **Lining Fields** Together, Sari and Rilee can mark the lines on the soccer fields at a recreation center 18 minutes faster than if Rilee lines the fields on her own. If it takes Sari 56 minutes to line the fields by herself, how long would it take Rilee to do the job alone?

41. **Filling a Hot Tub** A garden hose can fill a hot tub in 180 minutes, whereas the drain on a hot tub can empty it in 300 minutes. If the drain is accidentally left open, how long will it take to fill the tub?

42. **Emptying a Boat's Bilge** A leak in the hull will fill a boat's bilge in 8 hours. The boat's bilge pump can empty a full bilge in 3 hours. How long will it take the pump to empty a full bilge if the boat is leaking?

6.6 Variation

THINGS TO KNOW

Before working through this section, be sure you are familiar with the following concepts:

VIDEO   ANIMATION   INTERACTIVE

You Try It
1. Solve a Formula for a Given Variable
(Section 1.5, Objective 1)   ▶

You Try It
2. Use Linear Models to Solve Application Problems; Direct Variation
(Section 2.5, Objective 8)   ▶

OBJECTIVES

1 Solve Problems Involving Direct Variation

2 Solve Problems Involving Inverse Variation

3 Solve Problems Involving Combined Variation

....

OBJECTIVE 1   SOLVE PROBLEMS INVOLVING DIRECT VARIATION

In **Section 2.5**, we introduced **direct variation** in the context of linear **models**. Often in application problems, we need to know how one quantity varies with respect to other quantities. A variation equation allows us to describe how one quantity changes with respect to one or more additional quantities. In this section, we discuss direct variation, inverse variation, and combined variation.

Direct variation means that one **variable** is a constant multiple of another variable.

Definition   Direct Variation

For an equation of the form

$$y = kx,$$

we say that y **varies directly** with x, or y is **directly proportional** to x. The nonzero constant k is called the **constant of variation** or the **proportionality constant**.

My video summary ▶ **Example 1 Direct Variation**

Suppose y varies directly with x, and $y = 20$ when $x = 8$.

a. Find the equation that relates x and y.

b. Find y when $x = 12$.

Solutions Read the following, or watch this **video** for a complete solution.

a. Because y varies directly with x, the equation has the form $y = kx$. Use the fact that $y = 20$ when $x = 8$ to find k.

$$\text{Write the direct variation equation:} \quad y = kx$$

$$\text{Substitute } y = 20 \text{ and } x = 8: \quad 20 = k(8)$$

$$\text{Divide both sides by 8:} \quad \frac{20}{8} = \frac{k(8)}{8}$$

$$\text{Simplify:} \quad 2.5 = k$$

The constant of variation is 2.5, so the equation is $y = 2.5x$.

b. The variables y and x are related by the equation $y = 2.5x$. To find y when $x = 12$, substitute 12 for x in the equation and simplify:

$$y = 2.5(12) = 30$$

Therefore, $y = 30$ when $x = 12$.

You Try It Work through this You Try It problem.

Work Exercises 1 and 2 in this eText or in the MyMathLab Study Plan.

In Example 1, the model $y = kx$ is a **linear equation**. However, direct variation equations will not always be linear. For example, if y varies directly with the **cube of** x, then the equation will be $y = kx^3$, where k is the constant of variation.

My video summary ▶ **Example 2 Direct Variation**

Suppose y varies directly with the cube of x, and $y = 375$ when $x = 5$.

a. Find the equation that relates x and y.

b. Find y when $x = 2$.

Solutions Because y varies directly with the cube of x, or x^3, the variation equation has the form $y = kx^3$. Use the fact that $y = 375$ when $x = 5$ to find k, and then the equation. Try to complete this problem on your own. View the **answer**, or watch this **video** for a complete solution.

You Try It   Work through this **You Try It** problem.

Work Exercises 3–6 in this eText or in the MyMathLab Study Plan.

For direct variation, the **ratio** of the two quantities is constant (the **constant of variation**). For example, consider $y = kx$ and $y = kx^3$.

$\underbrace{y = kx}$	$\rightarrow$	$\underbrace{\dfrac{y}{x}}$	$=$	$\underbrace{k}$	$\underbrace{y = kx^3}$	$\rightarrow$	$\underbrace{\dfrac{y}{x^3}}$	$=$	$\underbrace{k}$
y varies directly with x		Ratio of the quantities y and x		Constant of variation	y varies directly with the cube of x		Ratio of the quantities y and x^3		Constant of variation

Problems involving variation can generally be solved using the following guidelines.

Solving Variation Problems

Step 1 Translate the problem into an equation that models the situation.

Step 2 Substitute given values for the variables into the equation and solve for the **constant of variation**, k.

Step 3 Substitute the value for k into the equation to form the general **model**.

Step 4 Use the general model to answer the question posed in the problem.

My video summary ▶ **Example 3 Kinetic Energy**

The kinetic energy of an object in motion varies directly with the square of its speed. If a van traveling at a speed of 30 meters per second has 945,000 joules of kinetic energy, how much kinetic energy does it have if it is traveling at a speed of 20 meters per second?

Solution Read the following, or watch this **video** for a complete solution.

Follow the guidelines for solving variation problems.

Step 1 We are told that the kinetic energy of an object in motion varies **directly** with the square of its speed. If we let $E =$ kinetic energy and $s =$ speed, we can translate the problem statement into the model

$$E = ks^2,$$

where k is the constant of variation.

Step 2 To determine the value of k, use the fact that the kinetic energy is 945,000 joules when the velocity is 30 meters per second.

Substitute 945,000 for E and 30 for s: $945,000 = k(30)^2$

Simplify 30^2: $945,000 = 900k$

Divide both sides by 900: $1050 = k$

Step 3 The constant of variation is 1050, so the general model is $E = 1050s^2$.

Step 4 We want to determine the kinetic energy of the van if its speed is 20 meters per second. Substituting 20 for s gives

$$E = 1050(20)^2 = 1050(400) = 420{,}000.$$

The van will have 420,000 joules of kinetic energy if it is traveling at a speed of 20 meters per second.

You Try It Work through this You Try It problem.

Work Exercises 7 and 8 in this eText or in the MyMathLab Study Plan.

 Example 4 Measuring Leanness

The Ponderal Index measure of leanness states that weight varies directly with the cube of height. If a "normal" person who is 1.2 m tall weighs 21.6 kg, how much will a "normal" person weigh if they are 1.8 m tall?

Solution Follow the guidelines for solving variation problems.

Let w = weight and h = height. Because weight varies directly with the cube of height, the model is

$$w = kh^3,$$

where k is the constant of variation.

Try to finish solving this problem on your own.

View the answer, or watch this video for a detailed solution.

You Try It Work through this You Try It problem.

Work Exercises 9 and 10 in this eText or in the MyMathLab Study Plan.

OBJECTIVE 2 SOLVE PROBLEMS INVOLVING INVERSE VARIATION

Inverse variation means that one variable is a constant multiple of the **reciprocal** of another variable.

Definition Inverse Variation

For equations of the form

$$y = \frac{k}{x} \quad \text{or} \quad y = k \cdot \frac{1}{x},$$

we say that y **varies inversely** with x, or y is **inversely proportional** to x. The constant k is called the **constant of variation**.

For inverse variation, the **product** of the two quantities is constant (the constant of variation). For example, consider $y = \frac{k}{x}$ and $y = \frac{k}{x^2}$.

$$y = \frac{k}{x} \quad \rightarrow \quad xy \quad = \quad k$$

$\underbrace{}$ y varies inversely with x $\underbrace{}$ Product of the quantities y and x $\underbrace{}$ Constant of variation

$$y = \frac{k}{x^2} \quad \rightarrow \quad x^2 y \quad = \quad k$$

$\underbrace{}$ y varies inversely with the square of x $\underbrace{}$ Product of the quantities y and x^2 $\underbrace{}$ Constant of variation

My video summary ▶ **Example 5 Inverse Variation**

Suppose y varies inversely with x, and $y = 72$ when $x = 50$.

a. Find the equation that relates x and y.

b. Find y when $x = 45$.

Solutions

a. Because y varies inversely with x, the equation has the form $y = \frac{k}{x}$. We use the fact that $y = 72$ when $x = 50$ to find k.

Write the inverse variation equation: $y = \dfrac{k}{x}$

Substitute $y = 72$ and $x = 50$: $72 = \dfrac{k}{50}$

Multiply both sides by 50: $3600 = k$

The constant of variation is 3600, so the equation is $y = \dfrac{3600}{x}$.

b. Try to work this part on your own. View the **answer**, or watch this **video** for a complete solution to both parts.

You Try It **Work through this You Try It problem.**

Work Exercises 11–14 in this eText or in the MyMathLab **Study Plan.**

Problems involving inverse variation can be solved using the same **guidelines for solving variation problems.**

My video summary ▶ **Example 6 Density of an Object**

For a given mass, the density of an object is **inversely proportional** to its volume. If 50 cubic centimeters (cm^3) of an object with a density of 28 g/cm^3 is compressed to 40 cm^3, what would be its new density?

Solution Read the following, or watch this **video** for a complete solution.

Step 1 We are told that the density of an object with a given mass varies inversely with its volume. If we let D = density and V = volume, we can translate the problem statement into the **model**

$$D = \frac{k}{V},$$

where k is the **constant of variation.**

Step 2 To determine the value of k, use the fact that the density is 28 g/cm^3 when the volume is 50 cm^3.

Substitute 28 for D and 50 for V: $28 = \dfrac{k}{50}$

Multiply both sides by 50: $1400 = k$

Step 3 The constant of variation is 1400, so the general model is $D = \dfrac{1400}{V}$.

Step 4 We want to determine the density of the object if the volume is compressed to 40 cm^3. Substituting 40 for V, gives

$$D = \frac{1400}{40} = 35 \text{ g/cm}^3.$$

The density of the compressed object would be 35 g/cm^3.

You Try It Work through this You Try It problem.

Work Exercises 15 and 16 in this eText or in the MyMathLab Study Plan.

My video summary ▶ **Example 7 Shutter Speed**

The shutter speed, S, of a camera varies inversely as the square of the aperture setting, f. If the shutter speed is 125 for an aperture of 5.6, what is the shutter speed if the aperture is 1.4?

Solution Follow the guidelines for solving variation problems.

Let S = shutter speed and f = aperture setting. Because shutter speed varies inversely with the square of the aperture setting, the **model** is

$$S = \frac{k}{f^2},$$

where k is the **constant of variation**.

Try to finish this problem on your own. View the **answer**, or watch this **video** for a detailed solution.

You Try It Work through this You Try It problem.

Work Exercises 17 and 18 in this eText or in the MyMathLab Study Plan.

OBJECTIVE 3 SOLVE PROBLEMS INVOLVING COMBINED VARIATION

When a variable is related to more than one other variable, we call this **combined variation**. Some examples of **combined variation** are

$$y = k\frac{x}{z} \qquad y = kxz \qquad y = k\frac{w^2 x}{z}.$$

In the first example, we say that y varies directly as x and inversely as z. In the second example, we say that y varies directly as x and z. In the third example, we say that y varies directly as x and the square of w, and inversely as z.

In general, variables **directly related** to the dependent variable occur in the numerator while variables **inversely related** occur in the denominator.

$$y = k\frac{x}{z}$$ ←— Directly related
←— Inversely related

$$y = kxz$$ ←— Directly related

$$y = k\frac{w^2x}{z}$$ ←— Directly related
←— Inversely related

When a variable is directly proportional to the product of two or more other variables, such as $y = kxz$, this is often called **joint variation**. In this case, we would say that y varies **jointly** as x and z.

My video summary ▶ **Example 8 Combined Variation**

Suppose y varies directly with w and inversely with x, and $y = 240$ when $w = 20$ and $x = 4$.

a. Find the equation that relates w, x, and y.

b. Find y when $w = 10$ and $x = 3$.

Solutions Read the following, or watch this **video** for a complete solution.

a. Because y varies directly with w and inversely with x, the equation has the form $y = k\dfrac{w}{x}$.

We use the fact that $y = 240$ when $w = 20$ and $x = 4$ to find k.

Write the combined variation equation: $y = k\dfrac{w}{x}$

Substitute $y = 240, w = 20$, and $x = 4$: $240 = k\dfrac{20}{4}$

Simplify: $240 = 5k$

Divide both sides by 5: $48 = k$

The **constant of variation** is 48, so the equation is $y = 48\dfrac{w}{x}$, or $y = \dfrac{48w}{x}$.

b. To find y when $w = 10$ and $x = 3$, substitute the values into the equation and simplify:

$$y = \frac{48(10)}{3} = 160$$

Therefore, $y = 160$ when $w = 10$ and $x = 3$.

You Try It Work through this You Try It problem.

Work Exercises 19–22 in this eText or in the MyMathLab Study Plan.

My video summary ▶ **Example 9 Volume of a Conical Tank**

The number of gallons of a liquid that can be stored in a conical tank is directly proportional to the area of the base of the tank and its height (joint variation). A tank with a base area of 1200 square feet and a height of 15 feet holds 45,000 gallons of liquid. How tall must the tank be to hold 75,000 gallons of liquid if its base area is 1500 square feet?

Solution Read the following, or watch this **video** for a complete solution.

Follow the **guidelines for solving variation problems.**

Step 1 We are told that the volume is **directly** related to the area of the base of the tank and its height. If we let V = volume, B = base area, and h = height, we can translate the problem statement into the model

$$V = kBh \leftarrow \text{Joint variation } (B \text{ and } h \text{ both direct}),$$

where k is the constant of variation.

Step 2 To determine the value of k, use the fact that a tank with a base area of 1200 square feet and a height of 15 feet holds 45,000 gallons.

Substitute 1200 for B, 15 for h, and 45,000 for V: $45,000 = k(1200)(15)$

Simplify: $45,000 = 18,000k$

Divide both sides by 18,000: $2.5 = k$

Step 3 The constant of variation is 2.5, so the general model is $V = 2.5Bh$.

Step 4 We want to determine the height of a tank with a base area of 1500 square feet that holds 75,000 gallons. Substituting 75,000 for V and 1500 for B, we then solve for h.

Substitute 1500 for B and 75,000 for V: $75,000 = 2.5(1500)h$

Simplify: $75,000 = 3750h$

Divide both sides by 3750: $20 = h$

The tank would need to be 20 feet tall.

You Try It **Work through this You Try It problem.**

Work Exercises 23 and 24 in this eText or in the My**Math**Lab **Study Plan.**

My video summary ▶ **Example 10 Electrical Resistance**

The resistance of a wire varies directly with the length of the wire and inversely with the square of its radius. A wire with a length of 500 cm and a radius of 0.5 cm has a resistance of 15 ohms. Determine the resistance of an 800 cm piece of similar wire with a radius of 0.8 cm.

Solution Follow the **guidelines for solving variation problems.**

Step 1 We are told that resistance varies **directly** with the length of the wire and **inversely** with the square of its radius. Letting R = resistance, L = length, and r = radius, we can translate the problem statement into the **model**

$$R = k\frac{L}{r^2} \quad \begin{array}{l} \leftarrow \text{Length (direct)} \\ \leftarrow \text{Square of the radius (inverse)} \end{array}$$

where k is the **constant of variation.**

Step 2 To determine the value of k, use the fact that a wire of length 500 cm and radius 0.5 cm has a resistance of 15 ohms.

Substitute 500 for L, 0.5 for r, and 15 for R: $15 = k\dfrac{(500)}{(0.5)^2}$

Simplify: $15 = 2000k$

Divide both sides by 2000: $0.0075 = k$

Try to finish solving this problem on your own. View the **answer**, or watch this **video** for a detailed solution.

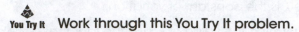 **Work through this You Try It problem.**

Work Exercises 25 and 26 in this eText or in the My Math**Lab Study Plan.**

✎ *My video summary* ▶ **Example 11 Burning Calories**

For a fixed speed, the number of calories burned while jogging varies **jointly** with the weight of the jogger (in kg) and the time spent jogging (in minutes). If a 100-kg man jogs for 40 minutes and burns 490 calories, how many calories will a 130-kg man burn if he jogs for 60 minutes at the same speed?

Solution Letting C = calories burned, W = weight in kg, and T = time jogging in minutes, gives the **model** $C = kWT$, where k is the **constant of variation**. Try to finish solving this problem on your own. View the **answer**, or watch this **video** for a complete solution.

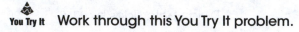 **Work through this You Try It problem.**

Work Exercises 27 and 28 in this eText or in the My Math**Lab Study Plan.**

6.6 Exercises

In Exercises 1–28, solve each variation problem.

1. Suppose y varies directly with x, and $y = 14$ when $x = 4$.

   a. Find the equation that relates x and y.

   b. Find y when $x = 7$.

2. Suppose S in directly proportional to t, and $S = 900$ when $t = 25$.

   a. Find the equation that relates t and S.

   b. Find S when $t = 4$.

3. Suppose y varies directly with the square of x, and $y = 216$ when $x = 3$.

   a. Find the equation that relates x and y.

   b. Find y when $x = 5$.

4. Suppose M is directly proportional to the square of r, and $M = 14$ when $r = \dfrac{1}{6}$.

   a. Find the equation that relates r and M.

   b. Find M when $r = \dfrac{2}{3}$.

5. Suppose p is directly proportional to the cube of n, and $p = 135$ when $n = 6$.

   a. Find the equation that relates n and p.

   b. Find p when $n = 10$.

6. Suppose A varies directly with the cube of m, and $A = 48$ when $m = 4$.

   a. Find the equation that relates m and A.

   b. Find A when $m = 8$.

7. **Scuba Diving** The water pressure on a scuba diver is directly proportional to the depth of the diver. If the pressure on a diver is 13.5 psi when she is 30 feet below the surface, how far below the surface will she be when the pressure is 18 psi?

8. **Pendulum Length** The length of a simple pendulum varies directly with the square of its period. If a pendulum of length 2.25 meters has a period of 3 seconds, how long is a pendulum with a period of 8 seconds?

9. **Water Flow Rate** For a fixed water flow rate, the amount of water that can be pumped through a pipe varies directly as the square of the diameter of the pipe. In one hour, a pipe with an 8-inch diameter can pump 400 gallons of water. Assuming the same water flow rate, how much water could be pumped through a pipe that is 12 inches in diameter?

10. **Falling Distance** The distance an object falls varies directly with the square of the time it spends falling. If a ball falls 19.6 meters after falling for 2 seconds, how far will it fall after 9 seconds?

11. Suppose y varies inversely with x, and $y = 24$ when $x = 25$.

    a. Find the equation that relates x and y.

    b. Find y when $x = 30$.

12. Suppose p is inversely proportional to q, and $p = 3.5$ when $q = 4.6$.

    a. Find the equation that relates q and p.

    b. Find p when $q = 1.4$.

13. Suppose n varies inversely with the square of m, and $n = 6.4$ when $m = 9$.

    a. Find the equation that relates m and n.

    b. Find n when $m = 4$.

14. Suppose d is inversely proportional to the cube of t, and $d = 76.8$ when $t = 2.5$.

    a. Find the equation that relates t and d.

    b. Find d when $t = 5$.

15. **Car Depreciation** The value of a car is inversely proportional to its age. If a car is worth $8100 when it is 4 years old, how old will it be when it is worth $3600?

16. **Electric Resistance** For a given voltage, the resistance of a circuit is inversely related to its current. If a circuit has a resistance of 5 ohms and a current of 12 amps, what is the resistance if the current is 20 amps?

17. **Weight of an Object** The weight of an object within Earth's atmosphere varies inversely with the square of the distance of the object from Earth's center. If a low Earth orbit satellite weighs 100 kg on Earth's surface (6400 km), how much will it weigh in its orbit 800 km above Earth?

18. **Light Intensity** The intensity of a light varies inversely as the square of the distance from the light source. If the intensity from a light source 3 feet away is 8 lumens, what is the intensity at a distance of 2 feet?

19. Suppose y varies directly with x and inversely with z, and $y = 48$ when $x = 8$ and $z = 2.5$.

    a. Find the equation that relates y, x, and z.

    b. Find y when $x = 24$ and $z = 18$.

20. Suppose A varies jointly with b and c, and $A = 15$ when $b = 10$ and $c = 3$.

    a. Find the equation that relates A, b, and c.

    b. Find b when $A = 27$ and $c = 9$.

21. Suppose P varies directly with n and the square of m, and $P = 43.2$ when $n = 4$ and $m = 1.5$.

    a. Find the equation that relates P, n and m.

    b. Find P when $n = 3.5$ and $m = 2$.

22. Suppose A varies directly with the square of t and inversely with the cube of r, and $A = 54$ when $t = 6$ and $r = 2$.

    a. Find the equation that relates A, t, and r.

    b. Find A when $t = 16$ and $r = 4$.

23. **Simple Interest** For a car loan using simple interest at a given rate, the amount of interest charged varies jointly with the loan amount and the time of the loan (in years). If a $15,000 car loan earns $2175 in simple interest over 5 years, how much interest will a $32,000 car loan earn over 6 years?

24. **Horsepower of a Pump** The horsepower of a water pump varies jointly with the weight of the water moved (in lb) and the length of the discharge head (in feet). A 3.1-horsepower pump can move 341 lb of water against a discharge head of 300 feet. What is the horsepower of a pump that moves 429 lb of water against a discharge head of 400 feet?

25. **Gas Pressure** The pressure of a gas in a container varies directly with its temperature and inversely with its volume. At a temperature of 90 Kelvin and a volume of 10 cubic meters, the pressure is 32.4 kilograms per square meter. What is the pressure if the temperature is increased to 100 Kelvin and the volume is decreased to 8 cubic meters?

26. **Spindle Speed** In milling operations, the spindle speed S (in revolutions per minute) is directly related to the cutting speed C (in feet per minute) and inversely related to the tool diameter D (in inches). A milling cut taken with a 2-inch high-speed drill and a cutting speed of 70 feet per minute has a spindle speed of 133.7 revolutions per minute. What is the spindle speed for a cut taken with a 4-inch high-speed drill and a cutting speed of 50 feet per minute?

27. **Load on a Beam** The load that can be supported by a rectangular beam varies jointly as the width of the beam and the square of its height, and inversely as the length of the beam. A beam 12 feet long, with a width of 6 inches and a height of 4 inches can support a maximum load of 900 pounds. If a similar board has a width of 8 inches and a height of 5 inches, how long must it be to support 1200 pounds?

28. **Windmill Power** The power produced by a windmill varies directly with the square of its diameter and the cube of the wind speed. A fan with a diameter of 2.5 meters produces 270 watts of power if the wind speed is 3 m/s. How much power would a similar fan produce if it had a diameter of 4 meters and the wind speed was 4 m/s?

CHAPTER SEVEN
Radicals and Rational Exponents

CHAPTER SEVEN CONTENTS

7.1 Radical Expressions

THINGS TO KNOW

Before working through this section, be sure you are familiar with the following concepts:

VIDEO   ANIMATION   INTERACTIVE

 You Try It
1. Compute the Absolute Value of a Real Number (Section R.1, Objective 5) ▶

 You Try It
2. Simplify Numeric Expressions Containing Exponents and Radicals (Section R.2, Objective 2) ▶

OBJECTIVES

1 Find Square Roots of Perfect Squares
2 Approximate Square Roots
3 Simplify Radical Expressions of the Form $\sqrt{a^2}$
4 Find Cube Roots
5 Find and Approximate nth Roots

....

OBJECTIVE 1   FIND SQUARE ROOTS OF PERFECT SQUARES

Recall from **Section R.2** that a **square root** of a **non-negative** real number a is a **real number** b that, when **squared**, results in a. So, b is a square root of a if $b^2 = a$.

Every positive real number has two square roots: one positive and one negative.

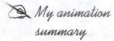 *My animation summary*

⊛ **Concept Animation** Watch this animation to review this concept.

Definition   Principal and Negative Square Roots

A non-negative real number b is the **principal square root** of a non-negative real number a, denoted as $b = \sqrt{a}$, if $b^2 = a$.

A negative real number b is the **negative square root** of a non-negative real number a, denoted as $b = -\sqrt{a}$, if $b^2 = a$.

The **expression** $\sqrt{4}$ is called a *radical expression*. A **radical expression** is an expression that contains a **radical sign**. The expression beneath the radical sign is called the **radicand**.

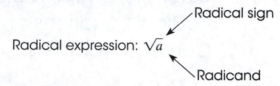

Radical sign

Radical expression: $\sqrt{a}$

Radicand

When the radicand is a **perfect square**, the **square root** will simplify to a **rational number**.

My video summary ▶ **Example 1 Finding Square Roots**

Evaluate.

a. $\sqrt{64}$ b. $-\sqrt{169}$ c. $\sqrt{-100}$

d. $\sqrt{\dfrac{9}{25}}$ e. $\sqrt{0.81}$ f. $\sqrt{0}$

Solutions

a. Find the **principal square root** of 64. Since $8^2 = 64$, $\sqrt{64} = 8$.

b. Find the **negative square root** of 169. Since $(-13)^2 = 169$, $-\sqrt{169} = -13$.

c. There is no **real number** that can be **squared** to result in -100 because the square of a real number will always be non-negative. So, $\sqrt{-100}$ is not a real number.

d.–f. Try to evaluate these square roots on your own. Check your **answers**, or watch this **video** to see complete solutions to all parts.

⚠ The square root of a negative number is not a real number.

 You Try It   Work through this You Try It problem.

Work Exercises 1–6 in this eText or in the MyMathLab Study Plan.

OBJECTIVE 2   APPROXIMATE SQUARE ROOTS

In Example 1, we saw that a **square root** simplifies to a **rational number** if the radicand is a **perfect square**. But what happens when the radicand is not a perfect square? In this case, the square root is an **irrational number**.

Consider $\sqrt{12}$. The radicand 12 is not a perfect square, so $\sqrt{12}$ is an irrational number. For such **radical expressions**, we can find decimal approximations using a calculator. Before doing this, consider what might be a **reasonable** result.

✎ *My animation summary* ✦ **Concept Animation** Watch this **animation** to explore this concept.

Figure 1 shows the TI-84 Plus C calculator display for $\sqrt{12}$. Rounding to three decimal places, we have $\sqrt{12} \approx 3.464$. This approximation is between 3 and 4 as expected.

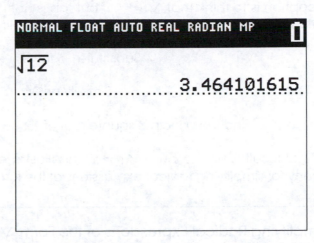

```
NORMAL FLOAT AUTO REAL RADIAN MP     🔋
√12
                        3.464101615
```

Figure 1
TI-84 Plus C Display Approximating $\sqrt{12}$

Example 2 Approximating Square Roots

Use a calculator to approximate each **square root**. Round to three decimal places. Check that the answer is **reasonable**.

a. $\sqrt{7}$ b. $\sqrt{40}$ c. $\sqrt{115}$

✎ *My video summary* ▶ **Solutions** Watch this **video**, or continue reading for the full solutions.

a. The radicand 7 is between the **perfect squares** $2^2 = 4$ and $3^2 = 9$, so $\sqrt{4} < \sqrt{7} < \sqrt{9}$ or $2 < \sqrt{7} < 3$. The square root is between 2 and 3. From Figure 2, we get $\sqrt{7} \approx 2.646$, which is between 2 and 3.

b. 40 is between $6^2 = 36$ and $7^2 = 49$, so $\sqrt{36} < \sqrt{40} < \sqrt{49}$ or $6 < \sqrt{40} < 7$. Figure 2 shows us $\sqrt{40} \approx 6.325$, which is between 6 and 7.

c. 115 is between $10^2 = 100$ and $11^2 = 121$, so $\sqrt{100} < \sqrt{115} < \sqrt{121}$ or $10 < \sqrt{115} < 11$. In Figure 2, we see $\sqrt{115} \approx 10.724$, which is reasonable.

```
NORMAL FLOAT AUTO REAL RADIAN MP     🔋
√7
                        2.645751311
√40
                         6.32455532
√115
                        10.72380529
```

Figure 2
TI-84 Plus C Display for Estimating $\sqrt{7}$, $\sqrt{40}$, and $\sqrt{115}$

 **You Try It** Work through this You Try It problem.

Work Exercises 7–12 in this eText or in the MyMathLab Study Plan.

OBJECTIVE 3 SIMPLIFY RADICAL EXPRESSIONS OF THE FORM $\sqrt{a^2}$

We now consider **square roots** with variables in the **radicand**. A common misconception is to think that $\sqrt{a^2} = a$, but this is not necessarily true. To see why, substitute $a = -5$ in the expression $\sqrt{a^2}$ and simplify:

$$\text{Substitute } -5 \text{ for } a: \quad \sqrt{a^2} = \sqrt{(-5)^2}$$

$$\text{Simplify } (-5)^2: \quad = \sqrt{25}$$

$$\text{Find the principal square root of 25:} \quad = 5$$

The final result is not -5 but $|-5| = 5$. This illustrates the following square root property for simplifying **radical expressions** of the form $\sqrt{a^2}$.

Simplifying Radical Expressions of the Form $\sqrt{a^2}$

For any real number a,

$$\sqrt{a^2} = |a|.$$

When taking the square root of a **base** raised to the second **power**, the result will be the **absolute value** of the base.

My video summary ▶ **Example 3 Simplifying Radical Expressions of the Form $\sqrt{a^2}$**

Simplify.

a. $\sqrt{(-12)^2}$ b. $\sqrt{(2x - 5)^2}$ c. $\sqrt{100x^2}$

d. $\sqrt{x^2 + 12x + 36}$ e. $\sqrt{9x^4}$ f. $\sqrt{y^6}$

Solutions

a. $\sqrt{(-12)^2} = |-12| = 12$

b. $\sqrt{(2x - 5)^2} = |2x - 5|$. Since $2x - 5$ could be negative, the **absolute value** symbol is required to ensure a non-negative result.

c. Since $100x^2 = (10x)^2$, then $\sqrt{100x^2} = \sqrt{(10x)^2} = |10x|$ or $10|x|$.

d.–f. Try to simplify each square root on your own by first writing the **radicand** as a **base** raised to the second **power**. Include absolute value notation if necessary. Check your answers, or watch this **video** for the complete solutions to all parts.

 You Try It Work through this You Try It problem.

Work Exercises 13–24 in this eText or in the MyMathLab Study Plan.

OBJECTIVE 4   FIND CUBE ROOTS

The process of finding **square roots** can be applied to other types of roots. For example, the **cube root** of a real number a is a real number b that, when **cubed**, results in a. So, b is the cube root of a if $b^3 = a$. For example, 2 is the cube root of 8 because $2^3 = 8$. To write the cube root of a, we use the notation $\sqrt[3]{a}$. The 3 in this **radical expression** indicates a cube root instead of a square root. Using this notation, we write $\sqrt[3]{8} = 2$.

> **Definition**   Cube Roots
>
> A real number b is the **cube root** of a real number a, denoted as $b = \sqrt[3]{a}$, if $b^3 = a$.

Recall that the cube of a negative number is a negative number. Unlike square roots, cube roots can have negative numbers in the **radicand**. For example, $\sqrt[3]{-64} = -4$ because $(-4)^3 = -64$. Every real number has one real cube root. If the radicand is positive, the cube root will be positive. If the radicand is negative, the cube root will be negative.

⚠ When simplifying expressions of the form $\sqrt[3]{a^3}$, do not use absolute value symbols. Doing so may lead to an incorrect result. For example, $\sqrt[3]{(-5)^3} = -5$, not $|-5| = 5$.

> **Simplifying Radical Expressions of the Form $\sqrt[3]{a^3}$**
>
> For any real number a,
> $$\sqrt[3]{a^3} = a.$$

 My video summary ▶ **Example 4 Finding Cube Roots**

Simplify.

a. $\sqrt[3]{125}$ b. $\sqrt[3]{-1000}$ c. $\sqrt[3]{x^{15}}$

d. $\sqrt[3]{0.064}$ e. $\sqrt[3]{\dfrac{8}{27}}$ f. $\sqrt[3]{-64y^9}$

Solutions

a. $\sqrt[3]{125} = \sqrt[3]{5^3} = 5$

b. $\sqrt[3]{-1000} = \sqrt[3]{(-10)^3} = -10$

c. $\sqrt[3]{x^{15}} = \sqrt[3]{(x^5)^3} = x^5$

d.–f. Try to simplify each **cube root** on your own by first writing each **radicand** as a **base** raised to the third **power**. Check your **answers**, or watch this **video** for the complete solutions to all parts.

You Try It   Work through this **You Try It** problem.

Work Exercises 25–32 in this eText or in the MyMathLab Study Plan.

OBJECTIVE 5   FIND AND APPROXIMATE nTH ROOTS

We know that 2 is a **square root** of 4 because $2^2 = 4$ and 2 is the **cube root** of 8 because $2^3 = 8$. Also, 2 is a **4th root** of 16 because $2^4 = 16$, and 2 is the **5th root** of 32 because $2^5 = 32$. We denote each of these roots using **radical signs** as follows: $\sqrt{4} = 2$, $\sqrt[3]{8} = 2$, $\sqrt[4]{16} = 2$, and $\sqrt[5]{32} = 2$. These are called *nth roots*.

Definition   Principal nth Roots

If a and b are **real numbers** and n is an **integer** such that $n \geq 2$, then b is the **principal nth root** of a, denoted as $b = \sqrt[n]{a}$, if $b^n = a$.

Note: b must be non-negative when n is even.

In the notation $\sqrt[n]{a}$, n is called the **index** of the **radical expression**. The index indicates the type of root. For example, a cube root has an index of $n = 3$. If no **index** is shown, then it is understood to be $n = 2$ for a square root.

If n is an odd integer, then the **radicand** a can be any real number. However, if n is an even integer, then the radicand a must be **non-negative**. To simplify nth roots, we use the following general rule:

Simplifying Radical Expressions of the Form $\sqrt[n]{a^n}$

If n is an integer such that $n \geq 2$ and a is a real number, then

$$\sqrt[n]{a^n} = a \quad \text{if } n \text{ is odd.}$$

$$\sqrt[n]{a^n} = |a| \quad \text{if } n \text{ is even.}$$

My video summary ▶ **Example 5 Finding nth Roots**

Simplify.

a. $\sqrt[4]{81}$ b. $\sqrt[5]{-32}$ c. $\sqrt[6]{\dfrac{1}{64}}$

d. $\sqrt[5]{x^{15}}$ e. $\sqrt[6]{(x-7)^6}$ f. $\sqrt[4]{-1}$

Solutions

a. $\sqrt[4]{81} = \sqrt[4]{3^4} = |3| = 3$

b. $\sqrt[5]{-32} = \sqrt[5]{(-2)^5} = -2$

c. $\sqrt[6]{\dfrac{1}{64}} = \sqrt[6]{\left(\dfrac{1}{2}\right)^6} = \left|\dfrac{1}{2}\right| = \dfrac{1}{2}$

d.–f. Try to simplify each **nth root** on your own by first writing each **radicand** as a base raised to the **nth power**. Check your **answers**, or watch this **video** for complete solutions to all parts.

When simplifying expressions of the form $\sqrt[n]{a^n}$, use absolute value symbols when n is even but not when n is odd.

 Work through this You Try It problem.

Work Exercises 33–38 in this eText or in the MyMathLab Study Plan.

In **Example 5**, we saw that $\sqrt[4]{81} = 3$ because $81 = 3^4$. Since 81 is a **perfect 4th power**, its 4th root is a **rational number**. But what if the **radicand** of an nth root is not a perfect nth power? Such expressions represent **irrational numbers**. As with square roots, we can use a calculator to approximate these types of nth roots.

My video summary ▶ **Example 6 Approximating nth Roots**

Use a calculator to approximate each root. Round to three decimal places. Check that the answer is **reasonable**.

a. $\sqrt[3]{7}$ b. $\sqrt[4]{215}$ c. $\sqrt[5]{176}$

Solutions

a. The radicand 7 is between the **perfect cubes** $1^3 = 1$ and $2^3 = 8$, so $\sqrt[3]{1} < \sqrt[3]{7} < \sqrt[3]{8}$ or $1 < \sqrt[3]{7} < 2$. From Figure 3, we get $\sqrt[3]{7} \approx 1.913$, which is between 1 and 2.

b. 215 is between $3^4 = 81$ and $4^4 = 256$, so $3 < \sqrt[4]{215} < 4$. In Figure 3, we see that $\sqrt[4]{215} \approx 3.829$, which is between 3 and 4.

c. 176 is between $2^5 = 32$ and $3^5 = 243$, so $2 < \sqrt[5]{176} < 3$. Figure 3 shows us that $\sqrt[5]{176} \approx 2.813$, which is reasonable.

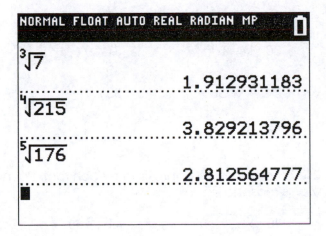

Figure 3
TI-84 Plus C Display for Estimating $\sqrt[3]{7}$, $\sqrt[4]{215}$, and $\sqrt[5]{176}$

 Work through this You Try It problem.

Work Exercises 39–46 in this eText or in the MyMathLab Study Plan.

7.1 Exercises

In Exercises 1–6, evaluate each square root. If the answer is not a real number, state so.

1. $\sqrt{100}$

2. $-\sqrt{49}$

3. $\sqrt{\dfrac{9}{121}}$

4. $\sqrt{0.09}$

5. $\sqrt{-16}$

6. $\sqrt{1.69}$

In Exercises 7–12, use a calculator to approximate each square root. Round to three decimal places. Check that the answer is reasonable.

7. $\sqrt{55}$

8. $\sqrt{637}$

9. $\sqrt{4.5}$

10. $\sqrt{\dfrac{7}{3}}$

11. $\sqrt{12\dfrac{3}{5}}$

12. $-\sqrt{74}$

In Exercises 13–24, simplify. Include the absolute value symbol if necessary.

13. $\sqrt{27^2}$

14. $\sqrt{(-17)^2}$

15. $\sqrt{(8x)^2}$

16. $\sqrt{(6x-5)^2}$

17. $\sqrt{(2-9x)^4}$

18. $\sqrt{16x^2}$

19. $\sqrt{81x^4}$

20. $\sqrt{w^{14}}$

21. $-\sqrt{121x^{10}}$

22. $-\sqrt{(-4m)^2}$

23. $\sqrt{x^2+16x+64}$

24. $\sqrt{x^2-20x+100}$

In Exercises 25–38, simplify. Include the absolute value symbol if necessary. If the answer is not a real number, state so.

25. $\sqrt[3]{343}$

26. $\sqrt[3]{-125}$

27. $\sqrt[3]{-\dfrac{27}{64}}$

28. $\sqrt[3]{0.216}$

29. $\sqrt[3]{w^{21}}$

30. $\sqrt[3]{-8x^{15}}$

31. $\sqrt[3]{(-5z^2)^3}$

32. $\sqrt[3]{(5n+2)^3}$

33. $\sqrt[5]{1024}$

34. $-\sqrt[4]{256}$

35. $\sqrt[4]{\dfrac{16}{625}}$

36. $\sqrt[5]{-243}$

37. $\sqrt[4]{x^{20}}$

38. $\sqrt[6]{(x-4)^6}$

In Exercises 39–46, use a calculator to approximate each root. Round to three decimal places. Check that the answer is reasonable.

39. $\sqrt[3]{42}$

40. $\sqrt[4]{232}$

41. $\sqrt[5]{50}$

42. $\sqrt[6]{348}$

43. $\sqrt[3]{98.6}$

44. $\sqrt[4]{\dfrac{127}{4}}$

45. $\sqrt[3]{-472}$

46. $\sqrt[4]{195.4}$

7.2 Radical Functions

THINGS TO KNOW

Before working through this section, be sure you are familiar with the following concepts:

| | VIDEO | ANIMATION | INTERACTIVE |

 1. Find the Domain and Range of a Relation
(Section 2.2, Objective 2)

 2. Evaluate Functions
(Section 2.3, Objective 2)

 3. Graph Simple Functions by Plotting Points
(Section 2.3, Objective 3)

 4. Find Square Roots of Perfect Squares
(Section 7.1, Objective 1)

 5. Approximate Square Roots
(Section 7.1, Objective 2)

 6. Simplify Radical Expressions of the Form $\sqrt{a^2}$
(Section 7.1, Objective 3)

 7. Find Cube Roots
(Section 7.1, Objective 4)

OBJECTIVES

1 Evaluate Radical Functions

2 Find the Domain of a Radical Function

3 Graph Functions That Contain Square Roots or Cube Roots

OBJECTIVE 1 EVALUATE RADICAL FUNCTIONS

A **radical function** is a function that contains one or more **radical expressions**. We evaluate radical functions in the same way that we **evaluated functions** in Section 2.3. Substitute the given value for the **variable** and simplify using the order of operations.

My video summary ▶ **Example 1 Evaluating Radical Functions**

For the radical functions $f(x) = \sqrt{2x - 5}$, $g(x) = \sqrt[3]{5x + 9}$, and $h(x) = -3\sqrt[4]{x} + 2$, evaluate the following.

a. $f(15)$ b. $g(-2)$ c. $h(625)$

d. $g\left(-\dfrac{1}{5}\right)$ e. $f(0.5)$ f. $h(1)$

Solutions

a. Recall that **radical signs** also serve as **grouping symbols**. So, after substituting the value for the variable, simplify the **radicand** before finding the root.

$$\text{Substitute 15 for } x \text{ in the function } f: \quad f(15) = \sqrt{2(15) - 5}$$

$$\text{Multiply:} \quad = \sqrt{30 - 5}$$

$$\text{Subtract:} \quad = \sqrt{25}$$

$$\text{Evaluate the square root:} \quad = 5 \quad \boxed{\text{because } 5^2 = 25}$$

b. Substitute -2 for x in the function g: $\quad g(-2) = \sqrt[3]{5(-2) + 9}$

$$\text{Multiply:} \quad = \sqrt[3]{-10 + 9}$$

$$\text{Add:} \quad = \sqrt[3]{-1}$$

$$\text{Evaluate the cube root:} \quad = -1 \quad \boxed{\text{because } (-1)^3 = -1}$$

c. Substitute 625 for x in the function h: $\quad h(625) = -3\sqrt[4]{625} + 2$

$$\text{Evaluate the 4th root:} \quad = -3(5) + 2 \quad \boxed{\text{because } 5^4 = 625}$$

$$\text{Multiply:} \quad = -15 + 2$$

$$\text{Add:} \quad = -13$$

d.–f. Try to evaluate each **radical function** on your own. Check your **answers**, or watch this **interactive video** for complete solutions to all parts.

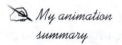 **You Try It** Work through this You Try It problem.

Work Exercises 1–12 in this eText or in the MyMathLab Study Plan.

OBJECTIVE 2 FIND THE DOMAIN OF A RADICAL FUNCTION

Recall that the **radicand** of a radical expression must be **non-negative** if the **index** of the radical expression is even. For example, $\sqrt[4]{-16}$ is not a **real number**. However, if the index of the radical expression is odd, then the radicand can be any real number. This information can be used to find the **domain of radical functions**.

My animation summary   ⊛ **Concept Animation** Watch this **animation** to learn more about finding the domain of a radical function.

> ### Guideline to Finding the Domain of a Radical Function
>
> If the **index** of a **radical function** is even, the **radicand** must be greater than or equal to zero.
>
> If the index of a radical function is odd, the radicand can be any real number.

Example 2 Finding the Domain of a Radical Function

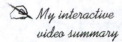 *My interactive video summary*

Find the **domain** for each radical function.

a. $F(x) = \sqrt[4]{12 - 4x}$ b. $h(x) = \sqrt[5]{3x + 5}$ c. $G(x) = \sqrt[6]{5x + 7}$

Solutions

a. The index is 4, which is even, so the radicand must be non-negative. Set the radicand greater than or equal to zero and solve.

Set the radicand greater than or equal to 0: $12 - 4x \geq 0$

Subtract 12 from both sides of the inequality: $-4x \geq -12$

Divide both sides by -4, reversing the inequality symbol: $x \leq 3$

The domain is $\{x | x \leq 3\}$ or, in interval notation, $(-\infty, 3]$.

b.–c. Try to find the domain for each radical function on your own. Check your answers, or watch this **interactive video** for complete solutions to all three parts.

You Try It Work through this You Try It problem.

Work Exercises 13–26 in this eText or in the MyMathLab Study Plan.

OBJECTIVE 3 GRAPH FUNCTIONS THAT CONTAIN SQUARE ROOTS OR CUBE ROOTS

In Section 2.3, simple functions were graphed by plotting points. The same strategy can be used to graph the **square root function** $f(x) = \sqrt{x}$.

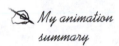 *My animation summary*

Concept Animation Watch this animation to see how to graph a square root function.

Other functions involving **square roots** will have a graph similar to the square root function, provided that the radicand is a **linear expression**.

Example 3 Graphing Functions That Contain Square Roots

Graph each function. Compare each graph to that of the square root function.

a. $F(x) = \sqrt{x + 1}$ b. $g(x) = \sqrt{x} + 1$ c. $h(x) = -\sqrt{x}$

Solutions

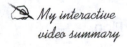 *My interactive video summary*

a. The domain of $F(x) = \sqrt{x + 1}$ is $\{x | x \geq -1\}$, or using **interval notation** $[-1, \infty)$. (Do you see why?) Evaluate the function for values of x greater than or equal to -1.

To make the computations work out nicely, choose values for x that will make the radicand a **perfect square**. Let $x = -1, 0, 3,$ and 8 (Table 1). Plotting the points and connecting them with a smooth curve gives the graph shown in Figure 4.

Table 1

x	$y = F(x) = \sqrt{x + 1}$	(x, y)
-1	$F(-1) = \sqrt{-1 + 1} = \sqrt{0} = 0$	$(-1, 0)$
0	$F(0) = \sqrt{0 + 1} = \sqrt{1} = 1$	$(0, 1)$
3	$F(3) = \sqrt{3 + 1} = \sqrt{4} = 2$	$(3, 2)$
8	$F(8) = \sqrt{8 + 1} = \sqrt{9} = 3$	$(8, 3)$

Figure 4

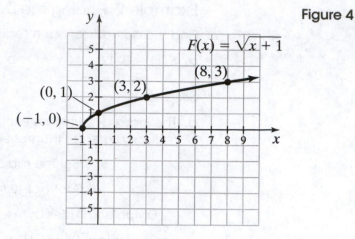

The graph of $F(x) = \sqrt{x + 1}$ looks like the graph of the **square root function,** but it is shifted one unit to the left. **View** these two graphs together on the same grid.

b. The **domain** of $g(x) = \sqrt{x} + 1$ is $\{x | x \geq 0\}$, or using **interval notation** $[0, \infty)$. **Evaluate** the function for non-negative values of x. For easier computations, substitute **perfect squares** for x (Table 2).

Table 2

x	$y = g(x) = \sqrt{x} + 1$	(x, y)
0	$g(0) = \sqrt{0} + 1 = 0 + 1 = 1$	$(0, 1)$
1	$g(1) = \sqrt{1} + 1 = 1 + 1 = 2$	$(1, 2)$
4	$g(4) = \sqrt{4} + 1 = 2 + 1 = 3$	$(4, 3)$
9	$g(9) = \sqrt{9} + 1 = 3 + 1 = 4$	$(9, 4)$

Plot the points and connect them with a smooth curve. Next, compare this graph to the graph of the square root function. Check your **answer,** or watch this **interactive video** for complete solutions to all three parts.

c. Try to work this problem on your own. Check your **answer,** or watch this **inter-active video** for complete solutions to all three parts.

You Try It Work through this You Try It problem.

Work Exercises 27–32 in this eText or in the MyMathLab Study Plan.

Now let's graph the **cube root function** $f(x) = \sqrt[3]{x}$. The **domain** of the cube root function is the set of all **real numbers** $\mathbb{R}$, or using **interval notation** $(-\infty, \infty)$. Therefore, **evaluate** the function for both positive and negative values. For easier computations, substitute **perfect cubes** for x (Table 3). Then plot the points and connect them with a smooth curve (Figure 5).

(eText Screens 7.2-1–7.2-15)

Table 3

x	$y = f(x) = \sqrt[3]{x}$	(x, y)
-8	$f(-8) = \sqrt[3]{-8} = -2$	$(-8, -2)$
-1	$f(-1) = \sqrt[3]{-1} = -1$	$(-1, -1)$
0	$f(0) = \sqrt[3]{0} = 0$	$(0, 0)$
1	$f(1) = \sqrt[3]{1} = 1$	$(1, 1)$
8	$f(8) = \sqrt[3]{8} = 2$	$(8, 2)$

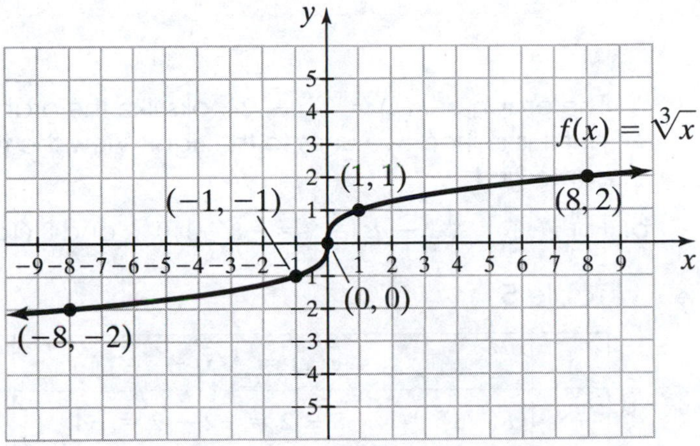

Figure 5

As with functions involving **square roots**, functions involving **cube roots** will have a graph similar to the cube root function if the **radicand** is a **linear expression**.

My interactive video summary

Example 4 Graphing Functions That Contain Cube Roots

Graph each function. Compare each graph to that of the **cube root function**.

a. $F(x) = \sqrt[3]{x - 2}$ **b.** $g(x) = \sqrt[3]{x} - 2$ **c.** $h(x) = -\sqrt[3]{x}$

Solutions The **domain** of each function is the set of all real numbers $\mathbb{R}$ or $(-\infty, \infty)$, so **evaluate** each function for positive and negative values. For easier computations, choose values for x that make the radicand a **perfect cube**.

a. Let $x = -6, 1, 2, 3,$ and 10 (Table 4). Plotting the points and connecting them with a smooth curve gives the graph shown in **Figure 6** on the next page.

Table 4

x	$y = F(x) = \sqrt[3]{x - 2}$	(x, y)
-6	$F(-6) = \sqrt[3]{-6 - 2} = \sqrt[3]{-8} = -2$	$(-6, -2)$
1	$F(1) = \sqrt[3]{1 - 2} = \sqrt[3]{-1} = -1$	$(1, -1)$
2	$F(2) = \sqrt[3]{2 - 2} = \sqrt[3]{0} = 0$	$(2, 0)$
3	$F(3) = \sqrt[3]{3 - 2} = \sqrt[3]{1} = 1$	$(3, 1)$
10	$F(10) = \sqrt[3]{10 - 2} = \sqrt[3]{8} = 2$	$(10, 2)$

Figure 6

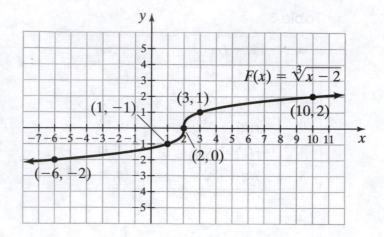

The graph of $F(x) = \sqrt[3]{x} - 2$ looks like the graph of the **cube root function**, but it is shifted two units to the right. **View the two graphs together on the same grid.**

b. For $g(x) = \sqrt[3]{x} - 2$, let $x = -8, -1, 0, 1,$ and 8 (Table 5).

Table 5

x	$y = g(x) = \sqrt[3]{x} - 2$	(x, y)
-8	$F(-8) = \sqrt[3]{-8} - 2 = -2 - 2 = -4$	$(-8, -4)$
-1	$F(-1) = \sqrt[3]{-1} - 2 = -1 - 2 = -3$	$(-1, -3)$
0	$F(0) = \sqrt[3]{0} - 2 = 0 - 2 = -2$	$(0, -2)$
1	$F(1) = \sqrt[3]{1} - 2 = 1 - 2 = -1$	$(1, -1)$
8	$F(8) = \sqrt[3]{8} - 2 = 2 - 2 = 0$	$(8, 0)$

Plot the points, connect them with a smooth curve, and compare the graph to the graph of the **cube root function**. Check your **answer**, or watch this **interactive video** for complete solutions to all three parts.

c. Try to work this problem on your own. Check your **answer**, or watch this **interactive video** for complete solutions to all three parts.

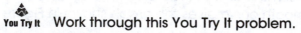 **You Try It** Work through this **You Try It** problem.

Work Exercises 33–38 in this eText or in the MyMathLab Study Plan.

7.2 Exercises

In Exercises 1–12, evaluate each radical function. If the function value is not a real number, state so.

1. If $f(x) = \sqrt{x - 7}$, find $f(88)$.

2. If $F(x) = \sqrt{9x + 10}$, find $F\left(\dfrac{2}{3}\right)$.

3. If $p(x) = \sqrt{5 - x^2}$, find $p(3)$.

4. If $F(x) = 8 - \sqrt{x - 4}$, find $F(40)$.

5. If $g(x) = \sqrt[3]{x} + 5$, find $g(3)$.

6. If $G(x) = \sqrt[3]{4x} - 5$, find $G(-5.5)$.

7. If $C(x) = 2x + \sqrt[3]{x^2 + 30x}$ find $C(2)$.

8. If $h(x) = \sqrt[4]{2x} - 3$, find $h(8)$.

9. If $H(x) = 2\sqrt[4]{x}$, find $H(1)$.

10. If $F(x) = \sqrt[4]{2x} - 14 + 7x$, find $F(5)$.

11. If $g(x) = \sqrt{3x} - \sqrt{x+9}$, find $g(27)$.

12. $q(x) = \dfrac{\sqrt{x+4}}{x^2+5}$, find $q(5)$

In Exercises 13–26, find the domain of each radical function.

13. $f(x) = \sqrt{x+4}$

14. $p(x) = \sqrt{5x-2}$

15. $F(x) = \sqrt{6-2x}$

16. $c(x) = \sqrt{x-4} + 2$

17. $D(x) = 3 - 7\sqrt{1-3x}$

18. $g(x) = \sqrt[3]{2x+1}$

19. $q(x) = \sqrt[3]{7-4x}$

20. $r(x) = \sqrt[3]{x^2 - 7x + 2x}$

21. $G(x) = \sqrt[4]{2x+5}$

22. $h(x) = \sqrt[4]{x} - 3$

23. $f(x) = \sqrt[4]{8x - \dfrac{3}{4}}$

24. $H(x) = \sqrt[5]{6x+7}$

25. $F(x) = 4x - \sqrt[5]{5-11x}$

26. $g(x) = \sqrt{2x+3} + \sqrt[3]{x-2}$

In Exercises 27–38, graph each radical function. Compare the graph to that of the square root function or cube root function.

27. $f(x) = \sqrt{x} - 2$

28. $F(x) = \sqrt{x} - 3$

29. $p(x) = \sqrt{x} + 5$

30. $q(x) = \sqrt{x} + 3$

31. $g(x) = 2\sqrt{x}$

32. $h(x) = -3\sqrt{x}$

33. $G(x) = \sqrt[3]{x+1}$

34. $c(x) = \sqrt[3]{x} - 1$

35. $h(x) = \sqrt[3]{x} + 2$

36. $d(x) = \sqrt[3]{x} - 4$

37. $r(x) = 3\sqrt[3]{x}$

38. $H(x) = -2\sqrt[3]{x}$

7.3 Rational Exponents and Simplifying Radical Expressions

THINGS TO KNOW

Before working through this section, be sure you are familiar with the following concepts:

| | | VIDEO | ANIMATION | INTERACTIVE |

1. Simplify Exponential Expressions Using the Product Rule (Section 4.1, Objective 1)

2. Simplify Exponential Expressions Using the Quotient Rule (Section 4.1, Objective 2)

3. Use the Negative-Power Rule (Section 4.1, Objective 4)

4. Use the Power-to-Power Rule (Section 4.1, Objective 5)

You Try It 5. Use the Product-to-Power and Quotient-to-Power Rules (Section 4.1, **Objective 6**)

 You Try It 6. Simplify Exponential Expressions Using a Combination of Rules (Section 4.1, **Objective 7**)

 You Try It 7. Find Square Roots of Perfect Squares (Section 7.1, **Objective 1**) ⏵

 You Try It 8. Simplify Radical Expressions of the Form $\sqrt{a^2}$ (Section 7.1, **Objective 3**) ⏵

 You Try It 9. Find and Approximate nth Roots (Section 7.1, **Objective 5**) ⏵

OBJECTIVES

1 Use the Definition for Rational Exponents of the Form $a^{\frac{1}{n}}$

2 Use the Definition for Rational Exponents of the Form $a^{\frac{m}{n}}$

3 Simplify Exponential Expressions Involving Rational Exponents

4 Use Rational Exponents to Simplify Radical Expressions

5 Simplify Radical Expressions Using the Product Rule

6 Simplify Radical Expressions Using the Quotient Rule

OBJECTIVE 1   USE THE DEFINITION FOR RATIONAL EXPONENTS OF THE FORM $a^{\frac{1}{n}}$

In **Section 4.1**, we defined an **exponential expression** as a **constant** or **algebraic expression** that is raised to a **power**. So far, we have only considered exponential expressions with **integer** powers. But what if the power is a **rational number** such as $\frac{1}{2}$?

 *My animation summary* ⊛ **Concept Animation** Watch this **animation** to see how to define an exponential expression when the exponent is the fraction $\frac{1}{2}$.

This idea can be used to define a **rational exponent of the form** $a^{\frac{1}{n}}$.

> **Definition**   Rational Exponent of the Form $a^{\frac{1}{n}}$
> If n is an integer such that $n \geq 2$ and if $\sqrt[n]{a}$ is a real number, then $a^{\frac{1}{n}} = \sqrt[n]{a}$.

The denominator n of the rational exponent is the **index** of the **root**. The **base** a of the exponential expression is the radicand of the root. If n is odd, then a can be any real number. If n is even, then a must be **non-negative**.

✎ *My video summary* ⊙ **Example 1** Converting Exponential Expressions to Radical Expressions

Write each exponential expression as a **radical expression**. Simplify if possible.

a. $25^{\frac{1}{2}}$ **b.** $(-64x^3)^{\frac{1}{3}}$ **c.** $-100^{\frac{1}{2}}$ **d.** $(-81)^{\frac{1}{4}}$ **e.** $(7x^3y)^{\frac{1}{5}}$

Solutions

a. The index of the root is 2, the denominator of the rational exponent. So, the radical expression is a **square root**. The radicand is 25.

$$25^{\frac{1}{2}} = \sqrt{25} = 5$$

b. From the denominator of the rational exponent, the index is 3. So, we have a **cube root**. The radicand is $-64x^3$.

$$(-64x^3)^{\frac{1}{3}} = \sqrt[3]{-64x^3} = \sqrt[3]{(-4x)^3} = -4x$$

c. As in part (a), the radical expression is a square root. The negative sign is not part of the **radicand** because it is not part of the **base**. We can tell this because there are no grouping symbols around -100.

$$-100^{\frac{1}{2}} = -\sqrt{100} = -10$$

d.–e. Try to convert each expression on your own. Check your **answers**, or watch this **video** for complete solutions to all parts.

You Try It Work through this **You Try It** problem.

Work Exercises 1–8 in this eText or in the MyMathLab Study Plan.

✎ *My video summary* ⊙ **Example 2** Converting Radical Expressions to Exponential Expressions

Write each radical expression as an **exponential expression**.

a. $\sqrt{5y}$ **b.** $\sqrt[3]{7x^2y}$ **c.** $\sqrt[4]{\dfrac{2m}{3n}}$

Solutions

a. The radical expression is a square root, so the **index** is 2 and the **denominator** of the **rational exponent** is 2. The base of the exponential expression is the radicand $5y$.

$$\sqrt{5y} = (5y)^{\frac{1}{2}}$$

b. The index is 3 and the radicand is $7x^2y$, so the denominator of the rational exponent is 3 and the base of the exponential expression is $7x^2y$. Finish writing the exponential expression. Check your **answer**, or watch this **video** for complete solutions to all three parts.

c. Try to convert this expression on your own. Check your **answer**, or watch this **video** for complete solutions to all three parts.

You Try It Work through this You Try It problem.

Work Exercises 17–19 in this eText or in the MyMathLab Study Plan.

OBJECTIVE 2 USE THE DEFINITION FOR RATIONAL EXPONENTS OF THE FORM $a^{\frac{m}{n}}$

What if the **numerator** of the **rational exponent** is not 1? For example, consider the **exponential expression** $a^{\frac{2}{3}}$. From the **power-to-power rule** for exponents, we write

$$a^{\frac{2}{3}} = a^{\frac{1}{3} \cdot 2} = \left(a^{\frac{1}{3}}\right)^2 = \left(\sqrt[3]{a}\right)^2 \text{ or } a^{\frac{2}{3}} = a^{2 \cdot \frac{1}{3}} = \left(a^2\right)^{\frac{1}{3}} = \sqrt[3]{a^2}.$$

This idea can be used to define a **rational exponent of the form** $a^{\frac{m}{n}}$.

Definition Rational Exponent of the Form $a^{\frac{m}{n}}$

If $\dfrac{m}{n}$ is a rational number in lowest terms, m and n are integers such that $n \geq 2$, and $\sqrt[n]{a}$ is a real number, then

$$a^{\frac{m}{n}} = \left(\sqrt[n]{a}\right)^m = \sqrt[n]{a^m}.$$

TIP To simplify $a^{\frac{m}{n}}$ using the form $\left(\sqrt[n]{a}\right)^m$, we find the **root** first and the **power** second. Using the form $\sqrt[n]{a^m}$, we find the **power** first and the **root** second. Both forms may be used, but the form $\left(\sqrt[n]{a}\right)^m$ is usually easier because it involves smaller numbers.

Note: For the rest of this chapter, we assume that all variable **factors** of the **radicand** of a **radical expression** with an even **index** will be **non-negative** real numbers. Likewise, all variable factors of the **base** of an **exponential expression** containing a rational exponent with an even **denominator** will be non-negative real numbers. This assumption allows us to avoid using **absolute value** symbols when simplifying radical expressions.

My video summary ▶ **Example 3** Converting Exponential Expressions to Radical Expressions

Write each exponential expression as a radical expression. **Simplify** if possible.

a. $16^{\frac{3}{2}}$ **b.** $\left(\dfrac{y^3}{1000}\right)^{\frac{2}{3}}$ **c.** $-81^{\frac{3}{4}}$ **d.** $(-36)^{\frac{5}{2}}$ **e.** $(x^2 y)^{\frac{2}{5}}$

Solutions

a. We use the form $a^{\frac{m}{n}} = \left(\sqrt[n]{a}\right)^m$. Since the denominator of the **rational exponent** is 2, the radical expression is a **square root**.

Begin with the original exponential expression: $16^{\frac{3}{2}}$

Use $a^{\frac{m}{n}} = \left(\sqrt[n]{a}\right)^m$ to rewrite as a radical expression: $= \left(\sqrt{16}\right)^3$

Simplify $\sqrt{16}$: $= (4)^3$

Simplify: $= 64$

b. $\left(\dfrac{y^3}{1000}\right)^{\frac{2}{3}} = \left(\sqrt[3]{\dfrac{y^3}{1000}}\right)^2 = \left(\sqrt[3]{\left(\dfrac{y}{10}\right)^3}\right)^2 = \left(\dfrac{y}{10}\right)^2 = \dfrac{y^2}{100}$

c. Note that the negative sign is not part of the **base**, so it goes in front of the radical expression.

$$-81^{\frac{3}{4}} = -\left(\sqrt[4]{81}\right)^3 = -(3)^3 = -27$$

d.–e. Try to convert each expression on your own. Check your **answers**, or watch this **video** for complete solutions.

You Try It Work through this You Try It problem.

Work Exercises 9–16 in this eText or in the MyMathLab Study Plan.

My video summary ▶ **Example 4 Converting Radical Expressions to Exponential Expressions**

Write each radical expression as an **exponential expression**.

a. $\sqrt[8]{x^5}$ **b.** $\left(\sqrt[5]{2ab^2}\right)^3$ **c.** $\sqrt[4]{(10x)^3}$

Solutions

a. The **index** is 8, so the **denominator** of the **rational exponent** is 8. The **radicand** is a **power** of 5, so the **numerator** is 5.

$$\sqrt[8]{x^5} = x^{\frac{5}{8}}$$

b. The index is 5, so the denominator of the rational exponent is 5. The expression is raised to a power of 3, so the numerator is 3. Finish writing the exponential expression. Check your **answer**, or watch this **video** for complete solutions to all three parts.

c. Try to convert this expression on your own. Check your **answer**, or watch this **video** for the complete solutions to all three parts.

You Try It Work through this You Try It problem.

Work Exercises 20–24 in this eText or in the MyMathLab Study Plan.

If a rational exponent is negative, we can first use the **negative-power rule** from Section 4.1 to rewrite the expression with a positive exponent.

My video summary ▶ **Example 5 Using the Negative-Power Rule with Negative Rational Exponents**

Write each exponential expression with positive exponents. **Simplify** if possible.

a. $1000^{-\frac{1}{3}}$ **b.** $\dfrac{1}{81^{-\frac{1}{4}}}$ **c.** $125^{-\frac{2}{3}}$ **d.** $\dfrac{1}{8^{-\frac{4}{3}}}$ **e.** $(-25)^{-\frac{3}{2}}$

Solutions

a. Begin with the original exponential expression: $1000^{-\frac{1}{3}}$

Use the negative-power rule $a^{-n} = \dfrac{1}{a^n}$: $= \dfrac{1}{1000^{\frac{1}{3}}}$

Rewrite the rational exponent as a radical expression: $= \dfrac{1}{\sqrt[3]{1000}}$

Evaluate the root: $= \dfrac{1}{10}$

b. Begin with the original exponential expression: $\dfrac{1}{81^{-\frac{1}{4}}}$

Use the negative-power rule $\dfrac{1}{a^{-n}} = a^n$: $= 81^{\frac{1}{4}}$

Rewrite the rational exponent as a radical expression: $= \sqrt[4]{81}$

Evaluate the root: $= 3$

c.–e. Try to rewrite and simplify each expression on your own. Check your **answers**, or watch this **video** for complete solutions to all parts.

You Try It Work through this You Try It problem.

Work Exercises 25–28 in this eText or in the MyMathLab Study Plan.

OBJECTIVE 3 SIMPLIFY EXPONENTIAL EXPRESSIONS INVOLVING RATIONAL EXPONENTS

We can apply the rules for exponents to **simplify exponential expressions** involving rational exponents.

Recall that an exponential expression is **simplified** when

- No parentheses or **grouping symbols** are present.
- No zero or **negative exponents** are present.
- No **powers** are raised to powers.
- Each **base** occurs only once.

My interactive video summary

Example 6 Simplifying Expressions Involving Rational Exponents

Use the rules for exponents to simplify each expression. Assume all **variables** represent **non-negative** values.

a. $x^{\frac{3}{8}} \cdot x^{\frac{1}{6}}$ **b.** $\dfrac{49^{\frac{7}{10}}}{49^{\frac{1}{5}}}$ **c.** $\left(64^{\frac{4}{9}}\right)^{\frac{3}{2}}$ **d.** $\left(32x^{\frac{5}{6}}y^{\frac{10}{9}}\right)^{\frac{3}{5}}$

e. $\left(\dfrac{125x^{\frac{5}{4}}}{y^{\frac{7}{8}}z^{\frac{9}{4}}}\right)^{\frac{4}{3}}, y \neq 0, z \neq 0$ **f.** $\left(4x^{\frac{1}{6}}y^{\frac{3}{4}}\right)^2\left(3x^{\frac{5}{9}}y^{-\frac{3}{2}}\right), y \neq 0$

Solutions

a. Begin with the original expression: $x^{\frac{3}{8}} \cdot x^{\frac{1}{6}}$

Use the **product rule** for exponents: $= x^{\frac{3}{8} + \frac{1}{6}}$

Add: $= x^{\frac{13}{24}}$ ⟵ $\boxed{\dfrac{3}{8} + \dfrac{1}{6} = \dfrac{9}{24} + \dfrac{4}{24} = \dfrac{13}{24}}$

b. Begin with the original expression: $\dfrac{49^{\frac{7}{10}}}{49^{\frac{1}{5}}}$

Use the **quotient rule** for exponents: $= 49^{\frac{7}{10} - \frac{1}{5}}$

Subtract: $= 49^{\frac{1}{2}}$ ⟵ $\boxed{\dfrac{7}{10} - \dfrac{1}{5} = \dfrac{7}{10} - \dfrac{2}{10} = \dfrac{5}{10} = \dfrac{1}{2}}$

Use $a^{\frac{1}{n}} = \sqrt[n]{a}$ to rewrite as a radical expression: $= \sqrt{49}$

Simplify: $= 7$

c. Begin with the original expression: $\left(64^{\frac{4}{9}}\right)^{\frac{3}{2}}$

Use the **power-to-power rule** for exponents: $= 64^{\frac{4}{9} \cdot \frac{3}{2}}$

Multiply: $= 64^{\frac{2}{3}}$ ⟵ $\boxed{\dfrac{4}{9} \cdot \dfrac{3}{2} = \dfrac{\overset{2}{\cancel{4}}}{\underset{3}{\cancel{9}}} \cdot \dfrac{\overset{1}{\cancel{3}}}{\underset{1}{\cancel{2}}} = \dfrac{2}{3}}$

Use $a^{\frac{m}{n}} = \left(\sqrt[n]{a}\right)^m$ to rewrite as a radical expression: $= \left(\sqrt[3]{64}\right)^2$

Simplify $\sqrt[3]{64}$: $= (4)^2$

Simplify: $= 16$

d.–f. Try to simplify each expression on your own. Check your **answers**, or watch this **interactive video** for complete solutions to all parts.

You Try It Work through this You Try It problem.

Work Exercises 29–38 in this eText or in the MyMathLab Study Plan.

OBJECTIVE 4 USE RATIONAL EXPONENTS TO SIMPLIFY RADICAL EXPRESSIONS

Some radical expressions can be simplified by first writing them with **rational exponents**. We can use the following process.

Using Rational Exponents to Simplify Radical Expressions

Step 1 Convert each radical expression to an **exponential expression** with rational exponents.

Step 2 Simplify by writing fractions in **lowest terms** or using the **rules of exponents**, as necessary.

Step 3 Convert any remaining rational exponents back to a radical expression.

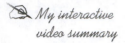
My interactive video summary

Example 7 Simplifying Radical Expressions

Use rational exponents to simplify each radical expression. Assume all **variables** represent **non-negative** values.

a. $\sqrt[6]{y} \cdot \sqrt[3]{y}$ **b.** $\sqrt{\sqrt[3]{x}}$ **c.** $\sqrt[6]{x^4}$

d. $\sqrt[8]{25x^2y^6}$ **e.** $\sqrt[4]{49}$ **f.** $\dfrac{\sqrt[3]{x}}{\sqrt[4]{x}}, x \neq 0$

Solutions

a.

Rewrite each radical using rational exponents: $\sqrt[6]{y} \cdot \sqrt[3]{y} = y^{\frac{1}{6}} \cdot y^{\frac{1}{3}}$

Use the **product rule** for exponents: $= y^{\frac{1}{6} + \frac{1}{3}}$

Add: $= y^{\frac{1}{2}} \longleftarrow \boxed{\dfrac{1}{6} + \dfrac{1}{3} = \dfrac{1}{6} + \dfrac{2}{6} = \dfrac{3}{6} = \dfrac{1}{2}}$

Convert back to a radical expression: $= \sqrt{y}$

b. Rewrite the **radicand** $\sqrt[3]{x}$ using a rational exponent: $\sqrt{\sqrt[3]{x}} = \sqrt{x^{\frac{1}{3}}}$

Rewrite the square root using a rational exponent: $= (x^{\frac{1}{3}})^{\frac{1}{2}}$

Use the **power-to-power** rule for exponents: $= x^{\frac{1}{3} \cdot \frac{1}{2}}$

Multiply: $= x^{\frac{1}{6}}$

Convert back to a radical expression: $= \sqrt[6]{x}$

c.

Convert to a rational exponent: $\sqrt[6]{x^4} = x^{\frac{4}{6}}$

Write the rational exponent in **lowest terms**: $= x^{\frac{2}{3}}$

Convert back to a radical expression: $= \sqrt[3]{x^2}$

d.–f. Try to simplify each radical expression on your own. Check your **answers**, or watch this **interactive video** for complete solutions to all parts.

You Try It   Work through this **You Try It** problem.

Work Exercises 39–44 in this eText or in the My MathLab **Study Plan.**

OBJECTIVE 5   SIMPLIFY RADICAL EXPRESSIONS USING THE PRODUCT RULE

We can develop a *product rule for radicals* that is similar to the **product-to-power** rule for exponents. Look at the radical expression $\sqrt[n]{ab}$.

Rewrite the radical as a rational exponent: $\sqrt[n]{ab} = (ab)^{\frac{1}{n}}$

Apply the product-to-power rule for exponents: $= a^{\frac{1}{n}}b^{\frac{1}{n}}$

Convert back to radical expressions: $= \sqrt[n]{a}\sqrt[n]{b}$

This **product rule for radicals** works in both directions: $\sqrt[n]{ab} = \sqrt[n]{a}\sqrt[n]{b}$ and $\sqrt[n]{a}\sqrt[n]{b} = \sqrt[n]{ab}$. So, we can use the rule to multiply radicals and to simplify radicals.

Product Rule for Radicals

If $\sqrt[n]{a}$ and $\sqrt[n]{b}$ are real numbers, then $\sqrt[n]{a}\sqrt[n]{b} = \sqrt[n]{ab}$.

⚠️ The **index** on each radical must be the same in order to use the product rule for radicals.

📝 *My video summary* ▶ **Example 8 Using the Product Rule to Multiply Radicals**

Multiply. Assume all **variables** represent **non-negative** values.

a. $\sqrt{3} \cdot \sqrt{7}$ b. $\sqrt{5} \cdot \sqrt{20}$ c. $\sqrt[5]{9x^2} \cdot \sqrt[5]{14y^3}$ d. $\sqrt[3]{-2} \cdot \sqrt[3]{4}$

Solutions Watch this **video**, or read through the following solutions.

In each case, use the **product rule for radicals** and then **simplify**.

	Original expression		Product rule for radicals		Simplify the radicand
a.	$\sqrt{3} \cdot \sqrt{7}$	$=$	$\sqrt{3 \cdot 7}$	$=$	$\sqrt{21}$

	Original expression		Product rule for radicals		Simplify the radicand		Evaluate
b.	$\sqrt{5} \cdot \sqrt{20}$	$=$	$\sqrt{5 \cdot 20}$	$=$	$\sqrt{100}$	$=$	10

	Original expression		Product rule for radicals		Simplify the radicand
c.	$\sqrt[5]{9x^2} \cdot \sqrt[5]{14y^3}$	$=$	$\sqrt[5]{9x^2 \cdot 14y^3}$	$=$	$\sqrt[5]{126x^2y^3}$

	Original expression		Product rule for radicals		Simplify the radicand		Evaluate
d.	$\sqrt[3]{-2} \cdot \sqrt[3]{4}$	$=$	$\sqrt[3]{-2 \cdot 4}$	$=$	$\sqrt[3]{-8}$	$=$	-2

🔺 **You Try It** Work through this You Try It problem.

Work Exercises 45–48 in this eText or in the MyMathLab **Study Plan.**

In Section 7.1, we simplified **radical expressions** of the form $\sqrt[n]{a^n}$. But when the radicand was not a **perfect** *n*th **power**, we approximated the **irrational number**. Now, we can use the **product rule** to simplify such radicals.

📝 *My animation summary* ⚙ **Concept Animation** Watch this animation to see how to simplify radicals when the radicand is not a perfect square.

Using the Product Rule to Simplify Radical Expressions of the Form $\sqrt[n]{a}$

Step 1 Write the radicand as a product of two factors, one being the largest possible perfect *n*th power.

Step 2 Use the product rule for radicals to take the *n*th root of each factor.

Step 3 Simplify the *n*th root of the perfect *n*th power.

Example 9 Using the Product Rule to Simplify Radicals

Use the **product rule** to **simplify**. Assume all **variables** represent non-negative values.

a. $\sqrt{700}$

b. $\sqrt[3]{40}$

c. $\sqrt[4]{x^8 y^5}$

d. $\sqrt{50x^4 y^3}$

Solutions Follow the **three-step process**.

a. The largest factor of 700 that is a **perfect square** is 100.

$$\begin{aligned} \text{Begin with the original expression:} \quad & \sqrt{700} \\ \text{Factor 700 into the product } 100 \cdot 7: \quad & = \sqrt{100 \cdot 7} \\ \text{Use the product rule for radicals:} \quad & = \sqrt{100} \cdot \sqrt{7} \\ \text{Evaluate } \sqrt{100}: \quad & = 10\sqrt{7} \end{aligned}$$

b. The largest factor of 40 that is a **perfect cube** is 8.

$$\begin{aligned} \text{Begin with the original expression:} \quad & \sqrt[3]{40} \\ \text{Factor 40 into the product } 8 \cdot 5: \quad & = \sqrt[3]{8 \cdot 5} \\ \text{Use the product rule for radicals:} \quad & = \sqrt[3]{8} \cdot \sqrt[3]{5} \\ \text{Evaluate } \sqrt[3]{8}: \quad & = 2\sqrt[3]{5} \end{aligned}$$

c. Note that x^8 is a **perfect 4th power** because $x^8 = (x^2)^4$. Also, $y^5 = y^4 \cdot y$. So, the largest factor of $x^8 y^5$ that is a perfect 4th power is $x^8 y^4$.

$$\begin{aligned} \text{Begin with the original expression:} \quad & \sqrt[4]{x^8 y^5} \\ \text{Factor } x^8 y^5 \text{ into the product } x^8 y^4 \cdot y: \quad & = \sqrt[4]{x^8 y^4 \cdot y} \\ \text{Use the product rule for radicals:} \quad & = \sqrt[4]{x^8 y^4} \cdot \sqrt[4]{y} \\ \text{Simplify } \sqrt[4]{x^8 y^4}: \quad & = x^2 y \sqrt[4]{y} \end{aligned}$$

My video summary 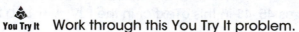 d. Try to simplify this **radical expression** on your own. Check your **answer**, or watch this **video** for a complete solution.

You Try It Work through this You Try It problem.

Work Exercises 49–58 in this eText or in the MyMathLab Study Plan.

Example 10 Using the Product Rule to Multiply and Simplify Radicals

Multiply and simplify. Assume all **variables** represent **non-negative** values.

a. $3\sqrt{10} \cdot 7\sqrt{2}$

b. $2\sqrt[3]{4} \cdot 5\sqrt[3]{6}$

c. $\sqrt[4]{18x^3} \cdot \sqrt[4]{45x^2}$

My interactive video summary

Solutions First, use the **product rule** to multiply **radicals**. Then, simplify using the three-step process.

a.

Begin with the original expression:	$3\sqrt{10} \cdot 7\sqrt{2}$
Rearrange factors to group radicals together:	$= 3 \cdot 7 \cdot \sqrt{10} \cdot \sqrt{2}$
Use the product rule for radicals:	$= 3 \cdot 7 \cdot \sqrt{10 \cdot 2}$
Multiply $3 \cdot 7$; multiply $10 \cdot 2$:	$= 21\sqrt{20}$
4 is the largest factor of 20 that is a perfect square:	$= 21\sqrt{4 \cdot 5}$
Use the product rule for radicals:	$= 21\sqrt{4} \cdot \sqrt{5}$
Simplify $\sqrt{4}$:	$= 21 \cdot 2\sqrt{5}$
Multiply $21 \cdot 2$:	$= 42\sqrt{5}$

b.–c. Try to work these problems on your own. Check your **answers**, or watch this interactive video for complete solutions to all three parts.

You Try It Work through this **You Try It** problem.

Work Exercises 59–64 in this eText or in the MyMathLab Study Plan.

My animation summary

Concept Animation Be careful not to confuse an **exponent** with the **index** of a **radical**, or vice versa. Watch this **animation** to explore this distinction in more detail.

OBJECTIVE 6 SIMPLIFY RADICAL EXPRESSIONS USING THE QUOTIENT RULE

We can now develop a *quotient rule for radicals*. Consider the **radical** expression $\sqrt[n]{\dfrac{a}{b}}$.

Rewrite the radical as a rational exponent:	$\sqrt[n]{\dfrac{a}{b}} = \left(\dfrac{a}{b}\right)^{\frac{1}{n}}$
Apply the **quotient-to-power rule** for exponents:	$= \dfrac{a^{\frac{1}{n}}}{b^{\frac{1}{n}}}$
Convert back to radical expressions:	$= \dfrac{\sqrt[n]{a}}{\sqrt[n]{b}}$

The **quotient rule for radicals** is used to simplify radical expressions involving fractions.

Quotient Rule for Radicals

If $\sqrt[n]{a}$ and $\sqrt[n]{b}$ are real numbers and $b \neq 0$, then $\sqrt[n]{\dfrac{a}{b}} = \dfrac{\sqrt[n]{a}}{\sqrt[n]{b}}$.

The **index** on each radical must be the same in order to use the quotient rule for radicals.

Like the **product rule**, the **quotient rule** works in both directions: $\sqrt[n]{\dfrac{a}{b}} = \dfrac{\sqrt[n]{a}}{\sqrt[n]{b}}$ and $\dfrac{\sqrt[n]{a}}{\sqrt[n]{b}} = \sqrt[n]{\dfrac{a}{b}}$.

We now add more requirements for a radical expression to be **simplified**.

Simplified Radical Expression

For a radical expression to be **simplified**, it must meet the following three conditions:

Condition 1. The radicand has no factor that is a **perfect power** of the **index** of the radical.

Condition 2. The radicand contains no fractions or **negative exponents**.

Condition 3. No **denominator** contains a radical.

As we saw in **Example 9**, the product rule for radicals can be used to help resolve issues with Condition 1. The quotient rule for radicals is used to resolve issues with Conditions 2 and 3.

Example 11 Using the Quotient Rule to Simplify Radicals

Use the quotient rule to simplify. Assume all **variables** represent **non-negative** values.

a. $\sqrt{\dfrac{25}{64}}$

b. $\sqrt[3]{\dfrac{16x^5}{27}}$

c. $\sqrt[4]{\dfrac{7x^4}{625}}$

d. $\sqrt{\dfrac{3x^9}{48x^3}}, x \neq 0$

Solutions In each case, we use the **quotient rule for radicals**.

$$
\underbrace{\sqrt{\dfrac{25}{64}}}_{\substack{\text{Original}\\\text{expression}}} \;=\; \underbrace{\dfrac{\sqrt{25}}{\sqrt{64}}}_{\substack{\text{Quotient rule}\\\text{for radicals}}} \;=\; \underbrace{\dfrac{5}{8}}_{\substack{\text{Evaluate each}\\\text{square root}}}
$$

a.

b.

$$
\underbrace{\sqrt[3]{\dfrac{16x^5}{27}}}_{\substack{\text{Original}\\\text{expression}}} = \underbrace{\dfrac{\sqrt[3]{16x^5}}{\sqrt[3]{27}}}_{\substack{\text{Quotient rule}\\\text{for radicals}}} = \underbrace{\dfrac{\sqrt[3]{8x^3 \cdot 2x^2}}{\sqrt[3]{27}}}_{\substack{\text{Write as product}\\\text{using perfect-cube}\\\text{factors}}} = \underbrace{\dfrac{\sqrt[3]{8x^3} \cdot \sqrt[3]{2x^2}}{\sqrt[3]{27}}}_{\substack{\text{Product rule}\\\text{for radicals}}} = \underbrace{\dfrac{2x\sqrt[3]{2x^2}}{3}}_{\substack{\text{Simplify the}\\\text{cube roots}}}
$$

My video summary ▶ **c.–d.** Try using the quotient rule to simplify these expressions on your own. Check your **answers**, or watch this **video** for complete solutions.

You Try It Work through this You Try It problem.

Work Exercises 65–70 in this eText or in the MyMathLab Study Plan.

In Example 11, we used the quotient rule for radicals to remove fractions from the **radicand**. In Example 12, we will use the quotient rule to remove **radicals** from a **denominator**.

My interactive video summary 🖎 **Example 12 Using the Quotient Rule to Simplify Radicals**

Use the quotient rule to **simplify**. Assume all **variables** represent positive numbers.

a. $\dfrac{\sqrt{240x^3}}{\sqrt{15x}}$ b. $\dfrac{\sqrt[3]{-500z^2}}{\sqrt[3]{4z^{-1}}}$

c. $\dfrac{\sqrt{150m^9}}{\sqrt{3m}}$ d. $\dfrac{\sqrt{45x^5y^{-3}}}{\sqrt{20xy^{-1}}}$

Solutions In each case, we use the **quotient rule for radicals.**

a. Begin with the original expression: $\dfrac{\sqrt{240x^3}}{\sqrt{15x}}$

Use the quotient rule for radicals: $= \sqrt{\dfrac{240x^3}{15x}}$

Simplify the radicand: $= \sqrt{16x^2}$

Simplify: $= 4x$

b.   Begin with the original expression: $\dfrac{\sqrt[3]{-500z^2}}{\sqrt[3]{4z^{-1}}}$

Use the quotient rule for radicals: $= \sqrt[3]{\dfrac{-500z^2}{4z^{-1}}}$

Divide factors and subtract exponents in radicand: $= \sqrt[3]{\dfrac{-500}{4} \cdot z^{2-(-1)}}$

Simplify radicand: $= \sqrt[3]{-125z^3}$

Simplify the cube root: $= -5z$

c.–d. Try to simplify each radical expression on your own. Check your **answers**, or watch this **interactive video** for complete solutions to all four parts.

You Try It Work through this You Try It problem.

Work Exercises 71–76 in this eText or in the MyMathLab Study Plan.

7.3 Exercises

In Exercises 1–16, write each exponential expression as a radical expression. Simplify if possible.

1. $36^{\frac{1}{2}}$

2. $(-125)^{\frac{1}{3}}$

3. $-81^{\frac{1}{4}}$

4. $-25^{\frac{1}{2}}$

5. $(27x^3)^{\frac{1}{3}}$

6. $(121m^4)^{\frac{1}{2}}$

7. $(7xy)^{\frac{1}{5}}$

8. $(13xy^2)^{\frac{1}{3}}$

9. $100^{\frac{3}{2}}$

10. $-8^{\frac{4}{3}}$

11. $(-16)^{\frac{3}{4}}$

12. $\left(\dfrac{x^3}{64}\right)^{\frac{2}{3}}$

13. $(-1024)^{\frac{2}{5}}$

14. $\left(\dfrac{1}{8}\right)^{\frac{5}{3}}$

15. $(-3ab)^{\frac{7}{3}}$

16. $(xy^2)^{\frac{3}{7}}$

In Exercises 17–24, write each radical expression as an exponential expression.

17. $\sqrt{10}$

18. $\sqrt[3]{7m}$

19. $\sqrt[5]{\dfrac{2x}{y}}$

20. $\sqrt{\left(\dfrac{m}{3n}\right)^3}$

21. $\sqrt[7]{x^4}$

22. $(\sqrt[5]{3xy^3})^4$

23. $\sqrt[9]{(2xy)^4}$

24. $\sqrt[4]{(7xy)^9}$

In Exercises 25–28, write each exponential expression with positive exponents. Simplify if possible.

25. $25^{-\frac{1}{2}}$

26. $\dfrac{1}{625^{-\frac{3}{4}}}$

27. $\left(\dfrac{27}{8}\right)^{-\frac{2}{3}}$

28. $-(-32)^{-\frac{3}{5}}$

For Exercises 29–76, assume that all variables represent non-negative values.

In Exercises 29–38, use the rules for exponents to simplify. Write final answers in exponential form when necessary.

29. $x^{\frac{1}{3}} \cdot x^{\frac{1}{2}}$

30. $9^{\frac{3}{10}} \cdot 9^{\frac{1}{5}}$

31. $\dfrac{36^{\frac{3}{4}}}{36^{\frac{1}{4}}}$

32. $(x^{\frac{5}{3}})^{\frac{9}{10}}$

33. $(p^{\frac{2}{5}}q^{-\frac{1}{3}})(p^{-\frac{3}{2}}q^{\frac{4}{3}})$

34. $(9m^4n^{-\frac{3}{2}})^{\frac{1}{2}}$

35. $(32x^5y^{-10})^{\frac{1}{5}}(xy^{-\frac{1}{2}})$

36. $\left(\dfrac{a^{\frac{1}{3}}b^{-\frac{3}{5}}}{8}\right)^{\frac{2}{3}}$

37. $\left(\dfrac{16x^{\frac{2}{3}}}{81x^{\frac{5}{4}}y^{\frac{2}{3}}}\right)^{\frac{3}{4}}$

38. $\left(\dfrac{x^{\frac{3}{4}}y}{x^3y^{-\frac{1}{5}}}\right)^{\frac{2}{3}}$

In Exercises 39–44, use rational exponents to simplify each radical expression. Write final answers in radical form when necessary.

39. $\sqrt[5]{x^2} \cdot \sqrt[4]{x}$

40. $\sqrt[4]{\sqrt[3]{x^2}}$

41. $\sqrt[6]{x^3}$

42. $\dfrac{\sqrt{5}}{\sqrt[6]{5}}$

43. $\sqrt[8]{81}$

44. $\sqrt[10]{49x^6y^8}$

In Exercises 45–48, multiply.

45. $\sqrt[4]{5} \cdot \sqrt[4]{7}$

46. $\sqrt{3} \cdot \sqrt{12}$

47. $\sqrt[3]{5x^2} \cdot \sqrt[3]{25x}$

48. $\sqrt[4]{5x} \cdot \sqrt[4]{15y^3}$

In Exercises 49–58, use the product rule to simplify.

49. $\sqrt{72}$

50. $\sqrt[3]{56}$

51. $-\sqrt{63}$

52. $\sqrt{48x^3}$

53. $\sqrt[3]{-125x^6}$

54. $3\sqrt{125y^9}$

55. $7\sqrt[3]{128x^4y^5}$

56. $\sqrt{a^{15}}$

57. $\sqrt[4]{x^5y^8z^9}$

58. $x^2y\sqrt[4]{x^5y^{11}}$

In Exercises 59–64, multiply and simplify.

59. $\sqrt{6} \cdot \sqrt{30}$

60. $4\sqrt{3x} \cdot 7\sqrt{6x}$

61. $\sqrt{20x^3y} \cdot \sqrt{18xy^4}$

62. $\sqrt[3]{9x^2} \cdot \sqrt[3]{6x^2}$

63. $2x\sqrt[3]{5y^2} \cdot 3x\sqrt[3]{25y}$

64. $\sqrt[4]{8x^2} \cdot \sqrt[4]{6x^3}$

In Exercises 65–76, use the quotient rule to simplify.

65. $\sqrt{\dfrac{49}{81}}$

66. $\sqrt{\dfrac{12x^3}{25y^2}}$

67. $\sqrt[3]{\dfrac{11x^3}{8}}$

68. $\sqrt[3]{\dfrac{16x^9}{125y^6}}$

69. $\sqrt[4]{\dfrac{21x^5}{16}}$

70. $\sqrt{\dfrac{6x^5}{54x}}$

71. $\dfrac{\sqrt{45}}{\sqrt{5}}$

72. $\dfrac{\sqrt{96x^6y^8}}{\sqrt{12x^2y^5}}$

73. $\dfrac{\sqrt{120x^9}}{\sqrt{3x^{-1}}}$

74. $\dfrac{\sqrt[3]{5000}}{\sqrt[3]{5}}$

75. $\dfrac{\sqrt[3]{-72m}}{\sqrt[3]{3m^{-2}}}$

76. $\dfrac{\sqrt{3x^5y^3}}{\sqrt{48xy^5}}$

7.4 Operations with Radicals

THINGS TO KNOW

Before working through this section, be sure you are familiar with the following concepts:

			VIDEO	ANIMATION	INTERACTIVE

You Try It 1. Find Square Roots of Perfect Squares
(Section 7.1, Objective 1) ▶

You Try It 2. Find Cube Roots
(Section 7.1, Objective 4) ▶

You Try It 3. Find and Approximate nth Roots
(Section 7.1, Objective 5) ▶

You Try It 4. Simplify Radical Expressions Using the
Product Rule (Section 7.3, Objective 5) ▶ 👆

You Try It 5. Simplify Radical Expressions Using the
Quotient Rule (Section 7.3, Objective 6) ▶ 👆

OBJECTIVES

1 Add and Subtract Radical Expressions

2 Multiply Radical Expressions

3 Rationalize Denominators of Radical Expressions

OBJECTIVE 1 ADD AND SUBTRACT RADICAL EXPRESSIONS

As we learned in Chapter 4, when simplifying **polynomial expressions** we combine like terms. Recall that **like terms** have the same **variables** raised to the same corresponding **exponents**. For example, $3x^2y^3$ and $5x^2y^3$ are like terms.

We can add or subtract these terms using the reverse of the **distributive property**.

$$3x^2y^3 + 5x^2y^3 = (3 + 5)x^2y^3 = 8x^2y^3$$
$$3x^2y^3 - 5x^2y^3 = (3 - 5)x^2y^3 = -2x^2y^3$$

$4xy^3$ and $2x^2y^2$ are not like terms, so we cannot add or subtract them. They have the same variables, but the variables in each term do not have the same corresponding exponents. This idea is true with **radicals** as well. We can only add and subtract **like radicals**.

> **Definition** Like Radicals
>
> Two radicals are **like radicals** if they have the same **index** and the same **radicand**.

My animation summary ⊛ **Concept Animation** Watch this animation to explore the concept of like radicals in more detail.

We add and subtract like radicals in the same way that we add and subtract like terms: reverse the **distributive property** to factor out the like radical and then simplify.

Example 1 Adding and Subtracting Radical Expressions

Add or subtract.

a. $\sqrt{15} + 7\sqrt{15}$ b. $10\sqrt{5} - 9\sqrt[4]{5}$ c. $\sqrt[3]{\dfrac{4}{27}} + 8\sqrt[3]{4}$

My video summary ▶ **Solutions** Watch this **video** for the complete solutions, or read through the following.

a. Begin with the original expression: $\sqrt{15} + 7\sqrt{15}$

$\qquad\qquad$ Identify like radicals: $= \sqrt{15} + 7\sqrt{15}$

$\qquad$ Reverse distributive property: $= (1 + 7)\sqrt{15}$

$\qquad\qquad\qquad\qquad$ Simplify: $= 8\sqrt{15}$

b. The indices are different, so $\sqrt{5}$ and $\sqrt[4]{5}$ are not **like radicals**. We cannot simplify $10\sqrt{5} - 9\sqrt[4]{5}$ further since $\sqrt{5}$ and $\sqrt[4]{5}$ are not like radicals.

c. Begin with the original expression: $\sqrt[3]{\dfrac{4}{27}} + 8\sqrt[3]{4}$

$\qquad\qquad\qquad$ Quotient rule: $= \dfrac{\sqrt[3]{4}}{\sqrt[3]{27}} + 8\sqrt[3]{4}$

$\qquad\qquad\qquad\qquad$ Simplify: $= \dfrac{\sqrt[3]{4}}{3} + 8\sqrt[3]{4}$

$\qquad\qquad$ Identify like radicals: $= \dfrac{\sqrt[3]{4}}{3} + 8\sqrt[3]{4}$

$\qquad$ Reverse distributive property: $= \left(\dfrac{1}{3} + 8\right)\sqrt[3]{4}$

$\qquad\qquad\qquad\qquad$ Simplify: $= \dfrac{25}{3}\sqrt[3]{4}$ or $\dfrac{25\sqrt[3]{4}}{3}$

You Try It Work through this **You Try It** problem.

Work Exercises 1–4 in this eText or in the MyMathLab Study Plan.

My video summary ▶ **Example 2 Adding and Subtracting Radical Expressions**

Add or subtract. Assume **variables** represent non-negative values.

a. $12\sqrt[3]{7x^2} + 4\sqrt[3]{7x^2}$ b. $-3\sqrt{5x} - \sqrt{\dfrac{x}{9}}$

Solutions Read through the following solutions, or watch this **video**.

a. Begin with the original expression: $12\sqrt[3]{7x^2} + 4\sqrt[3]{7x^2}$

Identify like radicals: $= 12\sqrt[3]{7x^2} + 4\sqrt[3]{7x^2}$

Reverse distributive property: $= (12 + 4)\sqrt[3]{7x^2}$

Simplify: $= 16\sqrt[3]{7x^2}$

b. Begin with the original expression: $-3\sqrt{5x} - \sqrt{\dfrac{x}{9}}$

Quotient rule: $= -3\sqrt{5x} - \dfrac{\sqrt{x}}{\sqrt{9}}$

Simplify: $= -3\sqrt{5x} - \dfrac{\sqrt{x}}{3}$

The **radicands** $5x$ and x are different, so $-3\sqrt{5x}$ and $\dfrac{\sqrt{x}}{3}$ are not like radicals. We cannot simplify further.

You Try It   Work through this You Try It problem.

Work Exercises 5–13 in this eText or in the MyMathLab Study Plan.

!   Sometimes it is necessary to simplify radicals first before adding or subtracting.

Example 3 Adding and Subtracting Radical Expressions

Add or subtract.

a. $\sqrt{54} + 6\sqrt{72} - 3\sqrt{24}$ b. $\sqrt[3]{24} - \sqrt[3]{192} + 4\sqrt[3]{250}$

Solutions

a. Begin with the original expression: $\sqrt{54} + 6\sqrt{72} - 3\sqrt{24}$

Factor: $= \sqrt{9 \cdot 6} + 6\sqrt{36 \cdot 2} - 3\sqrt{4 \cdot 6}$

Product rule: $= \sqrt{9} \cdot \sqrt{6} + 6\sqrt{36} \cdot \sqrt{2} - 3\sqrt{4} \cdot \sqrt{6}$

Simplify radicals: $= 3\sqrt{6} + 6 \cdot 6\sqrt{2} - 3 \cdot 2\sqrt{6}$

Multiply factors: $= 3\sqrt{6} + 36\sqrt{2} - 6\sqrt{6}$

Collect like radicals: $= (3 - 6)\sqrt{6} + 36\sqrt{2}$

Simplify: $= -3\sqrt{6} + 36\sqrt{2}$

My video summary ▶ b. Work the problem on your own. Check your **answer**, or watch this **video** for a detailed solution.

You Try It   Work through this You Try It problem.

Work Exercises 14–24 in this eText or in the MyMathLab Study Plan.

Example 4 Adding and Subtracting Radical Expressions

Add or subtract. Assume **variables** represent non-negative values.

a. $\sqrt[3]{27m^5n^4} + 2mn\sqrt[3]{m^2n} - m\sqrt[3]{m^2n^4}$ **b.** $2a\sqrt{16ab^3} + 4\sqrt{9a^2b} - 5\sqrt{4a^3b^3}$

Solutions

a. Original expression: $\sqrt[3]{27m^5n^4} + 2mn\sqrt[3]{m^2n} - m\sqrt[3]{m^2n^4}$

Factor: $= \sqrt[3]{27m^3n^3 \cdot m^2n} + 2mn\sqrt[3]{m^2n} - m\sqrt[3]{n^3 \cdot m^2n}$

Product rule: $= \sqrt[3]{27m^3n^3} \cdot \sqrt[3]{m^2n} + 2mn\sqrt[3]{m^2n} - m\sqrt[3]{n^3} \cdot \sqrt[3]{m^2n}$

Simplify radicals: $= 3mn\sqrt[3]{m^2n} + 2mn\sqrt[3]{m^2n} - mn\sqrt[3]{m^2n}$

Collect like radicals: $= (3 + 2 - 1)mn\sqrt[3]{m^2n}$

Simplify: $= 4mn\sqrt[3]{m^2n}$

My video summary ▶ **b.** Work the problem on your own. Check your **answer**, or watch this **video** for a detailed solution.

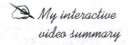 **You Try It** Work through this You Try It problem.

Work Exercises 25–30 in this eText or in the MyMathLab Study Plan.

Example 5 Adding and Subtracting Radical Expressions

Add or subtract. Assume **variables** represent non-negative values.

a. $\dfrac{\sqrt{45}}{6x} - \dfrac{4\sqrt{20}}{5x}$ **b.** $\dfrac{\sqrt[4]{a^5}}{3} + \dfrac{a\sqrt[4]{a}}{12}$ **c.** $\dfrac{3x^3\sqrt{24x^3y^3}}{2x\sqrt{3x^2y}} - \dfrac{x^2\sqrt{10xy^4}}{\sqrt{5y^2}}$

My interactive video summary **Solutions** Try to work the problems on your own. Remember to find a **common denominator** before adding or subtracting **fractions**. Check your **answers**, or watch this **interactive** video for detailed solutions to all three parts.

You Try It Work through this You Try It problem.

Work Exercises 31–34 in this eText or in the MyMathLab Study Plan.

OBJECTIVE 2 MULTIPLY RADICAL EXPRESSIONS

In **Section 7.3**, we saw how to multiply two **radical expressions** when each had only one **term**. Here we extend this idea to radical expressions with more than one term.

To multiply radical expressions, follow the same approach as when multiplying **polynomial** expressions. Use the **distributive property** to multiply each term in the first expression by each term in the second. Then simplify the resulting products and combine like terms and like radicals.

Example 6 Multiplying Radical Expressions

Multiply. Assume **variables** represent non-negative values.

a. $5\sqrt{2x}(3\sqrt{2x} - \sqrt{3})$ **b.** $\sqrt[3]{2n^2}(\sqrt[3]{4n} + \sqrt[3]{5n})$

Solutions

a. Begin with the original expression: $5\sqrt{2x}\left(3\sqrt{2x} - \sqrt{3}\right)$

$$\text{Distributive property:} \quad = 5\sqrt{2x} \cdot 3\sqrt{2x} - 5\sqrt{2x} \cdot \sqrt{3}$$

$$\text{Rearrange factors:} \quad = 5 \cdot 3 \cdot \sqrt{2x} \cdot \sqrt{2x} - 5 \cdot \sqrt{2x} \cdot \sqrt{3}$$

$$\text{Multiply:} \quad = 15\sqrt{4x^2} - 5\sqrt{6x}$$

$$\text{Simplify radical:} \quad = 15 \cdot 2x - 5\sqrt{6x}$$

$$\text{Simplify:} \quad = 30x - 5\sqrt{6x}$$

My video summary ▶ **b.** Try this problem on your own. Check your **answer**, or watch this video for a detailed solution.

You Try It   Work through this You Try It problem.

Work Exercises 35–38 in this eText or in the MyMathLab Study Plan.

Example 7 Multiplying Radical Expressions

Multiply. Assume variables represent non-negative values.

a. $\left(7\sqrt{2} - 2\sqrt{3}\right)\left(\sqrt{2} - 5\right)$ b. $\left(\sqrt{m} - 4\right)\left(3\sqrt{m} + 7\right)$

Solutions

a. Use the FOIL method to multiply the radical expressions. The two first terms are $7\sqrt{2}$ and $\sqrt{2}$. The two outside terms are $7\sqrt{2}$ and -5. The two inside terms are $-2\sqrt{3}$ and $\sqrt{2}$. The two last terms are $-2\sqrt{3}$ and -5.

$$\left(7\sqrt{2} - 2\sqrt{3}\right)\left(\sqrt{2} - 5\right) = \overbrace{7\sqrt{2} \cdot \sqrt{2}}^{F} + \overbrace{7\sqrt{2} \cdot (-5)}^{O} - \overbrace{2\sqrt{3} \cdot \sqrt{2}}^{I} - \overbrace{2\sqrt{3} \cdot (-5)}^{L}$$

$$= 7\sqrt{4} - 35\sqrt{2} - 2\sqrt{6} + 10\sqrt{3}$$

$$= 7 \cdot 2 - 35\sqrt{2} - 2\sqrt{6} + 10\sqrt{3}$$

$$= 14 - 35\sqrt{2} - 2\sqrt{6} + 10\sqrt{3}$$

My video summary ▶ **b.** Try this problem on your own. Check your **answer**, or watch this video for a detailed solution.

You Try It   Work through this You Try It problem.

Work Exercises 39–44 in this eText or in the MyMathLab Study Plan.

Example 8 Using Special Products to Multiply Radical Expressions

Multiply. Assume variables represent non-negative values.

a. $\left(\sqrt{y} + 3\right)\left(\sqrt{y} - 3\right)$ b. $\left(3\sqrt{x} - 2\right)^2$

Solutions

a. $(\sqrt{y} + 3)(\sqrt{y} - 3)$ is a **product of conjugates.** We use the rule for $(A + B)(A - B)$ with $A = \sqrt{y}$ and $B = 3$.

Write the product of conjugates rule: $(A + B)(A - B) = A^2 - B^2$

Substitute $\sqrt{y}$ for A and 3 for B: $(\sqrt{y} + 3)(\sqrt{y} - 3) = (\sqrt{y})^2 - (3)^2$

Simplify: $= y - 9$

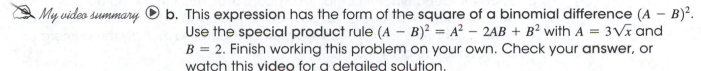 b. This **expression** has the form of the **square of a binomial difference** $(A - B)^2$. Use the **special product** rule $(A - B)^2 = A^2 - 2AB + B^2$ with $A = 3\sqrt{x}$ and $B = 2$. Finish working this problem on your own. Check your **answer,** or watch this **video** for a detailed solution.

You Try It Work through this You Try It problem.

Work Exercises 45–50 in this eText or in the MyMathLab **Study Plan.**

TIP In part (a) of Example 8, the product of **conjugates** involving **square roots** resulted in an expression without any **radicals.** This result is true in general and occurs because the product of conjugates equals the **difference of two squares.**

OBJECTIVE 3   RATIONALIZE DENOMINATORS OF RADICAL EXPRESSIONS

In **Section 7.3,** we saw that a **simplified radical expression** has no **radicals** in the **denominator.** We used the **quotient rule** to eliminate radicals from the denominator of a radical expression. However, the quotient rule is not always enough. For example, simplifying $\sqrt{\dfrac{3}{32}}$ using the quotient rule and **product rule** yields the following:

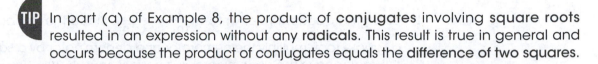

$$\sqrt{\frac{3}{32}} = \underbrace{\frac{\sqrt{3}}{\sqrt{32}}}_{\substack{\text{Quotient} \\ \text{rule}}} = \underbrace{\frac{\sqrt{3}}{\sqrt{16} \cdot \sqrt{2}}}_{\substack{\text{Product} \\ \text{rule}}} = \frac{\sqrt{3}}{4\sqrt{2}}$$

We still have a radical in the denominator. Although there is nothing wrong with this, it is common practice to remove radicals from the denominator so that the denominator is a **rational number.** This process is called **rationalizing the denominator.**

Rationalizing a Denominator with One Term

To rationalize a denominator with a single radical of **index** n, multiply the **numerator** and denominator by a radical of index n so that the radicand in the denominator is a **perfect** nth **power.**

Multiplying the numerator and denominator by the same radical is equivalent to multiplying by 1 since $\dfrac{\sqrt{2}}{\sqrt{2}} = 1$, $\dfrac{\sqrt{11}}{\sqrt{11}} = 1$, $\dfrac{\sqrt{x}}{\sqrt{x}} = 1$, and so on. We are simply writing an **equivalent expression** without radicals in the denominator. This idea is like writing **equivalent fractions** with a **least common denominator.**

My video summary ▶ **Example 9** Rationalizing Denominators with Square Roots

Rationalize the denominator.

a. $\dfrac{\sqrt{7}}{\sqrt{5}}$ b. $\sqrt{\dfrac{3}{10x}}$

Solutions Watch this video, or read through the following solutions.

a. Since the **denominator** contains a **square root**, multiply the **numerator** and denominator by a square root so that the **radicand** in the denominator is a **perfect square**. $5 \cdot 5 = 5^2 = 25$ is a perfect square, so multiply the numerator and denominator by $\sqrt{5}$.

$$\overset{\text{Original expression}}{\overbrace{\dfrac{\sqrt{7}}{\sqrt{5}}}} = \overset{\text{Multiplying by 1}}{\overbrace{\dfrac{\sqrt{7}}{\sqrt{5}} \cdot \dfrac{\sqrt{5}}{\sqrt{5}}}} = \overset{\text{Product rule}}{\overbrace{\dfrac{\sqrt{35}}{\sqrt{25}}}} = \overset{\text{Simplify}}{\overbrace{\dfrac{\sqrt{35}}{5}}}$$

b. Using the **quotient rule**, write $\sqrt{\dfrac{3}{10x}} = \dfrac{\sqrt{3}}{\sqrt{10x}}$. Since the denominator contains a square root, multiply the numerator and denominator by a **square root** so that the radicand in the denominator is a **perfect square**. $10x \cdot 10x = (10x)^2 = 100x^2$ is a perfect square, so multiply the numerator and denominator by $\sqrt{10x}$.

$$\overset{\text{Original expression}}{\overbrace{\sqrt{\dfrac{3}{10x}}}} = \overset{\text{Quotient rule}}{\overbrace{\dfrac{\sqrt{3}}{\sqrt{10x}}}} = \overset{\text{Multiplying by 1}}{\overbrace{\dfrac{\sqrt{3}}{\sqrt{10x}} \cdot \dfrac{\sqrt{10x}}{\sqrt{10x}}}} = \overset{\text{Product rule}}{\overbrace{\dfrac{\sqrt{30x}}{\sqrt{100x^2}}}} = \overset{\text{Simplify}}{\overbrace{\dfrac{\sqrt{30x}}{10x}}}$$

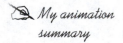

You Try It Work through this You Try It problem.

Work Exercises 51–54 in this eText or in the MyMathLab Study Plan.

My animation summary ✸ **Concept Animation** Why would we want to rationalize a denominator? Watch this animation to find out.

Example 10 Rationalizing Denominators with Cube Roots or Fourth Roots

Rationalize the denominator.

a. $\sqrt[3]{\dfrac{11}{25x}}$ b. $\dfrac{\sqrt[4]{7x}}{\sqrt[4]{27y^2}}$

Solutions

a. Using the **quotient rule**, write $\sqrt[3]{\dfrac{11}{25x}} = \dfrac{\sqrt[3]{11}}{\sqrt[3]{25x}}$. For roots greater than 2, it is

helpful to write the **radicand** of the denominator in **exponential form**.

$$\sqrt[3]{25x} = \sqrt[3]{5^2 x^1}$$

Since the denominator contains a **cube root**, we multiply the numerator and denominator by a cube root so that the radicand in the denominator is a **perfect cube**. To do this, we need the **factors** in the radicand to have

exponents of 3 or **multiples** of 3. We have two 5's and one x, so we need one more 5 and two more x's to get $5^2x^1 \cdot 5^1x^2 = 5^3x^3 = (5x)^3$, which is a perfect cube. Therefore, multiply the numerator and denominator by $\sqrt[3]{5^1x^2} = \sqrt[3]{5x^2}$.

$$\underbrace{\sqrt[3]{\dfrac{11}{25x}}}_{\substack{\text{Original} \\ \text{expression}}} = \underbrace{\dfrac{\sqrt[3]{11}}{\sqrt[3]{25x}}}_{\substack{\text{Quotient} \\ \text{rule}}} = \underbrace{\dfrac{\sqrt[3]{11}}{\sqrt[3]{25x}} \cdot \dfrac{\sqrt[3]{5x^2}}{\sqrt[3]{5x^2}}}_{\substack{\text{Multiplying} \\ \text{by 1}}} = \underbrace{\dfrac{\sqrt[3]{55x^2}}{\sqrt[3]{125x^3}}}_{\substack{\text{Product} \\ \text{rule}}} = \underbrace{\dfrac{\sqrt[3]{55x^2}}{5x}}_{\text{Simplify}}$$

My video summary ▶ **b.** Try this problem on your own. Check your **answer**, or watch this **video** for a detailed solution.

You Try It Work through this You Try It problem.

Work Exercises 55 and 56 in this eText or in the MyMathLab Study Plan.

⚠ Though not required, it is best to **rationalize the denominator** last when simplifying radical expressions since the **radicand** in the denominator will then involve simpler expressions.

My video summary ▶ **Example 11 Rationalizing Denominators**

Simplify each expression first and then rationalize the denominator.

a. $\sqrt{\dfrac{3x}{50}}$ **b.** $\dfrac{\sqrt{18x}}{\sqrt{27xy}}$ **c.** $\sqrt[3]{\dfrac{-4x^5}{16y^5}}$

Solutions Try these problems on your own. Check your **answers**, or watch this **video** for detailed solutions to all three parts.

You Try It Work through this You Try It problem.

Work Exercises 57 and 58 in this eText or in the MyMathLab Study Plan.

The goal of **rationalizing a denominator** is to write an equivalent expression without **radicals** in the denominator. Earlier in this section, we noted that multiplying **conjugates** involving **square roots** results in an expression without any radicals. So, we can use the **product of conjugates** to rationalize the denominator of a radical expression whose **denominator** contains two **terms** involving one or more square roots.

Rationalizing a Denominator with Two Terms

To rationalize a denominator with two terms involving one or more square roots, multiply the numerator and denominator by the conjugate of the denominator.

Example 12 Rationalizing Denominators with Two Terms

Rationalize the denominator.

a. $\dfrac{2}{\sqrt{3} + 5}$ **b.** $\dfrac{7}{3\sqrt{x} - 4}$ **c.** $\dfrac{\sqrt{y} - 3}{\sqrt{y} + 2}$

Solutions

a. Since the **denominator** has two **terms** and involves a **square root**, multiply the **numerator** and denominator by $\sqrt{3} - 5$, which is the **conjugate** of the denominator.

Multiply numerator and denominator by $\sqrt{3} - 5$: $\quad \dfrac{2}{\sqrt{3} + 5} = \dfrac{2}{\sqrt{3} + 5} \cdot \dfrac{\sqrt{3} - 5}{\sqrt{3} - 5}$

Multiply numerators and multiply denominators: $\qquad = \dfrac{2(\sqrt{3} - 5)}{(\sqrt{3})^2 - (5)^2}$

Simplify the denominator: $\qquad = \dfrac{2(\sqrt{3} - 5)}{3 - 25}$

$\qquad = \dfrac{2(\sqrt{3} - 5)}{-22}$

Divide out a common factor of 2: $\qquad = \dfrac{\overset{1}{2}(\sqrt{3} - 5)}{\underset{11}{-22}}$

Simplify: $\qquad = -\dfrac{\sqrt{3} - 5}{11} \text{ or } \dfrac{5 - \sqrt{3}}{11}$

b. Since the **denominator** has two **terms** and involves a **square root**, multiply the **numerator** and denominator by the **conjugate** of the denominator, $3\sqrt{x} + 4$.

$$\dfrac{7}{3\sqrt{x} - 4} = \overbrace{\dfrac{7}{3\sqrt{x} - 4} \cdot \dfrac{3\sqrt{x} + 4}{3\sqrt{x} + 4}}^{\substack{\text{Multiply numerator} \\ \text{and denominator by} \\ 3\sqrt{x} + 4}} = \overbrace{\dfrac{7(3\sqrt{x} + 4)}{(3\sqrt{x})^2 - (4)^2}}^{\substack{\text{Multiply numerators} \\ \text{and multiply} \\ \text{denominators}}} = \overbrace{\dfrac{7(3\sqrt{x} + 4)}{9x - 16}}^{\text{Simplify}} \text{ or } \overbrace{\dfrac{21\sqrt{x} + 28}{9x - 16}}^{\text{Distribute}}$$

The result can be left in factored form, or we can distribute in the final step. Leaving the numerator factored until the end helps with dividing out **common factors**, if they are present.

 *My video summary* ▶ **c.** Work this problem on your own. Check your **answer**, or watch this **video** for a detailed solution.

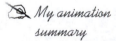 **You Try It** Work through this You Try It problem.

Work Exercises 59–64 in this eText or in the MyMathLab Study Plan.

My animation summary ⊛ **Concept Animation** Would we ever want to rationalize the **numerator**? Watch this **animation** to find out.

7.4 Exercises

In Exercises 1–50, assume that all variables represent non-negative values and be sure to simplify your answer.

In Exercises 1–34, add or subtract, if possible.

1. $4\sqrt{7} + 9\sqrt{7}$

2. $\sqrt[3]{\dfrac{32}{27}} - \sqrt[3]{108}$

3. $3\sqrt{2} + 5\sqrt{6}$

4. $3\sqrt{7} - 4\sqrt[3]{6} + 2\sqrt{7} - 8\sqrt[3]{6}$

5. $\sqrt[3]{6x} - 7\sqrt[3]{6x}$

6. $2\sqrt[3]{4y} - 7\sqrt[3]{4y}$

7. $\sqrt{3x} + 3\sqrt{x} - 9\sqrt{3x}$

8. $6x\sqrt{11} - 13x\sqrt{11}$

9. $\dfrac{\sqrt{x}}{4} + \dfrac{3}{8}\sqrt{x}$

10. $\dfrac{4\sqrt{3x}}{9} + \dfrac{2\sqrt{3x}}{15} - \dfrac{7\sqrt{3x}}{6}$

11. $10\sqrt{5} - 3\sqrt{x} - 4\sqrt{5} + 9\sqrt{x}$

12. $6\sqrt{a} + 2\sqrt{b} - a\sqrt{3}$

13. $4xy\sqrt[3]{2} + 3xy\sqrt[3]{2} - 5x\sqrt[3]{2}$

14. $\sqrt{18} + \sqrt{98}$

15. $\sqrt{45} - \sqrt{80}$

16. $3\sqrt[3]{32} + \sqrt[3]{500}$

17. $3\sqrt{48} - \sqrt{24}$

18. $\sqrt{\dfrac{3}{16}} + \sqrt{\dfrac{27}{4}}$

19. $5\sqrt{12} - \sqrt{75} + 3\sqrt{20}$

20. $9\sqrt{7} - 6\sqrt{72} + 4\sqrt{18} - 5\sqrt{112}$

21. $\sqrt[3]{72} - \sqrt[3]{9}$

22. $6\sqrt[3]{54} + 2\sqrt{8} - 4\sqrt[3]{16}$

23. $7\sqrt[3]{54} + 3\sqrt[3]{8} - 5\sqrt[3]{16}$

24. $\sqrt[3]{\dfrac{3}{125}} + \sqrt[3]{81}$

25. $2\sqrt[4]{81x^5} - 3\sqrt[4]{x^5}$

26. $6\sqrt{20x^3} + 4x\sqrt{45x}$

27. $\sqrt{100t^3} + 6\sqrt{t^2} - \sqrt{25t^3}$

28. $9\sqrt[3]{y^4} - 3y\sqrt[3]{125y} + \sqrt[3]{64y^4}$

29. $3a^2\sqrt{ab^3} + \sqrt{16a^5b^3} - 5b\sqrt{a^5b}$

30. $\sqrt[3]{81m^3n} - 4\sqrt[3]{3m^3n} + 3m\sqrt[3]{162n}$

31. $\dfrac{\sqrt{72}}{3y} + \dfrac{5\sqrt{18}}{4y}$

32. $\dfrac{\sqrt[3]{8a^4}}{4a} - \dfrac{\sqrt[3]{27a}}{7}$

33. $\dfrac{\sqrt[4]{81a^7}}{3} - \dfrac{5a\sqrt[4]{16a^3}}{2}$

34. $\dfrac{5x^2\sqrt{128x^5y^3}}{\sqrt{64x^4y}} + \dfrac{2xy\sqrt{18x^3y}}{3\sqrt{y}}$

In Exercises 35–50, multiply.

35. $\sqrt{5}\left(2x + \sqrt{3}\right)$

36. $\sqrt{2x}\left(\sqrt{6x} - 5\right)$

37. $2\sqrt{3m}\left(m - 3\sqrt{5m}\right)$

38. $\sqrt[3]{x}\left(\sqrt[3]{54x^2} - \sqrt[3]{x}\right)$

39. $\left(8\sqrt{3} - 5\right)\left(\sqrt{3} - 1\right)$

40. $\left(3 + \sqrt{5}\right)\left(2 + \sqrt{7}\right)$

41. $\left(\sqrt{x} + 4\right)\left(\sqrt{x} + 6\right)$

42. $\left(4 - 2\sqrt{3x}\right)\left(5 + \sqrt{3x}\right)$

43. $\left(3 - \sqrt{x}\right)\left(4 - \sqrt{y}\right)$

44. $\left(\sqrt[3]{y} + 1\right)\left(\sqrt[3]{y} - 2\right)$

45. $\left(\sqrt{7z} + 3\right)\left(\sqrt{7z} - 3\right)$

46. $\left(\sqrt{5} + 3y\right)^2$

47. $(\sqrt{x} + 3)(3 - \sqrt{x})$

48. $(\sqrt{3a} - \sqrt{2b})(\sqrt{3a} + \sqrt{2b})$

49. $(2\sqrt{x} + 5)^2$

50. $(2\sqrt{m} - 3\sqrt{n})^2$

In Exercises 51–64, rationalize the denominator.

51. $\dfrac{4}{\sqrt{6}}$

52. $\dfrac{\sqrt{3}}{\sqrt{5}}$

53. $\sqrt{\dfrac{9}{2x}}$

54. $\dfrac{5}{\sqrt{11x}}$

55. $\sqrt[3]{\dfrac{9}{4x^2}}$

56. $\dfrac{\sqrt[4]{3b^5}}{\sqrt[4]{25a}}$

57. $\dfrac{8}{\sqrt{12x^3y^4}}$

58. $\dfrac{\sqrt[3]{2x^2}}{\sqrt[3]{36y^5}}$

59. $\dfrac{5}{2 - \sqrt{3}}$

60. $\dfrac{-4x}{5 + \sqrt{6}}$

61. $\dfrac{-3}{\sqrt{x} + 1}$

62. $\dfrac{\sqrt{a}}{3\sqrt{a} - \sqrt{b}}$

63. $\dfrac{\sqrt{m} - 4}{\sqrt{m} - 7}$

64. $\dfrac{5\sqrt{3} + 2\sqrt{6}}{4\sqrt{6} - \sqrt{3}}$

7.5 Radical Equations and Models

THINGS TO KNOW

Before working through this section, be sure you are familiar with the following concepts:

| | | VIDEO | ANIMATION | INTERACTIVE |

 You Try It 1. Solve Linear Equations in One Variable (Section 1.1, Objective 2) ▶ ✸

 You Try It 2. Solve a Formula for a Given Variable (Section 1.5, Objective 1) ▶

 You Try It 3. Solve Application Problems Involving Geometric Formulas (Section 1.5, Objective 2) ▶

 You Try It 4. Use the Power-to-Power Rule (Section 4.1, Objective 5)

 You Try It 5. Multiply Two Binomials Using Special Product Rules (Section 4.3, Objective 4) ▶

 You Try It 6. Solve Polynomial Equations by Factoring (Section 5.4, Objective 1) ▶

 You Try It 7. Multiply Radical Expressions (Section 7.4, Objective 2) ▶

OBJECTIVES

1 Solve Equations Involving One Radical Expression

2 Solve Equations Involving Two Radical Expressions

3 Use Radical Equations and Models to Solve Application Problems

OBJECTIVE 1 SOLVE EQUATIONS INVOLVING ONE RADICAL EXPRESSION

Recall that when **solving an equation in one variable**, we find all values of the **variable** that make the equation true. All of these values together form the **solution set** of the equation. Previously, we have solved **linear equations**, **polynomial equations**, and **rational equations**. Now we learn how to solve *radical equations*.

> **Definition** Radical Equation
>
> A **radical equation** is an equation that contains at least one **radical expression** with a variable in the **radicand**.

Some examples of radical equations are

$$\sqrt{2x+1} = 3, \quad \sqrt[3]{5x-4} + 7 = 10, \quad \sqrt[4]{19x-2} = 2\sqrt[4]{x+1}, \text{ and } \sqrt{x+9} - \sqrt{x-6} = 3.$$

The first two equations contain one radical expression, whereas the last two equations contain two radical expressions.

 Not all equations that contain radical expressions are radical equations. If no radical expression contains a variable in the radicand, then the equation is not a radical equation. For example, $2x + 1 = \sqrt{3}$ is not a radical equation.

Consider the equation $\sqrt{x} = 5$, which has one **radical expression** $\sqrt{x}$. The radical expression $\sqrt{x}$ is an **isolated radical expression** because it stands alone on one side of the equal sign.

The key to solving a radical equation is to eliminate all of the radicals from the equation. To do this, we can use the following **theorem**.

> **Theorem** The Power Principle of Equality
>
> If A and B represent **algebraic expressions** and n is a positive **real number**, then any **solution** to the equation $A = B$ is also a solution to the equation $A^n = B^n$.

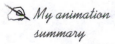 *My animation summary*

⊗ **Concept Animation** Watch this **animation** to see how the Power Principle of equality can be used to eliminate an isolated radical.

Sometimes when we raise both sides of an **equation** to an even **power**, the new equation will have **solutions** that are not solutions to the original equation. Such "solutions" are called **extraneous solutions**. They must be identified and excluded from the **solution set**.

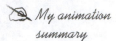

 Concept Animation Watch this **animation** for a more in-depth explanation of extraneous solutions.

We can use the following steps to **solve** equations involving one **radical expression.**

Solving Equations Involving One Radical Expression

Step 1 Isolate the radical expression. Use the **properties of equality** to get the radical expression by itself on one side of the equal sign.

Step 2 Eliminate the radical. Identify the **index** of the radical expression and raise both sides of the equation to the index power.

Step 3 Solve the resulting equation.

Step 4 Check each solution from Step 3 in the original equation. Disregard any **extraneous solutions.**

 Extraneous solutions do not result from raising both sides of an equation to an odd power, but it is still good practice to check your answers anyway.

When solving **radical equations,** answers can fail to check for two reasons: 1) An answer may be an extraneous solution, or 2) an error may have been made while solving. When an answer is disregarded as an extraneous solution, it is important to make sure that no errors were made while solving.

 ▶ **Example 1 Solving an Equation Involving One Radical Expression**

Solve $\sqrt{5x + 6} - 2 = 7$.

Solution We follow the four-step process. Read through the solution below, or watch this **video.**

Step 1 Isolate the radical expression by adding 2 to both sides.

Begin with the original equation: $\sqrt{5x + 6} - 2 = 7$

Add 2 to both sides: $\sqrt{5x + 6} - 2 + 2 = 7 + 2$

Simplify: $\sqrt{5x + 6} = 9$

Step 2 The radical expression is a **square root,** so **square** both sides.

Square both sides: $\left(\sqrt{5x + 6}\right)^2 = (9)^2$

Simplify: $5x + 6 = 81$

Step 3 Solve the resulting **linear equation.**

Subtract 6 from both sides: $5x = 75$

Divide both sides by 5: $x = 15$

Step 4 Check $x = 15$ in the original equation.

Substitute 15 for x: $\sqrt{5(15) + 6} - 2 \stackrel{?}{=} 7$

Simplify beneath the radical: $\sqrt{81} - 2 \stackrel{?}{=} 7$

Evaluate the square root: $9 - 2 \stackrel{?}{=} 7$

Add: $7 = 7$ True

The answer checks in the original equation, so the **solution set** is $\{15\}$.

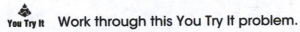 **You Try It** Work through this You Try It problem.

Work Exercises 1–9 in this eText or in the MyMathLab Study Plan.

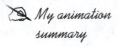 *My animation summary* ✺ **Concept Animation** When solving a radical equation, it is important to **isolate** the radical before raising both sides to the index power. If this is not done, then the radical expression will not be eliminated. Watch this **animation** for an illustration of this concept.

Example 2 Solving an Equation Involving One Radical Expression

Solve $\sqrt{5 + 4x} + 12 = 5$.

My video summary ▶ **Solution** Read through the solution below, or watch this **video**.

Step 1 Begin with the original equation: $\sqrt{5 + 4x} + 12 = 5$

Subtract 12 from both sides: $\sqrt{5 + 4x} + 12 - 12 = 5 - 12$

Simplify: $\sqrt{5 + 4x} = -7$

Step 2 Square both sides: $\left(\sqrt{5 + 4x}\right)^2 = (-7)^2$

Simplify: $5 + 4x = 49$

Step 3 Subtract 5 from both sides: $4x = 44$

Divide both sides by 4: $x = 11$

Step 4 Check $x = 11$ in the original equation: $\sqrt{5 + 4(11)} + 12 \stackrel{?}{=} 5$

Simplify beneath the radical: $\sqrt{49} + 12 \stackrel{?}{=} 5$

Evaluate the square root: $7 + 12 \stackrel{?}{=} 5$

Add: $19 = 5$ False

The answer $x = 11$ does not check in the original equation. It is an **extraneous** solution. Since this is the only possible solution, the equation has no real solution. The **solution set** is $\{\}$ or $\varnothing$.

You Try It Work through this You Try It problem.

Work Exercises 10–12 in this eText or in the MyMathLab Study Plan.

We might have noticed sooner that the equation in Example 2 would have no real solution. In Step 1, when we isolated the radical, the resulting equation was $\sqrt{5 + 4x} = -7$. This equation states that a principal square root equals a negative value, but this is impossible since principal square roots must be non-negative. So, this equation has no real solution.

My video summary ▶ **Example 3 Solving an Equation Involving One Radical Expression**

Solve $\sqrt{3x + 7} - x = 1$.

Solution We follow the four-step process.

Step 1 Isolate the radical expression by adding x to both sides.

$$\text{Begin with the original equation:} \qquad \sqrt{3x + 7} - x = 1$$

$$\text{Add } x \text{ to both sides:} \quad \sqrt{3x + 7} - x + x = 1 + x$$

$$\text{Simplify each side:} \qquad \sqrt{3x + 7} = x + 1$$

Step 2 The radical expression is a square root, so square both sides of the equation to eliminate the radical.

$$\text{Square both sides:} \quad \left(\sqrt{3x + 7}\right)^2 = (x + 1)^2$$

$$\text{Simplify each side:} \qquad 3x + 7 = x^2 + 2x + 1$$

Try to solve this **polynomial equation** on your own by using the four-step process from Section 5.4. Then finish solving the original equation by checking for extraneous solutions. Check your **answer**, or watch this **video** for the complete solution.

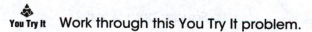 **You Try It** Work through this You Try It problem.

Work Exercises 13–19 in this eText or in the MyMathLab Study Plan.

So far, all of the radical equations solved have involved square roots. In Example 4, we look at an equation involving a cube root.

My video summary ▶ **Example 4 Solving an Equation Involving One Radical Expression**

Solve $\sqrt[3]{3x^2 + 23x} + 10 = 12$.

Solution

Step 1 Begin with the original equation: $\qquad \sqrt[3]{3x^2 + 23x} + 10 = 12$

$$\text{Subtract 10 from both sides:} \quad \sqrt[3]{3x^2 + 23x} + 10 - 10 = 12 - 10$$

$$\text{Simplify:} \qquad \sqrt[3]{3x^2 + 23x} = 2$$

Step 2 The radical expression is a cube root, so cube both sides of the equation to eliminate the radical.

$$\text{Cube both sides:} \quad \left(\sqrt[3]{3x^2 + 23x}\right)^3 = (2)^3$$

$$\text{Simplify:} \qquad 3x^2 + 23x = 8$$

The result is a **polynomial equation**. Try to finish solving this equation on your own. Check your **answer**, or watch this **video** for the complete solution.

You Try It Work through this You Try It problem.

Work Exercises 20–24 in this eText or in the MyMathLab Study Plan.

Recall from Section 7.3 that $a^{\frac{1}{n}} = \sqrt[n]{a}$. Sometimes equations containing one **rational exponent** can be solved using the same process for solving equations with one **radical expression**. We solve such an equation in Example 5.

My video summary ⊙ **Example 5 Solving an Equation Involving a Rational Exponent**

Solve $(x^2 - 9)^{\frac{1}{4}} + 3 = 5$.

Solution This equation is equivalent to $\sqrt[4]{x^2 - 9} + 3 = 5$. Try solving it on your own by using the **four-step process**. Check your **answer**, or watch this **video** for a complete solution.

You Try It Work through this You Try It problem.

Work Exercises 25–28 in this eText or in the MyMathLab Study Plan.

OBJECTIVE 2 SOLVE EQUATIONS INVOLVING TWO RADICAL EXPRESSIONS

Solving an equation involving two **radical expressions** is similar to solving an equation involving one radical expression. However, it may be necessary to repeat the process for eliminating a radical in order to eliminate both radicals from the equation.

Solving Equations Involving Two Radical Expressions

Step 1 Isolate one of the radical expressions. Use the **properties of equality** to get a radical expression by itself on one side of the equal sign.

Step 2 Eliminate the radical from the isolated radical expression. Identify the **index** of the isolated radical expression and raise both sides of the equation to this index **power**.

Step 3 If all the radicals have been eliminated, then solve the resulting equation. Otherwise, repeat Steps 1 and 2.

Step 4 Check each solution from Step 3 in the original equation. Disregard any **extraneous solutions**.

My video summary ⊙ **Example 6 Solving an Equation Involving Two Radical Expressions**

Solve $\sqrt{x + 9} - \sqrt{x} = 1$.

Solution Follow the four-step process.

Step 1 Choose to isolate $\sqrt{x+9}$.

$$\text{Begin with the original equation:} \quad \sqrt{x+9} - \sqrt{x} = 1$$
$$\text{Add } \sqrt{x} \text{ to both sides:} \quad \sqrt{x+9} - \sqrt{x} + \sqrt{x} = 1 + \sqrt{x}$$
$$\text{Simplify:} \quad \sqrt{x+9} = 1 + \sqrt{x}$$

Step 2 The isolated **radical expression** is a square root, so **square** both sides.

$$\text{Square both sides:} \quad (\sqrt{x+9})^2 = (1 + \sqrt{x})^2$$

The right side has the form of the **square of a binomial sum** $(A+B)^2$, so use the **special product rule** $(A+B)^2 = A^2 + 2AB + B^2$ with $A = 1$ and $B = \sqrt{x}$.

$$\text{Use } (A+B)^2 = A^2 + 2AB + B^2$$
$$\text{on the right side:} \quad (\sqrt{x+9})^2 = (1)^2 + 2(1)(\sqrt{x}) + (\sqrt{x})^2$$
$$\text{Simplify:} \quad x + 9 = 1 + 2\sqrt{x} + x$$

Step 3 The equation still contains a radical expression, so repeat Steps 1 and 2. Try to finish solving this equation on your own. Remember to check for **extraneous solutions**. Check your **answer**, or watch this **video** for the complete solution.

You Try It Work through this You Try It problem.

Work Exercises 29–31 in this eText or in the MyMathLab Study Plan.

My video summary ▶ **Example 7 Solving an Equation Involving Two Radical Expressions**
Solve $\sqrt{z+5} + \sqrt{2z+1} = 6$.

Solution Try solving this equation on your own by using the four-step process. Check your **answer**, or watch this **video** for the complete solution.

You Try It Work through this You Try It problem.

Work Exercises 32–34 in this eText or in the MyMathLab Study Plan.

My video summary ▶ **Example 8 Solving an Equation Involving Two Radical Expressions**
Solve $\sqrt[3]{2x^2 - 9} + \sqrt[3]{3x - 11} = 0$.

Solution Follow the four-step process.

Step 1 Begin with the original equation: $\quad \sqrt[3]{2x^2 - 9} + \sqrt[3]{3x - 11} = 0$
Subtract $\sqrt[3]{3x - 11}$ from both sides: $\quad \sqrt[3]{2x^2 - 9} = -\sqrt[3]{3x - 11}$

Step 2 The two radical expressions are cube roots, so cube both sides.

$$\text{Cube both sides:} \quad (\sqrt[3]{2x^2-9})^3 = (-\sqrt[3]{3x-11})^3$$
$$\text{Simplify:} \quad 2x^2 - 9 = -(3x - 11)$$
$$\text{Distribute the negative:} \quad 2x^2 - 9 = -3x + 11$$

Step 3 Both radicals are eliminated. Try to finish solving on your own. Check your answer, or watch this **video** for the complete solution.

You Try It   Work through this You Try It problem.

Work Exercises 35–38 in this eText or in the MyMathLab Study Plan.

OBJECTIVE 3   USE RADICAL EQUATIONS AND MODELS TO SOLVE APPLICATION PROBLEMS

Often, real-world situations are **modeled** by formulas that involve **radical expressions**. For example, the formula $t = \dfrac{\sqrt{d}}{4}$ gives the time t, in seconds, that it takes a **free-falling** object to fall a distance of d feet.

In **Section 1.5**, we learned to solve a formula for a given **variable**. If the given variable is beneath a radical, then we must remove the radical by **isolating the radical** and raising both sides of the equation to the appropriate **power**.

Example 9 Solving a Formula for a Variable beneath a Radical
Solve each formula for the given variable.

a. Free-falling object: $t = \dfrac{\sqrt{d}}{4}$ for d.

b. Radius of a sphere: $r = \sqrt[3]{\dfrac{3V}{4\pi}}$ for V.

Solutions

a.
Begin with the original formula:	$t = \dfrac{\sqrt{d}}{4}$
Multiply both sides by 4 to isolate the radical:	$4(t) = 4\left(\dfrac{\sqrt{d}}{4}\right)$
Simplify:	$4t = \sqrt{d}$
Square both sides:	$(4t)^2 = (\sqrt{d})^2$
Simplify:	$16t^2 = d$

Solving for d, the formula for a free-falling object is $d = 16t^2$.

My video summary ⊙ b. The **variable** V is already contained within the **isolated radical expression**. The radical expression is a **cube root**, so eliminate the radical by **cubing** both sides. Try to finish solving the formula for V. Check your **answer**, or watch this **video** for a complete solution.

You Try It   Work through this You Try It problem.

Work Exercises 39–44 in this eText or in the MyMathLab Study Plan.

We can use **radical equations** and **models** to solve a variety of application problems in many different disciplines. In Example 10, a radical equation is used to model the readability of written text.

📝 *My video summary* ▶ **Example 10 Assessing the Readability of Written Text**

A **SMOG** grade for written text is a minimum reading grade level G that a reader must possess in order to fully understand the written text being graded. If w is the number of words that have three or more syllables in a sample of 30 sentences from a given text, then the SMOG grade for that text is given by the **formula** $G = \sqrt{w} + 3$. Use the SMOG grade formula to answer the following questions. (*Source*: readabilityformulas.com).

a. If a sample of 30 sentences contains 18 words with three or more syllables, then what is the SMOG grade for the text? If necessary, round to a **whole number** for the grade level.

b. If a text must have a tenth-grade reading level, then how many words with three or more syllables would be needed in the sample of 30 sentences?

Solutions

a. There are 18 words with three or more syllables, so substitute 18 for w in the given **formula**.

Begin with the original formula: $G = \sqrt{w} + 3$

Substitute 18 for w: $G = \sqrt{18} + 3$

Approximate the square root: $G \approx 4.24 + 3$

Simplify: $G \approx 7.24$

Rounding to the nearest **whole number**, we get $G = 7$. So, the **SMOG** grade for this text is a seventh-grade reading level.

b. The text must have a tenth-grade reading level, so substitute 10 for G in the formula and solve for w.

Begin with the original formula: $G = \sqrt{w} + 3$

Substitute 10 for G: $10 = \sqrt{w} + 3$

Try to finish solving this problem on your own. Check your **answer**, or watch this **video** for complete solutions to both parts.

🔺 **You Try It** **Work through this You Try It problem.**

Work Exercises 45–48 in this eText or in the MyMathLab Study Plan.

📝 *My video summary* ▶ **Example 11 Punting a Football**

An important component of a good punt in football is **hang time**, which is the length of time that the punted ball remains in the air. If wind resistance is ignored, the relationship between the hang time t, in seconds, and the vertical height h, in feet, that the ball reaches can be **modeled** by the **formula** $t = \dfrac{\sqrt{h}}{2}$. Use this formula to answer the following questions.

a. If the average hang time for an NFL punt is 4.6 seconds, then what is the vertical height for an average NFL punt? Round to the nearest foot.

b. Cowboys Stadium in Arlington, Texas, has a huge high-definition screen centered over most of the football field. The bottom of the screen is 90 feet above the field. What hang time would result in the ball hitting the screen? Round to the nearest hundredth of a second.

Solutions Read through the following, or watch this **video** for complete solutions to both parts.

a. Substitute the hang time 4.6 for t in the formula and solve for h.

$$\text{Begin with the original formula:} \qquad t = \frac{\sqrt{h}}{2}$$

$$\text{Substitute 4.6 for } t: \qquad 4.6 = \frac{\sqrt{h}}{2}$$

$$\text{Multiply both sides by 2:} \qquad 9.2 = \sqrt{h}$$

$$\text{Square both sides:} \qquad (9.2)^2 = (\sqrt{h})^2$$

$$\text{Simplify:} \qquad 84.64 = h$$

Rounding, the average NFL punt reaches a vertical height of about 85 feet.

b. Substitute the vertical height, 90 feet, for h and **simplify**.

$$\text{Begin with the original formula:} \qquad t = \frac{\sqrt{h}}{2}$$

$$\text{Substitute 90 for } h: \qquad t = \frac{\sqrt{90}}{2}$$

$$\text{Approximate the square root:} \qquad t \approx \frac{9.4868}{2}$$

$$\text{Divide:} \qquad t \approx 4.7434$$

Rounding, a hang time of about 4.74 seconds will result in a punt that hits the screen.

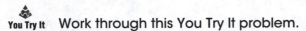 **You Try It** **Work through this You Try It problem.**

Work Exercises 49 and 50 in this eText or in the MyMathLab **Study Plan.**

7.5 Exercises

In Exercises 1–38, solve each radical equation.

1. $\sqrt{x + 15} = 3$

2. $\sqrt{2x - 1} = 3$

3. $\sqrt{r - 3} = -1$

4. $3\sqrt{m} + 5 = 7$

5. $\sqrt{3 - x} = 4$

6. $\sqrt{z + 5} + 2 = 9$

7. $\sqrt{1 - x} + 1 = 4$

8. $\sqrt{4p + 1} - 3 = 2$

9. $5\sqrt{w - 9} - 2 = 8$

10. $\sqrt{x + 6} = \sqrt{2 - 3x}$

11. $\sqrt{q + 10} = -2$

12. $\sqrt{2x - 1} + 6 = 1$

13. $\sqrt{2m + 1} - m = -1$

14. $p - \sqrt{4p + 9} + 3 = 0$

15. $\sqrt{32 - 4x} - x = 0$

16. $\sqrt{7z + 2} + 3z = 2$

17. $\sqrt{x^2 + 4x - 5} = x$

18. $2\sqrt{4y + 1} + 3y = 4y + 4$

19. $2q + 1 = \sqrt{q^3 + 17}$

20. $\sqrt[3]{y^2 - 2y + 8} = 2$

21. $\sqrt[3]{2x - 1} = -2$

22. $\sqrt[3]{7 - 5w} + 1 = 4$

23. $\sqrt[3]{2n^2 + 15n} + 1 = 4$

24. $\sqrt[4]{x^2 - 6x} - 2 = 0$

25. $(x - 5)^{\frac{1}{2}} - 1 = 4$

26. $(2y + 3)^{\frac{1}{3}} + 7 = 9$

27. $(2w + 5)^{\frac{1}{4}} + 1 = 4$

28. $(p^2 - 19)^{\frac{1}{4}} + 2 = 5$

29. $\sqrt{x + 15} - \sqrt{x} = 3$

30. $\sqrt{x - 1} = \sqrt{x - 5}$

31. $\sqrt{x - 64} + 8 = \sqrt{x}$

32. $\sqrt{6 - x} + \sqrt{5x + 6} = 6$

33. $\sqrt{3x + 1} - \sqrt{x + 4} = 1$

34. $\sqrt{6 + 5x} + \sqrt{3x + 4} = 2$

35. $\sqrt{7x - 4} = \sqrt{4x + 11}$

36. $\sqrt[3]{4y - 3} = \sqrt[3]{6y + 9}$

37. $\sqrt[3]{2x^2 + 5x} = \sqrt[3]{2x + 14}$

38. $\sqrt[4]{6w + 1} = \sqrt[4]{2w + 17}$

In Exercises 39–44, solve each formula for the given variable.

39. $T = \dfrac{\pi \sqrt{2L}}{4}$ for L

40. $d = k\sqrt[3]{E}$ for E

41. $r = \sqrt[3]{\dfrac{3V}{\pi h}}$ for h

42. $V = \sqrt{\dfrac{FR}{m}}$ for m

43. $A = P\sqrt{1 + r}$ for r

44. $T = \left(\dfrac{LH^2}{25}\right)^{\frac{1}{4}}$ for L

In Exercises 45–50, use the given model to solve each application problem.

45. **SMOG Grade** Recall from Example 10 that the formula $G = \sqrt{w} + 3$ gives the minimum reading grade level G needed to fully understand written text containing w words with three or more syllables in a sample of 30 sentences.

   a. If a sample of 30 sentences contains 51 words with three or more syllables, then what is the SMOG grade for the text? If necessary, round your answer to a whole number to find the grade level.

   b. If a text must have a ninth-grade reading level, then how many words with three or more syllables are needed in a sample of 30 sentences?

46. **Measure of Leanness** The *ponderal index* is a measure of the "leanness" of a person. A person who is h inches tall and weighs w pounds has a ponderal index I given by $I = \dfrac{h}{\sqrt[3]{w}}$.

   a. Compute the ponderal index for a person who is 75 inches tall and weighs 190 pounds. Round to the nearest hundredth.

   b. What is a man's weight if he is 70 inches tall and has a ponderal index of 12.35? Round to the nearest whole number.

47. **Body Surface Area** The Mosteller formula is used in the medical field to estimate a person's body surface area. The formula is $A = \dfrac{\sqrt{hw}}{60}$, where A is body surface area in square meters, h is height in centimeters, and w is weight in kilograms.

    a. Compute the body surface area of a person who is 178 cm tall and weighs 90.8 kg. Round to the nearest hundredth.

    b. If a woman is 165 cm tall and has body surface area of 1.72 m², how much does she weigh? Round to the nearest tenth.

48. **Skid Marks** Under certain road conditions, the length of a skid mark S, in feet, is related to the velocity v, in miles per hour, by the formula $v = \sqrt{10S}$. Assuming these same road conditions, answer the following:

    a. Compute a car's velocity if it leaves a skid mark of 360 feet.

    b. If a car is traveling at 50 miles per hour when it skids, what will be the length of the skid mark?

49. **Hang Time** Recall from Example 11 that hang time t, in seconds, and the vertical height h, in feet, can be modeled by $t = \dfrac{\sqrt{h}}{2}$. This same formula can be used to model an athlete's hang time when jumping.

    a. If LeBron James has a vertical leap of 3.7 feet, what is his hang time? Round to the nearest hundredth.

    b. When Mark "Wild Thing" Wilson of the Harlem Globetrotters slam-dunked a regulation basketball on a 12-foot rim in front of an Indianapolis crowd, his hang time for the shot was approximately 1.16 seconds. What was the vertical distance of his jump? Round to the nearest tenth.

50. **Distance to the Horizon** From a boat, the distance d, in miles, that a person can see to the horizon is modeled by the formula $d = \dfrac{3\sqrt{h}}{2}$, where h is the height, in feet, of eye level above the sea.

    a. From his ship, how far can a sailor see to the horizon if his eye level is 36 feet above the sea?

    b. How high is the eye level of a sailor who can see 12 miles to the horizon?

7.6 Complex Numbers

THINGS TO KNOW

Before working through this section, be sure you are familiar with the following concepts:

| | | VIDEO | ANIMATION | INTERACTIVE |

1. Find Square Roots of Perfect Squares (Section 7.1, Objective 1)

2. Simplify Radical Expressions Using the Product Rule (Section 7.3, Objective 5)

OBJECTIVES

1 Simplify Powers of i

2 Add and Subtract Complex Numbers

3 Multiply Complex Numbers

4 Divide Complex Numbers

5 Simplify Radicals with Negative Radicands

OBJECTIVE 1 SIMPLIFY POWERS OF i

So far, we have learned about **real number** solutions to **equations**, but not every equation has a real number **solution**.

 *My animation summary*

⊛ **Concept Animation** Watch this animation to see why some equations do not have real number solutions.

To find the solution to equations without real number solutions, such as $x^2 + 1 = 0$, we introduce a new number called the **imaginary unit** i.

Definition Imaginary Unit i

The **imaginary unit** i is defined as

$$i = \sqrt{-1}, \text{ where } i^2 = -1.$$

When working with the **imaginary unit**, we will encounter various powers of i. Let's consider some powers of i and look for patterns that will help us simplify them.

$$\underbrace{i^1 = i = \sqrt{-1}}_{\text{Defined}} \qquad\qquad \underbrace{i^2 = -1}_{\text{Defined}}$$

$$i^3 = \underbrace{i^2 \cdot i}_{\substack{\text{Product rule} \\ \text{for exponents}}} = \underbrace{(-1)}_{i^2 = -1} \cdot i = -i \qquad i^4 = \underbrace{i^2 \cdot i^2}_{\substack{\text{Product rule} \\ \text{for exponents}}} = \underbrace{(-1)}_{i^2 = -1} \cdot \underbrace{(-1)}_{i^2 = -1} = 1$$

$$i^5 = \underbrace{i^4 \cdot i}_{\substack{\text{Product rule} \\ \text{for exponents}}} = \underbrace{(1)}_{i^4 = 1} \cdot i = i \qquad i^6 = \underbrace{i^4 \cdot i^2}_{\substack{\text{Product rule} \\ \text{for exponents}}} = \underbrace{(1)}_{i^4 = 1} \cdot \underbrace{(-1)}_{i^2 = -1} = -1$$

$$i^7 = \underbrace{i^4 \cdot i^3}_{\substack{\text{Product rule} \\ \text{for exponents}}} = \underbrace{(1)}_{i^4 = 1} \cdot \underbrace{(-i)}_{i^3 = -i} = -i \qquad i^8 = \underbrace{i^4 \cdot i^4}_{\substack{\text{Product rule} \\ \text{for exponents}}} = \underbrace{(1)}_{i^4 = 1} \cdot \underbrace{(1)}_{i^4 = 1} = 1$$

$$\vdots$$

Notice that the powers of i follow the pattern $i, -1, -i, 1$. Based on this pattern, what is the value of i^0? View this **popup** to find out.

We can use the following procedure to simplify powers of i.

Simplifying i^n for $n > 4$

Step 1 Divide n by 4 and find the **remainder** r.

Step 2 Replace the exponent (power) on i by the remainder, $i^n = i^r$.

Step 3 Use the results $i^0 = 1$, $i^1 = i$, $i^2 = -1$, and $i^3 = -i$ to simplify if necessary.

My video summary ▶ **Example 1 Simplifying Powers of i**

Simplify.

a. i^{17} b. i^{60} c. i^{39} d. $-i^{90}$ e. $i^{14} + i^{29}$

Solutions

a. **Step 1** Divide the exponent by 4 and find the remainder:

$$
\begin{array}{r}
4 \\
4{\overline{)17}} \quad \longleftarrow \text{ Exponent, } n \\
\underline{16} \\
1 \quad \longleftarrow \text{ Remainder, } r
\end{array}
$$

Step 2 Replace n by r: $i^{\overset{n}{17}} = i^{\overset{r}{1}}$

Step 3 Simplify: $i^{17} = i^1 = i$

b. Divide the **exponent** 60 by 4 and find the **remainder**.

$$
\begin{array}{r}
15 \\
4{\overline{)60}} \quad \longleftarrow \text{ Exponent, } n \\
\underline{4} \\
20 \\
\underline{20} \\
0 \quad \longleftarrow \text{ Remainder, } r
\end{array}
$$

Replace n by r and simplify if necessary.

$$
i^{\overset{n}{60}} = i^{\overset{r}{0}} = 1
$$

c.–e. Try to simplify these **powers of i** on your own. Check your **answers**, or watch this **video** to see detailed solutions to all five parts.

You Try It Work through this **You Try It** problem.

Work **Exercises 1–7** in this eText or in the MyMathLab Study Plan.

With the **imaginary unit**, we can now expand our number system from the set of real numbers to the set of **complex numbers**.

> ### Complex Numbers
>
> The set of all numbers of the form
>
> $$a + bi,$$
>
> where a and b are **real numbers** and i is the **imaginary unit**, is called the set of **complex numbers**. The number a is called the **real part**, and the number b is called the **imaginary part**.

If $b = 0$, then the complex number is a purely real number. If $a = 0$, then the complex number is a purely **imaginary number**. **Figure 7** on the next page illustrates the relationships between complex numbers.

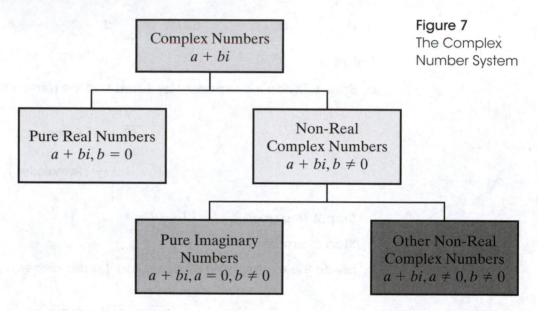

Figure 7
The Complex Number System

Figure 7 shows us that all **real numbers** are **complex numbers**, but not all complex numbers are real numbers. This distinction is important, particularly when solving equations. For example, $x^2 + 1 = 0$ has no *real* solutions, but it does have two *complex* solutions.

A complex number of the form $a + bi$ is written in **standard form**, and is the typical way to write complex numbers. Below are examples of complex numbers written in standard form.

	Standard form		Real part, $a = 7$		Imaginary part, $b = 0$	
Real Number:	7	$=$	7	$+$	0	i

	Standard form		Real part, $a = 0$		Imaginary part, $b = 3$	
Imaginary Number:	$3i$	$=$	0	$+$	3	i

	Standard form		Real part, $a = 4$		Imaginary part, $b = 9$	
Non-Real Complex Number:	$4 + 9i$	$=$	4	$+$	9	i

	Standard form		Real part, $a = \frac{1}{3}$		Imaginary part, $b = \frac{2}{3}$	
	$\frac{1}{3} + \frac{2}{3}i$	$=$	$\frac{1}{3}$	$+$	$\frac{2}{3}$	i

OBJECTIVE 2 ADD AND SUBTRACT COMPLEX NUMBERS

To add or subtract **complex numbers**, we combine the **real parts** and combine the **imaginary parts**.

Adding and Subtracting Complex Numbers

To add complex numbers, add the real parts and add the imaginary parts.

$$(a + bi) + (c + di) = (a + c) + (b + d)i$$

To subtract complex numbers, subtract the real parts and subtract the imaginary parts.

$$(a + bi) - (c + di) = (a - c) + (b - d)i$$

My video summary ⏵ **Example 2 Adding and Subtracting Complex Numbers**

Perform the indicated operations.

a. $(3 + 5i) + (2 - 7i)$　　　　　　　　　　**b.** $(3 + 5i) - (2 - 7i)$

c. $(-3 - 4i) + (2 - i) - (3 + 7i)$

Solutions

a.　　　　Original expression:　$(3 + 5i) + (2 - 7i)$

　　　Remove parentheses:　$= 3 + 5i + 2 - 7i$

　　　　　　　　　　　　　　Real　Imaginary

　　　Collect like terms:　$= 3 + 2 + 5i - 7i$

　　　　　　　　　　　　　　$3 + 2$

　　　Combine real parts:　$= 5 + 5i - 7i$

　　　　　　　　　　　　　　$5i - 7i$

　　Combine imaginary parts:　$= 5 - 2i$

b.　　　　Original expression:　$(3 + 5i) - (2 - 7i)$

　　Change to add the opposite:　$= (3 + 5i) + (-2 + 7i)$

　　　Remove parentheses:　$= 3 + 5i - 2 + 7i$

Finish working the problem on your own. Check your **answer**, or watch this **video** to see the full solutions to all three parts.

c. Try to work the problem on your own. Check your **answer**, or watch this **video** for detailed solutions to all three parts.

You Try It Work through this **You Try It** problem.

Work Exercises 8–17 in this eText or in the MyMathLab **Study Plan.**

OBJECTIVE 3 MULTIPLY COMPLEX NUMBERS

When multiplying **complex numbers**, use the **distributive property** and the **FOIL method** just as when multiplying **polynomials**. Remember that $i^2 = -1$ when simplifying.

🖎 *My video summary* ▶ **Example 3 Multiplying Complex Numbers**

Multiply.

a. $-4i(3 - 8i)$ b. $(3 - 4i)(6 + 11i)$

Solutions Read through the following solutions, or watch this **video**.

a. Multiply using the distributive property.

$$\begin{aligned}
\text{Original expression:} \quad & -4i(3 - 8i) \\
\text{Distribute:} \quad = \; & -4i \cdot 3 + (-4i)(-8i) \\
\text{Multiply:} \quad = \; & -12i + 32i^2 \\
\text{Replace } i^2 \text{ with } -1: \quad = \; & -12i + 32\underbrace{(-1)}_{i^2 = -1} \\
\text{Simplify:} \quad = \; & -12i - 32 \\
\text{Write in standard form:} \quad = \; & -32 - 12i
\end{aligned}$$

b. Multiply using the **FOIL** method.

$$(3 - 4i)(6 + 11i) = \overbrace{3 \cdot 6}^{F} + \overbrace{3 \cdot 11i}^{O} - \overbrace{4i \cdot 6}^{I} - \overbrace{4i \cdot 11i}^{L}$$

$$= 18 + \underbrace{33i - 24i}_{\text{Collect like terms}} - 44i^2$$

$$= 18 + 9i - \underbrace{44i^2}_{i^2 = -1}$$

$$= 18 + 9i - 44(-1)$$

$$= \underbrace{18 + 9i + 44}_{\text{Collect like terms}}$$

$$= 62 + 9i$$

You Try It Work through this You Try It problem.

Work Exercises 18–23 in this eText or in the MyMathLab Study Plan.

🖎 *My video summary* ▶ **Example 4 Multiplying Complex Numbers**

Multiply $\left(\dfrac{2}{3} - \dfrac{1}{5}i \right)\left(-\dfrac{6}{5} + \dfrac{7}{2}i \right)$.

Solution Work the problem on your own. Check your **answer**, or watch this **video** for a detailed solution.

You Try It Work through this You Try It problem.

Work Exercises 24–26 in this eText or in the MyMathLab Study Plan.

Example 5 Special Products Involving Complex Numbers

Multiply.

a. $(4 + 2i)^2$ **b.** $(\sqrt{5} - 7i)^2$

Solutions

a. We recognize this expression as the square of a **binomial sum**, where $A = 4$ and $B = 2i$. Use the **special product rule** to multiply.

Write the square of a binomial sum rule: $(A + B)^2 = A^2 + 2AB + B^2$

Substitute 4 for A and $2i$ for B: $\underbrace{(4}_{A} + \underbrace{2i)^2}_{B} = \underbrace{(4)^2}_{A^2} + \underbrace{2(4)(2i)}_{2AB} + \underbrace{(2i)^2}_{B^2}$

Simplify: $= 16 + 16i + 4i^2$

Replace i^2 with -1: $= 16 + 16i + 4(-1)$

Multiply: $= 16 + 16i - 4$

Simplify: $= 12 + 16i$

We can also find this product using the **FOIL method.**

Begin with the original expression: $(4 + 2i)^2$

Write the expression as a product: $= (4 + 2i)(4 + 2i)$

Multiply using the FOIL method: $= \overset{F}{\overbrace{4 \cdot 4}} + \overset{O}{\overbrace{4 \cdot 2i}} + \overset{I}{\overbrace{2i \cdot 4}} + \overset{L}{\overbrace{2i \cdot 2i}}$

Simplify: $= 16 + 8i + 8i + 4i^2$

Collect like terms: $= 16 + 16i + 4i^2$

At this point, we would continue simplifying as before to find $(4 + 2i)^2 = 12 + 16i$.

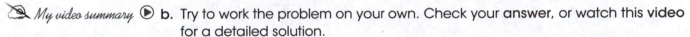 **b.** Try to work the problem on your own. Check your **answer**, or watch this **video** for a detailed solution.

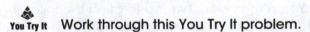 Work through this **You Try It** problem.

Work Exercises 27–30 in this eText or in the MyMathLab Study Plan.

In Section 4.3, we learned that a **binomial sum** and a **binomial difference** made from the same two terms are **conjugates** of each other. We also saw that the product of conjugates resulted in the **difference of two squares**. These results extend to our discussion of **complex numbers** as follows.

Complex Conjugates

The complex numbers $(a + bi)$ and $(a - bi)$ are called **complex conjugates** of each other. A complex conjugate is obtained by changing the sign of the imaginary part in a complex number. Also, $(a + bi)(a - bi) = a^2 + b^2$.

Notice that the product of complex conjugates is a *sum* of two **squares** rather than a difference and is always a **real number**. See **why**.

My video summary ▶ **Example 6 Multiplying Complex Conjugates**

Multiply $(-2 + 7i)(-2 - 7i)$.

Solution Read through the following, or watch this **video** for the complete solution.

Since the two **complex numbers** are conjugates, find the product using the result for the **product of complex conjugates**.

Identifying $a = -2$ and $b = 7$ gives

$$\overbrace{(a + bi)(a - bi)}^{} \quad \overbrace{a^2}^{} \quad \overbrace{b^2}^{}$$
$$(-2 + 7i)(-2 - 7i) = (-2)^2 + (7)^2 = 4 + 49 = 53.$$

We could also find the same result using the **FOIL method**. See how.

🔺
You Try It Work through this **You Try It** problem.

Work Exercises 31–40 in this eText or in the MyMathLab **Study Plan**.

OBJECTIVE 4 DIVIDE COMPLEX NUMBERS

When dividing **complex numbers**, the goal is to eliminate the imaginary part from the denominator and to express the **quotient in standard form**, $a + bi$. To do this, multiply the numerator and denominator by the **complex conjugate** of the denominator.

My video summary ▶ **Example 7 Dividing Complex Numbers**

Divide. Write the quotient in standard form.

$$\frac{1 - 3i}{5 - 2i}$$

Solution Watch this **video**, or read through the following solution. The denominator is $5 - 2i$, so its complex conjugate is $5 + 2i$. Multiply both the **numerator** and denominator by the complex conjugate and simplify to standard form.

Multiply numerator and denominator by $5 + 2i$: $\quad \dfrac{1 - 3i}{5 - 2i} = \dfrac{1 - 3i}{5 - 2i} \cdot \dfrac{5 + 2i}{5 + 2i}$

Multiply numerators and multiply denominators:
(remember that $(a + bi)(a - bi) = a^2 + b^2$) $\quad = \dfrac{5 + 2i - 15i - 6i^2}{(5)^2 + (2)^2}$

Simplify exponents: (remember that $i^2 = -1$) $\quad = \dfrac{5 + 2i - 15i - 6(-1)}{25 + 4}$

Simplify the numerator and denominator: $\quad = \dfrac{5 - 13i + 6}{29}$

$$= \dfrac{11 - 13i}{29}$$

Write in standard form, $a + bi$: $\quad = \dfrac{11}{29} - \dfrac{13}{29}i$

You Try It Work through this You Try It problem.

Work Exercises 41–44 in this eText or in the MyMathLab Study Plan.

TIP Remember that multiplying the numerator and denominator of an expression by the same quantity is the same as multiplying the expression by 1.

 *My video summary* **Example 8 Dividing Complex Numbers**

Divide. Write the quotient in standard form.

$$\frac{5 + 7i}{2i}$$

Solution First, multiply the numerator and denominator by the **complex conjugate** of the denominator. The denominator is $2i = 0 + 2i$, so its complex conjugate is $0 - 2i = -2i$.

$$\frac{5 + 7i}{2i} = \overbrace{\frac{5 + 7i}{2i} \cdot \frac{-2i}{-2i}}^{\substack{\text{Multiply numerator} \\ \text{and denominator} \\ \text{by } -2i}} = \overbrace{\frac{-10i - 14i^2}{-4i^2}}^{\substack{\text{Multiply numerators} \\ \text{and multiply} \\ \text{denominators}}}$$

Finish simplifying on your own, and write the answer in standard form. Check your **answer**, or watch this **video** for the complete solution.

 You Try It Work through this You Try It problem.

Work Exercises 45–48 in this eText or in the MyMathLab Study Plan.

In Example 8, we would get the same result if we multiplied both the numerator and denominator by $2i$. Find out why.

OBJECTIVE 5   SIMPLIFY RADICALS WITH NEGATIVE RADICANDS

In the next chapter, we will solve equations involving solutions with radicals having a negative **radicand**. Thus, we must first learn how to simplify a **radical** with a negative radicand such as $\sqrt{-49}$. By remembering that $\sqrt{-1} = i$, we can use the following rule to simplify this expression.

Square Root of a Negative Number

For any positive real number a,

$$\sqrt{-a} = \sqrt{-1} \cdot \sqrt{a} = i\sqrt{a}$$

So, $\sqrt{-49} = \sqrt{-1} \cdot \sqrt{49} = i \cdot 7 = 7i$.

At this point, you might want to review how to simplify radicals using the **product rule** in Section 7.3.

My video summary ▶ **Example 9** Simplifying a Square Root with a Negative Radicand

Simplify.

a. $\sqrt{-81}$ 　　　　　　　　　　**b.** $\sqrt{-48}$ 　　　　　　　　　　**c.** $\sqrt{-108}$

Solutions

a. $\sqrt{-81} = \underbrace{\sqrt{-1}}_{i} \cdot \sqrt{81} = i \cdot 9 = 9i$

b. $\sqrt{-48} = \underbrace{\sqrt{-1}}_{i} \cdot \sqrt{48} = i \cdot \sqrt{16} \cdot \sqrt{3}$

$$= i \cdot 4\sqrt{3} = 4i\sqrt{3} \longleftarrow$$

> **Note:** Since $4\sqrt{3}i$ can be confused with $4\sqrt{3i}$, we write $4\sqrt{3}i$ as $4i\sqrt{3}$.

c. Try this problem on your own. Check your **answer**, or watch this **video** for detailed solutions to all three parts.

You Try It Work through this You Try It problem.

Work Exercise 49 in this eText or in the MyMathLab Study Plan.

⚠ When simplifying or performing operations involving **radicals** with a negative **radicand** and an even **index**, it is important to first write the numbers in terms of the **imaginary unit** i if possible.

The property $\sqrt{a} \cdot \sqrt{b} = \sqrt{ab}$ is only true when $a \geq 0$ and $b \geq 0$ so that $\sqrt{a}$ and $\sqrt{b}$ are real numbers. This property does not apply to non-real numbers. To find the correct answer if a or b are negative, we must first write each number in terms of the imaginary unit i.

$$\sqrt{-3} \cdot \sqrt{-12} = \sqrt{(-3)(-12)} = \sqrt{36} = 6 \quad \text{False}$$

$$\sqrt{-3} \cdot \sqrt{-12} = \underbrace{\sqrt{-1}}_{i} \cdot \sqrt{3} \cdot \underbrace{\sqrt{-1}}_{i} \cdot \sqrt{12} = i\sqrt{3} \cdot i\sqrt{12} = \underbrace{i^2\sqrt{36}}_{i^2 = -1} = -6 \quad \text{True}$$

We can use a graphing calculator (in complex mode) to check the result. See Figure 8.

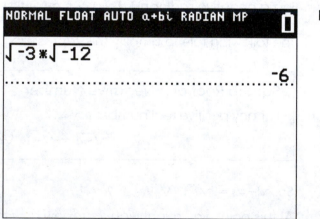

Figure 8

Notice that in order to get the correct answer, we had to first write each number in terms of the **imaginary unit** i.

My video summary ▶ **Example 10** Simplifying Expressions with Negative Radicands

Simplify.

a. $\sqrt{-8} + \sqrt{-18}$

b. $\sqrt{-8} \cdot \sqrt{-18}$

c. $\dfrac{6 + \sqrt{(6)^2 - 4(2)(5)}}{2}$

d. $\dfrac{4 - \sqrt{-12}}{4}$

Solutions

a. $\sqrt{-8} + \sqrt{-18} = \underbrace{\sqrt{-1} \cdot \sqrt{8}}_{i} + \underbrace{\sqrt{-1} \cdot \sqrt{18}}_{i}$

$= i \cdot 2\sqrt{2} + i \cdot 3\sqrt{2}$

$= 2i\sqrt{2} + 3i\sqrt{2}$

$= 5i\sqrt{2}$

b. $\sqrt{-8} \cdot \sqrt{-18} = \left(\underbrace{\sqrt{-1} \cdot \sqrt{8}}_{i}\right) \cdot \left(\underbrace{\sqrt{-1} \cdot \sqrt{18}}_{i}\right)$

$= (i \cdot 2\sqrt{2}) \cdot (i \cdot 3\sqrt{2})$

$= \underbrace{i^2}_{i^2 = -1} \cdot 6\sqrt{4}$

$= -1 \cdot 6 \cdot 2$

$= -12$

c.–d. Try these problems on your own. Check your answers, or watch this **video** for detailed solutions to all four parts.

You Try It Work through this You Try It problem.

Work Exercises 50–57 in this eText or in the MyMathLab Study Plan.

7.6 Exercises

In Exercises 1–7, write each power of i as i, -1, $-i$, or 1.

1. i^{41} 2. i^{28} 3. $-i^{19}$ 4. $(-i)^7$ 5. $i^{22} + i^{13}$ 6. $i^4 \cdot i^7$ 7. $\dfrac{i^{24}}{i^{-22}}$

In Exercises 8–17, find the sum or difference. Write each answer in standard form, $a + bi$.

8. $(3 - 2i) + (-7 + 9i)$

9. $(3 - 2i) - (-7 + 9i)$

10. $i - (1 + i)$

11. $5 + (2 - 3i)$

12. $(2 + 5i) - (2 - 5i)$

13. $(2 + 5i) + (2 - 5i)$

14. $\left(\sqrt{2} - 3i\right) + \left(2\sqrt{2} + 3i\right)$

15. $\left(\sqrt{7} + 4i\right) - \left(3\sqrt{7} - 5i\right)$

16. $[(-1 + 8i) - (3 - 4i)] + (9 - 4i)$

17. $(6 + 3i) - [(2 + 4i) + (5 - 2i)]$

In Exercises 18–30, perform the indicated operations. Write each answer in standard form.

18. $3i(7i)$

19. $2i(4 - 3i)$

20. $-i(1 - i)$

21. $(3 - 2i)(6 + i)$

22. $(-2 - i)(3 - 4i)$

23. $(5 + i)(2 + 3i)$

24. $\left(\dfrac{2}{5} - 6i\right)\left(\dfrac{1}{3} + 10i\right)$

25. $\left(\dfrac{1}{4} + i\right)\left(\dfrac{1}{2} - 4i\right)$

26. $\left(\dfrac{3}{5} + \dfrac{1}{2}i\right)\left(\dfrac{10}{3} + \dfrac{4}{5}i\right)$

27. $(2 + 7i)^2$

28. $(6 - 2i)^2$

29. $\left(\sqrt{2} - i\right)^2$

30. $\left(4 + \sqrt{3}i\right)^2$

In Exercises 31–40, find the product of the complex number and its conjugate.

31. $5 - 2i$

32. $1 - i$

33. $\dfrac{1}{2} - 3i$

34. $\sqrt{5} + i$

35. $4i$

36. $\dfrac{\sqrt{3}}{2} + \dfrac{1}{2}i$

37. $10 + 4i$

38. $\dfrac{3}{5} - \dfrac{1}{4}i$

39. $2\sqrt{3} + 5i$

40. $-8i$

In Exercises 41–48, write each quotient in standard form.

41. $\dfrac{2 - i}{3 + 4i}$

42. $\dfrac{1}{2 - i}$

43. $\dfrac{3i}{2 + 2i}$

44. $\dfrac{5 + i}{5 - i}$

45. $\dfrac{2 - 3i}{5i}$

46. $\dfrac{12 + 7i}{3i}$

47. $\dfrac{\sqrt{3} + 2i}{\sqrt{3} - 2i}$

48. $\dfrac{-8i}{4 - 11i}$

In Exercises 49–57, write each expression in standard form.

49. $\sqrt{-320}$

50. $\sqrt{-36} - \sqrt{49}$

51. $\sqrt{-1} + 3 - \sqrt{-64}$

52. $\sqrt{-2} \cdot \sqrt{-18}$

53. $\left(\sqrt{-8}\right)^2$

54. $\left(i\sqrt{-4}\right)^2$

55. $\dfrac{-4 - \sqrt{-20}}{2}$

56. $\dfrac{-3 - \sqrt{-81}}{6}$

57. $\dfrac{4 + \sqrt{-8}}{4}$

CHAPTER EIGHT

Quadratic Equations and Functions; Circles

CHAPTER EIGHT CONTENTS

8.1 Solving Quadratic Equations

THINGS TO KNOW

Before working through this section, be sure you are familiar with the following concepts:

		VIDEO	ANIMATION	INTERACTIVE
You Try It	1. Factor Trinomials Using Substitution (Section 5.2, Objective 4)	▶		
You Try It	2. Factor Polynomials Completely (Section 5.3, Objective 4)			🖱
You Try It	3. Solve Polynomial Equations by Factoring (Section 5.4, Objective 1)	▶		🖱
You Try It	4. Simplify Radical Expression Using the Product Rule (Section 7.3, Objective 5)	▶		🖱

OBJECTIVES

1 Solve Quadratic Equations Using Factoring

2 Solve Quadratic Equations Using the Square Root Property

3 Solve Quadratic Equations by Completing the Square

4 Solve Quadratic Equations Using the Quadratic Formula

5 Use the Discriminant to Determine the Number and Type of Solutions to a Quadratic Equation

6 Solve Equations That Are Quadratic in Form

OBJECTIVE 1   SOLVE QUADRATIC EQUATIONS USING FACTORING

In Section 1.1, we studied **linear equations** of the form $ax + b = c$, with $a \neq 0$. These equations were called first-degree **polynomial** equations. In this section, we learn how to solve second-degree polynomial equations. These equations are called **quadratic equations.**

Definition   Quadratic Equation in One Variable

A **quadratic equation in one variable** is an equation that can be written in the form $ax^2 + bx + c = 0$, where a, b, and c are real numbers, and $a \neq 0$. Quadratic equations in this form are said to be in *standard form.*

As we have seen with **polynomial equations,** some quadratic equations can be solved quickly by **factoring** and by using the **zero product property.** Remember that the zero product property states that if two **factors** multiplied together are equal to zero, then at least one of the factors must be zero.

Solving Quadratic Equations by Factoring

Step 1 Write the quadratic equation in the **standard form** $ax^2 + bx + c = 0$.

Step 2 Factor the left-hand side.

Step 3 Set each factor from Step 2 equal to zero (zero product property), and solve the resulting linear equations.

Step 4 Check each potential **solution** in the original equation and form the solution set.

My video summary ▶ **Example 1 Solving a Quadratic Equation Using Factoring**

Solve $6x^2 - 17x = -12$.

Solution   Original equation: $6x^2 - 17x = -12$

Write in standard form: $6x^2 - 17x + 12 = 0$

Factor the left-hand side. Then set each factor equal to zero and solve the resulting equations. The solution set is $\left\{ \dfrac{4}{3}, \dfrac{3}{2} \right\}$. View the **check,** or watch this **video** for the complete solution.

You Try It   Work through this **You Try It** problem.

Work Exercises 1–6 in this eText or in the MyMathLab Study Plan.

OBJECTIVE 2 SOLVE QUADRATIC EQUATIONS USING THE SQUARE ROOT PROPERTY

Consider the **quadratic equation** $x^2 = 100$. We can solve this equation by **factoring** and then using the **zero-product** property as follows:

Write the original equation:	$x^2 = 100$
Subtract 100 from both sides:	$x^2 - 100 = 0$
Factor the **difference of squares**:	$(x - 10)(x + 10) = 0$
Set each **factor** equal to 0:	$x - 10 = 0$ or $x + 10 = 0$
Solve each resulting equation:	$x = 10$ or $x = -10$

Both values **check**, so the **solution set** is $\{-10, 10\}$.

Some quadratic equations are not easy to solve by factoring, and other quadratic equations cannot be solved by factoring at all. Therefore, in this chapter, we discuss other methods for solving quadratic equations, beginning with the *square root property*.

To understand this property, look again at the equation $x^2 = 100$. Notice that the two solutions, -10 and 10, are the negative and positive **square roots** of the number 100. Looking at a similar equation, $x^2 = 36$, do you see that the solutions will be the negative and positive square roots of 36? Therefore, $x = -\sqrt{36} = -6$ or $x = \sqrt{36} = 6$. This illustrates the **square root property**.

Square Root Property

If u is an algebraic expression and k is a real number, then $u^2 = k$ is equivalent to $u = -\sqrt{k}$ or $u = \sqrt{k}$. Equivalently, if $u^2 = k$, then $u = \pm\sqrt{k}$.

 *My video summary* ▶ **Concept Video** Watch this **video** for further explanation of the square root property.

Note: If $k \geq 0$, then the solutions to the equation $u^2 = k$ will be **real numbers**. However, if $k < 0$, then the solutions to $u^2 = k$ will be non-real **complex numbers**.

My video summary ▶ **Example 2 Using the Square Root Property**

Use the square root property to solve each **quadratic equation**. Write each answer in **simplest form**.

a. $x^2 = 144$ **b.** $x^2 = 48$

Solutions For each equation, we apply the square root property and simplify. Work through each of the following, or watch this **video** for fully worked solutions.

a.

Write the original equation:	$x^2 = 144$
Apply the square root property:	$x = \pm\sqrt{144}$
Evaluate:	$x = \pm 12$

These results **check**, so the solution set is $\{-12, 12\}$.

b.   Write the original equation: $x^2 = 48$

Apply the square root property: $x = \pm\sqrt{48}$

Factor the radicand: $x = \pm\sqrt{16 \cdot 3}$

Simplify: $x = \pm 4\sqrt{3}$

View the check. The solution set is $\{-4\sqrt{3}, 4\sqrt{3}\}$.

 You Try It   Work through this You Try It problem.

Work Exercises 7–10 in this eText or in the MyMathLab **Study Plan.**

When solving equations of the form $u^2 = k$, we often simply say that we are taking the square root of both sides. However, the square root of a number yields only one value, the **principal root**. Remember that applying the square root property for $k \neq 0$ will result in *two* values: the positive and negative square roots of k. Be sure to include the $\pm$.

To solve quadratic equations using the square root property use the following guidelines.

Solving Quadratic Equations Using the Square Root Property

Step 1 Write the equation in the form $u^2 = k$ to isolate the quantity being squared.

Step 2 Apply the square root property.

Step 3 Solve the resulting equations.

Step 4 Check the solutions in the original equation.

Example 3 Solving a Quadratic Equation Using the Square Root Property

Solve.

a. $x^2 - 16 = 0$ **b.** $2x^2 + 72 = 0$

c. $(x - 1)^2 = 9$ **d.** $2(x + 1)^2 - 17 = 23$

Solutions

a. Isolate x^2, then apply the square root property.

Original equation: $x^2 - 16 = 0$

Add 16 to both sides: $x^2 = 16$ ← $\boxed{\text{Square is isolated}}$

Apply the square root property: $x = \pm\sqrt{16}$

Simplify the radical: $x = \pm 4$

The solution set is $\{-4, 4\}$. The check is left to you.

b. Isolate x^2, then apply the square root property.

Original equation: $2x^2 + 72 = 0$

Subtract 72 from both sides: $2x^2 = -72$

Divide both sides by 2:	$x^2 = -36$ ←$\boxed{\text{Square is isolated}}$
Apply the square root property:	$x = \pm\sqrt{-36}$
Simplify the **radical** with a negative **radicand**:	$x = \pm\sqrt{-1} \cdot \sqrt{36}$
	$x = \pm 6i$

The solution set is $\{-6i, 6i\}$. The check is left to you.

📝 *My video summary* ▶ **c.** In this case, an **algebraic expression** is being squared. However, this will not change the process. Isolate the square and apply the square root property.

Original equation:	$(x-1)^2 = 9$ ←$\boxed{\text{Square is isolated}}$
Apply the square root property:	$x - 1 = \pm\sqrt{9}$

Try to finish solving this equation on your own. View the **answer**, or watch this **video** for a complete solution to part c.

📝 *My video summary* ▶ **d.** Try to work this problem on your own. Remember to isolate the square first, then apply the square root property. Simplify radicals if possible. View the **answer**, or watch this **video** for the complete solution to part d.

🔺 **You Try It** Work through this **You Try It** problem.

Work Exercises 11–36 in this eText or in the My Math Lab **Study Plan.**

In Example 3c, the **quadratic equation** $(x-1)^2 = 9$ can be solved using the square root property because the left side of the equation is a **perfect square** and the right side is a **constant**. But what about quadratic equations such as $x^2 - 5x + 3 = 0$? Even if we get the square term by itself

$$x^2 = 5x - 3,$$

we cannot apply the **square root property** because the right-hand side is not a constant. However, every quadratic equation can be written in the form $(x-h)^2 = k$ (as in Example 3c) by using a method known as *completing the square*.

OBJECTIVE 3 **SOLVE QUADRATIC EQUATIONS BY COMPLETING THE SQUARE**

Consider the following **perfect square trinomials**:

$$x^2 + 2x + 1 = (x+1)^2 \qquad x^2 - 6x + 9 = (x-3)^2 \qquad x^2 - 7x + \frac{49}{4} = \left(x - \frac{7}{2}\right)^2$$

$$\left(\frac{1}{2} \cdot 2\right)^2 = 1 \qquad\qquad \left(\frac{1}{2} \cdot (-6)\right)^2 = 9 \qquad\qquad \left(\frac{1}{2} \cdot (-7)\right)^2 = \frac{49}{4}$$

In each case, notice the relationship between the **coefficient** of the **linear term** (*x*-term) and the **constant** term. The constant term of a perfect square trinomial is equal to the square of $\frac{1}{2}$ the **linear coefficient**.

To **complete the square** means to add an appropriate constant so that a binomial of the form $x^2 + bx$ becomes a perfect square trinomial. The appropriate constant is the square of half the linear coefficient, $\left(\frac{1}{2} \cdot b\right)^2$. For example, to complete the square given $x^2 + 10x$, we add $\left(\frac{1}{2} \cdot 10\right)^2 = 5^2 = 25$ so we can write $x^2 + 10x + 25 = (x + 5)^2$.

Example 4 Completing the Square

What number must be added to make the binomial a perfect square trinomial?

a. $x^2 - 12x$ **b.** $x^2 + 5x$ **c.** $x^2 - \dfrac{3}{2}x$

Solutions

a. The linear coefficient is -12, so we must add $\left(\frac{1}{2}(-12)\right)^2 = (-6)^2 = 36$ to complete the square. Thus, the expression $x^2 - 12x + 36$ is a perfect square trinomial and $x^2 - 12x + 36 = (x - 6)^2$.

b. The linear coefficient is 5, so we must add $\left(\frac{1}{2} \cdot 5\right)^2 = \left(\frac{5}{2}\right)^2 = \dfrac{25}{4}$ to complete the square: $x^2 + 5x + \dfrac{25}{4} = \left(x + \dfrac{5}{2}\right)^2$.

My video summary ▶ **c.** Try to work this problem on your own. View the **answer**, or watch this **video** for the complete solution.

You Try It Work through this **You Try It** problem.

Work Exercises 37–40 in this eText or in the MyMathLab Study Plan.

When writing the perfect square trinomial as a binomial squared, note that the first term of the binomial is x (the variable) and the second term is $\dfrac{1}{2}$ the linear coefficient from the trinomial. Consider the perfect square trinomials we saw earlier:

$$x^2 + 2x + 1 = (x + 1)^2 \qquad x^2 - 6x + 9 = (x - 3)^2 \qquad x^2 - 7x + \frac{49}{4} = \left(x - \frac{7}{2}\right)^2$$

$$\left(\frac{1}{2} \cdot 2\right) = 1 \qquad\qquad \left(\frac{1}{2} \cdot (-6)\right) = -3 \qquad\qquad \left(\frac{1}{2} \cdot (-7)\right) = -\frac{7}{2}$$

To solve a **quadratic equation** of the form $ax^2 + bx + c = 0$ by **completing the square**, where a, b, and c are real numbers, and $a \neq 0$, use the following guidelines:

Solving $ax^2 + bx + c = 0$, $a \neq 0$, by Completing the Square

Step 1 If $a \neq 1$, divide both sides of the equation by a.

Step 2 Move all constants to the right-hand side.

Step 3 Find $\dfrac{1}{2}$ times the coefficient of the x-term, square it, and add the result to both sides of the equation.

Step 4 The left-hand side is now a **perfect square**. Rewrite it as a **binomial squared**.

Step 5 Use the **square root property** and solve for x.

📝 *My video summary* ▶ **Example 5 Solving a Quadratic Equation by Completing the Square**

Solve $x^2 - 8x + 2 = 0$ by completing the square.

Solution Watch this video, or read the following for a complete solution.

Step 1 The leading coefficient is 1, so proceed to Step 2.

Step 2 Move all constants to the right-hand side.

$$x^2 - 8x = -2$$

Step 3 Multiply $\frac{1}{2}$ times the coefficient of the x-term, square the result, and add this to both sides of the equation.

$$\left(\frac{1}{2} \cdot (-8)\right)^2 = (-4)^2 = 16 \quad \rightarrow \quad x^2 - 8x + 16 = -2 + 16$$

$$x^2 - 8x + 16 = 14$$

Step 4 Rewrite the left-hand side as a perfect square: $(x - 4)^2 = 14$

Step 5 Use the square root property and solve for x.

Use the square root property: $\quad x - 4 = \pm\sqrt{14}$

Add 4 to both sides: $\quad x = 4 \pm \sqrt{14}$

The solution set is $\left\{4 - \sqrt{14},\, 4 + \sqrt{14}\right\}$.

🔺 **You Try It** Work through this **You Try It** problem.

Work Exercises 41–46 in this eText or in the MyMathLab **Study Plan.**

📝 *My video summary* ▶ **Example 6 Solving a Quadratic Equation by Completing the Square**

Solve $2x^2 - 10x - 6 = 0$ by completing the square.

Solution Read the following, or watch this video for a complete solution.

Step 1   Original equation: $\quad 2x^2 - 10x - 6 = 0$

Divide both sides by 2: $\quad \dfrac{2x^2}{2} - \dfrac{10x}{2} - \dfrac{6}{2} = \dfrac{0}{2}$

Simplify: $\quad x^2 - 5x - 3 = 0$

Step 2 Add 3 to both sides: $\quad x^2 - 5x = 3$

Step 3 $\left(\frac{1}{2} \cdot (-5)\right)^2 = \left(-\frac{5}{2}\right)^2 = \frac{25}{4} \quad \rightarrow \quad x^2 - 5x + \frac{25}{4} = 3 + \frac{25}{4}$

$$x^2 - 5x + \frac{25}{4} = \frac{12}{4} + \frac{25}{4}$$

$$x^2 - 5x + \frac{25}{4} = \frac{37}{4}$$

Step 4 Write the left-hand side as a binomial squared: $\left(x - \dfrac{5}{2}\right)^2 = \dfrac{37}{4}$

Step 5 Use the square root property: $x - \dfrac{5}{2} = \pm\sqrt{\dfrac{37}{4}}$

Apply $\sqrt{\dfrac{a}{b}} = \dfrac{\sqrt{a}}{\sqrt{b}}$: $x - \dfrac{5}{2} = \pm\dfrac{\sqrt{37}}{\sqrt{4}}$

Simplify: $x - \dfrac{5}{2} = \pm\dfrac{\sqrt{37}}{2}$

Add $\dfrac{5}{2}$ to both sides and simplify: $x = \dfrac{5}{2} \pm \dfrac{\sqrt{37}}{2} = \dfrac{5 \pm \sqrt{37}}{2}$

The solution set is $\left\{\dfrac{5 - \sqrt{37}}{2}, \dfrac{5 + \sqrt{37}}{2}\right\}$.

You Try It Work through this You Try It problem.

Work Exercises 47–52 in this eText or in the MyMathLab Study Plan.

My video summary ▶ **Example 7 Solving a Quadratic Equation by Completing the Square**

Solve $9x^2 + 12x + 5 = 0$ by completing the square.

Solution Try working this example on your own. View the **answer**, or watch this video for the complete solution.

You Try It Work through this You Try It problem.

Work Exercises 53–61 in this eText or in the MyMathLab Study Plan.

OBJECTIVE 4 SOLVE QUADRATIC EQUATIONS USING THE QUADRATIC FORMULA

Any **quadratic equation** can be solved by **completing the square**. However, this process can be very time consuming. Solving the general quadratic equation $ax^2 + bx + c = 0$ by completing the square, where a, b, and c are real numbers, and $a \neq 0$, provides a useful result known as the **quadratic formula**.

The quadratic formula can be used to solve any quadratic equation and is often less time-consuming than completing the square.

My animation summary ✱ **Concept Animation** Watch this animation to see how to derive the quadratic formula.

Quadratic Formula

The **solutions** to the quadratic equation $ax^2 + bx + c = 0$, $a \neq 0$, are given by the following formula:

$$x = \frac{-b \pm \sqrt{b^2 - 4ac}}{2a}$$

⚠ Remember to write the quadratic equation in **standard form** before identifying the coefficients a, b, and c.

🖎 *My video summary* ▶ **Example 8 Solving a Quadratic Equation Using the Quadratic Formula**

Use the quadratic formula to solve $x^2 + 4x = -1$.

Solution Work through the following, or watch this **video** for the complete solution.

First, we write the equation in **standard form** by adding 1 to both sides. The standard form is $x^2 + 4x + 1 = 0$. Next, we identify the **coefficients**. No coefficient is written on the x^2-term, so it is understood to be 1.

$$1x^2 + 4x + 1 = 0$$
$$a = 1, \ b = 4, \ c = 1$$

Substitute 1 for a, 4 for b, and 1 for c in the quadratic formula.

Write down the quadratic formula: $\quad x = \dfrac{-b \pm \sqrt{b^2 - 4ac}}{2a}$

Substitute values for a, b, and c: $\quad = \dfrac{-(4) \pm \sqrt{(4)^2 - 4(1)(1)}}{2(1)}$

Simplify: $\quad = \dfrac{-4 \pm \sqrt{12}}{2} \leftarrow \boxed{\sqrt{(4)^2 - 4(1)(1)} = \sqrt{16 - 4} = \sqrt{12}}$

Use the **product rule for radicals**: $\quad = \dfrac{-4 \pm 2\sqrt{3}}{2} \leftarrow \boxed{\sqrt{12} = \sqrt{4} \cdot \sqrt{3} = 2\sqrt{3}}$

Factor out 2 in the numerator: $\quad = \dfrac{2(-2 \pm \sqrt{3})}{2}$

Divide out the common factor: $\quad = -2 \pm \sqrt{3} \quad \leftarrow \boxed{\dfrac{\cancel{2}(-2 \pm \sqrt{3})}{\cancel{2}}}$

View the check. The solution set is $\{-2 - \sqrt{3}, -2 + \sqrt{3}\}$.

You Try It Work through this **You Try It** problem.

Work Exercises 62–67 in this eText or in the MyMathLab Study Plan.

🖎 *My video summary* ▶ **Example 9 Solving a Quadratic Equation Using the Quadratic Formula**

Solve $3x^2 + 2x - 2 = 0$ using the quadratic formula.

Solution The equation is in standard form, so we can identify the **coefficients**.

$$3x^2 + 2x - 2 = 0$$
$$a = 3, \ b = 2, \ c = -2$$

Substitute 3 for a, 2 for b, and -2 for c in the quadratic formula.

Write down the quadratic formula: $\quad x = \dfrac{-b \pm \sqrt{b^2 - 4ac}}{2a}$

Substitute values for a, b, and c: $\quad = \dfrac{-(2) \pm \sqrt{(2)^2 - 4(3)(-2)}}{2(3)}$

Try to finish finding the two solutions on your own. Remember to **simplify the radical** and **divide out** any common factors. View the **answer**, or watch this **video** for a complete solution.

You Try It Work through this You Try It problem.

Work Exercises 68–72 in this eText or in the MyMathLab Study Plan.

My video summary ▶ **Example 10 Solving a Quadratic Equation Using the Quadratic Formula**

Solve $14x^2 - 5x = 5x^2 + 7x - 4$ using the **quadratic formula**.

Solution Try to work this example on your own. Remember to write the equation in **standard form** first. View the **answer**, or watch this **video** for the complete solution.

You Try It Work through this You Try It problem.

Work Exercise 73 in this eText or in the MyMathLab Study Plan.

When a **quadratic equation** contains fractions, it is good practice to **clear the fractions** before using the quadratic formula. Doing so makes the work more manageable.

My video summary ▶ **Example 11 Solving a Quadratic Equation Using the Quadratic Formula**

Solve $\frac{1}{12}x^2 + \frac{1}{6} = \frac{1}{2}x$ using the **quadratic formula**.

Solution

Write the original equation: $\quad \frac{1}{12}x^2 + \frac{1}{6} = \frac{1}{2}x$

Multiply both sides by the **LCD**, 12: $\quad 12\left(\frac{1}{12}x^2 + \frac{1}{6}\right) = 12\left(\frac{1}{2}x\right)$

Simplify: $\quad x^2 + 2 = 6x$

We now have an equation without fractions. Try to finish solving this equation on your own using the quadratic formula. Remember to write it in **standard form** first. View the **answer**, or watch this **video** for the complete solution.

You Try It Work through this You Try It problem.

Work Exercises 74–76 in this eText or in the MyMathLab Study Plan.

My video summary ▶ **Example 12 Solving a Quadratic Equation Using the Quadratic Formula**

Solve $16x^2 - 8x + 5 = 0$ using the **quadratic formula**.

Solution Work through the following, or watch this **video** for a complete solution. The equation is in **standard form**, so we can identify the **coefficients**.

$$16x^2 - 8x + 5 = 0$$
$$a = 16 \quad b = -8 \quad c = 5$$

Substitute 16 for a, -8 for b, and 5 for c in the quadratic formula.

$$x = \frac{-b \pm \sqrt{b^2 - 4ac}}{2a} = \frac{-(-8) \pm \sqrt{(-8)^2 - 4(16)(5)}}{2(16)} = \frac{8 \pm \sqrt{64 - 320}}{32} = \frac{8 \pm \sqrt{-256}}{32}$$

The **radicand** is negative, so the solutions will be non-real **complex numbers**.

$$x = \frac{8 \pm 16i}{32} = \frac{8}{32} \pm \frac{16i}{32} = \frac{1}{4} \pm \frac{1}{2}i$$

The solution set, in standard form $a + bi$, is $\left\{ \frac{1}{4} - \frac{1}{2}i, \frac{1}{4} + \frac{1}{2}i \right\}$.

You Try It Work through this You Try It problem.

Work Exercises 77–82 in this eText or in the MyMathLab Study Plan.

OBJECTIVE 5 USE THE DISCRIMINANT TO DETERMINE THE NUMBER AND TYPE OF SOLUTIONS TO A QUADRATIC EQUATION

In **Example 12**, the quadratic equation $16x^2 - 8x + 5 = 0$ had two non-real solutions. The solutions were non-real **complex numbers** because the expression $b^2 - 4ac$ under the **radical** of the **quadratic formula** had a negative value. This expression $b^2 - 4ac$ is called the **discriminant**.

$$\text{Quadratic Formula: } x = \frac{-b \pm \sqrt{\boxed{b^2 - 4ac}}}{2a} \leftarrow \text{Discriminant}$$

The value of the discriminant can be used to determine the number and type of solutions to a quadratic equation.

Discriminant

Given a quadratic equation $ax^2 + bx + c = 0$, $a \neq 0$, the expression $D = b^2 - 4ac$ is called the **discriminant**.

If $D > 0$, then the quadratic equation has two real solutions.
If $D < 0$, then the quadratic equation has two non-real solutions.
If $D = 0$, then the quadratic equation has exactly one real solution.

My video summary ▶ **Example 13 Using the Discriminant**

Use the **discriminant** to determine the number and type of solutions to each quadratic equation.

a. $2x^2 - 5x + 1 = 0$ **b.** $25x^2 + 9 = 30x$ **c.** $x^2 = 4x - 5$

Solutions

a. The equation $2x^2 - 5x + 1 = 0$ is written in **standard form**, so $a = 2$, $b = -5$, and $c = 1$. Then

$$D = b^2 - 4ac = (-5)^2 - 4(2)(1) = 25 - 8 = 17.$$

The discriminant is positive ($D > 0$), so the equation has two real solutions.

b. The equation $25x^2 + 9 = 30x$ is not written in standard form. Subtract $30x$ from both sides to result in the standard form $25x^2 - 30x + 9 = 0$. Use the coefficients $a = 25$, $b = -30$, and $c = 9$ to find the discriminant. Then use the discriminant to determine the number and type of solutions. View the **answer** to part b, or watch this **video** for complete solutions to all three parts.

c. Try working this problem on your own. View the **answer** to part c, or watch this **video** for complete solutions to all three parts.

You Try It Work through this You Try It problem.

Work Exercises 83–88 in this eText or in the MyMathLab Study Plan.

OBJECTIVE 6 SOLVE EQUATIONS THAT ARE QUADRATIC IN FORM

We have now seen several different techniques for solving **quadratic equations**. View this **popup** for a summary along with some advantages and disadvantages for each approach.

Sometimes equations that are not quadratic can be changed into a quadratic equation by using **substitution**. In **Section 5.2**, we used substitution to rewrite an **algebraic expression** into an easily factorable form. We use this same approach to rewrite some equations so that they look like quadratic equations. Equations of this type are called **quadratic in form**. These equations will have the form $au^2 + bu + c = 0$, $a \neq 0$, after an appropriate substitution.

For example, the equation $3x^4 - 13x^2 + 4 = 0$ is quadratic in form. Letting $u = x^2$, then $u^2 = x^4$, and the equation becomes $3u^2 - 13u + 4 = 0$, which is quadratic. We solve this equation in the next example.

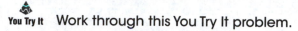 **Example 14 Solving Equations That Are Quadratic in Form**

Solve $3x^4 - 13x^2 + 4 = 0$.

Solution Watch this **video**, or continue reading for a complete solution.

To solve the equation $3x^4 - 13x^2 + 4 = 0$, begin by looking for an appropriate substitution.

Note that the middle term contains x^2, and the first term can be written as $3(x^2)^2$. Therefore, an appropriate substitution is $u = x^2$. Let $u = x^2$, then $u^2 = (x^2)^2 = x^4$. Substituting u for x^2 and u^2 for x^4, the equation becomes

$$3u^2 - 13u + 4 = 0,$$

which is a **quadratic equation**, in terms of u. We can solve this quadratic equation for u by using any of the approaches previously discussed. For this example, let's solve the quadratic equation by factoring.

$$\text{Quadratic equation:} \quad 3u^2 - 13u + 4 = 0$$

$$\text{Factor:} \quad (3u - 1)(u - 4) = 0$$

Now apply the **zero product property** and solve the resulting equations for u.

$$3u - 1 = 0 \quad \text{or} \quad u - 4 = 0$$
$$3u = 1 \qquad\qquad u = 4$$
$$u = \frac{1}{3}$$

Here, we may be tempted to form a **solution set**. However, we have only solved for u. Since the original equation involved the variable x, we still need to solve for x. Because $u = x^2$, this means

$$x^2 = \frac{1}{3} \quad \text{or} \quad x^2 = 4.$$

Solve these two equations for x by applying the **square root property**.

$$x^2 = \frac{1}{3} \qquad\qquad\qquad \text{or} \quad x^2 = 4$$

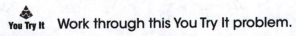

$$x = \pm\sqrt{\frac{1}{3}} = \pm\frac{\sqrt{3}}{3} \qquad\qquad x = \pm\sqrt{4} = \pm 2$$

$$\underbrace{}$$
Rationalize the
denominator

The solution set is $\left\{ -\dfrac{\sqrt{3}}{3}, \dfrac{\sqrt{3}}{3}, -2, 2 \right\}$.

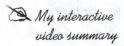 **You Try It** Work through this You Try It problem.

Work Exercises 89 and 90 in this eText or in the MyMathLab Study Plan.

My interactive video summary

✍ **Example 15 Solving Equations That Are Quadratic in Form**

Solve each equation.

a. $x^{\frac{2}{3}} - 4x^{\frac{1}{3}} - 5 = 0$ **b.** $4x^{-2} + 5x^{-1} - 6 = 0$ **c.** $x - 3\sqrt{x} - 10 = 0$

Solutions

Try to solve these equations on your own. View the **answers**, or watch this interactive video for complete solutions to all three parts.

You Try It Work through this You Try It problem.

Work Exercises 91–98 in this eText or in the MyMathLab Study Plan.

8.1 Exercises

In Exercises 1–6, solve each equation by factoring.

1. $x^2 - 8x = 0$

2. $x^2 - x = 6$

3. $x^2 - 10x + 24 = 0$

4. $3x^2 + 8x - 3 = 0$

5. $8m^2 - 15 = 14m$

6. $28z^2 - 13z - 6 = 0$

In Exercises 7–36, solve each equation using the square root property.

7. $x^2 = 256$

8. $y^2 = 80$

9. $m^2 = \dfrac{25}{64}$

10. $p^2 = \dfrac{20}{49}$

11. $x^2 - 121 = 0$

12. $z^2 - 45 = -13$

13. $3x^2 = 84$

14. $5p^2 = 7$

15. $(x - 7)^2 = 625$

16. $(n + 5)^2 = 13$

17. $(z + 3)^2 = 54$

18. $\left(y - \dfrac{1}{2}\right)^2 = \dfrac{9}{4}$

19. $7x^2 - 35 = 0$

20. $16t^2 + 3 = 4$

21. $5x^2 - 7 = -4$

22. $\dfrac{1}{3}x^2 + 8 = 14$

23. $(x + 3)^2 - 16 = 0$

24. $(n - 13)^2 + 11 = 79$

25. $3(t - 11)^2 + 19 = 544$

26. $(2m + 3)^2 - 5 = 76$

27. $x^2 = -4$

28. $y^2 + 45 = 0$

29. $2w^2 + 23 = 7$

30. $(z + 9)^2 = -196$

31. $(n - 1)^2 = -5$

32. $-\dfrac{3}{4}t^2 = \dfrac{1}{75}$

33. $(p - 2)^2 + 32 = 0$

34. $(2x - 14)^2 = -100$

35. $25 - (t - 15)^2 = 169$

36. $(3x - 15)^2 = -162$

In Exercises 37–40, add the appropriate constant to complete the square, then factor the resulting perfect square trinomial.

37. $x^2 + 8x$

38. $y^2 - 20y$

39. $z^2 + 3z$

40. $x^2 - \dfrac{5}{3}x$

In Exercises 41–61, solve each quadratic equation by completing the square.

41. $x^2 + 2x = 24$

42. $x^2 - 14x = -33$

43. $z^2 - 14z + 9 = 0$

44. $x^2 - 8x - 2 = 0$

45. $x^2 + 7x + 14 = 0$

46. $x^2 - 3x = -1$

47. $3x^2 + 6x - 3 = 0$

48. $2x^2 - 2x + 1 = 0$

49. $3y^2 + 12y + 3 = 0$

50. $4z^2 + 10 = 16z$

51. $2x^2 - 3x = 7$

52. $3x^2 = 7 - 24x$

53. $x^2 - 12x + 40 = 0$

54. $x^2 + 14 = -6x$

55. $x^2 - 4x + 13 = 0$

56. $5y^2 - 10y + 7 = 0$

57. $5x^2 + 8x + 4 = 0$

58. $16x^2 + 73 = 56x$

59. $4y^2 + 1 = 9y$

60. $x(x + 6) = 2$

61. $(x - 4)(x - 6) = 31$

In Exercises 62–82, solve each quadratic equation using the quadratic formula.

62. $x^2 - 7x - 144 = 0$

63. $3x^2 + 8x - 3 = 0$

64. $2x^2 - 21x + 40 = 0$

65. $x^2 - 8x - 2 = 0$

66. $2x^2 + 6x = 5$

67. $3x^2 = 1 + 4x$

68. $x^2 + x - 3 = 0$

69. $2x^2 + 8x + 5 = 0$

70. $x^2 + 4x + 1 = 0$

71. $2x^2 - 7x = x^2 + x - 11$

72. $3x^2 + 5x + 1 = 7x + 4$

73. $9x^2 - 6x = -1$

74. $\frac{1}{2}x^2 - x - \frac{1}{3} = 0$

75. $\frac{3}{16}x^2 - \frac{1}{8} = \frac{1}{4}x$

76. $2x^2 + 5.5x - 1.5 = 0$

77. $x^2 - 4x + 29 = 0$

78. $x^2 + 20 = 5x$

79. $4x^2 - x + 8 = 0$

80. $5x^2 + 3x + 1 = 0$

81. $9x^2 = 12x - 5$

82. $\frac{1}{10}x^2 + \frac{1}{2} = \frac{2}{5}x$

In Exercises 83–88, find the discriminant. Then use it to determine the number and type of solutions to each quadratic equation. Do not solve the equations.

83. $x^2 + 3x + 1 = 0$

84. $4x^2 + 4x + 1 = 0$

85. $2x^2 + x = 5$

86. $4x^2 + 25 = 12x$

87. $9x^2 + 49 = 42x$

88. $25x - 6x^2 = 24$

In Exercises 89–98, solve the equation after making an appropriate substitution.

89. $x^4 - 6x^2 + 8 = 0$

90. $2x^4 + 7x^2 - 30 = 0$

91. $2x^{\frac{2}{3}} - 5x^{\frac{1}{3}} + 2 = 0$

92. $x^6 - x^3 = 6$

93. $2x^{-2} - 3x^{-1} - 2 = 0$

94. $\left(\frac{1}{x}\right)^2 + 14\left(\frac{1}{x}\right) + 40 = 0$

95. $x - 6\sqrt{x} - 27 = 0$

96. $\sqrt{x} - 3\sqrt[4]{x} - 4 = 0$

97. $(13x - 1)^2 - 2(13x - 1) - 3 = 0$

98. $3\left(\frac{1}{x-1}\right)^2 - \frac{5}{x-1} - 2 = 0$

8.2 Quadratic Functions and Their Graphs

THINGS TO KNOW

Before working through this section, be sure you are familiar with the following concepts:

			VIDEO	ANIMATION	INTERACTIVE

You Try It 1. Find *x*- and *y*-Intercepts (Section 2.1, Objective 5) ▶

You Try It 2. Find the Domain and Range of a Relation (Section 2.2, Objective 2) 👆

You Try It 3. Solve Quadratic Equations Using Factoring (Section 8.1, Objective 1) ▶

You Try It 4. Solve Quadratic Equations by Completing the Square (Section 8.1, Objective 3) ▶

You Try It 5. Solve Quadratic Equations Using the Quadratic Formula (Section 8.1, Objective 4) ▶

OBJECTIVES

1 Identify the Characteristics of a Quadratic Function from Its Graph

2 Graph Quadratic Functions by Using Translations

3 Graph Quadratic Functions of the Form $f(x) = a(x - h)^2 + k$

4 Find the Vertex of a Quadratic Function by Completing the Square

5 Graph Quadratic Functions of the Form $f(x) = ax^2 + bx + c$ by Completing the Square

6 Find the Vertex of a Quadratic Function by Using the Vertex Formula

7 Graph Quadratic Functions of the Form $f(x) = ax^2 + bx + c$ by Using the Vertex Formula

....

OBJECTIVE 1   IDENTIFY THE CHARACTERISTICS OF A QUADRATIC FUNCTION FROM ITS GRAPH

In the previous section, we learned how to solve **quadratic equations**. In this section, we learn about *graphing quadratic functions*.

Definition   Quadratic Function

A **quadratic function** is a second-degree **polynomial function** of the form $f(x) = ax^2 + bx + c$, where a, b, and c are real numbers and $a \neq 0$. Every quadratic function has a "u-shaped" graph called a **parabola**.

The function $f(x) = x^2$ is a quadratic function with $a = 1$, $b = 0$, and $c = 0$. Its graph is shown in Figure 1a. The function $g(x) = -x^2$ is a quadratic function with $a = -1$, $b = 0$, and $c = 0$. Its graph is shown in Figure 1b. Notice that both graphs are **parabolas** and have the characteristic "u-shape."

Figure 1
Graphs of $f(x) = x^2$ and $g(x) = -x^2$

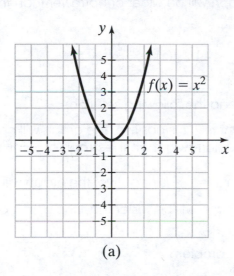

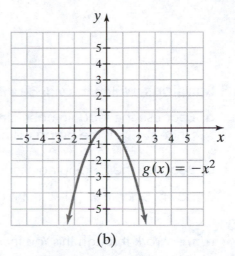

(a) (b)

A **parabola** either opens up (called **concave up**) or opens down (called **concave down**) depending on the **leading coefficient**, a. If $a > 0$, as in Figure 1a, the parabola will "open up." If $a < 0$, as in Figure 1b, the parabola will "open down."

The **leading coefficient** also affects the shape of the **parabola**. The function $f(x) = 2x^2$ is a **quadratic function** with $a = 2$, $b = 0$, and $c = 0$. Its graph is shown in Figure 2a. The function $g(x) = \frac{1}{2}x^2$ is a quadratic function with $a = \frac{1}{2}$, $b = 0$, and $c = 0$. Its graph is shown in Figure 2b. The graph of $y = x^2$ is shown in gray.

Figure 2

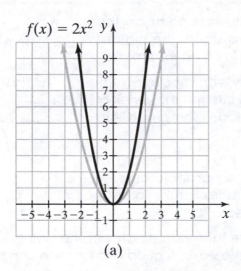

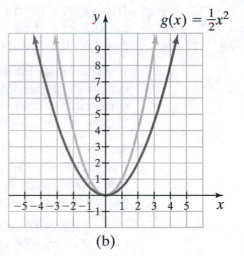

(a) (b)

If $|a| > 1$, as in Figure 2a, the graph will be narrower than the graph of $y = x^2$. If $0 < |a| < 1$, as in Figure 2b, the graph will be wider than the graph of $y = x^2$.

My video summary ▶ **Example 1** Determining the Shape of the Graph of a Quadratic Function

Without graphing, determine if the graph of each **quadratic function** opens up or down. Also determine if the graph will be wider or narrower than the graph of $y = x^2$.

a. $f(x) = 2x^2 - 3x + 5$ **b.** $g(x) = -\frac{2}{3}x^2 + 5x - 7$

Solutions Watch this **video**, or read the following solution.

a. Because the **leading coefficient** $a = 2$ is positive, the graph will open up. Since $|a| > 1$, the graph will be narrower than the graph of $y = x^2$.

b. Because the **leading coefficient** $a = -\frac{2}{3}$ is negative, the graph will open down. Since $0 < |a| < 1$, the graph will be wider than the graph of $y = x^2$.

You Try It Work through this **You Try It** problem.

Work Exercises 1–4 in this eText or in the MyMathLab Study Plan.

Before we can sketch graphs of **quadratic functions**, we must be able to identify the five basic characteristics of a **parabola**: *vertex, axis of symmetry, y-intercept, x-intercepts,* and *domain and range.*

My animation summary ✸ **Concept Animation** Watch this **animation** for an overview of these terms.

The following summarizes the characteristics of a parabola.

The **vertex** is the lowest point on the parabola if it opens up, or the highest point on the parabola if it opens down.

The **axis of symmetry** is an imaginary vertical line that passes through the vertex of a parabola and divides the graph into two mirror images.

The **y-intercept** is the y-coordinate of the point where the parabola crosses the y-axis. Every quadratic function has exactly one y-intercept.

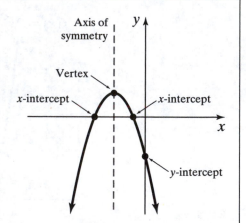

The **x-intercepts** are the x-coordinates of the points where the parabola crosses or touches the x-axis. A quadratic function can have two, one, or no x-intercepts.

The **domain** is the set of all x-values. The domain of every quadratic function is the set of all real numbers, or $(-\infty, \infty)$.

The **range** is the set of all y-values. The range of a quadratic function depends on whether the parabola opens us or down.

My video summary ▶ **Example 2 Finding the Characteristics of a Quadratic Function from Its Graph**

Use the given graph of a quadratic function to find the following:

a. Vertex   b. Axis of symmetry   c. *y*-intercept

d. *x*-intercept(s)   e. Domain and range

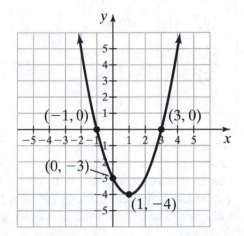

Solutions Read the following, or watch this **video** for a complete solution.

a. Because the graph "opens up," the vertex is the lowest point on the graph: $(h, k) = (1, -4)$.

b. Since the *x*-coordinate of the vertex is $h = 1$, the equation of the axis of symmetry is $x = 1$.

c. The graph crosses the *y*-axis at the point $(0, -3)$, so the **y-intercept** is -3.

d. The graph crosses the *x*-axis at the points $(-1, 0)$ and $(3, 0)$, so the **x-intercepts** are -1 and 3.

e. The **domain** of every quadratic function is all real numbers, or $(-\infty, \infty)$. To determine the **range**, first note that the graph has a **minimum value** because it opens up. The minimum value is the *y*-coordinate of the vertex, $k = -4$. So, the range is all real numbers greater than or equal to -4, or $[-4, \infty)$ in interval notation.

You Try It Work through this **You Try It** problem.

Work Exercises 5–8 in this eText or in the MyMathLab Study Plan.

OBJECTIVE 2 GRAPH QUADRATIC FUNCTIONS BY USING TRANSLATIONS

In mathematics, a **translation** is when every point on a graph is shifted the same distance in the same direction. We now examine translations of parabolas involving vertical or horizontal shifts.

In **Section 2.3**, we **graphed** simple functions by plotting points. We can use the same technique to graph the basic **quadratic function** $f(x) = x^2$. Table 1 gives some **ordered pairs**, and Figure 3 shows the resulting graph.

Table 1

x	$y = x^2$	(x, y)
-3	$(-3)^2 = 9$	$(-3, 9)$
-2	$(-2)^2 = 4$	$(-2, 4)$
-1	$(-1)^2 = 1$	$(-1, 1)$
0	$(0)^2 = 0$	$(0, 0)$
1	$(1)^2 = 1$	$(1, 1)$
2	$(2)^2 = 4$	$(2, 4)$
3	$(3)^2 = 9$	$(3, 9)$

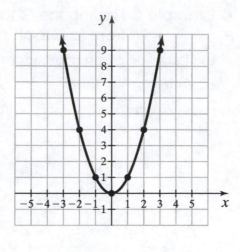

Figure 3
Graph of $f(x) = x^2$

Notice that the graph in **Figure 3** is "u-shaped" and opens up. The point $(0, 0)$, the lowest point on the graph, is the **vertex** of this **parabola**. The graph is **symmetric** about the vertical line $x = 0$. The **domain** is $(-\infty, \infty)$. From the graph, we see that y can be any real number greater than or equal to 0, so the **range** is $[0, \infty)$. Similarly, we can use point plotting to graph other **quadratic functions**.

My video summary ⏵ **Example 3 Graphing a Quadratic Function with a Vertical Shift**

Sketch the graph of $g(x) = x^2 + 2$. Compare the graph to the graph of $f(x) = x^2$.

Solution Watch this video, or read through the following solution.

Table 2 shows that for every value of x, the y-coordinate of the function g is always 2 greater than the y-coordinate for f.

Table 2

x	$f(x) = x^2$	$g(x) = x^2 + 2$
-2	$(-2)^2 = 4$	$(-2)^2 + 2 = 4 + 2 = 6$
-1	$(-1)^2 = 1$	$(-1)^2 + 2 = 1 + 2 = 3$
0	$(0)^2 = 0$	$(0)^2 + 2 = 0 + 2 = 2$
1	$(1)^2 = 1$	$(1)^2 + 2 = 1 + 2 = 3$
2	$(2)^2 = 4$	$(2)^2 + 2 = 4 + 2 = 6$

The two functions are graphed in Figure 4. The graph of $g(x) = x^2 + 2$ is exactly the same as the graph of $f(x) = x^2$, except the graph of g is shifted *up* two units.

The point $(0, 2)$, now the lowest point on the graph, is the **vertex** of this parabola. The graph of g is symmetric about the vertical line $x = 0$. The domain is $(-\infty, \infty)$. From the graph, we see that y can be any real number greater than or equal to 2, so the **range** is $[2, \infty)$. Notice how the vertical shift affected the y-coordinate of the **vertex** and the range, but did not affect the x-coordinate of the vertex.

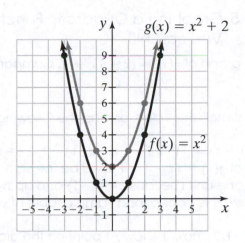

Figure 4
Graph of $g(x) = x^2 + 2$

You Try It Work through this You Try It problem.

Work Exercises 9 and 10 in this eText or in the MyMathLab Study Plan.

We see from **Example 3** that if $k > 0$, the graph of $y = x^2 + k$ is the graph of $y = x^2$ shifted *up* k units. It follows that for $k < 0$, the graph of $y = x^2 + k$ is the graph of $y = x^2$ shifted *down* k units.

Vertical Shifts of Quadratic Functions

The graph of $y = x^2 + k$ is a **parabola** with the same shape as the graph of $y = x^2$, but it is shifted vertically $|k|$ units. The graph is shifted up if $k > 0$ and down if $k < 0$. The **vertex** of the parabola is $(0, k)$.

My video summary ▶ **Example 4 Graphing a Quadratic Function with a Horizontal Shift**

Sketch the graph of $g(x) = (x - 2)^2$. Compare this graph to the graph of $f(x) = x^2$.

Solution Construct a table of values to find points on the graph. Use the points to sketch the graph. View this **popup** for a sample table of values and to check your graph. Watch this **video** for a detailed solution. Confirm that the graph of $g(x) = (x - 2)^2$ is the same as the graph of $f(x) = x^2$ but shifted two units to the right.

You Try It Work through this You Try It problem.

Work Exercises 11 and 12 in this eText or in the MyMathLab Study Plan.

We see from Example 4 that if $h > 0$, the graph of $y = (x - h)^2$ is the graph of $y = x^2$ shifted *right* h units. It follows that for $h < 0$, the graph of $y = (x - h)^2$ is the graph of $y = x^2$ shifted *left* h units.

Horizontal Shifts of Quadratic Functions

The graph of $y = (x - h)^2$ is a **parabola** with the same shape as the graph of $y = x^2$, but it is shifted horizontally $|h|$ units. The graph is shifted right if $h > 0$ and left if $h < 0$. The **vertex** of the parabola is $(h, 0)$.

📖 *My video summary* ▶ **Example 5** Graphing a Quadratic Function with a Horizontal and Vertical Shift

Sketch the graph of $g(x) = (x + 1)^2 + 2$. Compare this graph to the graph of $f(x) = x^2$.

Solution Watch this **video**, or read the following solution.

Writing $g(x) = (x + 1)^2 + 2 = (x - (-1))^2 + 2$ we see that $h = -1$ and $k = 2$. The graph of $g(x) = (x + 1)^2 + 2$ will be the same as the graph of $f(x) = x^2$ but shifted left one unit (because $h = -1$) and up two units (because $k = 2$). See **Figure 5**. View this **popup** for a sample table of values.

The point $(-1, 2)$, now the lowest point on the graph, is the **vertex** of this **parabola**. The graph of g is **symmetric** about the vertical line $x = -1$. The **domain** is $(-\infty, \infty)$. From the graph, we see that y can be any real number greater than or equal to 2, so the **range** is $[2, \infty)$.

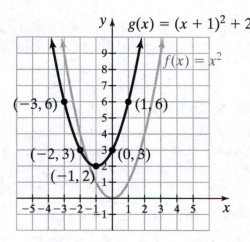

Figure 5
Graph of $g(x) = (x + 1)^2 + 2$

🔺 **You Try It** Work through this You Try It problem.

Work Exercises 13 and 14 in this eText or in the MyMathLab Study Plan.

OBJECTIVE 3 GRAPH QUADRATIC FUNCTIONS OF THE FORM $f(x) = a(x - h)^2 + k$

The quadratic function $f(x) = \frac{1}{2}(x - 3)^2 - 2$ has vertex $(3, -2)$ and is written in *standard form*.

Standard Form of a Quadratic Function

A quadratic function is in **standard form** if it is written as $f(x) = a(x - h)^2 + k$. The graph is a **parabola** with vertex (h, k). The parabola "opens up" if $a > 0$ or "opens down" if $a < 0$.

Since we can easily determine the coordinates of the vertex (h, k) when written this way, standard form is sometimes called **vertex form**.

🖋 *My video summary* ▶ **Example 6** Graphing a Quadratic Function in the Form
$$f(x) = a(x - h)^2 + k$$

Given the **quadratic function** $f(x) = \frac{1}{2}(x + 3)^2 - 2$, answer the following:

a. What are the coordinates of the **vertex**?

b. Does the graph "open up" or "open down"?

c. What is the equation of the **axis of symmetry**?

d. Find any **x-intercepts**.

e. Find the **y-intercept**.

f. Sketch the graph.

g. State the **domain** and **range** in interval notation.

Solutions Work through the following, or watch this **video** for a complete solution.

a. First change $x + 3$ to $x - (-3)$ to write the function in **standard form**. Then find the **vertex** by determining the values for h and k.

$$f(x) = \frac{1}{2}(x + 3)^2 - 2 \quad \rightarrow \quad f(x) = \frac{1}{2}(x - (-3))^2 + (-2)$$

$$a = \frac{1}{2} \qquad h = -3 \qquad k = -2$$

Because $h = -3$ and $k = -2$, the vertex for this **parabola** is $(-3, -2)$.

b. Since the **leading coefficient** is $a = \frac{1}{2} > 0$, the parabola opens up.

c. The x-coordinate of the vertex is -3, so the equation of the **axis of symmetry** is $x = -3$.

d. The **x-intercepts** (if any) are the real solutions to the equation $f(x) = 0$.

Write the original function:	$f(x) = \frac{1}{2}(x + 3)^2 - 2$
Set $f(x)$ equal to 0:	$\frac{1}{2}(x + 3)^2 - 2 = 0$
Add 2 to both sides:	$\frac{1}{2}(x + 3)^2 = 2$
Multiply both sides by 2:	$(x + 3)^2 = 4$
Use the square root property:	$x + 3 = \pm 2$
Subtract 3 from both sides:	$x = -3 \pm 2$

The x-intercepts are $-3 - 2 = -5$ and $-3 + 2 = -1$.

e. The *y*-intercept is found by evaluating $f(0)$.

$$f(0) = \frac{1}{2}(0 + 3)^2 - 2$$

$$= \frac{1}{2}(3)^2 - 2 = \frac{9}{2} - 2 = \frac{5}{2}$$

The *y*-intercept is $\frac{5}{2}$.

f. Let's summarize what we know:
The **vertex** is $(h, k) = (-3, -2)$; the graph opens up; the equation of the **axis of symmetry** is $x = -3$; there are two *x*-intercepts, -5 and -1; the *y*-intercept is $\frac{5}{2}$. From the vertex and intercepts, we can plot the points $(-3, -2)$, $(-5, 0)$, $(-1, 0)$, and $\left(0, \frac{5}{2}\right)$. The *y*-intercept lies three units to the right of the axis of symmetry. Because of the symmetry of the graph, we can move three units to left of the axis of symmetry to obtain another point on the graph: $\left(-6, \frac{5}{2}\right)$. Using this information, we can sketch the **parabola**, as shown in Figure 6.

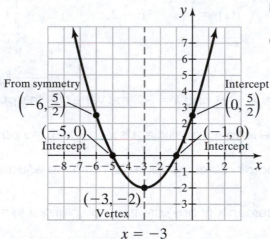

Figure 6

Graph of $f(x) = \frac{1}{2}(x + 3)^2 - 2$

g. The **domain** of every **quadratic function** is all real numbers $(-\infty, \infty)$. Because the parabola opens up, the graph has a **minimum** value at the vertex. The minimum value is the *y*-coordinate of the vertex, $k = -2$. So, the **range** is all real numbers greater than or equal to -2, or $[-2, \infty)$ in **interval notation**.

You Try It Work through this You Try It problem.

Work Exercises 15 and 16 in this eText or in the MyMathLab Study Plan.

My video summary ▶ **Example 7** Graphing a Quadratic Function in the Form
$$f(x) = a(x - h)^2 + k$$

Given that the quadratic function $f(x) = -(x - 2)^2 - 4$ is in standard form, answer the following:

a. What are the coordinates of the **vertex**?

b. Does the graph "open up" or "open down"?

c. What is the equation of the **axis of symmetry**?

d. Find any x-intercepts.

e. Find the y-intercept.

f. Sketch the graph.

g. State the **domain** and **range** in **interval notation**.

Solutions Try to answer these questions on your own. View the **answers**, or watch this **video** for a complete solution.

You Try It Work through this **You Try It** problem.

Work Exercises 17–20 in this eText or in the **MyMathLab** Study Plan.

OBJECTIVE 4 FIND THE VERTEX OF A QUADRATIC FUNCTION BY COMPLETING THE SQUARE

In **Section 8.1**, we saw how to use completing the square to **solve a quadratic equation**. The process involved adding the same quantity to both sides of the equation. In this section, we use **completing the square** to write a quadratic function in **standard form**. Since we are dealing with a **function** instead of an **equation**, we take a slightly different approach. Rather than adding the same quantity to both sides, we will *add and subtract* the same quantity, which is the same as adding 0 to the function.

Writing $f(x) = ax^2 + bx + c$ in Standard Form by Completing the Square

Step 1 Group the variable terms together within parentheses.

Step 2 If $a \neq 1$, factor a out of the **variable terms**.

Step 3 Take half the **coefficient** of the x-term inside the parentheses, **square** it, and *add* it *inside* the parentheses. Multiply this value by a, then *subtract* from c.

Step 4 The expression inside the parentheses is now a **perfect square**. Rewrite it as a **binomial** squared and simplify the **constant** term outside of the parentheses.

My video summary ▶ **Example 8 Writing a Quadratic Function in Standard Form**

Write the function $f(x) = 2x^2 - 8x + 7$ in **standard form** and find the **vertex**.

Solution Watch this **video**, or use the following steps to **write the function in standard form** by completing the square.

Step 1 $f(x) = (2x^2 - 8x) + 7$

Step 2 $f(x) = 2(x^2 - 4x) + 7$

Factor out a, the coefficient of
x^2, from the variable terms.

Step 3 $\left(\dfrac{1}{2}(-4)\right)^2 = (-2)^2 = 4$

Coefficient of x
inside parentheses

Add inside
parentheses

We are really
adding $2(4) = 8$
to the function.

So we must
subtract 8 as
well.

$f(x) = 2(x^2 - 4x + 4) + 7 - 8$

Notice that although we added 4 inside the parentheses, we had to subtract 8 on the outside. Don't forget that a **distributes** to each term inside the parentheses so we must take it into account when determining what to subtract.

Step 4 $f(x) = 2\underbrace{(x^2 - 4x + 4)}_{\text{Perfect square trinomial}} + \underbrace{7 - 8}_{\text{Simplify}} \Rightarrow f(x) = 2(x - 2)^2 - 1$

Now compare the function to the **standard form** $f(x) = a(x - h)^2 + k$, to determine the vertex. The vertex is $(h, k) = (2, -1)$.

You Try It Work through this You Try It problem.

Work Exercises 21–24 in this eText or in the MyMathLab Study Plan.

My video summary ▶ **Example 9 Writing a Quadratic Function in Standard Form**

Write the function $f(x) = -3x^2 - 24x$ in standard form and find the vertex.

Solution The constant term is $c = 0$ so we can write $f(x) = -3x^2 - 24x + 0$. Complete the square to write the function in standard form and determine the vertex. View the **answer**, or watch this **video** for a complete solution.

You Try It Work through this You Try It problem.

Work Exercises 25 and 26 in this eText or in the MyMathLab Study Plan.

OBJECTIVE 5 GRAPH QUADRATIC FUNCTIONS OF THE FORM $f(x) = ax^2 + bx + c$ BY COMPLETING THE SQUARE

Once we have the function in standard form, we can sketch its graph.

My video summary ▶ **Example 10 Graphing a Quadratic Function in the Form**
$$f(x) = ax^2 + bx + c$$

Rewrite the quadratic function $f(x) = 3x^2 - 6x - 1$ in standard form, and then answer the following:

a. What are the coordinates of the **vertex**?

b. Does the graph "open up" or "open down"?

c. What is the equation of the **axis of symmetry**?

d. Find any x-intercepts.

 e. Find the *y*-intercept.

 f. Sketch the graph.

 g. State the **domain** and **range** in interval notation.

Solutions Write the function in standard form by completing the square then try answering the questions as in **Example 6**. View the **answers**, or watch this **video** for a detailed solution.

You Try It   Work through this You Try It problem.

Work Exercises 27–32 in this eText or in the MyMathLab **Study Plan.**

OBJECTIVE 6   FIND THE VERTEX OF A QUADRATIC FUNCTION BY USING THE VERTEX FORMULA

Just as we saw how the **quadratic formula** comes from solving a general **quadratic equation** by completing the square, we can use completing the square to establish a **formula** for the **vertex**.

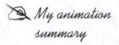 *My animation summary*

❊ **Concept Animation** Watch this **animation** to verify that the function $f(x) = ax^2 + bx + c$ is equivalent to $f(x) = a\left(x + \dfrac{b}{2a}\right)^2 + c - \dfrac{b^2}{4a}$, and the

coordinates of the vertex must be $\left(-\dfrac{b}{2a}, c - \dfrac{b^2}{4a}\right)$.

It is not really necessary to memorize the formulas for both coordinates. Since we can find the value of the *y*-coordinate by evaluating the function at the *x*-coordinate, it is common to write the **vertex** as $\left(-\dfrac{b}{2a}, f\left(-\dfrac{b}{2a}\right)\right)$.

Formula for the Vertex of a Parabola

Given a quadratic function of the form $f(x) = ax^2 + bx + c$, $a \neq 0$, the vertex of the parabola is given by

$$(h, k) = \left(-\frac{b}{2a}, f\left(-\frac{b}{2a}\right)\right).$$

The axis of symmetry is the vertical line $x = -\dfrac{b}{2a}$.

My video summary ▶ **Example 11 Using the Vertex Formula**

Use the **vertex formula** to find the vertex for each **quadratic function**.

 a. $f(x) = 3x^2 - 12x - 4$ **b.** $f(x) = -\dfrac{1}{2}x^2 - 10x + 5$

Solutions

a. Start by identifying the coefficients of the function: $a = 3$, $b = -12$, and $c = -4$.

The x-coordinate is $h = -\dfrac{b}{2a} = -\dfrac{(-12)}{2(3)} = \dfrac{12}{6} = 2$. We find the y-coordinate by evaluating $k = f(h) = f(2)$.

$$f(2) = 3(2)^2 - 12(2) - 4$$
$$= 12 - 24 - 4 = -16$$

The vertex is $(h, k) = (2, -16)$.

b. Determine the vertex on your own. View the **answer**, or watch this **video** for a complete solution to both parts.

You Try It Work through this You Try It problem.

Work Exercises 33–36 in this eText or in the MyMathLab Study Plan.

OBJECTIVE 7 GRAPH QUADRATIC FUNCTIONS OF THE FORM $f(x) = ax^2 + bx + c$ BY USING THE VERTEX FORMULA

My video summary ▶ **Example 12** Graphing a Quadratic Function Using the Vertex Formula

Given the quadratic function $f(x) = x^2 - 4x - 1$, answer the following:

a. What are the coordinates of the **vertex**?

b. Does the graph "open up" or "open down"?

c. What is the equation of the **axis of symmetry**?

d. Find any x-intercepts.

e. Find the y-intercept.

f. Sketch the graph.

g. State the **domain and range** in interval notation.

Solutions Find the vertex by using the **vertex formula**. Then try to answer the remaining questions as in **Example 6**. View the answers, or watch this **video** for a complete solution.

You Try It Work through this You Try It problem.

Work Exercises 37–42 in this eText or in the MyMathLab Study Plan.

8.2 Exercises

In Exercises 1–4, without graphing, determine if the graph of each quadratic function opens up or down. Also determine if the graph will be wider or narrower than the graph of $y = x^2$.

1. $f(x) = \frac{1}{4}x^2 - 3$

2. $f(x) = -2x^2 + 4x + 1$

3. $f(x) = -4 + 5x^2 - 2x$

4. $f(x) = -\frac{2}{3}x^2 + x + 7$

In Exercises 5–8, use the given graph of a quadratic function to find the following:

a. Vertex   b. Axis of symmetry   c. y-intercept

d. x-intercept(s)   e. Domain and range

5.

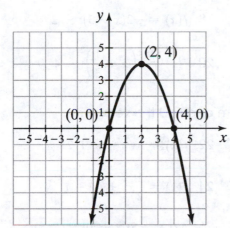

6.

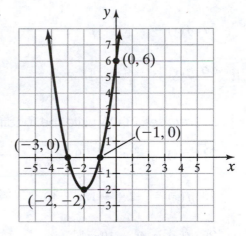

7.

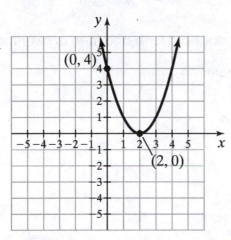

8.

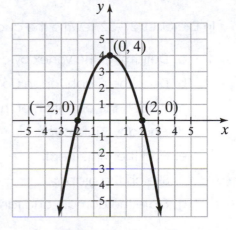

In Exercises 9–14, sketch the graph of each quadratic function by using translations (vertical or horizontal shifts). Compare each graph to the graph of $y = x^2$.

9. $f(x) = x^2 - 4$

10. $f(x) = x^2 + \frac{3}{2}$

11. $f(x) = (x + 3)^2$

12. $f(x) = (x - 4)^2$

13. $f(x) = (x - 1)^2 - 2$

14. $f(x) = (x + 2)^2 - 3$

In Exercises 15–20, use the given quadratic function to answer the following:

a. What are the coordinates of the vertex?

b. Does the graph "open up" or "open down"?

c. What is the equation of the axis of symmetry?

d. Find any x-intercepts.

e. Find the y-intercept.

f. Sketch the graph.

g. State the domain and range in interval notation.

15. $f(x) = (x - 2)^2 - 4$

16. $f(x) = \dfrac{1}{2}(x + 2)^2 + 2$

17. $f(x) = -(x + 1)^2 - 9$

18. $f(x) = -2(x - 3)^2 + 2$

19. $f(x) = -\dfrac{1}{4}(x - 4)^2 + 2$

20. $f(x) = 3\left(x + \dfrac{1}{3}\right)^2 - 4$

In Exercises 21–26, write the function in standard form and find the vertex.

21. $f(x) = x^2 + 8x + 9$

22. $f(x) = 3x^2 - 6x + 2$

23. $f(x) = -2x^2 - 12x - 10$

24. $f(x) = \dfrac{1}{2}x^2 + 6x + 1$

25. $f(x) = x^2 + 10x$

26. $f(x) = -5x^2 + 20x$

In Exercises 27–32, rewrite the quadratic function in standard form and then answer the following:

a. Determine the vertex by completing the square.

b. Does the graph "open up" or "open down"?

c. What is the equation of the axis of symmetry?

d. Find any x-intercepts.

e. Find the y-intercept.

f. Sketch the graph.

g. State the domain and range in interval notation.

27. $f(x) = x^2 + 6x - 7$

28. $f(x) = -x^2 - 3x + 4$

29. $f(x) = 4x^2 - 7x + 8$

30. $f(x) = -3x^2 - 6x + 1$

31. $f(x) = \dfrac{1}{4}x^2 - 2x + 1$

32. $f(x) = 2x^2 - 16x$

In Exercises 33–36, use the vertex formula to find the vertex of the quadratic function.

33. $f(x) = x^2 + 5x - 3$

34. $f(x) = -x^2 + 2x - 5$

35. $f(x) = -4x^2 + 12x$

36. $f(x) = \frac{1}{2}x^2 + 7x + 4$

In Exercises 37–42, use the quadratic function to answer the following:

    a. Determine the vertex by using the vertex formula.

    b. Does the graph "open up" or "open down"?

    c. What is the equation of the axis of symmetry?

    d. Find any x-intercepts.

    e. Find the y-intercept.

    f. Sketch the graph.

    g. State the domain and range in interval notation.

37. $f(x) = x^2 - 4x - 60$

38. $f(x) = 3x^2 + 6x - 4$

39. $f(x) = -x^2 + 2x - 6$

40. $f(x) = \frac{1}{2}x^2 + 6x + 1$

41. $f(x) = -3x^2 + 7x + 5$

42. $f(x) = x^2 - 8x$

8.3 Applications and Modeling of Quadratic Functions

THINGS TO KNOW

Before working through this section, be sure you are familiar with the following concepts:

                                            VIDEO   ANIMATION   INTERACTIVE

You Try It

1. Use Linear Equations to Solve Application Problems (Section 1.1, Objective 4) ▶

You Try It

2. Solve Application Problems Involving Geometric Formulas (Section 1.5, Objective 2) ▶

You Try It

3. Solve Application Problems Involving Uniform Motion (Section 1.5, Objective 4) ▶

You Try It

4. Use Linear Models to Solve Application Problems; Direct Variation (Section 2.5, Objective 8) ▶

You Try It 5. Use Polynomial Equations and Models
to Solve Application Problems
(Section 5.4, **Objective 3**)

You Try It 6. Use Rational Equations to Solve Application
Problems (Section 6.5, **Objective 5**)

You Try It 7. Solve Quadratic Equations Using Factoring
(Section 8.1, **Objective 1**)

You Try It 8. Solve Quadratic Equations Using
the Quadratic Formula
(Section 8.1, **Objective 4**)

You Try It 9. Graph Quadratic Functions of the
Form $f(x) = ax^2 + bx + c$ by Using
the Vertex Formula (Section 8.2, **Objective 7**)

OBJECTIVES

1 Solve Applications Involving Unknown Numbers

2 Solve Applications Involving Projectile Motion

3 Solve Applications Involving Geometric Formulas

4 Solve Applications Involving Distance, Rate, and Time

5 Solve Applications Involving Work

6 Maximize Quadratic Functions to Solve Application Problems

7 Minimize Quadratic Functions to Solve Application Problems

OBJECTIVE 1 SOLVE APPLICATIONS INVOLVING UNKNOWN NUMBERS

In **Section 1.1**, we solved application problems involving **linear equations**. In this section, we follow the same six-step process to solve application problems involving **quadratic equations**.

Problem-Solving Strategy for Applications

Step 1 Define the Problem. Read the problem carefully, or multiple times if necessary. Identify what you are trying to find and determine what information is available to help you find it.

Step 2 Assign Variables. Choose a variable to assign to an unknown quantity in the problem. If other unknown quantities exist, express them in terms of the selected variable.

Step 3 Translate into an Equation. Use the relationships among the known and unknown quantities to form an equation.

Step 4 Solve the Equation. Determine the value of the variable and use the result to find any other unknown quantities in the problem.

Step 5 Check the Reasonableness of Your Answer. Check to see if your answer makes sense within the context of the problem. If not, check your work for errors and try again.

Step 6 Answer the Question. Write a clear statement that answers the question(s) posed.

My video summary ▶ **Example 1 Finding Two Numbers**

The **product** of a number and 1 more than twice the number is 36. Find the two numbers.

Solution Work through the following, or watch this **video** for a complete solution.

Step 1 We are looking for two numbers. We know that the product is 36 and the second number is 1 more than twice the first.

Step 2 Let $x =$ the first number. Then $2x + 1 =$ the second number.

Step 3 Because the product of the two numbers is 36, we get the equation $x(2x + 1) = 36$, which simplifies to the **quadratic equation** $2x^2 + x - 36 = 0$.

Step 4 Solve:

$$\text{Original equation:} \quad 2x^2 + x - 36 = 0$$

$$\text{Factor:} \quad (2x + 9)(x - 4) = 0$$

Apply the **zero product property** and solve for x:

$$2x + 9 = 0 \quad \text{or} \quad x - 4 = 0$$

$$2x = -9 \qquad \qquad x = 4$$

$$x = -\frac{9}{2}$$

Step 5 If $x = -\dfrac{9}{2}$, then the other number is $2\left(-\dfrac{9}{2}\right) + 1 = -8$. Since $\left(-\dfrac{9}{2}\right)(-8) = 36$, this result is reasonable.

If $x = 4$, then the other number is $2(4) + 1 = 9$. Since $(4)(9) = 36$, this result is also reasonable.

Step 6 The two numbers are $-\dfrac{9}{2}$ and -8, or 4 and 9.

You Try It Work through this **You Try It** problem.

Work Exercises 1 and 2 in this eText or in the MyMathLab Study Plan.

✎ *My video summary* ▶ **Example 2** Finding Consecutive Even Integers

Three consecutive *positive* even integers are such that the square of the third is 20 less than the sum of the squares of the first two. Find the positive integers.

Solution Watch this **video**, or work through the following solution.

Step 1 We are looking for three consecutive positive even integers. We know that the square of the third is 20 less than the sum of the squares of the first two.

Step 2 Let $x =$ the first positive even integer. Then $x + 2 =$ the second, and $x + 4 =$ the third.

Step 3 Because the square of the third integer is 20 less than the sum of the squares of the first two integers, we get the equation

$$\underbrace{(x + 4)^2}_{\substack{\text{Square of} \\ \text{the third}}} = \underbrace{\underbrace{x^2}_{\substack{\text{Square of} \\ \text{the first}}} + \underbrace{(x + 2)^2}_{\substack{\text{Square} \\ \text{of the} \\ \text{second}}}}_{\text{Sum}} \underbrace{- \ 20.}_{\substack{\text{20 less} \\ \text{than the} \\ \text{sum}}}$$

Step 4 Solve:

$$\begin{aligned}
\text{Original equation:} && (x + 4)^2 &= x^2 + (x + 2)^2 - 20 \\
\text{Expand:} && x^2 + 8x + 16 &= x^2 + x^2 + 4x + 4 - 20 \\
\text{Combine like terms:} && x^2 + 8x + 16 &= 2x^2 + 4x - 16 \\
\text{Write in standard form:} && 0 &= x^2 - 4x - 32 \\
\text{Factor:} && 0 &= (x - 8)(x + 4)
\end{aligned}$$

Apply the **zero product property** and solve for x:

$$x - 8 = 0 \quad \text{or} \quad x + 4 = 0$$
$$x = 8 \qquad\qquad x = -4$$

Step 5 Because the consecutive even integers must be positive, discard the negative solution. The only feasible solution remaining is $x = 8$ as the first positive even integer which would give 10 and 12 as the next two consecutive even integers. Since $8^2 + 10^2 - 20 = 12^2$, this result is reasonable.

Step 6 The three consecutive positive even integers are 8, 10, and 12.

▲
You Try It Work through this **You Try It** problem.

Work Exercises 3 and 4 in this eText or in the MyMathLab Study Plan.

OBJECTIVE 2 SOLVE APPLICATIONS INVOLVING PROJECTILE MOTION

An object launched, thrown, or shot vertically into the air with an initial velocity of v_0 meters per second (m/s) from an initial height of h_0 meters above the ground can be **modeled** by the equation $h = -4.9t^2 + v_0 t + h_0$. The **variable h** is the height above the ground (in meters) of the object (also known as a *projectile*) t seconds after its departure.

🖋 *My video summary* ▶ **Example 3 Launching a Toy Rocket**

A toy rocket is launched at an initial velocity of 14.7 meters per second from a platform that sits 49 meters above the ground. The height h of the rocket above the ground at any time t seconds after launch is given by the equation $h = -4.9t^2 + 14.7t + 49$. When will the rocket hit the ground?

Solution Watch this video, or read through the following solution.

Step 1 We want to find the time when $h = 0$ (the rocket hits the ground) using the equation $h = -4.9t^2 + 14.7t + 49$.

Step 2 We know that h = height in meters and t = time in seconds.

Step 3 Using the equation $h = -4.9t^2 + 14.7t + 49$, set $h = 0$ and solve for t.

Step 4 Solve:

$$\text{Original equation:} \quad 0 = -4.9t^2 + 14.7t + 49$$

$$\text{Divide both sides by } -4.9: \quad 0 = t^2 - 3t - 10$$

$$\text{Factor:} \quad 0 = (t - 5)(t + 2)$$

Apply the **zero product property** and solve for t:

$$t - 5 = 0 \quad \text{or} \quad t + 2 = 0$$
$$t = 5 \qquad\qquad t = -2$$

Step 5 Because t represents the time (in seconds) after launch, its value cannot be negative. So, $t = -2$ seconds does not make sense. The only reasonable solution is $t = 5$ seconds.

Step 6 The rocket will hit the ground 5 seconds after launch.

🔺 **You Try It** Work through this You Try It problem.

Work Exercises 5 and 6 in this eText or in the My Math Lab Study Plan.

OBJECTIVE 3 SOLVE APPLICATIONS INVOLVING GEOMETRIC FORMULAS

Now let's explore application problems that involve geometric formulas and quadratic equations.

🖋 *My video summary* ▶ **Example 4 Dimensions of a Rectangle**

The length of a rectangle is 7 inches less than six times the width. Find the dimensions of the rectangle if the area of the rectangle is 49 square inches.

Solution Try to solve this problem on your own. View the answer, or watch this video for a complete solution.

🔺 **You Try It** Work through this You Try It problem.

Work Exercises 7 and 8 in this eText or in the My Math Lab Study Plan.

📝 *My video summary* ▶ **Example 5 Dimensions of a Television**

The size of a television is the length of the **diagonal** of its rectangular screen. If a 55-inch television is 21 inches longer than it is wide, find the length and width of the screen, rounded to the nearest inch.

Solution

Step 1 We want to find the length and width of the television. We know that the length is 21 inches more than the width, and the diagonal of the rectangular screen is 55 inches.

Step 2 Let w = the width of the television. Then $w + 21$ = the length.

Step 3 The diagonal forms a **right triangle** with legs w and $w + 21$, and hypotenuse 55. Use the **Pythagorean Theorem** to form a **quadratic equation**:

$$(\text{leg})^2 + (\text{leg})^2 = (\text{hypotenuse})^2$$
$$w^2 + (w + 21)^2 = 55^2$$

Step 4 Begin with the original equation: $w^2 + (w + 21)^2 = 55^2$

Square the binomial: $w^2 + w^2 + 42w + 441 = 3025$

Combine like terms: $2w^2 + 42w - 2584 = 0$

Divide both sides by 2: $w^2 + 21w - 1292 = 0$

Because this equation does not factor, use the **quadratic formula** to finish solving it. Then use the result to find the dimensions of the television. View the **answer**, or watch this **video** for the complete solution.

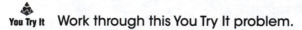

You Try It Work through this **You Try It** problem.

Work Exercises 9–12 in this eText or in the MyMathLab Study Plan.

OBJECTIVE 4 SOLVE APPLICATIONS INVOLVING DISTANCE, RATE, AND TIME

In the next example, we use quadratic equations to solve an application problem involving distance, rate, and time.

📝 *My video summary* ▶ **Example 6 Speed of an Airplane**

Kevin flew his new Cessna O-2A airplane from Jonesburg to Mountainview, a distance of 2560 miles. The average speed for the return trip was 64 mph faster than the average outbound speed. If the total flying time for the round trip was 18 hours, what was the plane's average speed on the outbound trip from Jonesburg to Mountainview?

Solution

Step 1 We are asked to find the average outbound speed of the plane from Jonesburg to Mountainview. We know that the distance traveled in each direction is 2560 miles and the total time of the trip is 18 hours. We also know that the speed on the return trip is 64 mph more than on the outbound trip. Because the problem involves distance, rate, and time, we will need the distance formula $d = r \cdot t$.

Step 2 Let $r =$ speed of plane on the outbound trip. Then $r + 64 =$ speed of plane on the return trip from Mountainview to Jonesburg.

Step 3 Since the total time of the trip was 18 hours, we write

$$\text{time}_{\text{outbound}} + \text{time}_{\text{return}} = \text{time}_{\text{total}}$$

$$\text{time}_{\text{outbound}} + \text{time}_{\text{return}} = 18$$

Because $\text{distance} = \text{rate} \cdot \text{time}$, we also know that $\text{time} = \dfrac{\text{distance}}{\text{rate}}$. Using the given distance and rate expressions, we get the following equation:

$$\frac{2560}{r} + \frac{2560}{r + 64} = 18$$

Solve this equation on your own. Be sure to check the reasonableness of your answer. View the **answer**, or watch this **video** for the complete solution.

You Try It Work through this You Try It problem.

Work Exercises 13 and 14 in this eText or in the MyMathLab Study Plan.

OBJECTIVE 5   SOLVE APPLICATIONS INVOLVING WORK

We can also use quadratic equations to solve application problems involving work.

My video summary ▶ **Example 7 Monthly Sales Reports**

Dawn can finish the monthly sales reports in 2 hours less time than it takes Adam. Working together, they were able to finish the sales reports in 8 hours. How long does it take each person to finish the monthly sales reports alone? (Round to the nearest minute.)

Solution

Step 1 We must find the time it takes for each person to finish the sales reports alone. We know that it takes Dawn 2 hours less than Adam to do the job alone, and it takes them 8 hours to do the job together. Because this problem involves combining **rates of work**, we use the equation $\dfrac{1}{t_1} + \dfrac{1}{t_2} = \dfrac{1}{t}$.

Step 2 Let t_1 = the time it takes Dawn to complete the reports alone (in hours). Then $t_2 = t_1 + 2$ = the time it takes Adam to complete the reports alone. We know that it takes $t = 8$ hours to complete the job when they work together.

Step 3 Substituting the expressions for the individual times and 8 for the total time, we get the following equation:

$$\frac{1}{t_1} + \frac{1}{t_1 + 2} = \frac{1}{8}$$

Solve this equation for t_1. Confirm that it will take Dawn approximately 15 hours and 4 minutes to complete the reports on her own and it will take Adam 17 hours and 4 minutes to complete the same job alone. Be sure to check that the sum of their rates is equal to $\frac{1}{8}$ (within rounding error). To see this problem worked out in detail, watch this **video**.

You Try It Work through this You Try It problem.

Work Exercises 15 and 16 in this eText or in the MyMathLab Study Plan.

OBJECTIVE 6 MAXIMIZE QUADRATIC FUNCTIONS TO SOLVE APPLICATION PROBLEMS

Sometimes with application problems involving **functions**, we often need to find the *maximum* or *minimum value* of the function. For example, a builder with a fixed amount of fencing may wish to maximize the area enclosed, or an economist may want to minimize **cost** or maximize **profit**. **Quadratic functions** are relatively easy to maximize or minimize. Recall that if $f(x) = ax^2 + bx + c, a \neq 0$, we know that the **coordinates** of the **vertex** are $\left(-\frac{b}{2a}, f\left(-\frac{b}{2a}\right)\right)$. In addition, we know that if $a > 0$, the **parabola** opens *up* and the function has a **minimum value** at the vertex. If $a < 0$, the parabola opens *down* and the function has a **maximum value** at the vertex. See Figure 7.

Figure 7

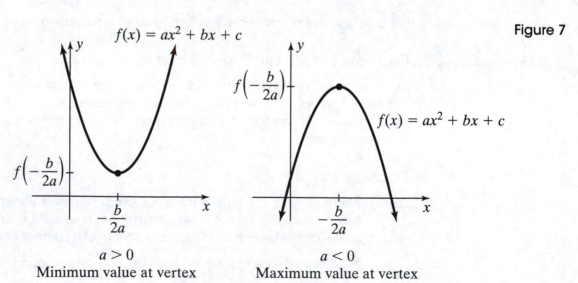

$a > 0$
Minimum value at vertex

$a < 0$
Maximum value at vertex

✑ *My video summary* ▶ **Example 8** Launching a Toy Rocket

A toy rocket is launched with an initial **velocity** of 44.1 meters per second from a platform located 1 meter above the ground. The height h of the object above the ground at any time t seconds after launch is given by the function $h(t) = -4.9t^2 + 44.1t + 1$. How long after launch did it take the rocket to reach its maximum height? What is the rocket's **maximum** height?

Solution Watch the video, or read the following solution.

The function is **quadratic** with $a = -4.9 < 0$, so the graph is a **parabola** that opens down. This means, the function has a maximum value at the **vertex**. The t-coordinate of the vertex is $t = -\dfrac{b}{2a} = -\dfrac{44.1}{2(-4.9)} = \dfrac{-44.1}{-9.8} = 4.5$ seconds. Therefore, the rocket reaches its maximum height at 4.5 seconds after launch. The rocket's maximum height is $h(4.5) = -4.9(4.5)^2 + 44.1(4.5) + 1 = 100.225$ meters.

🔺
You Try It   Work through this You Try It problem.

Work Exercises 17–19 in this eText or in the MyMathLab **Study Plan.**

Revenue and *profit* applications often involve maximizing quadratic functions.

Revenue is the dollar amount received from selling x items at a price of p dollars per item. **Total revenue** is found by multiplying the number of units sold by the price per unit, $R = xp$. For example, if a child sells 50 cups of lemonade at a price of \$0.25 per cup, then the revenue generated is $R = \underbrace{(50)}_{x}\underbrace{(0.25)}_{p} = \$12.50$. A **demand equation** relates the quantity sold to the price per item, and we can write this equation as quantity in terms of price, such as $x = 2p - 6$, or as price in terms of quantity, such as $p = 0.5x + 3$. This allows us to write the revenue function in terms of a single variable, either x or p:

$$R = xp \text{ and } x = 2p - 6 \text{ (or equivalently, } p = 0.5x + 3)$$

$$R(x) = x(0.5x + 3) = 0.5x^2 + 3x \leftarrow \text{Revenue in terms of } x$$

or

$$R(p) = (2p - 6)p = 2p^2 - 6p \leftarrow \text{Revenue in terms of } p$$

✑ *My video summary* ▶ **Example 9** Maximizing Revenue

Records can be kept on the price of clothing and the number of items sold in order to gather enough data to reasonably **model** shopping trends for a particular type of shirt. Suppose the marketing and research department of a clothing company determined the price of a certain shirt obeys the demand equation $p = -\dfrac{1}{80}x + 50$.

a. According to the demand equation, how much should the shirt sell for if 1000 shirts are sold? 2500 shirts?

b. What is the revenue if 1000 shirts are sold? 2500 shirts?

c. How many shirts should be sold in order to maximize revenue? What is the maximum revenue?

d. What price should be charged in order to maximize revenue?

Solutions Watch this **video**, or read the following for a complete solution.

a. If 1000 shirts are sold, the price should be

$$p(1000) = -\frac{1}{80}(1000) + 50 = -12.50 + 50 = \$37.50.$$

If 2500 shirts are sold, the price should be

$$p(2500) = -\frac{1}{80}(2500) + 50 = -31.25 + 50 = \$18.75.$$

b. Because $R = xp$, we can substitute the **demand equation**, $p = -\frac{1}{80}x + 50$, for p to get R as a function of x.

$$R(x) = x\underbrace{\left(-\frac{1}{80}x + 50\right)}_{p} \quad \text{or} \quad R(x) = -\frac{1}{80}x^2 + 50x$$

The **revenue** generated by selling 1000 shirts is

$$R(1000) = 1000\left(-\frac{1}{80}(1000) + 50\right) = 1000(37.50) = \$37,500.$$

The revenue from selling 2500 shirts is

$$R(2500) = 2500\left(-\frac{1}{80}(2500) + 50\right) = 2500(18.75) = \$46,875$$

c. $R(x) = -\frac{1}{80}x^2 + 50x$ is a **quadratic function** with $a = -\frac{1}{80} < 0$ and $b = 50$. Because $a < 0$, the function has a **maximum** value at the **vertex**. Therefore, the value of x that produces the maximum **revenue** is

$$x = -\frac{b}{2a} = -\frac{50}{2\left(-\frac{1}{80}\right)} = \frac{-50}{-\frac{1}{40}} = \underbrace{\frac{-50}{1} \cdot \left(-\frac{40}{1}\right)}_{\substack{\text{Change to} \\ \text{multiplication} \\ \text{by the reciprocal}}} = 2000 \text{ shirts.}$$

The maximum revenue is

$$R(2000) = 2000\left(-\frac{1}{80}(2000) + 50\right) = 2000(25) = \$50,000.$$

d. Using the **demand equation**, the price that should be charged to maximize revenue when selling 2000 shirts is

$$p(2000) = -\frac{1}{80}(2000) + 50 = \$25.$$

You Try It Work through this **You Try It** problem.

Work Exercises 20 and 21 in this eText or in the MyMathLab Study Plan.

My video summary ▶ **Example 10 Maximizing Profit**

To sell x waterproof CD alarm clocks, WaterTime, LLC, has determined that the price in dollars must be $p = 250 - 2x$, which is the **demand equation**. Each clock costs \$2 to produce, with fixed costs of \$4000 per month, producing the **cost** function $C(x) = 2x + 4000$.

a. Express the **revenue** R as a function of x.

b. Express the **profit** P as a function of x.

c. Find the value of x that maximizes profit. What is the **maximum** profit?

d. What is the price of the alarm clock that will maximize profit?

Solutions

a. The equation for revenue is $R = xp$. So, we substitute the demand equation $p = 250 - 2x$ for p to obtain the function:

$$R(x) = \underbrace{x(250 - 2x)}_{p} \quad \text{or} \quad R(x) = -2x^2 + 250x.$$

b. Profit is equal to revenue minus cost.

$$P(x) = R(x) - C(x)$$
$$= -2x^2 + 250x - (2x + 4000)$$
$$= -2x^2 + 250x - 2x - 4000$$
$$= -2x^2 + 248x - 4000$$

c.–d. Try to finish the remaining parts on your own. Find the **vertex** to answer the questions in part (c), and use the **demand equation** to answer part (d). View the **answers**, or watch this **video** for a complete solution.

▲
You Try It Work through this You Try It problem.

Work Exercises 22–25 in this eText or in the MyMathLab Study Plan.

Area problems sometimes involve maximizing quadratic functions.

My video summary ▶ **Example 11 Maximizing Area**

A rancher has 2500 feet of fencing available to create a rectangular enclosure bordering a river. If no fencing is required along the river, find the dimensions of the enclosure that maximizes the area. What is the **maximum area**?

Figure 8

Solution From Figure 8, let x = width of the enclosure and y = length of the enclosure. Since the enclosure is rectangular, the area is found by multiplying the length and the width:

$$A = xy$$

Ideally, the area should depend on only one variable. So, we need to write x in terms of y, or write y in terms of x. The 2500 feet of fencing will be used for two widths and one length, giving us the equation

$$2x + y = 2500.$$

Solving this equation for y gives $y = -2x + 2500$. Substitute $-2x + 2500$ for y to get the area as a **function** of x alone.

$$A(x) = x\underbrace{(-2x + 2500)}_{y} \quad \text{or} \quad A(x) = -2x^2 + 2500x$$

The area function is **quadratic** with $a = -2 < 0$. Therefore, the function has a **maximum** value at its **vertex**. Try to finish working this problem on your own. View the **answer**, or watch this **video** for a complete solution.

You Try It Work through this **You Try It** problem.

Work Exercises 26 and 27 in this eText or in the MyMathLab **Study Plan.**

OBJECTIVE 7 MINIMIZE QUADRATIC FUNCTIONS TO SOLVE APPLICATION PROBLEMS

Let's now consider some situations in which we **minimize a function.**

My video summary ▶ **Example 12 Minimizing Costs**

A fabric manufacturer has daily production costs of $C = 7000 - 100x + 0.5x^2$, where C is the total **cost** (in dollars) and x is the number of units produced. How many units should be produced each day in order to minimize costs? What is the minimum daily cost?

Solution Watch this **video**, or read the following for a complete solution. Writing the cost function in standard form, we have

$$C(x) = 0.5x^2 - 100x + 7000.$$

This is a **quadratic function** with $a = 0.5 > 0$, $b = -100$, and $c = 7000$. Because $a > 0$, the graph of the function is a **parabola** that *opens up*, and the function has a minimum value at the **vertex**. The x-coordinate of the vertex tells us the number of units to produce in order to minimize costs, and the y-coordinate tells us the minimum cost. To find these coordinates, we use the **vertex formula**

$$\left(-\frac{b}{2a}, f\left(-\frac{b}{2a}\right)\right).$$

$$x = -\frac{(-100)}{2(0.5)} \quad \text{and} \quad C(100) = 0.5(100)^2 - 100(100) + 7000$$

$$= \frac{100}{1} \qquad\qquad\qquad = 0.5(10{,}000) - 100(100) + 7000$$

$$= 100 \qquad\qquad\qquad = 5000 - 10{,}000 + 7000$$

$$\qquad\qquad\qquad = 2000$$

The manufacturer should produce 100 units each day to minimize costs. The minimum daily cost will be $2000.

You Try It Work through this You Try It problem.

Work Exercises 28–30 in this eText or in the MyMathLab **Study Plan.**

In economics, **long run average cost (LRAC)**, is the total long run cost divided by the number of units produced. Companies sometimes use the *LRAC* to determine if they should merge resources such as sales regions or franchises. If the *LRAC* is decreasing, then it might be best for the company to merge resources, provided that the merger does not cause the *LRAC* to increase. If the *LRAC* is increasing, a merger may not be beneficial since it would lead to higher costs.

My video summary ⊙ **Example 13 Merging Resources**

An account rep in one territory oversees $N = 20$ accounts and a second account rep in a nearby territory manages $N = 8$ accounts. The **long run average cost** function for their industry is $C = N^2 - 70N + 1400$.

a. Determine the long run average cost for $N = 20$ accounts and $N = 8$ accounts.

b. What number of accounts minimizes the long run average cost? What is the **minimum** long run average cost?

c. Should the two territories be merged into a single territory?

Solutions Try to work this problem on your own. View the **answers**, or watch this video for a detailed solution.

You Try It Work through this You Try It problem.

Work Exercises 31 and 32 in this eText or in the MyMathLab **Study Plan.**

8.3 Exercises

1. **Finding a Number** The product of some negative number and 5 less than three times that number is 12. Find the number.

2. **Finding a Number** The square of a number plus the number is 132. What is the number?

3. **Finding Integers** Three consecutive odd integers are such that the square of the third integer is 15 more than the sum of the squares of the first two. Find the integers.

4. **Finding a Number** The sum of the square of a number and the square of 7 more than a number is 169. What is the number?

5. **Rocket Launch** A toy rocket is launched from a platform 2.8 meters above the ground in such a way that its height, h (in meters), after t seconds is given by the equation $h = -4.9t^2 + 18.9t + 2.8$. How long will it take for the rocket to hit the ground?

6. **Projectile Motion** Shawn threw a rock straight up from a cliff that was 24 feet above the water. If the height of the rock h, in feet, after t seconds is given by the equation $h = -16t^2 + 20t + 24$, how long will it take for the rock to hit the water?

7. **Dimensions of a Rectangle** The length of a rectangle is 1 cm less than three times the width. If the area of the rectangle is 30 cm², find the dimensions of the rectangle.

8. **Dimensions of a Rectangle** The length of a rectangle is 1 inch less than twice the width. If the diagonal is 2 inches more than the length, find the dimensions of the rectangle.

9. **Computer Screen** The size of a computer screen is the length of its diagonal. If the 12-inch screen of a computer is 3.5 inches longer than it is wide, find the length and width of the screen, rounded to the nearest hundredth of an inch.

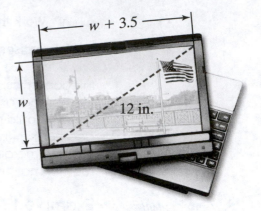

10. **Washing Windows** A homeowner leans a 15-foot ladder against his house to wash windows. If the base of the ladder is 6 feet from the house, how high does the ladder reach? Assume the ground is level. Find the height, rounded to the nearest tenth of a foot.

11. **Loading Ramp** A loading ramp in a steel yard has a horizontal run that is 26 feet longer than its vertical rise. If the ramp is 30 feet long, what is the vertical rise? (Round to two decimal places if necessary.)

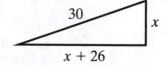

12. **Building a Walkway** A 35- by 20-feet rectangular swimming pool is surrounded by a walkway of uniform width. If the total area of the walkway is 434 ft^2, how wide is the walkway?

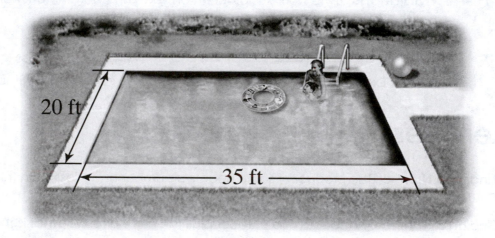

13. **Speed of a Boat** Logan rowed her boat upstream a distance of 9 miles and then rowed back to the starting point. The total time of the trip was 10 hours. If the rate of the current was 4 mph, find the average speed of the boat in still water.

14. **Speed of a Car** Imogene's car traveled 280 miles averaging a certain speed. If the car had gone 5 mph faster, the trip would have taken 1 hour less. Find the average speed.

15. **Mowing Lawns** Twin brothers, Billy and Bobby, can mow their grandparent's lawn together in 56 minutes. Billy could mow the lawn by himself in 15 minutes less time than it would take Bobby. How long would it take Bobby to mow the lawn by himself?

16. **Working Together** Jeff and Kirk can build a 75-foot. retaining wall together in 12 hours. Because Jeff has more experience, he could build the wall himself 4 hours quicker than Kirk. How long would it take Kirk (to the nearest minute) to build the wall by himself?

17. **Maximum Height** A baseball player swings and hits a pop fly straight up in the air to the catcher. The height of the baseball in meters t seconds after it is hit is given by the quadratic function $h(t) = -4.9t^2 + 34.3t + 1$. How long does it take for the baseball to reach its maximum height? What is the baseball's maximum height?

18. **Maximum Height** An object is launched vertically in the air at 36.75 meters per second from a platform 10 meters above the ground. The height of the object above the ground (in meters) t seconds after launch is given by $h(t) = -4.9t^2 + 36.75t + 10$. How long does it take the object to reach its maximum height? What is the object's maximum height?

19. **Maximum Height of a Toy Rocket** A toy rocket is shot vertically into the air from a launching pad 5 feet above the ground with an initial velocity of 112 feet per second. The height h, in feet, of the rocket above the ground at t seconds after launch is given by the function $h(t) = -16t^2 + 112t + 5$. How long will it take the rocket to reach its maximum height? What is the maximum height?

20. **Maximum Revenue** The price p and the quantity x sold of a small TV obeys the demand equation $p = -0.15x + 300$.

    a. How much should be charged for the TV if there are 50 TVs in stock?

    b. What quantity x will maximize revenue? What is the maximum revenue?

    c. What price should be charged per TV in order to maximize revenue?

21. **Maximum Revenue** The dollar price for a barrel of oil sold at a certain oil refinery tends to obey the demand equation $p = -\dfrac{1}{10}x + 72$, where x is the number of barrels of oil on hand (in millions).

    a. How much should be charged per barrel of oil if there are 4 million barrels on hand?

    b. What quantity x will maximize revenue? What is the maximum revenue?

    c. What price should be charged per barrel in order to maximize revenue?

In Exercises 22–25, use the fact that profit is defined as revenue minus cost, or $P(x) = R(x) - C(x)$.

22. **Maximum Profit** Rite-Cut riding lawnmowers obey the demand equation $p = -\dfrac{1}{20}x + 1000$.

    The cost of producing x lawnmowers is given by the function $C(x) = 100x + 5000$.

    a. Express the revenue R as a function of x.

    b. Express the profit P as a function of x.

    c. Find the value of x that maximizes profit. What is the maximum profit?

    d. What price should be charged per lawnmower to maximize profit?

23. **Maximum Profit** The CarryItAll minivan, a popular vehicle among soccer moms, obeys the demand equation $p = -\dfrac{1}{40}x + 8000$. The cost of producing x vans is given by the function $C(x) = 4000x + 20{,}000$.

    a. Express the revenue R as a function of x.

    b. Express the profit P as a function of x.

c. Find the value of x that maximizes profit. What is the maximum profit?

d. What price should be charged to maximize profit?

24. **Maximum Profit** Silver Scooter, Inc., finds that it costs $200 to produce each motorized scooter and that the fixed costs are $1500 per day yielding the cost function $C(x) = 200x + 1500$. The price is given by $p = 600 - 5x$, where p is the price in dollars at which exactly x scooters will be sold. Find the quantity of scooters that Silver Scooter, Inc., should produce and the price it should charge to maximize profit. Find the maximum profit.

25. **Maximum Profit** Amy, the owner of Amy's Pottery, can produce china pitchers at a cost of $5 each. She estimates her price function to be $p = 17 - 5x$, where p is the price at which exactly x pitchers will be sold per week. Find the number of pitchers that she should produce and the price that she should charge in order to maximize profit. Also find the maximum profit.

26. **Maximum Area** A farmer has 1800 feet of fencing available to enclose a rectangular area bordering a river. If no fencing is required along the river, find the dimensions of the fence that will maximize the area. What is the maximum area?

27. **Maximum Area** Jim wants to build a rectangular parking lot along a busy street but has only 2500 feet of fencing available. If no fencing is required along the street, find the maximum area of the parking lot.

28. **Minimizing Cost** A manufacturer of denim jeans has daily production costs of $C = 0.2x^2 - 90x + 10{,}700$, where C is the total cost (in dollars) and x is the number of jeans produced. How many jeans should be produced each day in order to minimize costs? What is the minimum daily cost?

29. **Minimizing Loss Function** Quality control analysts often use a *quadratic loss function* to determine costs associated with failing to meet product specifications. For a certain machine part, the loss function relating cost to diameter is $L = 4x^2 - 4x + 2.5$, where L is the total loss (in dollars) per part and x is the diameter of the machined part. Find the diameter that minimizes total loss per part. What is the minimum total loss?

30. **Minimum Bankruptcies** Based on data from the Federal Judiciary, the number of business bankruptcy filings per year can be approximated by the model $B(x) = 3.9x^2 - 50.8x + 192.9$, where x is the number of years since 2000 and B is the number of business bankruptcy filings (in 1000s). During what year was the number of filings at a minimum? What was the approximate minimum number of filings to the nearest thousand?

31. **Merging Account Territories** An account manager in one territory oversees $N = 14$ accounts, and a second account manager in a nearby territory manages $N = 18$ accounts. The long run average cost function for their industry is $C = 484 - 44N + N^2$.

a. Find the long run average cost for $N = 14$ accounts and for $N = 18$ accounts.

b. How many accounts will minimize the long run average cost? What is the minimum long run average cost?

c. Should the two territories be merged into a single territory?

32. **Bank Merger** One bank services $N = 150$ customers, and a second bank services $N = 80$ customers. The long run average cost function for their industry is $C = N^2 - 700N + 110{,}000$.

a. Find the long run average cost for $N = 150$ customers and for $N = 80$ customers.

b. What number of customers will minimize the long run average cost? What is the minimum long run average cost?

c. Should the two banks merge?

8.4 Circles

THINGS TO KNOW

Before working through this section, be sure you are familiar with the following concepts:

VIDEO   ANIMATION   INTERACTIVE

 You Try It 1. Plot Ordered Pairs in the Rectangular Coordinate System (Section 2.1, Objective 1)   ▶

 You Try It 2. Simplify Radical Expressions Using the Product Rule (Section 7.3, Objective 5)   ▶

 You Try It 3. Solve Quadratic Equations Using the Square Root Property (Section 8.1, Objective 2)   ▶

 You Try It 4. Solve Quadratic Equations by Completing the Square (Section 8.1, Objective 3)   ▶

OBJECTIVES

1 Find the Distance between Two Points

2 Find the Midpoint of a Line Segment

3 Write the Standard Form of an Equation of a Circle

4 Sketch the Graph of a Circle Given in Standard Form

5 Write the General Form of a Circle in Standard Form and Sketch Its Graph

....

OBJECTIVE 1   FIND THE DISTANCE BETWEEN TWO POINTS

We can use the **Pythagorean theorem** to develop a **formula** for finding the distance between two points in a **plane**.

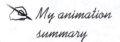 *My animation summary*   ✸ **Concept Animation** Watch this **animation** to explore this concept.

The **distance formula** is used to find the distance $d(A, B)$ between the **points** $A(x_1, y_1)$ and $B(x_2, y_2)$ (see Figure 9).

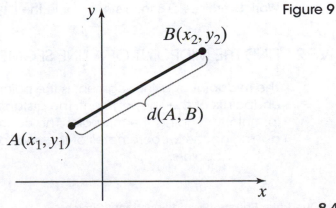

Figure 9

Distance Formula

The distance $d(A, B)$ between two points $A(x_1, y_1)$ and $B(x_2, y_2)$ is given by

$$d(A, B) = \sqrt{(x_2 - x_1)^2 + (y_2 - y_1)^2}.$$

TIP When using the distance formula, it does not matter which point is (x_1, y_1) and which is (x_2, y_2).

 My video summary ▶ **Example 1 Using the Distance Formula**

Find the distance between the points $(-1, 5)$ and $(4, -5)$.

Solution Watch this **video**, or read the following for a complete solution.

Let $(x_1, y_1) = (-1, 5)$ and $(x_2, y_2) = (4, -5)$ and use the **distance formula**.

Distance formula:	$d(A, B) = \sqrt{(x_2 - x_1)^2 + (y_2 - y_1)^2}$
Substitute the coordinates:	$= \sqrt{(4 - (-1))^2 + (-5 - 5)^2}$
Subtract:	$= \sqrt{(5)^2 + (-10)^2}$
Square:	$= \sqrt{25 + 100}$
Add:	$= \sqrt{125}$
Simplify the radical:	$= 5\sqrt{5}$

Figure 10

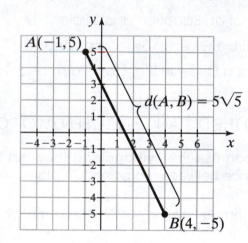

You Try It Work through this **You Try It** problem.

Work Exercises 1–8 in this eText or in the MyMathLab Study Plan.

OBJECTIVE 2   FIND THE MIDPOINT OF A LINE SEGMENT

The **midpoint** of a line segment is the point that lies exactly halfway between the endpoints of the line segment. The **distance formula** can be used to develop a formula for finding the midpoint. The **coordinates** of the midpoint are found by averaging the **x-coordinates** of the endpoints and averaging the **y-coordinates** of the endpoints.

Midpoint of a Line Segment

The **midpoint** of the line segment from $A(x_1, y_1)$ to $B(x_2, y_2)$ is the point with coordinates

$$\left(\frac{x_1 + x_2}{2}, \frac{y_1 + y_2}{2} \right).$$

⚠ The distance formula involves finding the **difference** between coordinates, but the midpoint formula involves finding the **sum** of the coordinates.

Because the **midpoint** lies halfway between the endpoints of a line segment (see Figure 11), the distance from the midpoint to either endpoint is half the distance between the endpoints.

$$d(A, M) = d(M, B) = \frac{d(A, B)}{2}$$

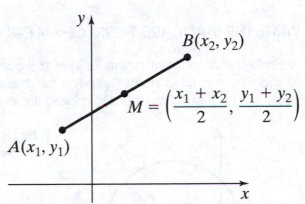

Figure 11

⚠ The distance between two points is a number, but the midpoint of a line segment is a point. When writing a midpoint, be sure to use an **ordered pair**.

✎ *My video summary* ▶ **Example 2 Finding the Midpoint of a Line Segment**

Find the **midpoint** of the line segment with endpoints $(-3, 2)$ and $(4, 6)$.

Solution Read the following, or watch this **video** for a complete solution.

Let $(x_1, y_1) = (-3, 2)$ and $(x_2, y_2) = (4, 6)$ and use the midpoint formula.

$$\text{Midpoint formula:} \quad \left(\frac{x_1 + x_2}{2}, \frac{y_1 + y_2}{2} \right)$$

$$\text{Substitute the coordinates:} \quad = \left(\frac{-3 + 4}{2}, \frac{2 + 6}{2} \right)$$

$$\text{Simplify:} \quad = \left(\frac{1}{2}, \frac{8}{2} \right)$$

$$= \left(\frac{1}{2}, 4 \right)$$

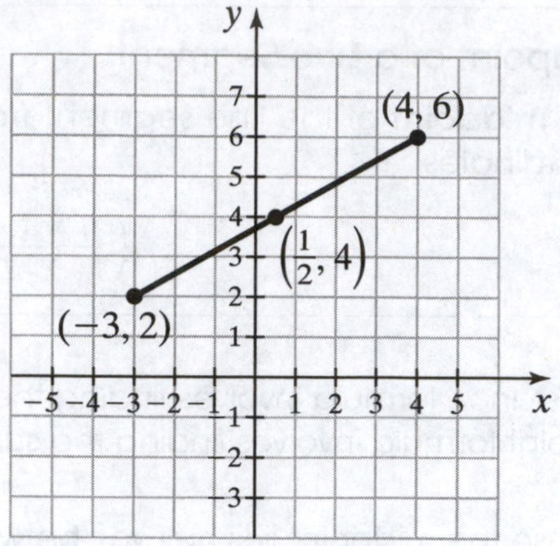

Figure 12

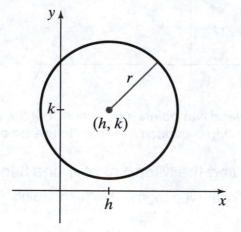

The graph shows a line segment with endpoints $(-3, 2)$ and $(4, 6)$, with midpoint $\left(\frac{1}{2}, 4\right)$.

🔺 **You Try It** Work through this You Try It problem.

Work Exercises 9–16 in this eText or in the MyMathLab Study Plan.

OBJECTIVE 3 WRITE THE STANDARD FORM OF AN EQUATION OF A CIRCLE

A **circle** is the set of all points (x, y) in the **coordinate plane** that are a fixed distance r from a fixed point (h, k). The fixed distance r is called the **radius** of the circle, and the fixed point (h, k) is called the **center** of the circle. See Figure 13.

Figure 13
Circle with Center (h, k) and Radius, r

The **distance formula** can be used to develop the standard form of the equation of a circle.

✎ *My animation summary*

⊛ **Concept Animation** Watch this animation to explore how this is done.

Standard Form of the Equation of a Circle

The **standard form of the equation of a circle** with center (h, k) and radius r is

$$(x - h)^2 + (y - k)^2 = r^2.$$

My video summary ⊙ **Example 3** Writing the Standard Form of the Equation of a Circle

Write the standard form of the equation of a **circle** with center $(0, 0)$ and radius 3.

Solution Watch this video, or read through the following solution.

The **center** is $(0, 0)$. Since the center is represented by (h, k) in the standard form equation, this means that $h = 0$ and $k = 0$. The **radius** is 3, so $r = 3$. Substitute these values into the equation.

$$(x - (0))^2 + (y - (0))^2 = 3^2$$

$$x^2 + y^2 = 9$$

The graph of the circle is shown in Figure 14.

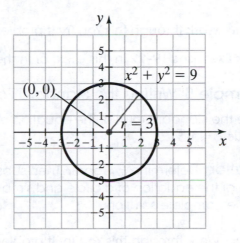

Figure 14

👤 **You Try It** Work through this You Try It problem.

Work Exercises 17 and 18 in this eText or in the MyMathLab Study Plan.

Example 3 shows that the equation of a **circle** centered at the **origin** takes on a special form. The **standard form** of the equation of a circle centered at the origin with **radius** r is

$$x^2 + y^2 = r^2.$$

My video summary ⊙ **Example 4** Writing the Standard Form of the Equation of a Circle

Write the standard form of the equation of the circle with center $(1, -2)$ and radius 4.

Solution The **center** is $(1, -2)$. Since the center is (h, k) in the standard form equation, this means that $h = 1$ and $k = -2$. The radius is 4, so $r = 4$. Substitute these values into the equation $(x - h)^2 + (y - k)^2 = r^2$ and simplify. View the **answer**, or watch this **video** for the complete solution.

The graph of the circle is shown in Figure 15.

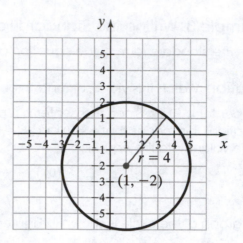

Figure 15

▶ **You Try It** Work through this You Try It problem.

Work Exercises 19–22 in this eText or in the MyMathLab Study Plan.

📝 *My video summary* ▶ **Example 5 Writing the Standard Form of the Equation of a Circle**

Write the standard form of the equation of the circle with center $(0, -4)$ and radius $r = \sqrt{5}$.

Solution Determine h, k, and r. Then substitute these values into the **standard form** of the equation of a circle and simplify. View the **answer**, or watch this **video** for the complete solution.

You Try It Work through this You Try It problem.

Work Exercises 23 and 24 in this eText or in the MyMathLab Study Plan.

OBJECTIVE 4 SKETCH THE GRAPH OF A CIRCLE GIVEN IN STANDARD FORM

📝 *My video summary* ▶ **Example 6 Sketching the Graph of a Circle**

Find the **center** and **radius** of the circle $x^2 + y^2 = 4$ and sketch its graph.

Solution Read the following, or watch this **video** for a complete solution.

Start by finding the center (h, k) and radius, r. The equation $x^2 + y^2 = 4$ has the form $(x - 0)^2 + (y - 0)^2 = 2^2$. Therefore, the circle is centered at the **origin**, $(h, k) = (0, 0)$, and has radius $r = 2$.

To sketch the graph of this circle, locate the center and then plot a few **points** on the circle. The easiest points to plot are located two units left and right from the center and two units up and down from the center.

$$\underbrace{(0 \overbrace{- 2}^{\text{2 units}}, 0)}_{r} \to (-2, 0) \qquad \underbrace{(0 \overbrace{+ 2}^{\text{2 units}}, 0)}_{r} \to (2, 0)$$

$$\underbrace{(0, 0 \overbrace{+ 2}^{\text{2 units}})}_{r} \to (0, 2) \qquad \underbrace{(0, 0 \overbrace{- 2}^{\text{2 units}})}_{r} \to (0, -2)$$

Complete the graph by drawing the **circle** through these points, as shown in Figure 16.

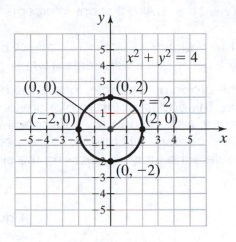

Figure 16
Graph of $x^2 + y^2 = 4$

You Try It   Work through this You Try It problem.

Work Exercises 25 and 26 in this eText or in the MyMathLab Study Plan.

My video summary ▶ **Example 7 Sketching the Graph of a Circle**

Find the **center** and **radius** of the circle $(x + 1)^2 + (y - 4)^2 = 25$ and sketch its graph.

Solution The equation $(x + 1)^2 + (y - 4)^2 = 25$ can be written as $(x - (-1))^2 + (y - 4)^2 = 5^2$. Comparing this equation to the **standard form** of the equation of a circle, we find that the circle has center $(h, k) = (-1, 4)$ and radius $r = 5$.

Locate the center, then plot a few points on the circle. Start with the points that are five units left and right of the center and five units up and down from the center.

$$\overbrace{5 \text{ units left}} \qquad\qquad \overbrace{5 \text{ units right}}$$
$$(\underbrace{-1 - 5}_{r}, 4) \rightarrow (-6, 4) \qquad (\underbrace{-1 + 5}_{r}, 4) \rightarrow (4, 4)$$

$$\overbrace{5 \text{ units up}} \qquad\qquad \overbrace{5 \text{ units down}}$$
$$(-1, \underbrace{4 + 5}_{r}) \rightarrow (-1, 9) \qquad (-1, \underbrace{4 - 5}_{r}) \rightarrow (-1, -1)$$

Use these points to sketch the graph of the circle. View the **answer**, or watch this **video** for a complete solution.

You Try It   Work through this You Try It problem.

Work Exercises 27–30 in this eText or in the MyMathLab Study Plan.

Recall that each point on the graph of an equation is an **ordered pair solution** to the equation. We can improve our sketch by finding additional points. For example, we may want to find the **intercepts** (if any) and then plot those corresponding points.

My video summary ▶ **Example 8 Sketching the Graph of a Circle**

Find the center and the radius, and sketch the graph of the circle $(x - 1)^2 + (y + 2)^2 = 9$. Also find any intercepts.

Solution The equation $(x - 1)^2 + (y + 2)^2 = 9$ can be written in **standard form** as $(x - 1)^2 + (y - (-2))^2 = 3^2$ with center $(h, k) = (1, -2)$ and radius $r = 3$.

To sketch the graph of this circle, locate the center, then plot a few points on the circle. The easiest points to plot are located three units left and right from the center, $(-2, -2)$ and $(4, -2)$ respectively, and three units up and down from the center, $(1, 1)$ and $(1, -5)$ respectively. After plotting these four points, complete the graph by drawing the circle through the points as shown in **Figure 17**.

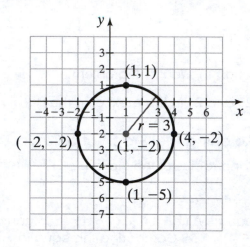

Figure 17
Graph of $(x - 1)^2 + (y + 2)^2 = 9$

To find the *x*-intercepts, set $y = 0$ and solve the resulting **quadratic equation** for *x*. To find the *y*-intercepts, set $x = 0$ and solve the resulting **quadratic equation** for *y*. View the **answer**, or watch this **video** for the complete solution.

You Try It Work through this **You Try It** problem.

Work Exercises 31 and 32 in this eText or in the **My**Math**Lab** Study Plan.

OBJECTIVE 5 WRITE THE GENERAL FORM OF A CIRCLE IN STANDARD FORM AND SKETCH ITS GRAPH

Since a **circle** is completely defined by its **center** and **radius**, we only need these two pieces of information to write the equation of a circle or to sketch its **graph**. We saw in Examples 3–5 that given the center and radius, we can write the **standard form** of an equation of a circle. Then we used this standard form in Examples 6–8 to find the center and radius and then to sketch the graph of a circle.

Not all equations of circles are given in standard form. Consider the circle $(x + 3)^2 + (y - 1)^2 = 49$. This equation is in standard form, so we can quickly see that the center is $(h, k) = (-3, 1)$ and the radius is $r = \sqrt{49} = 7$. However, let's see what happens if we expand the equation.

Original equation:	$(x + 3)^2 + (y - 1)^2 = 49$
Expand to remove parentheses:	$x^2 + 6x + 9 + y^2 - 2y + 1 = 49$
Collect like terms:	$x^2 + 6x + y^2 - 2y + 10 = 49$
Subtract 49 from both sides:	$x^2 + 6x + y^2 - 2y - 39 = 0$
Rearrange terms:	$x^2 + y^2 + 6x - 2y - 39 = 0$

The equation $x^2 + y^2 + 6x - 2y - 39 = 0$ is written in **general form**. In this form, the center and radius are less obvious.

General Form of the Equation of a Circle

The **general form** of the equation of a circle is

$$Ax^2 + By^2 + Cx + Dy + E = 0,$$

where A, B, C, D, and E are real numbers and $A = B \neq 0$.

To find the center and radius of a circle from its equation in general form, we need to **complete the square** in both x and y to write the equation in **standard form**. Before working through Example 9, you might want to review completing the square from **Section 8.1**.

✎ *My video summary* ▶ **Example 9 Writing the General Form of a Circle in Standard Form and Sketching Its Graph**

Write the equation $x^2 + y^2 + 10x + 9 = 0$ in **standard form**. Find the **center**, **radius**, and **intercepts**, then sketch the graph.

Solution Read the following, or watch this **video** for a complete solution.

The equation contains a **linear term** in x, but no linear term in y. So, we only need to **complete the square** in x. Start by rearranging terms to group the x terms together and move any constants to the right-hand side of the equation.

No linear y term so we already have a perfect square in y	Linear x term so we must complete the square in x

Original equation:	$x^2 + y^2 + 10x + 9 = 0$
Rearrange terms:	$x^2 + 10x + y^2 + 9 = 0$
Move constants to the right-hand side:	$x^2 + 10x + y^2 = -9$

To complete the square in x, we divide the **coefficient** of the linear term by 2, **square** the result, and add this to both sides of the equation. Since $\left(\frac{1}{2} \cdot 10\right)^2 = 5^2 = 25$, we need to add 25 to both sides, and then factor the left-hand side.

Complete the square in x: $x^2 + 10x + 25 + y^2 = -9 + 25$

$$\left(\frac{1}{2} \cdot 10\right)^2 = 25$$

Factor the left-hand side: $(x + 5)^2 + y^2 = 16$

Now the equation is written in **standard form**. Comparing to $(x - h)^2 + (y - k)^2 = r^2$, the **center** is $(h, k) = (-5, 0)$, and the **radius** is $r = \sqrt{16} = 4$.

To find the x-intercepts, we let $y = 0$ and solve for x.

Standard form: $(x + 5)^2 + y^2 = 16$

Substitute 0 for x: $(x + 5)^2 + (0)^2 = 16$

Simplify: $(x + 5)^2 = 16$

Square root property: $x + 5 = \pm\sqrt{16}$

Simplify: $x + 5 = \pm 4$

Subtract 5 from both sides: $x = -5 \pm 4$

The x-intercepts are $-5 - 4 = -9$ and $-5 + 4 = -1$. So, the points $(-9, 0)$ and $(-1, 0)$ are on the graph.

To find the y-intercepts, we let $x = 0$ and solve for y.

Original equation: $x^2 + y^2 + 10x + 9 = 0$

Substitute 0 for x: $(0)^2 + y^2 + 10(0) + 9 = 0$

Simplify: $y^2 + 9 = 0$

Subtract 9 from both sides: $y^2 = -9$

There is no **real number** whose **square** is -9, so the graph has no y-intercepts. The graph of the circle is shown in Figure 18.

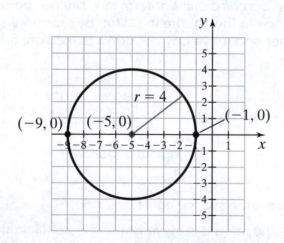

Figure 18
Graph of $x^2 + 10x + y^2 + 9 = 0$

You Try It Work through this You Try It problem.

Work Exercises 33 and 34 in this eText or in the MyMathLab Study Plan.

My video summary ▶ **Example 10** **Writing the General Form of a Circle in Standard Form and Sketching Its Graph**

Write the equation $x^2 + y^2 - 8x + 6y + 16 = 0$ in standard form. Find the **center**, **radius**, and **intercepts**, then sketch the graph.

Solution The equation contains linear terms in both x and y, so we will need to **complete the square** in both x and y. We rearrange the terms, complete the square, and move any **constants** to the right-hand side of the equation.

Rearrange the terms: $x^2 - 8x \quad + y^2 + 6y \quad = -16$

Complete the square in x and y. Remember to add 16 and 9 to both sides: $x^2 - 8x + 16 + y^2 + 6y + 9 = -16 + 16 + 9$

$$\left(\frac{1}{2} \cdot (-8)\right)^2 = 16 \quad \left(\frac{1}{2} \cdot 6\right)^2 = 9$$

Factor the left: $(x - 4)^2 + (y + 3)^2 = 9$

Now we have converted the **general form** of the circle into **standard form** with center $(h, k) = (4, -3)$ and radius $r = 3$. Watch this **video** to verify that there are no y-intercepts. The only x-intercept is $x = 4$. The graph of the circle is shown in Figure 19.

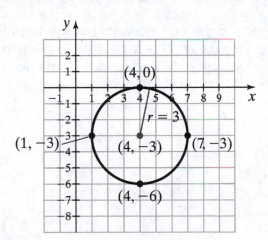

Figure 19
Graph of $x^2 + y^2 - 8x + 6y + 16 = 0$

You Try It Work through this You Try It problem.

Work Exercises 35–40 in this eText or in the MyMathLab Study Plan.

My animation summary ✳ **Example 11** **Writing the General Form of a Circle in Standard Form, Where $A \neq 1$ and $B \neq 1$**

Write the equation $4x^2 + 4y^2 + 4x - 8y + 1 = 0$ in standard form. Find the center, radius, and intercepts. Then sketch the graph.

Solution Work through this **animation** to verify that this equation is equivalent to the following equation in standard form:

$$\left(x + \frac{1}{2}\right)^2 + (y - 1)^2 = 1; \text{ center } (h, k) = \left(-\frac{1}{2}, 1\right) \text{ and } r = \sqrt{1} = 1$$

Verify that the one x-intercept is found at $\left(-\frac{1}{2}, 0\right)$, and the y-intercepts are found at $\left(0, 1 - \frac{\sqrt{3}}{2}\right)$ and $\left(0, 1 + \frac{\sqrt{3}}{2}\right)$. The graph of the circle is shown in **Figure 20**.

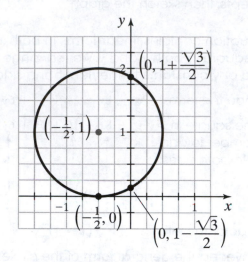

Figure 20
Graph of $4x^2 + 4y^2 + 4x - 8y + 1 = 0$

 You Try It Work through this **You Try It** problem.

Work Exercises 41–44 in this eText or in the MyMathLab Study Plan.

> Every equation of a **circle** can be written in **general form**. However, not every equation of the form $Ax^2 + By^2 + Cx + Dy + E = 0$ is the equation of a circle. For example, the equation $2x^2 + 2y^2 + 1 = 0$ has no graph, and the graph of $x^2 + y^2 + 2x - 4y + 5 = 0$ is a single point.

8.4 Exercises

In Exercises 1–8, find the distance $d(A, B)$ between points A and B.

1. $A(2, 7); B(5, 11)$

2. $A(1, 5); B(-2, 1)$

3. $A(3, 5); B(-2, -2)$

4. $A(2, -4); B(-3, 6)$

5. $A\left(\frac{2}{3}, 5\right), B\left(-1, \frac{1}{2}\right)$

6. $A(0, -\sqrt{2}); B(\sqrt{3}, 0)$

7. $A(2, -3); B(2, 5)$

8. $A(-1, 4); B(5, 4)$

In Exercises 9–16, find the midpoint of the line segment with endpoints A and B.

9. $A(3, 7); B(5, 9)$

10. $A(2, -5); B(4, 1)$

11. $A(-1, 4); B(-2, -2)$

12. $A(-3, 0); B(0, 7)$

13. $A(0, 1); B\left(-3, \frac{1}{2}\right)$

14. $A(\sqrt{27}, 4); B(\sqrt{3}, 2)$

15. $A(1, 3); B(1, -5)$

16. $A(4, -2); B(-3, -2)$

In Exercises 17–24, write the standard form of the equation of each circle described.

17. Center $(0, 0)$, $r = 1$

18. Center $(0, 0)$, $r = \sqrt{3}$

19. Center $(-2, 3)$, $r = 4$

20. Center $(0, 4)$, $r = 8$

21. Center $(1, -4)$, $r = \dfrac{3}{4}$

22. Center $\left(\dfrac{1}{5}, 3\right)$, $r = 2$

23. Center $(3, 0)$, $r = \sqrt{2}$

24. Center $\left(-\dfrac{1}{4}, -\dfrac{1}{3}\right)$, $r = \dfrac{3}{4}$

In Exercises 25–30, find the center and radius of each circle and sketch its graph.

25. $x^2 + y^2 = 9$

26. $(x + 3)^2 + y^2 = 4$

27. $(x - 1)^2 + (y + 5)^2 = 16$

28. $(x + 2)^2 + (y + 4)^2 = 36$

29. $\left(x - \dfrac{1}{4}\right)^2 + \left(y + \dfrac{1}{2}\right)^2 = 4$

30. $(x + 4)^2 + (y - 5)^2 = 10$

In Exercises 31 and 32, find the center, radius, and intercepts of each circle and then sketch its graph.

31. $(x - 4)^2 + (y + 7)^2 = 12$

32. $(x + 1)^2 + (y - 3)^2 = 20$

In Exercises 33–44, write the equation of each circle in standard form. Find the center, radius, and intercepts. Then sketch the graph.

33. $x^2 + y^2 + 6x + 5 = 0$

34. $x^2 + y^2 + 2y - 8 = 0$

35. $x^2 + y^2 + 2x - 4y + 1 = 0$

36. $x^2 + y^2 - 10x + 6y + 18 = 0$

37. $x^2 + y^2 - 4x - 8y + 19 = 0$

38. $x^2 + y^2 - 6x - 12y - 5 = 0$v

39. $x^2 + y^2 - 3x - y - \dfrac{1}{2} = 0$

40. $x^2 + y^2 + \dfrac{2}{3}x - \dfrac{1}{2}y - \dfrac{7}{18} = 0$

41. $2x^2 + 2y^2 - 4x + 8y + 2 = 0$

42. $16x^2 + 16y^2 - 16x - 8y - 11 = 0$

43. $144x^2 + 144y^2 - 72x - 96y - 551 = 0$

44. $36x^2 + 36y^2 + 12x + 72y - 35 = 0$

8.5 Polynomial and Rational Inequalities

THINGS TO KNOW

Before working through this section, be sure you are familiar with the following concepts:

VIDEO   ANIMATION   INTERACTIVE

 You Try It

1. Factor by Grouping
   (Section 5.1, Objective 3)

 You Try It

2. Factor Trinomials of the Form $x^2 + bx + c$
   (Section 5.2, Objective 1)

3. Factor Trinomials of the Form $ax^2 + bx + c$ Using Trial and Error (Section 5.2, Objective 2)

4. Factor Trinomials of the Form $ax^2 + bx + c$ Using the ac Method (Section 5.2, Objective 3)

5. Solve Polynomial Equations by Factoring (Section 5.4, Objective 1)

OBJECTIVES

1 Solve Polynomial Inequalities

2 Solve Rational Inequalities

OBJECTIVE 1 SOLVE POLYNOMIAL INEQUALITIES

In **Section 1.2**, we learned how to solve **linear inequalities**. In this objective, we learn how to solve **polynomial inequalities**.

> **Definition** Polynomial Inequality
>
> A **polynomial inequality** is an inequality that can be written as
>
> $$P(x) < 0, P(x) > 0, P(x) \le 0, \text{ or } P(x) \ge 0,$$
>
> where $P(x)$ is a **polynomial function**.

When solving polynomial inequalities, **x-intercepts** will play an important part. To understand why, consider the polynomial functions graphed in **Figure 21**.

Figure 21

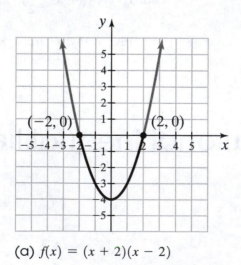

(a) $f(x) = (x + 2)(x - 2)$

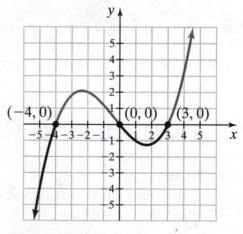

(b) $g(x) = 0.1x(x + 4)(x - 3)$

From **Figure 21**, note that for some x values, the value of the polynomial function will be positive; for others, it will be negative, and for still other x values, the resulting value of the polynomial function will be zero. The x values where the function value is zero are the **x-intercepts**, which are also **boundary points** that divide the **real number line** into **intervals**. Notice that in each interval, the function is always positive or always negative. This is true for all **polynomials** and is central to our approach to solving **polynomial inequalities**.

My video summary ▶ **Example 1 Solving a Polynomial Inequality**

Solve $x^3 - 3x^2 + 2x \geq 0$.

Solution Read the following solution, or watch this video for a complete solution.

First, **factor** the left-hand side to get $x(x - 1)(x - 2) \geq 0$. Second, find all real values of x that make the left-hand side *equal to* zero. These values are the **boundary points**. To find these boundary points, set the **factored polynomial** on the left equal to zero and solve for x.

Set the factored polynomial equal to 0: $x(x - 1)(x - 2) = 0$

Use the zero product property: $x = 0$ or $x - 1 = 0$ or $x - 2 = 0$

The boundary points are $x = 0$, $x = 1$, and $x = 2$.

Next, plot each boundary point on a **number line**. Because the expression $x(x - 1)(x - 2)$ is equal to zero at our three boundary points, we use a solid circle ● at each boundary point to indicate that the inequality $x(x - 1)(x - 2) \geq 0$ is satisfied at these points. (**Note:** If the inequality had been a **strict inequality**, such as > or <, we would have used an open circle ○ to represent that the boundary points were *not* part of the solution. We will look at this in Example 2.) Notice that in **Figure 22**, we have naturally divided the number line into four **intervals**.

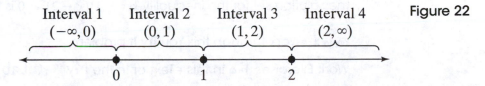

Interval 1 $(-\infty, 0)$ Interval 2 $(0, 1)$ Interval 3 $(1, 2)$ Interval 4 $(2, \infty)$ **Figure 22**

The expression $x(x - 1)(x - 2)$ is equal to zero *only* at the three **boundary points**: 0, 1, and 2. So, in any of the four intervals shown in Figure 22, the expression $x(x - 1)(x - 2)$ must be either always *positive* or always *negative* throughout the entire interval. To check whether this expression is positive or negative on each interval, pick a number called a **test value** from each interval. The test value can be any point in the interval but not a boundary point. Possible test values are plotted in Figure 23.

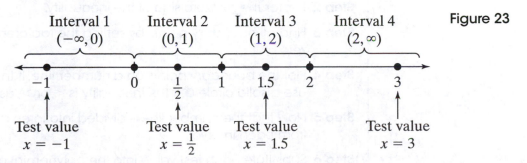

Interval 1 $(-\infty, 0)$ Interval 2 $(0, 1)$ Interval 3 $(1, 2)$ Interval 4 $(2, \infty)$ **Figure 23**

Test value $x = -1$ Test value $x = \frac{1}{2}$ Test value $x = 1.5$ Test value $x = 3$

Substitute the **test value** into the expression $x(x - 1)(x - 2)$, and check to see if it yields a positive or negative value.

Interval	Test Value	Substitute Test Value into $x(x - 1)(x - 2)$	Comment
1. $(-\infty, 0)$	$x = -1$	$(-1)(-1 - 1)(-1 - 2) \Rightarrow (-)(-)(-) = -$	Expression is negative on $(-\infty, 0)$
2. $(0, 1)$	$x = \dfrac{1}{2}$	$\left(\dfrac{1}{2}\right)\left(\dfrac{1}{2} - 1\right)\left(\dfrac{1}{2} - 2\right) \Rightarrow (+)(-)(-) = +$	Expression is positive on $(0, 1)$
3. $(1, 2)$	$x = 1.5$	$(1.5)(1.5 - 1)(1.5 - 2) \Rightarrow (+)(+)(-) = -$	Expression is negative on $(1, 2)$
4. $(2, \infty)$	$x = 3$	$(3)(3 - 1)(3 - 2) \qquad \Rightarrow (+)(+)(+) = +$	Expression is positive on $(2, \infty)$

If the expression $x(x - 1)(x - 2)$ is positive on an interval, we place a "+" above the interval on the **number line**. If the expression $x(x - 1)(x - 2)$ is negative, we place a "−" above the interval. See Figure 24.

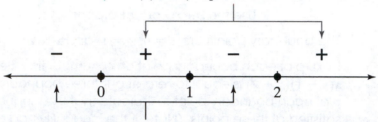

$x(x - 1)(x - 2)$ is positive on these intervals   **Figure 24**

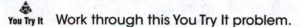

$x(x - 1)(x - 2)$ is negative on these intervals

In **Figure 24**, we see that $x(x - 1)(x - 2)$ is greater than zero on the intervals $(0, 1)$ and $(2, \infty)$ and is equal to zero at the **boundary points** $x = 0$, $x = 1$, and $x = 2$. So, the solution set for the inequality $x(x - 1)(x - 2) \geq 0$ is the interval $[0, 1] \cup [2, \infty)$.

You Try It   Work through this You Try It problem.

Work Exercises 1–6 in this eText or in the MyMathLab Study Plan.

Example 1 illustrates the following approach for solving a factorable **polynomial inequality.**

Solving Polynomial Inequalities

Step 1 Move all terms to one side of the inequality, leaving zero on the other side.

Step 2 Factor the nonzero side of the inequality.

Step 3 Find all boundary points by setting the factored polynomial equal to zero.

Step 4 Plot the boundary points on a **number line.** If the inequality is $\leq$ or $\geq$, use a solid circle ●. If the inequality is $<$ or $>$, use an open circle ○.

Step 5 Now that the number line is divided into **intervals**, pick a **test value** from each interval.

Step 6 Substitute each test value into the **polynomial** and determine whether the expression is positive or negative on the corresponding interval.

Step 7 Find the intervals that **satisfy** the inequality.

My video summary ▶ **Example 2 Solving a Polynomial Inequality**

Solve $x^2 + 5x < 3 - x^2$.

Solution Work through the following, or watch this **video** for a complete solution. Use the steps for **solving polynomial inequalities**.

Step 1 Move all terms to one side of the inequality.

$$2x^2 + 5x - 3 < 0$$

Step 2 Factor.

$$(2x - 1)(x + 3) < 0$$

Step 3 Find boundary points by setting the factored polynomial equal to zero.

$$(2x - 1)(x + 3) = 0$$

Zero product property: $\quad 2x - 1 = 0 \quad$ or $\quad x + 3 = 0$

$$2x = 1 \qquad\qquad x = -3$$

$$x = \frac{1}{2}$$

The boundary points are $x = -3$ and $x = \frac{1}{2}$.

Step 4 Plot the boundary points.

We use open circles to plot our **boundary points** because these points are *not* part of the solution. (**Note:** We are only looking for values that make the expression strictly less than zero.)

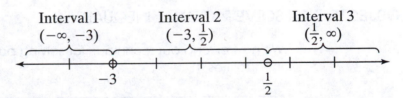

Step 5 We have three intervals. Pick a **test value** from each interval.

Interval 1: Test value $x = -4$

Interval 2: Test value $x = 0$

Interval 3: Test value $x = 1$

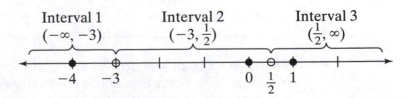

Step 6 Substitute each test value into the polynomial.

$x = -4$: $(2(-4) - 1)(-4 + 3) \Rightarrow (-)(-) = +$,
expression is positive on $(-\infty, -3)$.

$$x = 0: (2(0) - 1)(0 + 3) \Rightarrow (-)(+) = -,$$

expression is negative on $\left(-3, \dfrac{1}{2}\right)$.

$$x = 1: (2(1) - 1)(1 + 3) \Rightarrow (+)(+) = +,$$

expression is positive on $\left(\dfrac{1}{2}, \infty\right)$.

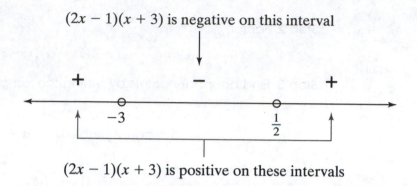

$(2x - 1)(x + 3)$ is negative on this interval

$(2x - 1)(x + 3)$ is positive on these intervals

Step 7 Find the intervals that satisfy the inequality.
Because we want x-values that result in a polynomial value less than zero (negative values), the solution set is the interval $\left(-3, \dfrac{1}{2}\right)$.

You Try It Work through this You Try It problem.

Work Exercises 7–10 in this eText or in the MyMathLab Study Plan.

OBJECTIVE 2 SOLVE RATIONAL INEQUALITIES

We now extend our technique for solving polynomial inequalities to solve **rational inequalities.**

Definition   Rational Inequality

A **rational inequality** is an inequality that can be written as

$$R(x) < 0, R(x) > 0, R(x) \le 0, \text{ or } R(x) \ge 0,$$

where $R(x) = \dfrac{P(x)}{Q(x)}$ is a rational function and $P(x)$ and $Q(x)$ are polynomial functions.

We solve rational inequalities in much the same way that we solve polynomial inequalities. The only differences are that we need the left-hand side to be a single rational expression and we find the boundary points by setting both the numerator and denominator equal to zero.

(eText Screens 8.5-1–8.5-21)

Solving Rational Inequalities

Step 1 Move all terms to one side of the inequality, leaving zero on the other side.

Step 2 Combine the terms into a single **rational expression** and factor the nonzero side of the inequality (both the **numerator** and **denominator**).

Step 3 Find all **boundary points** by setting the factored polynomials in the numerator and denominator equal to zero.

Step 4 Plot the boundary points on a **number line**. If the inequality is ≤ or ≥, use a solid circle ● for the boundary points from the numerator. If the inequality is < or >, use an open circle ○ . Boundary points from the denominator always have an open circle. If the same boundary point is obtained from both the numerator and denominator, then it is plotted with an open circle.

Step 5 Pick a test value from each interval.

Step 6 Substitute each test value into the rational expression and determine whether the expression is positive or negative on the corresponding interval.

Step 7 Find the intervals that satisfy the inequality.

When solving rational inequalities, do not multiply both sides of the inequality by the **LCD**. Doing so may remove boundary points and lead to incorrect results.

My video summary ▶ **Example 3 Solving a Rational Inequality**

Solve $\frac{x-4}{x+1} \le 0$.

Solution Using the steps for **solving a rational inequality**, work through the following, or watch this **video** for a complete solution.

Because the left-hand side of the inequality is already a single **rational expression** in completely factored form, we can skip Steps 1 and 2 and start with Step 3.

Step 3 Find the **boundary points** by setting the factored polynomials in the numerator and the denominator equal to zero.

Numerator: $x - 4 = 0$, so $x = 4$ is a boundary point.

Denominator: $x + 1 = 0$, so $x = -1$ is a boundary point.

Step 4 Plot the boundary points.

Since the inequality is a *less than or equal to* inequality, we use a closed circle ● to represent the boundary point $x = 4$. Because we cannot divide by zero, we use an open circle ○ to represent the boundary point from the denominator, $x = -1$.

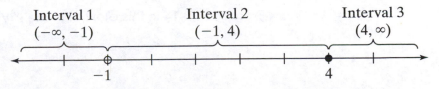

Step 5 Pick a test value from each interval.

Interval 1: Test value $x = -2$

Interval 2: Test value $x = 0$

Interval 3: Test value $x = 5$

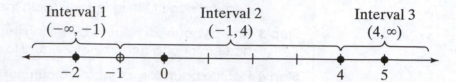

Step 6 Substitute each **test value** into the rational expression, and determine whether the expression is positive or negative on the corresponding interval.

$$x = -2: \frac{(-2 - 4)}{(-2 + 1)} \Rightarrow \frac{(-)}{(-)} = +,$$

expression is positive on the interval $(-\infty, -1)$.

$$x = 0: \frac{(-0 - 4)}{(0 + 1)} \Rightarrow \frac{(-)}{(+)} = -,$$

expression is negative on the interval $(-1, 4)$.

$$x = 5: \frac{(5 - 4)}{(5 + 1)} \Rightarrow \frac{(+)}{(+)} = +,$$

expression is positive on the interval $(4, \infty)$.

$\dfrac{x - 4}{x + 1}$ is negative on this interval.

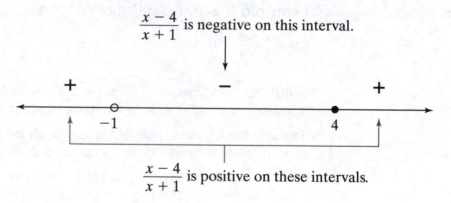

$\dfrac{x - 4}{x + 1}$ is positive on these intervals.

Step 7 Find the intervals that satisfy the inequality.

We look for x values that make the **rational expression** less than or equal to zero. From the last step, we see that the **solution set** to the inequality is $(-1, 4]$.

Note that 4 is included as a solution but not -1 because $x = -1$ makes the denominator equal to zero.

You Try It Work through this **You Try It** problem.

Work Exercises 11–14 in this eText or in the MyMathLab Study Plan.

My video summary ▶ **Example 4 Solving a Rational Inequality**

Solve $x \geq \dfrac{3}{x-2}$.

Solution

Step 1 Subtract $\dfrac{3}{x-2}$ from both sides of the inequality.

$$x - \frac{3}{x-2} \geq 0$$

All nonzero terms are on the left-hand side, so we combine the terms by finding a **common denominator**.

Rewrite the inequality: $\quad x - \dfrac{3}{x-2} \geq 0$

Rewrite using a common
denominator of $x-2$: $\quad \dfrac{x(x-2)}{x-2} - \dfrac{3}{x-2} \geq 0$

Combine terms: $\quad \dfrac{x(x-2)-3}{x-2} \geq 0$

Distribute: $\quad \dfrac{x^2 - 2x - 3}{x-2} \geq 0$

Step 2 Combine the terms into a single rational expression and factor the nonzero side of the inequality (both the numerator and denominator).

$$\frac{(x-3)(x+1)}{x-2} \geq 0$$

Finish Steps 3–7 on your own. View the **answer**, or watch this **video** to see the entire solution.

You Try It Work through this You Try It problem.

Work Exercises 15–20 in this eText or in the MyMathLab Study Plan.

My video summary ▶ **Example 5 Solving a Rational Inequality**

Solve $\dfrac{x+1}{x-2} > \dfrac{7x+1}{x^2+x-6}$.

Solution Complete the **seven-step process** to solve the rational inequality. View the **answer**, or watch this **video** for the complete solution.

You Try It Work through this You Try It problem.

Work Exercises 21 and 22 in this eText or in the MyMathLab Study Plan.

8.5 Exercises

In Exercises 1–10, solve each polynomial inequality. Express each solution using interval notation.

1. $(x - 1)(x + 3) \geq 0$

2. $x^2 + 4x - 21 < 0$

3. $x(3x + 2) \leq 0$

4. $2x^2 - 4x > 0$

5. $x^2 \leq 1$

6. $(x - 1)(x + 4)(x - 3) \geq 0$

7. $3x^2 + x < 3x + 1$

8. $2x^3 > 24x - 2x^2$

9. $x^3 + x^2 - x \leq 1$

10. $x^3 \geq -2x^2 - x$

In Exercises 11–22, solve each rational inequality. Express each solution using interval notation.

11. $\dfrac{x + 3}{x - 1} \leq 0$

12. $\dfrac{2 - x}{3x + 9} \geq 0$

13. $\dfrac{x}{x - 1} > 0$

14. $\dfrac{x^2 - 9}{x + 2} \leq 0$

15. $\dfrac{4}{x + 1} \geq 2$

16. $\dfrac{x + 5}{2x - 3} > 1$

17. $\dfrac{x^2 - 8}{x + 4} \geq x$

18. $\dfrac{x}{2 - x} \leq \dfrac{1}{x}$

19. $\dfrac{x - 1}{x - 2} \geq \dfrac{x + 2}{x + 3}$

20. $\dfrac{x - 1}{x + 1} + \dfrac{x + 1}{x - 1} \leq \dfrac{x + 5}{x^2 - 1}$

21. $\dfrac{x + 1}{x + 4} < \dfrac{-11x - 17}{x^2 - x - 20}$

22. $\dfrac{x - 8}{x^2 + 5x + 4} < \dfrac{-21x}{(x + 1)(x - 3)(x + 4)}$

Exponential and Logarithmic Functions and Equations

CHAPTER NINE CONTENTS

9.1 Transformations of Functions

THINGS TO KNOW

Before working through this section, be sure you are familiar with the following concepts:

		VIDEO	ANIMATION	INTERACTIVE
You Try It	1. Graph Equations by Plotting Points (Section 2.1, Objective 4)			☞
You Try It	2. Find the Domain and Range of a Relation (Section 2.2, Objective 2)		✸	☞
You Try It	3. Graph Simple Functions by Plotting Points (Section 2.3, Objective 3)		✸	
You Try It	4. Graph Functions That Contain Square Roots or Cube Roots (Section 7.2, Objective 3)		✸	☞
You Try It	5. Graph Quadratic Functions by Using Translations (Section 8.2, Objective 2)	▶		

OBJECTIVES

1 Use Vertical Shifts to Graph Functions

2 Use Horizontal Shifts to Graph Functions

3 Use Reflections to Graph Functions

4 Use Vertical Stretches and Compressions to Graph Functions

5 Use Horizontal Stretches and Compressions to Graph Functions

6 Use Combinations of Transformations to Graph Functions

In this section, we learn how to sketch the graphs of new **functions** using the graphs of known functions. Starting with the graph of a known function, we *transform* it into a new function by applying various transformations.

Before we begin our discussion about transformations, it is critical that you know the graphs of some basic functions that have been explored earlier. Take a moment to review these basic functions. Select a function name to review its properties.

REVIEW OF THE BASIC FUNCTIONS

Select any function to review its graph.

The **identity function**	$f(x) = x$
The **absolute value function**	$f(x) = \|x\|$
The **square function**	$f(x) = x^2$
The **cube function**	$f(x) = x^3$
The **square root function**	$f(x) = \sqrt{x}$
The **cube root function**	$f(x) = \sqrt[3]{x}$

OBJECTIVE 1   USE VERTICAL SHIFTS TO GRAPH FUNCTIONS

My video summary ▶ **Example 1 Vertically Shifting a Function**

Sketch the graphs of $f(x) = |x|$ and $g(x) = |x| + 2$.

Solution Work through the following solution, or watch this **video**.

Table 1 shows that for every value of x, the y-coordinate of the function g is always two greater than the y-coordinate of the function f. The two functions are sketched in Figure 1. The graph of $g(x) = |x| + 2$ is exactly the same as the graph of $f(x) = |x|$, except the graph of g is shifted *up* two units.

Table 1

x	$f(x) = \lvert x\rvert$	$g(x) = \lvert x\rvert + 2$
−3	3	5
−2	2	4
−1	1	3
0	0	2
1	1	3
2	2	4
3	3	5

Figure 1

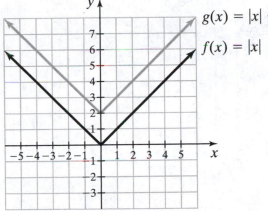

$g(x) = \lvert x\rvert + 2$

$f(x) = \lvert x\rvert$

You Try It Work through this You Try It problem.

Work Exercises 1–7 in this eText or in the MyMathLab Study Plan.

We see from Example 1 that if c is a positive number, then $y = f(x) + c$ is the graph of f shifted *up* c units. It follows that for $c > 0$, the graph of $y = f(x) - c$ is the graph of f shifted *down* c units.

My animation summary   ⊛ **Concept Animation** Watch this animation to learn more about graphing vertical shifts.

Vertical Shifts of Functions

If c is a positive real number,

The graph of $y = f(x) + c$ is obtained by shifting the graph of $y = f(x)$ vertically upward c units.

The graph of $y = f(x) - c$ is obtained by shifting the graph of $y = f(x)$ vertically downward c units.

OBJECTIVE 2 USE HORIZONTAL SHIFTS TO GRAPH FUNCTIONS

My video summary ▶ **Example 2 Horizontally shifting a Function**

Sketch the graphs of $f(x) = x^2$ and $g(x) = (x + 2)^2$.

Solution Tables 2 and 3 show tables of values for f and g, respectively. The graphs of f and g are sketched in Figure 2. The graph of g is the graph of f shifted to the *left* two units.

Table 2

x	$f(x) = x^2$
-2	4
-1	1
0	0
1	1
2	4

Table 3

x	$g(x) = (x + 2)^2$
-4	4
-3	1
-2	0
-1	1
0	4

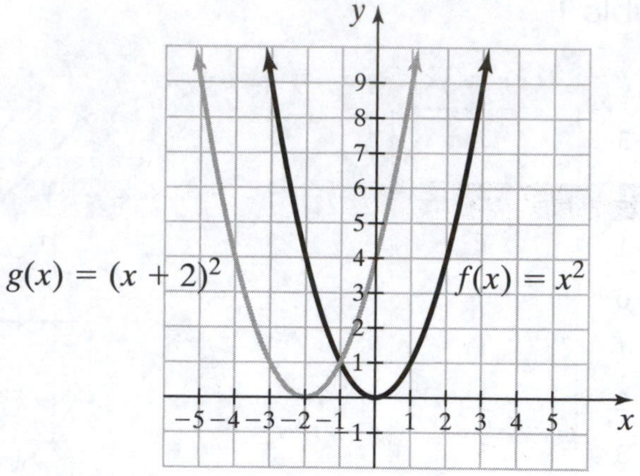

Figure 2

$g(x) = (x + 2)^2$ $f(x) = x^2$

You Try It Work through this You Try It problem.

Work Exercises 8–14 in this eText or in the MyMathLab Study Plan.

It follows that if c is a positive number, then $y = f(x + c)$ is the graph of f shifted to the *left* c units. For $c > 0$, the graph of $y = f(x - c)$ is the graph of f shifted to the *right* c units.

My animation summary

⊛ **Concept Animation** Watch this animation to learn more about graphing horizontal shifts.

At first glance, it appears that the rule for horizontal shifts is the opposite of what seems natural. Substituting $x + c$ for x causes the graph of $y = f(x)$ to be shifted to the *left* c units, whereas substituting $x - c$ for x causes the graph to shift to the *right* c units.

Horizontal Shifts of Functions

If c is a positive real number,

The graph of $y = f(x + c)$ is obtained by shifting the graph of $y = f(x)$ horizontally to the left c units.

The graph of $y = f(x - c)$ is obtained by shifting the graph of $y = f(x)$ horizontally to the right c units.

My animation summary

⊛ **Example 3 Combining Horizontal and Vertical Shifts**

Use the graph of $y = x^3$ to sketch the graph of $g(x) = (x - 1)^3 + 2$.

Solution The graph of g is obtained by shifting the graph of the basic function $y = x^3$ first horizontally to the right one unit and then vertically upward two units. When doing a problem with multiple transformations, it is good practice to always perform the vertical transformation last. Watch this animation to see the sketch of $g(x)$.

You Try It Work through this You Try It problem.

Work Exercises 15–21 in this eText or in the MyMathLab Study Plan.

OBJECTIVE 3   USE REFLECTIONS TO GRAPH FUNCTIONS

Using a graphing utility with $y_1 = x^2$ and $y_2 = -x^2$, we can see that the graph of $y_2 = -x^2$ is the graph of $y_1 = x^2$ reflected about the **x-axis**.

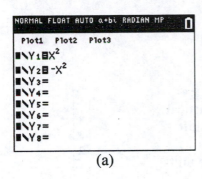 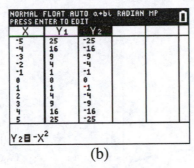 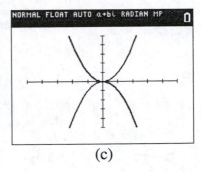   **Figure 3**

        (a)   (b)   (c)

Functions can also be reflected about the y-axis. Given the graph of $y = f(x)$, the graph of $y = f(-x)$ will be the graph of $y = f(x)$ reflected about the y-axis. Using a graphing utility, we illustrate a y-axis reflection by letting $y_1 = \sqrt{x}$ and $y_2 = \sqrt{-x}$. You can see that the functions are mirror images of each other about the y-axis.

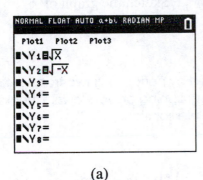 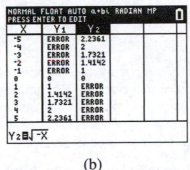 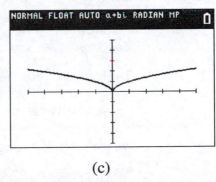   **Figure 4**

        (a)   (b)   (c)

Reflections of Functions about the x-Axis

The graph of $y = -f(x)$ is obtained by reflecting the graph of $y = f(x)$ about the x-axis.

Reflections of Functions about the y-Axis

The graph of $y = f(-x)$ is obtained by reflecting the graph of $y = f(x)$ about the y-axis.

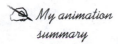 *My animation summary*   ✺ **Concept Animation** Watch this animation to learn more about graphing reflections.

Example 4 Sketching Functions Using Reflections and Shifts

Use the graph of the basic function $y = \sqrt[3]{x}$ to sketch each graph.

    a. $g(x) = -\sqrt[3]{x} - 2$ **b.** $h(x) = \sqrt[3]{1 - x}$.

Solutions

a. Starting with the graph of $y = \sqrt[3]{x}$, we can obtain the graph of $g(x) = -\sqrt[3]{x} - 2$ by performing two transformations:

1. Reflect about the x-axis.

2. Vertically shift down two units.

See Figure 5.

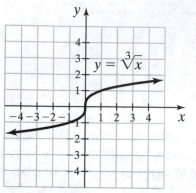 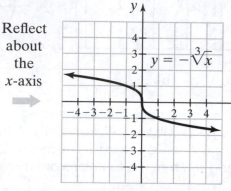 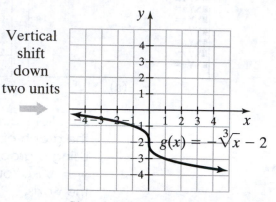

Start with the graph of the basic function $y = \sqrt[3]{x}$.

Sketch the graph of $y = -\sqrt[3]{x}$.

Sketch the graph of $g(x) = -\sqrt[3]{x} - 2$.

Figure 5

b. The graph of this function can also be obtained by performing two transformations: a horizontal shift and a reflection. Try graphing on your own. Check your **answer**, or watch this **video** for the complete solution.

You Try It Work through this You Try It problem.

Work Exercises 22–33 in this eText or in the MyMathLab Study Plan.

OBJECTIVE 4 USE VERTICAL STRETCHES AND COMPRESSIONS TO GRAPH FUNCTIONS

My video summary ▶ **Example 5** Vertically Stretching and Compressing

Use the graph of $f(x) = x^2$ to sketch the graph of $g(x) = 2x^2$.

Solution Work through the following solution, or watch this **video**.

In Table 4, for each value of x, the y-coordinate of g is two times as large as the corresponding y-coordinate of f. In Figure 6, the graph of $f(x) = x^2$ is vertically stretched by a factor of two to obtain the graph of $g(x) = 2x^2$.

Table 4

x	$f(x) = x^2$	$g(x) = 2x^2$
-2	4	8
-1	1	2
0	0	0
1	1	2
2	4	8

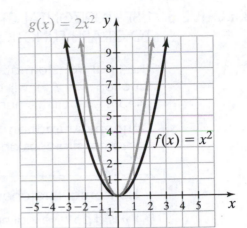

Figure 6

You Try It Work through this You Try It problem.

Work Exercises 34–40 in this eText or in the MyMathLab Study Plan.

It follows that if $a > 1$, the graph of $y = af(x)$ is a **vertical stretch** of the graph of $y = f(x)$ and is obtained by multiplying each y-coordinate on the graph of f by a factor of a. If $0 < a < 1$, then the graph of $y = af(x)$ is a **vertical compression** of the graph of $y = f(x)$.

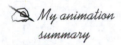 *My animation summary*

⊛ **Concept Animation** Watch this animation to learn more about graphing vertical stretches and compressions.

Table 5 and Figure 7 show the relationship between the graphs of the functions $f(x) = x^2$ and $h(x) = \dfrac{1}{2}x^2$.

Table 5

x	$f(x) = x^2$	$h(x) = \dfrac{1}{2}x^2$
-2	4	2
-1	1	$\dfrac{1}{2}$
0	0	0
1	1	$\dfrac{1}{2}$
2	4	2

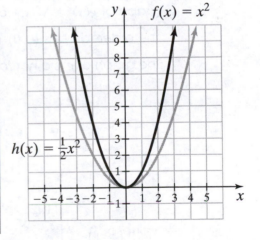

Figure 7

Vertical Stretches and Compressions of Functions

Suppose a is a positive real number:

The graph of $y = af(x)$ is obtained by multiplying each y-coordinate of $y = f(x)$ by a. If $a > 1$, the graph of $y = af(x)$ is a vertical stretch of the graph of $y = f(x)$. If $0 < a < 1$, the graph of $y = af(x)$ is a vertical compression of the graph of $y = f(x)$.

OBJECTIVE 5 USE HORIZONTAL STRETCHES AND COMPRESSIONS TO GRAPH FUNCTIONS

The final transformation to discuss is a horizontal stretch or compression. A function, $y = f(x)$, will be horizontally stretched or compressed when x is multiplied by a positive number, $a \neq 1$, to obtain the new function, $y = f(ax)$.

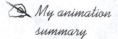 *My animation summary*

⊛ **Concept Animation** Watch this animation to learn more about graphing horizontal stretches and compressions.

Horizontal Stretches and Compressions of Functions

If a is a positive real number,

For $a > 1$, the graph of $y = f(ax)$ is obtained by dividing each x-coordinate of $y = f(x)$ by a. The resultant graph is a **horizontal compression**.

For $0 < a < 1$, the graph of $y = f(ax)$ is obtained by dividing each x-coordinate of $y = f(x)$ by a. The resultant graph is a **horizontal stretch**.

My video summary ▶ **Example 6 Horizontally Stretching and Compressing**

Use the graph of $f(x) = \sqrt{x}$ to sketch the graphs of $g(x) = \sqrt{4x}$ and $h(x) = \sqrt{\frac{1}{4}x}$.

Solution Read through the following, or watch this **video** for the complete solution.

The graph of $f(x) = \sqrt{x}$ contains the ordered pairs $(0, 0), (1, 1), (4, 2)$. To sketch the graph of $g(x) = \sqrt{4x}$, we must divide each previous x-coordinate by 4. Therefore, the ordered pairs $(0, 0), \left(\frac{1}{4}, 1\right), (1, 2)$ must lie on the graph of g. You can see that the graph of g is a horizontal compression of the graph of f. See Figure 8.

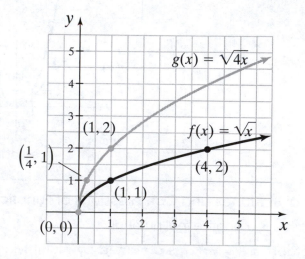

Figure 8
Graphs of $f(x) = \sqrt{x}$ and $g(x) = \sqrt{4x}$

Similarly, to sketch the graph of $h(x) = \sqrt{\frac{1}{4}x}$, divide the x-coordinates of the ordered pairs of f by $\frac{1}{4}$ to get the ordered pairs $(0, 0), (4, 1), (16, 2)$. You can see that the graph of h is a horizontal stretch of the graph of f. See Figure 9.

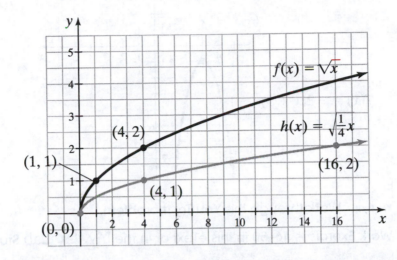

Figure 9

Graphs of $f(x) = \sqrt{x}$

and $g(x) = \sqrt{\frac{1}{4}x}$

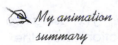

You Try It   Work through this **You Try It** problem.

Work Exercises 41–45 in this eText or in the MyMathLab Study Plan.

OBJECTIVE 6   USE COMBINATIONS OF TRANSFORMATIONS TO GRAPH FUNCTIONS

You may encounter functions that combine many (if not all) of the transformations discussed in this section. When sketching a function that involves multiple transformations, it is important to follow a certain "order of operations." Following is the order in which each transformation is performed in this eText:

1. Horizontal shifts

2. Horizontal stretches/compressions

3. Reflection about y-axis

4. Vertical stretches/compressions

5. Reflection about x-axis

6. Vertical shifts

Different ordering is possible for transformations 2 through 5, but you should always perform the horizontal shift first and the vertical shift last.

My animation summary

⊛ **Example 7 Combining Transformations**

Use transformations to sketch the graph of $f(x) = -2(x + 3)^2 - 1$.

Solution Watch the **animation** to see how to sketch the function
$f(x) = -2(x + 3)^2 - 1$ as seen in Figure 10.

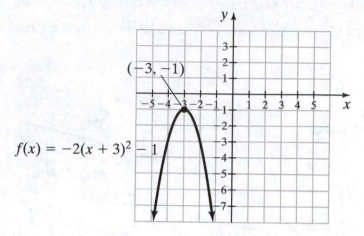

$(-3, -1)$

$f(x) = -2(x + 3)^2 - 1$

Figure 10
Graph of
$f(x) = -2(x + 3)^2 - 1$

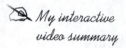

You Try It   Work through this You Try It problem.

Work Exercises 46–50 in this eText or in the MyMathLab Study Plan.

My interactive video summary   **Example 8** Combining Transformations

Use the graph of $y = f(x)$ to sketch each of the following functions.

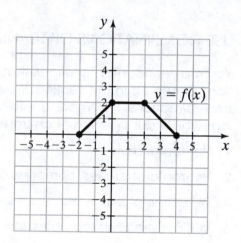

$y = f(x)$

a. $y = -f(2x)$ **b.** $y = 2f(x - 3) - 1$ **c.** $y = -\dfrac{1}{2}f(2 - x) + 3$

Solutions Watch the interactive video to see any one of the solutions worked
out in detail.

a. The graph of $y = -f(2x)$ can be obtained from the graph of $y = f(x)$ using two
transformations: (1) a **horizontal compression** and (2) a **reflection about the
x-axis**. The resultant graph is shown in Figure 11.

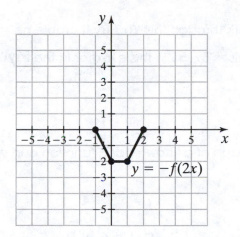

Figure 11

Graph of $y = -f(2x)$

b. The graph of $y = 2f(x - 3) - 1$ can be obtained from the graph of $y = f(x)$ using three transformations: (1) a **horizontal shift** to the right three units, (2) a **vertical stretch** by a factor of 2, and (3) a **vertical shift** down one unit. The resultant graph is shown in Figure 12.

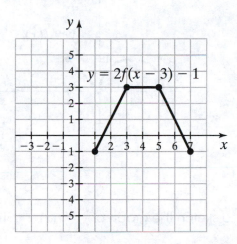

Figure 12

Graph of $y = 2f(x - 3) - 1$

c. The graph of $y = -\dfrac{1}{2}f(2 - x) + 3$ can be obtained from the graph of $y = f(x)$

using five transformations: (1) a **horizontal shift** to the left two units,

(2) a **reflection** about the **y-axis**, (3) a **vertical compression** by a factor of $\dfrac{1}{2}$

(4) a **reflection** about the **x-axis**, and (5) a **vertical shift** up three units. The resultant graph is shown in Figure 13.

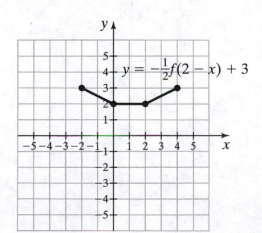

Figure 13

Graph of $y = -\dfrac{1}{2}f(2 - x) + 3$

△
You Try It   Work through this You Try It problem.

Work Exercises 51–55 in this eText or in the MyMathLab Study Plan.

9.1 Exercises

In Exercises 1–5, use the graph of a known basic function and vertical shifting to sketch each function.

1. $f(x) = x^2 - 1$

2. $y = \sqrt{x} + 2$

3. $h(x) = \sqrt[3]{x} - 2$

4. $y = |x| - 3$

5. $g(x) = x^3 + 1$

6. Use the graph of $y = f(x)$ to sketch the graph of $y = f(x) - 1$. Label at least three points on the new graph.

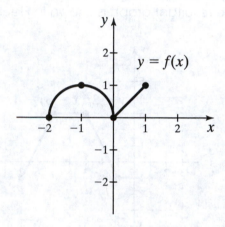

7. Use the graph of $y = f(x)$ to sketch the graph of $y = f(x) + 2$. Label at least three points on the new graph.

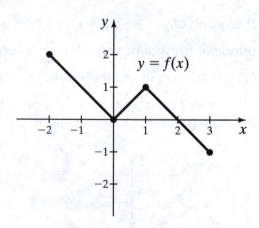

In Exercises 8–12, use the graph of a known basic function and horizontal shifts to sketch each function.

8. $f(x) = \sqrt[3]{x} - 2$

9. $y = \sqrt{x} - 4$

10. $h(x) = (x + 1)^3$

11. $k(x) = |x - 1|$

12. $y = (x - 3)^2$

13. Use the graph of $y = f(x)$ to sketch the graph of $y = f(x - 2)$. Label at least three points on the new graph.

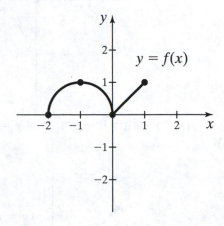

14. Use the graph of $y = f(x)$ to sketch the graph of $y = f(x + 2)$. Label at least three points on the new graph.

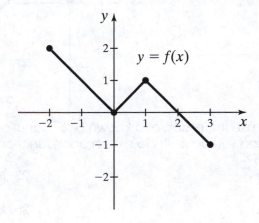

In Exercises 15–19, use the graph of a known basic function and a combination of horizontal and vertical shifts to sketch each function.

15. $y = (x + 1)^2 - 2$

16. $f(x) = (x - 3)^2 + 1$

17. $y = \sqrt{x + 3} + 2$

18. $f(x) = |x + 2| + 2$

19. $y = \sqrt[3]{x + 1} - 1$

20. Use the graph of $y = f(x)$ to sketch the graph of $y = f(x - 2) - 1$. Label at least three points on the new graph.

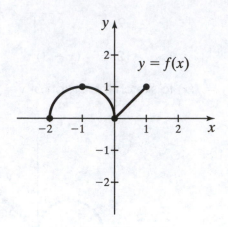

21. Use the graph of $y = f(x)$ to sketch the graph of $y = f(x + 1) + 2$. Label at least three points on the new graph.

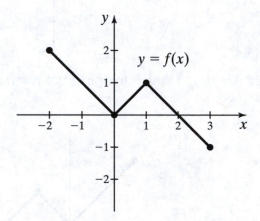

In Exercises 22–31, use the graph of a known basic function and a combination of horizontal shifts, reflections, and vertical shifts to sketch each function.

22. $g(x) = -x^2 - 2$

23. $h(x) = \sqrt{2 - x}$

24. $f(x) = |-1 - x|$

25. $h(x) = \sqrt[3]{-x - 2}$

26. $g(x) = -x^3 + 1$

27. $h(x) = -\sqrt[3]{x} + 2$

28. $g(x) = (3 - x)^2$

29. $h(x) = -\sqrt{x} - 1$

30. $f(x) = -|x| + 1$

31. $g(x) = (1 - x)^3$

32. Use the graph of $y = f(x)$ to sketch the graph of $y = -f(x)$. Label at least three points on the new graph.

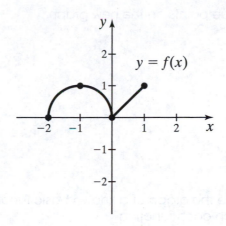

33. Use the graph of $y = f(x)$ to sketch the graph of $y = f(-x)$. Label at least three points on the new graph.

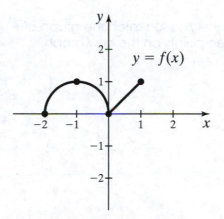

In Exercises 34–38, use the graph of a known basic function and a vertical stretch or vertical compression to sketch each function.

34. $f(x) = \dfrac{1}{4}|x|$

35. $g(x) = 2\sqrt{x}$

36. $f(x) = \dfrac{1}{3}x^3$

37. $f(x) = \dfrac{1}{2}\sqrt[3]{x}$

38. $g(x) = 3x^2$

39. Use the graph of $y = f(x)$ to sketch the graph of $y = 3f(x)$. Label at least three points on the new graph.

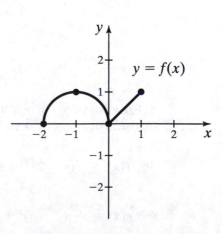

40. Use the graph of $y = f(x)$ to sketch the graph of $y = \dfrac{1}{2}f(x)$.

    Label at least three points on the new graph.

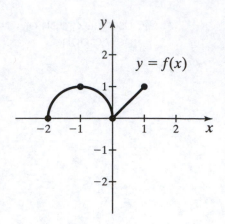

In Exercises 41–43, use the graph of a known basic function and a horizontal stretch or horizontal compression to sketch each function.

41. $y = \left|\dfrac{1}{4}x\right|$

42. $f(x) = \sqrt{2x}$

43. $g(x) = \sqrt[3]{3x}$

44. Use the graph of $y = f(x)$ to sketch the graph of $y = f(2x)$.
    Label at least three points on the new graph.

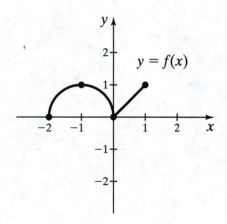

45. Use the graph of $y = f(x)$ to sketch the graph of $y = f\left(\dfrac{1}{2}x\right)$.

    Label at least three points on the new graph.

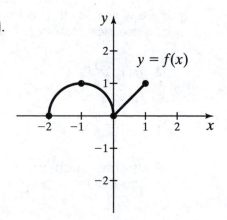

In Exercises 46–50, use the graph of a known basic function and a combination of transformations to sketch each function.

46. $f(x) = -(x - 2)^2 + 3$

47. $g(x) = \dfrac{1}{2}|x + 1| - 1$

48. $f(x) = 2\sqrt[3]{x} - 1$

49. $g(x) = -\dfrac{1}{2}(2 - x)^3 + 1$

50. $h(x) = 2\sqrt{4 - x} + 5$

In Exercises 51–55, use the graph of $y = f(x)$ to sketch each function. Label at least three points on each graph.

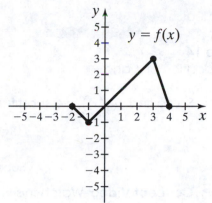

51. $y = -f(-x) - 1$

52. $y = \frac{1}{2}f(2 - x)$

53. $y = -2f(x + 1) + 2$

54. $y = 3 - 3f(x + 3)$

55. $y = -f(1 - x) - 2$

9.2 Composite and Inverse Functions

THINGS TO KNOW

Before working through this section, be sure you are familiar with the following concepts:

VIDEO   ANIMATION   INTERACTIVE

 1. Find the Domain and Range of a Relation
(Section 2.2, Objective 2)

 2. Evaluate Functions
(Section 2.3, Objective 2)

 3. Graph Simple Functions by Plotting Points
(Section 2.3, Objective 3)

OBJECTIVES

1 Form and Evaluate Composite Functions

2 Determine the Domain of Composite Functions

3 Determine If a Function Is One-to-One Using the Horizontal Line Test

4 Verify Inverse Functions

5 Sketch the Graphs of Inverse Functions

6 Find the Inverse of a One-to-One Function

....

OBJECTIVE 1   FORM AND EVALUATE COMPOSITE FUNCTIONS

My video summary ▶ Consider the functions $f(x) = x^2$ and $g(x) = 2x + 1$. How could we find $f(g(x))$?

The diagram in Figure 14 shows that given a number x, we first apply it to the function g to obtain $g(x)$. We then substitute $g(x)$ into f to get the result. The function $f(g(x))$ is called a *composite function* because one function is "composed" of another function.

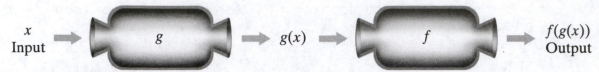

Figure 14
Composition of f and g

So, to find $f(g(x))$, we substitute $g(x)$ for x in the function f to get

$$f(g(x)) = f(2x + 1) = (2x + 1)^2.$$

Substitute $g(x)$ into f

My video summary ▶ **Concept Video** Watch this video to further explore this concept.

Definition   Composite Function

Given functions f and g, the **composite function**, $f \circ g$ (also called the **composition of f and g**), is defined by

$$(f \circ g)(x) = f(g(x)),$$

provided $g(x)$ is in the domain of f.

⚠ The composition of f and g does not equal the product of f and g: $(f \circ g)(x) \neq f(x)g(x)$. Also, the composition of f and g does not necessarily equal the composition of g and f, although this equality does exist for certain pairs of functions.

My interactive video summary 🕮 **Example 1 Forming and Evaluating Composite Functions**

Let $f(x) = 4x + 1$, $g(x) = \dfrac{x}{x - 2}$ and $h(x) = \sqrt{x + 3}$.

a. Find the function $f \circ g$.

b. Find the function $g \circ h$.

c. Find the function $h \circ f \circ g$.

d. Evaluate $(f \circ g)(4)$, or state that it is undefined.

e. Evaluate $(g \circ h)(1)$, or state that it is undefined.

f. Evaluate $(h \circ f \circ g)(6)$, or state that it is undefined.

Solutions

a. Writing $(f \circ g)(x)$ as $f(g(x))$, we find the composition by substituting the function g for x in the function f. Since $g(x) = \dfrac{x}{x - 2}$, this means we substitute $\dfrac{x}{x - 2}$ for x in f.

$$(f \circ g)(x) = f(g(x))$$

$$g(x) = \frac{x}{x-2}: \quad = f\left(\frac{x}{x-2}\right)$$

Substitute $\dfrac{x}{x-2}$ for x in f: $\quad = 4\left(\dfrac{x}{x-2}\right) + 1$

Simplify: $\quad = \dfrac{4x}{x-2} + \dfrac{x-2}{x-2}$

$$= \frac{5x-2}{x-2}$$

b.–c. Try finding these compositions on your own. Check your **answers**, or watch this **interactive video** to see solutions to all parts.

d. Rewriting $(f \circ g)(4)$ as $f\big(g(4)\big)$, we first need to find $g(4)$.

$$g(4) = \frac{4}{4-2} = \frac{4}{2} = 2$$

Next, we substitute this value for x in $f(x)$.

$$f(2) = 4(2) + 1 = 8 + 1 = 9$$

Therefore, $(f \circ g)(4) = 9$.

Alternatively, we could have used the result from part a to find $(f \circ g)(4)$ directly by substituting 4 for x in that result.

$$(f \circ g)(4) = \frac{5(4)-2}{4-2} = \frac{18}{2} = 9$$

e.–f. Try finding these compositions on your own. Check your **answers**, or watch this **interactive video** to see solutions to all parts.

 You Try It Work through this You Try It problem.

Work Exercises 1–16 in this eText or in the MyMathLab Study Plan.

My interactive video summary 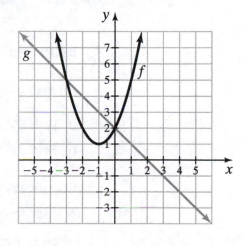 **Example 2 Evaluating Composite Functions Using a Graph**

Use the graph to evaluate each expression:

a. $(f \circ g)(4)$

b. $(g \circ f)(-3)$

c. $(f \circ f)(-1)$

d. $(g \circ g)(4)$

e. $(f \circ g \circ f)(1)$

Solutions

a. To evaluate $(f \circ g)(4) = f(g(4))$, we first need to find the value of $g(4)$. Using the graph of g, we locate the input 4 along the x-axis and then move down to the graph to see that $g(4) = -2$. Next, use this result as the input for $f(x)$. Locating the input -2 along the x-axis and moving up to the graph of f, we find that $f(-2) = 2$. Therefore,

$$(f \circ g)(4) = f(g(4)) = 2$$

b. To evaluate $(g \circ f)(-3) = g(f(-3))$, we first need to find the value of $f(-3)$. Using the graph of f, locate the input -3 along the x-axis and then move up to the graph to see that $f(-3) = 5$. Next, use this result as the input for $g(x)$. Locating the input 5 along the x-axis and moving down to the graph of g, we find that $g(5) = -3$. Therefore,

$$(g \circ f)(-3) = g(f(-3)) = -3$$

c.–e. Try evaluating these expressions on your own. Check your **answers**, or watch this **interactive video** for complete solutions to all parts.

You Try It Work through this You Try It problem.

Work Exercises 17 and 18 in this eText or in the MyMathLab **Study Plan.**

OBJECTIVE 2 DETERMINE THE DOMAIN OF COMPOSITE FUNCTIONS

Suppose f and g are **functions**. For a number x to be in the **domain** of $f \circ g$, x must be in the domain of g *and* $g(x)$ must be in the domain of f.

My video summary ▶ **Concept Video** Watch this **video** to further explore the domain of a composite function.

Finding the Domain of a Composite Function

Step 1 Find the domain of g.

Step 2 Exclude from the domain of g all values of x for which $g(x)$ is not in the domain of f.

My interactive video summary **Example 3 Finding the Domain of a Composite Function**

Let $f(x) = \dfrac{-10}{x - 4}$ and $g(x) = \sqrt{5 - x}$.

a. Find the domain of $f \circ g$.

b. Find the domain of $g \circ f$.

Solutions

a. First, form the composite function $(f \circ g)(x) = \dfrac{-10}{\sqrt{5 - x} - 4}$. To find the domain of $f \circ g$, follow these two steps:

Step 1 The domain of $g(x) = \sqrt{5-x}$ is $(-\infty, 5]$. The domain of $f \circ g$ cannot contain any values of x that are not in this interval. In other words, the domain of $f \circ g$ is a **subset** of $(-\infty, 5]$.

Step 2 All real numbers except 4 are in the domain of f. This implies that $g(x)$ cannot equal 4 because $g(x)$ equal to 4 would make the denominator of f, $x - 4$, equal to zero. Thus, we must exclude all values of x such that $g(x) = 4$.

$$g(x) = 4$$

Substitute $\sqrt{5-x}$ for $g(x)$: $\quad \sqrt{5-x} = 4$

Square both sides: $\quad 5 - x = 16$

Solve for x: $\quad x = -11$

We must *exclude* $x = -11$ from the domain of $f \circ g$. Therefore, the domain of $f \circ g$ is all values of x less than 5 such that $x \neq -11$, or the interval $(-\infty, -11) \cup (-11, 5]$. Watch the **interactive video** to view the solution in more detail.

b. Carefully work through the **interactive video** to verify that the domain of

$$(g \circ f)(x) = \sqrt{5 + \frac{10}{x-4}} = \sqrt{\frac{5x - 10}{x - 4}} \text{ is the interval } (-\infty, 2] \cup (4, \infty).$$

You Try It Work through this You Try It problem.

Work Exercises 19–24 in this eText or in the MyMathLab Study Plan.

OBJECTIVE 3 DETERMINE IF A FUNCTION IS ONE-TO-ONE USING THE HORIZONTAL LINE TEST

Later in this section, we examine a process for finding the inverse of a function. Note that we can find the inverse of many functions using this process, but that inverse will not always be a function. In this text, we are only interested in inverses that are functions, so we first develop a test to determine whether a function has an inverse *function*. When the word *inverse* is used throughout the remainder of this section, we assume that we are referring to the inverse that is a function.

My video summary ▶ **Concept Video** Watch this video to learn about **one-to-one functions**.

Definition One-to-One Function

A function f is **one-to-one** if for any values $a \neq b$ in the domain of f, $f(a) \neq f(b)$.

This definition suggests that a function is one-to-one if for any two *different* input values (domain values), the corresponding output values (range values) must be different.

An alternate definition says that if two range values are the same, $f(u) = f(v)$, then the domain values must be the same; that is, $u = v$.

> **Alternate Definition of a One-to-One Function**
>
> A function f is **one-to-one** if for any two range values $f(u)$ and $f(v), f(u) = f(v)$ implies that $u = v$.

The function sketched in Figure 15 is one-to-one because for any two distinct x-values in the domain ($a \neq b$), the function values or range values are not equal ($f(a) \neq f(b)$). In Figure 16, we see that the function $y = g(x)$ is *not* one-to-one because we can easily find two different domain values that correspond to the same range value.

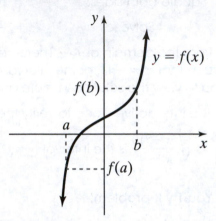

Figure 15
Example of a One-to-
One Function

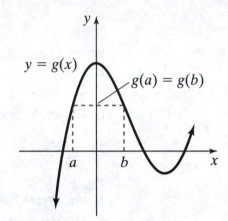

Figure 16
Example of a Function That Is Not
One-to-One

My video summary ⏵ **Concept Video** Watch this **video** to learn how horizontal lines can be used to determine if a function is one-to-one.

> **Horizontal Line Test**
>
> If every horizontal line intersects the graph of a function f at most once, then f is one-to-one.

My animation summary ✴ **Example 4 Determining Whether a Function Is One-to-One**

Determine whether each function is one-to-one.

a.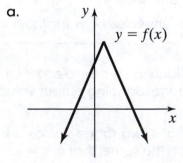

b.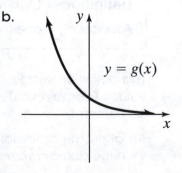

c. $f(x) = x^2 + 1, x \le 0$

d. $f(x) = \begin{cases} 2x + 4 \text{ for } x \le -1 \\ 2x - 6 \text{ for } x \ge 4 \end{cases}$

Solutions Work through this problem on your own, or watch this **animation**. Note that the functions in parts b and c are one-to-one, while the functions in parts a and d are not one-to-one.

You Try It Work through this You Try It problem.

Work Exercises 25–39 in this eText or in the MyMathLab Study Plan.

OBJECTIVE 4 VERIFY INVERSE FUNCTIONS

We are now ready to ask ourselves the question, "Why should we be concerned with one-to-one functions?"

Answer: Every one-to-one function has an inverse function!

My video summary ▶ **Concept Video** Watch this video to learn more about inverse functions.

> **Definition** Inverse Function
>
> Let f be a one-to-one function with domain A and range B. Then f^{-1} is the **inverse function of** f with domain B and range A. Furthermore, if $f(a) = b$, then $f^{-1}(b) = a$.

⚠ Do not confuse f^{-1} with $\dfrac{1}{f(x)}$. The -1 in f^{-1} is *not* an exponent! It is simply notation to indicate an inverse function.

According to the definition of an inverse function, the domain of f is exactly the same as the range of f^{-1}, and the range of f is the same as the domain of f^{-1}. Figure 17 illustrates that if the function f assigns a number a to b, then the inverse function will assign b back to a. In other words, if the point (a, b) is an ordered pair on the graph of f, then the point (b, a) must be on the graph of f^{-1}.

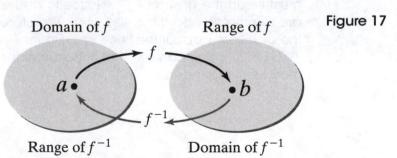

Figure 17

My video summary ▶ **Concept Video** Watch this video to see how this "undoing" nature can be used to verify two functions are inverses of each other.

As with our opening example using an inverse function to convert a Fahrenheit temperature back into a Celsius temperature (see the Concept Video above related to inverse functions), Figure 17 illustrates that inverse functions "undo" each other.

Composition Cancellation Equations

$f\left(f^{-1}(x)\right) = x$ for all x in the domain of f^{-1}

and

$f^{-1}\left(f(x)\right) = x$ for all x in the domain of f

These cancellation equations can be used to show whether two functions are inverses of each other. For example, if $f(x) = x^3$ and $f^{-1}(x) = \sqrt[3]{x}$, then

$$f\left(f^{-1}(x)\right) = f\left(\sqrt[3]{x}\right) = \left(\sqrt[3]{x}\right)^3 = x \quad \text{and} \quad f^{-1}\left(f(x)\right) = f^{-1}(x^3) = \sqrt[3]{x^3} = x.$$

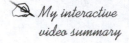 *My interactive video summary*

Example 5 Verify Inverse Functions

Show that $f(x) = \dfrac{x}{2x + 3}$ and $g(x) = \dfrac{3x}{1 - 2x}$ are inverse functions using the composition cancellation equations.

Solution To show that f and g are inverses of each other, we must show that $(f \circ g)(x) = x$ and $(g \circ f)(x) = x$. Work through this example on your own. Then watch this **interactive video** to verify that both composition cancellation equations are satisfied.

You Try It Work through this **You Try It** problem.

Work Exercises 40–44 in this eText or in the MyMathLab Study Plan.

OBJECTIVE 5   SKETCH THE GRAPHS OF INVERSE FUNCTIONS

If f is a **one-to-one** function, then we know that it must have an **inverse function**, f^{-1}. Given the graph of a one-to-one function f, we can obtain the graph of f^{-1} by simply interchanging the coordinates of each ordered pair that lies on the graph of f. In other words, for any point (a, b) on the graph of f, the point (b, a) must lie on the graph of f^{-1}. Notice in Figure 18 that the points (a, b) and (b, a) are **symmetric** about the line $y = x$. Therefore, the graph of f^{-1} is a reflection of the graph of f about the line $y = x$. Figure 19 shows the graphs of $f(x) = x^3$ and $f^{-1}(x) = \sqrt[3]{x}$. You can see that if the functions have any points in common, they must lie along the line $y = x$.

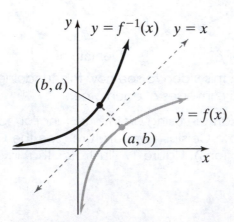

Figure 18
Graph of a One-to-One Function and Its Inverse

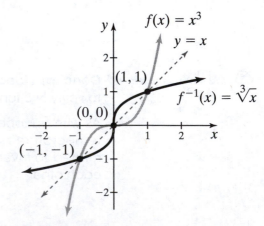

Figure 19
Graph of $f(x) = x^3$ and $f^{-1}(x) = \sqrt[3]{x}$

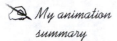 *My animation summary*

⊛ **Example 6** Sketch the Graph of a One-to-One Function and Its Inverse

Sketch the graph of $f(x) = x^2 + 1, x \leq 0$, and its inverse. Also state the domain and range of f and f^{-1}.

Solution Work through this **animation**, or read the following solution

The graphs of f and f^{-1} are sketched in Figure 20. Notice how the graph of f^{-1} is a reflection of the graph of f about the line $y = x$. Also notice that the domain of f is the same as the range of f^{-1}. Likewise, the domain of f^{-1} is equivalent to the range of f.

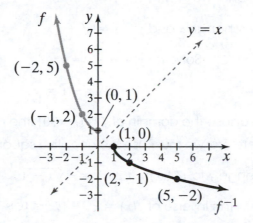

Figure 20
Graph of
$f(x) = x^2 + 1, x \leq 0$,
and Its Inverse

Domain of f: $(-\infty, 0]$ Domain of f^{-1}: $[1, \infty)$
Range of f: $[1, \infty)$ Range of f^{-1}: $(-\infty, 0]$

Using a TI-84 Plus C, we can sketch the functions by letting $y_1 = (x^2 + 1)/(x \leq 0)$. We can draw the inverse function by typing the command **DrawInv Y_1** in the calculator's main viewing window. The graph of $y = x$ is included to show the symmetry. See Figure 21.

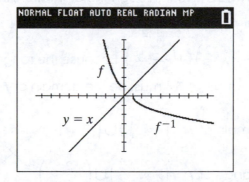

Figure 21

You Try It Work through this You Try It problem.

Work Exercises 45–50 in this eText or in the MyMathLab Study Plan.

OBJECTIVE 6   FIND THE INVERSE OF A ONE-TO-ONE FUNCTION

My video summary ▶ We are now ready to find the inverse of a **one-to-one function** algebraically. We know that if a point (x, y) is on the graph of a one-to-one function, then the point (y, x) is on the graph of its inverse function. We can use this information to develop a process for finding the inverse of a function algebraically simply by switching x and y in the original function to produce its inverse function.

We use as a motivating example the function $f(x) = x^2 + 1, x \le 0$, discussed in Example 6. To find the inverse of a one-to-one function, we follow the four-step process outlined here.

Step 1   Change $f(x)$ to y: $y = x^2 + 1$

Step 2   Interchange x and y: $x = y^2 + 1$

Step 3   Solve for y: $x - 1 = y^2$

$$\pm\sqrt{x - 1} = y$$

(Because the domain of f is $(-\infty, 0]$, the range of f^{-1} must be $(-\infty, 0]$. Therefore, we must use the negative square root or $y = -\sqrt{x - 1}$.)

Step 4 Change y to $f^{-1}(x)$: $f^{-1}(x) = -\sqrt{x - 1}$

Thus, the inverse of $f(x) = x^2 + 1, x \le 0$, is $f^{-1}(x) = -\sqrt{x - 1}$.

My animation summary   ✳ **Example 7 Find the Inverse of a Function**

Find the **inverse** of the function $f(x) = \dfrac{2x}{1 - 5x}$, and state the **domain** and **range** of f and f^{-1}.

Solution Work through the animation, and follow the four-step process to verify that $f^{-1}(x) = \dfrac{x}{5x + 2}$. The domain of f is $\left(-\infty, \dfrac{1}{5}\right) \cup \left(\dfrac{1}{5}, \infty\right)$, while the domain of f^{-1} is $\left(-\infty, -\dfrac{2}{5}\right) \cup \left(-\dfrac{2}{5}, \infty\right)$. Because the range of f must be the domain of f^{-1} and the range of f^{-1} must be the domain of f, we get the following result:

Domain of f: $\left(-\infty, \dfrac{1}{5}\right) \cup \left(\dfrac{1}{5}, \infty\right)$ Domain of f^{-1}: $\left(-\infty, -\dfrac{2}{5}\right) \cup \left(-\dfrac{2}{5}, \infty\right)$

Range of f: $\left(-\infty, -\dfrac{2}{5}\right) \cup \left(-\dfrac{2}{5}, \infty\right)$ Range of f^{-1}: $\left(-\infty, \dfrac{1}{5}\right) \cup \left(\dfrac{1}{5}, \infty\right)$

You Try It   Work through this **You Try It** problem.

Work Exercises 51–59 in this eText or in the MyMathLab Study Plan.

Summary of Key Points About Inverse Functions

1. f^{-1} exists if and only if the function f is one-to-one.

2. The **domain** of f is the same as the **range** of f^{-1}, and the range of f is the same as the domain of f^{-1}.

3. To verify that two one-to-one functions, f and g, are **inverses of each other**, we must use the **composition cancellation equations** to show that $f(g(x)) = g(f(x)) = x$.

4. The graph of f^{-1} is a reflection of the graph of f about the line $y = x$. That is, for any point (a, b) that lies on the graph of f, the point (b, a) must lie on the graph of f^{-1}.

5. To find the inverse of a one-to-one function, replace $f(x)$ with y, interchange the variables x and y, and solve for y. This is the function $f^{-1}(x)$.

9.2 Exercises

In Exercises 1–8, let $f(x) = 3x + 1$, $g(x) = \dfrac{2}{x + 1}$, and $h(x) = \sqrt{x + 3}$.

1. Find the function $f \circ g$.

2. Find the function $g \circ f$.

3. Find the function $f \circ h$.

4. Find the function $g \circ h$.

5. Find the function $h \circ f$.

6. Find the function $h \circ g$.

7. Find the function $f \circ f$.

8. Find the function $g \circ g$.

In Exercises 9–16, evaluate the following composite functions given that

$f(x) = 3x + 1$, $g(x) = \dfrac{2}{x + 1}$, and $h(x) = \sqrt{x + 3}$.

9. $(f \circ g)(0)$

10. $(f \circ h)(6)$

11. $(g \circ f)(1)$

12. $(g \circ h)(-2)$

13. $(h \circ f)(0)$

14. $(h \circ g)(3)$

15. $(f \circ f)(-1)$

16. $(g \circ g)(4)$

In Exercises 17 and 18, use the graph to evaluate each expression.

17. a. $(f \circ g)(1)$ b. $(g \circ f)(-1)$ 18. a. $(f \circ g)(1)$ b. $(g \circ f)(-1)$

    c. $(g \circ g)(0)$ d. $(f \circ f)(1)$ c. $(g \circ g)(0)$ d. $(f \circ f)(1)$

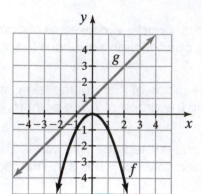
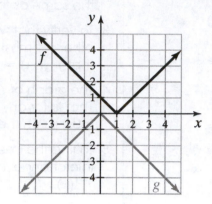

In Exercises 19–24, find the domain of $(f \circ g)(x)$ and $(g \circ f)(x)$.

19. $f(x) = x^2, g(x) = 2x - 1$ 20. $f(x) = 3x - 5, g(x) = 2x^2 + 1$

21. $f(x) = x^2, g(x) = \sqrt{x}$ 22. $f(x) = \dfrac{1}{x}, g(x) = x^2 - 4$

23. $f(x) = \dfrac{3}{x + 1}, g(x) = \dfrac{x}{x - 2}$ 24. $f(x) = \dfrac{2x}{x - 3}, g(x) = \dfrac{x + 1}{x - 1}$

In Exercises 25–39, determine whether each function is one-to-one.

25. $f(x) = 3x - 1$ 26. $f(x) = 2x^2$ 27. $f(x) = (x - 1)^2, x \geq 1$

28. $f(x) = (x - 1)^2, x \geq -1$ 29. $f(x) = \dfrac{1}{x} - 2$ 30. $f(x) = 4\sqrt{x}$

31. $f(x) = -2|x|$ 32. $f(x) = 2$ 33. $f(x) = (x + 1)^3 - 2$

34.

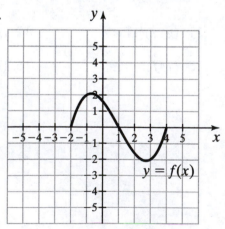

35.

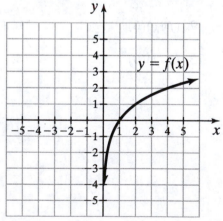

36.

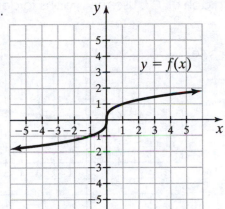

37.

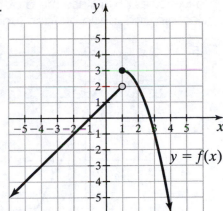

38.

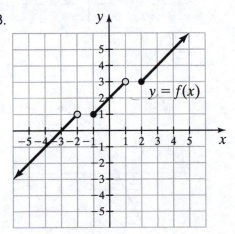

39.

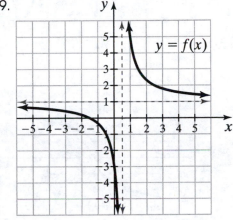

In Exercises 40–44, use the composition cancellation equations to verify that f and g are inverse functions.

40. $f(x) = \dfrac{3}{2}x - 4$ and $g(x) = \dfrac{2x + 8}{3}$

41. $f(x) = (x - 1)^2, x \geq 1$, and $g(x) = \sqrt{x} + 1$

42. $f(x) = \dfrac{7}{x + 1}$ and $g(x) = \dfrac{7 - x}{x}$

43. $f(x) = \dfrac{x}{5 + 3x}$ and $g(x) = \dfrac{5x}{1 - 3x}$

44. $f(x) = 2\sqrt[3]{x - 1} + 3$ and $g(x) = \dfrac{(x - 3)^3}{8} + 1$

In Exercises 45–50, use the graph of f to sketch the graph of f^{-1}. Use the graphs to determine the domain and range of each function.

45.

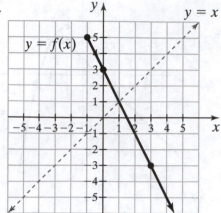

46.

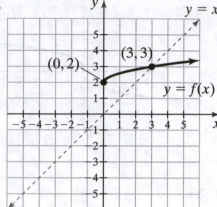

47.

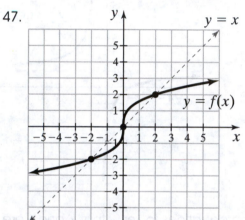

48.

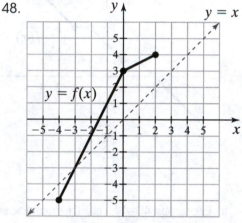

49.

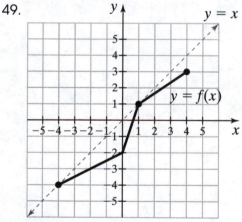

50.

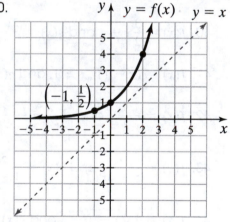

In Exercises 51–59, write an equation for the inverse function, and then state the domain and range of f and f^{-1}.

51. $f(x) = \dfrac{1}{3}x - 5$

52. $f(x) = \dfrac{3x + 9}{7}$

53. $f(x) = \sqrt[3]{2x - 3}$

54. $f(x) = 1 - \sqrt[5]{x + 4}$

55. $f(x) = -x^2 - 2, x \geq 0$

56. $f(x) = (x + 3)^2 - 5, x \leq -3$

57. $f(x) = \dfrac{3}{x}$

58. $f(x) = \dfrac{1 - x}{2x}$

59. $f(x) = \dfrac{8x - 1}{7 - 5x}$

9.3 Exponential Functions

THINGS TO KNOW

Before working through this section, be sure you are familiar with the following concepts:

VIDEO   ANIMATION   INTERACTIVE

 You Try It

1. Use Combinations of Transformations to Graph Functions (Section 9.1, Objective 6)

 You Try It

2. Determine If a Function Is One-to-One Using the Horizontal Line Test (Section 9.2, Objective 3)

OBJECTIVES

1 Use the Characteristics of Exponential Functions

2 Sketch the Graphs of Exponential Functions Using Transformations

3 Solve Exponential Equations by Relating the Bases

4 Solve Applications of Exponential Functions

OBJECTIVE 1   USE THE CHARACTERISTICS OF EXPONENTIAL FUNCTIONS

Many natural phenomena and real-life applications can be modeled using *exponential functions*. Before we define the exponential function, it is important to remember how to manipulate exponential expressions because this skill is necessary when solving certain equations involving exponents. Previously, expressions of the form b^r were evaluated for **rational numbers** r. For example,

$$3^2 = 9, \quad 4^{-2} = \frac{1}{4^2} = \frac{1}{16}, \quad \text{and} \quad 27^{-2/3} = \frac{1}{27^{2/3}} = \frac{1}{\left(\sqrt[3]{27}\right)^2} = \frac{1}{(3)^2} = \frac{1}{9}.$$

In this section, we extend the meaning of b^r to include *all* real values of r by defining the exponential function $f(x) = b^x$.

Definition   Exponential Function

An **exponential function** is a function of the form $f(x) = b^x$, where x is any real number and $b > 0$ such that $b \neq 1$.

The constant, b, is called the base of the exponential function.

Notice in the definition that the base, b, must be positive and must not equal 1. View this **explanation** to see why.

My video summary ▶ **Concept Video** Before we generalize the graph of $f(x) = b^x$, watch this video to explore some basic properties of exponential functions.

Figure 22 shows the graph of $y = b^x$ for $b = 2, 3, \frac{1}{2}$, and $\frac{1}{3}$. You can see from the graphs that all four functions intersect the y-axis at the point $(0, 1)$. This is true because $b^0 = 1$ for all nonzero values of b. For values of $b > 1$, the graph of $y = b^x$ increases rapidly as the values of x approach positive infinity ($b^x \to \infty$ as $x \to \infty$). In fact, the larger the base, the faster the graph will grow. Also, for $b > 1$, the graph of $y = b^x$ decreases quickly, approaching 0 as the values of x approach negative infinity ($b^x \to 0$ as $x \to -\infty$). Thus, the x-axis (the line $y = 0$) is a **horizontal asymptote**.

However, for $0 < b < 1$, the graph decreases quickly, approaching the horizontal asymptote $y = 0$ as the values of x approach positive infinity ($b^x \to 0$ as $x \to \infty$), while the graph increases rapidly as the values of x approach negative infinity ($b^x \to \infty$ as $x \to -\infty$). The preceding statements, along with some other characteristics of the graphs of exponential functions, are summarized on the following page.

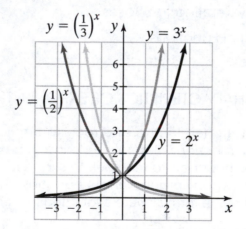

Figure 22
Graphs of $y = 2^x, y = 3^x$,

$y = \left(\frac{1}{2}\right)^x$, and $y = \left(\frac{1}{3}\right)^x$

Characteristics of Exponential Functions

For $b > 0, b \neq 1$, the **exponential function with base** b is defined by $f(x) = b^x$.

The **domain** of $f(x) = b^x$ is $(-\infty, \infty)$, and the **range** is $(0, \infty)$. The graph of $f(x) = b^x$ has one of the following two shapes:

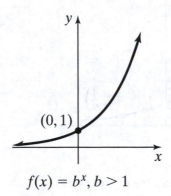

$f(x) = b^x, b > 1$

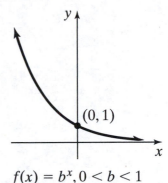

$f(x) = b^x, 0 < b < 1$

The graph intersects the y-axis at $(0, 1)$.

$b^x \to \infty$ as $x \to \infty$

$b^x \to 0$ as $x \to -\infty$

The line $y = 0$ is a **horizontal asymptote**.

The function is **one-to-one**.

The graph intersects the y-axis at $(0, 1)$.

$b^x \to 0$ as $x \to \infty$

$b^x \to \infty$ as $x \to -\infty$

The line $y = 0$ is a **horizontal asymptote**.

The function is **one-to-one**.

My video summary ▶ **Example 1 Sketching the Graph of an Exponential Function**

Sketch the graph of $f(x) = \left(\dfrac{2}{3}\right)^x$.

Solution Read the following solution, or watch this **video**.

Because the base of the **exponential function** is $\dfrac{2}{3}$, which is between 0 and 1, the graph must approach the x-axis as the value of x approaches positive infinity. The graph intersects the y-axis at $(0, 1)$. We can find a few more points by choosing some negative and positive values of x:

$$f(-2) = \left(\frac{2}{3}\right)^{-2} = \left(\frac{3}{2}\right)^{2} = \frac{9}{4}$$

$$f(-1) = \left(\frac{2}{3}\right)^{-1} = \left(\frac{3}{2}\right)^{1} = \frac{3}{2}$$

$$f(1) = \left(\frac{2}{3}\right)^{1} = \frac{2}{3}$$

$$f(2) = \left(\frac{2}{3}\right)^{2} = \frac{4}{9}$$

We can complete the graph by connecting the points with a smooth curve. See Figure 23.

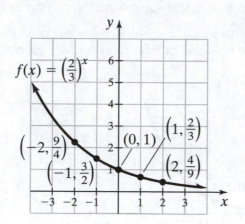

Figure 23

Graph of $f(x) = \left(\dfrac{2}{3}\right)^x$

You Try It Work through this You Try It problem.

Work Exercises 1–5 in this eText or in the MyMathLab Study Plan.

My video summary ▶ **Example 2 Determining an Exponential Function Given the Graph**

Find the exponential function $f(x) = b^x$ with the graph shown.

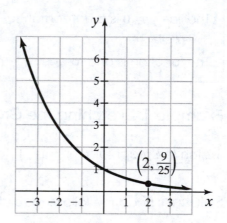

Solution Read through the following solution, or watch this **video**.

From the point $\left(2, \dfrac{9}{25}\right)$, we see that $f(2) = \dfrac{9}{25}$. Thus,

Write the exponential function $f(x) = b^x$: $f(x) = b^x$

Evaluate $f(2)$: $f(2) = b^2$

The graph contains the point $\left(2, \dfrac{9}{25}\right)$: $f(2) = \dfrac{9}{25}$

Equate the two expressions for $f(2)$: $b^2 = \dfrac{9}{25}$

Therefore, we are looking for a constant b such that $b^2 = \dfrac{9}{25}$. Using the **square**

root property, we get

$$\sqrt{b^2} = \pm\sqrt{\dfrac{9}{25}}$$

$$b = \pm\dfrac{3}{5}.$$

By definition of an **exponential function**, $b > 0$; thus, $b = \dfrac{3}{5}$. Therefore, this is the

graph of $f(x) = \left(\dfrac{3}{5}\right)^x$.

You Try It Work through this You Try It problem.

Work Exercises 6–12 in this eText or in the MyMathLab Study Plan.

OBJECTIVE 2   SKETCH THE GRAPHS OF EXPONENTIAL FUNCTIONS USING TRANSFORMATIONS

We can use the **transformation techniques** discussed in **Section 9.1** to sketch the graphs of exponential functions. For example, the graph of $f(x) = 3^x - 1$ can be obtained by vertically shifting the graph of $y = 3^x$ down one unit. Note in Figure 24 that the y-intercept of $f(x) = 3^x - 1$ is $(0, 0)$ and the **horizontal asymptote** is the line $y = -1$.

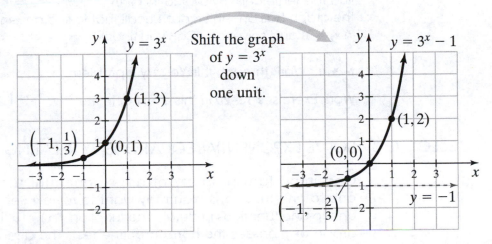

Figure 24
The graph of $f(x) = 3^x - 1$ can be obtained by vertically shifting the graph of $y = 3^x$ down one unit.

My video summary ▶ **Example 3** Using Transformations to Sketch an Exponential Function

Use transformations to sketch the graph of $f(x) = -2^{x+1} + 3$.

Solution Read through the following solution, or watch this **video**.

Starting with the graph of $y = 2^x$, we can obtain the graph of $f(x) = -2^{x+1} + 3$ through a series of three transformations:

1. Horizontally shift the graph of $y = 2^x$ to the left one unit, producing the graph of $y = 2^{x+1}$.

2. Reflect the graph of $y = 2^{x+1}$ about the x-axis, producing the graph of $y = -2^{x+1}$.

3. Vertically shift the graph of $y = -2^{x+1}$ up three units, producing the graph of $f(x) = -2^{x+1} + 3$.

The graph of $f(x) = -2^{x+1} + 3$ is shown in Figure 25.

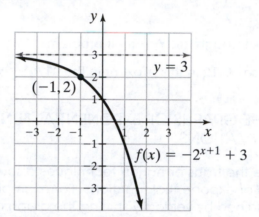

Figure 25
Graph of $f(x) = -2^{x+1} + 3$

Notice that the graph of $f(x) = -2^{x+1} + 3$ in **Figure 25** has a y-intercept. We can find the y-intercept by evaluating $f(0) = -2^{0+1} + 3 = -2 + 3 = 1$. Also notice that the graph has an x-intercept. Recall that to find an x-intercept, we need to set $f(x) = 0$ and solve for x. Can you find it?

You Try It Work through this You Try It problem.

Work Exercises 13–20 in this eText or in the MyMathLab Study Plan.

OBJECTIVE 3 SOLVE EXPONENTIAL EQUATIONS BY RELATING THE BASES

One important property of all **exponential functions** is that they are **one-to-one functions**. You may want to review **Section 9.2**, which discusses one-to-one functions in detail. The function $f(x) = b^x$ is one-to-one because the graph of f passes the **horizontal line test**. The **alternate definition of one-to-one** can be used along with letting $f(x) = b^x$, to say that if $b^u = b^v$, then $u = v$. In other words, if the bases of an exponential equation of the form $b^u = b^v$ are the same, then the exponents must be the same. Solving exponential equations with this property is known as the **method of relating the bases** for solving exponential equations.

Method of Relating the Bases

If an exponential equation can be written in the form $b^u = b^v$, then $u = v$.

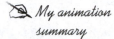
My animation summary

⊛ **Example 4** Using the Method of Relating the Bases to Solve Exponential Equations

Solve the following equations.

a. $8 = \dfrac{1}{16^x}$　　　　b. $\dfrac{1}{27^x} = (\sqrt[4]{3})^{x-2}$

Solutions

a.–b. Try solving these equations on your own. Check your **answers**, or watch this **animation** for complete solutions to both parts.

You Try It Work through this **You Try It** problem.

Work Exercises 21–30 in this eText or in the MyMathLab Study Plan.

OBJECTIVE 4　SOLVE APPLICATIONS OF EXPONENTIAL FUNCTIONS

Exponential functions are used to describe many real-life situations and natural phenomena.

My video summary ▶ **Example 5** Learn to Hit a 3-Wood on a Golf Driving Range

Most golfers find that their golf skills improve dramatically at first and then level off rather quickly. For example, suppose that the distance (in yards) that a typical beginning golfer can hit a 3-wood after t weeks of practice on the driving range is given by the exponential function $d(t) = 225 - 100(2.7)^{-0.7t}$. This function has been developed after many years of gathering data on beginning golfers.

How far can a typical beginning golfer initially hit a 3-wood? How far can a typical beginning golfer hit a 3-wood after 1 week of practice on the driving range? After 5 weeks? After 9 weeks? Round to the nearest hundredth yard.

Solution Read through the following solution, or watch this **video**. Initially, when $t = 0$, $d(0) = 225 - 100(2.7)^0 = 225 - 100 = 125$ yards. Therefore, a typical beginning golfer can hit a 3-wood 125 yards.

After 1 week of practice on the driving range, a typical beginning golfer can hit a 3-wood $d(1) = 225 - 100(2.7)^{-0.7(1)} \approx 175.11$ yards. After 5 weeks, $d(5) = 225 - 100(2.7)^{-0.7(5)} \approx 221.91$ yards. After 9 weeks, $d(9) = 225 - 100(2.7)^{-0.7(9)} \approx 224.81$ yards.

Using a graphing utility, we can sketch the graph of $d(t) = 225 - 100(2.7)^{-0.7t}$. See Figure 26. The distance increases rather quickly and then tapers off toward a horizontal asymptote of 225 yards.

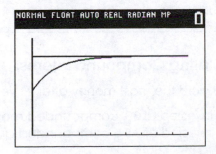

Figure 26
A TI-84 Plus C was used to sketch the function $d(t) = 225 - 100(2.7)^{-0.7t}$ and the horizontal asymptote $y = 225$.

You Try It   Work through this You Try It problem.

Work Exercises 31–34 in this eText or in the MyMathLab Study Plan.

COMPOUND INTEREST

A real-life application of exponential functions is the concept of **compound interest**, or interest that is paid on *both principal and interest*. First, we take a look at how **simple interest** is accrued. If an investment of P dollars is invested at r percent annually (written as a decimal) using simple interest, then the interest earned after 1 year is Pr dollars. Adding this interest to the original investment yields a total amount, A, of

$$A = \underbrace{P}_{\substack{\text{Original} \\ \text{investment}}} + \underbrace{Pr}_{\substack{\text{Interest} \\ \text{earned}}} = P(1 + r).$$

If this amount is reinvested at the same interest rate, then the total amount after 2 years becomes

$$A = \underbrace{P(1 + r)}_{\substack{\text{Total investment} \\ \text{after 1 year}}} + \underbrace{P(1 + r)r}_{\substack{\text{Interest} \\ \text{earned}}} = P(1 + r)(1 + r) = P(1 + r)^2.$$

Reinvesting this amount for a third year gives an amount of $P(1 + r)^3$. Continuing this process for k years, we can see that the amount becomes $A = P(1 + r)^k$. This is an exponential function with base $1 + r$.

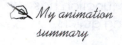 *My animation summary*

⊛ **Concept Animation** Watch this **animation** to see how the **compounded** simple interest formula can be modified for general situations involving periodic compound interest.

Periodic Compound Interest Formula

Periodic compound interest can be calculated using the formula

$$A = P\left(1 + \frac{r}{n}\right)^{nt}$$

where

A = Total amount after t years

P = Principal (original investment)

r = Interest rate per year

n = Number of times interest is compounded per year

t = Number of years

Example 6 Calculating Compound Interest

Which investment will yield the most money after 25 years?

Investment A: $12,000 invested at 3% compounded monthly
Investment B: $10,000 invested at 3.9% compounded quarterly

📝 *My video summary* ▶ **Solution** Work through the following solution, or watch this **video**.

Investment A: $P = 12,000, r = 0.03, n = 12, t = 25$:

$$A = 12,000\left(1 + \frac{0.03}{12}\right)^{12(25)} \approx \$25,380.23$$

Investment B: $P = 10,000, r = 0.039, n = 4, t = 25$:

$$A = 10,000\left(1 + \frac{0.039}{4}\right)^{4(25)} \approx \$26,386.77$$

Investment B will yield the most money after 25 years.

🔺 **You Try It** Work through this You Try It problem.

Work Exercises 35–37 in this eText or in the MyMathLab Study Plan.

PRESENT VALUE

Sometimes, investors want to know how much money to invest now in order to reach a certain investment goal in the future. This amount of money, P, is known as the **present value** of A dollars. To find a formula for present value, start with the formula for periodic compound interest and solve the formula for P. This formula is known as the **present value formula**. View the **derivation**.

Present Value Formula

Present value can be calculated using the formula

$$P = A\left(1 + \frac{r}{n}\right)^{-nt}$$

where

$P =$ Principal (original investment)

$A =$ Total amount after t years

$r =$ Interest rate per year

$n =$ Number of times interest is compounded per year

$t =$ Number of years

📝 *My video summary* ▶ **Example 7 Determining Present Value**

Find the present value of \$8000 if interest is paid at a rate of 5.6% compounded quarterly for 7 years. Round to the nearest cent.

Solution Work through the following solution, or watch this **video**.

Using the present value formula $P = A\left(1 + \frac{r}{n}\right)^{-nt}$ with $A = \$8000, r = 0.056, n = 4,$ and $t = 7$, we get

$$P = A\left(1 + \frac{r}{n}\right)^{-nt}$$

$$P = 8000\left(1 + \frac{0.056}{4}\right)^{-(4)(7)} \approx 5420.35.$$

Therefore, the present value of $8000 in 7 years at 5.6% compounded quarterly is $5420.35.

 You Try It **Work through this You Try It problem.**

Work Exercises 38–40 in this eText or in the MyMathLab **Study Plan.**

9.3 Exercises

In Exercises 1–5, sketch the graph of each exponential function. Label the y-intercept and at least two other points on the graph using both positive and negative values of x.

1. $f(x) = 4^x$

2. $f(x) = \left(\dfrac{1}{4}\right)^x$

3. $f(x) = \left(\dfrac{3}{2}\right)^x$

4. $f(x) = (0.4)^x$

5. $f(x) = (2.7)^x$

In Exercises 6–12, determine the correct exponential function of the form $f(x) = b^x$ whose graph is given.

6.

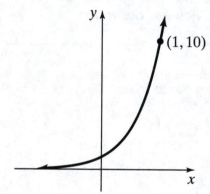

$(1, 10)$

7.

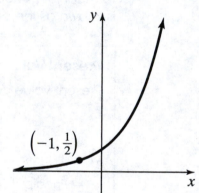

$\left(-1, \dfrac{1}{2}\right)$

8.

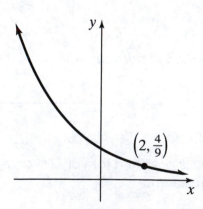

$\left(2, \dfrac{4}{9}\right)$

9.

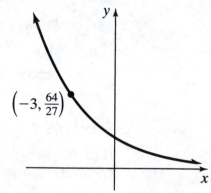

$\left(-3, \dfrac{64}{27}\right)$

10.

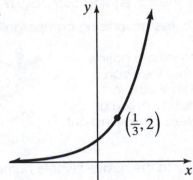

$\left(\frac{1}{3}, 2\right)$

11.

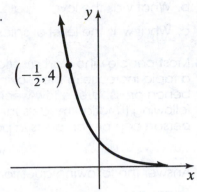

$\left(-\frac{1}{2}, 4\right)$

12.

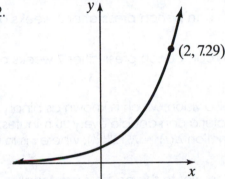

$(2, 7.29)$

In Exercises 13–20, use the graph of $y = 2^x$ or $y = 3^x$ and transformations to sketch each exponential function. Determine the domain and range. Also, determine the y-intercept, and find the equation of the horizontal asymptote.

13. $f(x) = 2^{x-1}$

14. $f(x) = 3^x - 1$

15. $f(x) = -3^{x+2}$

16. $f(x) = -2^{x+1} - 1$

17. $f(x) = \left(\frac{1}{2}\right)^{x+1}$

18. $f(x) = \left(\frac{1}{3}\right)^x - 3$

19. $f(x) = 2^{-x} + 1$

20. $f(x) = 3^{1-x} - 2$

In Exercises 21–30, solve each exponential equation using the method of "relating the bases" by first rewriting the equation in the form $b^u = b^v$.

21. $2^x = 16$

22. $3^{x-1} = \frac{1}{9}$

23. $\sqrt{5} = 25^x$

24. $\left(\sqrt[3]{3}\right)^x = 9$

25. $\frac{1}{\sqrt[5]{8}} = 2^x$

26. $\frac{9}{\sqrt[4]{3}} = \left(\frac{1}{27}\right)^x$

27. $(49)^x = \left(\frac{1}{7}\right)^{x-1}$

28. $\frac{125}{\sqrt[3]{5^x}} = \left(\frac{1}{25^x}\right)$

29. $\frac{3^{x^2}}{9^x} = 27$

30. $2^{x^3} = \frac{4^x}{2^{-x^2}}$

31. Typically, weekly sales will drop off rather quickly after the end of an advertising campaign. This drop in sales is known as *sales decay*. Suppose that the gross sales S, in hundreds of dollars, of a certain product is given by the exponential function $S(t) = 3000(1.5^{-0.3t})$, where t is the number of weeks after the end of the advertising campaign.

Answer the following questions, rounding each answer to the nearest whole number:

a. What was the level of sales immediately after the end of the advertising campaign when $t = 0$?

b. What was the level of sales 1 week after the end of the advertising campaign?

c. What was the level of sales 5 weeks after the end of the advertising campaign?

32. Most people who start a serious weight-lifting regimen initially notice a rapid increase in the maximum amount of weight that they can bench press. After a few weeks, this increase starts to level off. The following function models the maximum weight, w, that a particular person can bench press in pounds at the end of t weeks of working out.

$$w(t) = 250 - 120(2.7)^{-0.3t}$$

Answer the following questions, rounding each answer to the nearest whole number:

a. What is the maximum weight that this person can bench press initially?

b. What is the maximum weight that this person can bench press after 3 weeks of weight lifting?

c. What is the maximum weight that this person can bench press after 7 weeks of weight lifting?

33. *Escherichia coli* bacteria reproduce by simple cell division, which is known as binary fission. Under ideal conditions, a population of E. coli bacteria can double every 20 minutes. This behavior can be modeled by the exponential function $N(t) = N_0(2^{0.05t})$, where t is in minutes and N_0 is the initial number of E. coli bacteria.

Answer the following questions, rounding each answer to the nearest bacteria:

a. If the initial number of E. coli bacteria is five, how many bacteria will be present in 3 hours?

b. If the initial number of E. coli bacteria is eight, how many bacteria will be present in 3 hours?

c. If the initial number of E. coli bacteria is eight, how many bacteria will be present in 10 hours?

34. A wildlife-management research team noticed that a certain forest had no rabbits, so they decided to introduce a rabbit population into the forest for the first time. The rabbit population will be controlled by wolves and other predators. This rabbit population can

be modeled by the function $R(t) = \dfrac{960}{0.6 + 23.4(2.7)^{-0.045t}}$, where t is the number of weeks

after the research team first introduced the rabbits into the forest.

Answer the following questions, rounding each answer to the nearest whole number:

a. How many rabbits did the wildlife-management research team bring into the forest?

b. How many rabbits can be expected after 10 weeks?

c. How many rabbits can be expected after the first year?

d. What is the expected rabbit population after 4 years? 5 years? What can the expected rabbit population approach as time goes on?

Use the **periodic compound interest formula** to solve Exercises 35–37.

35. Suppose that $9000 is invested at 3.5% compounded quarterly. Find the total amount of this investment after 10 years. Round to the nearest cent.

36. Suppose that you have $5000 to invest. Which investment yields the greater return over a 10-year period: 7.35% compounded daily or 7.4% compounded quarterly?

37. Which investment yields the greatest return?

Investment A: $4000 invested for 5 years compounded semiannually (twice per year) at 8%

Investment B: $5000 invested for 4 years compounded quarterly at 4.5%

Use the **present value formula** to solve Exercises 38–40.

38. Find the present value of $10,000 if interest is paid at a rate of 4.5% compounded semiannually for 12 years. Round to the nearest cent.

39. Find the present value of $1,000,000 if interest is paid at a rate of 9.5% compounded monthly for 8 years. Round to the nearest cent.

40. How much money would you have to invest at 10% compounded semiannually so that the total investment had a value of $2205 after 1 year? Round to the nearest cent.

9.4 The Natural Exponential Function

THINGS TO KNOW

Before working through this section, be sure you are familiar with the following concepts:

| | VIDEO | ANIMATION | INTERACTIVE |

You Try It 1. Sketch the Graphs of Exponential Functions Using Transformations (Section 9.3, Objective 2) ▶

You Try It 2. Solve Exponential Equations by Relating the Bases (Section 9.3, Objective 3) ✦

You Try It 3. Solve Applications of Exponential Functions (Section 9.3, Objective 4) ▶

OBJECTIVES

1 Use the Characteristics of the Natural Exponential Function

2 Sketch the Graphs of Natural Exponential Functions Using Transformations

3 Solve Natural Exponential Equations by Relating the Bases

4 Solve Applications of the Natural Exponential Function

....

OBJECTIVE 1 USE THE CHARACTERISTICS OF THE NATURAL EXPONENTIAL FUNCTION

We learned in the previous section that any positive number b, where $b \neq 1$, can be used as the base of an **exponential function**. However, there is one number that appears as the base in exponential applications more than any other number. This number is called the **natural base** and is symbolized using the letter e.

(eText Screens 9.4-1–9.4-21)

✎ *My video summary* ▶ **Concept Video** Watch this **video** to learn more about the natural base, then continue reading.

The number e is an **irrational number** that is defined as the value of the expression $\left(1 + \dfrac{1}{n}\right)^n$ as n approaches infinity.

As the values of n get large, the value e (rounded to six decimal places) is 2.718282.

The function $f(x) = e^x$ is called the **natural exponential function**. Because $2 < e < 3$, it follows that the graph of $f(x) = e^x$ lies between the graphs of $y = 2^x$ and $y = 3^x$, as seen in Figure 27.

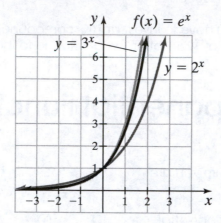

Figure 27
Graph of the Natural
Exponential Function $f(x) = e^x$

Characteristics of the Natural Exponential Function

The natural exponential function is the exponential function with base e and is defined as $f(x) = e^x$.

The domain of $f(x) = e^x$ is $(-\infty, \infty)$, and the range is $(0, \infty)$. The graph of $f(x) = e^x$ and some of its characteristics are stated here.

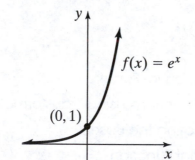

The graph of $f(x) = e^x$ intersects the y-axis at $(0, 1)$.

$e^x \to \infty$ as $x \to \infty$

$e^x \to 0$ as $x \to -\infty$

The line $y = 0$ is a **horizontal asymptote**.

The function $f(x) = e^x$ is **one-to-one**.

It is important that you are able to use your calculator to evaluate various powers of e. Most calculators have an $\boxed{e^x}$ key (sometimes as a 2nd function). Find this special key on your calculator and evaluate the expressions in the following example.

✎ *My video summary* ▶ **Example 1 Evaluating the Natural Exponential Function**

Evaluate each expression correctly to six decimal places.

a. e^2 **b.** $e^{-0.534}$ **c.** $1000e^{0.013}$

Solutions Work through the solutions, or watch this video.

Using the $\boxed{e^x}$ key on a calculator, we get

a. $e^2 \approx 7.389056$

b. $e^{-0.534} \approx 0.586255$

c. $1000e^{0.013} \approx 1013.084867$

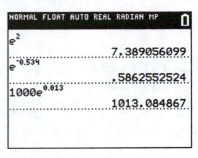

Figure 28
Screen Shot from TI-84 Plus C

🔺 **You Try It** Work through this You Try It problem.

Work Exercises 1–5 in this eText or in the My Math Lab **Study Plan.**

OBJECTIVE 2 SKETCH THE GRAPHS OF NATURAL EXPONENTIAL FUNCTIONS USING TRANSFORMATIONS

We can use the transformation techniques discussed in **Section 9.1** to sketch variations of the natural exponential function.

✎ *My video summary* ▶ **Example 2 Using Transformations to Sketch Natural Exponential Functions**

Use transformations to sketch the graph of $f(x) = -e^x + 2$. Determine the domain, range, and y-intercept, and find the equation of any asymptotes.

Solution Work through the following, or watch this video.

We can sketch the graph of $f(x) = -e^x + 2$ through a series of the following two transformations.

Start with the graph of $y = e^x$.

1. Reflect the graph of $y = e^x$ about the x-axis, producing the graph of $y = -e^x$.

2. Vertically shift the graph of $y = -e^x$ up two units, producing the graph of $f(x) = -e^x + 2$.

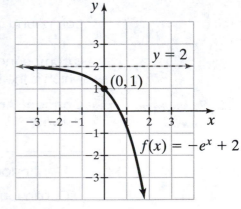

Figure 29
Graph of $f(x) = -e^x + 2$

The graph of $f(x) = -e^x + 2$ is shown in Figure 29. The domain of f is the interval $(-\infty, \infty)$. The range of f is the interval $(-\infty, 2)$. The y-intercept is 1, and the equation of the horizontal asymptote is $y = 2$.

🔺 **You Try It** Work through this You Try It problem.

Work Exercises 6–10 in this eText or in the My Math Lab **Study Plan.**

OBJECTIVE 3 SOLVE NATURAL EXPONENTIAL EQUATIONS BY RELATING THE BASES

Recall the **method of relating the bases** for solving exponential equations from **Section 9.3**. If we can write an exponential equation in the form of $b^u = b^v$, then $u = v$. This method for solving exponential equations also holds true for the natural base as illustrated in the following example.

My interactive video summary

Example 3 Using the Method of Relating the Bases to Solve Natural Exponential Equations

Use the method of relating the bases to solve each exponential equation:

a. $e^{3x-1} = \dfrac{1}{\sqrt{e}}$

b. $\dfrac{e^{x^2}}{e^{10}} = (e^x)^3$

Solutions

a.–b. Try to solve these problems on your own. Check your **answers**, or watch this **interactive video** for complete solutions to both parts.

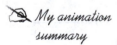 **You Try It** Work through this You Try It problem.

Work Exercises 11–18 in this eText or in the MyMathLab Study Plan.

OBJECTIVE 4 SOLVE APPLICATIONS OF THE NATURAL EXPONENTIAL FUNCTION

CONTINUOUS COMPOUND INTEREST

Recall the **periodic compound interest formula** that is introduced in **Section 9.3**. Some banks use **continuous compounding**; that is, they compound the interest every fraction of a second every day! If we start with the formula for periodic compound interest, $A = P\left(1 + \dfrac{r}{n}\right)^{nt}$, and let n (the number of times the interest is compounded each year) approach infinity, we can derive the formula $A = Pe^{rt}$, which is the formula for continuous compound interest.

My animation summary

Concept Animation Watch this **animation** to see exactly how this formula is derived.

Continuous Compound Interest Formula

Continuous compound interest can be calculated using the formula

$$A = Pe^{rt},$$

where

A = Total amount after t years

P = Principal

r = Interest rate per year

t = Number of years

My video summary ▶ **Example 4 Calculating Continuous Compound Interest**

How much money would be in an account after 5 years if an original investment of $6000 was compounded continuously at 4.5%? Compare this amount to the same investment that was compounded daily. Round to the nearest cent.

Solution Read through this solution, or watch this **video**.

First, the amount after 5 years compounded continuously is
$A = Pe^{rt} = 6000e^{0.045(5)} \approx \$7513.94.$

The same investment compounded daily yields an amount of
$A = P\left(1 + \dfrac{r}{n}\right)^{nt} = 6000\left(1 + \dfrac{0.045}{365}\right)^{365(5)} \approx \$7513.83.$ Continuous compound interest yields only $0.11 more interest after 5 years!

You Try It Work through this **You Try It** problem.

Work Exercises 19–21 in this eText or in the MyMathLab Study Plan.

PRESENT VALUE

Recall that the present value P is the amount of money to be invested now to obtain A dollars in the future. To find a formula for present value on money that is compounded continuously, we start with the formula for continuous compound interest and solve for P.

Write the continuous compound interest formula: $A = Pe^{rt}$

Divide both sides by e^{rt}: $\dfrac{A}{e^{rt}} = P\dfrac{e^{rt}}{e^{rt}}$

Rewrite $\dfrac{1}{e^{rt}}$ as e^{-rt}: $Ae^{-rt} = P$

Present Value Formula—Continuous Compounding

The present value of A dollars after t years of continuous compound interest, with interest rate r, is given by the formula

$$P = Ae^{-rt}.$$

My video summary ▶ **Example 5 Calculating Present Value**

Find the present value of $18,000 if interest is paid at a rate of 8% compounded continuously for 20 years. Round to the nearest cent.

Solution Read the following solution, or watch this **video**.

Using the present value formula $P = Ae^{-rt}$ with $A = \$18,000$, $r = 0.08$, and $t = 20$, we get

$$P = Ae^{-rt}$$
$$P = (18,000)e^{-(0.08)(20)} \approx \$3634.14.$$

You Try It Work through this You Try It problem.

Work Exercises 22 and 23 in this eText or in the MyMathLab Study Plan.

EXPONENTIAL GROWTH MODEL

You probably know that some populations grow exponentially. Most populations grow at a rate proportional to the size of the population. In other words, the larger the population, the faster the population grows. With this in mind, it can be shown in a more advanced math course that the **mathematical model** that can describe population growth is given by the function $P(t) = P_0 e^{kt}$.

Exponential Growth

A model that describes the population, P, after a certain time, t, is

$$P(t) = P_0 e^{kt},$$

where $P_0 = P(0)$ is the initial population and $k > 0$ is a constant called the **relative growth rate**. (**Note:** k may be given as a percent.)

The graph of the exponential growth model is shown in Figure 30. Notice that the graph has a y-intercept of P_0.

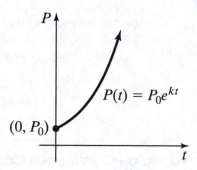

Figure 30
Graph of the Exponential
Growth Model $P(t) = P_0 e^{kt}$

$P(t) = P_0 e^{kt}$

$(0, P_0)$

My video summary ▶ **Example 6 Population Growth**

The population of a small town follows the exponential growth model $P(t) = 900 e^{0.015t}$, where t is the number of years after 2000.

Answer the following questions, rounding each answer to the nearest whole number:

a. What was the population of this town in 2000?

b. What was the population of this town in 2014?

c. Use this model to predict the population of this town in 2035.

Solutions Work through the following solutions, or watch this **video**.

a. The initial population was $P(0) = 900 e^{0.015(0)} = 900$.

b. Because 2014 is 14 years after 2000, we must evaluate $P(14)$.
$P(14) = 900 e^{0.015(14)} \approx 1110$.

c. In the year 2035, we can predict that the population will be
$P(35) = 900 e^{0.015(35)} \approx 1521$.

You Try It   Work through this You Try It problem.

Work Exercises 24 and 25 in this eText or in the MyMathLab Study Plan.

My video summary ▶ **Example 7 Determining the Initial Population**

Twenty years ago, the State of Idaho Fish and Game Department introduced a new breed of wolf into a certain Idaho forest. The current wolf population in this forest is now estimated at 825, with a relative growth rate of 12%.

Answer the following questions, rounding each answer to the nearest whole number:

a. How many wolves did the Idaho Fish and Game Department initially introduce into this forest?

b. How many wolves can be expected after another 20 years?

Solutions Work through the following solutions, or watch this **video**.

a. The relative growth rate is 0.12, so we use the exponential growth model $P(t) = P_0 e^{0.12t}$. Because $P(20) = 825$, we get

$$\text{Substitute 20 for } t: \quad P(20) = P_0 e^{0.12(20)}$$

$$\text{Substitute 825 for } P(20): \quad 825 = P_0 e^{0.12(20)}$$

$$\text{Solve for } P_0: \quad P_0 = \frac{825}{e^{0.12(20)}} \approx 75$$

Therefore, the Idaho Fish and Game Department initially introduced 75 wolves into the forest.

b. Because $P_0 = 75$, we can use the exponential growth model $P(t) = 75e^{0.12t}$. In another 20 years, the value of t will be 40. Thus, we must evaluate $P(40)$.

$$P(40) = 75e^{0.12(40)} \approx 9113$$

Therefore, we can expect the wolf population to be approximately 9113 in another 20 years.

You Try It   Work through this You Try It problem.

Work Exercises 26 and 27 in this eText or in the MyMathLab Study Plan.

9.4 Exercises

In Exercises 1–5, use a calculator to approximate each exponential expression to six decimal places.

1. e^3 2. $e^{-0.2}$ 3. $e^{1/3}$ 4. $100e^{-.123}$ 5. $\sqrt{2}e^{\pi}$

In Exercises 6–10, use transformations to sketch each exponential function. Determine the domain and range. Also, label the y-intercept, and find the equation of the horizontal asymptote.

6. $f(x) = e^{x-1}$ 7. $f(x) = e^x - 1$ 8. $f(x) = -e^{x+2}$

9. $f(x) = -e^{x+1} - 1$ 10. $f(x) = e^{-x} - 2$

In Exercises 11–18, solve each exponential equation using the method of relating the bases by first rewriting the equation in the form $e^u = e^v$.

11. $e^x = \dfrac{1}{e^2}$

12. $e^{5x+2} = \sqrt[3]{e}$

13. $\dfrac{1}{e^x} = \dfrac{\sqrt{e}}{e^{1-x}}$

14. $(e^{x^2})^2 = e^8$

15. $e^{x^2} = (e^x) \cdot e^{12}$

16. $e^{x^2} = \dfrac{e^3}{(e^x)^5}$

17. $\dfrac{e^{x^3}}{e^x} = \dfrac{e^{2x^2}}{e^2}$

18. $e^{x^3} = \dfrac{(e^{2x^2})^2 \cdot e^x}{e^4}$

19. An original investment of $6000 earns 6.25% interest compounded continuously. What will the investment be worth in 2 years? In 20 years? Round to the nearest cent.

20. How much more will an investment of $10,000 earning 5.5% compounded continuously for 9 years earn compared to the same investment at 5.5% compounded quarterly for 9 years? Round to the nearest cent.

21. Suppose your great-great grandfather invested $500 earning 6.5% interest compounded continuously 100 years ago. How much would his investment be worth today? Round to the nearest cent.

22. Find the present value of $16,000 if interest is paid at a rate of 4.5% compounded continuously for 10 years. Round to the nearest cent.

23. Which has the lower present value: (a) $20,000 if interest is paid at a rate of 5.18% compounded continuously for 2 years or (b) $25,000 if interest is paid at a rate of 3.8% compounded continuously for 30 months?

24. The population of a rural city follows the exponential growth model $P(t) = 2000e^{0.035t}$, where t is the number of years after 2005.

   a. What was the population of this city in 2005?

   b. What is the relative growth rate as a percent?

   c. Use this model to approximate the population in 2040, rounding to the nearest whole number.

25. The relative growth rate of a certain bacteria colony is 25%. Suppose there are 10 bacteria initially.

   a. Find a function that describes the population of the bacteria after t hours.

   b. How many bacteria should we expect after 1 day? Round to the nearest whole number.

26. In 2014, the population of a certain American city was 18,221. If the relative growth rate has been 6% since 1994, what was the population of this city in 1994? Round to the nearest whole number.

27. In 1980, a wildlife resource management team introduced a certain rabbit species into a forest for the first time. In 2014, the rabbit population had grown to 7183. The relative growth rate for this rabbit species is 20%.

   Answer the following questions, rounding each answer to the nearest whole number:

   a. How many rabbits did the wildlife resource management team introduce into the forest in 1980?

   b. How many rabbits can be expected in the year 2035?

9.5 Logarithmic Functions

THINGS TO KNOW

Before working through this section, be sure you are
familiar with the following concepts:

		VIDEO	ANIMATION	INTERACTIVE

You Try It 1. Solve Polynomial Inequalities
(Section 8.5, Objective 1) ▶

You Try It 2. Solve Rational Inequalities
(Section 8.5, Objective 2) ▶

You Try It 3. Determine If a Function Is One-to-One
Using the Horizontal Line Test
(Section 9.2, Objective 3) ▶ ✸

You Try It 4. Verify Inverse Functions
(Section 9.2, Objective 4) ▶ ☞

You Try It 5. Sketch the Graphs of Inverse Functions
(Section 9.2, Objective 5) ✸

You Try It 6. Find the Inverse of a One-to-One
Function (Section 9.2, Objective 6) ▶ ✸

You Try It 7. Use the Characteristics of Exponential
Functions (Section 9.3, Objective 1) ▶

You Try It 8. Sketch the Graphs of Exponential
Functions Using Transformations
(Section 9.3, Objective 2) ▶

You Try It 9. Solve Exponential Equations by Relating
the Bases (Section 9.3, Objective 3) ✸

OBJECTIVES

1 Use the Definition of a Logarithmic Function

2 Evaluate Logarithmic Expressions

3 Use the Properties of Logarithms

4 Use the Common and Natural Logarithms

5 Use the Characteristics of Logarithmic Functions

6 Sketch the Graphs of Logarithmic Functions Using Transformations

7 Find the Domain of Logarithmic Functions

OBJECTIVE 1 USE THE DEFINITION OF A LOGARITHMIC FUNCTION

Every exponential function of the form $f(x) = b^x$, where $b > 0$ and $b \neq 1$, is **one-to-one** and thus has an **inverse function**. (You may want to refer to **Section 9.2** to review one-to-one functions and inverse functions.) Remember, given the graph of a one-to-one function f, the graph of the inverse function is a reflection about the line $y = x$. That is, for any point (a, b) that lies on the graph of f, the point (b, a) must lie on the graph of f^{-1}. In other words, the graph of f^{-1} can be obtained by simply switching the x and y coordinates of the ordered pairs of $f(x) = b^x$.

My video summary ▶ **Concept Video** Watch this **video** to see how to sketch the graph of $f(x) = b^x$ and its inverse.

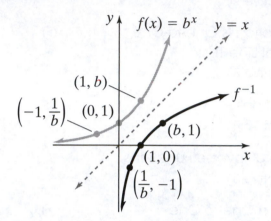

Figure 31
Graph of $f(x) = b^x, b > 1,$
and Its Inverse Function

The graphs of f and f^{-1} are sketched in **Figure 31**, but what is the equation of the inverse of $f(x) = b^x$? To find the equation of f^{-1}, we follow the **four-step process** for finding inverse functions that is discussed in Section 9.2.

Step 1 Change $f(x)$ to y: $\quad y = b^x$

Step 2 Interchange x and y: $\quad x = b^y$

Step 3 Solve for y: $\qquad\qquad$??

Before we can solve for y, we must introduce the following definition:

Definition Logarithmic Function

For $x > 0, b > 0$, and $b \neq 1$, the **logarithmic function** with base b is defined by

$$y = \log_b x \quad \text{if and only if} \quad x = b^y$$

The equation $y = \log_b x$ is said to be in **logarithmic form**, while the equation $x = b^y$ is in **exponential form**. We can now continue to find the inverse of $f(x) = b^x$ by completing Steps 3 and 4.

Step 3 Solve for y: $\qquad\quad x = b^y$ can be written as $y = \log_b x$

Step 4 Change y to $f^{-1}(x)$: $\quad f^{-1}(x) = \log_b x$

In general, if $f(x) = b^x$ for $b > 0$ and $b \neq 1$, then the inverse function is $f^{-1}(x) = \log_b x$. For example, the inverse of $f(x) = 2^x$ is $f^{-1}(x) = \log_2 x$, which is read as "the log base 2 of x." We revisit the graphs of logarithmic functions later on in this section, but first it is very important to understand the definition of the logarithmic function and practice how to go back and forth writing exponential equations as logarithmic equations and vice versa.

My video summary ▶ **Example 1** Changing from Exponential Form to Logarithmic Form

Write each exponential equation as an equation involving a logarithm.

a. $2^3 = 8$ **b.** $5^{-2} = \dfrac{1}{25}$ **c.** $1.1^M = z$

Solutions Read the following solutions, or watch this **video**.

We use the fact that the equation $x = b^y$ is equivalent to the equation $y = \log_b x$.

a. $2^3 = 8$ is equivalent to $\log_2 8 = 3$.

b. $5^{-2} = \dfrac{1}{25}$ is equivalent to $\log_5 \dfrac{1}{25} = -2$.

c. $1.1^M = z$ is equivalent to $\log_{1.1} z = M$.

 You Try It Work through this You Try It problem.

Work Exercises 1–6 in this eText or in the MyMathLab Study Plan.

TIP Note that the exponent of the original (exponential) equation ends up by itself on the right side of the second (logarithmic) equation. Therefore, a logarithmic expression can be thought of as describing the exponent of a certain exponential equation.

 My video summary ▶ **Example 2** Changing from Logarithmic Form to Exponential Form

Write each logarithmic equation as an equation involving an exponent.

a. $\log_3 81 = 4$ **b.** $\log_4 16 = y$ **c.** $\log_{3/5} x = 2$

Solutions Read the following, or watch this **video** to see this example worked out in more detail.

We use the fact that the equation $y = \log_b x$ is equivalent to the equation $x = b^y$.

a. $\log_3 81 = 4$ is equivalent to $3^4 = 81$.

b. $\log_4 16 = y$ is equivalent to $4^y = 16$.

c. $\log_{3/5} x = 2$ is equivalent to $\left(\dfrac{3}{5}\right)^2 = x$.

You Try It Work through this You Try It problem.

Work Exercises 7–11 in this eText or in the MyMathLab Study Plan.

OBJECTIVE 2   EVALUATE LOGARITHMIC EXPRESSIONS

Because a logarithmic expression represents the exponent of an exponential equation, it is possible to evaluate many logarithms by inspection or by creating the corresponding exponential equation. Remember that the expression $\log_b x$ is the exponent to which b must be raised in order get x. For example, suppose we are to evaluate the expression $\log_4 64$. To evaluate this expression, we must ask ourselves, "4 raised to what power is 64?" Because $4^3 = 64$, we conclude that $\log_4 64 = 3$. For some logarithmic expressions, it is often convenient to create an exponential equation and use the **method of relating the bases** for solving exponential equations.

My interactive video summary

Example 3 Evaluating Logarithmic Expressions

Evaluate each logarithm:

a. $\log_5 25$ **b.** $\log_3 \dfrac{1}{27}$ **c.** $\log_{\sqrt{2}} \dfrac{1}{4}$

Solutions

a. To evaluate $\log_5 25$, we must ask, "5 raised to what exponent is 25?" Because $5^2 = 25$, $\log_5 25 = 2$.

b. The expression $\log_3 \dfrac{1}{27}$ requires more analysis. In this case, we ask, "3 raised to what exponent is $\dfrac{1}{27}$?" Suppose we let y equal this exponent. Then $3^y = \dfrac{1}{27}$. To solve for y, we can use the **method of relating the bases** for solving exponential equations.

$$\text{Write the exponential equation:} \quad 3^y = \frac{1}{27}$$

$$\text{Rewrite 27 as } 3^3: \quad 3^y = \frac{1}{3^3}$$

$$\text{Use } \frac{1}{b^n} = b^{-n}: \quad 3^y = 3^{-3}$$

$$\text{Use the method of relating the bases:} \quad y = -3$$
$$(\text{If } b^u = b^v, \text{ then } u = v.)$$

Thus, $\log_3 \dfrac{1}{27} = -3$.

c. Try this problem on your own. Check your **answer**, or watch the **interactive video** to see all three solutions worked out in detail.

You Try It Work through this You Try It problem.

Work Exercises 12–18 in this eText or in the MyMathLab Study Plan.

OBJECTIVE 3 USE THE PROPERTIES OF LOGARITHMS

Because $b^1 = b$ for any real number b, we can use the definition of the **logarithmic function** to rewrite this expression as $\log_b b = 1$. Similarly, because $b^0 = 1$ for any nonzero real number b, we can rewrite this expression as $\log_b 1 = 0$. These two general properties are summarized as follows.

General Properties of Logarithms

For $b > 0$ and $b \neq 1$,

1. $\log_b b = 1$ and
2. $\log_b 1 = 0$.

In **Section 9.2**, we saw that a function f and its inverse function f^{-1} satisfy the composition cancellation equations. If $f(x) = b^x$, then $f^{-1}(x) = \log_b x$. Applying the two composition cancellation equations, we get

$$f\left(f^{-1}(x)\right) = b^{\log_b x} = x \quad \text{and}$$

$$f^{-1}\left(f(x)\right) = \log_b b^x = x.$$

Cancellation Properties of Exponentials and Logarithms

For $b > 0$ and $b \neq 1$,

1. $b^{\log_b x} = x$ and
2. $\log_b b^x = x$.

My video summary ▶ **Example 4 Using the Properties of Logarithms**

Use the properties of logarithms to evaluate each expression.

a. $\log_3 3^4$ **b.** $\log_{12} 12$ **c.** $7^{\log_7 13}$ **d.** $\log_8 1$

Solutions Read the following solutions, or watch this **video**.

a. By the second cancellation property, $\log_3 3^4 = 4$.

b. Because $\log_b b = 1$ for all $b > 0$ and $b \neq 1$, $\log_{12} 12 = 1$.

c. By the first cancellation property, $7^{\log_7 13} = 13$.

d. Because $\log_b 1 = 0$ for all $b > 0$ and $b \neq 1$, $\log_8 1 = 0$.

You Try It Work through this You Try It problem.

Work Exercises 19–26 in this eText or in the MyMathLab Study Plan.

OBJECTIVE 4 USE THE COMMON AND NATURAL LOGARITHMS

There are two bases that are used more frequently than any other base. They are base 10 and **base e**. Because our counting system is based on the number 10, the base 10 logarithm is known as the **common logarithm**. Instead of using the notation $\log_{10} x$ to denote the common logarithm, it is usually abbreviated without the subscript 10 as simply $\log x$. The base e logarithm is called the **natural logarithm** and is abbreviated as $\ln x$ instead of $\log_e x$. Most scientific calculators are equipped with a (log) key and a (ln) key. We can apply the definition of the logarithmic function for the base 10 and for the base e logarithm as follows.

Definition Common Logarithmic Function

For $x > 0$, the **common logarithmic function** is defined by

$$y = \log x \quad \text{if and only if} \quad x = 10^y.$$

Definition Natural Logarithmic Function

For $x > 0$, the **natural logarithmic function** is defined by

$$y = \ln x \quad \text{if and only if} \quad x = e^y.$$

My video summary ▶ **Example 5 Changing from Exponential Form to Logarithmic Form**

Write each exponential equation as an equation involving a common logarithm or natural logarithm.

a. $e^0 = 1$ 　b. $10^{-2} = \dfrac{1}{100}$ 　c. $e^K = w$

Solutions Read the following, or watch this **video** to see the example worked out in more detail.

a. $e^0 = 1$ is equivalent to $\ln 1 = 0$.

b. $10^{-2} = \dfrac{1}{100}$ is equivalent to $\log\left(\dfrac{1}{100}\right) = -2$.

c. $e^K = w$ is equivalent to $\ln w = K$.

You Try It Work through this You Try It problem.

Work Exercises 27–31 in this eText or in the MyMathLab Study Plan.

My video summary ▶ **Example 6 Changing from Logarithmic Form to Exponential Form**

Write each logarithmic equation as an equation involving an exponent.

a. $\log 10 = 1$ 　b. $\ln 20 = Z$ 　c. $\log(x - 1) = T$

Solutions Read the following, or watch this **video** to see the example worked out in more detail.

a. $\log 10 = 1$ is equivalent to $10^1 = 10$.

b. $\ln 20 = Z$ is equivalent to $e^Z = 20$.

c. $\log(x - 1) = T$ is equivalent to $10^T = x - 1$.

You Try It Work through this You Try It problem.

Work Exercises 32–35 in this eText or in the MyMathLab Study Plan.

My video summary ▶ **Example 7 Evaluating Common and Natural Logarithmic Expressions**

Evaluate each expression without the use of a calculator.

a. $\log 100$ 　b. $\ln\sqrt{e}$ 　c. $e^{\ln 51}$ 　d. $\log 1$

Solutions Read the following, or watch this **video** to see the example worked out in more detail.

a. $\log 100 = 2$ because $10^2 = 100$ by the definition of the logarithmic function or $\log 100 = \log 10^2 = 2$ by cancellation property **(2)**.

b. $\ln\sqrt{e} = \ln e^{1/2} = \dfrac{1}{2}$ by cancellation property **(2)**.

c. $e^{\ln 51} = 51$ by cancellation property **(1)**.

d. $\log 1 = 0$ by general property **(1)**.

You Try It   Work through this You Try It problem.

Work Exercises 36–43 in this eText or in the MyMathLab Study Plan.

OBJECTIVE 5   USE THE CHARACTERISTICS OF LOGARITHMIC FUNCTIONS

To sketch the graph of a **logarithmic function** of the form $f(x) = \log_b x$, where $b > 0$ and $b \neq 1$, follow these three steps:

Step 1   Start with the graph of the exponential function $y = b^x$, labeling several ordered pairs.

Step 2   Because $f(x) = \log_b x$ is the inverse of $y = b^x$, we can find several points on the graph of $f(x) = \log_b x$ by reversing the coordinates of the ordered pairs of $y = b^x$.

Step 3   Plot the ordered pairs from Step 2, and complete the graph of $f(x) = \log_b x$ by connecting the ordered pairs with a smooth curve. The graph of $f(x) = \log_b x$ is a reflection of the graph of $y = b^x$ about the line $y = x$.

My video summary ▶ **Example 8 Sketching the Graph of a Logarithmic Function**

Sketch the graph of $f(x) = \log_3 x$.

Solution   Work through the following solution, or watch this **video**.

Step 1   The graph of $y = 3^x$ passes through the points $\left(-1, \dfrac{1}{3}\right)$, $(0, 1)$, and $(1, 3)$. See Figure 32.

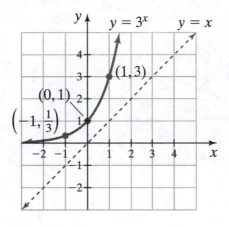

Figure 32
Graph of $y = 3^x$

Step 2   We reverse the three ordered pairs from Step 1 to obtain the following three points: $\left(\dfrac{1}{3}, -1\right)$, $(1, 0)$, and $(3, 1)$.

Step 3   Plot the points $\left(\dfrac{1}{3}, -1\right)$, $(1, 0)$, and $(3, 1)$, and connect them with a smooth curve to obtain the graph of $f(x) = \log_3 x$. See Figure 33.

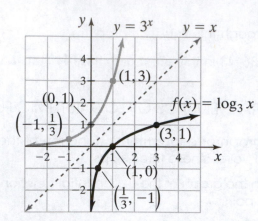

Figure 33
Graph of $y = 3^x$ and $f(x) = \log_3 x$

You Try It Work through this You Try It problem.

Work Exercises 44 and 45 in this eText or in the My MathLab Study Plan.

Notice in Figure 33 that the y-axis is a **vertical asymptote** of the graph of $f(x) = \log_3 x$. Every logarithmic function of the form $y = \log_b x$, where $b > 0$ and $b \neq 1$ has a vertical asymptote at the y-axis.

The graphs and the characteristics of logarithmic functions are outlined as follows.

Characteristics of Logarithmic Functions

For $b > 0, b \neq 1$, the logarithmic function with base b is defined by $y = \log_b x$. The **domain** of $f(x) = \log_b x$ is $(0, \infty)$, and the **range** is $(-\infty, \infty)$. The graph of $f(x) = \log_b x$ has one of the following two shapes.

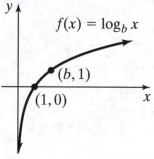

$f(x) = \log_b x, b > 1$

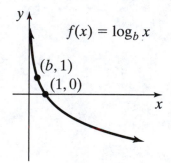

$f(x) = \log_b x, 0 < b < 1$

The graph intersects the x-axis at $(1, 0)$.
The graph contains the point $(b, 1)$.
The graph is **increasing** on the interval $(0, \infty)$.
The y-axis ($x = 0$) is a **vertical asymptote**.
The function is **one-to-one**.

The graph intersects the x-axis at $(1, 0)$.
The graph contains the point $(b, 1)$.
The graph is **decreasing** on the interval $(0, \infty)$.
The y-axis ($x = 0$) is a **vertical asymptote**.
The function is **one-to-one**.

OBJECTIVE 6 SKETCH THE GRAPHS OF LOGARITHMIC FUNCTIONS USING TRANSFORMATIONS

We can use the transformation techniques discussed in **Section 9.1** to sketch the graph of logarithmic functions.

My video summary ▶ **Example 9** Using Transformations to Sketch the Graph of a Logarithmic Function

Sketch the graph of $f(x) = -\ln(x + 2) - 1$.

Solution Read through the solution, or watch this **video** for more details. Recall that the function $y = \ln x$ has a base of e, where $2 < e < 3$. This means that the graph of $y = \ln x$ is increasing on the interval $(0, \infty)$ and contains the points $(1, 0)$ and $(e, 1)$. Starting with the graph of $y = \ln x$, we can obtain the graph of $f(x) = -\ln(x + 2) - 1$ through the following series of transformations:

1. Shift the graph of $y = \ln x$ horizontally to the left two units to obtain the graph of $y = \ln(x + 2)$.

2. Reflect the graph of $y = \ln(x + 2)$ about the x-axis to obtain the graph of $y = -\ln(x + 2)$.

3. Shift the graph of $y = -\ln(x + 2)$ vertically down one unit to obtain the final graph of $f(x) = -\ln(x + 2) - 1$.

The graph of $f(x) = -\ln(x + 2) - 1$ is sketched in Figure 34. You can see from the graph that the domain of $f(x) = -\ln(x + 2) - 1$ is $(-2, \infty)$. The vertical asymptote is $x = -2$, and the x-intercept is $\left(\dfrac{1}{e} - 2, 0\right)$.

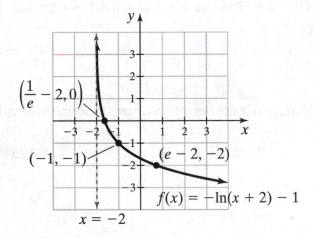

Figure 34
Graph of $f(x) = -\ln(x + 2) - 1$

You Try It Work through this **You Try It** problem.

Work Exercises 46–51 in this eText or in the MyMathLab **Study Plan.**

OBJECTIVE 7 FIND THE DOMAIN OF LOGARITHMIC FUNCTIONS

In Example 9, we sketched the function $f(x) = -\ln(x + 2) - 1$ and observed that the domain was $(-2, \infty)$. We do not have to sketch the graph of a **logarithmic function** to determine the domain.

My animation summary

⊕ **Concept Animation** Watch this animation to see how to find the domain of a logarithmic function without graphing.

Example 10 is a bit more challenging because the **argument** of the logarithm is a rational expression.

Example 10 Finding the Domain of a Logarithmic Function with a Rational Argument

Find the domain of $f(x) = \log_5\left(\dfrac{2x - 1}{x + 3}\right)$.

My interactive video summary

✎ **Solution** Read the following solution, or watch the **interactive video** to see this problem worked out in detail.

To find the domain of f, we must find all values of x for which the argument $\dfrac{2x - 1}{x + 3}$ is greater than zero. That is, you must solve the rational inequality $\left(\dfrac{2x - 1}{x + 3}\right) > 0$. See **Section 8.5** if you need help remembering how to solve this inequality. By the techniques discussed in Section 8.5, we find that the solution to $\left(\dfrac{2x - 1}{x + 3}\right) > 0$ is $x < -3$ or $x > \dfrac{1}{2}$. Therefore, the domain of $f(x) = \log_5\left(\dfrac{2x - 1}{x + 3}\right)$ in set notation is $\left\{x \mid x < -3 \text{ or } x > \dfrac{1}{2}\right\}$. In interval notation, the domain is $(-\infty, -3) \cup \left(\dfrac{1}{2}, \infty\right)$.

You Try It Work through this You Try It problem.

Work Exercises 52–58 in this eText or in the MyMathLab Study Plan.

9.5 Exercises

In Exercises 1–6, write each exponential equation as an equation involving a logarithm.

1. $3^2 = 9$

2. $16^{1/2} = 4$

3. $2^{-3} = \dfrac{1}{8}$

4. $\left(\sqrt{2}\right)^\pi = W$

5. $\left(\dfrac{1}{3}\right)^t = 27$

6. $7^{5k} = L$

In Exercises 7–11, write each logarithmic equation as an exponential equation.

7. $\log_5 1 = 0$

8. $\log_7 343 = 3$

9. $\log_{\sqrt{2}} 8 = 6$

10. $\log_4 K = L$

11. $\log_a (x - 1) = 3$

In Exercises 12–18, evaluate each logarithm without the use of a calculator.

12. $\log_2 8$

13. $\log_6 \sqrt{6}$

14. $\log_3 \dfrac{1}{9}$

15. $\log_{\sqrt{5}} 25$

16. $\log_4 \left(\dfrac{1}{\sqrt[5]{64}} \right)$

17. $\log_{1/7} \sqrt[3]{7}$

18. $\log_{0.1} 100$

In Exercises 19–26, use the properties of logarithms to evaluate each expression without the use of a calculator.

19. $2^{\log_2 11}$

20. $\log_4 4$

21. $\log_9 1$

22. $\log_7 7^{-3}$

23. $\log_a a, a > 1$

24. $5^{\log_5 M}, M > 0$

25. $\log_y 1, y > 0$

26. $\log_x x^{20}, x > 1$

In Exercises 27–31, write each exponential equation as an equation involving a common logarithm or a natural logarithm.

27. $10^3 = 1000$

28. $e^{-1} = \dfrac{1}{e}$

29. $e^k = 2$

30. $10^e = M$

31. $e^{10} = Z$

In Exercises 32–35, write each logarithmic equation as an exponential equation.

32. $\ln 1 = 0$

33. $\log (1{,}000{,}000) = 6$

34. $\log K = L$

35. $\ln Z = 4$

In Exercises 36–43, evaluate each expression without the use of a calculator, and then verify your answer using a calculator.

36. $\log 10{,}000$

37. $\log \left(\dfrac{1}{1000} \right)$

38. $\ln 1$

39. $\ln \sqrt[3]{e^2}$

40. $10^{\log e}$

41. $e^{\ln 49}$

42. $\log 10^6$

43. $\ln e + \ln e^3$

In Exercises 44–51, sketch each logarithmic function. Label at least two points on the graph, and determine the domain and the equation of any vertical asymptotes.

44. $h(x) = \log_4 x$

45. $g(x) = \log_{\frac{1}{3}} x$

46. $f(x) = \log_2 (x) - 1$

47. $f(x) = \log_5 (x - 1)$

48. $f(x) = -\ln (x)$

49. $y = \log_{1/2} (x + 1) + 2$

50. $y = \log_3 (1 - x)$

51. $h(x) = -\dfrac{1}{2} \log_3 (x + 3) + 1$

In Exercises 52–58, find the domain of each logarithmic function.

52. $f(x) = \log(-x)$

53. $f(x) = \log_{1/4}(2x + 6)$

54. $f(x) = \ln(1 - 3x)$

55. $f(x) = \log_2(x^2 - 9)$

56. $f(x) = \log_7(x^2 - x - 20)$

57. $f(x) = \ln\left(\dfrac{x + 5}{x - 8}\right)$

58. $f(x) = \log\left(\dfrac{x^2 - x - 6}{x + 10}\right)$

9.6 Properties of Logarithms

THINGS TO KNOW

Before working through this section, be sure you are familiar with the following concepts:

| | VIDEO | ANIMATION | INTERACTIVE |

 1. Solve Exponential Equations by Relating the Bases (Section 9.3, **Objective 3**)

 2. Change from Exponential Form to Logarithmic Form (Section 9.5, **Objective 1**)

 3. Change from Logarithmic Form to Exponential Form (Section 9.5, **Objective 1**)

 4. Evaluate Logarithmic Expressions (Section 9.5, **Objective 2**)

 5. Use the Common and Natural Logarithms (Section 9.5, **Objective 4**)

 6. Find the Domain of Logarithmic Functions (Section 9.5, **Objective 7**)

OBJECTIVES

1 Use the Product Rule, Quotient Rule, and Power Rule for Logarithms

2 Expand and Condense Logarithmic Expressions

3 Solve Logarithmic Equations Using the Logarithm Property of Equality

4 Use the Change of Base Formula

OBJECTIVE 1 USE THE PRODUCT RULE, QUOTIENT RULE, AND POWER RULE FOR LOGARITHMS

In this section, we learn how to manipulate logarithmic expressions using properties of logarithms. Understanding how to use these properties will help us solve exponential and logarithmic equations that are encountered in the next section. Recall from **Section 9.5** the **general properties** and **cancellation properties** of logarithms. We now look at three additional properties of logarithms.

My video summary ▶

Properties of Logarithms

If $b > 0, b \neq 1$, u and v represent positive numbers and r is any real number, then

$$\log_b uv = \log_b u + \log_b v \quad \text{Product rule for logarithms} \quad ▶$$

$$\log_b \frac{u}{v} = \log_b u - \log_b v \quad \text{Quotient rule for logarithms} \quad ▶$$

$$\log_b u^r = r \log_b u \quad \text{Power rule for logarithms} \quad ▶$$

Concept Video Select the appropriate video proof link above to see a proof for each of these properties.

My video summary ▶ ## Example 1 Using the Product Rule

Use the product rule for logarithms to expand each expression. Assume $x > 0$.

a. $\ln (5x)$

b. $\log_2 (8x)$

Solutions Read the following solutions, or watch this **video**.

a. Use the product rule for logarithms: $\ln (5x) = \ln 5 + \ln x$

b. Use the product rule for logarithms: $\log_2 (8x) = \log_2 8 + \log_2 x$

Use the **definition of the logarithmic function** to rewrite $\log_2 8$ as 3 because $2^3 = 8$: $\log_2 (8x) = 3 + \log_2 x$

 You Try It Work through this You Try It problem.

Work Exercises 1–4 in this eText or in the MyMathLab Study Plan.

 $\log_b (u + v)$ is *not* equivalent to $\log_b u + \log_b v$.

My video summary ▶ ## Example 2 Using the Quotient Rule

Use the **quotient rule** for logarithms to expand each expression. Assume $x > 0$.

a. $\log_5 \left(\dfrac{12}{x}\right)$

b. $\ln \left(\dfrac{x}{e^5}\right)$

Solutions Read through the following solutions, or watch this **video**.

a. Use the quotient rule for logarithms: $\log_5\left(\dfrac{12}{x}\right) = \log_5 12 - \log_5 x$

b. Use the quotient rule for logarithms: $\ln\left(\dfrac{x}{e^5}\right) = \ln x - \ln e^5$

       Use cancellation property (2) to
rewrite $\ln e^5$ as 5: $= \ln x - 5$

 You Try It Work through this You Try It problem.

Work Exercises 5–7 in this eText or in the MyMathLab Study Plan.

⚠ $\log_b (u - v)$ is *not* equivalent to $\log_b u - \log_b v$, and $\dfrac{\log_b u}{\log_b v}$ is *not* equivalent to

$\log_b u - \log_b v$.

My video summary ▶ **Example 3 Using the Power Rule**

Use the **power rule** for logarithms to rewrite each expression. Assume $x > 0$.

a. $\log 6^3$ b. $\log_{1/2} \sqrt[4]{x}$

Solutions Read through the following solutions, or watch this **video**.

a. Use the power rule for logarithms: $\log 6^3 = 3 \log 6$

b.   Rewrite the fourth root of x
using a rational exponent: $\log_{1/2} \sqrt[4]{x} = \log_{1/2} x^{1/4}$

     Use the power rule for logarithms: $= \dfrac{1}{4} \log_{1/2} x$

The process of using the power rule to simplify a logarithmic expression is often casually referred to as "bringing down the exponent."

 You Try It Work through this You Try It problem.

Work Exercises 8–10 in this eText or in the MyMathLab Study Plan.

⚠ $(\log_b u)^r$ is *not* equivalent to $r \log_b u$.

OBJECTIVE 2   EXPAND AND CONDENSE LOGARITHMIC EXPRESSIONS

Sometimes it is necessary to combine several **properties of logarithms** to expand a logarithmic expression into the sum and/or difference of logarithms or to condense several logarithms into a single logarithm.

Example 4 Expanding a Logarithmic Expression

Use properties of logarithms to expand each logarithmic expression as much as possible.

a. $\log_7\left(49x^3\sqrt[5]{y^2}\right)$ b. $\ln\left(\dfrac{x^2 - 4}{9e^{x^3}}\right)$

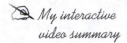 *My interactive video summary*

Solutions Read through the following solutions, or watch the **interactive video** to see the solutions worked out in detail.

a.

Write the original expression: $\quad \log_7\left(49x^3\sqrt[5]{y^2}\right)$

Use the **product rule**: $\quad = \log_7 49 + \log_7 x^3\sqrt[5]{y^2}$

Use the product rule again: $\quad = \log_7 49 + \log_7 x^3 + \log_7 \sqrt[5]{y^2}$

Rewrite $\sqrt[5]{y^2}$ using a rational exponent: $\quad = \log_7 49 + \log_7 x^3 + \log_7 y^{2/5}$

Rewrite $\log_7 49$ as 2 and use the **power rule**: $\quad = 2 + 3\log_7 x + \dfrac{2}{5}\log_7 y$

b. Factor the expression in the numerator: $\quad \ln\left(\dfrac{x^2-4}{9e^{x^3}}\right) = \ln\left(\dfrac{(x-2)(x+2)}{9e^{x^3}}\right)$

Use the **quotient rule**: $\quad = \ln(x-2)(x+2) - \ln 9e^{x^3}$

Use the **product rule** twice: $\quad = \ln(x-2) + \ln(x+2) - \left[\ln 9 + \ln e^{x^3}\right]$

Use cancellation property (2)
to rewrite $\ln e^{x^3}$ as x^3: $\quad = \ln(x-2) + \ln(x+2) - \left[\ln 9 + x^3\right]$

Distribute: $\quad = \ln(x-2) + \ln(x+2) - \ln 9 - x^3$

You Try It Work through this You Try It problem.

Work Exercises 11–20 in this eText or in the MyMathLab Study Plan.

 *My interactive video summary*

Example 5 Condensing a Logarithmic Expression

Use **properties of logarithms** to rewrite each expression as a single logarithm.

a. $\dfrac{1}{2}\log(x-1) - 3\log z + \log 5$

b. $\dfrac{1}{3}(\log_3 x - 2\log_3 y) + \log_3 10$

Solutions Read through the following solutions, or watch the **interactive video** to see the solutions worked out in detail.

a. Write the original expression: $\quad \dfrac{1}{2}\log(x-1) - 3\log z + \log 5$

Use the **power rule** twice: $\quad = \log(x-1)^{1/2} - \log z^3 + \log 5$

Use the **quotient rule**: $\quad = \log\dfrac{(x-1)^{1/2}}{z^3} + \log 5$

Use the **product rule**: $\quad = \log\dfrac{5(x-1)^{1/2}}{z^3} \ \text{or}\ \log\dfrac{5\sqrt{x-1}}{z^3}$

b. Write the original expression: $\quad \dfrac{1}{3}(\log_3 x - 2\log_3 y) + \log_3 10$

Use the **power rule**: $\quad = \dfrac{1}{3}(\log_3 x - \log_3 y^2) + \log_3 10$

Use the **quotient rule**: $\quad = \dfrac{1}{3}\log_3\dfrac{x}{y^2} + \log_3 10$

Use the power rule: $= \log_3 \left(\dfrac{x}{y^2}\right)^{1/3} + \log_3 10$

Use the product rule: $= \log_3 \left[10\left(\dfrac{x}{y^2}\right)^{1/3}\right]$ or $\log_3 \left[10\sqrt[3]{\dfrac{x}{y^2}}\right]$

You Try It Work through this You Try It problem.

Work Exercises 21–30 in this eText or in the MyMathLab Study Plan.

OBJECTIVE 3 SOLVE LOGARITHMIC EQUATIONS USING THE LOGARITHM PROPERTY OF EQUALITY

Remember that all logarithmic functions of the form $f(x) = \log_b x$ for $b > 0$ and $b \neq 1$ are **one-to-one**. Using the alternate definition of one-to-one and letting $f(x) = \log_b x$, we can say that if $\log_b u = \log_b v$, then $u = v$. In other words, if the bases of a logarithmic equation of the form $\log_b u = \log_b v$ are equal, then the arguments (the quantities "inside" the logs) must be equal. This is known as the **logarithm property of equality**.

Logarithm Property of Equality

If a logarithmic equation can be written in the form $\log_b u = \log_b v$, then $u = v$.

Furthermore, if $u = v$, then $\log_b u = \log_b v$.

The second statement of the logarithm property of equality says that if we start with the equation $u = v$, then we can rewrite the equation as $\log_b u = \log_b v$. This process is often casually referred to as "taking the log of both sides."

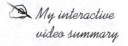 *My interactive video summary*

Example 6 Using the Logarithm Property of Equality to Solve Logarithmic Equations

Solve the following equations.

a. $\log_7 (x - 1) = \log_7 12$

b. $2 \ln x = \ln 16$

Solutions Read through the following solutions, or watch this **interactive video** for full solutions to both parts.

a. Because the base of each logarithm is 7, we can use the logarithm property of equality to eliminate the logarithms.

Write the original equation: $\log_7 (x - 1) = \log_7 12$

If $\log_b u = \log_b v$, then $u = v$: $(x - 1) = 12$

Solve for x: $x = 13$

b. Write the original expression: $2 \ln x = \ln 16$

Use the power rule: $\ln x^2 = \ln 16$

If $\log_b u = \log_b v$, then $u = v$: $x^2 = 16$

Use the square root property: $x = \pm 4$

The domain of $\ln x$ is $x > 0$; this implies that $x = -4$ is an **extraneous solution**, and hence, we must discard it. Therefore, this equation has only one solution, $x = 4$.

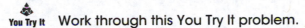 **You Try It** Work through this You Try It problem.

Work Exercises 31–36 in this eText or in the MyMathLab Study Plan.

⚠ Remember to check potential solutions in the *original* equation before forming the solution set.

OBJECTIVE 4   USE THE CHANGE OF BASE FORMULA

Most scientific calculators are equipped with a (log) key and a (ln) key to evaluate common logarithms and natural logarithms. But how do we use a calculator to evaluate logarithmic expressions having a base other than 10 or e? The answer is to use the following **change of base formula**.

Change of Base Formula

For any positive base $b \neq 1$ and for any positive real number u, then

$$\log_b u = \frac{\log_a u}{\log_a b}$$

where a is any positive number such that $a \neq 1$.

✍ *My video summary* ▶ **Concept Video** Watch this video proof of the change of base formula.

The change of base formula allows us to change the base of a logarithmic expression into a ratio of two logarithms using any base we choose. For example, suppose we are given the logarithmic expression $\log_3 10$. We can use the change of base formula to write this logarithm as a logarithm involving any base we choose:

$$\log_3 10 = \frac{\log_7 10}{\log_7 3} \quad \text{or} \quad \log_3 10 = \frac{\log_2 10}{\log_2 3} \quad \text{or} \quad \log_3 10 = \frac{\log 10}{\log 3} \quad \text{or} \quad \log_3 10 = \frac{\ln 10}{\ln 3}$$

In each of the previous four cases, we introduced a new base (7, 2, 10, and e, respectively). However, if we want to use a calculator to get a numerical approximation of $\log_3 10$, then it really only makes sense to change $\log_3 10$ into an expression involving base 10 or base e because these are the only two bases many calculators can handle.

Note: $\log_3 10 = \dfrac{\log 10}{\log 3} \approx 2.0959 \quad \text{or} \quad \log_3 10 = \dfrac{\ln 10}{\ln 3} \approx 2.0959$

✍ *My video summary* ▶ **Example 7** Using the Change of Base Formula

Approximate the following expressions. Round each to four decimal places.

a. $\log_9 200$ b. $\log_{\sqrt{3}} \pi$

Solutions Read these solutions, or watch this **video**.

a. $\log_9 200 = \dfrac{\log 200}{\log 9} \approx 2.4114$ or $\log_9 200 = \dfrac{\ln 200}{\ln 9} \approx 2.4114$

b. $\log_{\sqrt{3}} \pi = \dfrac{\log \pi}{\log \sqrt{3}} \approx 2.0840$ or $\log_{\sqrt{3}} \pi = \dfrac{\ln \pi}{\ln \sqrt{3}} \approx 2.0840$

See Figure 35.

You Try It Work through this You Try It problem.

Work Exercises 37–40 in this eText or in the MyMathLab Study Plan.

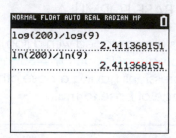
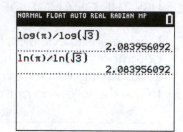

Figure 35
Approximating
Logarithms

⚠ Be careful to close parentheses around the argument of each logarithm.

My video summary ▶ **Example 8 Using the Change of Base Formula and Properties of Logarithms**

Use the **change of base formula** and the **properties of logarithms** to rewrite as a single logarithm involving base 2.

$$\log_4 x + 3 \log_2 y$$

Solution To use properties of logarithms, the base of each logarithmic expression must be the same. We use the change of base formula to rewrite $\log_4 x$ as a logarithmic expression involving base 2:

$$\boxed{\log_4 x = \dfrac{\log_2 x}{\log_2 4}}$$

Use the change of base formula: $\log_4 x + 3 \log_2 y = \dfrac{\log_2 x}{\log_2 4} + 3 \log_2 y$

Rewrite $\log_2 4$ as 2 because $2^2 = 4$: $= \dfrac{\log_2 x}{2} + 3 \log_2 y$

Rewrite $\dfrac{\log_2 x}{2}$ as $\dfrac{1}{2}\log_2 x$: $= \dfrac{1}{2}\log_2 x + 3 \log_2 y$

Use the power rule: $= \log_2 x^{1/2} + \log_2 y^3$

Use the product rule: $= \log_2 x^{1/2} y^3$ or $\log_2 \sqrt{x}\, y^3$

Therefore, the expression $\log_4 x + 3 \log_2 y$ is equivalent to $\log_2 \sqrt{x}\, y^3$. Note that we could have chosen to rewrite the original expression as a single logarithm involving base 4.

Check your **answer**, or watch this **video** to see the complete solution.

You Try It Work through this You Try It problem.

Work Exercises 41–44 in this eText or in the MyMathLab Study Plan.

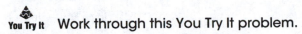 *My video summary* ▶ **Example 9 Using the Change of Base Formula to Solve Logarithmic Equations**

Use the change of base formula and the properties of logarithms to solve the equation

$$2 \log_3 x = \log_9 16.$$

Solution Try this problem on your own. Check your **answer**, or watch this **video** for a complete solution.

You Try It Work through this You Try It problem.

Work Exercises 45–48 in this eText or in the MyMathLab Study Plan.

9.6 Exercises

In Exercises 1–10, use the product rule, quotient rule, or power rule to expand each logarithmic expression. Wherever possible, evaluate logarithmic expressions.

1. $\log_4 (xy)$

2. $\ln 5e^2$

3. $\log 100P$

4. $\log_{\sqrt{2}} 8x$

5. $\log \left(\dfrac{9}{t} \right)$

6. $\ln \left(\dfrac{e^5}{r} \right)$

7. $\log_2 \left(\dfrac{M}{32} \right)$

8. $\log_5 y^3$

9. $\log_3 w^{-9}$

10. $\log_9 \sqrt[4]{k}$

In Exercises 11–20, use the properties of logarithms to expand each logarithmic expression. Wherever possible, evaluate logarithmic expressions.

11. $\log_7 x^2 y^3$

12. $\ln \dfrac{a^2 b^3}{c^4}$

13. $\log \dfrac{\sqrt{x}}{10y^3}$

14. $\log_3 9(x^2 - 25)$

15. $\log_2 \sqrt{4xy}$

16. $\log_5 \dfrac{\sqrt{5x^5}}{\sqrt[3]{25y^4}}$

17. $\ln \dfrac{\sqrt[5]{ez}}{\sqrt{x-1}}$

18. $\log_3 \sqrt[4]{\dfrac{x^3 y^5}{9}}$

19. $\ln \left[\dfrac{x+1}{(x^2-1)^3} \right]^{2/3}$

20. $\log \dfrac{(10x)^3 \sqrt{x-4}}{(x^2-16)^5}$

In Exercises 21–30, use properties of logarithms to rewrite each expression as a single logarithm. Wherever possible, evaluate logarithmic expressions.

21. $\log_b A + \log_b C$

22. $\log_4 M - \log_4 N$

23. $2 \log_8 x + \dfrac{1}{3} \log_8 y$

24. $\log 20 + \log 5$

25. $\log_2 80 - \log_2 5$

26. $\ln \sqrt{x} - \dfrac{1}{3} \ln x + \ln \sqrt[4]{x}$

27. $\log_5 (x-2) + \log_5 (x+2)$

28. $\log_9 (x^2 - 5x + 6) - \log_9 (x^2 - 4) + \log_9 (x + 2)$

29. $\log (x - 3) + 2 \log (x + 3) - \log (x^3 + x^2 - 9x - 9)$

30. $\dfrac{1}{2} [\ln (x - 1)^2 - \ln (2x^2 - x - 1)^4] + 2 \ln (2x + 1)$

In Exercises 31–36, use the properties of logarithms and the logarithm property of equality to solve each logarithmic equation.

31. $\log_3 (2x + 1) = \log_3 11$

32. $\log_{11} \sqrt{x} = \log_{11} 6$

33. $2 \log (x + 5) = \log 12 + \log 3$

34. $\ln 5 + \ln x = \ln 7 + \ln (3x - 2)$

35. $\log_7 (x + 6) - \log_7 (x + 2) = \log_7 x$

36. $\log_2 (x + 3) + \log_2 (x - 4) = \log_2 (x + 12)$

In Exercises 37–40, use the change of base formula and a calculator to approximate each logarithmic expression. Round your answers to four decimal places.

37. $\log_4 51$

38. $\log_7 0.8$

39. $\log_{1/5} 72$

40. $\log_{\sqrt{7}} 100$

In Exercises 41–44, use the change of base formula and the properties of logarithms to rewrite each expression as a single logarithm in the indicated base.

41. $\log_3 x + 4 \log_9 w$, base 3

42. $\log_5 x + \log_{1/5} x^3$, base 5

43. $\log_{16} x^4 + \log_8 y^3 + \log_4 w^2$, base 2

44. $\log_{e^2} x^5 + \log_{e^3} x^6 + \log_{e^4} x^{12}$, base e

In Exercises 45–48, solve each logarithmic equation.

45. $\log_2 x = \log_4 25$

46. $\log_{1/3} x = \log_3 20$

47. $\log_5 x = \log_{\sqrt{5}} 6$

48. $2 \ln x = \log_{e^3} 125$

9.7 Exponential and Logarithmic Equations

THINGS TO KNOW

Before working through this section, be sure you are familiar with the following concepts:

| | VIDEO | ANIMATION | INTERACTIVE |

 You Try It

1. Solve Exponential Equations by Relating the Bases (Section 9.3, Objective 3)

 You Try It

2. Change from Exponential Form to Logarithmic Form (Section 9.5, Objective 1)

 You Try It

3. Change from Logarithmic Form to Exponential Form (Section 9.5, Objective 1)

 You Try It

4. Use the Cancellation Properties of Exponentials and Logarithms (Section 9.5, Objective 3)

You Try It   5.   Expand and Condense Logarithmic Expressions (Section 9.6, Objective 2)

 You Try It   6.   Solve Logarithmic Equations Using the Logarithm Property of Equality (Section 9.6, Objective 3)

OBJECTIVES

1 Solve Exponential Equations

2 Solve Logarithmic Equations

....

OBJECTIVE 1   SOLVE EXPONENTIAL EQUATIONS

We have already solved exponential equations using the **method of relating the bases**. For example, we can solve the equation $4^{x+3} = \dfrac{1}{2}$ by converting the base on both sides of the equation to base 2. View the **details** of how to solve this equation.

But suppose we are given an exponential equation in which the bases cannot be related, such as $2^{x+1} = 3$. Remember in **Section 9.3, Example 3**, we wanted to find the one x-intercept of the graph of $f(x) = -2^{x+1} + 3$ (Figure 36).

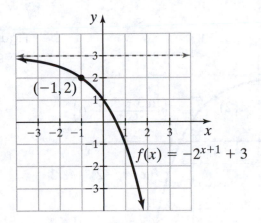

Figure 36
Graph of $f(x) = -2^{x+1} + 3$

To find the x-intercept of $f(x) = -2^{x+1} + 3$, we need to set $f(x) = 0$ and solve for x.

$$f(x) = -2^{x+1} + 3$$
$$0 = -2^{x+1} + 3$$
$$2^{x+1} = 3$$

In **Section 9.3**, we could not solve this equation for x because we had not yet defined the logarithm. We can now use some properties of logarithms to solve this equation. Recall the following logarithmic properties.

If $u = v$, then $\log_b u = \log_b v$.	Logarithm Property of Equality
$\log_b u^r = r \log_b u$	Power Rule for Logarithms

We can use these two properties to solve the equation $2^{x+1} = 3$ and thus determine the x-intercept of $f(x) = -2^{x+1} + 3$. We solve the equation $2^{x+1} = 3$ in Example 1.

My video summary ▶ **Example 1 Solving an Exponential Equation**

Solve $2^{x+1} = 3$.

Solution Read through the following solution, or watch this **video**.

Write the original equation: $\qquad\qquad 2^{x+1} = 3$

Use the **logarithm property of equality**: $\qquad \ln 2^{x+1} = \ln 3$

Use the power rule for logarithms: $\quad (x + 1)\ln 2 = \ln 3$

Use the distributive property: $\quad x \ln 2 + \ln 2 = \ln 3$

Subtract $\ln 2$ from both sides: $\qquad\quad x \ln 2 = \ln 3 - \ln 2$

Divide both sides by $\ln 2$: $\qquad\qquad\qquad x = \dfrac{\ln 3 - \ln 2}{\ln 2}$

The solution to Example 1 verifies that the **x-intercept** of $f(x) = -2^{x+1} + 3$ is $x = \dfrac{\ln 3 - \ln 2}{\ln 2} \approx 0.5850$. See Figure 37.

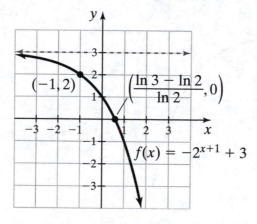

Figure 37
Graph of $f(x) = -2^{x+1} + 3$

You Try It Work through this You Try It problem.

Work Exercises 1 and 2 in this eText or in the MyMathLab Study Plan.

When we cannot easily relate the bases of an exponential equation, as in Example 1, we use logarithms and their properties to solve them. The methods used to solve exponential equations are outlined as follows.

Solving Exponential Equations

- If the equation can be written in the form $b^u = b^v$, then solve the equation $u = v$.
- If the equation cannot easily be written in the form $b^u = b^v$,
  1. Use the logarithm property of equality to "take the log of both sides" (typically using base 10 or base e).
  2. Use the product rule of logarithms to "bring down" any exponents.
  3. Solve for the given variable.

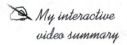 *My interactive video summary*

 **Example 2 Solving Exponential Equations**

Solve each equation. For part b, round to four decimal places.

a. $3^{x-1} = \left(\dfrac{1}{27}\right)^{2x+1}$

b. $7^{x+3} = 4^{2-x}$

Solutions Watch the interactive video, or read through the following solutions.

a. The expression $\dfrac{1}{27}$ can be rewritten as 3^{-3}. Therefore, we can use the method of relating the bases to solve.

Write the original equation:	$3^{x-1} = \left(\dfrac{1}{27}\right)^{2x+1}$
Rewrite $\dfrac{1}{27}$ as 3^{-3}:	$3^{x-1} = (3^{-3})^{2x+1}$
Use $(b^m)^n = b^{mn}$:	$3^{x-1} = 3^{-6x-3}$
Use relating the bases (if $b^u = b^v$, then $u = v$):	$x - 1 = -6x - 3$
Add $6x$ to both sides, and add 1 to both sides:	$7x = -2$
Divide both sides by 7:	$x = -\dfrac{2}{7}$

b. We cannot easily use the method of relating the bases because we cannot easily write both 7 and 4 using a common base. Therefore, we use logarithms to solve.

Write the original equation:	$7^{x+3} = 4^{2+x}$
Use the logarithm property of equality:	$\ln 7^{x+3} = \ln 4^{2-x}$
Use the power rule for logarithms:	$(x + 3)\ln 7 = (2 - x)\ln 4$
Use the distributive property:	$x \ln 7 + 3 \ln 7 = 2 \ln 4 - x \ln 4$
Add $x \ln 4$ to both sides, and subtract $3 \ln 7$ from both sides:	$x \ln 7 + x \ln 4 = 2 \ln 4 - 3 \ln 7$
Factor out an x from the left-hand side:	$x(\ln 7 + \ln 4) = 2 \ln 4 - 3 \ln 7$

Divide both sides by $\ln 7 + \ln 4$: $\quad x = \dfrac{2 \ln 4 - 3 \ln 7}{\ln 7 + \ln 4}$

Use the power rule for logarithms in the numerator, and use the product rule for logarithms in the denominator: $\quad = \dfrac{\ln 16 - \ln 343}{\ln 28}$

Use the quotient rule for logarithms to rewrite $\ln 16 - \ln 343$ as $\ln\left(\dfrac{16}{343}\right)$: $\quad = \dfrac{\ln\left(\dfrac{16}{343}\right)}{\ln 28}$

Use a calculator to round to four decimal places: $\quad \approx -0.9199$

You Try It Work through this You Try It problem.

Work Exercises 3–10 in this eText or in the MyMathLab Study Plan.

My interactive video summary

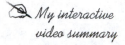

Example 3 Solving Exponential Equations Involving the Natural Exponential Function

Solve each equation. Round to four decimal places.

a. $25e^{x-5} = 17$ 　　　　 b. $e^{2x-1} \cdot e^{x+4} = 11$

Solutions Read through the following, or watch the **interactive video** to see the solutions worked out in detail.

a. Isolate the exponential term on the left by dividing both sides of the equation by 25.

Write the original equation: $\quad 25e^{x-5} = 17$

Divide both sides by 25: $\quad e^{x-5} = \dfrac{17}{25}$

Use the logarithm property of equality: $\quad \ln e^{x-5} = \ln \dfrac{17}{25}$

Use cancellation property (2) to rewrite $\ln e^{x-5}$ as $x - 5$: $\quad x - 5 = \ln \dfrac{17}{25}$

Add 5 to both sides: $\quad x = \ln \dfrac{17}{25} + 5$

Use a calculator to round to four decimal places: $\quad \approx 4.6143$

b.　Write the original equation: $\quad e^{2x-1} \cdot e^{x+4} = 11$

Use $b^m \cdot b^n = b^{m+n}$: $\quad e^{(2x-1)+(x+4)} = 11$

Combine like terms in the exponent: $\quad e^{3x+3} = 11$

Use the logarithm property of equality: $\quad \ln e^{3x+3} = \ln 11$

Use cancellation property (2) to rewrite $\ln e^{3x+3}$ as $3x + 3$: $\quad 3x + 3 = \ln 11$

Subtract 3 from both sides: $\quad 3x = \ln 11 - 3$

Divide both sides by 3: $\qquad x = \dfrac{\ln 11 - 3}{3}$

Use a calculator to round to four
decimal places: $\qquad \approx -0.2007$

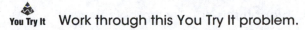 **You Try It** Work through this You Try It problem.

Work Exercises 11–15 in this eText or in the MyMathLab Study Plan.

OBJECTIVE 2   SOLVE LOGARITHMIC EQUATIONS

We now turn our attention to solving logarithmic equations. In **Section 9.6**, we learned how to solve certain logarithmic equations by using the **logarithm property of equality**. That is, if we can write a logarithmic equation in the form $\log_b u = \log_b v$, then $u = v$. Before we look at an example, let's review three of the properties of logarithms.

Properties of Logarithms

If $b > 0, b \neq 1, u$ and v represent positive numbers and r is any real number, then

$$\log_b uv = \log_b u + \log_b v \qquad \textbf{Product Rule for Logarithms}$$

$$\log_b \frac{u}{v} = \log_b u - \log_b v \qquad \textbf{Quotient Rule for Logarithms}$$

$$\log_b u^r = r \log_b u \qquad \textbf{Power Rule for Logarithms}$$

My video summary ▶ **Example 4** Solving a Logarithmic Equation Using the Logarithm Property of Equality

Solve $2 \log_5 (x - 1) = \log_5 64$.

Solution Read through the following solution, or watch this **video**.

We can use the **power rule for logarithms** and the **logarithmic property of equality** to solve.

Write the original equation: $2 \log_5 (x - 1) = \log_5 64$

Use the power rule: $\log_5 (x - 1)^2 = \log_5 64$

Use the logarithm property of equality: $(x - 1)^2 = 64$

Use the square root property: $x - 1 = \pm 8$

Solve for x: $x = 1 \pm 8$

Simplify: $x = 9$ or $x = -7$

Recall that the **domain** of a logarithmic function must contain only positive numbers; thus, $x - 1$ must be positive. Therefore, the solution of $x = -7$ must be discarded. The only solution is $x = 9$. You may want to review how to determine the domain of a logarithmic function, which is discussed in **Section 9.5**.

 You Try It Work through this You Try It problem.

Work Exercises 16–19 in this eText or in the MyMathLab Study Plan.

When solving logarithmic equations, it is important to always verify the solutions. Logarithmic equations often lead to **extraneous solutions**, as in Example 4.

When a logarithmic equation cannot be written in the form $\log_b u = \log_b v$, we adhere to the steps outlined as follows:

Solving Logarithmic Equations

1. Determine the domain of the variable.

2. Use properties of logarithms to combine all logarithms, and write as a single logarithm, if needed.

3. Eliminate the logarithm by rewriting the equation in exponential form. (Review how to change from logarithmic form to exponential form, which is discussed in **Section 9.5**.)

4. Solve for the given variable.

5. Check for any extraneous solutions. Verify that each solution is in the domain of the variable.

My video summary ▶ **Example 5 Solving a Logarithmic Equation**

Solve $\log_4 (2x - 1) = 2$.

Solution Read through the following solution, or watch this **video**.

The domain of the variable in this equation is the solution to the inequality

$2x - 1 > 0$ or $x > \dfrac{1}{2}$.

Thus, our solution must be greater than $\dfrac{1}{2}$. Because the equation involves a single logarithm, proceed to the third step.

$$
\begin{aligned}
\text{Write the original equation:} && \log_4 (2x - 1) &= 2 \\
\text{Rewrite in exponential form:} && 4^2 &= 2x - 1 \\
\text{Simplify:} && 16 &= 2x - 1 \\
\text{Add 1 to both sides:} && 17 &= 2x \\
\text{Divide by 2:} && x &= \frac{17}{2}
\end{aligned}
$$

Because the solution satisfies the inequality, there are no extraneous solutions.

Verify the solution by substituting $x = \dfrac{17}{2}$ into the original equation.

Check: Write the original equation: $\log_4(2x - 1) = 2$

Substitute $x = \dfrac{17}{2}$: $\log_4\left(2\left(\dfrac{17}{2}\right) - 1\right) \overset{?}{=} 2$

Simplify: $\log_4(17 - 1) \overset{?}{=} 2$

$\log_4(16) = 2$

This is a true statement because $4^2 = 16$.

You Try It   Work through this **You Try It** problem.

Work Exercises 20–24 in this eText or in the MyMathLab Study Plan.

 My interactive video summary   **Example 6 Solving a Logarithmic Equation**

Solve $\log_2(x + 10) + \log_2(x + 6) = 5$.

Solution Read through the following solution, or watch this **interactive video**.

The domain of the variable in this equation is the solution to the compound inequality $x + 10 > 0$ and $x + 6 > 0$. The solution to this **compound inequality** is $x > -6$. (You may want to review compound inequalities from **Section 1.3**.)

Write the original equation: $\log_2(x + 10) + \log_2(x + 6) = 5$

Use the product rule: $\log_2(x + 10)(x + 6) = 5$

Rewrite in exponential form: $(x + 10)(x + 6) = 2^5$

Simplify: $x^2 + 16x + 60 = 32$

Subtract 32 from both sides: $x^2 + 16x + 28 = 0$

Factor: $(x + 14)(x + 2) = 0$

Use the **zero product property** to solve: $x = -14$ or $x = -2$

Because the domain of the variable is $x > -6$, we must *exclude* the solution $x = -14$. Therefore, the only solution to this logarithmic equation is $x = -2$.

You Try It   Work through this **You Try It** problem.

Work Exercises 25–30 in this eText or in the MyMathLab Study Plan.

My video summary   ▶ **Example 7 Solving a Logarithmic Equation**

Solve $\ln(x - 4) - \ln(x - 5) = 2$. Round to four decimal places.

Solution Read through the following solution, or watch this **video**.

The domain of the variable is the solution to the **compound inequality** $x - 4 > 0$ and $x - 5 > 0$. The solution to this compound inequality is $x > 5$. (You may want to review compound inequalities from **Section 1.3**.)

Write the original equation: $\ln(x-4) - \ln(x-5) = 2$

Use the **quotient rule**: $\ln\left(\dfrac{x-4}{x-5}\right) = 2$

Rewrite in exponential form: $e^2 = \dfrac{x-4}{x-5}$

Multiply both sides by $x - 5$: $e^2(x-5) = x - 4$

Use the distributive property: $e^2 x - 5e^2 = x - 4$

Add $5e^2$ to both sides and subtract x from both sides: $e^2 x - x = 5e^2 - 4$

Factor out an x from the left-hand side: $x(e^2 - 1) = 5e^2 - 4$

Solve for x. Use a calculator to round to four decimal places: $x = \dfrac{5e^2 - 4}{e^2 - 1} \approx 5.1565$

We approximate the exact answer $x = \dfrac{5e^2 - 4}{e^2 - 1}$ in order to verify that the solution is in the domain of the variable. In this example, we see that 5.1565 is clearly greater than 5. In some cases, we may need to use the exact answer to verify a solution to a logarithmic equation. View the **verification** for Example 7.

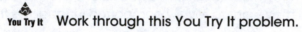

You Try It Work through this You Try It problem.

Work Exercises 31 and 32 in this eText or in the MyMathLab Study Plan.

9.7 Exercises

In Exercises 1–15, solve each exponential equation. For **irrational** solutions, round to four decimal places.

1. $3^x = 5$

2. $2^{x/3} = 19$

3. $4^{x^2 - 2x} = 64$

4. $3^{x+7} = -20$

5. $(1.52)^{-3x/7} = 11$

6. $\left(\dfrac{1}{5}\right)^{x-1} = 25^x$

7. $8^{4x-7} = 11^{5+x}$

8. $3(9)^{x-1} = (81)^{2x+1}$

9. $(3.14)^x = \pi^{1-2x}$

10. $7(2 - 10^{4x-2}) = 8$

11. $e^x = 2$

12. $150e^{x-4} = 5$

13. $e^{x-3} \cdot e^{3x+7} = 24$

14. $2(e^{x-1})^2 \cdot e^{3-x} = 80$

15. $8e^{-x/3} \cdot e^x = 1$

In Exercises 16–30, solve each logarithmic equation.

16. $\log_4(x+1) = \log_4(6x-5)$

17. $\log_3(x^2 - 21) = \log_3 4x$

18. $2\log_5(3-x) - \log_5 2 = \log_5 18$

19. $2\ln x - \ln(2x-3) = \ln 2x - \ln(x-1)$

20. $\log_2 (4x - 7) = 3$

21. $\log (1 - 5x) = 2$

22. $\log_3 (2x - 5) = -2$

23. $\log_x 3 = -1$

24. $\log_{x/2} 16 = 2$

25. $\log_2 (x - 2) + \log_2 (x + 2) = 5$

26. $\log_7 (x + 9) + \log_7 (x + 15) = 1$

27. $\log_2 (3x + 1) - \log_2 (x - 2) = 2$

28. $\log_4 (x - 7) + \log_4 x = \dfrac{3}{2}$

29. $\ln 3 + \ln \left(x^2 + \dfrac{2x}{3} \right) = 0$

30. $\log_2 (x - 4) + \log_2 (x + 6) = 2 + \log_2 x$

In Exercises 31 and 32, solve each logarithmic equation. Round to four decimal places.

31. $\ln x - \ln (x + 6) = 1$

32. $\ln (x + 3) - \ln (x - 2) = 4$

9.8 Applications of Exponential and Logarithmic Functions

THINGS TO KNOW

Before working through this section, be sure you are familiar with the following concepts:

VIDEO   ANIMATION   INTERACTIVE

 You Try It
1. Solve Applications of Exponential Functions (Section 9.3, Objective 4)

 You Try It
2. Solve Applications of the Natural Exponential Function (Section 9.4, Objective 4)

 You Try It
3. Change from Exponential Form to Logarithmic Form (Section 9.5, Objective 1)

OBJECTIVES

1 Solve Compound Interest Applications

2 Solve Exponential Growth and Decay Applications

3 Solve Logistic Growth Applications

4 Use Newton's Law of Cooling

OBJECTIVE 1   SOLVE COMPOUND INTEREST APPLICATIONS

In this section, we revisit some applications that are discussed previously in this chapter and then introduce several new applications. The difference between these applications is that we can now solve for variables that appear as exponents. We start with applications involving **compound interest**. You may want to review **periodic compound interest** from **Section 9.3** and **continuous compound interest** from **Section 9.4** before proceeding.

From Sections 9.3 and 9.4, the formulas for compound interest and continuous compound interest are presented as follows for convenience.

Compound Interest Formulas

Periodic Compound Interest Formula

$$A = P\left(1 + \frac{r}{n}\right)^{nt}$$

Continuous Compound Interest Formula

$$A = Pe^{rt}$$

where

A = Total amount after t years
P = Principal (original investment)
r = Interest rate per year
n = Number of times interest is compounded per year
t = Number of years

My video summary ▶ **Example 1 Finding the Doubling Time**

How long will it take (in years and months) for an investment to double if it earns 7.5% compounded monthly?

Solution Watch this video, or read through the following solution.

We use the **periodic compound interest formula** with $r = 0.075$ and $n = 12$ and solve for t. Notice that the **principal** is not given. As it turns out, any value of P will suffice. If the principal is P, then the amount needed to double the investment is $A = 2P$. Now we can solve for t:

Use the periodic compound interest formula: $\quad A = P\left(1 + \frac{r}{n}\right)^{nt}$

Substitute the appropriate values: $\quad 2P = P\left(1 + \frac{0.075}{12}\right)^{12t}$

Divide both sides by P: $\quad 2 = \left(1 + \frac{0.075}{12}\right)^{12t}$

Simplify within the parentheses: $\quad 2 = (1.00625)^{12t}$

Use the logarithm property of equality: $\quad \ln 2 = \ln (1.00625)^{12t}$

Use the power rule: $\log_b u^r = r \log_b u$: $\quad \ln 2 = 12t \ln (1.00625)$

Divide both sides by $12 \ln (1.00625)$: $\quad t = \dfrac{\ln 2}{12 \ln (1.00625)}$

Round to two decimal places: $\quad t \approx 9.27$ years

Note that 0.27 years = 0.27 years × $\dfrac{12 \text{ months}}{1 \text{ year}}$ = 3.24 months. Because the interest is compounded at the end of each month, the investment will not double until 9 years and 4 months.

You Try It Work through this You Try It problem.

Work Exercises 1–3 in this eText or in the MyMathLab Study Plan.

My video summary ▶ **Example 2 Continuous Compound Interest**

Suppose an investment of \$5000 compounded continuously grew to an amount of \$5130.50 in 6 months. Find the interest rate, and then determine how long it will take for the investment to grow to \$6000. Round the interest rate to the nearest hundredth of a percent and the time to the nearest hundredth of a year.

Solution Because the investment is compounded continuously, we use the formula $A = Pe^{rt}$.

We are given that $P = 5000$, so $A = 5000e^{rt}$. In 6 months, or when $t = 0.5$ years, we know that $A = 5130.50$. Substituting these values into our formula will enable us to solve for r:

Substitute the appropriate values: $\qquad 5130.50 = 5000e^{r(0.5)}$

Divide by 5000: $\qquad \dfrac{5130.50}{5000} = e^{0.5r}$

Use the logarithm property of equality: $\qquad \ln\left(\dfrac{5130.50}{5000}\right) = \ln e^{0.5r}$

Use cancellation property (2) to rewrite $\ln e^{0.5r}$ as $0.5r$: $\qquad \ln\left(\dfrac{5130.50}{5000}\right) = 0.5r$

Divide by 0.5: $\quad r = \dfrac{\ln\left(\dfrac{5130.50}{5000}\right)}{0.5} \approx 0.05153$

Therefore, the interest rate is 5.15%. To find the time that it takes for the investment to grow to \$6000, we use the formula $A = Pe^{rt}$, with $A = 6000$, $P = 5000$, and $r = 0.0515$, and solve for t. Check your **answer**, or watch the **video** to see the full solution.

You Try It Work through this You Try It problem.

Work Exercises 4–7 in this eText or in the MyMathLab Study Plan.

OBJECTIVE 2 SOLVE EXPONENTIAL GROWTH AND DECAY APPLICATIONS

In **Section 9.4**, the exponential growth model is introduced. This model is used when a population grows at a rate proportional to the size of its current population, so is often called the uninhibited growth model. We review this exponential growth model and sketch the graph in Figure 38.

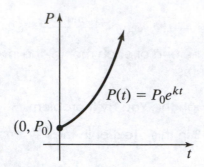

Figure 38
Graph of $P(t) = P_0e^{kt}$ for $k > 0$

$P(t) = P_0e^{kt}$

$(0, P_0)$

Exponential Growth

A model that describes the exponential uninhibited growth of a population, P, after a certain time, t, is

$$P(t) = P_0e^{kt},$$

where $P_0 = P(0)$ is the initial population and $k > 0$ is a constant called the **relative growth rate**. (**Note:** k is sometimes given as a percent.)

My video summary ▶ **Example 3 Population Growth**

The population of a small town grows at a rate proportional to its current size. In 1900, the population was 900. In 1920, the population had grown to 1600. What was the population of this town in 1950? Round to the nearest whole number.

Solution Watch this video, or read the following solution.

Using the model $P(t) = P_0e^{kt}$, we must first determine the constants P_0 and k. The initial population was 900 in 1900 so $P_0 = 900$. Therefore, $P(t) = 900e^{kt}$. To find k, we use the fact that in 1920, or when $t = 20$, the population was 1600.

$$\text{Substitute } P(20) = 1600: \quad P(20) = 900e^{k(20)} = 1600$$

$$900e^{20k} = 1600$$

$$\text{Divide by 900 and simplify:} \quad e^{20k} = \frac{16}{9}$$

$$\text{Use the logarithm property of equality:} \quad \ln e^{20k} = \ln \frac{16}{9}$$

Use **cancellation property (2)** to rewrite
$\ln e^{20k}$ as $20k$:
$$20k = \ln \frac{16}{9}$$

$$\text{Divide by 20:} \quad k = \frac{\ln\left(\frac{16}{9}\right)}{20}$$

The function that models the population of this town at any time t is given by
$P(t) = 900e^{\frac{\ln(16/9)}{20}t}$. To determine the population in 1950, or when $t = 50$, we evaluate $P(50)$:

$$P(50) = 900e^{\frac{\ln(16/9)}{20}(50)} \approx 3793$$

The population of this town in 1950 was 3793.

(eText Screens 9.8-1–9.8-28)

You Try It Work through this You Try It problem.

Work Exercises 8–12 in this eText or in the MyMathLab Study Plan.

EXPONENTIAL DECAY

Some populations exhibit *negative exponential growth*. In other words, the population, quantity, or amount *decreases* over time. Such models are called **exponential decay** models. The only difference between an exponential growth model and an exponential decay model is that the constant, k, is less than zero in an exponential decay model. See Figure 39.

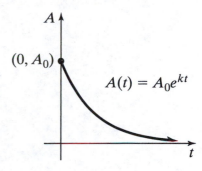

Figure 39
Graph of $A(t) = A_0 e^{kt}$ for $k < 0$

Exponential Decay

A model that describes the exponential decay of a population, quantity, or amount A, after a certain time, t, is

$$A(t) = A_0 e^{kt},$$

where $A_0 = A(0)$ is the initial quantity and $k < 0$ is a constant called the **relative decay constant**. (**Note:** k is sometimes given as a percent.)

HALF-LIFE

 *My animation summary*

Every radioactive element has a **half-life**, which is the required time for a given quantity of that element to decay to half of its original mass. For example, the half-life of Cesium-137 is 30 years. Thus, it takes 30 years for any amount of Cesium-137 to decay to $\frac{1}{2}$ of its original mass. It takes an additional 30 years to decay to $\frac{1}{4}$ of its original mass and so on. See Figure 40, and view the **animation** that illustrates the half-life of Cesium-137.

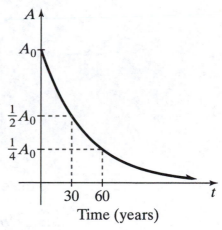

Figure 40
Half-Life of Cesium-137

My video summary ▶ **Example 4 Radioactive Decay**

Suppose that a meteorite is found containing 4% of its original Krypton-99. If the half-life of Krypton-99 is 80 years, how old is the meteorite? Round to the nearest year.

Solution Watch this video, or read the following solution.

We use the formula $A(t) = A_0 e^{kt}$, where A_0 is the original amount of Krypton-99. We first must find the constant k. To find k, we use the fact that the half-life of Krypton-99 is 80 years. Therefore, $A(80) = \frac{1}{2}A_0$. Because $A(80) = A_0 e^{k(80)}$, we can set $\frac{1}{2}A_0 = A_0 e^{k(80)}$ and solve for k.

$$\frac{1}{2}A_0 = A_0 e^{k(80)}$$

Divide both sides by A_0: $\quad \frac{1}{2} = e^{80k}$

Use the **logarithm property of equality**: $\quad \ln \frac{1}{2} = \ln e^{80k}$

Use **cancellation property (2)** to rewrite $\ln e^{80k}$ as $80k$: $\quad \ln \frac{1}{2} = 80k$

Divide both sides by 80: $\quad \dfrac{\ln \frac{1}{2}}{80} = k$

$\ln \frac{1}{2} = \ln 1 - \ln 2 = 0 - \ln 2 = -\ln 2$: $\quad \dfrac{-\ln 2}{80} = k$

Now that we know $k = \dfrac{-\ln 2}{80}$, our function becomes $A(t) = A_0 e^{\frac{-\ln 2}{80}t}$. To find out the age of the meteorite, we set $A(t) = 0.04A_0$ because the meteorite now contains 4% of the original amount of Krypton-99.

Substitute $0.04A_0$ for $A(t)$: $\quad 0.04A_0 = A_0 e^{\frac{-\ln 2}{80}t}$

Divide both sides by A_0: $\quad 0.04 = e^{\frac{-\ln 2}{80}t}$

Use the **logarithm property of equality**: $\quad \ln 0.04 = \ln e^{\frac{-\ln 2}{80}t}$

Use **cancellation property (2)** to rewrite $\ln e^{\frac{-\ln 2}{80}t}$ as $\dfrac{-\ln 2}{80}t$: $\quad \ln 0.04 = \dfrac{-\ln 2}{80}t$

Divide both sides by $\dfrac{-\ln 2}{80}$: $\quad \dfrac{\ln 0.04}{\frac{-\ln 2}{80}} = t \approx 372 \text{ years}$

The meteorite is about 372 years old.

You Try It Work through this You Try It problem.

Work Exercises 13–16 in this eText or in the MyMathLab Study Plan.

OBJECTIVE 3 SOLVE LOGISTIC GROWTH APPLICATIONS

The **uninhibited exponential growth model** $P(t) = P_0 e^{kt}$ for $k > 0$ is used when there are no outside limiting factors such as predators or disease that affect the population growth. When such outside factors exist, scientists often use a **logistic model** to describe the population growth. One such logistic model is described and sketched in Figure 41.

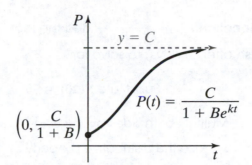

Figure 41

Graph of $P(t) = \dfrac{C}{1 + Be^{kt}}$

Definition   Logistic Growth

A model that describes the logistic growth of a population P at any time t is given by the function

$$P(t) = \frac{C}{1 + Be^{kt}},$$

where B, C, and k are constants with $C > 0$ and $k < 0$.

The number C is called the **carrying capacity**. In the logistic model, the population will approach the value of the carrying capacity over time but never exceed it. You can see in the graph sketched in Figure 41 that the graph of the logistic growth model approaches the **horizontal asymptote** $y = C$.

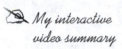
My interactive video summary

Example 5 Logistic Growth

Ten goldfish were introduced into a small pond. Because of limited food, space, and oxygen, the carrying capacity of the pond is 400 goldfish. The goldfish population at any time t, in days, is modeled by the logistic growth function $F(t) = \dfrac{C}{1 + Be^{kt}}$. If 30 goldfish are in the pond after 20 days,

a. Find B.

b. Find k.

c. When will the pond contain 250 goldfish? Round to the nearest whole number.

Solutions Work through this interactive video, or read the following solutions.

a. The carrying capacity is 400; thus, $C = 400$. Also, initially (at $t = 0$), there were 10 goldfish, so $F(0) = 10$. Therefore,

Substitute $C = 400$ and $F(0) = 10$: $\quad 10 = \dfrac{400}{1 + Be^{k(0)}}$

Evaluate $e^0 = 1$: $\quad 10 = \dfrac{400}{1 + B}$

Multiply both sides by $1 + B$: $\quad 10 + 10B = 400$

Solve for B: $\quad B = 39$

b. Use the function $F(t) = \dfrac{400}{1 + 39e^{kt}}$ and the fact that $F(20) = 30$ (there are 30 goldfish after 20 days) to solve for k.

Substitute $F(20) = 30$:	$30 = \dfrac{400}{1 + 39e^{k(20)}}$
Multiply both sides by $1 + 39e^{20k}$:	$30(1 + 39e^{20k}) = 400$
Use the distributive property:	$30 + 1170e^{20k} = 400$
Subtract 30 from both sides:	$1170e^{20k} = 370$
Divide both sides by 1170:	$e^{20k} = \dfrac{370}{1170}$
Simplify:	$e^{20k} = \dfrac{37}{117}$
Use the logarithm property of equality:	$\ln e^{20k} = \ln \dfrac{37}{117}$
Use cancellation property **(2)** to rewrite $\ln e^{20k}$ as $20k$:	$20k = \ln \dfrac{37}{117}$
Divide both sides by 20:	$k = \dfrac{\ln \dfrac{37}{117}}{20}$

c. Use the function $F(t) = \dfrac{400}{1 + 39e^{\frac{\ln(37/117)}{20}t}}$, and then find t when $F(t) = 250$.

By repeating the exact same process as in part (b), we find that it will take approximately 73 days until there are 250 goldfish in the pond.

You Try It Work through this You Try It problem.

Work Exercises **17–19** in this eText or in the MyMathLab Study Plan.

OBJECTIVE 4   USE NEWTON'S LAW OF COOLING

Newton's law of cooling states that the temperature of an object changes at a rate proportional to the difference between its temperature and that of its surroundings. The function describing Newton's law of cooling is given by the following.

Newton's Law of Cooling

The temperature T of an object at any time t is given by

$$T(t) = S + (T_0 - S)e^{kt},$$

where T_0 is the original temperature of the object, S is the constant temperature of the surroundings, and k is the cooling constant.

My animation summary ✣ **Concept Animation** View the animation to learn more about Newton's Law of Cooling.

Example 6 Newton's Law of Cooling

Suppose that the temperature of a cup of hot tea obeys Newton's law of cooling. If the tea has a temperature of 200°F when it is initially poured and 1 minute later has cooled to 189°F in a room that maintains a constant temperature of 69°F, determine when the tea reaches a temperature of 146°F. Round to the nearest minute.

My video summary ▶ **Solution** Watch this **video**, or read through the following solution.

Use the formula for Newton's law of cooling with $T_0 = 200$ and $S = 69$.

Use Newton's law of cooling formula: $\quad T(t) = S + (T_0 - S)e^{kt}$

Substitute $T_0 = 200$ and $S = 69$: $\quad T(t) = 69 + (200 - 69)e^{kt}$

Simplify: $\quad T(t) = 69 + 131e^{kt}$

We now proceed to find k. The object cools to 189°F in 1 minute; thus, $T(1) = 189$. Therefore,

Substitute $T(1) = 189$: $\qquad 189 = 69 + 131e^{k(1)}$

Subtract 69 from both sides: $\qquad 120 = 131e^{k}$

Divide both sides by 131: $\qquad \dfrac{120}{131} = e^{k}$

Use the logarithm property of equality: $\quad \ln\dfrac{120}{131} = \ln e^{k}$

Use cancellation property **(2)** to rewrite $\ln e^k$ as k: $\quad \ln\dfrac{120}{131} = k$

Now that we know the cooling constant $k = \ln\dfrac{120}{131}$, we can use the function $T(t) = 69 + 131e^{\ln\frac{120}{131}t}$ to determine the value of t when $T(t) = 146$.

Set $T(t) = 146$: $\qquad 146 = 69 + 131e^{\left(\ln\frac{120}{131}\right)t}$

Subtract 69 from both sides: $\qquad 77 = 131e^{\left(\ln\frac{120}{131}\right)t}$

Divide both sides by 131: $\qquad \dfrac{77}{131} = e^{\left(\ln\frac{120}{131}\right)t}$

Use the logarithm property of equality: $\quad \ln\dfrac{77}{131} = \ln e^{\left(\ln\frac{120}{131}\right)t}$

Use cancellation property **(2)** to rewrite $\ln e^{\left(\ln\frac{120}{131}\right)t}$ as $\ln\dfrac{120}{131}t$: $\quad \ln\dfrac{77}{131} = \left(\ln\dfrac{120}{131}\right)t$

Divide both sides by $\ln\dfrac{120}{131}$: $\qquad \dfrac{\ln\dfrac{77}{131}}{\ln\dfrac{120}{131}} = t$

Use a calculator to approximate the time rounded to the nearest minute: $\qquad t \approx 6 \text{ minutes}$

So, it takes approximately 6 minutes for the tea to cool to 146°F.

The graph of $T(t) = 69 + 131e^{\ln\frac{120}{131}t}$, which describes the temperature of the tea t minutes after being poured, was created using a graphing utility. See Figure 42. Note that the line $y = 69$, which represents the temperature of the surroundings, is a horizontal asymptote.

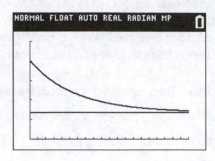

Figure 42
Graph of $T(t) = 69 + 131e^{\left(\ln\frac{120}{131}\right)t}$

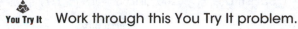

You Try It   Work through this You Try It problem.

Work Exercises 20–22 in this eText or in the MyMathLab Study Plan.

9.8 Exercises

1. How long (in years and months) will it take for an investment to double at 9% compounded monthly?

2. How long will it take for an investment to triple if it is compounded continuously at 8%? Round to two decimal places.

3. What is the interest rate necessary for an investment to quadruple after 8 years of continuous compound interest? (Round to the nearest hundredth of a percent.)

4. Jimmy invests $15,000 in an account that pays 6.25% compounded quarterly. How long (in years and months) will it take for his investment to reach $20,000?

5. How long will it take an investment of $38,500 to grow to $80,000 if interest is compounded continuously at an annual rate of 4.2%? Round to the nearest tenth of a year.

6. Marsha and Jan both invested money on March 1, 2012. Marsha invested $5000 at Bank A, where the interest was compounded quarterly. Jan invested $3000 at Bank B, where the interest was compounded continuously. On March 1, 2014, Marsha had a balance of $5468.12, whereas Jan had a balance of $3289.09. What was the interest rate at each bank? (Round to the nearest tenth of a percent.)

7. The **effective interest rate** is the annual interest rate that takes compounding into effect. That is, for compound interest, the effective interest rate is the simple interest rate that would result in the same amount of interest after one year. If $18,000 is invested at 6.3% per year, compounded continuously, what is the effective interest rate? Write your answer as a percent, rounded to two decimal places.

8. The population of Adamsville grew from 9000 to 15,000 in 6 years. Assuming uninhibited exponential growth, what is the expected population in an additional 4 years? Round to the nearest whole number.

9. During a research experiment, it was found that the number of bacteria in a culture grew at a rate proportional to its size. At 8:00 AM, there were 2000 bacteria present in the culture.

At noon, the number of bacteria grew to 2400. How many bacteria will there be at midnight? Round to the nearest whole number.

10. A skull cleaning factory cleans animal skulls such as deer, buffalo, and other types of animal skulls using flesh-eating beetles to clean the skulls. The factory owner started with only 10 adult beetles. After 40 days, the beetle population grew to 30 adult beetles. Assuming uninhibited exponential growth, how long did it take before the beetle population reached 10,000 beetles? Round to the nearest whole number.

11. The 2014 Ebola outbreak in West Africa was the first Ebola epidemic the world had ever known. The number of cases in Guinea was 86 when first reported on March 25th and had grown to 197 by April 16th. Assuming exponential growth, how many days would it take for there to be at least 1000 cases?

12. Across the globe, use of solar power has been growing exponentially. At the end of 2002, 1.6 terawatt-hours of solar power had been installed. This amount had risen to 29.9 terawatt-hours by the end of 2010. Assuming exponential growth, in what year will 5000 terawatt-hours of solar power be installed? Round to the nearest year.

13. The population of a Midwest industrial town decreased from 210,000 to 205,000 in just 3 years. Assuming negative exponential growth and that this trend continues, what will the population be after an additional 3 years? Round to the nearest whole number.

14. A certain radioactive isotope is leaked into a small stream. Three hundred days after the leak, 2% of the original amount of the substance remained. Determine the half-life of this radioactive isotope. Round to the nearest whole number.

15. Radioactive Iodine-131 is a by-product of certain nuclear reactors. On April 26, 1986, one of the nuclear reactors in Chernobyl, Ukraine, a republic of the former Soviet Union, experienced a massive release of radioactive iodine. Fortunately, Iodine-131 has a very short half-life of 8 days. Estimate the percentage of the original amount of Iodine-131 released by the Chernobyl explosion on May 1, 1986, 5 days after the explosion. Round to two decimal places.

16. Superman is rendered powerless when exposed to 50 or more grams of kryptonite. A 500-year-old rock that originally contained 300 grams of kryptonite was recently stolen from a rock museum by Superman's enemies. The half-life of kryptonite is known to be 200 years.

    a. How many grams of kryptonite are still contained in the stolen rock? Round to two decimal places.

    b. For how many years can this rock be used by Superman's enemies to render him powerless? Round to the nearest whole number.

17. The logistic growth model $H(t) = \dfrac{6000}{1 + 2e^{-0.65t}}$ represents the number of families that own a home in a certain small (but growing) Idaho city t years after 1980.

    a. What is the maximum number of families that will own a home in this city?

    b. How many families owned a home in 1980?

    c. In what year did 5920 families own a home?

18. The number of students that hear a rumor on a small college campus t days after the rumor starts is modeled by the logistic function $R(t) = \dfrac{3000}{1 + Be^{kt}}$. Determine the following if 8 students initially heard the rumor and 100 students heard the rumor after 1 day.

    a. What is the carrying capacity for the number of students who will hear the rumor?

    b. Find B.

    c. Find k.

    d. How long will it take for 2900 students to hear the rumor?

19. In 1999, 1500 runners entered the inaugural Run-for-Your-Life marathon in Joppetown, USA. In 2005, 21,500 runners entered the race. Because of the limited number of hotels, restaurants, and portable toilets in the area, the carrying capacity for the number of racers is 61,500. The number of racers at any time, t, in years, can be modeled by the logistic function $P(t) = \dfrac{C}{1 + Be^{kt}}$.

    a. What is the value of C?

    b. Find B.

    c. Find k.

    d. In what year should at least 49,500 runners be expected to have run in the race? Round to the nearest year.

20. Estabon poured himself a hot beverage that had a temperature of 198°F and then set it on the kitchen table to cool. The temperature of the kitchen was a constant 75°F. If the drink cooled to 180°F in 2 minutes, how long will it take for the drink to cool to 100°F?

21. Police arrive at a murder scene at 1:00 AM and immediately record the body's temperature, which was 92°F. At 2:30 AM, after thoroughly inspecting and fingerprinting the area, they again took the temperature of the body, which had dropped to 85°F. The temperature of the crime scene had remained at a constant 60°F. Determine when the person was murdered. (Assume that the victim was healthy at the time of death. That is, assume that the temperature of the body at the time of death was 98.6°F.)

22. Jodi poured herself a cold soda that had an initial temperature of 40°F and immediately went outside to sunbathe where the temperature was a steady 99°F. After 5 minutes, the temperature of the soda was 47°F. Jodi had to run back into the house to answer the phone. What is the expected temperature of the soda after an additional 10 minutes?

CHAPTER TEN
Conic Sections

CHAPTER TEN CONTENTS

10.1 Introduction to Conic Sections; The Parabola

THINGS TO KNOW

Before working through this section, be sure you are familiar with the following concepts:

			VIDEO	ANIMATION	INTERACTIVE
You Try It	1.	Write the Equation of a Line from Given Information (Section 2.5, Objective 6)	▶		
You Try It	2.	Solve Quadratic Equations by Completing the Square (Section 8.1, Objective 3)	▶		
You Try It	3.	Find the Distance between Two Points (Section 8.4, Objective 1)	▶	✸	
You Try It	4.	Write the General Form of a Circle in Standard Form and Sketch Its Graph (Section 8.4, Objective 5)	▶	✸	

OBJECTIVES

1 Identify Conic Sections

2 Work with the Equation of a Parabola with a Vertical Axis of Symmetry

3 Work with the Equation of a Parabola with a Horizontal Axis of Symmetry

4 Find the Equation of a Parabola Given Information about the Graph

5 Complete the Square to Find the Equation of a Parabola in Standard Form

6 Solve Applications Involving Parabolas

OBJECTIVE 1 IDENTIFY CONIC SECTIONS

In this chapter, we focus on the geometric study of *conic sections*. **Conic sections (or conics)** are curves formed when a plane intersects a pair of **right circular cones**. The surface of the cones comprises the set of all line segments that intersect the outer edges of the circular bases of the cones and pass through a fixed point. The fixed point is called the **vertex** of the cone, and the line segments are called the **elements**. See Figure 1.

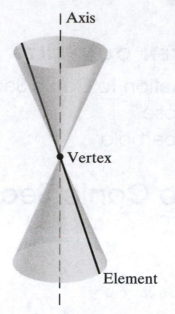

Figure 1
Pair of Right Circular Cones

The four **conic sections** that can be formed are **circles**, **parabolas**, **ellipses**, and **hyperbolas**.

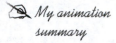 *My animation summary*

Concept Animation Watch the following four animations to see how each conic section is formed.

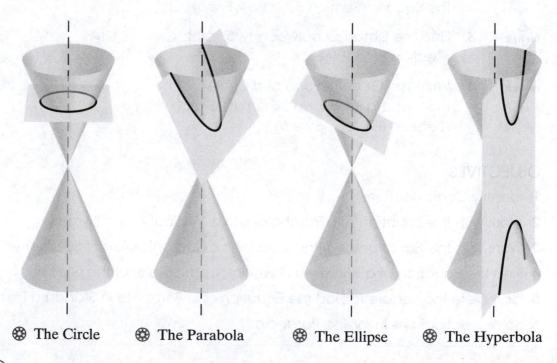

⊛ **The Circle**　⊛ **The Parabola**　⊛ **The Ellipse**　⊛ **The Hyperbola**

My video summary ▶ Because we studied circles in **Section 8.4**, they are not covered again in this chapter. It may, however, be useful to review circles before going on. Watch this **video** to review how to write the equation of a circle in standard form by **completing the square**. We also need to be able to complete the square to write each of the other conic sections in standard form.

It is worth noting that when the circle, ellipse, parabola, or hyperbola is formed, the intersecting plane does not pass through the vertex of the cones. When a plane does intersect the vertex of the cones, a **degenerate conic section** is formed. The three degenerate conic sections are a **point**, a **line**, and **a pair of intersecting lines**. See Figure 2. We do not concern ourselves with degenerate conic sections in this chapter.

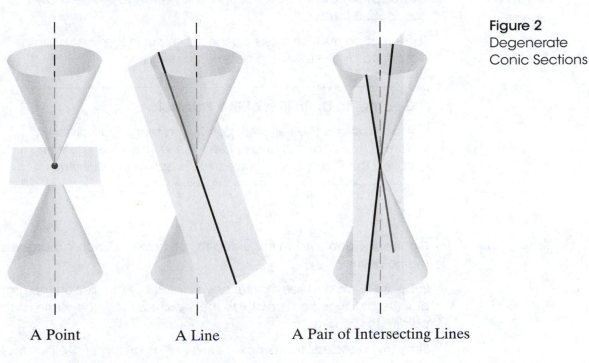

Figure 2
Degenerate
Conic Sections

A Point   A Line   A Pair of Intersecting Lines

You Try It Work through this You Try It problem.

Work Exercises 1–4 in this eText or in the MyMathLab Study Plan.

OBJECTIVE 2   WORK WITH THE EQUATION OF A PARABOLA WITH A VERTICAL AXIS OF SYMMETRY

In Section 8.2, we studied quadratic functions of the form $f(x) = ax^2 + bx + c, a \neq 0$. We learned that every quadratic function has a u-shaped graph called a *parabola*. You may want to review the different characteristics of a parabola.

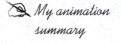 *My animation summary*

⚙ **Concept Animation** Work through the following **animation**, and view each characteristic listed below for a detailed description.

Characteristics of a Parabola

1. Vertex

2. Axis of symmetry

3. y-Intercept

4. x-Intercept(s) or real zeros

5. Domain and range

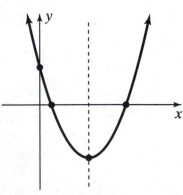

In **Section 8.2**, we studied **quadratic functions** and **parabolas** from an algebraic point of view. We now look at parabolas from a geometric perspective. Recall that when a plane is parallel to an **element** of the cone, the plane will intersect the cone in a parabola.

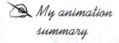 *My animation summary*

⊛ **Concept Animation** Watch this **animation** to review how a parabola is formed.

The set of points that make up the parabola in this illustration is described by the following geometric definition.

Geometric Definition of the Parabola

A **parabola** is the set of all points in a plane equidistant from a fixed point F and a fixed line D. The fixed point is called the **focus**, and the fixed line is called the **directrix**.

The Parabola

 *My video summary* ▶ **Concept video** Watch this **video** to see how to sketch a parabola using the geometric definition.

In Figure 3, we can see that for any point $P(x, y)$ that lies on the graph of the parabola, the distance from point P to the **focus** is exactly the same as the distance from point P to the **directrix**.

Similarly, because the **vertex**, V, lies on the graph of the parabola, the distance from V to the focus must also be the same as the distance from V to the directrix. Therefore, if the distance from V to F is p units, then the distance from V to the directrix is also p units.

If the coordinates of the vertex in Figure 3 are (h, k), then the coordinates of the focus must be $(h, k + p)$ and the equation of the directrix is $y = k - p$. We can use this information and the fact that the distance from $P(x, y)$ to the focus is equal to the distance from $P(x, y)$ to the directrix to derive the equation of a parabola.

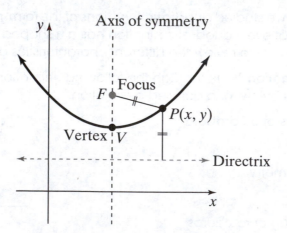

Figure 3

The distance from any point P to the focus is the same as the distance from point P to the directrix.

Equation of a Parabola in Standard Form with a Vertical Axis of Symmetry

The equation of a parabola with a vertical **axis of symmetry** is $(x - h)^2 = 4p(y - k)$, where the **vertex** is $V(h, k)$, $|p|$ = distance from the vertex to the **focus** = distance from the vertex to the **directrix**, the focus is $F(h, k + p)$, and the equation of the directrix is $y = k - p$.

The parabola opens *upward* if $p > 0$ or *downward* if $p < 0$.

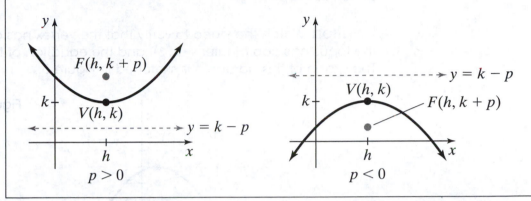

My video summary ▶ **Example 1 Finding the Vertex, Focus, and Directrix of a Parabola and Sketching Its Graph**

Find the vertex, focus, and directrix of the parabola $x^2 = 8y$ and sketch its graph.

Solution Read the following, or watch this **video** for a complete solution. Rewrite the equation $x^2 = 8y$ as $(x - 0)^2 = 8(y - 0)$. Now compare the equation $(x - 0)^2 = 8(y - 0)$ to the standard form equation $(x - h)^2 = 4p(y - k)$ to see that $h = 0$ and $k = 0$; hence, the vertex is at the origin $(0, 0)$. To find the focus and directrix, find p.

$$4p = 8$$

Divide both sides by 4: $\quad p = 2$

Because the value of p is positive, the parabola opens upward and the focus is located two units vertically *above* the vertex, while the directrix is the horizontal line located two units vertically *below* the vertex. The focus has coordinates $(0, 2)$, and the equation of the directrix is $y = -2$. The graph is shown in **Figure 4**.

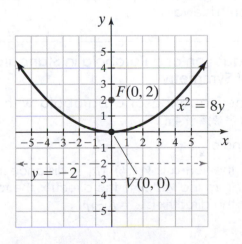

Figure 4

👤 **You Try It** Work through this You Try It problem.

Work Exercises 5 and 6 in this eText or in the MyMathLab Study Plan.

🖊 *My video summary* ▶ **Example 2 Finding the Vertex, Focus, and Directrix of a Parabola and Sketching Its Graph**

Find the vertex, focus, and directrix of the parabola $(x + 1)^2 = -4(y - 3)$ and sketch its graph.

Solution Watch the video to verify that the vertex has coordinates $(-1, 3)$, the focus has coordinates $(-1, 2)$, and the equation of the directrix is $y = 4$. The graph of this parabola is sketched in Figure 5.

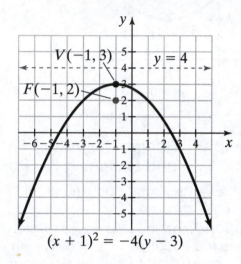

Figure 5

$$(x + 1)^2 = -4(y - 3)$$

👤 **You Try It** Work through this You Try It problem.

Work Exercises 7–10 in this eText or in the MyMathLab Study Plan.

OBJECTIVE 3   WORK WITH THE EQUATION OF A PARABOLA WITH A HORIZONTAL AXIS OF SYMMETRY

In Examples 1 and 2, the graphs of both parabolas had vertical axes of symmetry. The graph of a parabola could also have a horizontal axis of symmetry and open "sideways."

Equation of a Parabola in Standard Form with a Horizontal Axis of Symmetry

The equation of a parabola with a horizontal axis of symmetry is $(y - k)^2 = 4p(x - h)$,

where

the vertex is $V(h, k)$, $|p| =$ distance from the vertex to the focus $=$ distance from the vertex to the directrix, the focus is $F(h + p, k)$, and the equation of the directrix is $x = h - p$.

The parabola opens *right* if $p > 0$ or *left* if $p < 0$.

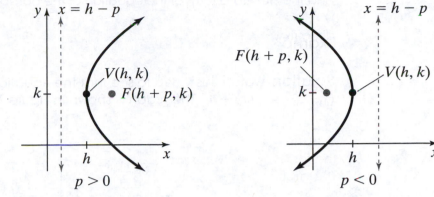

$p > 0$ $p < 0$

📝 *My video summary* ▶ **Example 3 Finding the Vertex, Focus, and Directrix of a Parabola and Sketching Its Graph**

Find the vertex, focus, and directrix of the parabola $(y - 3)^2 = 8(x + 2)$ and sketch its graph.

Solution Watch this video to verify that the vertex has coordinates $(-2, 3)$, the focus has coordinates $(0, 3)$, and the equation of the directrix is $x = -4$. The graph of this parabola is sketched in Figure 6.

Figure 6

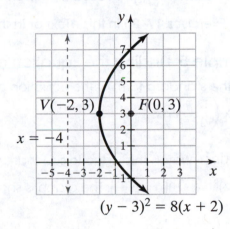

$$(y - 3)^2 = 8(x + 2)$$

🔺 **You Try It** Work through this You Try It problem.

Work Exercises 11–16 in this eText or in the MyMathLab Study Plan.

OBJECTIVE 4 FIND THE EQUATION OF A PARABOLA GIVEN INFORMATION ABOUT THE GRAPH

Sometimes we must determine the equation of a parabola given certain information. It is always useful to first determine whether the parabola has a vertical axis of symmetry or a horizontal axis of symmetry.

My video summary ▶ **Example 4 Finding the Equation of a Parabola**

Find the standard form of the equation of the parabola with focus $\left(-3, \frac{5}{2}\right)$ and directrix $y = \frac{11}{2}$.

Solution Watch this video to see that the equation of this parabola is $(x + 3)^2 = -6(y - 4)$. The graph is shown in Figure 7.

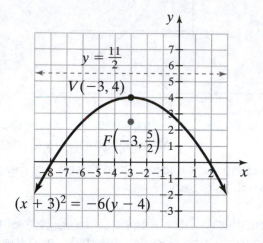

Figure 7

$$(x + 3)^2 = -6(y - 4)$$

You Try It Work through this You Try It problem.

Work Exercises 17–20 in this eText or in the MyMathLab Study Plan.

My video summary ▶ **Example 5 Finding the Equation of a Parabola**

Find the standard form of the equation of the parabola with focus $(4, -2)$ and vertex $\left(\frac{13}{2}, -2\right)$.

Solution Watch this video to see that the equation of this parabola is $(y + 2)^2 = -10\left(x - \frac{13}{2}\right)$. The graph is shown in Figure 8.

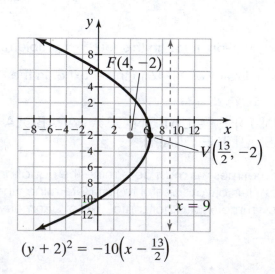

Figure 8

$$(y + 2)^2 = -10\left(x - \frac{13}{2}\right)$$

We can use a graphing utility to graph the parabola from **Example 5** by solving the equation for y.

$$(y + 2)^2 = -10\left(x - \frac{13}{2}\right)$$

$$y + 2 = \pm\sqrt{-10\left(x - \frac{13}{2}\right)}$$

$$y = -2 \pm \sqrt{-10\left(x - \frac{13}{2}\right)}$$

Using $y_1 = -2 + \sqrt{-10\left(x - \frac{13}{2}\right)}$, and $y_2 = -2 - \sqrt{-10\left(x - \frac{13}{2}\right)}$, we obtain the graph seen in Figure 9. **Note:** The directrix was created using the calculator's DRAW feature.

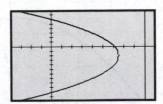

Figure 9

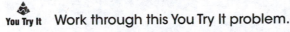 **You Try It**   Work through this You Try It problem.

Work Exercises 21–26 in this eText or in the MyMathLab Study Plan.

OBJECTIVE 5   COMPLETE THE SQUARE TO FIND THE EQUATION OF A PARABOLA IN STANDARD FORM

If the equation of a parabola is in the standard form of $(x - h)^2 = 4p(y - k)$ or $(y - k)^2 = 4p(x - h)$, it is not too difficult to determine the vertex, focus, and directrix and sketch its graph. However, the equation might not be given in standard form. If this is the case, we **complete the square** on the variable that is squared to rewrite the equation in standard form as in Example 6.

My video summary ▶ **Example 6   Writing the Equation of a Parabola in Standard Form by Completing the Square**

Find the vertex, focus, and directrix and sketch the graph of the parabola $x^2 - 8x + 12y = -52$.

Solution Read the following, or watch this **video** for a complete solution.

Because x is squared, we will complete the square on the variable x.

$$x^2 - 8x + 12y = -52$$

$$x^2 - 8x = -12y - 52$$

$$x^2 - 8x + 16 = -12y - 52 + 16$$

$$(x - 4)^2 = -12y - 36$$

$$(x - 4)^2 = -12(y + 3)$$

The equation is now in standard form with vertex $(4, -3)$ and $4p = -12$ so $p = -3$. The parabola must open down because the variable x is squared and $p < 0$. The focus is located three units below the vertex, while the directrix is three units above the vertex. Thus, the focus has coordinates $(4, -6)$, and the equation of the directrix is $y = 0$ or the x-axis. The graph is shown in Figure 10.

The directrix is the x-axis or $y = 0$.

Figure 10

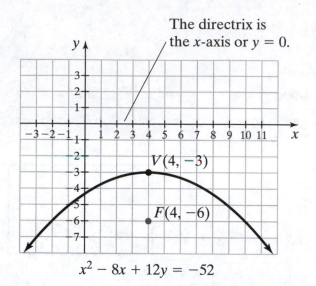

$$x^2 - 8x + 12y = -52$$

You Try It   Work through this You Try It problem.

Work Exercises 27–30 in this eText or in the MyMathLab Study Plan.

OBJECTIVE 6   SOLVE APPLICATIONS INVOLVING PARABOLAS

The Romans were one of the first civilizations to use the engineering properties of parabolic structures in their creation of **arch bridges**. The cables of many suspension bridges, such as the **Golden Gate Bridge** in San Francisco, span from tower to tower in the shape of a parabola.

Parabolic surfaces are used in the manufacture of many satellite dishes, search lights, car headlights, telescopes, lamps, heaters, and other objects. This is because parabolic surfaces have the property that incoming rays of light or radio waves traveling parallel to the **axis of symmetry** of a parabolic reflector or receiver will reflect off the parabolic surface and travel directly toward the antenna that is placed at the focus. See Figure 11. When a light source such as the headlight of a car is placed at the focus of a parabolic reflector, the light reflects off the surface outward, producing a narrow beam of light and thus maximizing the output of illumination.

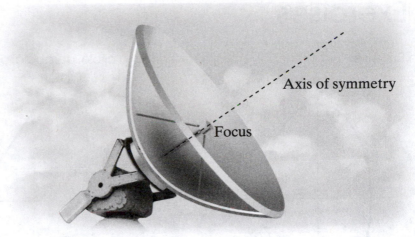

Figure 11 Incoming rays reflect off the parabolic surface toward the antenna placed at the focus.

My video summary ▶ **Example 7 Finding the Focus of a Parabolic Microphone**

Parabolic microphones can be seen on the sidelines of professional sporting events so that television networks can capture audio sounds from the players on the field. If the surface of a parabolic microphone is 27 centimeters deep and has a diameter of 72 centimeters at the top, where should the microphone be placed relative to the **vertex** of the **parabola**?

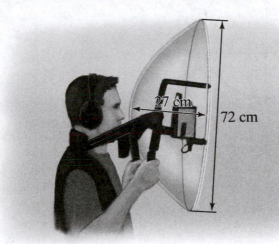

Solution Watch this **video**, or read through the following solution.

We can draw a **parabola** with the **vertex** at the origin representing the center cross section of the parabolic microphone. The equation of this parabola in standard form is $x^2 = 4py$. Substitute the point $(36, 27)$ into the equation to get

$$x^2 = 4py$$

$$(36)^2 = 4p(27)$$

$$1{,}296 = 108p$$

$$p = 12.$$

The microphone must be placed 12 centimeters from the vertex.

You Try It Work through this You Try It problem.

Work Exercises 31–35 in this eText or in the MyMathLab Study Plan.

10.1 Exercises

In Exercises 1–4, identify the conic section as a circle, parabola, ellipse, or hyperbola.

1.

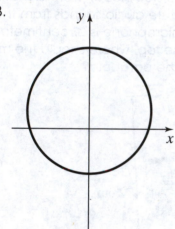

2.

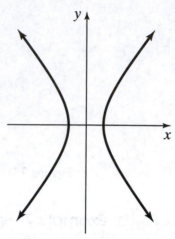

3.

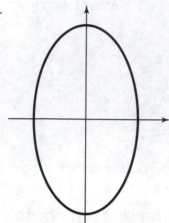

4.

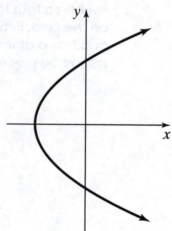

In Exercises 5–16, determine the vertex, focus, and directrix of the parabola and sketch its graph.

5. $x^2 = 16y$

6. $x^2 = -8y$

7. $(x - 1)^2 = -12(y - 4)$

8. $(x + 3)^2 = 6(y - 1)$

9. $(x + 2)^2 = 5(y + 6)$

10. $(x - 3)^2 = -8(y + 2)$

11. $y^2 = -8x$

12. $y^2 = 4x$

13. $(y - 5)^2 = -4(x - 2)$

14. $(y + 3)^2 = 20(x - 4)$

15. $(y + 4)^2 = 9(x + 3)$

16. $(y - 5)^2 = -6(x + 1)$

In Exercises 17–26, find the equation in standard form of the parabola described.

17. The focus has coordinates $(2, 0)$, and the equation of the directrix is $x = -2$.

18. The focus has coordinates $\left(0, -\dfrac{1}{2}\right)$, and the equation of the directrix is $y = \dfrac{1}{2}$.

19. The focus has coordinates $(3, -5)$, and the equation of the directrix is $y = -1$.

20. The focus has coordinates $(2, 4)$, and the equation of the directrix is $x = -4$.

21. The vertex has coordinates $\left(-\dfrac{11}{4}, -2\right)$, and the focus has coordinates $(-3, -2)$.

22. The vertex has coordinates $\left(4, -\dfrac{1}{4}\right)$, and the focus has coordinates $\left(4, \dfrac{1}{4}\right)$.

23. The vertex has coordinate $(1, 2)$, and the equation of the directrix is $y = 5$.

24. The vertex has coordinates $(-3, 4)$, and the equation of the directrix is $x = -7$.

25. Find the equations of the two parabolas in standard form that have a horizontal axis of symmetry and a focus at the point $(0, 4)$, and that pass through the origin.

26. Find the equations of the two parabolas in standard form that have a vertical axis of symmetry and a focus at the point $\left(2, \dfrac{3}{2}\right)$, and that pass through the origin.

In Exercises 27–30, determine the vertex, focus, and directrix of the parabola and sketch its graph.

27. $x^2 + 10x = 5y - 10$

28. $y^2 - 12y + 6x + 30 = 0$

29. $x^2 - 2x = -y + 1$

30. $y^2 + x + 6y = -10$

31. A parabolic eavesdropping device is used by a CIA agent to record terrorist conversations. The parabolic surface measures 120 centimeters in diameter at the top and is 90 centimeters deep at its center. How far from the vertex should the microphone be located?

32. A parabolic space heater is 18 inches in diameter and 8 inches deep. How far from the vertex should the heat source be located to maximize the heating output?

33. A large NASA parabolic satellite dish is 22 feet across and has a receiver located at the focus 4 feet from its base. The satellite dish should be how deep?

34. A parabolic arch bridge spans 160 feet at the base and is 40 feet above the water at the center. Find the equation of the parabola if the vertex is placed at the point $(0, 40)$. Can a sailboat that is 35 feet tall fit under the bridge 30 feet from the center?

35. The cable between two 40-meter towers of a suspension bridge is in the shape of a parabola that just touches the bridge halfway between the towers. The two towers are 100 meters apart. Vertical cables are spaced every 10 meters along the bridge. What are the lengths of the vertical cables located 30 meters from the center of the bridge?

10.2 The Ellipse

THINGS TO KNOW

Before working through this section, be sure you are familiar with the following concepts:

		VIDEO	ANIMATION	INTERACTIVE

You Try It 1. Find the Distance Between Two Points (Section 8.4, Objective 1) — VIDEO ▶, ANIMATION ✦

You Try It 2. Identify Conic Sections (Section 10.1, Objective 1) — ANIMATION ✦

You Try It 3. Complete the Square to Find the Equation of a Parabola in Standard Form (Section 10.1, Objective 5) — VIDEO ▶

OBJECTIVES

1 Sketch the Graph of an Ellipse

2 Find the Equation of an Ellipse Given Information about the Graph

3 Complete the Square to Find the Equation of an Ellipse in Standard Form

4 Solve Applications Involving Ellipses

OBJECTIVE 1   SKETCH THE GRAPH OF AN ELLIPSE

Recall that when a plane intersects a **right circular cone** at an angle between 0 and 90 degrees to the axis of the cone, the conic section formed is an *ellipse*.

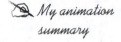 *My animation summary*

⊛ **Concept Animation** Watch this **animation** to review how an ellipse is formed. The set of points that make up the ellipse in this illustration is described by the following geometric definition.

The Ellipse

Geometric Definition of the Ellipse

An **ellipse** is the set of all points in a plane, the sum of whose distances from two fixed points is a positive constant. The two fixed points, F_1 and F_2, are called the *foci*.

The **geometric definition** of an ellipse implies that for any two points P and Q that lie on the graph of the ellipse, the sum of the distance between P and F_1 plus the distance between P and F_2 is equal to the sum of the distance between Q and F_1 plus the distance between Q and F_2. In symbols, we write $d(P, F_1) + d(P, F_2) = d(Q, F_1) + d(Q, F_2)$. See Figure 12.

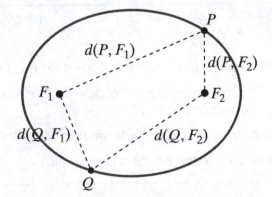

Figure 12
For any points P and Q on an ellipse, $d(P, F_1) + d(P, F_2) = d(Q, F_1) + d(Q, F_2)$.

An ellipse has two **axes of symmetry**. The longer axis, which is the line segment that connects the two **vertices**, is called the **major axis**. The foci are always located along the major axis. The shorter axis is called the **minor axis**. It is the line segment perpendicular to the major axis that passes through the center having endpoints that lie on the ellipse.

✎ *My video summary* ▶ **Concept Video** Watch this video to see how to sketch an ellipse using the geo-metric definition.

In Figure 13(a), the ellipse has a **horizontal major axis**. The ellipse in Figure 13(b) has a **vertical major axis**.

Figure 13

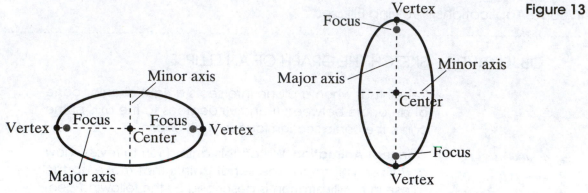

(a) Ellipse with Horizontal Major Axis   (b) Ellipse with Vertical Major Axis

Consider the ellipse with a horizontal major axis centered at (h, k), as shown in Figure 14, where $c > 0$ is the distance between the center and a focus, $a > 0$ is the distance between the center and one of the vertices, and $b > 0$ is the distance from the center to an endpoint of the **minor axis**.

Figure 14

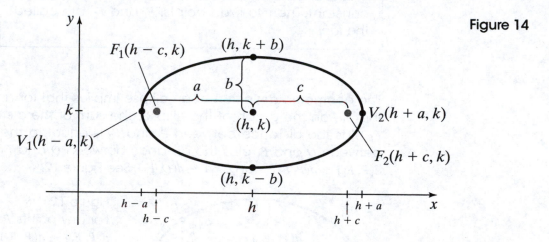

Before we can derive the equation of this ellipse, we must establish two facts:

Fact 1: The sum of the distances from any point on the ellipse to the two foci is $2a$.
Fact 2: $b^2 = a^2 - c^2$, or equivalently, $c^2 = a^2 - b^2$.

Once we have established these two facts, we can derive the equation of the ellipse.

We now state the two standard equations of an ellipse.

Equation of an Ellipse in Standard Form with Center (h, k); Horizontal Major Axis

$$\frac{(x - h)^2}{a^2} + \frac{(y - k)^2}{b^2} = 1$$

- $a > b > 0$

- Foci: $F_1(h - c, k)$ and $F_2(h + c, k)$

- Vertices: $V_1(h - a, k)$ and $V_2(h + a, k)$
- Endpoints of minor axis: $(h, k - b)$ and $(h, k + b)$
- $c^2 = a^2 - b^2$

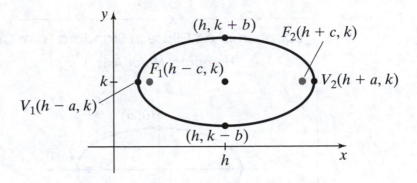

Equation of an Ellipse in Standard Form with Center (*h*, *k*); Vertical Major Axis

$$\frac{(x - h)^2}{b^2} + \frac{(y - k)^2}{a^2} = 1$$

- $a > b > 0$
- Foci: $F_1(h, k - c)$ and $F_2(h, k + c)$
- Vertices: $V_1(h, k - a)$ and $V_2(h, k + a)$
- Endpoints of minor axis: $(h - b, k)$ and $(h + b, k)$
- $c^2 = a^2 - b^2$

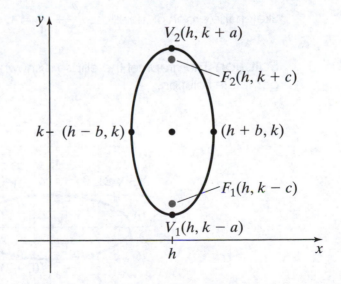

Notice that the two equations are identical except for the placement of a^2. If the equation of an **ellipse** is in standard form, we can quickly determine whether the ellipse has a **horizontal major axis** or a **vertical major axis** by looking at the denominator. If the larger denominator, a^2, appears under the x-term, then the ellipse has a horizontal major axis. If the larger denominator appears under the y-term, then the ellipse has a vertical major axis. If the denominators are equal ($a^2 = b^2$), then the ellipse is a **circle**.

If $h = 0$ and $k = 0$, then the ellipse is centered at the **origin**. Ellipses centered at the origin have the following equations.

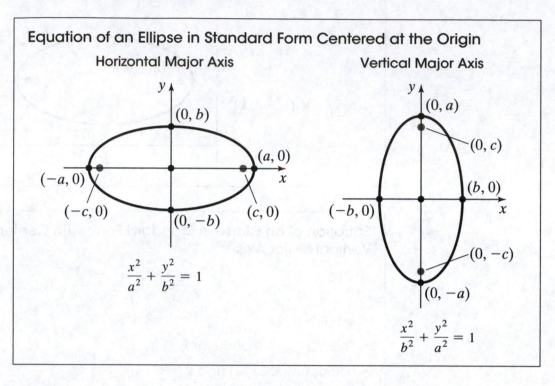

Equation of an Ellipse in Standard Form Centered at the Origin

Horizontal Major Axis

$$\frac{x^2}{a^2} + \frac{y^2}{b^2} = 1$$

Vertical Major Axis

$$\frac{x^2}{b^2} + \frac{y^2}{a^2} = 1$$

My video summary ▶ **Example 1 Sketching the Graph of an Ellipse Centered at the Origin**

Sketch the graph of the ellipse $\dfrac{x^2}{25} + \dfrac{y^2}{4} = 1$, and label the **center**, **foci**, and **vertices**.

Solution The graph of the ellipse is shown in Figure 15. Watch this **video** for a complete solution.

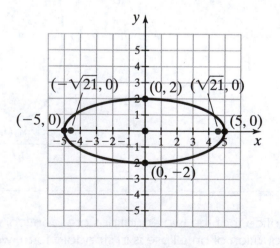

Figure 15

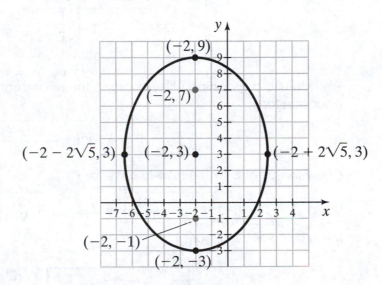

You Try It Work through this You Try It problem.

Work Exercises 1–4 in this eText or in the MyMathLab Study Plan.

My video summary ▶ **Example 2 Sketching the Graph of an Ellipse**

Sketch the graph of the ellipse $\dfrac{(x+2)^2}{20} + \dfrac{(y-3)^2}{36} = 1$, and label the **center, foci,** and **vertices.**

Solution Note that the larger denominator appears under the y-term. This indicates that the ellipse has a **vertical major axis.** Watch this **video** to verify that the center of the ellipse is $(-2, 3)$. The foci have coordinates $(-2, -1)$ and $(-2, 7)$, the vertices have coordinates $(-2, -3)$ and $(-2, 9)$, and the coordinates of the minor axis are $(-2 - 2\sqrt{5}, 3)$ and $(-2 + 2\sqrt{5}, 3)$. The graph is shown in Figure 16.

Figure 16

You Try It Work through this You Try It problem.

Work Exercises 5–8 in this eText or in the MyMathLab Study Plan.

OBJECTIVE 2 **FIND THE EQUATION OF AN ELLIPSE GIVEN INFORMATION ABOUT THE GRAPH**

Sometimes we must determine the equation of an **ellipse** given certain information. Then it is useful to first determine whether the ellipse has a **horizontal major axis** or a **vertical major axis.**

My video summary ▶ **Example 3 Finding the Equation of an Ellipse**

Find the standard form of the equation of the ellipse with **foci** at $(-6, 1)$ and $(-2, 1)$ such that the length of the **major axis** is eight units.

Solution Watch the **video** to verify that this is an ellipse centered at $(-4, 1)$ with a horizontal major axis such that $a = 4$ and $b = \sqrt{12}$. The equation in standard form is $\dfrac{(x + 4)^2}{4^2} + \dfrac{(y - 1)^2}{(\sqrt{12})^2} = 1$ or $\dfrac{(x + 4)^2}{16} + \dfrac{(y - 1)^2}{12} = 1$. The graph of this ellipse is shown in Figure 17.

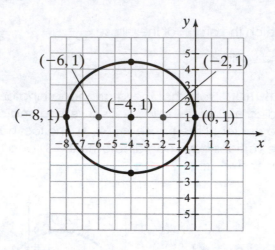

Figure 17

We can use a graphing utility to graph the ellipse from **Example 3** by solving the equation for y:

$$y = 1 \pm \sqrt{12\left(1 - \frac{(x + 4)^2}{16}\right)}$$

(View this **popup** to see how to solve for y.)

Using $y_1 = 1 + \sqrt{12\left(1 - \dfrac{(x + 4)^2}{16}\right)}$ and $y_2 = 1 - \sqrt{12\left(1 - \dfrac{(x + 4)^2}{16}\right)}$, we obtain the graph seen in Figure 18.

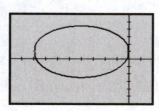

Figure 18

You Try It Work through this You Try It problem.

Work Exercises 9–18 in this eText or in the MyMathLab Study Plan.

My video summary ▶ **Example 4 Finding the Equation of an Ellipse**

Determine the equation of the ellipse with foci located at $(0, 6)$ and $(0, -6)$ that passes through the point $(-5, 6)$.

Solution To find the equation, we can use **Fact 1** and **Fact 2** of ellipses. Watch the **video** to see how we can use these two facts to determine that the equation of this ellipse is $\dfrac{x^2}{45} + \dfrac{y^2}{81} = 1$.

You Try It Work through this You Try It problem.

Work Exercises **19** and **20** in this eText or in the MyMathLab Study Plan.

OBJECTIVE 3 COMPLETE THE SQUARE TO FIND THE EQUATION OF AN ELLIPSE IN STANDARD FORM

If the equation of an **ellipse** is in the standard form of $\dfrac{(x-h)^2}{a^2} + \dfrac{(y-k)^2}{b^2} = 1$ or

$\dfrac{(x-h)^2}{b^2} + \dfrac{(y-k)^2}{a^2} = 1$, it is not too difficult to determine the **center** and **foci** and sketch its graph. However, the equation might not be given in standard form. If this is the case, we **complete the square** on both variables to rewrite the equation in standard form.

My video summary ▶ **Example 5 Writing the Equation of an Ellipse in Standard Form by Completing the Square**

Find the center and foci and sketch the ellipse $36x^2 + 20y^2 + 144x - 120y - 396 = 0$.

Solution Read the following, or watch this **video** for a complete solution. Rearrange the terms, leaving some room to complete the square and move any constants to the right-hand side:

Rearrange the terms: $36x^2 + 144x + 20y^2 - 120y = 396$

Then **factor** and **complete the square**.

Factor out 36 and 20: $36(x^2 + 4x \quad) + 20(y^2 - 6y \quad) = 396$

Complete the square on x and y. Remember to add $36 \cdot 4 = 144$ and $20 \cdot 9 = 180$ to the right side: $36(x^2 + 4x + 4) + 20(y^2 - 6y + 9) = 396 + 144 + 180$

Factor the left side, and simplify the right side: $36(x + 2)^2 + 20(y - 3)^2 = 720$

Divide both sides by 720: $\dfrac{36(x+2)^2}{720} + \dfrac{20(y-3)^2}{720} = \dfrac{720}{720}$

Simplify: $\dfrac{(x+2)^2}{20} + \dfrac{(y-3)^2}{36} = 1$

The equation is now in standard form. Notice that this is the exact same ellipse that we sketched in **Example 2**.

You Try It Work through this You Try It problem.

Work Exercises **21–24** in this eText or in the MyMathLab Study Plan.

OBJECTIVE 4 SOLVE APPLICATIONS INVOLVING ELLIPSES

Ellipses have many applications. The planets of our solar system travel around the Sun in an elliptical orbit with the Sun at one focus. See Figure 19. Some comets, such as Halley's comet, also travel in elliptical orbits. (Some comets are only seen in our solar system once because they travel in hyperbolic orbits.) We saw in **Section 10.1** that the **parabola** had a special reflecting property where incoming rays of light or radio waves traveling parallel to the axis of symmetry of a parabolic reflector or receiver will reflect off the parabolic surface and travel directly toward the antenna that is placed at the focus.

Figure 19
Planets travel around the Sun in an elliptical orbit.

Ellipses have a similar reflecting property. When light or sound waves originate from one **focus** of an ellipse, the waves will reflect off the surface of the ellipse and travel directly toward the other focus. See Figure 20. This reflecting property is used in a medical procedure called *sound wave lithotripsy* in which the patient is placed in an elliptical tank with the kidney stone placed at one focus.

An ultrasound wave emitter is positioned at the other focus. The sound waves reflect off the walls of the tank directly to the kidney stone, thus obliterating the stone into fragments that are passed naturally through the patient's body.

Figure 20
Sound waves or light rays emitted from one focus of an ellipse reflect off the surface directly to the other focus.

🖎 *My video summary* ▶ **Example 6 Position a Patient During Kidney Stone Treatment**

A patient is placed in an elliptical tank that is 280 centimeters long and 250 centimeters wide to undergo sound wave lithotripsy treatment for kidney stones. Determine where the sound emitter and the stone should be positioned relative to the **center** of the **ellipse**.

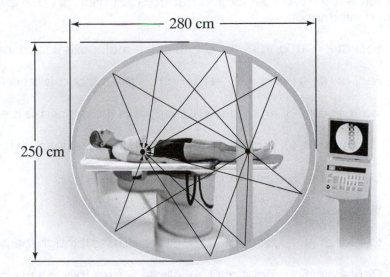

Solution Read the following, or watch this **video** for a complete solution. The kidney stone and the sound emitter must be placed at the **foci** of the ellipse. We know that the **major axis** is 280 centimeters. Therefore, the **vertices** must be 140 centimeters from the center. Similarly, the endpoints of the minor axis are located 125 centimeters from the center.

To find c, we use the fact that $c^2 = a^2 - b^2$.

$$c^2 = 140^2 - 125^2$$
$$c^2 = 3975$$
$$c = \sqrt{3975} \approx 63.05$$

The stone and the sound emitter should be positioned approximately 63.05 centimeters from the center of the tank on the major axis.

 You Try It   Work through this You Try It problem.

Work Exercises 25–31 in this eText or in the MyMathLab **Study Plan.**

10.2 Exercises

In Exercises 1–8, determine the center, foci, and vertices of the ellipse and sketch its graph.

1. $\dfrac{x^2}{16} + \dfrac{y^2}{4} = 1$

2. $\dfrac{x^2}{12} + \dfrac{y^2}{25} = 1$

3. $4x^2 + 9y^2 = 36$

4. $20x^2 + 5y^2 = 100$

5. $\dfrac{(x-2)^2}{36} + \dfrac{(y-4)^2}{25} = 1$

6. $\dfrac{(x-1)^2}{9} + \dfrac{(y+4)^2}{49} = 1$

7. $9(x-5)^2 + 25(y+1)^2 = 225$

8. $(x+7)^2 + 9(y-2)^2 = 9$

In Exercises 9–18, determine the standard equation of each ellipse using the given graph or the stated information.

9. Foci at $(-5, 0)$ and $(5, 0)$; length of the major axis is fourteen units.

10. Foci at $(8, -1)$ and $(-2, -1)$; length of the major axis is twelve units.

11. Vertices at $(-6, 10)$ and $(-6, 0)$; length of the minor axis is eight units.

12. Center at $(4, 5)$; vertical minor axis with length sixteen units; $c = 6$.

13. Center at $(-1, 4)$; vertex at $(-1, 8)$; focus at $(-1, 7)$.

14. Vertices at $(2, -7)$ and $(2, 5)$; focus at $(2, 3)$.

15. Center at $(4, 1)$; focus at $(4, 8)$; ellipse passes through the point $(6, 1)$.

16. Center at $(1, 3)$; focus at $(1, 6)$; ellipse passes through the point $(2, 3)$.

17.

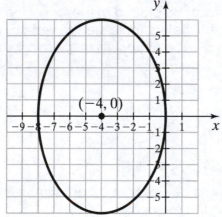

18.

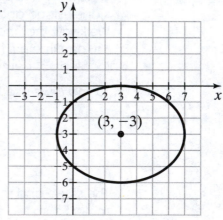

In Exercises 19 and 20, determine the standard equation of each ellipse by first using Fact 1 to find a and then by using Fact 2 to find b.

19. The foci have coordinates $(-2, 0)$ and $(2, 0)$. The ellipse contains the point $(2, 3)$.

20. The foci have coordinates $(0, -8)$ and $(0, 8)$. The ellipse contains the point $(-2, 6)$.

In Exercises 21–24, complete the square to write each equation in the form $\dfrac{(x-h)^2}{a^2} + \dfrac{(y-k)^2}{b^2} = 1$

or $\dfrac{(x-h)^2}{b^2} + \dfrac{(y-k)^2}{a^2} = 1$.

Determine the center, foci, and vertices of the ellipse and sketch its graph.

21. $16x^2 + 20y^2 + 64x - 40y - 236 = 0$

22. $9y^2 + 16x^2 + 224x + 54y + 721 = 0$

23. $x^2 + 4x + 16y^2 - 32y + 4 = 0$

24. $50x^2 + y^2 + 20y = 0$

25. An elliptical arch railroad tunnel 16 feet high at the center and 30 feet wide is cut through the side of a mountain. Find the equation of the ellipse if a vertex is represented by the point $(0, 16)$.

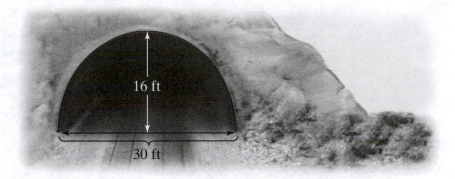

26. A rectangular playing field lies in the interior of an elliptical track that is 50 yards wide and 140 yards long. What is the width of the rectangular playing field if the width is located 10 yards from either vertex?

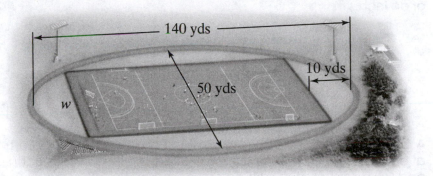

27. A 78-inch by 36-inch door contains a decorative glass elliptical pattern. One vertex of the ellipse is located 9 inches from the top of the door. The other vertex is located 27 inches from the bottom of the door. The endpoints of the minor axis are located 9 inches from each side. Find the equation of the elliptical pattern. (Assume that the center of the elliptical pattern is at the origin.)

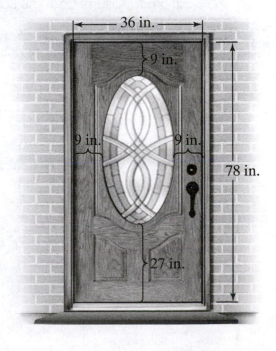

28. A patient is placed in an elliptical tank that is 180 centimeters long and 170 centimeters wide to undergo sound wave lithotripsy treatment for kidney stones. Determine where the sound emitter and the stone should be positioned relative to the center of the tank. (Round to the nearest hundredth of a centimeter.)

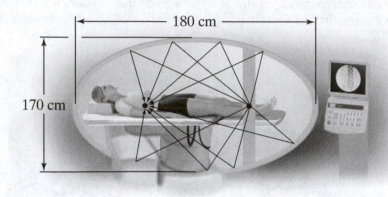

29. A window is constructed with the top half of an ellipse on top of a square. The square portion of the window has a 36-inch base. If the window is 50 inches tall at its highest point, find the height, h, of the window 14 inches from the center of the base. (Round the height to the nearest hundredth of an inch.)

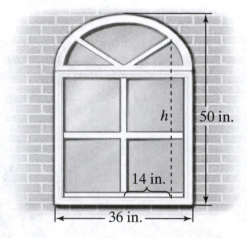

30. A government spy satellite is in an elliptical orbit around the Earth with the center of the Earth at a focus. If the satellite is 150 miles from the surface of the Earth at one vertex of the orbit and 500 miles from the surface of the Earth at the other vertex, find the equation of the elliptical orbit. (Assume that the Earth is a sphere with a diameter of 8000 miles.)

31. An elliptical arch bridge spans 160 feet. The elliptical arch has a maximum height of 40 feet. What is the height of the arch at a distance of 10 feet from the center?

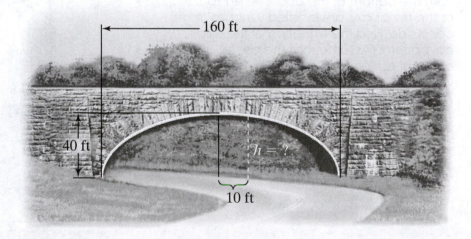

10.3 The Hyperbola

THINGS TO KNOW

Before working through this section, be sure you are familiar with the following concepts:

VIDEO   ANIMATION   INTERACTIVE

 You Try It
1. Write the Equation of a Line from Given Information (Section 2.5, Objective 6)

 You Try It
2. Find the Distance between Two Points (Section 8.4, Objective 1)

 You Try It
3. Identify Conic Sections (Section 10.1, Objective 1)

 You Try It
4. Complete the Square to Find the Equation of an Ellipse in Standard Form (Section 10.2, Objective 3)   ▶

OBJECTIVES

1 Sketch the Graph of a Hyperbola

2 Find the Equation of a Hyperbola in Standard Form

3 Complete the Square to Find the Equation of a Hyperbola in Standard Form

4 Solve Applications Involving Hyperbolas

....

OBJECTIVE 1   SKETCH THE GRAPH OF A HYPERBOLA

Recall that when a plane intersects two **right circular cones** at the same time, the **conic section** formed is a *hyperbola*.

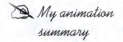 *My animation summary*

⊛ **Concept Animation** Watch this **animation** to review how a hyperbola is formed. The set of points that make up the hyperbola in this illustration is described by the following geometric definition.

> ### Geometric Definition of the Hyperbola
>
> A **hyperbola** is the set of all points in a plane, the difference of whose distances from two fixed points is a positive constant. The two fixed points, F_1 and F_2, are called the *foci*.

The Hyperbola

The **geometric definition of a hyperbola** is very similar to the **geometric definition of the ellipse**. Recall that for a point to lie on the graph of an ellipse, the **sum** of the distances from the point to the two foci must be constant. For a point to lie on the graph of a hyperbola, the **difference** between the distances from the point to the two foci must be constant. Because subtraction is not commutative, we consider the absolute value of the difference in the distances between a point on the hyperbola and the foci to ensure that the constant is positive. Thus, for any two points P and Q that lie on the graph of a hyperbola, $|d(P, F_1) - d(P, F_2)| = |d(Q, F_1) - d(Q, F_2)|$. See Figure 21.

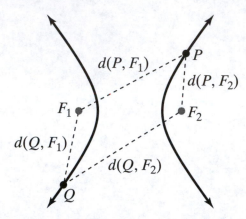

Figure 21
For any points P and Q that lie on the graph of a hyperbola,
$$|d(P, F_1) - d(P, F_2)| = |d(Q, F_1) - d(Q, F_2)|.$$

The graph of a **hyperbola** has two branches. These branches look somewhat like **parabolas**, but they are certainly not because the branches do not satisfy the **geometric definition of the parabola**.

Every hyperbola has a *center*, two *vertices*, and two *foci*. The vertices are located at the endpoints of an invisible line segment called the **transverse axis**. The transverse axis is either parallel to the x-axis (horizontal transverse axis) or parallel to the y-axis (vertical transverse axis). The **center** of a hyperbola is located midway between the two vertices (or two foci). The hyperbola has another invisible line segment called the **conjugate axis** that passes through the center and lies perpendicular to the transverse axis.

Each branch of the hyperbola approaches (but never intersects) a pair of lines called *asymptotes*. A **reference rectangle** is typically used as a guide to help sketch the asymptotes. The reference rectangle is a rectangle whose midpoints of each side are the vertices of the hyperbola or the endpoints of the conjugate axis. The asymptotes pass diagonally through opposite corners of the reference rectangle. See **Figure 22**.

My video summary ▶ **Concept Video** Watch this video to see how to sketch a hyperbola using the geometric definition.

Figure 22

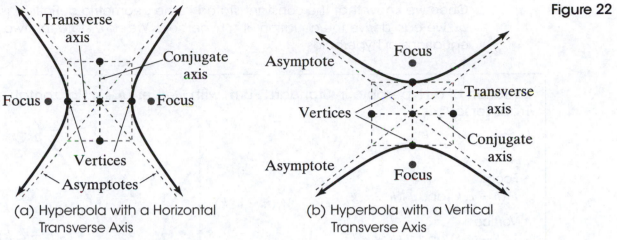

(a) Hyperbola with a Horizontal
Transverse Axis

(b) Hyperbola with a Vertical
Transverse Axis

To derive the equation of a **hyperbola**, consider a hyperbola with a horizontal **transverse axis** centered at (h, k). If the distance between the **center** and either **vertex** is $a > 0$, then the coordinates of the vertices are $V_1(h - a, k)$ and $V_2(h + a, k)$, and the length of the transverse axis is equal to $2a$. If $c > 0$ is the distance between the center and either foci, then the foci have coordinates $F_1(h - c, k)$ and $F_2(h + c, k)$. See Figure 23.

Figure 23

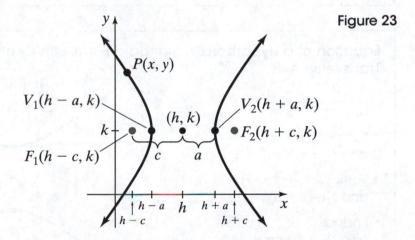

By the **geometric definition of a hyperbola**, we know that for any point $P(x, y)$ that lies on the hyperbola, the difference of the distances from P to the two **foci** is a constant. It can be shown that this constant is equal to $2a$, the length of the **transverse axis**. We now state this fact and denote it as Fact 1 for Hyperbolas.

Fact 1 for Hyperbolas

Given the foci of a hyperbola F_1 and F_2 and any point P that lies on the graph of the hyperbola, the difference of the distance between P and the foci is equal to $2a$. In other words, $|d(P, F_1) - d(P, F_2)| = 2a$. The constant $2a$ represents the length of the transverse axis.

Once we know that the constant stated in the geometric definition is equal to $2a$, we can derive the equation of a hyperbola. We now state the two standard equations of hyperbolas.

Equation of a Hyperbola in Standard Form with Center (h, k); Horizontal Transverse Axis

$$\frac{(x - h)^2}{a^2} - \frac{(y - k)^2}{b^2} = 1$$

- Foci:
  $F_1(h - c, k)$ and $F_2(h + c, k)$

- Vertices:
  $V_1(h - a, k)$ and $V_2(h + a, k)$

- Endpoints of conjugate axis:
  $(h, k - b)$ and $(h, k + b)$

- $b^2 = c^2 - a^2$

- Asymptotes: $y - k = \pm\dfrac{b}{a}(x - h)$

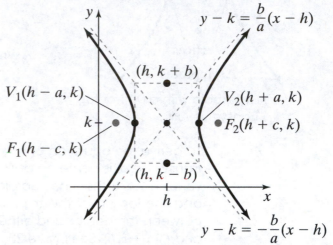

Equation of a Hyperbola in Standard Form with Center (h, k); Vertical Transverse Axis

$$\frac{(y - k)^2}{a^2} - \frac{(x - h)^2}{b^2} = 1$$

- Foci: $F_1(h, k - c)$
  and $F_2(h, k + c)$

- Vertices: $V_1(h, k - a)$
  and $V_2(h, k + a)$

- Endpoints of conjugate
  axis: $(h - b, k)$ and
  $(h + b, k)$

- $b^2 = c^2 - a^2$

- Asymptotes: $y - k = \pm\dfrac{a}{b}(x - h)$

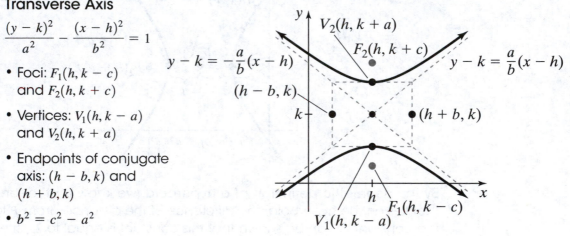

The two hyperbola equations are nearly identical except for two differences. The x-terms are positive and the y-terms are negative in the equation with a horizontal transverse axis. The signs are reversed in the equation with a vertical transverse axis. Also note that a^2 appears in the denominator of the positive squared term in each equation. We can derive the equations of the asymptotes by using the **point-slope form** of the equation of a line.

If $h = 0$ and $k = 0$, then the hyperbola is centered at the **origin**. Hyperbolas centered at the origin have the following equations.

Standard Equations of a Hyperbola with the Center at the Origin

$$\frac{x^2}{a^2} - \frac{y^2}{b^2} = 1 \qquad\qquad \frac{y^2}{a^2} - \frac{x^2}{b^2} = 1$$

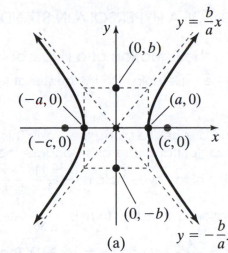

(a)

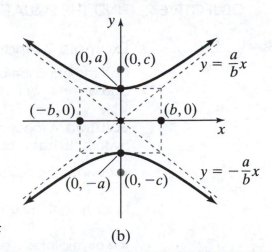

(b)

🖋 *My interactive video summary*

🔖 **Example 1 Sketching the Graph of a Hyperbola in Standard Form**

Sketch the following hyperbolas. Determine the **center, transverse axis, vertices,** and **foci** and find the equations of the **asymptotes.**

a. $\dfrac{(y-4)^2}{36} - \dfrac{(x+5)^2}{9} = 1$

b. $25x^2 - 16y^2 = 400$

Solutions Watch this **interactive video** to see how to sketch each hyperbola shown in Figure 24(a) and 24(b).

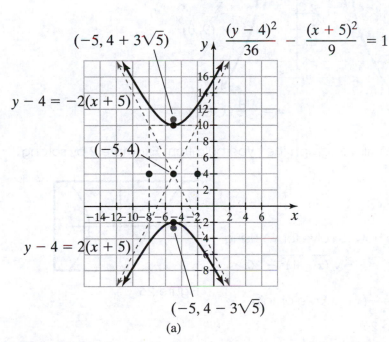

(a)

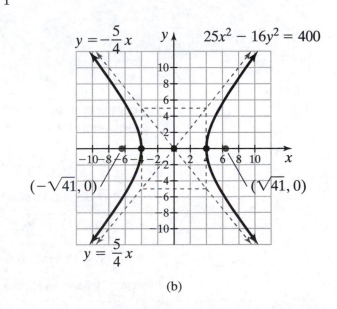

(b)

Figure 24

You Try It Work through this You Try It problem.

Work Exercises 1–6 in this eText or in the MyMathLab Study Plan.

OBJECTIVE 2 FIND THE EQUATION OF A HYPERBOLA IN STANDARD FORM

My video summary ▶ **Example 2** Finding the Equation of a Hyperbola

Find the equation of the hyperbola with the **center** at $(-1, 0)$, a **focus** at $(-11, 0)$, and a **vertex** at $(5, 0)$.

Solution A focus and a vertex lie along the x-axis. This indicates that the hyperbola has a horizontal transverse axis. Because the center is at $(-1, 0)$, we know that the equation of the hyperbola is $\dfrac{(x + 1)^2}{a^2} - \dfrac{y^2}{b^2} = 1$. Watch this **video** to verify that the equation in standard form is $\dfrac{(x + 1)^2}{36} - \dfrac{y^2}{64} = 1$. The equations of the **asymptotes** are $y = -\dfrac{4}{3}(x + 1)$ and $y = \dfrac{4}{3}(x + 1)$. The graph of this hyperbola is shown in Figure 25.

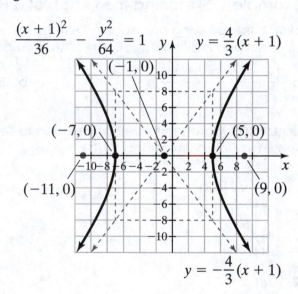

Figure 25

We can use a graphing utility to graph the hyperbola from **Example 2** by solving the equation for y:

$$y = \pm\frac{4}{3}\sqrt{(x + 1)^2 - 36}$$

(View this **popup** to see how to solve for y.) Using $y_1 = \dfrac{4}{3}\sqrt{(x + 1)^2 - 36}$, $y_2 = -\dfrac{4}{3}\sqrt{(x + 1)^2 - 36}$, $y_3 = -\dfrac{4}{3}(x + 1)$, and $y_4 = \dfrac{4}{3}(x + 1)$, we obtain the graph seen in Figure 26.

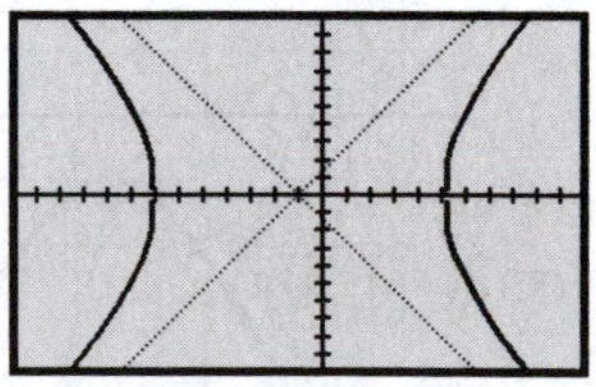

Figure 26

You Try It Work through this You Try It problem.

Work Exercises 7–14 in this eText or in the MyMathLab Study Plan.

OBJECTIVE 3 COMPLETE THE SQUARE TO FIND THE EQUATION OF A HYPERBOLA IN STANDARD FORM

If the equation of a hyperbola is in the standard form of $\dfrac{(x-h)^2}{a^2} - \dfrac{(y-k)^2}{b^2} = 1$ or $\dfrac{(y-k)^2}{a^2} - \dfrac{(x-h)^2}{b^2} = 1$, then it is not too difficult to determine the **center, vertices, foci,** and **asymptotes** and sketch its graph. However, the equation might not be given in standard form. If this is the case, we may need to **complete the square** on both variables.

✎ *My video summary* ▶ **Example 3 Writing the Equation of a Hyperbola in Standard Form by Completing the Square**

Find the center, vertices, foci, and equations of asymptotes and sketch the hyperbola $12x^2 - 4y^2 - 72x - 16y + 140 = 0$.

Solution Read through the following, or watch this **video** for a complete solution.

Rearrange the terms leaving some room to **complete the square,** and move any constants to the right-hand side:

Rearrange the terms: $12x^2 - 72x \quad - 4y^2 - 16y \quad = -140$

Then **factor** and complete the square.

Factor out 12 and -4: $12(x^2 - 6x \quad) - 4(y^2 + 4y \quad) = -140$

Complete the square on x and y. Remember to add $12 \cdot 9 = 108$ and $-4 \cdot 4 = -16$ to the right side: $12(x^2 - 6x + 9) - 4(y^2 + 4y + 4) = -140 + 108 - 16$

Factor the left side, and simplify the right side: $12(x-3)^2 - 4(y+2)^2 = -48$

Divide both sides by -48: $\dfrac{12(x-3)^2}{-48} - \dfrac{4(y+2)^2}{-48} = \dfrac{-48}{-48}$

Simplify: $-\dfrac{(x-3)^2}{4} + \dfrac{(y+2)^2}{12} = 1$

Rewrite the equation: $\dfrac{(y+2)^2}{12} - \dfrac{(x-3)^2}{4} = 1$

The equation is now in standard form. You should verify that this is the equation of a hyperbola with a vertical **transverse axis** with center $(3, -2)$. The vertices have coordinates $(3, -2 - 2\sqrt{3})$ and $(3, -2 + 2\sqrt{3})$. The foci have coordinates $(3, -6)$ and $(3, 2)$. The equations of the asymptotes are $y + 2 = -\sqrt{3}(x - 3)$ and $y + 2 = \sqrt{3}(x - 3)$.

You Try It Work through this You Try It problem.

Work Exercises 15–20 in this eText or in the MyMathLab Study Plan.

OBJECTIVE 4   SOLVE APPLICATIONS INVOLVING HYPERBOLAS

Hyperbolas have many applications. Some comets are only seen once in our solar system because they travel on the **path of a hyperbola** with the Sun at a focus. As an airplane moves faster than the speed of sound, a **cone-shaped shock wave** is produced. The cone intersects the ground in the shape of a hyperbola. When two rocks are simultaneously tossed into a calm pool of water, ripples move outward in the form of **concentric circles**. These circles intersect in points that form a hyperbola.

Hyperbolas can be used to locate ships by sending radio signals simultaneously from radio transmitters placed at some fixed distance apart. A device measures the difference in the time it takes the radio signals to reach the ship. The equation of a hyperbola can then be determined to describe the current path of the ship. If three transmitters are used, two hyperbolic equations can be determined. The precise location of the ship can be determined by finding the intersection of the two hyperbolas. This system of locating ships is known as **long-range navigation** or **LORAN**.

My video summary ▶ **Example 4 Use a Hyperbola to Locate a Ship**

One transmitting station is located 100 miles due east from another transmitting station. Each station simultaneously sends out a radio signal. The signal from the west tower is received by a ship $\dfrac{1600}{3}$ microseconds after the signal from the east tower. If the radio signal travels at 0.18 miles per microsecond, find the equation of the **hyperbola** on which the ship is presently located.

Solution Read the following, or watch this **video** for a complete solution.

Start by plotting the two foci of the hyperbola at points $F_1(-50, 0)$ and $F_2(50, 0)$. These two points represent the position of the two transmitting towers. Note that $c = 50$. Because the hyperbola is centered at the origin with a horizontal transverse axis,

the equation must be of the form $\dfrac{x^2}{a^2} - \dfrac{y^2}{b^2} = 1$. The difference in the distances

from the two transmitters to the ship is $\left(\dfrac{1600}{3} \text{ microseconds} \right) \cdot \left(0.18 \dfrac{\text{miles}}{\text{microsecond}} \right) =$

96 miles. This distance represents the constant stated in the **Fact 1 for Hyperbolas**.

Therefore, $2a = 96$, so $a = 48$ or $a^2 = 2304$. To find b^2, we use the fact that $b^2 = c^2 - a^2$.

$$b^2 = c^2 - a^2$$

$$b^2 = 50^2 - 48^2$$

$$b^2 = 196$$

We now substitute the values of $a^2 = 2304$ and $b^2 = 196$ into the previous equation to obtain the equation $\dfrac{x^2}{2304} - \dfrac{y^2}{196} = 1$.

You Try It   Work through this You Try It problem.

Work Exercises 21–24 in this eText or in the MyMathLab Study Plan.

10.3 Exercises

In Exercises 1–6, determine the center, transverse axis, vertices, foci, and the equations of the asymptotes and sketch the hyperbola.

1. $\dfrac{x^2}{16} - \dfrac{y^2}{9} = 1$

2. $\dfrac{y^2}{9} - \dfrac{x^2}{16} = 1$

3. $\dfrac{(y-4)^2}{25} - \dfrac{(x-2)^2}{36} = 1$

4. $\dfrac{(x+1)^2}{9} - \dfrac{(y+3)^2}{49} = 1$

5. $20x^2 - 5y^2 = 100$

6. $20(x-1)^2 - 16(y-3)^2 = -320$

In Exercises 7–12, determine the standard equation of the hyperbola with the given characteristics and sketch the graph.

7. The center is at $(0,0)$, a focus is at $(5,0)$, and a vertex is at $(3,0)$.

8. The center is at $(0,0)$, a focus is at $(0,10)$, and a vertex is at $(0,-6)$.

9. The center is at $(4,-4)$, a focus is at $(6,-4)$, and a vertex is at $(5,-4)$.

10. The center is at $(-6,-1)$, a focus is at $(-6,-9)$, and a vertex is at $(-6,-5)$.

11. The foci are at $(9,3)$ and $(9,9)$; a vertex is at $(9,8)$.

12. The vertices are at $(-2,-3)$ and $(10,-3)$; an asymptote has equation $y + 3 = \dfrac{7}{6}(x-4)$.

In Exercises 13 and 14, determine the equation of the hyperbola with the given characteristics and sketch the graph. **Hint:** $\left| d(P, F_1) - d(P, F_2) \right| = 2a$.

13. The hyperbola has foci with coordinates $F_1(0,-6)$ and $F_2(0,6)$ and passes through the point $P(8, 10)$.

14. The hyperbola has foci with coordinates $F_1(-6, 1)$ and $F_2(10, 1)$ and passes through the point $P(10, 13)$.

In Exercises 15–20, complete the square to write each equation in the form $\dfrac{(x-h)^2}{a^2} - \dfrac{(y-k)^2}{b^2} = 1$ or $\dfrac{(y-k)^2}{a^2} - \dfrac{(x-h)^2}{b^2} = 1$. Determine the center, vertices, foci, endpoints of the conjugate axis, and the equations of the asymptotes of the hyperbola, and sketch its graph.

15. $x^2 - y^2 - 4x + 2y - 1 = 0$

16. $x^2 - y^2 + 8x - 6y + 9 = 0$

17. $y^2 - 9x^2 - 12y - 36x - 9 = 0$

18. $x^2 - 16y^2 + 10x + 64y - 55 = 0$

19. $25y^2 - 144x^2 + 1728x + 400y + 16 = 0$

20. $49y^2 - 576x^2 - 98y - 1152x - 28{,}751 = 0$

21. A light on a wall produces a shadow in the shape of a hyperbola. If the distance between the two vertices of the hyperbola is 14 inches and the distance between the two foci is 16 inches, find the equation of the hyperbola.

22. This figure shows the hyperbolic orbit of a comet with the center of the Sun positioned at a focus, 100 million miles from the origin. (The units are in millions of miles.) The comet will be 50 million miles from the center of the Sun at its nearest point during the orbit. Find the equation of the hyperbola describing the comet's orbit assuming it is centered at the origin.

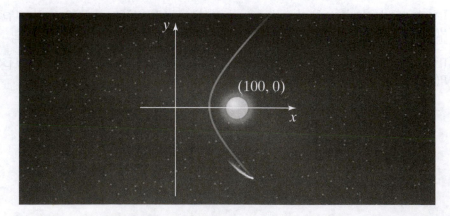

23. A nuclear power plant has a large cooling tower with sides curved in the shape of a hyperbola. The radius of the base of the tower is 60 meters. The radius of the top of the tower is 50 meters. The sides of the tower are 60 meters apart at the closest point located 90 meters above the ground.

    a. Find the equation of the hyperbola that describes the sides of the cooling tower. (Assume that the center is at the origin.)

    b. Determine the height of the tower. (Round your answer to the nearest meter.)

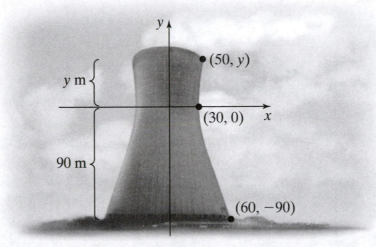

24. One transmitting station is located 80 miles north of another transmitting station. Each station simultaneously sends out a radio signal. The signal from the north station is received 200 microseconds after the signal from the south station. If the radio signal travels at 0.18 miles per microsecond, find the equation of the hyperbola on which the ship is presently located.

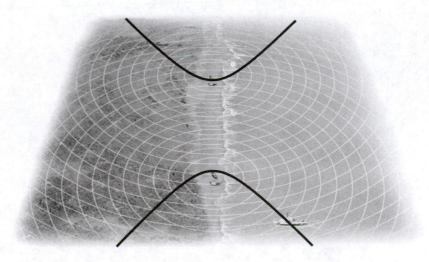

CHAPTER ELEVEN
Sequences and Series

CHAPTER ELEVEN CONTENTS

11.1 Introduction to Sequences and Series

THINGS TO KNOW

Before working through this section, be sure you are familiar with the following concept:

VIDEO   ANIMATION   INTERACTIVE

 You Try It

1. Determine If Graphs Are Functions (Section 2.2, Objective 4)

OBJECTIVES

1 Write the Terms of a Sequence
2 Write the Terms of a Recursive Sequence
3 Write the General Term for a Given Sequence
4 Compute Partial Sums of a Series
5 Determine the Sum of a Finite Series Written in Summation Notation
6 Write a Series Using Summation Notation

. . . .

OBJECTIVE 1   WRITE THE TERMS OF A SEQUENCE

Consider the **function** $f(n) = 2n - 1$, where n is a **natural number**. The graph of this function consists of infinitely many ordered pairs of the form $(n, 2n - 1)$, where $n \geq 1$. Therefore, the ordered pairs that lie on the graph of this function are $(1, 1), (2, 3), (3, 5), (4, 7)$, and so on. A portion of the graph of this function can be seen in Figure 1. Notice that this graph passes the **vertical line test**.

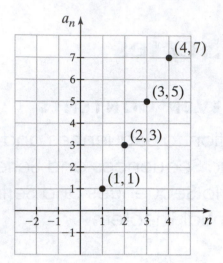

Figure 1
A Portion of the Graph of the Function $f(n) = 2n - 1$, Where n Is a Natural Number.

Any function whose **domain** is the set of **natural numbers** is called an infinite sequence. Instead of using the conventional function notation $f(n)$ to name a sequence, we will use a subscript notation, such as a_n (read as "a sub n"), to name sequences. Let's now formally define a sequence.

Definition Sequence

A **finite sequence** is a function whose domain is the finite set $\{1, 2, 3, \ldots, n\}$, where n is a natural number.

An **infinite sequence** is a function whose domain is the set of all natural numbers.

The **range** values of a sequence are called the **terms** of the sequence.

We now rename the sequence $f(n) = 2n - 1$ as $a_n = 2n - 1$. The first four terms of this sequence are $a_1 = 1, a_2 = 3, a_3 = 5$, and $a_4 = 7$.

A graphing utility set to *sequence mode* can be used to sketch the graph of a sequence. Figure 2 shows a portion of the graph of the sequence $a_n = 2n - 1$.

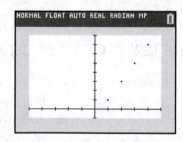

Figure 2

Before we start to find the terms of a sequence, it is important to introduce factorial notation.

Definition The Factorial of a Non-Negative Integer

The **factorial** of a non-negative integer n, denoted as $n!$, is the product of all positive integers less than or equal to n. Thus, $n! = n(n - 1) \cdot \cdots \cdot 3 \cdot 2 \cdot 1$.

Note: By definition, we say that zero factorial is equal to 1, or $0! = 1$.

Some examples of factorial notation are $5! = 5 \cdot 4 \cdot 3 \cdot 2 \cdot 1 = 120$ and $8! = 8 \cdot 7 \cdot 6 \cdot 5 \cdot 4 \cdot 3 \cdot 2 \cdot 1 = 40{,}320$. As you can see, the value of $n!$ gets large rather quickly. In fact, the value of $13!$ is over 6 billion, which is a rough estimate of the population of Earth in 2000!

My interactive video summary

Example 1 Writing the Terms of a Sequence

Write the first four terms of each sequence whose nth term is given.

a. $a_n = 2n - 1$ **b.** $b_n = n^2 - 1$

c. $c_n = \dfrac{3^n}{(n-1)!}$ **d.** $d_n = (-1)^n 2^{n-1}$

Solutions

a. To find the first four terms of the sequence, evaluate $a_n = 2n - 1$ when n is 1, 2, 3, and 4.

$$a_1 = 2(1) - 1 = 2 - 1 = 1, \; a_2 = 2(2) - 1 = 4 - 1 = 3,$$

$$a_3 = 2(3) - 1 = 6 - 1 = 5, \text{ and } a_4 = 2(4) - 1 = 8 - 1 = 7$$

Therefore, the first four terms of the sequence $a_n = 2n - 1$ are 1, 3, 5, and 7.

b. Work through this **interactive video** to verify that the first four **terms** of the sequence $b_n = n^2 - 1$ are 0, 3, 8, and 15.

c. To find the first four terms of the sequence $c_n = \dfrac{3^n}{(n-1)!}$, we evaluate c_n when n is 1, 2, 3, and 4.

$$c_1 = \frac{3^1}{(1-1)!} = \frac{3}{0!} = \frac{3}{1} = 3, \qquad c_2 = \frac{3^2}{(2-1)!} = \frac{9}{1!} = \frac{9}{1} = 9,$$

$$c_3 = \frac{3^3}{(3-1)!} = \frac{27}{2!} = \frac{27}{2}, \text{ and } \quad c_4 = \frac{3^4}{(4-1)!} = \frac{81}{3!} = \frac{81}{6} = \frac{27}{2}$$

Therefore, the first four terms of the sequence $c_n = \dfrac{3^n}{(n-1)!}$ are

$$3, 9, \frac{27}{2}, \text{and } \frac{27}{2}.$$

d. Work through this **interactive video** to verify that the first four terms of the sequence $d_n = (-1)^n 2^{n-1}$ are $-1, 2, -4,$ and 8.

Note: The sequence $d_n = (-1)^n 2^{n-1}$ is an example of an **alternating sequence** because the successive terms alternate in sign.

You Try It Work through this **You Try It** problem.

Work Exercises 1–10 in this eText or in the MyMathLab Study Plan.

OBJECTIVE 2 WRITE THE TERMS OF A RECURSIVE SEQUENCE

Some sequences are defined recursively. A **recursive sequence** is a sequence in which each term is defined using one or more of its previous terms. Typically, the first term of a recursive sequence is given, followed by the formula for the nth term of the sequence. The following example illustrates two recursive sequences.

My interactive video summary

Example 2 Writing the Terms of a Recursive Sequence

Write the first four terms of each of the following recursive sequences.

a. $a_1 = -3, a_n = 5a_{n-1} - 1$ for $n \geq 2$

b. $b_1 = 2, b_n = \dfrac{(-1)^{n-1}n}{b_{n-1}}$ for $n \geq 2$

Solutions

a. The first four terms of this recursive sequence are $-3, -16, -81$, and -406. Work through this **interactive video** to verify.

b. The first four terms of this recursive sequence are $2, -1, -3$, and $\dfrac{4}{3}$. Work through this **interactive video** to verify.

You Try It Work through this You Try It problem.

Work Exercises 11–14 in this eText or in the MyMathLab Study Plan.

Arguably the most famous recursively defined sequence is the **Fibonacci sequence** named after the 13th-century Italian mathematician Leonardo of Pisa, also known as Fibonacci. The Fibonacci sequence is defined in Example 3.

My video summary

Example 3 Writing the Terms of the Fibonacci Sequence

The Fibonacci sequence is defined recursively by $a_n = a_{n-1} + a_{n-2}$, where $a_1 = 1$ and $a_2 = 1$. Write the first eight terms of the Fibonacci sequence.

Solution Watch this **video**, or read through the following solution.

We are given that $a_1 = 1$ and $a_2 = 1$. We use the recursive formula $a_n = a_{n-1} + a_{n-2}$ to find the next six terms starting with $n = 3$.

$$a_3 = a_2 + a_1 = 1 + 1 = 2$$
$$a_4 = a_3 + a_2 = 2 + 1 = 3$$
$$a_5 = a_4 + a_3 = 3 + 2 = 5$$
$$a_6 = a_5 + a_4 = 5 + 3 = 8$$
$$a_7 = a_6 + a_5 = 8 + 5 = 13$$
$$a_8 = a_7 + a_6 = 13 + 8 = 21$$

You can see in Example 3 that given the first two terms, each term of the Fibonacci sequence is the sum of the preceding two terms. We now write the first 12 terms of the Fibonacci sequence.

$$1, 1, 2, 3, 5, 8, 13, 21, 34, 55, 89, 144, \ldots$$

The numbers of this sequence are known as **Fibonacci numbers**. The Fibonacci sequence and Fibonacci numbers occur in many natural phenomena such as the spiral formation of seeds of various plants, the number of petals of a flower, and the formation of the branches of a tree.

You Try It Work through this You Try It problem.

Work Exercises 15, 16, and 46 in this eText or in the MyMathLab Study Plan.

OBJECTIVE 3 WRITE THE GENERAL TERM FOR A GIVEN SEQUENCE

Sometimes the first several terms of a sequence are given without listing the nth term. When this occurs, we must try to determine a pattern and use **deductive reasoning** to establish a rule that describes the general term, or nth term, of the sequence. Example 4 illustrates two such sequences.

Example 4 Finding the General Term of a Sequence

Write a formula for the nth term of each infinite sequence, then use this formula to find the eighth term of the sequence.

a. $\dfrac{1}{1}, \dfrac{1}{2}, \dfrac{1}{3}, \dfrac{1}{4}, \dfrac{1}{5}, \cdots$

b. $-\dfrac{2}{1}, \dfrac{4}{2}, -\dfrac{8}{6}, \dfrac{16}{24}, -\dfrac{32}{120}, \cdots$

Solutions

a. The nth term of the sequence is $a_n = \dfrac{1}{n}$. Thus, the eighth term of this sequence is $a_8 = \dfrac{1}{8}$.

My video summary ▶ b. Watch this **video**, or read through the following solution. For this sequence, notice that the first term is negative and that terms alternate in sign. We can therefore represent the sign of each term as $(-1)^n$. Also, notice that the numerators are successive powers of 2. We now have the following pattern:

$$
\begin{array}{ccccc}
a_1 & a_2 & a_3 & a_4 & a_5 \\
\downarrow & \downarrow & \downarrow & \downarrow & \downarrow \\
-\dfrac{2}{1}, & \dfrac{4}{2}, & -\dfrac{8}{6}, & \dfrac{16}{24}, & -\dfrac{32}{120}, \cdots \\
\downarrow & \downarrow & \downarrow & \downarrow & \downarrow \\
\dfrac{(-1)^1 2^1}{1}, & \dfrac{(-1)^2 2^2}{2}, & \dfrac{(-1)^3 2^3}{6}, & \dfrac{(-1)^4 2^4}{24}, & \dfrac{(-1)^5 2^5}{120}, \cdots
\end{array}
$$

Finally, if we factor each successive denominator we get $1 = 1$, $2 = 2 \cdot 1$, $6 = 3 \cdot 2 \cdot 1$, $24 = 4 \cdot 3 \cdot 2 \cdot 1$, and $120 = 5 \cdot 4 \cdot 3 \cdot 2 \cdot 1$. This suggests that the denominator of the nth term can be represented by $n!$.

Therefore, the nth term of the sequence is $a_n = \dfrac{(-1)^n 2^n}{n!}$. The eighth term of this sequence is

$$a_8 = \dfrac{(-1)^8 2^8}{8!} = \dfrac{256}{40,320} = \dfrac{2}{315}.$$

You Try It Work through this You Try It problem.

Work Exercises 17–24 in this eText or in the MyMathLab Study Plan.

OBJECTIVE 4 COMPUTE PARTIAL SUMS OF A SERIES

Suppose that we wanted to find the sum of the first four terms of the sequence $a_n = 2n - 1$. From **Example 1**, we saw that the first four terms of this sequence were $a_1 = 1, a_2 = 3, a_3 = 5$, and $a_4 = 7$. Therefore, the sum of the first four terms is $a_1 + a_2 + a_3 + a_4 = 1 + 3 + 5 + 7 = 16$. The expression $1 + 3 + 5 + 7$ is called a **series**.

> **Definition   Series**
>
> Let $a_1, a_2, a_3, \ldots$ be a sequence. The expression of the form $a_1 + a_2 + a_3 + \cdots + a_n$ is called a **finite series**.
>
> The expression of the form $a_1 + a_2 + a_3 + \cdots + a_n + a_{n+1} + \cdots$ is called an **infinite series**.
>
> The sum of the first n terms of a series is called the nth **partial sum** of the series and is denoted as S_n.

For the series $1 + 3 + 5 + 7 + 9 + \cdots + 2n - 1$, the first five **partial sums** are as follows:

$$S_1 = 1$$
$$S_2 = 1 + 3 = 4$$
$$S_3 = 1 + 3 + 5 = 9$$
$$S_4 = 1 + 3 + 5 + 7 = 16$$
$$S_5 = 1 + 3 + 5 + 7 + 9 = 25$$

It appears that $S_n = n^2$. In fact, it can be shown that for any positive integer n, the sum of the series $1 + 3 + 5 + 7 + 9 + \cdots + 2n - 1$ is equal to n^2.

My video summary ▶ **Example 5 Computing Partial Sums of a Series**

Given the general term of each sequence, find the indicated **partial sum**.

a. $a_n = \dfrac{1}{n}$, find S_3.   b. $b_n = (-1)^n 2^{n-1}$, find S_5.

Solutions Watch this video, or read through the following solutions.

a. The first three terms are $a_1 = 1, a_2 = \dfrac{1}{2}$, and $a_3 = \dfrac{1}{3}$. Therefore, the partial sum, S_3, is $S_3 = 1 + \dfrac{1}{2} + \dfrac{1}{3} = \dfrac{11}{6}$.

b. The first five terms are $b_1 = -1, b_2 = 2, b_3 = -4, b_4 = 8$, and $b_5 = -16$. Therefore, the partial sum, S_5, is $S_5 = -1 + 2 + (-4) + 8 + (-16) = -11$.

You Try It Work through this You Try It problem.

Work Exercises 25–30 in this eText or in the MyMathLab Study Plan.

OBJECTIVE 5 DETERMINE THE SUM OF A FINITE SERIES WRITTEN IN SUMMATION NOTATION

Writing out an entire finite series of the form $a_1 + a_2 + a_3 + \cdots + a_n$ can be quite tedious, especially if n is fairly large. Fortunately, there is a convenient way to express a finite series using a short-hand notation called *summation notation* (also called *sigma notation*). This notation involves the use of the uppercase Greek letter sigma, which is written as Σ.

My video summary ▶ **Concept Video** Watch this video for an explanation of summation notation.

Definition Summation Notation

If $a_1, a_2, a_3, \ldots$ is a sequence, then the finite series $a_1 + a_2 + a_3 + \cdots + a_n$

can be written in **summation notation** as $\sum\limits_{i=1}^{n} a_i$. The infinite series

$a_1 + a_2 + \cdots + a_n + a_{n+1} + \cdots$ can be written as $\sum\limits_{i=1}^{\infty} a_i$.

The variable i is called the **index of summation**. The number 1 is the **lower limit of summation** and n is the **upper limit of summation**.

The lower limit of summation, $i = 1$, below the sigma tells us which term to start with. The upper limit of summation, n, that appears above the sigma tells us which term of the sequence will be the last term to add. There is nothing special about the letter i to represent the index of summation. We will often use different letters such as j or k. Also, it is not necessary for the lower limit of summation to start at 1. In Examples 6b and 6c, the lower limits of summation are 2 and 0, respectively.

My interactive video summary

Example 6 Determining the Sum of a Series Written in Summation Notation

Find the sum of each finite series.

a. $\sum\limits_{i=1}^{5} i^2$ b. $\sum\limits_{j=2}^{5} \dfrac{j-1}{j+1}$ c. $\sum\limits_{k=0}^{6} \dfrac{1}{k!}$

(Round the sum to three decimal places.)

Solutions

a. $\sum\limits_{i=1}^{5} i^2 = 1^2 + 2^2 + 3^2 + 4^2 + 5^2$

$= 1 + 4 + 9 + 16 + 25 = 55$

b. $\sum\limits_{j=2}^{5} \dfrac{j-1}{j+1} = \dfrac{2-1}{2+1} + \dfrac{3-1}{3+1} + \dfrac{4-1}{4+1} + \dfrac{5-1}{5+1}$

$= \dfrac{1}{3} + \dfrac{2}{4} + \dfrac{3}{5} + \dfrac{4}{6} = \dfrac{21}{10}$

c. $\sum\limits_{k=0}^{6} \dfrac{1}{k!} = \dfrac{1}{0!} + \dfrac{1}{1!} + \dfrac{1}{2!} + \dfrac{1}{3!} + \dfrac{1}{4!} + \dfrac{1}{5!} + \dfrac{1}{6!}$

$= 1 + 1 + \dfrac{1}{2} + \dfrac{1}{6} + \dfrac{1}{24} + \dfrac{1}{120} + \dfrac{1}{720} = \dfrac{1957}{720} \approx 2.718$

Work through this **interactive video** to see these solutions worked out in detail.

Using a TI-84 Plus C, we can calculate the sum obtained in **Example 6c**. Notice that this number is a good approximation of the **number e**. In fact, it can be

shown that the exact value of e is $e = \sum\limits_{n=0}^{\infty} \dfrac{1}{n!}$.

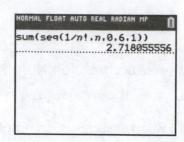

Figure 3

$$\sum_{n=0}^{6} \frac{1}{n!}$$

You Try It Work through this You Try It problem.

Work Exercises 31–37 in this eText or in the MyMathLab Study Plan.

OBJECTIVE 6 WRITE A SERIES USING SUMMATION NOTATION

Given the first several terms of a series, it is important to be able to rewrite the series using summation notation as in Example 7.

My video summary ▶ **Example 7 Writing a Series Using Summation Notation**

Rewrite each series using summation notation. Use 1 as the lower limit of summation.

a. $2 + 4 + 6 + 8 + 10 + 12$ 　　 b. $1 + 2 + 6 + 24 + 120 + 720 + \cdots + 3{,}628{,}800$

Solutions Read through the following solutions, or watch this **video**.

a. This series is the sum of six terms. Therefore, the lower limit of summation is 1 and the upper limit of summation is 6. Each term is a successive multiple of 2.

So, one possible series is $\displaystyle\sum_{i=1}^{6} 2i$.

b. Notice that $1 = 1!, 2 = 2!, 6 = 3!, 24 = 4!, 120 = 5!, 720 = 6!$, and $3{,}628{,}800 = 10!$

Thus, a possible series is $\displaystyle\sum_{n=1}^{10} n!$.

You Try It Work through this You Try It problem.

Work Exercises 38–45 in this eText or in the MyMathLab Study Plan.

11.1 Exercises

In Exercises 1–10, write the first four terms of each sequence.

1. $a_n = 3n + 1$ 　　 2. $a_n = 4^n$ 　　 3. $a_n = \dfrac{4n}{n+3}$ 　　 4. $a_n = (-4)^n$

5. $a_n = 5(n+2)!$ 　　 6. $a_n = \dfrac{n^3}{(n+1)!}$ 　　 7. $a_n = (-1)^n(5n)$

8. $a_n = \dfrac{3^n}{(-1)^{n+1}+5}$ 　　 9. $a_n = \dfrac{(-1)^n}{(n+5)(n+6)}$ 　　 10. $a_n = \dfrac{(-1)^n(3)^{2n+1}}{(2n+1)!}$

In Exercises 11–14, write the first four terms of each recursive sequence.

11. $a_1 = 7, a_n = 3 + a_{n-1}$ for $n \geq 2$

12. $a_1 = -1, a_n = n - a_{n-1}$ for $n \geq 2$

13. $a_1 = 6, a_n = \dfrac{a_{n-1}}{n^2}$ for $n \geq 2$

14. $a_1 = -4, a_n = 1 - \dfrac{1}{a_{n-1}}$ for $n \geq 2$

In Exercises 15 and 16, write the first six terms of each recursive sequence.

15. $a_1 = 3, a_2 = 0, a_n = 2 \cdot a_{n-1} + a_{n-2}$

16. $a_1 = -2, a_2 = 0, a_3 = 1, a_n = a_{n-3} - a_{n-1}$

In Exercises 17–24, write a formula for the general term, or nth term, for the given sequence. Then find the indicated term.

17. $-1, 1, 3, 5, 7, \ldots \; ; a_{11}.$

18. $\dfrac{1}{5}, \dfrac{2}{6}, \dfrac{3}{7}, \dfrac{4}{8}, \dfrac{5}{9}, \ldots \; ; a_8.$

19. $1 \cdot 6, 2 \cdot 7, 3 \cdot 8, 4 \cdot 9, \ldots \; ; a_7.$

20. $-2, 4, -8, 16, \ldots \; ; a_7.$

21. $\dfrac{2}{5}, \dfrac{2}{25}, \dfrac{2}{125}, \dfrac{2}{625}, \ldots \; ; a_6.$

22. $-6, 12, -24, 48, -96, \ldots \; ; a_9.$

23. $-6, 24, -120, 720, \ldots \; ; a_5.$

24. $\dfrac{3}{2}, \dfrac{9}{6}, \dfrac{27}{24}, \dfrac{81}{120}, \ldots \; ; a_5.$

In Exercises 25–27, the first several terms of a sequence are given. Find the indicated partial sum.

25. $2, 4, 6, 8, 10, \ldots \; ; S_4$

26. $3, -6, 9, -12, 15, -18, \ldots \; ; S_9$

27. $\dfrac{1}{2}, -\dfrac{1}{4}, \dfrac{1}{8}, -\dfrac{1}{16}, \ldots \; ; S_5$

In Exercises 28–30, the general term of a sequence is given. Find the indicated partial sum.

28. $a_n = 3n + 8; S_6$

29. $a_n = (-1)^n \cdot (4n); S_6$

30. $a_1 = 4, a_n = a_{n-1} - 8$ for $n \geq 2; S_8$

In Exercises 31–37, find the sum of each series.

31. $\displaystyle\sum_{i=1}^{9} i$

32. $\displaystyle\sum_{i=1}^{6} (4i + 3)$

33. $\displaystyle\sum_{i=1}^{7} i(i + 2)$

34. $\displaystyle\sum_{i=1}^{21} 7$

35. $\displaystyle\sum_{j=0}^{5} (j + 4)^2$

36. $\displaystyle\sum_{k=2}^{7} \dfrac{k!}{(k - 2)!}$

37. $\displaystyle\sum_{j=0}^{5} (j - 2)^3$

In Exercises 38–45, rewrite each series using summation notation. Use 1 as the lower limit of summation.

38. $1 + 2 + 3 + \cdots + 29$

39. $5 + 10 + 15 + \cdots + 50$

40. $1^2 + 2^2 + 3^2 + \cdots + 11^2$

41. $\dfrac{4}{5} + \dfrac{5}{6} + \dfrac{6}{7} + \cdots + \dfrac{12}{13}$

11.1 Introduction to Sequences and Series   11-9

42. $2 + (-4) + 8 + (-16) + \cdots + (-256)$

43. $-\dfrac{1}{9} + \dfrac{1}{18} - \dfrac{1}{27} + \cdots + \dfrac{1}{54}$

44. $5 + \dfrac{5^2}{2} + \dfrac{5^3}{3} + \cdots + \dfrac{5^n}{n}$

45. $1 + 7 + \dfrac{7^2}{2!} + \dfrac{7^3}{3!} + \dfrac{7^4}{4!} + \cdots + \dfrac{7^n}{n!}$

46. The figure shown gives the progression of the branching of a tree during each stage of development. Notice that the number of branches formed during a given stage is a Fibonacci number. Assuming that this branching pattern continues, how many branches will form during the 10th stage of development?

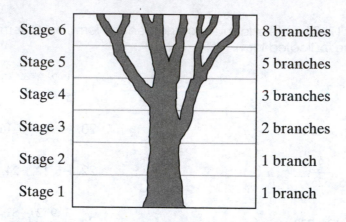

Stage 6	8 branches
Stage 5	5 branches
Stage 4	3 branches
Stage 3	2 branches
Stage 2	1 branch
Stage 1	1 branch

11.2 Arithmetic Sequences and Series

THINGS TO KNOW

Before working through this section, be sure you are familiar with the following concepts:

VIDEO   ANIMATION   INTERACTIVE

 1. Solve Systems of Linear Equations in Two Variables by Substitution (Section 3.1, Objective 3)   ▶

 2. Solve Systems of Linear Equations in Two Variables by Elimination (Section 3.1, Objective 4)   ▶

 3. Determine the Sum of a Finite Series Written in Summation Notation (Section 11.1, Objective 5)

OBJECTIVES

1 Determine If a Sequence Is Arithmetic

2 Find the General Term or a Specific Term of an Arithmetic Sequence

3 Compute the nth Partial Sum of an Arithmetic Series

4 Solve Applications of Arithmetic Sequences and Series

OBJECTIVE 1   DETERMINE IF A SEQUENCE IS ARITHMETIC

In this section, we will work exclusively with a specific type of sequence known as an **arithmetic sequence**. A sequence is arithmetic if the difference in any two successive terms is constant. For example, the sequence

$$5, 9, 13, 17, \ldots$$

is arithmetic because the difference of any two successive terms is 4. The first term of this sequence is $a_1 = 5$ and the common difference is $d = 4$. Notice that we can rewrite the terms of this sequence as $5, 5 + 4, 5 + 2(4), 5 + 3(4), \ldots$

In general, given an **arithmetic sequence** with a first term of a_1 and a common difference of d, then the first n terms of the sequence are as follows:

$$a_1$$
$$a_2 = a_1 + d$$
$$a_3 = a_2 + d = \underbrace{(a_1 + d)}_{a_2} + d = a_1 + 2d$$
$$a_4 = a_3 + d = \underbrace{(a_1 + 2d)}_{a_3} + d = a_1 + 3d$$
$$\vdots$$
$$a_n = a_1 + (n - 1)d$$

Definition   Arithmetic Sequence

An **arithmetic sequence** is a sequence of the form $a_1, a_1 + d, a_1 + 2d,$ $a_1 + 3d, a_1 + 4d, \ldots$, where a_1 is the first term of the sequence and d is the common difference. The general term, or nth term, of an arithmetic sequence has the form $a_n = a_1 + (n - 1)d$.

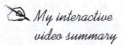 *My interactive video summary*

Example 1 Determining If a Sequence Is Arithmetic

For each of the following sequences, determine if it is arithmetic. If the sequence is arithmetic, find the common difference.

a. $1, 4, 7, 10, 13, \ldots$
b. $b_n = n^2 - n$

c. $a_n = -2n + 7$
d. $a_1 = 14, a_n = 3 + a_{n-1}$

Solutions Watch this interactive video to verify that the sequences in parts (a), (c), and (d) are arithmetic. The sequence in part (b) is not arithmetic. Notice that the arithmetic sequence in part (d) is a **recursive sequence**.

You Try It Work through this You Try It problem.

Work Exercises 1–6 in this eText or in the 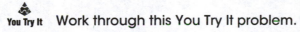 Study Plan.

It is worth noting that every arithmetic sequence is a **linear function** whose **domain** is the **natural numbers**. A portion of the graphs of the arithmetic sequences from Example 1a and Example 1c are seen in Figure 4. Notice that the ordered pairs of each sequence are **collinear**.

Figure 4

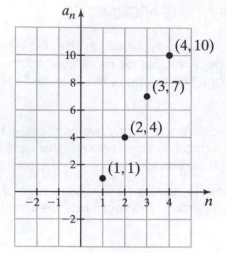

(a) A Portion of the Graph of the
Sequence $1, 4, 7, 10, 13, \ldots$

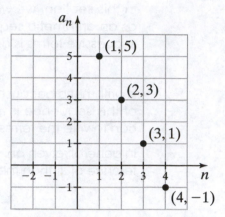

(b) A Portion of the Graph of the
Sequence $a_n = -2n + 7$

The graph of every arithmetic sequence is represented by a set of ordered pairs that lies on a straight line.

Note: When the **common difference** of an **arithmetic sequence** is positive, the **terms** of the sequence *increase* and the graph is represented by a set of ordered pairs that lie along a line with positive slope. When the common difference of an arithmetic sequence is negative, the terms of the sequence *decrease* and the graph is represented by a set of ordered pairs that lies along a line with negative slope.

OBJECTIVE 2 FIND THE GENERAL TERM OR A SPECIFIC TERM OF AN ARITHMETIC SEQUENCE

By the definition of an arithmetic sequence, the **general term** of an arithmetic sequence has the form $a_n = a_1 + (n - 1)d$. We can use this formula to find any term of an arithmetic sequence.

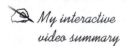 *My interactive video summary*

Example 2 Finding the General Term of an Arithmetic Sequence

Find the general term of each arithmetic sequence, then find the indicated term of the sequence. (In part (c), only a portion of the graph is given. Assume that the **domain** of this sequence is all **natural numbers**.)

a. $11, 17, 23, 29, 35, \ldots$; a_{50}

b. $2, 0, -2, -4, -6, \ldots$; a_{90}

c. Find a_{31}.

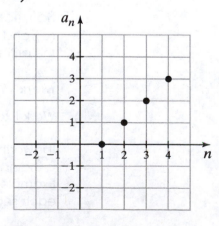

Solutions

a. The first term of the sequence is $a_1 = 11$ and the common difference is $d = 6$. The general term is given by $a_n = 11 + (n-1)(6) = 11 + 6n - 6 = 6n + 5$. Therefore, $a_{50} = 6(50) + 5 = 305$.

b. The first term of the sequence is $a_1 = 2$ and the common difference is $d = -2$. The general term is given by $a_n = 2 + (n-1)(-2) = 2 - 2n + 2 = 4 - 2n$. Therefore, $a_{90} = 4 - 2(90) = -176$.

c. The first four terms of this sequence are $a_1 = 0, a_2 = 1, a_3 = 2$, and $a_4 = 3$. Thus, $a_1 = 0$ and the common difference is $d = 1$. The general term is given by $a_n = 0 + (n-1)(1) = n - 1$. Thus, $a_{31} = 31 - 1 = 30$.

You may also watch this interactive video to see each of these solutions worked out in detail.

You Try It Work through this You Try It problem.

Work Exercises 7–12 in this eText or in the MyMathLab Study Plan.

My interactive video summary

Example 3 Finding a Specific Term of an Arithmetic Sequence

a. Given an arithmetic sequence with $d = -4$ and $a_3 = 14$, find a_{50}.

b. Given an arithmetic sequence with $a_4 = 12$ and $a_{15} = -10$, find a_{41}.

Solutions Read through the following solutions, or watch this interactive video.

a. We are given that $d = -4$ and $a_3 = 14$. We can use this information to solve for a_1.

Use the formula for the general term of an arithmetic sequence: $a_n = a_1 + (n-1)d$

Substitute $n = 3$ and $d = -4$: $a_3 = a_1 + (3-1)(-4)$

Simplifying, we get $a_3 = a_1 - 8$. Now substitute $a_3 = 14$ to solve for a_1.

Start with the formula for a_3: $a_3 = a_1 - 8$

Substitute $a_3 = 14$: $14 = a_1 - 8$

Add 8 to both sides: $22 = a_1$

Using the formula $a_n = a_1 + (n-1)d$ with $a_1 = 22$ and $d = -4$, simplify to get $a_n = 26 - 4n$. Therefore, $a_{50} = 26 - 4(50) = -174$.

b. Using the fact that $a_n = a_1 + (n-1)d$, we get $a_4 = a_1 + (4-1)d = 12$ and $a_{15} = a_1 + (15-1)d = -10$. This leads to the following **system of linear equations**:

$$\begin{cases} a_1 + 3d = 12 \\ a_1 + 14d = -10 \end{cases}$$

Using the **substitution method** or **elimination method** to solve this system, we get $a_1 = 18$ and $d = -2$. Using the formula $a_n = a_1 + (n-1)d$ with $a_1 = 18$ and $d = -2$, we can find the general term.

Use the formula for the general term of an arithmetic sequence: $a_n = a_1 + (n-1)d$

Substitute $a_1 = 18$ and $d = -2$: $= 18 + (n-1)(-2)$

Use the **distributive property:** $= 18 - 2n + 2$

Simplify: $= 20 - 2n$

The general term is $a_n = 20 - 2n$. Therefore, $a_{41} = 20 - 2(41) = -62$.

 **You Try It** Work through this **You Try It** problem.

Work Exercises 13–18 in this eText or in the MyMathLab Study Plan.

OBJECTIVE 3 COMPUTE THE nTH PARTIAL SUM OF AN ARITHMETIC SERIES

If $a_1, a_2, a_3, \ldots$ is an arithmetic sequence, then the expression $a_1 + a_2 + a_3 + \cdots + a_n + a_{n+1} + \cdots$ is called an **infinite arithmetic series** and can be written using summation notation as $\displaystyle\sum_{i=1}^{\infty} a_i$. Recall that the sum of the first n terms of a series is called the **nth partial sum** of the series and is given by $S_n = a_1 + a_2 + a_3 + \cdots + a_n$. We can also represent the nth partial sum using summation notation as $S_n = \displaystyle\sum_{i=1}^{n} a_i$. The nth partial sum is simply the sum of a finite arithmetic series.

My video summary ▶ **Concept Video** Fortunately, there is a convenient formula for computing the nth partial sum of an arithmetic series. Watch this **video** to see the derivation of this formula.

Formula for the nth Partial Sum of an Arithmetic Series

The sum of the first n terms of an arithmetic series is called the **nth partial sum** of the series and is given by $S_n = \displaystyle\sum_{i=1}^{n} a_i = a_1 + a_2 + a_3 + \cdots + a_n$. This sum can be computed using the formula $S_n = \dfrac{n(a_1 + a_n)}{2}$.

Note: The **nth partial sum** of an arithmetic series is simply the sum of a finite arithmetic series. An arithmetic series *must* be **finite** in order to compute the sum. This is not true for some other types of series. You will see how to find the sum of a special type of infinite series in **Section 11.3**.

My interactive video summary 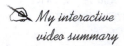 **Example 4 Finding the Sum of an Arithmetic Series**

Find the sum of each arithmetic series.

a. $\displaystyle\sum_{i=1}^{20} (2i - 11)$

b. $-5 + (-1) + 3 + 7 + \cdots + 39$

Solutions

a. We can use the formula $S_{20} = \dfrac{20(a_1 + a_{20})}{2}$ to compute the sum of the first 20 terms of this series.

$$\text{Substitute } i = 1 \text{ in the}$$
$$\text{formula } 2i - 11 \text{ to find } a_1: \quad a_1 = 2(1) - 11 = -9$$

$$\text{Substitute } i = 20 \text{ in the}$$
$$\text{formula } 2i - 11 \text{ to find } a_{20}: \quad a_{20} = 2(20) - 11 = 29$$

Now substitute $a_1 = -9$ and $a_{20} = 29$ into the formula $S_{20} = \dfrac{20(a_1 + a_{20})}{2}$.

$$\text{Use the formula for 20th partial}$$
$$\text{sum of an arithmetic series:} \quad S_{20} = \frac{20(a_1 + a_{20})}{2}$$

$$\text{Substitute } a_1 = -9 \text{ and } a_{20} = 29: \quad = \frac{20(-9 + 29)}{2}$$

$$\text{Simplify:} \quad = 200$$

Therefore, $\displaystyle\sum_{i=1}^{20}(2i - 11) = 200$. You may also watch this **interactive video** to see this solution worked out in detail.

b. Try this problem on your own, or work through this **interactive video** to verify that the sum of this arithmetic series is 204.

You Try It Work through this **You Try It** problem.

Work Exercises 19–27 in this eText or in the MyMathLab Study Plan.

OBJECTIVE 4 SOLVE APPLICATIONS OF ARITHMETIC SEQUENCES AND SERIES

My video summary ▶ **Example 5 Selling Digital Media Subscriptions**

A digital media provider has hired teenagers to go door-to-door to try to solicit new members. The teenagers receive $10 for selling the first membership. For each additional membership sold, the provider will pay the teenagers 20 cents more than what was paid for the previous membership. How much will the teenagers get paid for selling the 100th membership? How much money will the teenagers earn by selling 100 memberships?

Solution The amount of money earned by selling one membership can be represented by $a_1 = 10$. The money earned by selling the second membership is $a_2 = 10.20$. The money earned by selling the third membership is $a_3 = 10.40$. We see that the amount of money earned by selling n memberships is an arithmetic sequence with $a_1 = 10$ and $d = 0.20$. This sequence is defined by $a_n = 10 + (n - 1)(0.20) = 10 + (0.20)n - 0.20 = 0.20n + 9.80$.

The cash earned by selling the 100th membership is the 100th term of this sequence, or $a_{100} = 0.20(100) + 9.80 = 29.80$. Therefore, the teenagers are paid $29.80 for selling the 100th membership.

To find the total amount earned by selling 100 memberships, we must find the sum of the series $\displaystyle\sum_{i=1}^{100}[(0.20)i + 9.80]$.

Using the formula $S_n = \dfrac{n(a_1 + a_n)}{2}$ with $n = 100$, $a_1 = 10$, and $a_{100} = 29.80$, we get

$$S_{100} = \frac{100(a_1 + a_{100})}{2} = \frac{100(10 + 29.80)}{2} = 50(39.80) = 1990.$$

Thus, the teenagers will be paid $1990 if they sell 100 memberships.

You Try It Work through this You Try It problem.

Work Exercises 28–31 in this eText or in the MyMathLab Study Plan.

My video summary ▶ **Example 6 Seats in a Theater**

A large multiplex movie house has many theaters. The smallest theater has only 12 rows. There are six seats in the first row. Each row has two seats more than the previous row. How many total seats are there in this theater?

Solution Try solving this problem on your own. When you are done, watch this video to see if you are correct, then work through the following "You Try It" problem.

You Try It Work through this You Try It problem.

Work Exercises 32–35 in this eText or in the MyMathLab Study Plan.

11.2 Exercises

In Exercises 1–6, determine if the sequence is arithmetic. If the sequence is arithmetic, find the common difference.

1. 8, 14, 20, 26, 32, . . .

2. 8, 11, 13, 16, 18, . . .

3. $a_n = \dfrac{3n + 1}{2}$

4. $a_n = n(n + 1)$

5. $a_1 = 8, a_n = 2 + a_{n-1}$

6. $a_1 = 5, a_n = 3a_{n-1} + 1$

In Exercises 7–12, find the general term of each arithmetic sequence and then find the indicated term of the sequence. If the sequence is represented by a graph, assume that the domain of the sequence is all natural numbers.

7. 2, 7, 12, 17, . . . ; a_{10}

8. 5, 1, −3, −7, . . . ; a_{31}

9. $\dfrac{3}{2}, 3, \dfrac{9}{2}, 6, \dfrac{15}{2}, \ldots$; a_{50}

10. 5.0, 3.8, 2.6, 1.4, . . . ; a_{29}

11. Find a_{17}.

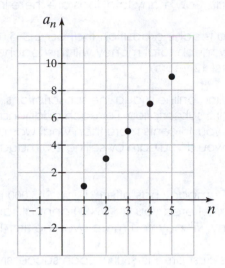

12. Find a_{11}.

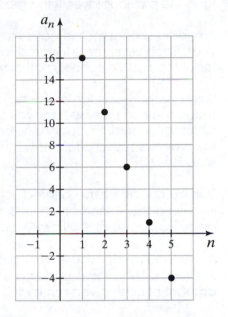

13. Given an arithmetic sequence with $d = 3$ and $a_8 = 5$, find a_{30}.

14. Given an arithmetic sequence with $d = -5$ and $a_7 = 11$, find a_{22}.

15. Given an arithmetic sequence with $a_5 = 4$ and $a_{22} = 55$, find a_{36}.

16. Given an arithmetic sequence with $a_6 = 4$ and $a_{20} = -52$, find a_{33}.

17. Given an arithmetic sequence with $a_{16} = 30$ and $a_{30} = 65$, find a_9.

18. Given an arithmetic sequence with $a_8 = -6$ and $a_{19} = -\dfrac{45}{2}$, find a_{34}.

In Exercises 19–27, find the indicated sum.

19. $\displaystyle\sum_{i=1}^{80} i$

20. $\displaystyle\sum_{j=1}^{10} (3j + 7)$

21. $7 + 10 + 13 + 16 + \cdots + 118$

22. $-14 + (-9) + (-4) + 1 + \cdots + 101$

23. $\displaystyle\sum_{i=3}^{14} (-7 - 9i)$

24. $1 + 11 + 21 + 31 + \cdots + a_{102}$

25. $6 + 14 + 22 + 30 + \cdots + (8n - 2)$

26. Find the sum of the first 100 odd integers.

27. Find the sum of the first 100 even positive integers.

28. A large multiplex movie house has many theaters. The largest theater has 40 rows. There are 12 seats in the first row. Each row has four seats more than the previous row. How many total seats are there in this theater?

29. A stack of logs has 47 logs on the bottom layer. Each subsequent layer has nine fewer logs than the previous layer. If the top layer has two logs, how many total logs are there in the pile?

30. A middle school mathematics teacher accepts a teaching position that pays $45,000 per year. Each year, the expected raise is $1400. How much total money will this teacher earn teaching middle school mathematics over the first 12 years?

31. Suppose that you plan on taking a summer job selling online magazine subscriptions. The magazine company will pay you $1 for selling the first subscription. For each additional subscription sold, the magazine company will pay you 15 cents more than what was paid for the previous subscription. How much total money would you earn by selling 200 magazine subscriptions?

32. Two companies have offered you a job. Alpha Company has offered you $45,000 per year with an annual raise of $2500. Beta Company has offered you a $56,000 annual salary with an annual raise of $900 per year. Which company will pay you more over the first 10 years?

33. A city fund-raiser raffle is raffling off 25 cash prizes. First prize is $5000. Each successive prize is $200 less than the preceding prize. What is the value of the 25th prize? What is the total amount of cash given out by this raffle?

34. Larry's Luxury Rental Car Company rents luxury cars for up to 18 days. The price is $300 for the first day, with the rental fee decreasing $7 for each additional day. How much will it cost to rent a luxury car for 18 days?

35. A ball thrown straight up in the air travels 48 inches in the first tenth of a second. In the next tenth of a second, the ball travels 44 inches. After each additional tenth of a second, the ball travels 4 inches less than it did during the preceding tenth of a second. How long will it take before the ball starts coming back down? What is the total distance that the ball has traveled when it has reached its maximum height?

11.3 Geometric Sequences and Series

THINGS TO KNOW

Before working through this section, be sure you are familiar with the following concepts:

		VIDEO	ANIMATION	INTERACTIVE

 You Try It 1. Solve Applications of Exponential Functions (Section 9.3, Objective 4)

 You Try It 2. Find the General Term or a Specific Term of an Arithmetic Sequence (Section 11.2, Objective 2)

 You Try It 3. Compute the nth Partial Sum of an Arithmetic Series (Section 11.2, Objective 3)

OBJECTIVES

1 Write the Terms of a Geometric Sequence

2 Determine If a Sequence Is Geometric

3 Find the General Term or a Specific Term of a Geometric Sequence

4 Compute the *n*th Partial Sum of a Geometric Series

5 Determine If an Infinite Geometric Series Converges or Diverges

6 Solve Applications of Geometric Sequences and Series

....

OBJECTIVE 1   WRITE THE TERMS OF A GEOMETRIC SEQUENCE

Suppose that you have agreed to work for Bill Gates on a particular job for 21 days. Mr. Gates gives you two choices of payment. You can be paid $100 for the first day and an additional $50 per day for each subsequent day. Or, you can choose to be paid 1 penny for the first day with your pay doubling each subsequent day. Which method of payment would you choose? (We will revisit this question later in this section. See **Example 7**.) Notice that each payment method can be represented by a sequence:

$$\text{Payment Method 1: } 100, 150, 200, 250, 300, \ldots$$

$$\text{Payment Method 2: } 0.01, 0.02, 0.04, 0.08, 0.16, \ldots$$

The first method of payment is an **arithmetic sequence** with $a_1 = 100$ and $d = 50$. The second method of payment is an example of a **geometric sequence**. Each **term** of this geometric sequence can be obtained by multiplying the previous term by 2. The number 2 in this case is called the **common ratio**. We can obtain this common ratio by dividing any term of the sequence (except the first term) by the previous term. That is, $r = \dfrac{a_2}{a_1} = \dfrac{a_3}{a_2} = \cdots = \dfrac{a_{n+1}}{a_n}$. The first term of the payment method 2 sequence is $a_1 = 0.01$ and the common ratio is $r = 2$.

Notice that we can rewrite the terms of this sequence as $0.01, (0.01)(2), (0.01)(2^2), (0.01)(2^3), \ldots$. In general, given a geometric sequence with a first term of a_1 and a **common ratio** of r, then the first n terms of the sequence are

$$a_1$$

$$a_2 = a_1 r$$

$$a_3 = a_2 r = \underbrace{(a_1 r)}_{a_2} r = a_1 r^2$$

$$a_4 = a_3 r = \underbrace{(a_1 r^2)}_{a_3} r = a_1 r^3$$

$$\vdots$$

$$a_n = a_1 r^{n-1}.$$

Definition   Geometric Sequence

A **geometric sequence** is a sequence of the form $a_1, a_1 r, a_1 r^2, a_1 r^3, a_1 r^4, \ldots,$ where a_1 is the first term of the sequence and r is the common ratio such that

$r = \dfrac{a_2}{a_1} = \dfrac{a_3}{a_2} = \cdots = \dfrac{a_{n+1}}{a_n}$ for all $n \geq 1$. The general term, or *n*th term, of a geometric sequence has the form $a_n = a_1 r^{n-1}$.

A portion of the two sequences representing the two payment methods are sketched in Figure 5. The sequence representing payment method 1 (Figure 5a) is **arithmetic**. The ordered pairs of this sequence are **collinear**. The sequence representing payment method 2 (Figure 5b) is **geometric**. Notice that the ordered pairs of this sequence do not lie along a common line but rather lie on an **exponential curve**.

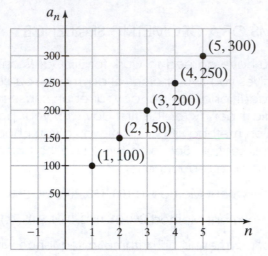

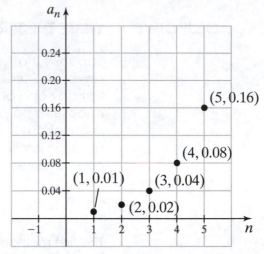

(a) A Portion of the Graph of the
    Sequence 100, 150, 200, 250, 300, . . .

(b) A Portion of the Graph of the
    Sequence 0.01, 0.02, 0.04, 0.08, 0.16, . . .

Figure 5

The graph of every arithmetic sequence is represented by a set of ordered pairs that lies on a straight line. The graph of a geometric sequence with $r > 0$ is represented by a set of ordered pairs that lies on an exponential curve.

If the first **term** and the **common ratio** of a **geometric sequence** are known, then we can determine the second term by multiplying the first term by the common ratio. The third term can be found by multiplying the second term by the common ratio. We continue this process to find the subsequent terms of the sequence.

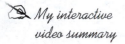

My interactive video summary

✎ Example 1 Writing a Geometric Sequence

a. Write the first five terms of the geometric sequence having a first term of 2 and a common ratio of 3.

b. Write the first five terms of the geometric sequence such that $a_1 = -4$ and $a_n = -5a_{n-1}$ for $n \geq 2$.

Solutions Read through the following solutions, or watch this **interactive video**.

a. We are given that the first term is $a_1 = 2$ and the common ratio is $r = 3$. The second term is $a_2 = 2 \cdot 3 = 6$. The third term is $a_3 = 6 \cdot 3 = 18$. The fourth term is $a_4 = 18 \cdot 3 = 54$. The fifth term is $a_5 = 54 \cdot 3 = 162$. Therefore, the first five terms of this sequence are 2, 6, 18, 54, and 162.

b. This sequence is defined **recursively**. The first term is $a_1 = -4$. To find a_2, substitute the value of 2 for n in the formula $a_n = -5a_{n-1}$ and simplify:

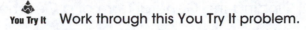

$$\text{Given recursive formula:}\quad a_n = -5a_{n-1}$$
$$\text{Substitute:}\quad a_2 = -5a_{2-1}$$
$$\text{Simplify:}\quad = -5a_1$$
$$\text{Substitute } a_1 = -4:\quad = -5(-4)$$
$$\text{Multiply:}\quad = 20$$

Therefore, $a_2 = 20$. We can follow this same procedure to find a_3, a_4, and a_5. You should verify that $a_3 = -100$, $a_4 = 500$, and $a_5 = -2500$ on your own.

You Try It Work through this **You Try It** problem.

Work Exercises 1–5 in this eText or in the MyMathLab Study Plan.

OBJECTIVE 2 DETERMINE IF A SEQUENCE IS GEOMETRIC

To determine if a given sequence is **geometric**, we must check to see if each term of the sequence can be obtained by multiplying the previous term by a common ratio r. That is, we must check to see if there exists a constant value r such that $r = \dfrac{a_{n+1}}{a_n}$ for any $n \geq 1$.

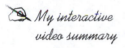
My interactive video summary

Example 2 Determining If a Sequence Is Geometric

For each of the following sequences, determine if it is geometric. If the sequence is geometric, find the common ratio.

a. $2, 4, 6, 8, 10, \ldots$

b. $\dfrac{2}{3}, \dfrac{4}{9}, \dfrac{8}{27}, \dfrac{16}{81}, \dfrac{32}{243}, \ldots$

c. $12, -6, 3, -\dfrac{3}{2}, \dfrac{3}{4}, \ldots$

Solutions

a. For this sequence, $\dfrac{a_2}{a_1} = \dfrac{4}{2} = 2$ and $\dfrac{a_3}{a_2} = \dfrac{6}{4} = \dfrac{3}{2}$. Since $\dfrac{a_2}{a_1} \neq \dfrac{a_3}{a_2}$, there does not exist a **common ratio**. Hence, this sequence is not geometric. (Note that this sequence is an **arithmetic** sequence.)

b. For this sequence, $\dfrac{a_2}{a_1} = \dfrac{\frac{4}{9}}{\frac{2}{3}} = \dfrac{4}{9} \cdot \dfrac{3}{2} = \dfrac{2}{3}$. Note that each term of this sequence

(other than the first term) can be obtained by multiplying the previous term

by $\dfrac{2}{3}$. Therefore, this sequence is geometric with a common ratio of $\dfrac{2}{3}$.

c. Try to determine if this sequence is geometric. Work through the **interactive video** to see if you are correct, or to view complete solutions to all parts.

You Try It Work through this You Try It problem.

Work Exercises 6–10 in this eText or in the MyMathLab Study Plan.

OBJECTIVE 3 FIND THE GENERAL TERM OR A SPECIFIC TERM OF A GEOMETRIC SEQUENCE

By the definition of a **geometric sequence**, the **general term**, or nth term, of a geometric sequence has the form $a_n = a_1 r^{n-1}$. We can use this formula to find the any term of a given geometric sequence.

My video summary **Example 3 Finding the General Term of a Geometric Sequence**

Find the general term of each geometric sequence.

a. $12, -6, 3, -\dfrac{3}{2}, \dfrac{3}{4}, \ldots$ b. $\dfrac{2}{3}, \dfrac{2}{9}, \dfrac{2}{27}, \dfrac{2}{81}, \dfrac{2}{243}, \ldots$

Solutions Read through the following solutions, or watch this **video**.

a. The first term of the sequence is $a_1 = 12$ and the common ratio is $r = \dfrac{a_2}{a_1} = \dfrac{-6}{12} = -\dfrac{1}{2}$. Therefore, $a_n = 12\left(-\dfrac{1}{2}\right)^{n-1}$.

b. The first term of the sequence is $a_1 = \dfrac{2}{3}$ and the common ratio is

$r = \dfrac{a_2}{a_1} = \dfrac{\frac{2}{9}}{\frac{2}{3}} = \dfrac{1}{3}$. Therefore, $a_n = \left(\dfrac{2}{3}\right)\left(\dfrac{1}{3}\right)^{n-1}$.

You Try It Work through this You Try It problem.

Work Exercises 11–13 in this eText or in the MyMathLab Study Plan.

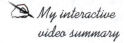 *My interactive video summary* **Example 4 Finding a Specific Term of a Geometric Sequence**

a. Find the seventh term of the geometric sequence whose first term is 2 and whose **common ratio** is -3.

b. Given a geometric sequence such that $a_6 = 16$ and $a_9 = 2$, find a_{13}.

Solutions

a. We can use the formula $a_n = a_1 r^{n-1}$ with $a_1 = 2$ and $r = -3$ to find the general term. The general term is $a_n = 2(-3)^{n-1}$. Therefore, $a_7 = 2(-3)^{7-1} = 2(-3)^6 = 2(729) = 1458$.

b. Since $a_6 = 16$, we can substitute $n = 6$ into the formula $a_n = a_1 r^{n-1}$ to get $a_6 = a_1 r^5 = 16$. Similarly, we can substitute $n = 9$ into the formula $a_n = a_1 r^{n-1}$ to get $a_9 = a_1 r^8 = 2$. This gives the following two equations.

$$(1)\quad a_1 r^5 = 16$$

$$(2)\quad a_1 r^8 = 2$$

Divide both sides of equation (1) by r^5 to get $a_1 = \dfrac{16}{r^5}$. Now, substitute $a_1 = \dfrac{16}{r^5}$ into equation (2) and solve for a_1.

$$\text{Start with equation (2):} \qquad a_1 r^8 = 2$$

$$\text{Substitute } a_1 = \frac{16}{r^5}: \qquad \left(\frac{16}{r^5}\right)r^8 = 2$$

$$\frac{r^8}{r^5} = r^3: \qquad 16r^3 = 2$$

$$\text{Divide both sides by 16:} \qquad r^3 = \frac{1}{8}$$

$$\text{Take the cube root of both sides:} \qquad r = \frac{1}{2}$$

Now substitute $r = \dfrac{1}{2}$ into equation (1) to solve for a_1.

$$\text{Start with equation (1):} \qquad a_1 r^5 = 16$$

$$\text{Substitute } r = \frac{1}{2}: \qquad a_1\left(\frac{1}{2}\right)^5 = 16$$

$$\left(\frac{1}{2}\right)^5 = \frac{1}{32}: \qquad a_1\left(\frac{1}{32}\right) = 16$$

$$\text{Multiply both sides by 32:} \qquad a_1 = 512$$

We can now use the formula $a_n = a_1 r^{n-1}$ with $a_1 = 512$ and $r = \dfrac{1}{2}$ to find the general term. The general term is $a_n = 512\left(\dfrac{1}{2}\right)^{n-1}$. Therefore,

$$a_{13} = 512\left(\frac{1}{2}\right)^{13-1} = 512\left(\frac{1}{2}\right)^{12} = \frac{512}{2^{12}} = \frac{512}{4096} = \frac{1}{8}.$$

You may wish to work through this **interactive video** to see these solutions worked out in detail.

You Try It Work through this **You Try It** problem.

Work Exercises 14–18 in this eText or in the MyMathLab Study Plan.

OBJECTIVE 4 COMPUTE THE nTH PARTIAL SUM OF A GEOMETRIC SERIES

If $a_1, a_1 r, a_1 r^2, a_1 r^3, \ldots$ is a geometric sequence, then the expression $a_1 + a_1 r + a_1 r^2 + a_1 r^3 + \cdots + a_1 r^{n-1} + \cdots$ is called an **infinite geometric series** and can be written in summation notation as $\displaystyle\sum_{i=1}^{\infty} a_1 r^{i-1}$. Recall that the sum of the first n terms of a series is called the **nth partial sum** of the series and is given by $S_n = a_1 + a_1 r + a_1 r^2 + a_1 r^3 + \cdots + a_1 r^{n-1}$. We can also represent the nth partial sum using summation notation as $S_n = \displaystyle\sum_{i=1}^{n} a_1 r^{i-1}$. This nth partial sum is simply the sum of a finite geometric series.

My video summary ▶ **Concept Video** Fortunately, there is a convenient formula for computing the nth partial sum of a geometric series. Watch this video to see the derivation of this formula.

> ### Formula for the nth Partial Sum of a Geometric Series
>
> The sum of the first n terms of a geometric series is called the **nth partial sum** of
>
> the series and is given by $S_n = \sum_{i=1}^{n} a_1 r^{i-1} = a_1 + a_1 r + a_1 r^2 + a_1 r^3 + \cdots + a_1 r^{n-1}$.
>
> This sum can be computed using the formula $S_n = \dfrac{a_1(1 - r^n)}{1 - r}$ for $r \neq 1$.

My interactive video summary 🖐 **Example 5 Computing the nth Partial Sum of a Geometric Series**

a. Find the sum of the series $\sum_{i=1}^{15} 5(-2)^{i-1}$.

b. Find the seventh partial sum of the geometric series $8 + 6 + \dfrac{9}{2} + \dfrac{27}{8} + \cdots$.

Solutions

a. Using the formula $S_n = \dfrac{a_1(1 - r^n)}{1 - r}$ with $n = 15$, $a_1 = 5$, and $r = -2$, we get

$$S_{15} = \frac{5(1 - (-2)^{15})}{1 - (-2)} = \frac{5(1 - (-32{,}768))}{3} = \frac{5(32{,}769)}{3} = \frac{163{,}845}{3} = 54{,}615.$$ You may also watch this interactive video to see this solution worked out in detail.

b. Try this problem on your own, or work through this **interactive video** to verify

that the seventh partial sum of this series is $S_7 = \dfrac{14{,}197}{512}$.

🔺 **You Try It** Work through this You Try It problem.

Work Exercises 19–23 in this eText or in the MyMathLab Study Plan.

OBJECTIVE 5   DETERMINE IF AN INFINITE GEOMETRIC SERIES CONVERGES OR DIVERGES

Consider the infinite geometric series $\sum_{n=1}^{\infty} a_1 r^{n-1} = a_1 + a_1 r + a_1 r^2 + \cdots + a_1 r^{n-1} + \cdots$.
Is it possible for a series of this form to have a finite sum? Is it possible to add infinitely many terms and get a finite sum? The answer is **yes**, it is possible, but it depends on the value of r.

My video summary ▶ **Concept Video** Watch this video to learn more about this concept and develop a formula to find such sums.

> ### Formula for the Sum of an Infinite Geometric Series
>
> Let $\sum_{n=1}^{\infty} a_1 r^{n-1} = a_1 + a_1 r + a_1 r^2 + a_1 r^3 + \cdots + a_1 r^{n-1} + \cdots$ be an infinite
>
> geometric series. If $|r| < 1$, then the sum of the series is given by $S = \dfrac{a_1}{1 - r}$.

Note: A formal proof of this formula requires calculus.

If $|r| < 1$, then the infinite geometric series has a finite sum and is said to **converge**. If $|r| \geq 1$, then the infinite geometric series does not have a finite sum and the series is said to **diverge**.

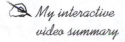

Example 6 Determining If an Infinite Geometric Series Converges or Diverges

Determine whether each of the following series converges or diverges. If the series converges, find the sum.

a. $\displaystyle\sum_{n=1}^{\infty} \frac{1}{2}\left(\frac{2}{3}\right)^{n-1}$ b. $3 - \dfrac{6}{5} + \dfrac{12}{25} - \dfrac{24}{125} + \cdots$

c. $12 + 18 + 27 + \dfrac{81}{2} + \dfrac{243}{4} + \cdots$

Solutions

a. This is an infinite geometric series with $|r| = \left|\dfrac{2}{3}\right| < 1$. Since $|r| < 1$, the infinite series must converge and, thus, must have a finite sum. The sum of the series is

$$S = \frac{a_1}{1-r} = \frac{\dfrac{1}{2}}{1 - \dfrac{2}{3}} = \frac{\dfrac{1}{2}}{\dfrac{1}{3}} = \frac{1}{2} \cdot \frac{3}{1} = \frac{3}{2}.$$

b. For this infinite geometric series, the common ratio is $r = \dfrac{-\dfrac{6}{5}}{3} = -\dfrac{6}{5} \cdot \dfrac{1}{3} = -\dfrac{2}{5}.$

Since $|r| = \left|-\dfrac{2}{5}\right| = \dfrac{2}{5} < 1$, the infinite series converges. The sum is

$$S = \frac{a_1}{1-r} = \frac{3}{1 - \left(-\dfrac{2}{5}\right)} = \frac{3}{1 + \dfrac{2}{5}} = \frac{3}{\dfrac{7}{5}} = 3 \cdot \frac{5}{7} = \frac{15}{7}.$$

c. This infinite series diverges. Try to show this on your own. Work through the interactive video to see why this series diverges.

 You Try It Work through this You Try It problem.

Work Exercises 24–28 in this eText or in the MyMathLab Study Plan.

OBJECTIVE 6 SOLVE APPLICATIONS OF GEOMETRIC SEQUENCES AND SERIES

In Example 7, we revisit the question that was presented at the beginning of this section.

Example 7 Choosing a Payment Method

Suppose that you have agreed to work for Bill Gates on a particular job for 21 days. Mr. Gates gives you two choices of payment. You can be paid $100 for the first day and an additional $50 per day for each subsequent day. Or, you can choose to be paid 1 penny for the first day with your pay doubling each subsequent day. Which method of payment yields the most income?

My video summary ▶ **Solution** Read through the following solution, or watch this **video**.

Each payment method can be represented by a sequence.

$$\text{Payment method 1: } 100, 150, 200, 250, \ldots$$

$$\text{Payment method 2: } 0.01, 0.02, 0.04, 0.08, \ldots$$

To find out which method of payment will yield the greatest income, we must find the sum of the first 21 terms of each sequence.

Payment method 1 is an **arithmetic sequence** with $a_1 = 100$ and $d = 50$. Using the formula for the **general term of an arithmetic sequence** we get $a_n = 100 + (n - 1)50 = 100 + 50n - 50 = 50n + 50$. Note that $a_{21} = 50(21) + 50 = 1100$. Using the formula for the **nth partial sum of an arithmetic series** with $n = 21$, we get

$$S_{21} = \frac{n(a_1 + a_{21})}{2} = \frac{21(100 + 1100)}{2} = \$12{,}600.$$

Payment method 2 is a geometric sequence with $a_1 = 0.01$ and $r = 2$. Using the formula for the **nth partial sum of a geometric series** with $n = 21$, we get

$$S_{21} = \frac{a_1(1 - r^{21})}{1 - r} = \frac{(0.01)(1 - 2^{21})}{1 - 2} = \$20{,}971.51.$$

Clearly, payment method 2 is the better choice.

You Try It Work through this **You Try It** problem.

Work Exercises 29 and 30 in this eText or in the MyMathLab **Study Plan.**

My video summary ▶ **Example 8 Total Amount Given to a Local Charity**

A local charity received \$8500 in charitable contributions during the month of January. Because of a struggling economy, it is projected that contributions will decline each month to 95% of the previous month's contributions. What are the expected contributions for the month of October? What is the total expected contributions that this charity can expect at the end of the year?

Solution Read the following solution, or watch this **video**.

The monthly contributions can be represented by a geometric sequence with $a_1 = 8500$ and $r = 0.95$. Thus, the contributions for the nth month is given by $a_n = 8500(0.95)^{n-1}$. The expected contributions for October, or when $n = 10$, are $a_{10} = 8500(0.95)^{10-1} \approx 5357.12$. Thus, the contributions for the month of October are expected to be about \$5357.12.

The total contributions for the year can be written as the following finite geometric series:

$$8500 + (8500)(0.95) + (8500)(0.95)^2 + \cdots + (8500)(0.95)^{11} = \sum_{i=1}^{12} 8500(0.95)^{i-1}$$

Using the formula for the **nth partial sum of a geometric series** with $n = 12$, we get

$$S_{12} = \frac{a_1(1 - r^{12})}{1 - r} = \frac{8500(1 - 0.95^{12})}{1 - 0.95} \approx 78{,}138.79.$$

Therefore, the charity can expect about $78,138.79 in donations for the year.

You Try It Work through this You Try It problem.

Work Exercises 31–33 in this eText or in the MyMathLab Study Plan.

Recall that every repeating decimal number is a **rational number** and can therefore be represented by the quotient of two integers.

My interactive video summary

Example 9 Expressing a Repeating Decimal as a Ratio of Two Integers

Write each of the following repeating decimal numbers as the quotient of two integers.

a. $0.\overline{4}$

b. $0.2\overline{13}$

Solutions

a. We can rewrite $0.\overline{4}$ as $0.44444\ldots = \dfrac{4}{10} + \dfrac{4}{100} + \dfrac{4}{1,000} + \dfrac{4}{10,000} + \dfrac{4}{100,000} + \cdots$.

This is an infinite geometric series with $a_1 = \dfrac{4}{10}$ and $r = \dfrac{1}{10}$. Because $|r| = \left|\dfrac{1}{10}\right| < 1$, we know that the series converges. Using the formula

$$S = \frac{a_1}{1-r}, \text{ we see that } 0.\overline{4} = \frac{\dfrac{4}{10}}{1 - \dfrac{1}{10}} = \frac{\dfrac{4}{10}}{\dfrac{9}{10}} = \frac{4}{9}.$$

b. Carefully work through the interactive video to see that $0.2\overline{13} = \dfrac{211}{990}$.

You Try It Work through this You Try It problem.

Work Exercises 34 and 35 in this eText or in the MyMathLab Study Plan.

ANNUITIES

In **Section 9.3**, we derived a formula for **periodic compound interest**. This formula is used to determine the future value of a *one-time* investment. Suppose that instead of investing one lump sum, you wish to invest equal amounts of money at steady intervals. An investment of equal amounts deposited at equal time intervals is called an **annuity**. If these equal deposits are made at the end of a compound period, the annuity is called an **ordinary annuity**.

For example, suppose that you want to invest $P at the end of each payment period at an annual rate r, in decimal form. Then the interest rate per payment period is $i = \dfrac{r}{\text{number of payment periods per year}}$.

We now summarize the total amount of the ordinary annuity after the first k payment periods.

End of 1st payment period: $\underbrace{P}_{\text{1st payment}}$

End of 2nd payment period: $\underbrace{P}_{\text{1st payment}} + \underbrace{Pi}_{\substack{\text{Interest earned} \\ \text{on 1st payment}}} + \underbrace{P}_{\text{2nd payment}} = \underbrace{P(1+i)}_{\substack{\text{Total amount} \\ \text{of 1st payment}}} + \underbrace{P}_{\text{2nd payment}}$

End of 3rd payment period:

$\underbrace{P(1+i)}_{\substack{\text{Amount of} \\ \text{1st payment}}} + \underbrace{Pi(1+i)}_{\substack{\text{Interest earned} \\ \text{on amount of} \\ \text{1st payment}}} + \underbrace{P}_{\text{2nd payment}} + \underbrace{Pi}_{\substack{\text{Interest earned} \\ \text{on 2nd payment}}} + \underbrace{P}_{\text{3rd payment}} = \underbrace{P(1+i)^2}_{\substack{\text{Total amount} \\ \text{of 1st payment}}} + \underbrace{P(1+i)}_{\substack{\text{Total amount} \\ \text{of 2nd payment}}} + \underbrace{P}_{\text{3rd payment}}$

End of kth payment period: $\underbrace{P(1+i)^{k-1}}_{\substack{\text{Total amount} \\ \text{of 1st payment}}} + \underbrace{P(1+i)^{k-2}}_{\substack{\text{Total amount} \\ \text{of 2nd payment}}} + \cdots + \underbrace{P(1+i)}_{\substack{\text{Total amount} \\ \text{of }(k-1)\text{st payment}}} + \underbrace{P}_{\text{kth payment}}$

The total amount of the ordinary annuity after k payment periods is

$$A = P + P(1+i) + \cdots + P(1+i)^{k-2} + P(1+i)^{k-1}.$$

This is a finite geometric series with $a_1 = P$ and a common ratio of $(1+i)$. Thus, the amount of the annuity after the kth payment is

$$A = \frac{P(1-(1+i)^k)}{1-(1+i)} = \frac{P(1-(1+i)^k))}{-i} = \frac{P((1+i)^k - 1)}{i}.$$

Amount of an Ordinary Annuity after the kth Payment

The total amount of an ordinary annuity after the kth payment is given by the formula

$$A = \frac{P((1+i)^k - 1)}{i}$$

where

A = Total amount of annuity after k payments

P = Deposit amount at the end of each payment period

i = Interest rate per payment period

My video summary ▶ **Example 10 Finding the Amount of an Ordinary Annuity**

Chie and Ben decided to save for their newborn son Jack's college education. They decided to invest \$200 every 3 months in an investment earning 8% interest compounded quarterly. How much is this investment worth after 18 years?

Solution This is an ordinary annuity with $P = \$200$ and $i = \dfrac{0.08}{4} = 0.02$. What is k?

See if you can determine k and use the formula $A = \dfrac{P((1+i)^k - 1)}{i}$ to determine the total amount of this annuity. When you are done, watch this video to see if you are correct.

You Try It **Work through this You Try It problem.**

Work Exercises 36–38 in this eText or in the MyMathLab Study Plan.

11.3 Exercises

In Exercises 1–5, write the first five terms of the geometric sequence with the given information.

1. The first term is 8 and the common ratio is 2.

2. The first term is 162 and the common ratio is $\frac{1}{3}$.

3. The first term is 25 and the common ratio is $-\frac{1}{5}$.

4. $a_n = 7a_{n-1}; a_1 = 3$

5. $a_n = -2a_{n-1}; a_1 = -4$

In Exercises 6–10, determine if the sequence is geometric. If the sequence is geometric, find the common ratio.

6. $4, 24, 144, 864, \ldots$

7. $-2, 2, -2, 2, \ldots$

8. $-3, 1, -1, -3, \ldots$

9. $2, -\frac{10}{3}, \frac{50}{9}, -\frac{250}{27}, \ldots$

10. $7.236, -5.7888, 4.63104, -3.704832, \ldots$

11. Determine the general term of the sequence $3, 6, 12, 24, \ldots$.

12. Determine the general term of the sequence $\frac{1}{2}, \frac{1}{8}, \frac{1}{32}, \frac{1}{128}, \ldots$.

13. Determine the general term of the sequence $\frac{1}{5}, -\frac{2}{15}, \frac{4}{45}, -\frac{8}{135}, \ldots$.

14. Find the seventh term of the geometric sequence whose first term is 5 and whose common ratio is 4.

15. Find the sixth term of the geometric sequence whose first term is 3804 and whose common ratio is $-\frac{1}{4}$.

16. Find the 11th term of the geometric sequence $\$5000, \$5050, \$5100.50, \ldots$.

17. Given a geometric sequence such that $a_4 = 108$ and $a_7 = 2916$, find a_{10}.

18. Given a geometric sequence such that $a_3 = 16$ and $a_8 = -\frac{1}{2}$, find a_{11}.

In Exercises 19–23, find the sum of each geometric series.

19. $\displaystyle\sum_{i=1}^{8} 7(-4)^{i-1}$

20. $\displaystyle\sum_{i=1}^{13} 2\left(\frac{3}{5}\right)^{i-1}$

21. $\displaystyle\sum_{i=1}^{10} 4(1.05)^{i-1}$

22. $2 + \dfrac{2}{3} + \dfrac{2}{9} + \dfrac{2}{27} + \cdots + \dfrac{2}{729}$

23. $1 - \dfrac{1}{2} + \dfrac{1}{4} - \dfrac{1}{8} + \cdots - \dfrac{1}{128}$

In Exercises 24–28, determine if each infinite geometric series converges or diverges. If the series converges, find the sum.

24. $-1 + \dfrac{1}{10} - \dfrac{1}{100} + \dfrac{1}{1000} - \cdots$

25. $343 + 49 + 7 + 1 + \cdots$

26. $\displaystyle\sum_{i=1}^{\infty} 7\left(\frac{1}{4}\right)^{i-1}$

27. $\displaystyle\sum_{i=1}^{\infty} \frac{1}{5}(3)^{i-1}$

28. $0.5 - 0.05 + 0.005 - 0.0005 + 0.00005 - \cdots$

29. Warren wanted to save money to purchase a new car. He started by saving \$1 on the first of January. On the first of February, he saved \$3. On the first of March he saved \$9. So, on the first of each month, he wanted to save three times as much as he did on the first day of the previous month. If Warren continues his savings pattern, how much will he need to save on the first day of September?

30. Suppose that you have accepted a job for 2 weeks that will pay \$0.07 for the first day, \$0.14 for the second day, \$0.28 for the third day, and so on. What will your total earnings be after 2 weeks?

31. Mary has accepted a teaching job that pays \$37,000 for the first year. According to the Teacher's Union, Mary will get guaranteed salary increases of 3 percent per year. If Mary plans to teach for 30 years, what will be her total salary earnings?

32. A child is given an initial push on a rope swing. On the first swing, the rope swings through an arc of 12 feet. On each successive swing, the length of the arc is 80% of the previous length. After 10 swings, what is the total length the rope will have swung? When the child stops swinging, what is the total length the rope will have swung?

33. Randy dropped a rubber ball from his apartment window from a height of 50 feet. The ball always bounces $\dfrac{3}{5}$ of the distance fallen. How far has the ball traveled once it is done bouncing?

34. Rewrite the number $0.\overline{7}$ as the quotient of two integers.

35. Rewrite the number $0.3\overline{25}$ as the quotient of two integers.

36. Kip contributes \$200 every month to his 401(k). What will the value of Kip's 401(k) be in 10 years if the yearly rate of return is assumed to be 12% compounded monthly?

37. Mark and Lisa decide to invest \$500 every 3 months in an education IRA to save for their son Beau's college education. What will the value of the IRA be after 10 years if the yearly assumed rate of return is 8% compounded quarterly?

38. Marv and Cindy decide to build a new home in 10 years. They will need \$80,000 to purchase the lot on which to build. How much should they save each month in an account that has an assumed yearly rate of return of 7% compounded monthly?

11.4 The Binomial Theorem

THINGS TO KNOW

Before working through this section, be sure you are familiar with the following concepts:

		VIDEO	ANIMATION	INTERACTIVE
You Try It	1. Multiply Two Binomials (Section 4.3, Objective 3)	▶	⊛	☞
You Try It	2. Multiply Two or More Polynomials (Section 4.3, Objective 5)	▶	⊛	☞

OBJECTIVES

1 Expand Binomials Raised to a Power Using Pascal's Triangle

2 Evaluate Binomial Coefficients

3 Expand Binomials Raised to a Power Using the Binomial Theorem

4 Find a Particular Term or a Particular Coefficient of a Binomial Expansion

OBJECTIVE 1 EXPAND BINOMIALS RAISED TO A POWER USING PASCAL'S TRIANGLE

In this section, we will focus on expanding algebraic expressions of the form $(a + b)^n$, where n is an integer greater than or equal to zero. Because $(a + b)$ is a **binomial**, we call the expansion of $(a + b)^n$ a *binomial expansion.*

My video summary ▶ **Concept Video** The goal in this section is to try to develop a method for expanding expressions of the form $(a + b)^n$ without actually performing all of the multiplication. Watch this **video** to see the development of such a method.

The coefficients of each expansion of $(a + b)^n$ are known as **binomial coefficients**. There are always $n + 1$ terms and the sum of the exponents of each term is always equal to n. The exponent of the first variable decreases by 1 and the exponent of the second variable increases by 1. Thus, the exponent pattern of the variables of each expansion is $a^nb^0, a^{n-1}b^1, a^{n-2}b^2, a^{n-3}b^3, \ldots, a^1b^{n-1}, a^0b^n$.

The pattern for the binomial coefficients is less obvious. To see the pattern for the coefficients, we start by rewriting the six expansions of $(a + b)^n$ again, but this time, we only write the coefficients. See Figure 6.

$$\begin{array}{lccccccccccc}
n = 0: & & & & & & 1 & & & & & \\
n = 1: & & & & & 1 & & 1 & & & & \\
n = 2: & & & & 1 & & 2 & & 1 & & & \\
n = 3: & & & 1 & & 3 & & 3 & & 1 & & \\
n = 4: & & 1 & & 4 & & 6 & & 4 & & 1 & \\
n = 5: & 1 & & 5 & & 10 & & 10 & & 5 & & 1
\end{array}$$

Figure 6
The Coefficients of the Expansions of $(a + b)^n$, Also Called Pascal's Triangle

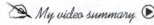

Notice that the coefficients in Figure 6 form a "triangle." This triangle is known as **Pascal's triangle**, named after the French mathematician, Blaise Pascal. The first and last number of each row of Pascal's triangle is 1. Every other number is equal to the sum of the two numbers directly above it. We can now write the next row of Pascal's triangle, which is the row corresponding to $n = 6$.

$$n = 5: \quad 1 \quad 5 \quad 10 \quad 10 \quad 5 \quad 1$$

$$n = 6: \quad 1 \quad 6 \quad 15 \quad 20 \quad 15 \quad 6 \quad 1$$

This new row of Pascal's triangle represents the coefficients of the expansion of $(a + b)^6$. Therefore,

$$(a + b)^6 = a^6 b^0 + 6a^5 b^1 + 15a^4 b^2 + 20a^3 b^3 + 15a^2 b^4 + 6a^1 b^5 + a^0 b^6$$

$$= a^6 + 6a^5 b + 15a^4 b^2 + 20a^3 b^3 + 15a^2 b^4 + 6ab^5 + b^6.$$

See if you can create Pascal's triangle for values of n up to 10. View this **popup box** to see if you are correct.

My interactive video summary

Example 1 Using Pascal's Triangle to Expand a Binomial Raised to a Power

Use Pascal's triangle to expand each binomial.

a. $(x + 2)^4$ **b.** $(x - 3)^5$ **c.** $(2x - 3y)^3$

Solutions

a. We start by looking at the row of Pascal's triangle corresponding with $n = 4$. We see that this row is 1 4 6 4 1. Using these coefficients and the exponent pattern, we get

$$(x + 2)^4 = 1(x^4 \cdot 2^0) + 4(x^3 \cdot 2^1) + 6(x^2 \cdot 2^2) + 4(x^1 \cdot 2^3) + 1(x^0 \cdot 2^4)$$

$$= x^4 + 8x^3 + 24x^2 + 32x + 16$$

b. The row of Pascal's triangle corresponding with $n = 5$ is 1 5 10 10 5 1. Using these coefficients and the exponent pattern, we get

$$(x - 3)^5 = 1(x^5 \cdot (-3)^0) + 5(x^4 \cdot (-3)^1) + 10(x^3 \cdot (-3)^2)$$

$$+ 10(x^2 \cdot (-3)^3) + 5(x^1 \cdot (-3)^4) + 1(x^0 \cdot (-3)^5)$$

$$= x^5 - 15x^4 + 90x^3 - 270x^2 + 405x - 243.$$

c. See if you can use Pascal's triangle to show that $(2x - 3y)^3 = 8x^3 - 36x^2 y + 54xy^2 - 27y^3$. Work through the **interactive video** for complete solutions to all three parts.

You Try It Work through this You Try It problem.

Work Exercises 1–6 in this eText or in the MyMathLab Study Plan.

The terms of the expansion of the form $(a - b)^n$ will *always* alternate in sign with the sign of the first term being positive.

OBJECTIVE 2   EVALUATE BINOMIAL COEFFICIENTS

Although **Pascal's triangle** is useful for determining the binomial coefficients of $(a + b)^n$ for fairly small values of n, it is not that useful for large values of n. For example, to find the binomial coefficients of the expansion of $(a + b)^{50}$ using Pascal's triangle, we would need to write the first 51 rows of the triangle to determine the coefficients. Fortunately, there is a convenient formula for the binomial coefficients. This formula requires the use of **factorials**. Recall, $n! = n \cdot (n - 1) \cdot (n - 2) \cdot \cdots \cdot 3 \cdot 2 \cdot 1$ and $0! = 1$.

To establish a formula for the binomial coefficients, let's take another look at the expansion for $(a + b)^6$.

$$(a + b)^6 = 1a^6b^0 + 6a^5b^1 + 15a^4b^2 + 20a^3b^3 + 15a^2b^4 + 6a^1b^5 + 1a^0b^6$$

Table 1 shows the relationship between the variable parts of the expansion of the form $a^{n-r}b^r$ and the corresponding coefficients. Select any of the coefficients in Table 1 to verify that the formula using factorial notation is true.

Table 1

Variables	Coefficient	Variables	Coefficient
a^6b^0	$1 = \dfrac{6!}{0! \cdot 6!}$	a^2b^4	$15 = \dfrac{6!}{4! \cdot 2!}$
a^5b^1	$6 = \dfrac{6!}{1! \cdot 5!}$	a^1b^5	$6 = \dfrac{6!}{5! \cdot 1!}$
a^4b^2	$15 = \dfrac{6!}{2! \cdot 4!}$	a^0b^6	$1 = \dfrac{6!}{6! \cdot 0!}$
a^3b^3	$20 = \dfrac{6!}{3! \cdot 3!}$		

You can see in Table 1 that for each pair of variables of the form $a^{n-r}b^r$, the corresponding binomial coefficient is of the form $\dfrac{n!}{r! \cdot (n - r)!}$. We will use the shorthand notation $\binom{n}{r}$, read as "n choose r," to denote a binomial coefficient.

Formula for a Binomial Coefficient

For non-negative integers n and r with $n \geq r$, the coefficient of the expansion of $(a + b)^n$ whose variable part is $a^{n-r}b^r$ is given by

$$\binom{n}{r} = \frac{n!}{r! \cdot (n - r)!}.$$

📖 *My video summary* ▶ **Example 2 Evaluating Binomial Coefficients**

Evaluate each of the following binomial coefficients.

a. $\binom{5}{3}$ b. $\binom{4}{1}$ c. $\binom{12}{8}$

Solutions Watch this **video**, or read through the following solutions.

a. $\binom{5}{3} = \dfrac{5!}{3!(5-3)!} = \dfrac{5!}{3! \cdot 2!} = \dfrac{5 \cdot 4 \cdot \cancel{3!}}{\cancel{3!} \cdot 2 \cdot 1} = \dfrac{20}{2} = 10$

b. $\binom{4}{1} = \dfrac{4!}{1!(4-1)!} = \dfrac{4!}{1! \cdot 3!} = \dfrac{4 \cdot \cancel{3!}}{1 \cdot \cancel{3!}} = \dfrac{4}{1} = 4$

c. $\binom{12}{8} = \dfrac{12!}{8!(12-8)!} = \dfrac{12!}{8! \cdot 4!} = \dfrac{12 \cdot 11 \cdot 10 \cdot 9 \cdot \cancel{8!}}{\cancel{8!} \cdot 4 \cdot 3 \cdot 2 \cdot 1} = \dfrac{11{,}880}{24} = 495$

A calculator can be used to compute binomial coefficients. Typically, the key $\boxed{nCr}$ is used. Figure 7 shows the computation of $\binom{12}{8}$ using a graphing utility.

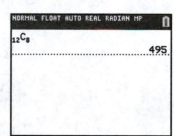

NORMAL FLOAT AUTO REAL RADIAN MP

$_{12}C_8$
495.

Figure 7

🔺 **You Try It** Work through this You Try It problem.

Work Exercises 7–10 in this eText or in the MyMathLab Study Plan.

OBJECTIVE 3 EXPAND BINOMIALS RAISED TO A POWER USING THE BINOMIAL THEOREM

Now that we know how to compute a binomial coefficient, we can state the Binomial Theorem.

Binomial Theorem

If n is a positive integer then,

$$(a+b)^n = \binom{n}{0}a^n + \binom{n}{1}a^{n-1}b + \binom{n}{2}a^{n-2}b^2 + \cdots + \binom{n}{n}b^n$$

$$= \sum_{i=0}^{n} \binom{n}{i}a^{n-i}b^i.$$

📖 *My interactive video summary* 🔖 **Example 3 Using the Binomial Theorem to Expand a Binomial Raised to a Power**

Use the Binomial Theorem to expand each binomial.

a. $(x-1)^8$ b. $\left(\sqrt{x} + y^2\right)^5$

Solutions Work through this **interactive video**, or read the following solutions.

a. Verify that $(x-1)^8 = x^8 - 8x^7 + 28x^6 - 56x^5 + 70x^4 - 56x^3 + 28x^2 - 8x + 1$ using the Binomial Theorem.

b. The expansion of $\left(\sqrt{x} + y^2\right)^5$ is as follows.

$$\left(\sqrt{x} + y^2\right)^5$$

$$= \binom{5}{0}\left(\sqrt{x}\right)^5 + \binom{5}{1}\left(\sqrt{x}\right)^4\left(y^2\right) + \binom{5}{2}\left(\sqrt{x}\right)^3\left(y^2\right)^2 + \binom{5}{3}\left(\sqrt{x}\right)^2\left(y^2\right)^3 + \binom{5}{4}\left(\sqrt{x}\right)\left(y^2\right)^4 + \binom{5}{5}\left(y^2\right)^5$$

$$= 1\cdot\left(\sqrt{x}\right)^5 + 5\cdot\left(\sqrt{x}\right)^4\left(y^2\right) + 10\cdot\left(\sqrt{x}\right)^3\left(y^2\right)^2 + 10\cdot\left(\sqrt{x}\right)^2\left(y^2\right)^3 + 5\cdot\left(\sqrt{x}\right)\left(y^2\right)^4 + 1\cdot\left(y^2\right)^5$$

$$= x^2\sqrt{x} + 5x^2y^2 + 10x\sqrt{x}y^4 + 10xy^6 + 5\sqrt{x}y^8 + y^{10}$$

You Try It Work through this You Try It problem.

Work Exercises 11–17 in this eText or in the MyMathLab Study Plan.

OBJECTIVE 4 FIND A PARTICULAR TERM OR A PARTICULAR COEFFICIENT OF A BINOMIAL EXPANSION

We may want to find a particular term of a binomial expansion. Fortunately, we can use the **Binomial Theorem** to develop a formula for a particular term. We start by writing out the first several terms of $(a + b)^n$ using the Binomial Theorem.

$$(a + b)^n = \binom{n}{0}a^n + \binom{n}{1}a^{n-1}b + \binom{n}{2}a^{n-2}b^2 + \binom{n}{3}a^{n-3}b^3 + \cdots + \binom{n}{n}b^n$$

The first term is $\binom{n}{0}a^n$. The second term is $\binom{n}{1}a^{n-1}b$. The third term is $\binom{n}{2}a^{n-2}b^2$. Following this pattern, we can see that the formula for the $(r+1)$st term (for $r \geq 0$) is given by $\binom{n}{r}a^{n-r}b^r$.

Formula for the $(r + 1)$st Term of a Binomial Expansion

If n is a positive integer and if $r \geq 0$, then the $(r + 1)$st term of the expansion of $(a + b)^n$ is given by

$$\binom{n}{r}a^{n-r}b^r = \frac{n!}{r! \cdot (n - r)!}a^{n-r}b^r.$$

My video summary ▶ **Example 4 Finding a Particular Term of a Binomial Expansion**

Find the third term of the expansion of $(2x - 3)^{10}$.

Solution Since we want to find the third term of this expansion, we will use the formula for the $(r + 1)$st term, which is equal to $\binom{n}{r}a^{n-r}b^r$ for $r = 2, n = 10, a = 2x$, and $b = -3$. Therefore, the third term is

$$\binom{10}{2}(2x)^{10-2}(-3)^2 = 45(256x^8)(9) = 103{,}680x^8.$$

Watch this **video** to see every step of this solution.

You Try It Work through this You Try It problem.

Work Exercises 18–21 in this eText or in the MyMathLab Study Plan.

My video summary ▶ **Example 5 Finding a Particular Coefficient of a Binomial Expansion**
Find the coefficient of x^7 in the expansion of $(x + 4)^{11}$.

Solution The formula for the $(r + 1)$st term of the expansion of $(x + 4)^{11}$ is given by the formula $\binom{11}{r}x^{11-r}4^r$. The term containing x^7 occurs when $11 - r = 7$. Solving this equation for r, we get $r = 4$. Therefore, the term involving x^7 is $\binom{11}{4}x^7 4^4$.

Simplifying this expression, we get $84{,}480x^7$. Thus, the coefficient of x^7 is $84{,}480$. Watch this **video** to see this solution worked out in detail.

You Try It Work through this You Try It problem.

Work Exercises 22–25 in this eText or in the MyMathLab Study Plan.

11.4 Exercises

In Exercises 1–6, use Pascal's triangle to expand each binomial.

1. $(m + n)^6$
2. $(x + 5)^4$
3. $(x - y)^7$

4. $(x - 3)^5$
5. $(2x + 3y)^6$
6. $(3x^2 - 4y^3)^4$

In Exercises 7–10, evaluate each binomial coefficient.

7. $\binom{7}{1}$
8. $\binom{10}{4}$
9. $\binom{7}{7}$
10. $\binom{23}{3}$

In Exercises 11–17, use the Binomial Theorem to expand each binomial.

11. $(x + 2)^7$
12. $(x - 3)^6$
13. $(4x + 1)^5$

14. $(x + 3y)^4$
15. $(5x - 3y)^5$
16. $(x^4 + y^5)^6$

17. $(\sqrt{x} - \sqrt{2})^4$

18. Find the sixth term of the expansion of $(x + 4)^9$.

19. Find the fifth term of the expansion of $(a - b)^8$.

20. Find the third term of the expansion of $(3c - d)^7$.

21. Find the seventh term of the expansion of $(3x + 2)^{10}$.

22. Find the coefficient of x^5 in the expansion of $(x - 3)^{11}$.

23. Find the coefficient of x^4 in the expansion of $(4x + 1)^{12}$.

24. Find the coefficient of x^0 in the expansion of $\left(x^2 + \dfrac{1}{x}\right)^{12}$.

25. Find the coefficient of x^{10} in the expansion of $\left(x - \dfrac{3}{\sqrt{x}}\right)^{19}$.

Review—Fractions, Decimals, Proportions, Percents

APPENDIX A CONTENTS

A.1 Fractions

TYPES OF FRACTIONS

Whole numbers are used to count whole units such as people, dogs, trees, and so on. To describe part of a whole unit, *fractions* are needed. For example, consider a pizza that is divided into twelve equally sized pieces. If you eat five of the pieces, then you have eaten five-twelfths, or $\frac{5}{12}$, of the pizza. The 12 is called the *denominator* and refers to the number of equal parts within the whole unit. The 5 is called the *numerator* and refers to the number of equal parts of interest, in this case 5.

Definition   Fraction

A **fraction** is a number of the form $\frac{a}{b}$, where a and b are **integers** and $b \neq 0$.

a is called the **numerator**, and b is called the **denominator**.

Example 1 Using Fractions to Describe a Part of a Whole

Write a fraction to represent the shaded area of each figure.

a.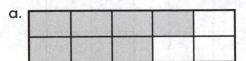

b.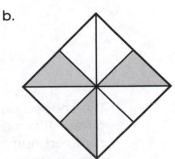

Solutions

a. Seven of the 10 equal parts are shaded, so the fraction is $\frac{7}{10}$.

b. Three of the 8 equal parts are shaded, so the fraction is $\frac{3}{8}$.

A **proper fraction** is a fraction in which the **numerator** is smaller than the **denominator**. Proper fractions represent quantities less than 1. The fractions $\frac{7}{10}$ and $\frac{3}{8}$ from Example 1 are proper fractions.

An **improper fraction** is a fraction in which the numerator is larger than or equal to the denominator. Improper fractions represent quantities greater than or equal to 1. The fractions $\frac{5}{4}$ and $\frac{5}{5}$ are examples of improper fractions.

Example 2 Using Improper Fractions to Describe a Figure

Write an improper fraction to represent the shaded area of each figure.

a. 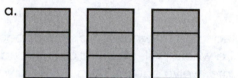   **b.**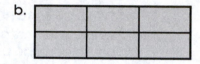

Solutions

a. Each whole object is divided into 3 equal parts. A total of 8 parts are shaded, so the **improper fraction** is $\frac{8}{3}$.

b. The whole object is divided into 6 equal parts. All 6 parts are shaded, so the improper fraction is $\frac{6}{6}$.

MIXED NUMBERS

Consider the shaded area in Figure 1, which represents the **improper fraction** $\frac{7}{4}$.

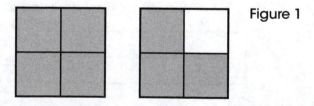

Figure 1

The shaded area represents one whole object and three-fourths of a second object. This represents the quantity *one and three-fourths*, which is known as a **mixed number**. A mixed number contains a **whole-number** part and a **proper-fraction** part. Symbolically, we denote "one and three-fourths" by $1\frac{3}{4}$, which means $1 + \frac{3}{4}$, but the $+$ symbol is not written.

Since Figure 1 can be written as both $1\frac{3}{4}$ and $\frac{7}{4}$, then $1\frac{3}{4} = \frac{7}{4}$. In fact, **mixed numbers** can *always* be converted into improper fractions, and improper fractions can always be converted into mixed numbers.

Converting a Mixed Number into an Improper Fraction

Step 1 Multiply the whole number by the denominator of the fraction.

Step 2 Add the product from Step 1 to the numerator of the fraction part.

Step 3 Write the sum from Step 2 over the denominator of the fraction part.

Example 3 Converting Mixed Numbers into Improper Fractions

Convert $4\frac{5}{8}$ into an improper fraction.

Solution Follow the steps for converting a mixed number to an improper fraction.

$$4\frac{5}{8} = \frac{\overbrace{4\cdot 8}^{\text{Step 1}} + \overbrace{5}^{\text{Step 2}}}{\underbrace{8}_{\text{Step 3}}} = \frac{32 + 5}{8} = \frac{37}{8}$$

To convert an improper fraction into a mixed number, divide.

Converting an Improper Fraction into a Mixed Number or Whole Number

Step 1 Divide the numerator by the denominator.

Step 2 If a remainder occurs in Step 1, use the **quotient** as the **whole-number** part of the mixed number. Place the remainder over the original denominator to form the fraction part of the mixed number. The mixed number will take the form

$$\text{Quotient}\,\frac{\text{Remainder}}{\text{Original Denominator}}.$$

If no remainder occurs in Step 1, then the result is a whole number, not a mixed number.

Example 4 Converting Improper Fractions into Mixed Numbers

Convert $\frac{47}{5}$ into a mixed number.

Solution 47 divided by 5 is 9 with a remainder of 2. So, $\frac{47}{5} = 9\frac{2}{5}$.

$$\frac{47}{5}: \text{Original denominator} \rightarrow 5\overline{)47} \begin{array}{l} \overset{\text{Step 1}}{9} \leftarrow \text{Quotient} \\ \underline{45} \\ 2 \leftarrow \text{Remainder} \end{array}$$

$$\overset{\text{Step 2}}{\frac{47}{5}} = 9\frac{2}{5} \begin{array}{l} \leftarrow \text{Remainder} \\ \leftarrow \text{Original denominator} \end{array}$$
$$\qquad\qquad\uparrow$$
$$\qquad\quad \text{Quotient}$$

EQUIVALENT FRACTIONS

Definition Equivalent Fraction

Two or more fractions are **equivalent fractions** if they represent the same value.

Consider the fractions shown in Figure 2. The shaded area is the same in all four figures, illustrating that the four fractions are equivalent. So, $\dfrac{1}{2} = \dfrac{2}{4} = \dfrac{4}{8} = \dfrac{8}{16}$.

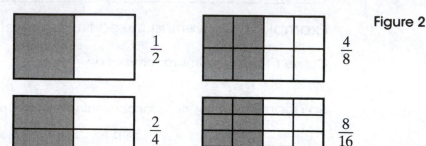

Figure 2

Multiplying or dividing the **numerator** and **denominator** of a fraction by the same nonzero number results in an equivalent fraction.

Property of Equivalent Fractions

If a, b, and c are numbers, then

$$\frac{a}{b} = \frac{a \cdot c}{b \cdot c} \quad \text{and} \quad \frac{a}{b} = \frac{a \div c}{b \div c}$$

as long as b and c are not equal to zero.

Example 5 Finding Equivalent Fractions

Write each fraction as an equivalent fraction with the given **denominator**.

a. $\dfrac{7}{8}$; denominator 40

b. $\dfrac{18}{24}$; denominator 4

Solutions

a. We multiply the current denominator 8 by 5 to obtain the new denominator 40, so we also multiply the **numerator** by 5.

$$\frac{7}{8} = \frac{7 \cdot 5}{8 \cdot 5} = \frac{35}{40}$$

b. To obtain the new denominator 4, divide the current denominator 24 by 6. So, we also divide the numerator by 6.

$$\frac{18}{24} = \frac{18 \div 6}{24 \div 6} = \frac{3}{4}$$

SIMPLIFYING FRACTIONS

A fraction can be written in infinitely many equivalent forms. Only one of these is the *simplest form*. For example, of all the equivalent fractions $\dfrac{1}{2} = \dfrac{2}{4} = \dfrac{4}{8} = \dfrac{8}{16}$ in Figure 2, the fraction $\dfrac{1}{2}$ is in simplest form.

Definition Simplest Form

A fraction is in **simplest form**, or **lowest terms**, when the numerator and denominator have no **common factors** other than 1.

Writing a Fraction in Simplest Form

Step 1 Write the prime factorizations of the numerator and denominator.

Step 2 Divide both the numerator and denominator by each common factor.

Step 3 Separately in the numerator and denominator, multiply any factors that are not divided out to obtain the simplified fraction. If all factors in the numerator divide out, the numerator will be 1.

Let's review finding the **prime factorization** of a number, and then we will simplify fractions.

Example 6 Finding the Prime Factorization of a Number

Find the prime factorization of each number.

a. 80   **b.** 72

Solutions

a.
$$
\begin{array}{c}
80 \\
8 \cdot 10 \\
2 \cdot 4 \cdot 2 \cdot 5 \\
2 \cdot 2 \cdot 2 \cdot 2 \cdot 5
\end{array}
$$

b.
$$
\begin{array}{c}
72 \\
8 \cdot 9 \\
2 \cdot 4 \cdot 3 \cdot 3 \\
2 \cdot 2 \cdot 2 \cdot 3 \cdot 3
\end{array}
$$

Example 7 Simplifying Fractions

Write each fraction in simplest form.

a. $\dfrac{8}{54}$ **b.** $\dfrac{40}{120}$

Solutions

a.
$$\frac{8}{54} = \frac{2 \cdot 2 \cdot 2}{2 \cdot 3 \cdot 3 \cdot 3} = \frac{\overset{1}{\cancel{2}} \cdot 2 \cdot 2}{\underset{1}{\cancel{2}} \cdot 3 \cdot 3 \cdot 3} = \frac{1 \cdot 2 \cdot 2}{1 \cdot 3 \cdot 3 \cdot 3} = \frac{4}{27}$$

b.
$$\frac{40}{120} = \frac{2 \cdot 2 \cdot 2 \cdot 5}{2 \cdot 2 \cdot 2 \cdot 3 \cdot 5} = \frac{\cancel{2} \cdot \cancel{2} \cdot \cancel{2} \cdot \cancel{5}}{\cancel{2} \cdot \cancel{2} \cdot \cancel{2} \cdot 3 \cdot \cancel{5}} = \frac{1 \cdot 1 \cdot 1 \cdot 1}{1 \cdot 1 \cdot 1 \cdot 3 \cdot 1} = \frac{1}{3}$$

MULTIPLYING FRACTIONS

To multiply two fractions, multiply the **numerators**, multiply the **denominators**, and simplify.

Multiplying Two Fractions

If a, b, c, and d are numbers and b and d are not equal to zero, then $\dfrac{a}{b} \cdot \dfrac{c}{d} = \dfrac{a \cdot c}{b \cdot d}$.

Example 8 Multiplying Fractions

Multiply and simplify.

a. $\dfrac{5}{7} \cdot \dfrac{8}{11}$

b. $\dfrac{4}{9} \cdot \dfrac{12}{20}$

Solutions

Multiply numerators

a. $\dfrac{5}{7} \cdot \dfrac{8}{11} = \dfrac{\overbrace{5 \cdot 8}}{\underbrace{7 \cdot 11}} = \dfrac{40}{77}$

Multiply denominators

b. $\dfrac{4}{9} \cdot \dfrac{12}{20} = \dfrac{4 \cdot 12}{9 \cdot 20} = \dfrac{48}{180} = \dfrac{2 \cdot 2 \cdot 2 \cdot 2 \cdot 3}{2 \cdot 2 \cdot 3 \cdot 3 \cdot 5} = \dfrac{\overset{1}{\cancel{2}} \cdot \overset{1}{\cancel{2}} \cdot 2 \cdot 2 \cdot \overset{1}{\cancel{3}}}{\underset{1}{\cancel{2}} \cdot \underset{1}{\cancel{2}} \cdot \underset{1}{\cancel{3}} \cdot 3 \cdot 5} = \dfrac{1 \cdot 1 \cdot 2 \cdot 2 \cdot 1}{1 \cdot 1 \cdot 1 \cdot 3 \cdot 5} = \dfrac{4}{15}$

 TIP When multiplying fractions, if a number in a **numerator** has a **factor** in common with a number in a **denominator**, then this **common factor** can be divided out from both numbers first before multiplying the fractions. This generally results in fewer shown steps. View this **alternative solution** for Example 8(b).

DIVIDING FRACTIONS

Before we divide fractions, let's review the definition of a *reciprocal* or *multiplicative inverse*.

Definition   Reciprocal or Multiplicative Inverse

Two nonzero numbers are **reciprocals** or **multiplicative inverses** if their **product** is 1.

For example, the reciprocal of $\dfrac{5}{8}$ is $\dfrac{8}{5}$ because $\dfrac{5}{8} \cdot \dfrac{8}{5} = 1$. Likewise, the reciprocal of 4 is $\dfrac{1}{4}$, and the reciprocal of $\dfrac{7}{10}$ is $\dfrac{10}{7}$ because $\dfrac{1}{4} \cdot 4 = 1$ and $\dfrac{10}{7} \cdot \dfrac{7}{10} = 1$, respectively.

Recall that we can find the reciprocal of a fraction by switching its numerator and denominator.

To divide two fractions, **multiply** the first fraction by the reciprocal of the second fraction.

Dividing Two Fractions

If a, b, c, and d are numbers and b, c, and d are not equal to zero, then

$$\frac{a}{b} \div \frac{c}{d} = \frac{a}{b} \cdot \frac{d}{c} = \frac{a \cdot d}{b \cdot c}.$$

Example 9 Dividing Fractions

Divide and simplify.

a. $\dfrac{4}{9} \div \dfrac{12}{20}$

b. $\dfrac{2}{9} \div \dfrac{5}{12}$

Solutions

a. $\dfrac{4}{9} \div \dfrac{12}{20} = \dfrac{4}{9} \cdot \dfrac{20}{12} = \dfrac{\overset{1}{4}}{9} \cdot \dfrac{20}{\underset{3}{12}} = \dfrac{1 \cdot 20}{9 \cdot 3} = \dfrac{20}{27}$

b. $\dfrac{2}{9} \div \dfrac{5}{12} = \dfrac{2}{9} \cdot \dfrac{12}{5} = \dfrac{2}{\underset{3}{9}} \cdot \dfrac{\overset{4}{12}}{5} = \dfrac{2 \cdot 4}{3 \cdot 5} = \dfrac{8}{15}$

ADDING AND SUBTRACTING FRACTIONS

Now let's add and subtract fractions. Consider $\dfrac{1}{2} + \dfrac{1}{3}$. Recall that fractions must be written with a **common denominator** before they can be added. This is illustrated in Figure 3.

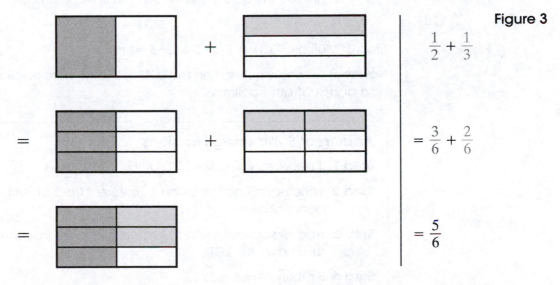

Figure 3

$$\frac{1}{2} + \frac{1}{3}$$

$$= \frac{3}{6} + \frac{2}{6}$$

$$= \frac{5}{6}$$

Typically, we use the *least common denominator* of a group of fractions as a common denominator.

Definition Least Common Denominator (LCD)

The **least common denominator (LCD)** of a group of fractions is the smallest number that is divisible by all the denominators in the group.

For example, the LCD of $\frac{1}{2}$ and $\frac{1}{3}$ is 6 because 6 is the smallest number that is divisible by both 2 and 3.

Sometimes the LCD is difficult to find, especially if the denominators are large numbers. The following steps will help:

Finding the LCD of a Group of Fractions

Step 1 Write out the **prime factorization** of each denominator in the group.

Step 2 Write down each distinct **prime factor** the greatest number of times it appears in any one factorization.

Step 3 Multiply the factors listed in Step 2 to find the LCD.

Example 10 Finding a Least Common Denominator

Find the **LCD** of two fractions with denominators 90 and 120.

Solution Use the steps for finding the LCD.

Step 1 Write out the **prime factorization** of each denominator:

$$90 = 2 \cdot 3 \cdot 3 \cdot 5$$
$$120 = 2 \cdot 2 \cdot 2 \cdot 3 \cdot 5$$

Step 2 There are three distinct prime **factors**: 2, 3, and 5. The greatest number of 2's in one factorization is three, so include three 2's for the LCD. Likewise, the greatest number of 3's is two, and the greatest number of 5's is one, so include two 3's and one 5 in the list for the LCD.

$$\text{LCD} = 2 \cdot 2 \cdot 2 \cdot 3 \cdot 3 \cdot 5$$

Step 3 Multiply. $\text{LCD} = 2 \cdot 2 \cdot 2 \cdot 3 \cdot 3 \cdot 5 = 360$

Now that we know how to find the **LCD** of two or more fractions, let's review how to add and subtract fractions.

Adding or Subtracting Fractions

Step 1 If necessary, find the LCD of the fractions.

Step 2 Write each fraction as an **equivalent fraction** with the LCD as its denominator.

Step 3 Add or subtract the numerators of the like fractions and write the result over the LCD.

Step 4 Simplify, if necessary.

Example 11 Adding and Subtracting Fractions

Add or subtract. Simplify if necessary.

a. $\frac{1}{6} + \frac{3}{4}$

b. $\frac{5}{8} - \frac{7}{12}$

Solutions

a. The LCD is 12.

Begin with the original problem: $\dfrac{1}{6} + \dfrac{3}{4}$

Write each fraction with a denominator of LCD = 12: $= \dfrac{2}{12} + \dfrac{9}{12}$

Add the numerators $2 + 9 = 11$; write the result over 12: $= \dfrac{11}{12}$

b. The LCD is 24.

Begin with the original problem: $\dfrac{5}{8} - \dfrac{7}{12}$

Write each fraction with a denominator of LCD = 24: $= \dfrac{15}{24} - \dfrac{14}{24}$

Subtract the numerators $15 - 14 = 1$; write the result over 24: $= \dfrac{1}{24}$

MULTIPLYING AND DIVIDING MIXED NUMBERS

To multiply or divide two or more **mixed numbers**, first convert each mixed number into an **improper fraction**.

Example 12 Multiplying and Dividing Mixed Numbers

Multiply or divide. Write the answer as a mixed number if possible.

a. $6\dfrac{1}{3} \cdot 4\dfrac{1}{2}$ **b.** $2\dfrac{3}{5} \div 3\dfrac{5}{7}$

Solutions

a.

Begin with the original problem: $6\dfrac{1}{3} \cdot 4\dfrac{1}{2}$

Convert to improper fractions: $= \dfrac{6 \cdot 3 + 1}{3} \cdot \dfrac{4 \cdot 2 + 1}{2} = \dfrac{19}{3} \cdot \dfrac{9}{2}$

Divide out common factors and multiply: $= \dfrac{19}{\cancel{3}_{1}} \cdot \dfrac{\cancel{9}^{3}}{2} = \dfrac{19 \cdot 3}{1 \cdot 2} = \dfrac{57}{2}$

Convert back to mixed number: $= 28\dfrac{1}{2}$

b.

Begin with the original problem: $2\dfrac{3}{5} \div 3\dfrac{5}{7}$

Convert to improper fractions: $= \dfrac{2 \cdot 5 + 3}{5} \div \dfrac{3 \cdot 7 + 5}{7} = \dfrac{13}{5} \div \dfrac{26}{7}$

Change to multiply by reciprocal: $= \dfrac{13}{5} \cdot \dfrac{7}{26}$

Divide of common factors and multiply: $= \dfrac{\overset{1}{\cancel{13}}}{5} \cdot \dfrac{7}{\underset{2}{\cancel{26}}} = \dfrac{1 \cdot 7}{5 \cdot 2} = \dfrac{7}{10}$

ADDING AND SUBTRACTING MIXED NUMBERS

To add two or more **mixed numbers**, first we can convert each mixed number into an improper fraction, just like we did to multiply and divide. However, because the equivalent improper fractions can involve very large numerators, it is usually easier to add the similar parts. This means that we can add the **proper-fraction** parts and add the **whole-number** parts.

Example 13 Adding Mixed Numbers

Add $1\dfrac{3}{8} + 4\dfrac{5}{12}$.

Solution It is usually easier to write the problem vertically rather than horizontally. Convert the **proper-fraction** parts to have **common denominators**. Then add the proper-fraction parts and the whole-number parts. In this case, the **LCD** of the fraction parts is 24.

Whole-number parts
$\downarrow$

$$1\dfrac{3}{8} = \quad 1\dfrac{9}{24}$$

↗ Proper-fraction parts

$$+4\dfrac{5}{12} = \quad +4\dfrac{10}{24}$$

$$5\dfrac{19}{24}$$

So, $1\dfrac{3}{8} + 4\dfrac{5}{12} = 5\dfrac{19}{24}$.

Sometimes when we add the proper-fraction parts of two **mixed numbers**, the result is an **improper fraction**. If this occurs, we must convert the improper-fraction part into a mixed number and "carry" the whole-number part of the result to the whole number parts of the original mixed numbers before finding the final sum.

This may sound confusing, so before we look at an example, consider the number $6\dfrac{8}{5}$.

This number is technically *not* a mixed number because the fraction part is not a proper fraction. However, we can convert this number into a mixed number as follows.

$$6\dfrac{8}{5} = 6 + \dfrac{8}{5} = 6 + 1\dfrac{3}{5} = 7\dfrac{3}{5}$$

Example 14 Adding Mixed Numbers

Add $1\frac{5}{6} + 4\frac{7}{9}$.

Solution Write the problem vertically and convert the **proper-fraction parts** to have **common denominators**. Next, add the fraction parts and the **whole-number parts**. In this case, the **LCD** of the fraction parts is 18.

$$
\begin{array}{r}
1\frac{5}{6} = 1\frac{15}{18} \\
+4\frac{7}{9} = +4\frac{14}{18} \\
\hline
5\frac{29}{18}
\end{array}
$$

Since $5\frac{29}{18}$ is not **simplified**, we must convert it into a mixed number.

$$5\frac{29}{18} = 5 + \frac{29}{18} = 5 + 1\frac{11}{18} = 6\frac{11}{18}$$

So, $1\frac{5}{6} + 4\frac{7}{9} = 6\frac{11}{18}$.

Similar to adding, when subtracting two or more **mixed numbers**, we subtract the **proper-fraction** parts and subtract the **whole-number** parts. This is shown in the next example.

Example 15 Subtracting Mixed Numbers

Subtract $6\frac{3}{4} - 4\frac{2}{3}$.

Solution Write the problem vertically and convert the proper-fraction parts to have **common denominators**. Then subtract the fraction parts and the whole-number parts. In this case, the **LCD** of the fraction parts is 12.

$$
\begin{array}{r}
6\frac{3}{4} = 6\frac{9}{12} \\
-4\frac{2}{3} = -4\frac{8}{12} \\
\hline
2\frac{1}{12}
\end{array}
$$

So, $6\frac{3}{4} - 4\frac{2}{3} = 2\frac{1}{12}$.

In some cases, when we try to subtract the **proper-fraction** parts of two **mixed numbers**, the **subtrahend** may be larger than the **minuend**. If this occurs, we must borrow from the **whole-number** part.

Example 16 Subtracting Mixed Numbers

Subtract $7\frac{1}{5} - 4\frac{3}{8}$.

Solution First, write the problem vertically and convert the proper-fraction parts to have **common denominators**. Then subtract the proper-fraction parts and the whole-number parts. In this case, the **LCD** of the fraction parts is 40.

$$7\frac{1}{5} = 7\frac{8}{40}$$

$$-4\frac{3}{8} = -4\frac{15}{40}$$

When we try to subtract the fraction parts, we are faced with $\frac{8}{40} - \frac{15}{40}$, which we cannot do because 15 is larger than 8. So we must borrow.

$$7\frac{8}{40} = 7 + \frac{8}{40} = 6 + 1\frac{8}{40} = 6 + \frac{1 \cdot 40 + 8}{40} = 6 + \frac{48}{40} = 6\frac{48}{40}$$

We can subtract $\frac{48}{40} - \frac{15}{40}$, so the problem becomes

$$7\frac{1}{5} = 7\frac{8}{40} = 6\frac{48}{40}$$

$$-4\frac{3}{8} = -4\frac{15}{40} = -4\frac{15}{40}$$

$$2\frac{33}{40}$$

Therefore, $7\frac{1}{5} - 4\frac{3}{8} = 2\frac{33}{40}$.

A.2 Decimals

DECIMAL NOTATION

As with **fractions**, a **decimal number**, or simply a **decimal**, can be used to describe a part of a whole. The position of each **digit** in a decimal determines its **place value**. Figure 4 shows a **place-value chart**. In the chart, the digit 4 represents a value of four-tenths, or $\frac{4}{10}$. The digit 9 represents a value of nine-hundredths, or $\frac{9}{100}$.

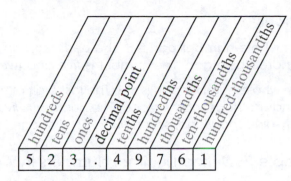

Figure 4
Place Value Chart

Digits appearing in front of the **decimal point** form the **whole-number part** of the decimal number. Digits appearing behind the decimal point form the **decimal part** of the number. In the number 82.1356, for example, 82 is the whole number part and .1356 is the decimal part.

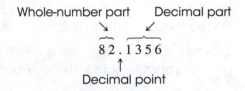

WRITING A DECIMAL NUMBER IN WORDS

We read or write **decimals** similar to the way we write **whole numbers**. For example, the whole number 1356 is written in words as *one thousand, three hundred fifty-six*. If these same four **digits** fall directly behind a **decimal point**, then we write the decimal number in the same way, but we follow it with the **place value** name of the digit in the last position. For the decimal number 0.1356, the last digit 6 has a "ten-thousandths" place value. So, 0.1356 is written as *one thousand three hundred fifty-six ten-thousandths*.

Writing a Decimal Number in Words

Step 1 Write the whole-number part in words.

Step 2 Write the word *and* for the decimal point.

Step 3 Write the decimal part as if it were a whole number and follow it with the place value name of the last digit.

The decimal number 82.1356 is written in words as *eighty-two and one thousand three hundred fifty-six ten-thousandths.*

82.1356

↗   ↑   ↖

eighty-two   and   one thousand three hundred fifty-six ten-thousandths

Example 1   Writing Decimal Numbers in Words

Write each decimal number in words.

a. 5.23   **b.** 523.49761

Solutions

a. The **whole-number part** is 5. The **decimal part** .23 has its last **digit** in the hundredths **place**. The number is *five and twenty-three hundredths.*

b. The whole-number part is 523. The last digit of the decimal part .49761 is in the hundred-thousandths place. The number is *five hundred twenty-three and forty-nine thousand, seven hundred sixty-one hundred-thousandths.*

Example 2 Writing Decimal Numbers in Words

Write each decimal number in words.

a. 0.407   b. 400.007

Solutions

a. This number has no whole-number part. The last digit of the decimal part is in the thousandths place. The number is *four hundred seven thousandths.*

b. The whole-number part is 400. The last digit of the decimal part .007 is in the thousandths place. The number is *four hundred and seven thousandths.*

 There is a big difference between the values of 0.407 and 400.007 from Example 2, but when reading or writing these numbers, the only difference is the word *and.*

Example 3 Converting from Word to Decimal Form

Write each number in **decimal** form.

a. *fifteen and seventy-eight hundredths*

b. *five hundred forty-one and two hundred seventeen thousandths*

Solutions

a. The word *and* separates the **whole-number part** from the **decimal part**. The last **digit** of the decimal part is in the hundredths **place**. The whole-number part is 15 and the decimal part .78, so the decimal is 15.78.
↑
Hundredths place

b. The last digit of the decimal part is in the thousandths place. The whole-number part is 541 and the decimal part is .217, so the decimal is 541.217.
↑
Thousandths place

Example 4 Converting from Word to Decimal Form

Write each number in decimal form.

a. *three hundred and ninety-six thousandths*

b. *three hundred ninety-six thousandths*

Solutions

a. The last **digit** of the **decimal part** is in the thousandths **place**. The word *and* separates the **whole-number** part 300 from the decimal **part** .096. The decimal is 300.096.

$$\underbrace{\text{three hundred}}_{300} \quad \underbrace{\text{and}}_{.} \quad \underbrace{\text{ninety-six thousandths}}_{096}$$

$$\uparrow$$
Thousandths place

b. The last digit of the decimal part is in the thousandths place. The word *and* does not appear in the written phrase, so there is no whole-number part. The decimal is 0.396.

WRITING DECIMALS AS FRACTIONS OR MIXED NUMBERS

If we know how to write decimals in words, then writing them as **fractions** or **mixed numbers** follows naturally. For example, we write the decimal 0.3 in words as *three tenths*, which is how we write the fraction $\frac{3}{10}$ in words. We write them the same way because they are equal in value: $0.3 = \frac{3}{10}$. Similarly, the decimal 5.371 and the mixed number $5\frac{371}{1000}$ are both written in words as *five and three hundred seventy-one thousandths* because $5.371 = 5\frac{371}{1000}$.

Example 5 Writing Decimals as Fractions or Mixed Numbers

Write each **decimal** as a fraction or mixed number. Be sure that your answer is written in **simplest form**.

a. 0.375   **b.** 7.8

Solutions

a. In words, 0.375 is *three hundred seventy-five thousandths*. As a fraction, this is $\frac{375}{1000}$, which simplifies to $\frac{3}{8}$.

b. In words, 7.8 is *seven and eight tenths*. As a mixed number, this is $7\frac{8}{10}$, which simplifies to $7\frac{4}{5}$.

ROUNDING DECIMALS

We can approximate a decimal number by rounding it to a given **place value**.

Rounding a Decimal to the Right of the Decimal Point

Step 1 Find the **digit** to the right of the given place value (the place that you are rounding to).

Step 2 If the digit from Step 1 is 5 or greater, then increase the digit in the given place value by 1 and drop all digits to its right.
If the digit from Step 1 is less than 5, then drop all digits to the right of the given place value.

Example 6 Rounding Decimals

a. Round 12.689 to the nearest tenth.

b. Round 7.256435 to the nearest thousandth.

Solutions Follow the steps for rounding a decimal.

a. 12.689
↗ ↖
Tenths Hundredths

The **digit** 6 is in the tenths **place**, where we want to round. The digit 8 is in the hundredths place, which is to the immediate right of the tenths place. Because 8 is larger than 5, increase the tenths place from 6 to 7. So, 12.689 rounds to 12.7.

b. 7.256435
↗ ↖
Thousandths Ten-thousandths

The digit 6 is in the thousandths place, where we want to round. The digit 4 is in the ten-thousandths place, which is to the immediate right of the thousandths place. Because 4 is smaller than 5, keep the thousandths place as 6 and drop all of the digits to its right. So, 7.256435 rounds to 7.256.

ADDING AND SUBTRACTING DECIMALS

Addition and subtraction of decimals is similar to the addition and subtraction of **whole numbers**. We add or subtract the **digits** in corresponding **place values**, working from right to left and carrying or borrowing as necessary. To ensure that digits in corresponding place values are added or subtracted, we line up the decimal points vertically.

Adding or Subtracting Decimals

Step 1 Write the numbers vertically, lining up the decimal points.

Step 2 Add or subtract the corresponding place values, working from right to left. Carry or borrow as needed.

Step 3 Write the decimal point in the answer so that it lines up with the decimal points in the problem.

 If one number has more decimal places than the other, fill in the "missing" place values with zeros for placeholders.

Example 7 Adding and Subtracting Decimals

Add or subtract.

a. $26.81 + 5.467$ 　　　　　　　**b.** $26.81 - 5.467$

Solutions Follow the steps for adding or subtracting decimals.

```
    1 1        ← Carrying                    7 10 10 ← Borrowing
  26.810  ← 0 place holder              2 6 . 8 1̶ 0̶  ← 0 place holder
a. +5.467                          b.  −   5.467
   32.277                                 21.343
    ↑                                       ↑
 Align decimal points                Align decimal points
```

MULTIPLYING DECIMALS

We multiply decimals like we multiply whole numbers. The only difference is that we must write the decimal point in the proper **place** in the answer. Consider the following:

$$0.2 \cdot 0.14 = \frac{2}{10} \cdot \frac{14}{100} = \frac{2 \cdot 14}{10 \cdot 100} = \frac{28}{1000} = 0.028.$$

In the **factors** of the original problem, notice that there are a total of three **digits** to the right of decimal points. The final **product** also has three digits to the right of the decimal point.

$$0.2 \quad \cdot \quad 0.14 \quad = \quad 0.028$$

1 decimal place 　　　2 decimal places 　　　3 decimal places

The total number of **digits** to the right of the **decimal points** in the **factors** will equal the number of digits to the right of the decimal point in the **product**. We can use this pattern to simplify the multiplication process.

Multiplying Decimals

Step 1 Multiply the numbers as though no decimal points were present.

Step 2 Place the decimal point in the product so that the number of digits to its right is the same as the total number of digits to the right of the decimal points in the factors.

Example 8 Multiplying Decimals

Multiply.

a. $2.3 \cdot 1.7$ 　　　　　　　**b.** $1.75 \cdot 0.05$

Solutions Follow the steps for multiplying decimals.

a. The factors have a total of two digits behind the decimal points, so the product will have two digits behind its decimal point.

$$
\begin{array}{r}
2.3 \leftarrow 1 \text{ decimal place} \\
\times\ 1.7 \leftarrow 1 \text{ decimal place} \\
\hline
1\,6\,1 \\
2\,3\,0 \\
\hline
3.9\,1 \leftarrow 2 \text{ decimal places}
\end{array}
$$

b. The factors have a total of three digits behind the decimal points, so the product will have three digits behind its decimal point.

$$
\begin{array}{r}
1.7\,5 \leftarrow 2 \text{ decimal places} \\
\times\ \ \ 0.5 \leftarrow 1 \text{ decimal place} \\
\hline
0.8\,7\,5 \leftarrow 3 \text{ decimal places}
\end{array}
$$

DIVIDING DECIMALS

Dividing decimals is similar to dividing whole numbers. However, we must write the decimal point in the proper place as we did when multiplying decimals. Consider the following:

$$
0.158 \div 0.2 = \frac{158}{1000} \div \frac{2}{10} = \frac{158}{1000} \cdot \frac{10}{2} = \frac{\overset{79}{\cancel{158}}}{\underset{100}{\cancel{1000}}} \cdot \frac{\overset{1}{\cancel{10}}}{\underset{1}{\cancel{2}}} = \frac{79 \cdot 1}{100 \cdot 1} = \frac{79}{100} = 0.79.
$$

Dividing Decimals

Step 1 Move the decimal point in the **divisor** to the right until it is behind the last digit.

Step 2 Move the decimal point in the **dividend** to the right the same number of places as the decimal point was moved in Step 1. Add zeros as placeholders if necessary.

Step 3 Divide. Place the decimal point in the **quotient** directly above the location of the moved decimal point in the dividend.

 The quotient will either be a terminating or repeating decimal.

Example 9 Dividing Decimals

Divide.

a. $4.325 \div 2.5$ **b.** $7.5 \div 0.33$

Solutions Follow the steps for dividing decimals.

a. The **decimal points** in the **divisor** and **dividend** must be moved one place to the right.

$$
2.5\overline{)4.325} \quad\Rightarrow\quad
\begin{array}{r}
1.73 \\
25.\overline{)43.25} \\
\underline{25} \\
182 \\
\underline{175} \\
75 \\
\underline{75} \\
0
\end{array}
$$

So, $4.325 \div 2.5 = 1.73$.

b. The decimal points in the divisor and dividend must be moved two places to the right. We can place zeros behind the decimal point to continue dividing.

$$
0.33\overline{)7.50} \quad\Rightarrow\quad
\begin{array}{r}
22.7272.... \\
33.\overline{)750.0000} \\
\underline{66} \\
90 \\
\underline{66} \\
240 \\
\underline{231} \\
90 \\
\underline{66} \\
240 \\
\underline{231} \\
90 \\
\underline{66} \\
24
\end{array}
$$

The pattern "72" keeps repeating, so $7.5 \div 0.33 = 22.\overline{72}$.

WRITING FRACTIONS OR MIXED NUMBERS AS DECIMALS

To write a fraction or mixed number as a decimal, use the following procedure:

Writing Fractions or Mixed Numbers as Decimals

To write a fraction as a decimal, divide the **numerator** by the **denominator**.

To write a mixed number as a decimal, divide the numerator of the **proper-fraction** part by the denominator, and add the **whole-number** part to the decimal result.

TIP The quotient will either be a terminating or repeating decimal.

Example 10 Writing Fractions or Mixed Numbers as Decimals

Write each fraction or mixed number as a decimal.

a. $\dfrac{5}{8}$ b. $\dfrac{3}{11}$ c. $13\dfrac{9}{25}$ d. $4\dfrac{2}{3}$

Solutions For parts a and b, divide the **numerator** by the **denominator**. Note that **decimal points** are understood to be behind the last digit of each **whole number**. We can place zeros behind the decimal point to continue dividing.

a.
```
    0.625
8.)5.000
   48
   20
   16
   40
   40
    0
```
So, $\dfrac{5}{8} = 0.625$.

b.
```
     0.2727...
11.)3.0000
    22
    80
    77
    30
    22
    80
    77
     3
```
The pattern "27" keeps repeating, so $\dfrac{3}{11} = 0.\overline{27}$.

For parts c and d, divide the numerator of the **proper-fraction** part by the denominator and add the result to the **whole-number** part.

c.
```
    0.36
25.)9.00
   75
  150
  150
    0
```
So, $\dfrac{9}{25} = 0.36$. Adding the whole-number part 13 to this decimal result, we get

$13\dfrac{9}{25} = 13 + 0.36 = 13.36$.

d.
```
    0.666...
3.)2.000
  18
  20
  18
  20
  18
   2
```
The pattern "6" keeps repeating, so $\dfrac{2}{3} = 0.\overline{6}$.

Adding the whole-number part 4 to this decimal result, we get $4\dfrac{2}{3} = 4 + 0.\overline{6} = 4.\overline{6}$.

A.3 Proportions

RATIOS

If there are 15 men and 20 women in an algebra class, then we can say that the *ratio* of men to women is 15 to 20. A **ratio** is a comparison of two quantities. There are three common notations for expressing ratios. One uses the word *to* as we just saw, but we can also use *fraction notation* or *colon notation*. So, we can write the ratio of 15 to 20 as

$$15 \text{ to } 20 \quad \text{or} \quad \frac{15}{20} \quad \text{or} \quad 15{:}20.$$

Colon notation

Fraction notation

⚠ The order of the quantities in a ratio is important. For example, the ratio of 15 to 20 is 15:20 or $\frac{15}{20}$, *not* ~~20:15~~ and *not* $\frac{20}{15}$.

Ratios behave like fractions, which is why fraction notation is often used. For example, the ratio of 15 to 20 is equivalent to the ratio of 3 to 4. We can see this using fraction notation and writing it in simplest form: $\frac{15}{20} = \frac{3 \cdot 5}{4 \cdot 5} = \frac{3 \cdot \cancel{5}}{4 \cdot \cancel{5}} = \frac{3}{4}$.

Simplifying a Ratio

To **simplify a ratio**, write the fraction notation in simplest form by dividing out all **common factors** and **common units**.

Example 1 Writing and Simplifying Ratios

a. A television has a length of 48 inches and a width of 27 inches. Write the ratio of length to width as a fraction in simplest form.

b. A recipe calls for $2\frac{1}{2}$ cups of milk and $\frac{3}{4}$ of a cup of sugar. Write the ratio of milk to sugar in simplest form.

Solutions

a. $\dfrac{\text{Length}}{\text{Width}} = \dfrac{48 \text{ inches}}{27 \text{ inches}} = \dfrac{16 \cdot \cancel{3} \text{ inches}}{9 \cdot \cancel{3} \text{ inches}} = \dfrac{16}{9}$

b. $\dfrac{\text{Milk}}{\text{Sugar}} = \dfrac{2\frac{1}{2} \text{ cups}}{\frac{3}{4} \text{ cups}} = \dfrac{2\frac{1}{2} \text{ cups}}{\frac{3}{4} \text{ cups}} = \dfrac{\frac{5}{2}}{\frac{3}{4}} = \dfrac{5}{2} \div \dfrac{3}{4} = \dfrac{5}{2} \cdot \dfrac{4}{3} = \dfrac{5}{\underset{1}{\cancel{2}}} \cdot \dfrac{\overset{2}{\cancel{4}}}{3} = \dfrac{10}{3}$

RATES

Rates are special kinds of **ratios** in which quantities of different types are compared. For example, consider a runner who ran 3 miles in 24 minutes. Writing this rate as a **fraction**, we have

$$\frac{3 \text{ miles}}{24 \text{ minutes}}, \text{ which simplifies to } \frac{1 \text{ mile}}{8 \text{ minutes}}.$$

Note that the units do not divide out.

A **unit rate** is a rate that has a **denominator** of 1 unit. A common unit rate is "miles per hour." For example, the speed of 65 miles per hour can be written as $\frac{65 \text{ miles}}{1 \text{ hour}}$.

Converting Rates to Unit Rates

To convert a rate to a unit rate, divide the quantity in the **numerator** by the quantity in the denominator.

 TIP The units can be written as "<units of numerator> per <units of denominator>." For example, "miles per hour" are the units for the unit rate of $\frac{65 \text{ miles}}{1 \text{ hour}}$.

Example 2 Converting Rates to Unit Rates

a. A jet travels 900 miles in 2 hours. What is the **unit rate** in miles per hour?

b. A 50-ounce bottle of detergent costs $7.59. What is the **unit price** in $/oz?

Solutions

a. $\dfrac{900 \text{ miles}}{2 \text{ hours}} = \dfrac{900 \div 2 \text{ miles}}{2 \div 2 \text{ hours}} = \dfrac{450 \text{ miles}}{1 \text{ hour}} = 450 \text{ miles per hour}$

b. $\dfrac{\$7.59}{50 \text{ ounces}} = \dfrac{\$7.59 \div 50}{50 \div 50 \text{ ounces}} = \dfrac{\$0.1518}{1 \text{ ounce}} = \$0.1518 \text{ per ounce}$

PROPORTIONS

Definition   Proportion

A **proportion** is an equation stating that two **ratios** (or rates) are equal, such as $\dfrac{P}{Q} = \dfrac{R}{S}$, where $R \neq 0$ and $S \neq 0$.

The proportion $\dfrac{15}{20} = \dfrac{6}{8}$ is true because both ratios **simplify** to $\dfrac{3}{4}$. Notice that $8 \cdot 15 = 120$ and $20 \cdot 6 = 120$. The products $8 \cdot 15$ and $20 \cdot 6$ are called the **cross products** of this proportion. When a proportion is true, the cross products are equal. Consider this cross product visually as follows:

$$\frac{15}{20} \diagup\!\!\!\!\!\diagdown \frac{6}{8}$$

$$8 \cdot 15 = 20 \cdot 6$$

We can use **cross-multiplication** to find an unknown quantity in a **proportion**.

Example 3 Solving Proportions

Cross-multiply to solve each proportion.

a. $\dfrac{x}{11} = \dfrac{7.1}{2}$

b. $\dfrac{y}{3\frac{1}{2}} = \dfrac{5\frac{1}{5}}{2\frac{1}{3}}$

Solutions

a. Begin with the original proportion: $\dfrac{x}{11} = \dfrac{7.1}{2}$

Cross-multiply: $2x = 11(7.1)$

Simplify: $2x = 78.1$

Divide both sides by 2: $\dfrac{2x}{2} = \dfrac{78.1}{2}$

Simplify: $x = 39.05$

b. Begin with the original proportion: $\dfrac{y}{3\frac{1}{2}} = \dfrac{5\frac{1}{5}}{2\frac{1}{3}}$

Cross-multiply: $\left(2\frac{1}{3}\right)y = \left(3\frac{1}{2}\right)\left(5\frac{1}{5}\right)$

Change the mixed numbers to improper fractions: $\dfrac{7}{3}y = \left(\dfrac{7}{2}\right)\left(\dfrac{26}{5}\right)$

Simplify: $\dfrac{7}{3}y = \dfrac{91}{5}$

Multiply both sides by $\dfrac{3}{7}$: $\dfrac{3}{7}\left(\dfrac{7}{3}y\right) = \dfrac{3}{7}\left(\dfrac{91}{5}\right)$

Simplify: $y = \dfrac{39}{5} = 7\dfrac{4}{5}$

Writing **proportions** is a powerful tool in solving application problems. Given a **ratio** (or **rate**) of two quantities, a proportion can be used to determine an unknown quantity.

Example 4 Using Proportions to Solve Application Problems

A quality-control inspector examined a sample of 200 lightbulbs and found 18 of them to be defective. At this ratio, how many defective bulbs can the inspector expect in a shipment of 22,000 lightbulbs?

Solution Let x represent the number of defective bulbs expected in the shipment. Since we know the ratio of defective bulbs, we can write a **proportion** by setting the ratio of defective bulbs to total bulbs in the shipment equal to the ratio of defective bulbs to total bulbs in the sample.

$$\underset{\text{Shipment}}{\downarrow} \qquad \underset{\text{Sample}}{\downarrow}$$

$$\frac{x \text{ defective bulbs}}{22{,}000 \text{ total bulbs}} = \frac{18 \text{ defective bulbs}}{200 \text{ total bulbs}}$$

Cross-multiply: $\qquad 200x = 22{,}000(18)$

Simplify: $\qquad 200x = 396{,}000$

Divide both sides by 200: $\qquad \dfrac{200x}{200} = \dfrac{396{,}000}{200}$

Simplify: $\qquad x = 1980$

The inspector can expect 1980 defective bulbs in a shipment of 22,000 lightbulbs.

Example 5 Using Proportions to Solve Application Problems

If 1 cup of skim milk contains 80 calories, how many calories are in an entire gallon of skim milk? Note that 1 gallon = 16 cups.

Solution Let x represent the number of calories in 1 gallon, or 16 cups, of milk. Since the **rate** of calories per cup is **constant**, we write a **proportion**.

$$\underset{\text{1 gallon}}{\downarrow} \qquad \underset{\text{1 cup}}{\downarrow}$$

$$\frac{x \text{ calories}}{16 \text{ cups}} = \frac{80 \text{ calories}}{1 \text{ cup}}$$

Cross-multiply: $\qquad 1x = 16(80)$

Simplify: $\qquad x = 1280$

There are 1280 calories in a gallon of skim milk.

SIMILAR TRIANGLES

Similar triangles have the same shape but not necessarily the same size. For similar triangles, **corresponding angles** are equal, and **corresponding sides** have lengths that are proportional. So, we can use proportions to find unknown lengths in similar triangles.

Example 6 Solving a Similar Triangle Problem

Find the unknown length n for the following **similar triangles**.

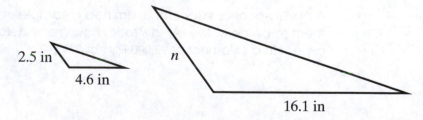

(eText Screens A.3-1–A.3-11)

Solution Since the triangles are similar, their **corresponding sides** are **proportional**. Side n on the larger triangle corresponds to the 2.5-inch side on the smaller triangle. The 16.1-inch side on the larger triangle corresponds to the 4.6-inch side on the smaller triangle. So, we have the following proportion:

$$\frac{n \text{ inches}}{2.5 \text{ inches}} = \frac{16.1 \text{ inches}}{4.6 \text{ inches}}$$

Cross-multiply:	$4.6n = 2.5(16.1)$
Simplify:	$4.6n = 40.25$
Divide both sides by 4.6:	$\dfrac{4.6n}{4.6} = \dfrac{40.25}{4.6}$
Simplify:	$n = 8.75$

The unknown side length n is 8.75 inches long.

Example 7 Using Proportions to Solve Application Problems

A forest ranger wants to determine the height of a tree. She measures the tree's shadow as 84 feet long. Her own shadow at the same time is 7.5 feet. If she is 5.5 feet tall, how tall is the tree?

Solution Since the shadows were measured at the same time, the **ratio** of the height-to-shadow length of the tree is **proportional** to the ratio of height-to-shadow length of the forest ranger. So, we have the following proportion.

$$\frac{x \text{ feet}}{84 \text{ feet}} = \frac{5.5 \text{ feet}}{7.5 \text{ feet}}$$

Cross-multiply:	$7.5x = 84(5.5)$
Simplify:	$7.5x = 462$
Divide both sides by 7.5:	$\dfrac{7.5x}{7.5} = \dfrac{462}{7.5}$
Simplify:	$x = 61.6$

The tree is 61.6 feet tall.

A.4 Percents

CONVERTING BETWEEN PERCENTS AND DECIMALS

The word **percent** means "per hundred" or "out of 100." A percent is denoted by the **percent symbol (%)**. If you score an 86% on a test, what does it mean? It means that you receive 86 points out of every 100 points possible on the test.

We can write percents as ratios. For example, 86% can be written as the ratio 86:100 or $\frac{86}{100}$. Since $\frac{86}{100} = 0.86$, we can also say that 86% = 0.86. In fact, any percent can be written as a **fraction** or a **decimal**.

Since $100\% = \frac{100}{100} = 1$, multiplying or dividing by 100% is equivalent to multiplying or dividing by 1. So, a value does not change when it is multiplied or divided by 100%.

Writing a Percent as a Decimal

To write a percent as a decimal, divide the percent by 100% and write the answer in decimal form.

Note: This is the same as dropping the percent symbol (%) and moving the decimal point two places to the left.

Example 1 Writing Percents as Decimals

Write each percent as a decimal.

a. 72.4%   **b.** 126%   **c.** 1.5%

Solutions

a. $72.4\% = \overbrace{\frac{72.4\%}{100\%}}^{\text{Divide by 100%}} = \frac{72.4\cancel{\%}}{100\cancel{\%}} = \frac{72.4}{100} = 0.724$

b. $126\% = \overbrace{\frac{126\%}{100\%}}^{\text{Divide by 100%}} = \frac{126\cancel{\%}}{100\cancel{\%}} = \frac{126}{100} = 1.26$

c. $1.5\% = \overbrace{\frac{1.5\%}{100\%}}^{\text{Divide by 100%}} = \frac{1.5\cancel{\%}}{100\cancel{\%}} = \frac{1.5}{100} = 0.015$

Writing a Decimal as a Percent

To write a decimal as a percent, multiply the decimal by 100%.

Note: This is the same as moving the **decimal point** two places to the right and including the percent symbol (%).

Example 2 Writing Decimals as Percents

Write each decimal as a percent.

a. 0.891 **b.** 2.47 **c.** 0.0678

Solutions

a. $0.891 = \overbrace{0.891(100\%)}^{\text{Multiply by 100\%}} = (0.891 \cdot 100)\% = 89.1\%$

b. $2.47 = \overbrace{2.47(100\%)}^{\text{Multiply by 100\%}} = (2.47 \cdot 100)\% = 247\%$

c. $0.0678 = \overbrace{0.0678(100\%)}^{\text{Multiply by 100\%}} = (0.0678 \cdot 100)\% = 6.78\%$

 TIP A decimal larger than 1 will yield a percent larger than 100%. Likewise, a percent larger than 100% will yield a decimal larger than 1.

CONVERTING BETWEEN PERCENTS AND FRACTIONS

Writing a Percent as a Fraction or Mixed Number

To write a percent as a **fraction** or **mixed number**, divide the percent by 100% and write the answer as a **simplified** fraction or mixed number.

Note: This is the same as dropping the percent symbol (%), writing the number over 100, and simplifying the fraction.

Example 3 Writing Percents as Fractions or Mixed Numbers

Write each percent as a fraction or mixed number.

a. 154% **b.** 10.8% **c.** $15\frac{2}{3}\%$

Solutions

a. $154\% = \overbrace{\dfrac{154\%}{100\%}}^{\text{Divide by 100\%}} = \dfrac{154\%}{100\%} = \dfrac{154}{100} = \dfrac{154 \div 2}{100 \div 2} = \dfrac{77}{50} = 1\dfrac{27}{50}$

b. $10.8\% = \overbrace{\dfrac{10.8\%}{100\%}}^{\text{Divide by 100\%}} = \dfrac{10.8\%}{100\%} = \dfrac{10.8}{100} = \dfrac{10.8 \cdot 10}{100 \cdot 10} = \dfrac{108}{1000} = \dfrac{108 \div 4}{1000 \div 4} = \dfrac{27}{250}$

c. $15\dfrac{2}{3}\% = \overbrace{\dfrac{15\frac{2}{3}\%}{100\%}}^{\text{Divide by 100\%}} = \dfrac{15\frac{2}{3}\%}{100\%} = \dfrac{\frac{47}{3}}{100} = \dfrac{47}{3} \cdot \dfrac{1}{100} = \dfrac{47}{300}$

Writing a Fraction as a Percent

To write a fraction or mixed number as a percent, multiply it by 100% and simplify.

Note: This is the same as multiplying by 100 and including the percent symbol (%).

Example 4 Writing Fractions or Mixed Numbers as Percents

Write each fraction or mixed number as a percent.

a. $\dfrac{3}{5}$ **b.** $5\dfrac{3}{8}$ **c.** $\dfrac{7}{9}$

Solutions

Multiply by 100%

a. $\dfrac{3}{5} = \dfrac{3}{5}(100\%) = \left(\dfrac{3}{5} \cdot \dfrac{100}{1}\right)\% = \left(\left(\dfrac{3}{\underset{1}{\cancel{5}}} \cdot \dfrac{\overset{20}{\cancel{100}}}{1}\right)\right)\% = 60\%$

Multiply by 100%

b. $5\dfrac{3}{8} = 5\dfrac{3}{8}(100\%) = \left(\dfrac{43}{8} \cdot 100\right)\% = \left(\dfrac{43}{\underset{2}{\cancel{8}}} \cdot \overset{25}{\cancel{100}}\right)\% = \dfrac{1075}{2}\% = 537\dfrac{1}{2}\%$

Multiply by 100%

c. $\dfrac{7}{9} = \dfrac{7}{9}(100\%) = \left(\dfrac{7}{9} \cdot 100\right)\% = \left(\dfrac{700}{9}\right)\% = 77\dfrac{7}{9}\%$

In **Example 4**, we found exact percent representations of fractions. Often, it is more practical to find approximations instead. For example,

$$77\dfrac{7}{9}\% = 77.7777\ldots\% \approx 77.78\%.$$

Example 5 Writing Fractions as Approximate Percents

Write $\dfrac{1}{12}$ as a percent rounded to the nearest hundredth of a percent.

Solution

Multiply by 100%

$$\dfrac{1}{12} = \dfrac{1}{12}(100\%) = \left(\dfrac{1}{12} \cdot 100\right)\% = 8.3333\ldots\% \approx 8.33\%$$

(eText Screens A.4-1–A.4-26)

USING EQUATIONS TO SOLVE PERCENT PROBLEMS

Recall that the word *of* indicates multiplication. For example, "three-fifths of forty is twenty-four" means

$$\frac{3}{5} \quad \text{of} \quad 40 \quad \text{is} \quad 24.$$

$$\frac{3}{5} \quad \cdot \quad 40 \quad = \quad 24$$

In **Example 4(a)**, we found that $\frac{3}{5} = 60\%$. So, the statement "three-fifths of forty is twenty-four" is equivalent to the statement "sixty percent of forty is twenty-four."

$$60\% \quad \text{of} \quad 40 \quad \text{is} \quad 24.$$

$$60\% \quad \cdot \quad 40 \quad = \quad 24$$

When we solve **percent** problems, the word *of* means multiplication, the word *is* means equals, and the words *what* or *what number* refer to the unknown number. We can use equations that include percents, called **percent equations**, to solve percent problems.

Example 6 Translating a Statement into a Percent Equation

Translate each statement into a **percent equation**. Let x be the number.

a. 18 is 30% of what number?

b. 123% of 62 is what number?

c. 8.2 is what percent of 12.5?

Solutions

a. 18   is   30%   of   what number?

$$18 \quad = \quad 30\% \quad \cdot \quad x$$

b. 123%   of   62   is   what number?

$$123\% \quad \cdot \quad 62 \quad = \quad x$$

c. 8.2   is   what percent   of   12.5?

$$8.2 \quad = \quad x \quad \cdot \quad 12.5$$

Each problem in **Example 6** contains three quantities. We name each of the quantities in order to develop a **general equation for percents**. Let's review the statement "60% of 40 is 24." In this statement, 60% is the **percent**, 40 is the **base**, and 24 is the **amount**.

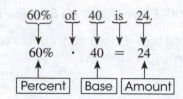

$$60\% \quad \text{of} \quad 40 \quad \text{is} \quad 24.$$
$$60\% \quad \cdot \quad 40 \quad = \quad 24$$
$$\boxed{\text{Percent}} \quad \boxed{\text{Base}} \quad \boxed{\text{Amount}}$$

Percent Equation

$$\text{Percent} \cdot \text{Base} = \text{Amount}$$

When we translate **percent** problems into **equations**, the percent can be identified by looking for the word *percent* or the % symbol. The **base** represents one whole and will usually follow the word *of*. The **amount** is the part compared to the whole and is normally isolated from the percent and the base on one side of the word *is*.

 When using the percent equation, the percent must be written as a **decimal** or **fraction** in order to solve the problem.

Example 7 Using Equations to Solve Percent Problems

Use equations to solve each percent problem from **Example 6**.

a. 18 is 30% of what number?

b. 123% of 62 is what number?

c. 8.2 is what percent of 12.5?

Solutions

a. Write the **percent equation** from Example 6: $\quad 18 = 30\% \cdot x$

Convert 30% to a decimal: $\quad 18 = 0.30 \cdot x$

Divide both sides by 0.30: $\quad \dfrac{18}{0.30} = \dfrac{0.30 \cdot x}{0.30}$

Simplify: $\quad 60 = x$

So, 18 is 30% of 60.

b. Rewrite the percent equation from Example 6: $\quad 123\% \cdot 62 = x$

Convert 123% to a decimal: $\quad 1.23 \cdot 62 = x$

Multiply: $\quad 76.26 = x$

123% of 62 is 76.26.

c. Rewrite the percent equation from Example 6: $\quad 8.2 = x \cdot 12.5$

Divide both sides of the equation by 12.5: $\quad \dfrac{8.2}{12.5} = \dfrac{x \cdot 12.5}{12.5}$

Simplify: $\quad 0.656 = x$

Convert 0.656 to a percent: $\quad 65.6\% = x$

8.2 is 65.6% of 12.5.

USING PROPORTIONS TO SOLVE PERCENT PROBLEMS

If the percent equation is true, we can say

$$\text{Percent} = \frac{\text{Amount}}{\text{Base}}.$$

If we then write the **percent** as a ratio over 100, we obtain the **percent proportion**.

Percent Proportion

$$\frac{\text{Amount}}{\text{Base}} = \frac{\text{Percent}}{100}$$

When we write the percent over 100%, the % symbols divide out. This is why the % symbols are left out of the percent proportion.

Example 8 Using Proportions to Solve Percent Problems

Use **proportions** to solve each percent problem.

a. 9% of 3250 is what number?

b. What percent of 47.32 is 106.47?

c. 74% of what number is 37?

Solutions

a. In this case, 9 is the **percent**, 3250 is the **base**, and we are looking for the **amount**. Let x represent the unknown number. The **proportion** is

$$\boxed{\text{Amount}} \mapsto \frac{x}{3250} = \frac{9}{100} \leftarrow \boxed{\text{Percent}}$$
$$\boxed{\text{Base}} \nearrow$$

Cross-multiply: $100x = 3250(9)$

Simplify: $100x = 29{,}250$

Divide both sides by 100: $\dfrac{100x}{100} = \dfrac{29{,}250}{100}$

Simplify: $x = 292.5$

9% of 3250 is 292.5.

b. Here, we are looking for the percent. 47.32 is the base and 106.47 is the amount. Let x represent the percent. The proportion is

$$\boxed{\text{Amount}} \mapsto \frac{106.47}{47.32} = \frac{x}{100} \leftarrow \boxed{\text{Percent}}$$
$$\boxed{\text{Base}} \nearrow$$

Cross-multiply: $100(106.47) = 47.32x$

Simplify: $10{,}647 = 47.32x$

Divide both sides by 47.32: $\dfrac{10{,}647}{47.32} = \dfrac{47.32x}{47.32}$

Simplify: $225 = x$

225% of 47.32 is 106.47.

c. 74 is the **percent**, 37 is the **amount**, and we are looking for the **base**. The proportion is

$$\boxed{\text{Amount}} \rightarrow \frac{37}{x} = \frac{74}{100} \leftarrow \boxed{\text{Percent}}$$
$$\boxed{\text{Base}} \rightarrow$$

Cross-multiply: $100(37) = x(74)$

Simplify: $3700 = 74x$

Divide both sides by 74: $\dfrac{3700}{74} = \dfrac{74x}{74}$

Simplify: $50 = x$

74% of 50 is 37.

 When using the **percent proportion** to find an unknown percent, be sure to label the answer with the % symbol and do not move the **decimal point**.

PERCENT APPLICATIONS

Now we can apply what we have learned about percents to real settings. We can solve percent problems in two ways—using the **percent equation** or using the percent proportion. The key to solving percent applications is to identify the percent, the amount, and the base for the problem.

Example 9 Finding the Percent of Field Goals Made

A basketball team scored 45 of 80 attempted field goals during one game. What **percent** of the field goal attempts were made?

Solution We want to find the percent of field goals made. The **base** is the 80 attempted field goals, and the **amount** is the 45 field goals that were actually made. Both methods are illustrated.

Using the **percent equation**:

$$\boxed{\text{Percent}} \quad \boxed{\text{Base}} \quad \boxed{\text{Amount}}$$
$$\downarrow \qquad \downarrow \qquad \downarrow$$
$$x \quad \cdot \quad 80 \quad = \quad 45$$

Divide by 80: $x = \dfrac{45}{80}$

Simplify: $x = 0.5625$

Convert 0.5625 to a percent: $x = 56.25\%$

Using the **percent proportion**:

$$\boxed{\text{Amount}} \rightarrow \frac{45}{80} = \frac{x}{100} \leftarrow \boxed{\text{Percent}}$$
$$\boxed{\text{Base}} \rightarrow$$

Cross-multiply: $100(45) = 80x$

Simplify: $4500 = 80x$

Divide by 80: $\dfrac{4500}{80} = \dfrac{80x}{80}$

Simplify: $56.25 = x$

56.25% of the field goal attempts were made.

In Examples 9–14, both methods will be shown for solving the percent problem. However, when solving percent problems on your own, practice both methods and then choose the method that you prefer.

Example 10 Finding the Percent of Questions Answered Correctly

Deon took a history exam that consisted of 75 multiple-choice questions. He answered 9 of the questions incorrectly. What **percent** of the questions did Deon answer correctly?

Solution We want to find the percent of questions answered correctly. The base is 75 questions, and the **amount** is the $75 - 9 = 66$ questions answered correctly. Both methods are illustrated.

Using the **percent equation**:

Percent	Base	Amount
x	$\cdot$ 75	$=$ 66

Divide by 75: $\quad x = \dfrac{66}{75}$

Simplify: $\quad x = 0.88$

Convert 0.88 to a percent: $\quad x = 88\%$

Using the **percent proportion**:

$$\text{Amount} \rightarrow \dfrac{66}{75} = \dfrac{x}{100} \leftarrow \text{Percent}$$
$$\text{Base} \rightarrow$$

Cross-multiply: $\quad 100(66) = 75x$

Simplify: $\quad 6600 = 75x$

Divide by 75: $\quad \dfrac{6600}{75} = \dfrac{75x}{75}$

Simplify: $\quad 88 = x$

Deon answered 88% of the questions correctly.

We could also solve Example 10 by first finding the percent of questions that Deon answered incorrectly. We could then subtract that percent from 100% to obtain the percent of questions answered correctly.

Example 11 Finding the Percent of a Fundraising Goal

At the beginning of a scholarship campaign, the college foundation set a goal to raise $25,000 in donations from businesses. By the end of the campaign, it had actually raised $32,800. What **percent** of the goal was actually raised?

Solution We need to find the percent of the goal that was actually raised. The **base** is the $25,000 goal, and the **amount** is the $32,800 that was actually raised. Both methods are illustrated.

Using the **percent equation**:

Percent	Base	Amount
x	$\cdot$ 25,000	$=$ 32,800

Divide by 25,000: $\quad x = \dfrac{32,800}{25,000}$

Simplify: $\quad x = 1.312$

Convert 1.312 to a percent: $\quad x = 131.2\%$

Using the **percent proportion**:

$$\text{Amount} \rightarrow \dfrac{32,800}{25,000} = \dfrac{x}{100} \leftarrow \text{Percent}$$
$$\text{Base} \rightarrow$$

Cross-multiply: $\quad 100(32,800) = 25,000x$

Simplify: $\quad 3,280,000 = 25,000x$

Divide by 25,000: $\quad \dfrac{3,280,000}{25,000} = \dfrac{25,000x}{25,000}$

Simplify: $\quad 131.2 = x$

The college foundation raised 131.2% of its goal.

Example 12 Finding the Number of Drivers Wearing Seatbelts

An observational study at a certain intersection found that of 650 drivers, 78% were wearing seatbelts. How many of the drivers were wearing seatbelts?

Solution We are looking for the **amount** of drivers who were wearing seatbelts. The **percent** is 78%, and the **base** is 650 drivers. Both methods are illustrated.

Using the **percent equation**:

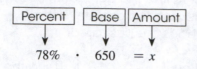

$$78\% \quad \cdot \quad 650 \quad = x$$

Convert 78% to a decimal: $\quad 0.78 \cdot 650 = x$

Multiply: $\qquad\qquad\qquad 507 = x$

Using the **percent proportion**:

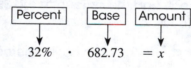

Cross-multiply: $\quad 100x = 650(78)$

Simplify: $\quad 100x = 50{,}700$

Divide by 100: $\quad \dfrac{100x}{100} = \dfrac{50{,}700}{100}$

Simplify: $\qquad\quad x = 507$

Of the 650 drivers, 507 were wearing seatbelts.

Example 13 Calculating Withholdings and Take Home Pay

Ben earns $682.73 per week at his job, but 32% of his pay is withheld for federal taxes, state taxes, social security, and so on.

a. How much money is withheld from Ben's weekly paycheck? **Round** to the nearest cent.

b. What is Ben's take home pay? Round to the nearest cent.

Solutions

a. For this problem, we need to find the **amount** of money that is withheld from Ben's weekly paycheck. The **percent** is 32%, and the **base** is the $682.73 gross. Both methods are shown.

Using the **percent equation**:

$$32\% \quad \cdot \quad 682.73 \quad = x$$

Convert 32% to a decimal: $\quad 0.32 \cdot 682.73 = x$

Multiply: $\qquad\qquad 218.4736 = x$

Round to hundredths: $\qquad 218.47 \approx x$

Using the **percent proportion**:

$$\frac{x}{682.73} = \frac{32}{100}$$

Cross-multiply: $\quad 100x = 682.73(32)$

Simplify: $\quad 100x = 21{,}847.36$

Divide by 100: $\quad \dfrac{100x}{100} = \dfrac{21{,}847.36}{100}$

Simplify: $\qquad\quad x = 218.4736$

Round: $\qquad\quad x \approx 218.47$

The amount withheld from Ben's weekly paycheck is $218.47.

b. Ben's take-home pay is $682.73 − $218.47 = $464.26.

Example 14 Calculating the Number of Calories in a Frozen Dinner

A frozen dinner advertises that 16% of its calories are from fat. If the dinner contains 65 calories from fat, what is the total number of calories in the dinner? Round to the nearest whole calorie.

Solution We are given the **percent** of the calories from fat (16%) and the **amount** of calories from fat (65 calories). We need to find the **base**, which is the total number of calories in the dinner. Both methods are shown.

Using the percent equation:

Percent	Base	Amount

$$16\% \cdot x = 65$$

Convert 16% to a decimal: $\quad 0.16x = 65$

Divide by 0.16: $\quad \dfrac{0.16x}{0.16} = \dfrac{65}{0.16}$

Simplify: $\quad x = 406.25$

Round to a whole number: $\quad x \approx 406$

Using the percent proportion:

Amount → $\dfrac{65}{x} = \dfrac{16}{100}$ ← Percent
Base →

Cross-multiply: $\quad 100(65) = x(16)$

Simplify: $\quad 6500 = 16x$

Divide by 16: $\quad \dfrac{6500}{16} = \dfrac{16x}{16}$

Simplify: $\quad 406.25 = x$

Round: $\quad 406 \approx x$

The dinner contains about 406 calories total.

PERCENT OF INCREASE AND PERCENT OF DECREASE

Percents are often used to show how much a quantity has increased or decreased. Let's discuss **percent of increase** and **percent of decrease**.

Percent of Increase

$$\text{Amount of increase} = \text{New amount} - \text{Original amount}$$

$$\frac{\text{Percent of increase}}{100} = \frac{\text{Amount of increase}}{\text{Original amount}}$$

Example 15 Calculating the Percent of Increase

Over the past four years, Ethan's rent increased from $875 per month to $1043 per month. What is the percent of increase?

Solution

Find the amount of increase: $\quad$ Amount of increase $= \$1043 - \$875 = \$168$.

Percent of increase → $\dfrac{x}{100} = \dfrac{168}{875}$ ← Amount of increase
← Original amount

Cross-multiply: $\quad 875x = 100(168)$

Simplify: $\quad 875x = 16{,}800$

Divide both sides by 875: $\quad \dfrac{875x}{875} = \dfrac{16{,}800}{875}$

Simplify: $\quad x = 19.2$

Ethan's rent has increased by 19.2%.

Example 16 Calculating the Amount of Increase

Last year, 16,528 students attended City Community College. This year, enrollment increased by 3.2%. How many students attend City Community College this year? **Round to the nearest whole student.**

Solution First, find the **amount of increase**. Then, add the amount of increase to last year's enrollment to find this year's enrollment.

Amount   Percent   Base

Amount of increase $= (3.2\%)(16{,}528) = (0.032)(16{,}528) = 528.896 \approx 529$

New enrollment $= 16{,}528 + 529 = 17{,}057$

This year's enrollment at City Community College is 17,057 students.

Percent of Decrease

$$\text{Amount of decrease} = \text{Original amount} - \text{New amount}$$

$$\frac{\text{Percent of decrease}}{100} = \frac{\text{Amount of decrease}}{\text{Original amount}}$$

Example 17 Calculating the Percent of Decrease

To attract more clients to her salon, Rachelle lowered the price of a haircut from $25 to $22.50. Find the **percent of decrease** for the price of a haircut at Rachelle's salon.

Solution Find the **amount of decrease**: Amount of decrease $= \$25 - \$22.50 = \$2.50$.

Percent of decrease $\rightarrow \dfrac{x}{100} = \dfrac{2.50 \leftarrow \text{Amount of decrease}}{25 \leftarrow \text{Original amount}}$

Cross-multiply: $25x = 100(2.50)$

Simplify: $25x = 250$

Divide both sides by 25: $\dfrac{25x}{25} = \dfrac{250}{25}$

Simplify: $x = 10$

The price of a haircut at Rachelle's salon decreased by 10%.

 When calculating the **percent of increase** or **percent of decrease**, be sure to divide by the original amount, not the new amount.

SIMPLE INTEREST

When we borrow money, the price that we pay for this privilege is called **interest**. Similarly, if we lend money to someone, we receive money for this generosity, which is also called **interest**. So, when we borrow money, we *pay* interest; when we loan or invest money, we *earn* interest. The amount of money that is borrowed or invested is called the **principal**. Usually, the amount of interest paid or earned is a **percent** of the principal over a given time period. The **interest rate** is the percent used to calculate the interest. Although there are different types of interest, we will look at only one type, called **simple interest**.

Simple Interest Formula

$$\text{Simple interest} = \text{Principal} \cdot \text{Rate} \cdot \text{Time}$$

$$I = P \cdot r \cdot t$$

 When entering the **interest rate** r into the simple interest formula, it must be entered as a **decimal**, not a percent. For example, an interest rate of 7% would be entered into the formula as 0.07.

 The interest rate r is given as a percent per unit of time. The time t must be entered in the same units. For example, if the interest rate is 7% per year, then the time must be in years. If the interest rate is 2% per month, then the time must be in months. Standard practice is to use annual (yearly) interest rates.

Example 18 Calculating Simple Interest

André borrowed $3000 for 9 months at a **simple interest** rate of 8.5% per year.

a. Find the **interest** on his loan.

b. Find the **total amount** that André had to pay back.

Solutions

a. The **interest rate** is per year, but the time is given in months. So we must change one or the other. We choose to change 9 months to years:

$$t = 9 \text{ months} = \frac{9}{12} \text{ year} = \frac{3}{4} \text{ year}.$$

The **principal** is $P = \$3000$, and the interest rate is $r = 8.5\%$, or 0.085 per year. The interest is

$$I = Prt = 3000 \cdot (0.085) \cdot \frac{3}{4} = 191.25.$$

André paid $191.25 in interest.

b. André had to pay back the principal, plus the interest, so he paid back a total of $3000 + \$191.25 = \$3191.25$.

APPENDIX B
Solving Systems of Linear Equations Using Matrices

THINGS TO KNOW

Before working through this section, be sure you are familiar with the following concepts:

VIDEO   ANIMATION   INTERACTIVE

  You Try It

1. Solve Systems of Linear Equations in Two Variables by Substitution (Section 3.1, Objective 3)

 You Try It

2. Solve Systems of Linear Equations in Two Variables by Elimination (Section 3.1, Objective 4)

 You Try It

3. Solve Inconsistent and Dependent Systems (Section 3.1, Objective 5)

You Try It

4. Solve Systems of Linear Equations in Three Variables (Section 3.2, Objective 2)

OBJECTIVES

1 Write an Augmented Matrix

2 Solve Systems of Two Equations Using Matrices

3 Solve Systems of Three Equations Using Matrices

....

OBJECTIVE 1   WRITE AN AUGMENTED MATRIX

In this section, we look at how to solve **systems of linear equations** by using a **matrix**. A matrix is a rectangular array of numbers arranged in rows and columns. The following are examples of **matrices** (the plural of *matrix*).

$$\begin{bmatrix} 3 & 7 \\ -2 & 5 \end{bmatrix} \quad \begin{bmatrix} 0 & 3 & 1 \\ 2 & -4 & 1 \end{bmatrix} \quad \begin{bmatrix} 1 & 0 & 0 \\ 0 & 1 & 0 \\ 0 & 0 & 1 \end{bmatrix} \quad \begin{bmatrix} 1 & 0 & -1 & 5 \\ 0 & 1 & 4 & -3 \\ 0 & 0 & 1 & \frac{1}{2} \end{bmatrix}$$

(eText Screens B.1-1–B.1-26)

The size of a matrix is determined by the number of rows and columns. The following matrix is a 2 × 3 (read *two by three*) matrix because it has 2 rows and 3 columns.

$$\text{Row 1} \rightarrow \begin{bmatrix} 4 & 9 & -6 \\ 3 & -1 & 2 \end{bmatrix}$$
$$\text{Row 2} \rightarrow$$
Column 1 ↑ Column 3
Column 2

When solving systems of equations using **elimination**, we wrote the equations in the system in **standard form** lining up the **variables**. Doing so gave a similar rectangular structure to our system. We then used the coefficients of the equations to eliminate variables because the solution to a system depends on the **coefficients** of the equations, not on the variables.

Once we have a system written in **standard form**, we can simplify the **elimination** process by first using the coefficients of the equations to form an **augmented matrix** that can be used to solve the system. Consider the following illustration.

Coefficients of y ─┐ ┌─Equal signs
Coefficients of x ─┐   ┌─Constants

Equation 1 $\begin{cases} 3x + 2y = 11 \longrightarrow \text{Row 1} \begin{bmatrix} 3 & 2 & | & 11 \\ 8 & -9 & | & 4 \end{bmatrix} \\ 8x - 9y = 4 \longrightarrow \text{Row 2} \end{cases}$
Equation 2

System in standard form   Augmented matrix

An augmented matrix for a system of equations in standard form is a **matrix** consisting of the **coefficients** of the variables and the constants from the right-hand side of the equations, separated by a vertical line. There is a column of coefficients for each variable and an additional column for the constants. Each row in the matrix corresponds to an equation in the system. Therefore, systems of two equations in two variables will produce a 2 × 3 augmented matrix, whereas systems of three equations in three variables will produce a 3 × 4 augmented matrix.

Example 1 Writing an Augmented Matrix

Write the corresponding **augmented matrix** for each system of equations.

a. $\begin{cases} -2x + 7y = 9 \\ 8x + 3y = 0 \end{cases}$

b. $\begin{cases} 6x - 3y = 11 \\ y = 2x - 5 \end{cases}$

c. $\begin{cases} 5x - 2y + z = 18 \\ x + 3z = -5 \\ 3y = 9 + z \end{cases}$

Solutions

a. The equations are both in **standard form**, so we can immediately write the augmented matrix.

$$\begin{cases} -2x + 7y = 9 \\ 8x + 3y = 0 \end{cases} \rightarrow \begin{bmatrix} -2 & 7 & | & 9 \\ 8 & 3 & | & 0 \end{bmatrix}$$

b. The first equation is in standard form, but the second is not. We can put it in standard form by subtracting $2x$ from both sides.

$$\begin{cases} 6x - 3y = 11 \\ y = 2x - 5 \end{cases} \rightarrow \begin{cases} 6x - 3y = 11 \\ -2x + 1y = -5 \end{cases} \rightarrow \left[\begin{array}{cc|c} 6 & -3 & 11 \\ -2 & 1 & -5 \end{array}\right]$$

Original system   Standard form   Augmented matrix

c. Start by writing the equations in **standard form**. If a variable is missing from an equation, it may be helpful to include the variable with a 0 coefficient.

$$\begin{cases} 5x - 2y + z = 18 \\ x + 3z = -5 \\ 3y = 9 + z \end{cases} \rightarrow \begin{cases} 5x - 2y + 1z = 18 & \leftarrow +z = +1z \\ 1x + 0y + 3z = -5 & \leftarrow \;\; x = 1x \\ 0x + 3y - 1z = 9 & \leftarrow -z = -1z \end{cases}$$

Original system   Standard form

Now write the **augmented matrix**. View the **answer**, or watch this **video** for a complete solution.

Notice that in the standard form we wrote $+z$ as $+1z$, x as $1x$, and $-z$ as $-1z$. This was to help make sure we entered the correct coefficient in the augmented matrix.

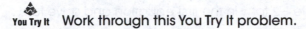 **You Try It**   Work through this **You Try It** problem.

Work Exercises 1–6 in this eText or in the MyMathLab Study Plan.

OBJECTIVE 2   SOLVE SYSTEMS OF TWO EQUATIONS USING MATRICES

To solve a **system of equations** using a **matrix** means to find an equivalent augmented matrix from which the solution can be easily obtained. Two augmented matrices are **equivalent matrices** if they represent systems that have the same solution set. We obtain equivalent matrices by applying **row operations**.

Row Operations

Applying any of the following operations to a matrix will result in an equivalent matrix.

1. Interchange any two rows.

2. Multiply the elements of a row by the same non-zero constant.

3. Add a multiple of one row to another row.

Note that these operations are the same as those applied to equations of a system when solving the system by the **elimination** method. We will use the following notation to describe the row operations.

Notation	**Meaning**
$R_i \Leftrightarrow R_j$	Interchange Rows i and j
$kR_i \rightarrow \text{New } R_i$	k times Row i becomes the new Row i
$kR_i + R_j \rightarrow \text{New } R_j$	k times Row i plus Row j becomes the new Row j

(eText Screens B.1-1–B.1-26)

View this **popup** to see an example of each row operation.

For a **system of two equations**, we apply **row operations** until we obtain a **matrix** of the form

$$\begin{bmatrix} 1 & a & | & b \\ 0 & 1 & | & c \end{bmatrix}.$$

This is called the **row-echelon form** of the matrix. In this form, there are 1's along the diagonal from the upper left to lower right, and 0's below the diagonal. From there we obtain a system of equations that can be solved by **back-substitution**. Consider the following.

Coefficients of y — ┌ Equal signs
Coefficients of x ┐ ↓↓ ┌ Constants

$$\begin{bmatrix} 1 & -3 & | & 7 \\ 0 & 1 & | & -2 \end{bmatrix} \qquad \begin{cases} 1x - 3y = 7 \\ 0x + 1y = -2 \end{cases} \longrightarrow \begin{cases} x - 3y = 7 \\ y = -2 \end{cases}$$

Augmented matrix   Corresponding
in row-echelon form   system

The second equation in the system tells us that $y = -2$. To find the value for x, we can substitute -2 for y in the first equation and solve for x.

Original equation: $x - 3y = 7$

Substitute -2 for y: $x - 3(-2) = 7$

Multiply: $x + 6 = 7$

Subtract 6 from both sides: $x = 1$

So the **solution** to the system is the ordered pair $(1, -2)$.

Example 2 Solving a System of Two Equations Using Matrices

Use **matrices** to solve each of the following systems.

a. $\begin{cases} 2x - 5y = 26 \\ 3x + 2y = 1 \end{cases}$ **b.** $\begin{cases} y = 3x + 2 \\ 6x + \dfrac{1}{2}y = 6 \end{cases}$

Solutions

a. Both equations are in **standard form**, so we can write the **augmented matrix**.

$$\begin{cases} 2x - 5y = 26 \\ 3x + 2y = 1 \end{cases} \longrightarrow \begin{bmatrix} 2 & -5 & | & 26 \\ 3 & 2 & | & 1 \end{bmatrix}$$

We now apply **row operations** to put the matrix in **row-echelon form**. Our first step is to obtain a 1 in row 1, column 1. We do this by multiplying each entry in the first row by $\dfrac{1}{2}$.

$$\begin{bmatrix} 2 & -5 & | & 26 \\ 3 & 2 & | & 1 \end{bmatrix} \xrightarrow{\frac{1}{2}R_1} \begin{bmatrix} \frac{1}{2}(2) & \frac{1}{2}(-5) & | & \frac{1}{2}(26) \\ 3 & 2 & | & 1 \end{bmatrix} = \begin{bmatrix} 1 & -\frac{5}{2} & | & 13 \\ 3 & 2 & | & 1 \end{bmatrix}$$

Next, we use row 1 to get a 0 in row 2, column 1. We do this by multiplying the entries in row 1 by -3 and adding the results to the entries in row 2.

$$\begin{bmatrix} 1 & -\dfrac{5}{2} & \Big| & 13 \\ 3 & 2 & \Big| & 1 \end{bmatrix} \xrightarrow{-3R_1 + R_2} \begin{bmatrix} 1 & -\dfrac{5}{2} & \Big| & 13 \\ -3(1)+3 & -3\left(-\dfrac{5}{2}\right)+2 & \Big| & -3(13)+1 \end{bmatrix} = \begin{bmatrix} 1 & -\dfrac{5}{2} & \Big| & 13 \\ 0 & \dfrac{19}{2} & \Big| & -38 \end{bmatrix}$$

Lastly, we get a 1 in row 2, column 2. We do this by multiplying the entries in row 2 by $\dfrac{2}{19}$.

$$\begin{bmatrix} 1 & -\dfrac{5}{2} & \Big| & 13 \\ 0 & \dfrac{19}{2} & \Big| & -38 \end{bmatrix} \xrightarrow{\frac{2}{19}R_2} \begin{bmatrix} 1 & -\dfrac{5}{2} & \Big| & 13 \\ \dfrac{2}{19}(0) & \dfrac{2}{19}\left(\dfrac{19}{2}\right) & \Big| & \dfrac{2}{19}(-38) \end{bmatrix} = \begin{bmatrix} 1 & -\dfrac{5}{2} & \Big| & 13 \\ 0 & 1 & \Big| & -4 \end{bmatrix}$$

The corresponding **equivalent system** is

$$\begin{cases} x - \dfrac{5}{2}y = 13 \\ y = -4 \end{cases}$$

The second equation tells us that $y = -4$. To find the value for x, we can **substitute** -4 for y in the first equation and solve for x.

$$\text{First equation:} \qquad x - \frac{5}{2}y = 13$$

$$\text{Substitute } -4 \text{ for } y\text{:} \qquad x - \frac{5}{2}(-4) = 13$$

$$\text{Multiply:} \qquad x + 10 = 13$$

$$\text{Subtract 10 from both sides:} \qquad x = 3$$

The **solution** to the system is the ordered pair $(3, -4)$.

My video summary ▶ **b.** Start by writing the equations in the system in **standard form**.

$$\begin{cases} y = 3x + 2 \\ 6x + \dfrac{1}{2}y = 6 \end{cases} \rightarrow \begin{cases} -3x + y = 2 \\ 6x + \dfrac{1}{2}y = 6 \end{cases}$$

Now write the corresponding **augmented matrix**.

$$\begin{cases} -3x + 1y = 2 \\ 6x + \dfrac{1}{2}y = 6 \end{cases} \rightarrow \begin{bmatrix} -3 & 1 & \Big| & 2 \\ 6 & \dfrac{1}{2} & \Big| & 6 \end{bmatrix}$$

Use **row operations** to solve the system on your own. View the **answer**, or watch this **video** for a complete solution.

You Try It Work through this You Try It problem.

Work Exercises 7–11 in this eText or in the MyMathLab Study Plan.

⚠ Recall that a system of equations may have one solution, no solution, or an infinite number of solutions.

(eText Screens B.1-1–B.1-26)

Example 3 Solving a System of Two Equations Using Matrices

Use matrices to solve the following system.

$$\begin{cases} 3x + \dfrac{1}{2}y = 4 \\ 2y = -12x + 16 \end{cases}$$

🖎 *My video summary* ▶ **Solution** Work through the following, or watch this **video** for the complete solution.

Start by writing the equations in the system in **standard form**.

$$\begin{cases} 3x + \dfrac{1}{2}y = 4 \\ 2y = -12x + 16 \end{cases} \quad \rightarrow \quad \begin{cases} 3x + \dfrac{1}{2}y = 4 \\ 12x + 2y = 16 \end{cases}$$

Now write the corresponding **augmented matrix**.

$$\begin{cases} 3x + \dfrac{1}{2}y = 4 \\ 12x + 2y = 16 \end{cases} \quad \rightarrow \quad \left[\begin{array}{cc|c} 3 & \dfrac{1}{2} & 4 \\ 12 & 2 & 16 \end{array}\right]$$

Apply **row operations** to put the matrix in **row-echelon form**.

$$\left[\begin{array}{cc|c} 3 & \dfrac{1}{2} & 4 \\ 12 & 2 & 16 \end{array}\right] \xrightarrow{\frac{1}{3}R_1} \left[\begin{array}{cc|c} 1 & \dfrac{1}{6} & \dfrac{4}{3} \\ 12 & 2 & 16 \end{array}\right] \xrightarrow{-12R_1 + R_2} \left[\begin{array}{cc|c} 1 & \dfrac{1}{6} & \dfrac{4}{3} \\ 0 & 0 & 0 \end{array}\right]$$

Notice that the last row is all zeros. This corresponds to the equation $0 = 0$, which is an **identity**.

The second equation in the system is just a multiple of the first equation, so it provides us with no new information. Therefore, this system is **dependent** and **consistent** and has an infinite number of solutions. Both equations in the system describe the same line so we can use either to obtain the **slope-intercept** form of the line.

$$\text{Second equation:} \quad 2y = -12x + 16$$

$$\text{Divide both sides by 2:} \quad \frac{2y}{2} = \frac{-12x + 16}{2}$$

$$\text{Simplify:} \quad y = -6x + 8$$

The solution set is $\{(x, y)\,|\,y = -6x + 8\}$

When solving a **system of two equations** using **matrices**, if we obtain a row of all zeros (as in this example), the system will have an infinite number of solutions. However, if we obtain a row that is all zeros to the left of the vertical bar, but non-zero to the right, then the system has no solution. Consider the following.

$$\left[\begin{array}{cc|c} 1 & 3 & 8 \\ 0 & 0 & 2 \end{array}\right]$$

The second row corresponds to the equation $0x + 0y = 2$, or $0 = 2$. Since this statement is false, the system has no solution.

You Try It Work through this **You Try It** problem.

Work Exercises 12–18 in this eText or in the MyMathLab Study Plan.

OBJECTIVE 3 SOLVE SYSTEMS OF THREE EQUATIONS USING MATRICES

For a **system of three equations**, we again apply **row operations** until we obtain a matrix in the following **row-echelon form**.

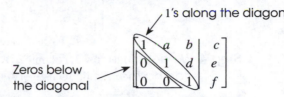

As with two equations, this form has 1's along the diagonal from the upper left to lower right, and 0's below the diagonal. In general, we can use the following steps to write the row-echelon form.

Writing Row-Echelon Form (Three Rows)

Step 1   Get a 1 in the row 1, column 1 entry.

Step 2   Use row 1 to obtain 0's in column 1 below the diagonal.

Step 3   Get a 1 in the row 2, column 2 entry.

Step 4   Use row 2 to obtain 0's in column 2 below the diagonal.

Step 5   Get a 1 in the row 3, column 3 entry.

Example 4 Solving a System of Three Equations Using Matrices

Use **matrices** to solve each of the following systems.

a. $\begin{cases} 2x + y - 3z = -3 \\ x + 2y + z = 4 \\ -3x + y - 4z = 1 \end{cases}$

b. $\begin{cases} 2x + 2y + z = 1 \\ 5x + 2y = 30 + 3z \\ 3x + 4z = -11 \end{cases}$

Solutions

a. All three equations are in **standard form**, so we can write the **augmented matrix**.

$$\begin{cases} 2x + 1y - 3z = -3 \\ 1x + 2y + 1z = 4 \\ -3x + 1y - 4z = 1 \end{cases} \rightarrow \left[\begin{array}{ccc|c} 2 & 1 & -3 & -3 \\ 1 & 2 & 1 & 4 \\ -3 & 1 & -4 & 1 \end{array}\right]$$

We now apply **row operations** to put the matrix in **row-echelon form**.

Step 1   We obtain a 1 in row 1, column 1 by interchanging row 1 and row 2.

$$\left[\begin{array}{ccc|c} 2 & 1 & -3 & -3 \\ 1 & 2 & 1 & 4 \\ -3 & 1 & -4 & 1 \end{array}\right] \xrightarrow{R_1 \Leftrightarrow R_2} \left[\begin{array}{ccc|c} 1 & 2 & 1 & 4 \\ 2 & 1 & -3 & -3 \\ -3 & 1 & -4 & 1 \end{array}\right]$$

Step 2   We get a 0 in row 2, column 1 by multiplying the entries in row 1 by -2 and adding the results to the entries in row 2. We get a 0 in row 3, column 1 by multiplying the entries in row 1 by 3 and adding the results to the entries in row 3.

$$\left[\begin{array}{ccc|c} 1 & 2 & 1 & 4 \\ 2 & 1 & -3 & -3 \\ -3 & 1 & -4 & 1 \end{array}\right] \xrightarrow[3R_1 + R_3]{-2R_1 + R_2} \left[\begin{array}{ccc|c} 1 & 2 & 1 & 4 \\ 0 & -3 & -5 & -11 \\ 0 & 7 & -1 & 13 \end{array}\right]$$

Step 3 We get a 1 in row 2, column 2 by multiplying the entries in row 2 by $-\frac{1}{3}$.

$$\begin{bmatrix} 1 & 2 & 1 & | & 4 \\ 0 & -3 & -5 & | & -11 \\ 0 & 7 & -1 & | & 13 \end{bmatrix} \xrightarrow{-\frac{1}{3}R_2} \begin{bmatrix} 1 & 2 & 1 & | & 4 \\ 0 & 1 & \frac{5}{3} & | & \frac{11}{3} \\ 0 & 7 & -1 & | & 13 \end{bmatrix}$$

Step 4 We get a 0 in row 3, column 2 by multiplying the entries in row 2 by -7 and adding the results to the entries in row 3.

$$\begin{bmatrix} 1 & 2 & 1 & | & 4 \\ 0 & 1 & \frac{5}{3} & | & \frac{11}{3} \\ 0 & 7 & -1 & | & 13 \end{bmatrix} \xrightarrow{-7R_2 + R_3} \begin{bmatrix} 1 & 2 & 1 & | & 4 \\ 0 & 1 & \frac{5}{3} & | & \frac{11}{3} \\ 0 & 0 & -\frac{38}{3} & | & -\frac{38}{3} \end{bmatrix}$$

Step 5 We get a 1 in row 3, column 3 by multiplying the entries in row 3 by $-\frac{3}{38}$.

$$\begin{bmatrix} 1 & 2 & 1 & | & 4 \\ 0 & 1 & \frac{5}{3} & | & \frac{11}{3} \\ 0 & 0 & -\frac{38}{3} & | & -\frac{38}{3} \end{bmatrix} \xrightarrow{-\frac{3}{38}R_3} \begin{bmatrix} 1 & 2 & 1 & | & 4 \\ 0 & 1 & \frac{5}{3} & | & \frac{11}{3} \\ 0 & 0 & 1 & | & 1 \end{bmatrix}$$

The corresponding **equivalent system** is

$$\begin{cases} x + 2y + z = 4 \\ y + \frac{5}{3}z = \frac{11}{3} \\ z = 1 \end{cases}$$

The third equation tells us that $z = 1$. To find the value for y, we can **substitute** 1 for z in the second equation and solve for y.

$$\text{Second equation:} \quad y + \frac{5}{3}z = \frac{11}{3}$$

$$\text{Substitute 1 for } z: \quad y + \frac{5}{3}(1) = \frac{11}{3}$$

$$\text{Multiply:} \quad y + \frac{5}{3} = \frac{11}{3}$$

$$\text{Subtract } \frac{5}{3} \text{ from both sides:} \quad y = \frac{6}{3} = 2$$

We now know that $y = 2$ and $z = 1$. We can find the value of x by **substituting** these values into the first equation and solving for x.

$$\text{First equation:} \quad x + 2y + z = 4$$

$$\text{Substitute 2 for } y \text{ and 1 for } z: \quad x + 2(2) + (1) = 4$$

$$\text{Multiply:} \quad x + 4 + 1 = 4$$

$$\text{Combine like terms:} \quad x + 5 = 4$$

$$\text{Subtract 5 from both sides:} \quad x = -1$$

The **solution** to the system is the ordered triple $(-1, 2, 1)$.

📄 *My video summary* ▶ **b.** Start by writing the equations in the system in **standard form**.

$$\begin{cases} 2x + 2y + z = 1 \\ 5x + 2y = 30 + 3z \\ 3x + 4z = -11 \end{cases} \rightarrow \begin{cases} 2x + 2y + z = 1 \\ 5x + 2y - 3z = 30 \\ 3x + 0y + 4z = -11 \end{cases}$$

Now write the corresponding **augmented matrix**.

$$\begin{cases} 2x + 2y + 1z = 1 \\ 5x + 2y - 3z = 30 \\ 3x + 0y + 4z = -11 \end{cases} \rightarrow \left[\begin{array}{ccc|c} 2 & 2 & 1 & 1 \\ 5 & 2 & -3 & 30 \\ 3 & 0 & 4 & -11 \end{array}\right]$$

Use **row operations** to solve the system on your own. View the **answer**, or watch this **video** for a complete solution.

🔺 **You Try It** Work through this **You Try It** problem.

Work Exercises 19–21 in this eText or in the MyMathLab Study Plan.

📄 *My video summary* ▶ **Example 5 Solving a Special System of Three Equations Using Matrices**

Use matrices to solve the following system.

$$\begin{cases} x + 2y - 3z = 5 \\ -2x - 3y + 4z = -6 \\ 2x + 4y - 6z = 10 \end{cases}$$

Solution Work through the following, or watch this **video** for the complete solution.

All three equations are in **standard form**, so we can write the **augmented matrix**.

$$\begin{cases} 1x + 2y - 3z = 5 \\ -2x - 3y + 4z = -6 \\ 2x + 4y - 6z = 10 \end{cases} \rightarrow \left[\begin{array}{ccc|c} 1 & 2 & -3 & 5 \\ -2 & -3 & 4 & -6 \\ 2 & 4 & -6 & 10 \end{array}\right]$$

We now apply **row operations** to put the matrix in **row-echelon form**.

Step 1 There is already a 1 in row 1, column 1.

Step 2 We get a 0 in row 2, column 1 by multiplying the entries in row 1 by 2 and adding the results to the entries in row 2. We get a 0 in row 3, column 1 by multiplying the entries in row 1 by −2 and adding the results to the entries in row 3.

$$\left[\begin{array}{ccc|c} 1 & 2 & -3 & 5 \\ -2 & -3 & 4 & -6 \\ 2 & 1 & -6 & 10 \end{array}\right] \xrightarrow[\substack{-2R_1 + R_3}]{-2R_1 + R_2} \left[\begin{array}{ccc|c} 1 & 2 & -3 & 5 \\ 0 & 1 & -2 & 4 \\ 0 & 0 & 0 & 0 \end{array}\right]$$

Notice that the last row is all zeros. This corresponds to the **identity** $0 = 0$ and indicates the system may have an infinite number of solutions.

The corresponding **equivalent system** is

$$\begin{cases} x + 2y - 3z = 5 \\ y - 2z = 4. \\ 0 = 0 \end{cases}$$

The third equation is an identity so provides no information about the solution. The second equation can be solved for y in terms of z.

$$\text{Second equation:} \quad y - 2z = 4$$

$$\text{Add } 2z \text{ to both sides:} \quad y = 2z + 4$$

We now know that $y = 2z + 4$. We can **substitute** $2z + 4$ for y in the first equation, then solve the equation for x in terms of z.

$$\text{First equation:} \quad x + 2y - 3z = 5$$

$$\text{Substitute } 2z + 4 \text{ for } y\text{:} \quad x + 2(2z + 4) - 3z = 5$$

$$\text{Distribute:} \quad x + 4z + 8 - 3z = 5$$

$$\text{Combine like terms:} \quad x + z + 8 = 5$$

$$\text{Subtract 8 from both sides:} \quad x + z = -3$$

$$\text{Subtract } z \text{ from both sides:} \quad x = -z - 3$$

Having expressed both x and y in terms of the third variable z, the system has an infinite number of solutions. We can express the **solutions** to the system as the **ordered triple** $(-z - 3, 2z + 4, z)$, where z is any real number.

 You Try It Work through this You Try It problem.

Work Exercises 22–30 in this eText or in the MyMathLab Study Plan.

Unlike the two equation case, if we obtain a row of all zeros when solving a system of three equations, there is not necessarily an infinite number of solutions. We must make sure that no other row yields a **contradiction**. Consider the following.

$$\begin{bmatrix} 1 & 3 & 0 & 2 \\ 0 & 0 & 0 & 3 \\ 0 & 0 & 0 & 0 \end{bmatrix}$$

The third row corresponds to the equation $0 = 0$, which is an **identity**. However, the second row corresponds to the equation $0 = 3$, which is a contradiction. If any row yields a contradiction, the system has no solution. Therefore, the system represented by this **matrix** has no solution.

B   Exercises

In Exercises 1–6, write the corresponding augmented matrix for each system.

1. $\begin{cases} 2x + 3y = 5 \\ 5x - 2y = 1 \end{cases}$

2. $\begin{cases} 5x - y = 3 \\ 2y = x + 1 \end{cases}$

3. $\begin{cases} \dfrac{1}{2}x - \dfrac{1}{3}y = 0 \\ \dfrac{2}{7}x + \dfrac{5}{9}y = 2 \end{cases}$

4. $\begin{cases} 2x + 3y - 2z = 7 \\ 5x - 2y + z = 4 \\ 3x + 2y - 4z = 1 \end{cases}$

5. $\begin{cases} 7x - 2y + 5z = 3 \\ 5y + 2z = -2 \\ 3x - 2z = 1 \end{cases}$

6. $\begin{cases} 2x = 1 - 3y \\ 5y = 2 + 2z \\ 3z = 2x - 5 \end{cases}$

In Exercises 7–30, solve each system of linear equations using matrices.

7. $\begin{cases} x + 3y = -6 \\ 3x + 2y = -4 \end{cases}$

8. $\begin{cases} x - 5y = 6 \\ 5x - y = 6 \end{cases}$

9. $\begin{cases} 3x + 2y = 4 \\ 6x + 5y = 7 \end{cases}$

10. $\begin{cases} \dfrac{1}{2}x + 2y = 8 \\ 3x - \dfrac{1}{3}y = 11 \end{cases}$

11. $\begin{cases} 2x - \dfrac{1}{2}y = 11 \\ 4x + 3y = 14 \end{cases}$

12. $\begin{cases} 3x - 2y = -1 \\ -6x + 4y = 2 \end{cases}$

13. $\begin{cases} 2x + 3y = 2 \\ 4x - 6y = 0 \end{cases}$

14. $\begin{cases} 3x - 2y = 3 \\ 6x + 6y = 1 \end{cases}$

15. $\begin{cases} 2x + \dfrac{1}{2}y = 3 \\ y = 7 - 4x \end{cases}$

16. $\begin{cases} 3y = 13 + 2x \\ 6x + 2y = -6 \end{cases}$

17. $\begin{cases} \dfrac{1}{2}x + \dfrac{1}{3}y = 3 \\ x + y = 7 \end{cases}$

18. $\begin{cases} 3x + 6y = 3 \\ 2y = 1 - x \end{cases}$

19. $\begin{cases} 2x - 3y + 4z = 4 \\ 6x + 4y - 2z = 0 \\ 4x - 2y - 3z = 11 \end{cases}$

20. $\begin{cases} x + 2y + 3z = -13 \\ 2x + y - 2z = 5 \\ 3x - 4y + z = 27 \end{cases}$

21. $\begin{cases} x + 2y + z = 9 \\ 2x - y + 3z = 5 \\ 3x + 2y - z = 7 \end{cases}$

22. $\begin{cases} x + 2y + 2z = 3 \\ 2x + y - 2z = -3 \\ 7x + 8y + 2z = 3 \end{cases}$

23. $\begin{cases} \dfrac{1}{2}x + y - 2z = -3 \\ 3x - \dfrac{1}{3}y + z = 15 \\ 2x - 5y + \dfrac{1}{4}z = -6 \end{cases}$

24. $\begin{cases} x + y + z = 0 \\ x + 2y + 2z = -1 \\ 2x + 2y - z = -3 \end{cases}$

25. $\begin{cases} x + 2y + z = 4 \\ 2x + y + 2z = 2 \\ x + 7y + 5z = 14 \end{cases}$

26. $\begin{cases} x + y + 2z = 2 \\ 3x - y - z = 1 \\ 5x + y + 3z = 3 \end{cases}$

27. $\begin{cases} x - z = 1 \\ -x + 2y - 3z = 12 \\ 2x - 4y = -6 \end{cases}$

28. $\begin{cases} 3x + 2y - 2z = -1 \\ x - 2y + 2z = 2 \\ x - y + 2z = 0 \end{cases}$

29. $\begin{cases} -2x - y + z = 7 \\ 4x - 6y + 8z = -4 \\ x - 2y + 3z = 0 \end{cases}$

30. $\begin{cases} 4x + 2y + 3z = 9 \\ 3x - y + z = 8 \\ x + y + z = 2 \end{cases}$

Determinants and Cramer's Rule

THINGS TO KNOW

Before working through this section, be sure you are familiar with the following concepts:

| | | VIDEO | ANIMATION | INTERACTIVE |

 1. Solve Systems of Linear Equations in Two Variables by Elimination (Section 3.1, Objective 4) ▶

 2. Solve Systems of Linear Equations in Three Variables (Section 3.2, Objective 2) ▶

 3. Write an Augmented Matrix (Appendix B, Objective 1) ▶

 4. Solve Systems of Two Equations Using Matrices (Appendix B, Objective 2) ▶

 5. Solve Systems of Three Equations Using Matrices (Appendix B, Objective 3) ▶

OBJECTIVES

1 Evaluate a 2 × 2 Determinant

2 Use Cramer's Rule to Solve a System of Linear Equations in Two Variables

3 Evaluate a 3 × 3 Determinant

4 Use Cramer's Rule to Solve a System of Linear Equations in Three Variables

....

OBJECTIVE 1   EVALUATE A 2 × 2 DETERMINANT

A **square matrix** has an equal number of rows and columns. For example, matrices A and B shown below are square matrices. A is a 2 × 2 matrix, while B is a 3 × 3 matrix.

$$A = \begin{bmatrix} 2 & -3 \\ 1 & 4 \end{bmatrix} \qquad B = \begin{bmatrix} 2 & 1 & -1 \\ 3 & 6 & -2 \\ 5 & 4 & 1 \end{bmatrix}$$

Every square matrix has a **real number** associated with it called its *determinant*. We will first define the determinant of a 2 × 2 matrix.

(eText Screens C.1-1–C.1-21)

Definition Determinant of a 2 × 2 Matrix

The **determinant** of a 2 × 2 matrix $\begin{bmatrix} a & b \\ c & d \end{bmatrix}$ is denoted by $\begin{vmatrix} a & b \\ c & d \end{vmatrix}$ and is defined

by $\begin{vmatrix} a & b \\ c & d \end{vmatrix} = ad - cb$.

To evaluate a 2 × 2 matrix, we find the **difference** between the **products** of the two diagonals. This is illustrated in the following diagram:

$$\begin{vmatrix} a & b \\ c & d \end{vmatrix} = ad - cb$$

Pay close attention to the notation for matrices and **determinants.** $\begin{bmatrix} a & b \\ c & d \end{bmatrix}$ is a

matrix, while $\begin{vmatrix} a & b \\ c & d \end{vmatrix}$ is the determinant of $\begin{bmatrix} a & b \\ c & d \end{bmatrix}$.

Example 1 Evaluating 2 × 2 Determinants

Evaluate each determinant.

a. $\begin{vmatrix} 5 & 2 \\ 8 & 7 \end{vmatrix}$
b. $\begin{vmatrix} -9 & 3 \\ -4 & 2 \end{vmatrix}$

Solutions

a. $\begin{vmatrix} 5 & 2 \\ 8 & 7 \end{vmatrix} = 5 \cdot 7 - 8 \cdot 2 = 35 - 16 = 19$

b. $\begin{vmatrix} -9 & 3 \\ -4 & 2 \end{vmatrix} = (-9)(2) - (-4)(3) = -18 + 12 = -6$

You Try It Work through this You Try It problem.

Work Exercises 1–4 in this eText or in the MyMathLab Study Plan.

OBJECTIVE 2 USE CRAMER'S RULE TO SOLVE A SYSTEM OF LINEAR EQUATIONS IN TWO VARIABLES

Consider the following system of linear equations in two variables:

$$\begin{cases} ax + by = m \\ cx + dy = n \end{cases}$$

where a, b, c, and d are numerical **coefficients,** and m and n are **constants.**

Cramer's Rule for solving a system of linear equations in two variables uses the following three **determinants** that are defined by the coefficients and constants from the equations.

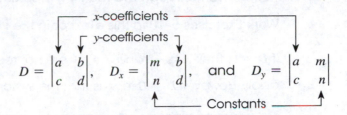

$$D = \begin{vmatrix} a & b \\ c & d \end{vmatrix}, \quad D_x = \begin{vmatrix} m & b \\ n & d \end{vmatrix}, \quad \text{and} \quad D_y = \begin{vmatrix} a & m \\ c & n \end{vmatrix}$$

Cramer's Rule for Solving Systems of Linear Equations in Two Variables

For the system of linear equations $\begin{cases} ax + by = m \\ cx + dy = n \end{cases}$, the solution is given by

$$x = \frac{D_x}{D} = \frac{\begin{vmatrix} m & b \\ n & d \end{vmatrix}}{\begin{vmatrix} a & b \\ c & d \end{vmatrix}} \quad \text{and} \quad y = \frac{D_y}{D} = \frac{\begin{vmatrix} a & m \\ c & n \end{vmatrix}}{\begin{vmatrix} a & b \\ c & d \end{vmatrix}} \quad \text{provided} \quad D = \begin{vmatrix} a & b \\ c & d \end{vmatrix} \neq 0.$$

My video summary ▶ **Concept Video** Watch this video to see Cramer's Rule derived.

Example 2 Using Cramer's Rule to Solve a Linear System in Two Variables

Use Cramer's Rule to solve each system.

a. $\begin{cases} 3x + 2y = 12 \\ 4x - 5y = -7 \end{cases}$
b. $\begin{cases} 6x + y = -2 \\ 9x - 2y = 11 \end{cases}$

Solutions

a. Begin by finding D, D_x, and D_y.

$$D = \begin{vmatrix} 3 & 2 \\ 4 & -5 \end{vmatrix} = 3(-5) - 4(2) = -15 - 8 = -23$$

$$D_x = \begin{vmatrix} 12 & 2 \\ -7 & -5 \end{vmatrix} = 12(-5) - (-7)(2) = -60 + 14 = -46$$

$$D_y = \begin{vmatrix} 3 & 12 \\ 4 & -7 \end{vmatrix} = 3(-7) - 4(12) = -21 - 48 = -69$$

Then $x = \dfrac{D_x}{D} = \dfrac{-46}{-23} = 2$ and $y = \dfrac{D_y}{D} = \dfrac{-69}{-23} = 3$. So, the solution to the system

is the ordered pair $(2, 3)$. View the **check**.

My video summary ▶ **b.** Try solving this system on your own. View the **answer**, or watch this **video** for a complete solution.

You Try It Work through this You Try It problem.

Work Exercises 5–10 in this eText or in the MyMathLab Study Plan.

If $D = 0$, then $x = \dfrac{D_x}{D}$ and $y = \dfrac{D_y}{D}$ are not real numbers, so Cramer's Rule does not apply. However, $D = 0$ tells us that the system is either inconsistent or dependent.

Cramer's Rule with Inconsistent and Dependent Systems

For the system of linear equations $\begin{cases} ax + by = m \\ cx + dy = n \end{cases}$, if $D = \begin{vmatrix} a & b \\ c & d \end{vmatrix} = 0$, then the system is either inconsistent or dependent.

1. If $D = \begin{vmatrix} a & b \\ c & d \end{vmatrix} = 0$ and at least one of the determinants $D_x = \begin{vmatrix} m & b \\ n & d \end{vmatrix}$ or $D_y = \begin{vmatrix} a & m \\ c & n \end{vmatrix}$ is not 0, then the system is inconsistent and the solution set is $\varnothing$.

2. If $D = \begin{vmatrix} a & b \\ c & d \end{vmatrix} = 0$ and the determinants $D_x = \begin{vmatrix} m & b \\ n & d \end{vmatrix}$ and $D_y = \begin{vmatrix} a & m \\ c & n \end{vmatrix}$ are both 0, then the system is dependent and has infinitely many solutions.

Example 3 Using Cramer's Rule to Solve a Linear System in Two Variables

Use Cramer's Rule to solve each system.

a. $\begin{cases} 2x - 5y = 7 \\ -4x + 10y = 11 \end{cases}$ 　　　　 **b.** $\begin{cases} 6x + 10y = -50 \\ -9x - 15y = 75 \end{cases}$

My video summary ▶ **Solutions**

a. Because $D = \begin{vmatrix} 2 & -5 \\ -4 & 10 \end{vmatrix} = 2(10) - (-4)(-5) = 20 - 20 = 0$, the system is either inconsistent or dependent. Because $D_x = \begin{vmatrix} 7 & -5 \\ 11 & 10 \end{vmatrix} = 7(10) - 11(-5) = 70 + 55 = 125 \neq 0$, the system is inconsistent. The solution set is $\varnothing$.

b. Try solving this system on your own. View the **answer**, or watch this **video** for a complete solution.

You Try It Work through this You Try It problem.

Work Exercises 11 and 12 in this eText or in the MyMathLab Study Plan.

OBJECTIVE 3   EVALUATE A 3 × 3 DETERMINANT

For every entry of a **3 × 3 determinant**, there is a **2 × 2 determinant** associated with it called its **minor**. The minor is the 2 × 2 determinant obtained by deleting the row and the column in which the entry appears. For example, consider the following 3 × 3 determinant.

$$\begin{vmatrix} a_1 & b_1 & c_1 \\ a_2 & b_2 & c_2 \\ a_3 & b_3 & c_3 \end{vmatrix}$$

Entry a_1 is located in the first row and first column of the determinant. To find the minor associated with a_1, delete the first row and the first column to obtain a 2 × 2 determinant.

$$\begin{vmatrix} a_1 & b_1 & c_1 \\ a_2 & b_2 & c_2 \\ a_3 & b_3 & c_3 \end{vmatrix}$$ The minor associated with a_1 is $\begin{vmatrix} b_2 & c_2 \\ b_3 & c_3 \end{vmatrix}$.

Similarly, the entry b_1 is located in the first row and second column. To find the minor associated with b_1, delete the first row and second column.

$$\begin{vmatrix} a_1 & b_1 & c_1 \\ a_2 & b_2 & c_2 \\ a_3 & b_3 & c_3 \end{vmatrix}$$ The minor associated with b_1 is $\begin{vmatrix} a_2 & c_2 \\ a_3 & c_3 \end{vmatrix}$.

To evaluate a 3 × 3 determinant, we use a technique called **expansion by minors**. In the definition below, we expand across the first row.

Definition   **Determinant of a 3 × 3 Matrix**

The **determinant** of a 3 × 3 matrix $\begin{bmatrix} a_1 & b_1 & c_1 \\ a_2 & b_2 & c_2 \\ a_3 & b_3 & c_3 \end{bmatrix}$ is denoted by $\begin{vmatrix} a_1 & b_1 & c_1 \\ a_2 & b_2 & c_2 \\ a_3 & b_3 & c_3 \end{vmatrix}$

and is defined by

$$\begin{vmatrix} a_1 & b_1 & c_1 \\ a_2 & b_2 & c_2 \\ a_3 & b_3 & c_3 \end{vmatrix} = a_1 \begin{vmatrix} b_2 & c_2 \\ b_3 & c_3 \end{vmatrix} \underset{\text{Subtract}}{-} b_1 \begin{vmatrix} a_2 & c_2 \\ a_3 & c_3 \end{vmatrix} \underset{\text{Add}}{+} c_1 \begin{vmatrix} a_2 & b_2 \\ a_3 & b_3 \end{vmatrix}.$$

Example 4   Evaluating 3 × 3 Determinants

Evaluate the determinant $\begin{vmatrix} 3 & 1 & -2 \\ -1 & 2 & 4 \\ 0 & 5 & 1 \end{vmatrix}$.

Solution To expand across the first row, begin by finding the minor associated with each entry in the first row.

$$\begin{vmatrix} 3 & 1 & -2 \\ -1 & 2 & 4 \\ 0 & 5 & 1 \end{vmatrix}$$ The minor associated with 3 is $\begin{vmatrix} 2 & 4 \\ 5 & 1 \end{vmatrix}$.

$$\begin{vmatrix} 3 & 1 & -2 \\ -1 & 2 & 4 \\ 0 & 5 & 1 \end{vmatrix}$$
The minor associated with 1 is $\begin{vmatrix} -1 & 4 \\ 0 & 1 \end{vmatrix}$.

$$\begin{vmatrix} 3 & 1 & -2 \\ -1 & 2 & 4 \\ 0 & 5 & 1 \end{vmatrix}$$
The minor associated with -2 is $\begin{vmatrix} -1 & 2 \\ 0 & 5 \end{vmatrix}$.

So, $\begin{vmatrix} 3 & 1 & -2 \\ -1 & 2 & 4 \\ 0 & 5 & 1 \end{vmatrix} = 3\begin{vmatrix} 2 & 4 \\ 5 & 1 \end{vmatrix} - 1\begin{vmatrix} -1 & 4 \\ 0 & 1 \end{vmatrix} + (-2)\begin{vmatrix} -1 & 2 \\ 0 & 5 \end{vmatrix}$

$$= 3(2 \cdot 1 - 5 \cdot 4) - 1(-1 \cdot 1 - 0 \cdot 4) + (-2)(-1 \cdot 5 - 0 \cdot 2)$$

$$= 3(2 - 20) - 1(-1 - 0) + (-2)(-5 - 0)$$

$$= 3(-18) - 1(-1) + (-2)(-5)$$

$$= -54 + 1 + 10$$

$$= -43$$

You Try It Work through this You Try It problem.

Work Exercise 13 in this eText or in the MyMathLab Study Plan.

In Example 4, we evaluated the **determinant** by expanding the minors across the first row. Actually, we can find the value of a determinant by expanding the minors across any row or down any column. However, we must be careful to properly add or subtract the product of each row or column entry and its minor.

If the **sum** of the row number and column number of an entry is **even**, then the **product** of the entry and its minor is "added." If the sum of the row number and column number of an entry is **odd**, then the product is "subtracted." These additions and subtractions are summarized below for 3×3 determinants:

$$\begin{vmatrix} + & - & + \\ - & + & - \\ + & - & + \end{vmatrix}.$$

When calculating determinants, it is typically easier to expand by minors for rows or columns that contain zeros. In Example 5, we evaluate a determinant by expanding the minors for different rows and columns.

My interactive video summary

Example 5 Evaluating 3×3 Determinants

Evaluate the **determinant** by expanding the minors for the given row or column.

$$\begin{vmatrix} 2 & 4 & -3 \\ 3 & 1 & 0 \\ -1 & -2 & 5 \end{vmatrix}$$

a. First column   **b.** Second row   **c.** Third column

Solutions

a. The entries down the first column are 2, 3, and -1. The corresponding minors are given below.

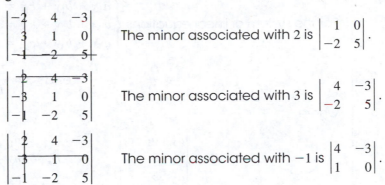

The minor associated with 2 is $\begin{vmatrix} 1 & 0 \\ -2 & 5 \end{vmatrix}$.

The minor associated with 3 is $\begin{vmatrix} 4 & -3 \\ -2 & 5 \end{vmatrix}$.

The minor associated with -1 is $\begin{vmatrix} 4 & -3 \\ 1 & 0 \end{vmatrix}$.

The additions and subtractions that correspond to the first-column entries are $+$, $-$, and $+$.

So, $\begin{vmatrix} 2 & 4 & -3 \\ 3 & 1 & 0 \\ -1 & -2 & 5 \end{vmatrix} = 2\begin{vmatrix} 1 & 0 \\ -2 & 5 \end{vmatrix} - 3\begin{vmatrix} 4 & -3 \\ -2 & 5 \end{vmatrix} + (-1)\begin{vmatrix} 4 & -3 \\ 1 & 0 \end{vmatrix}$

$$= 2(5-0) - 3(20-6) + (-1)(0+3)$$

$$= 10 - 42 - 3$$

$$= -35$$

b. The entries in the second row are 3, 1, and 0, and the corresponding minors are $\begin{vmatrix} 4 & -3 \\ -2 & 5 \end{vmatrix}$, $\begin{vmatrix} 2 & -3 \\ -1 & 5 \end{vmatrix}$, and $\begin{vmatrix} 2 & 4 \\ -1 & -2 \end{vmatrix}$. The additions and subtractions that correspond to the second-row entries are $-$, $+$, and $-$.

So, $\begin{vmatrix} 2 & 4 & -3 \\ 3 & 1 & 0 \\ -1 & -2 & 5 \end{vmatrix} = -3\begin{vmatrix} 4 & -3 \\ -2 & 5 \end{vmatrix} + 1\begin{vmatrix} 2 & -3 \\ -1 & 5 \end{vmatrix} - 0\begin{vmatrix} 2 & 4 \\ -1 & -2 \end{vmatrix}$.

Finish simplifying on your own to obtain the value -35, or watch this **interactive video** for complete solutions to all three parts.

c. Try working this expansion on your own to obtain the value -35, or watch this **interactive video** for complete solutions to all three parts.

You Try It Work through this **You Try It** problem.

Work Exercises 14–16 in this eText or in the MyMathLab Study Plan.

OBJECTIVE 4 USE CRAMER'S RULE TO SOLVE A SYSTEM OF LINEAR EQUATIONS IN THREE VARIABLES

We can use **Cramer's Rule** to solve a system of linear equation in three variables using **3 × 3 determinants** defined by the **coefficients** and **constants** from the equations.

Cramer's Rule for Solving Systems of Linear Equations in Three Variables

For the system of linear equations $\begin{cases} a_1x + b_1y + c_1z = n_1 \\ a_2x + b_2y + c_2z = n_2, \\ a_3x + b_3y + c_3z = n_3 \end{cases}$ the solution is given by

$$x = \frac{D_x}{D}, \quad y = \frac{D_y}{D} \quad \text{and} \quad z = \frac{D_z}{D},$$

where

$$D = \begin{vmatrix} a_1 & b_1 & c_1 \\ a_2 & b_2 & c_2 \\ a_3 & b_3 & c_3 \end{vmatrix}, \qquad D_x = \begin{vmatrix} n_1 & b_1 & c_1 \\ n_2 & b_2 & c_2 \\ n_3 & b_3 & c_3 \end{vmatrix},$$

$$D_y = \begin{vmatrix} a_1 & n_1 & c_1 \\ a_2 & n_2 & c_2 \\ a_3 & n_3 & c_3 \end{vmatrix}, \quad \text{and} \quad D_z = \begin{vmatrix} a_1 & b_1 & n_1 \\ a_2 & b_2 & n_2 \\ a_3 & b_3 & n_3 \end{vmatrix},$$

provided $D \neq 0$.

Example 6 Using Cramer's Rule to Solve a Linear System in Three Variables

Use Cramer's Rule to solve each system.

a. $\begin{cases} x - y + 4z = -5 \\ 6x - 2y + 5z = 1 \\ 3x + 2y + 2z = 10 \end{cases}$
b. $\begin{cases} x + y + z = 3 \\ 2x - 3y = 14 \\ 4y + 5z = -3 \end{cases}$

Solutions

a. Begin by finding $D, D_x, D_y,$ and D_z.

$$D = \begin{vmatrix} 1 & -1 & 4 \\ 6 & -2 & 5 \\ 3 & 2 & 2 \end{vmatrix} = 1\begin{vmatrix} -2 & 5 \\ 2 & 2 \end{vmatrix} - (-1)\begin{vmatrix} 6 & 5 \\ 3 & 2 \end{vmatrix} + 4\begin{vmatrix} 6 & -2 \\ 3 & 2 \end{vmatrix}$$

$$= 1(-4 - 10) - (-1)(12 - 15) + 4(12 - (-6))$$

$$= -14 - 3 + 72$$

$$= 55$$

$$D_x = \begin{vmatrix} -5 & -1 & 4 \\ 1 & -2 & 5 \\ 10 & 2 & 2 \end{vmatrix} = -5\begin{vmatrix} -2 & 5 \\ 2 & 2 \end{vmatrix} - (-1)\begin{vmatrix} 1 & 5 \\ 10 & 2 \end{vmatrix} + 4\begin{vmatrix} 1 & -2 \\ 10 & 2 \end{vmatrix}$$

$$= -5(-4 - 10) - (-1)(2 - 50) + 4(2 - (-20))$$

$$= 70 - 48 + 88$$

$$= 110$$

$$D_y = \begin{vmatrix} 1 & -5 & 4 \\ 6 & 1 & 5 \\ 3 & 10 & 2 \end{vmatrix} = 1\begin{vmatrix} 1 & 5 \\ 10 & 2 \end{vmatrix} - (-5)\begin{vmatrix} 6 & 5 \\ 3 & 2 \end{vmatrix} + 4\begin{vmatrix} 6 & 1 \\ 3 & 10 \end{vmatrix}$$

$$= 1(2 - 50) - (-5)(12 - 15) + 4(60 - 3)$$

$$= -48 - 15 + 228$$

$$= 165$$

$$D_z = \begin{vmatrix} 1 & -1 & -5 \\ 6 & -2 & 1 \\ 3 & 2 & 10 \end{vmatrix} = 1\begin{vmatrix} -2 & 1 \\ 2 & 10 \end{vmatrix} - (-1)\begin{vmatrix} 6 & 1 \\ 3 & 10 \end{vmatrix} + (-5)\begin{vmatrix} 6 & -2 \\ 3 & 2 \end{vmatrix}$$

$$= 1(-20 - 2) - (-1)(60 - 3) + (-5)(12 - (-6))$$

$$= -22 + 57 - 90$$

$$= -55$$

Then $x = \dfrac{D_x}{D} = \dfrac{110}{55} = 2$, $y = \dfrac{D_y}{D} = \dfrac{165}{55} = 3$, and $z = \dfrac{D_z}{D} = \dfrac{-55}{55} = -1$. So, the ordered-triple solution is $(2, 3, -1)$. View the **check**.

My video summary ▶ **b.** Try solving this system on your own. View the **answer**, or watch this **video** for a complete solution.

Like with the two-variable case, if $D = 0$, then $x = \dfrac{D_x}{D}$, $y = \dfrac{D_y}{D}$, and $z = \dfrac{D_z}{D}$ are not real numbers, and **Cramer's Rule** does not apply. But, $D = 0$ tells us that the system is either **inconsistent** or **dependent**.

Cramer's Rule with Inconsistent and Dependent Systems in Three Variables

If $D = 0$, then the system is either inconsistent or dependent.

1. If $D = 0$ and at least one of the determinants D_x, D_y, or D_z is not 0, then the system is inconsistent and the solution set is $\varnothing$.

2. If $D = 0$ and the determinants D_x, D_y, and D_z are all 0, then the system is dependent and has infinitely many solutions. The solutions to these systems can be found using the methods from **Section 3.2** or **Appendix B**.

You Try It Work through this You Try It problem.

Work Exercises 17–22 in this eText or in the MyMathLab Study Plan.

C   Exercises

In Exercises 1–4, evaluate each 2 × 2 determinant.

1. $\begin{vmatrix} 7 & 3 \\ 8 & 6 \end{vmatrix}$

2. $\begin{vmatrix} 9 & -7 \\ 4 & -2 \end{vmatrix}$

3. $\begin{vmatrix} 3 & -6 \\ -2 & 4 \end{vmatrix}$

4. $\begin{vmatrix} 10 & 1 \\ 6 & -5 \end{vmatrix}$

In Exercises 5–12, use Cramer's Rule to solve each system of linear equations in two variables.

5. $\begin{cases} x + y = 3 \\ -2x + y = 0 \end{cases}$

6. $\begin{cases} 2x - y = 1 \\ -3x + 3y = 0 \end{cases}$

7. $\begin{cases} 4x - y = -4 \\ -3x + 3y = -6 \end{cases}$

8. $\begin{cases} 4x + 5y = 51 \\ 3x - 2y = -2 \end{cases}$

9. $\begin{cases} 2x - 3y = 0 \\ 4x + 9y = 5 \end{cases}$

10. $\begin{cases} 3x + 4y = -2 \\ 6x - 3y = 7 \end{cases}$

11. $\begin{cases} -9x + 3y = -15 \\ 6x - 2y = 10 \end{cases}$

12. $\begin{cases} 10x - 15y = 12 \\ -4x + 6y = 9 \end{cases}$

In Exercises 13–16, evaluate each 3 × 3 determinant.

13. $\begin{vmatrix} 7 & -3 & 8 \\ 1 & -5 & -2 \\ 3 & 7 & 4 \end{vmatrix}$

14. $\begin{vmatrix} 2 & 5 & 1 \\ 4 & 0 & 3 \\ -6 & 1 & 0 \end{vmatrix}$

15. $\begin{vmatrix} 4 & 0 & 7 \\ -1 & 3 & -1 \\ 5 & 2 & 6 \end{vmatrix}$

16. $\begin{vmatrix} -5 & -1 & -2 \\ 8 & 10 & 0 \\ -7 & 5 & 0 \end{vmatrix}$

In Exercises 17–22, use Cramer's Rule to solve each system of linear equations in three variables.

17. $\begin{cases} x + y - z = 4 \\ 5x - y + 3z = 6 \\ x + y - 5z = 8 \end{cases}$

18. $\begin{cases} 3x + 3y + 4z = 6 \\ 4x - y + z = 8 \\ -x + y - 3z = -2 \end{cases}$

19. $\begin{cases} -2x + 5y + 5z = -4 \\ 5x + 6y + 5z = -23 \\ -3x + 4y + 5z = -3 \end{cases}$

20. $\begin{cases} x - z = -1 \\ 2y - z = 8 \\ 2x + 2y = 0 \end{cases}$

21. $\begin{cases} -x + 2y + z = 2 \\ 3x + y - z = -4 \\ x + 5y + z = 5 \end{cases}$

22. $\begin{cases} 4x - 3y + z = 4 \\ 10x - 9y = 8 \\ 6y + 2z = 0 \end{cases}$

Glossary

4th Root A real number b is the 4th root of a real number a provided that $b^4 = a$. Every positive real number a has two real 4th roots, one positive, $\sqrt[4]{a}$ (the principal 4th root), and one negative, $-\sqrt[4]{a}$.

5th Root A real number b is the 5th root of a real number a provided that $b^5 = a$.

Absolute Value The absolute value of a real number a, denoted as $|a|$, is the distance from 0 to a on the real number line.

ac Method A method of factoring a trinomial that involves splitting up the middle term in order to factor by grouping. Also known as the *grouping method* or *expansion method*.

Addends The numbers being added are called addends.

Addition Method See *Elimination Method*.

Addition Property of Equality Let a, b, and c be numbers or algebraic expressions. Then $a = b$, and $a + c = b + c$ are equivalent equations. Also, because subtraction is defined in terms of addition, $a = b$ and $a - c = b - c$ are equivalent equations.

Addition Property of Inequality Let a, b, and c be real numbers. If $a < b$, then $a + c < b + c$ and $a - c < b - c$. Adding or subtracting the same quantity to/from both sides of an inequality results in an equivalent inequality. The inequality symbol $<$ can be replaced with $>$, $\leq$, $\geq$, or $\neq$.

Additive Identity The number 0 is called the *additive identity* because if 0 is added to any number, the *sum* is the same identical number.

Additive Inverses Two numbers or expressions are additive inverses (also called *opposites*) if their sum is zero.

Algebraic Equation A statement that two quantities are equal where one or both quantities contain at least one variable.

Algebraic Expression A variable or combination of variables, constants, operations, and grouping symbols.

Alternating Sequence A sequence in which successive terms alternate signs.

Amount The amount is the part compared to the whole and is normally isolated from the percent and the base on one side of the word *is*.

Amount of Decrease The amount of decrease can be found by subtracting the new value from the original value. That is,
Amount of decrease = Original value − New value.

Amount of Increase The amount of increase can be found by subtracting the original value from the new value. That is, Amount of increase = New value − Original value.

Angle A figure formed by two rays that share a common endpoint.

Annually Happening once per year. In terms of compound interest, compounding *annually* means to compound interest once per year. Interest can also be compounded over other time intervals, such as *semiannually* (twice per year), *quarterly* (4 times per year), *monthly* (12 times per year), or *daily* (365 times per year).

Annuity An investment of equal amounts deposited at equal time intervals.

Apex See *Right circular cone*.

Area The amount of surface contained within the sides of a closed two-dimensional figure. Area is measured in square units such as square inches or square centimeters.

Area of a Circle If a circle has radius r, then its area A is given by the formula $A = \pi r^2$. The constant π is often approximated by the decimal 3.14 (or in some cases, the fraction $\frac{22}{7}$).

Area of a Rectangle If a rectangle has length l and width w, then its area A is given by the formula $A = l \cdot w$.

Area of a Square If a square has sides of length s, then its area A is given by the formula $A = s \cdot s = s^2$.

Area of a Triangle If a triangle has base b and height h, then its area A is given by the formula $A = \frac{1}{2}bh$.

Arithmetic Series The sum of the terms of an arithmetic sequence.

Arithmetic Sequence A sequence of the form $a_1, a_1 + d, a_1 + 2d, a_1 + 3d, a_1 + 4d, \ldots$, where a_1 is the first term of the sequence and d is the common difference. The general term, or nth term, of an arithmetic sequence has the form $a_n = a_1 + (n - 1)d$.

Associative Property of Addition Changing the grouping of terms does not affect the sum. That is, for any numbers, a, b, and c, $(a + b) + c = a + (b + c)$.

Associative Property of Multiplication Changing the grouping of factors does not affect the product. That is, for any numbers a, b, and c, $(a \cdot b) \cdot c = a \cdot (b \cdot c)$.

Augmented Matrix An augmented matrix for a system of equations in standard form is a matrix consisting of the coefficients of the variables and the constants from the right-hand side of the equations, separated by a vertical line.

Average A measure of central tendency found by adding the data values and then dividing by the number of values. Also called the *mean*.

Average Rate of Change The change in the value of a dependent variable divided by the corresponding change in the independent variable.

Axis of Symmetry The axis of symmetry is an imaginary line that divides a graph into two mirror images.

Back Substitution Taking known values of one or more variables and substituting them back into an equation involving the known variables and one unknown variable. The resulting equation is solved for the remaining variable.

Base The quantity being raised to a power in an exponential expression.

Binomial A simplified polynomial with two terms.

Binomial Coefficients The coefficients in the expansion of $(a + b)^n$. The coefficient of the term with variable part $a^{n-r}b^r$ is given by $\binom{n}{r} = \frac{n}{r!(n - r)!}$.

Binomial Difference   A binomial in which one term is subtracted from the other.

Binomial Sum   A binomial in which the two terms are added.

Book Value   The value of an asset as recorded in a balance sheet.

Boundary Line   When graphing a linear inequality, the *boundary line* is the graph of the corresponding linear equation. The boundary line divides the coordinate plane into two half-planes, only one of which contains solutions to the inequality. The line is dashed if the inequality is strict and points on the line are not part of the solution set. The line is solid if the inequality is non-strict and points on the line are included in the solution set.

Boundary Point (Polynomial Inequality)   A boundary point for a polynomial inequality $P(x) < 0$ is a value for x such that $P(x) = 0$. These are the x-intercepts for the graph of the polynomial. *Note:* The inequality symbol $<$ can be replaced by $>$, $\leq$, or $\geq$.

Boundary Point (Rational Inequality)   A boundary point for a rational inequality $R(x) < 0$ is a value for x such that $R(x) = 0$ or $R(x)$ is undefined. These are values that make either the numerator or denominator equal zero. *Note:* The inequality symbol $<$ can be replaced by $>$, $\leq$, or $\geq$.

Box Method   A method for multiplying two polynomials by using a rectangular array. Also known as the *grid method*.

Cartesian Coordinate System   See *Rectangular coordinate system*

Cartesian Plane   The plane represented by a rectangular coordinate system. Also known as the *xy-plane*.

Celsius Scale   A scale of temperature where 0°C is the freezing point of water (read "zero degrees Celsius") and 100°C is the boiling point of water. The raised circle indicates "degrees" and the "C" indicates we are using the *Celsius scale*.

Center of a Circle   The point within a circle, (h, k), that is the same distance from all points that lie on the circle.

Change of Base Formula   For any positive base $b \neq 1$ and for any positive real number u, then $\log_b u = \dfrac{\log_a u}{\log_a b}$, where a is any positive real number such that $a \neq 1$.

Circle   The set of all points (x, y) in the coordinate plane that are a fixed distance, r, from a fixed point (h, k). The fixed distance, r, is called the radius of the circle, and the fixed point (h, k) is called the center of the circle.

Circumference   The distance around a circle. Its value is found by using either $C = 2\pi r$ or $C = \pi d$, where r is the radius of the circle and d is the diameter. The constant π is often approximated by the decimal 3.14 (or in some cases, the fraction $\dfrac{22}{7}$).

Clear the Decimals   To *clear the decimals* from an equation, multiply both sides of the equation by an appropriate power of 10, such as 10, 100, or 1000. Determine which power of 10 to use by finding the largest number of decimal places in any of the constants.

Clear the Fractions   To *clear the fractions* from an equation, multiply both sides of the equation by the LCD of all the fractions in the equation.

Coefficient (or Numerical Coefficient)   The numeric factor of a term. For example, the coefficient in $4x^2$ is 4, the coefficient in $-3y$ is -3, and the coefficient in 13 is 13 (constant coefficient).

Coinciding Lines   Lines that have the same slope and the same y-intercept.

Collecting Like Terms   Reordering of terms in an algebraic expression so that like terms are grouped together.

Combine Like Terms   To add or subtract the like terms. To do this, reverse the distributive property to add or subtract the coefficients of the terms. The variable part remains unchanged.

Combined Variation   When a variable is related to more than one other variable.

Common Logarithm   The base 10 logarithm, which is abbreviated as $\log x$ instead of $\log_{10} x$.

Common Logarithmic Function   For $x > 0$, the common logarithmic function is defined by $y = \log x$ if and only if $x = 10^y$.

Common Multiple   A common multiple of two real numbers is a quantity that is evenly divisible by both numbers. For example, 24 is a common multiple of 6 and 12 because it is evenly divisible by both 6 and 12.

Common Ratio   In a geometric sequence, the ratio, r, of any term and the term right before it. Each term of the sequence can be obtained by multiplying the previous term by r.

Common Variable Factors for a GCF   If a variable is a common factor of a group of monomials, then the lowest power on that variable in the group will be the power on the variable in the GCF.

Commutative Property of Addition   The order of the terms in addition does not affect the sum. That is, for any numbers a and b, $a + b = b + a$.

Commutative Property of Multiplication   The order of the factors in multiplication does not affect the product. That is, for any numbers a and b, $a \cdot b = b \cdot a$.

Complementary Angles   Two angles are complementary angles, or *complements*, if the sum of their measures is 90°.

Complete the Square   To complete the square means to add an appropriate constant so that a binomial becomes a perfect square trinomial. The appropriate constant is the square of half the linear coefficient.

Complex Conjugates   The complex numbers $(a + bi)$ and $(a - bi)$ are called complex conjugates of each other. A complex conjugate is obtained by changing the sign of the imaginary part in a complex number.

Complex Fraction   A rational expression in which the numerator and/or denominator contain(s) rational expressions.

Complex Numbers   The set of all numbers of the form $a + bi$, where a and b are real numbers and i is the imaginary unit. The number a is called the *real part*, and the number b is called the *imaginary part*.

Complex Rational Expression   A rational expression in which the numerator and/or denominator contain(s) rational expressions.

Composite Function   Given functions f and g, the composite function, $f \circ g$ (also called the composition of f and g), is defined by $(f \circ g)(x) = f(g(x))$, provided $g(x)$ is in the domain of f.

Compound Inequality   Consists of two inequalities that are joined together using the words *and* or *or*.

Compound Interest   Interest that is paid or earned both on the original principal and on the interest that has been paid previously.

Compound Interest Formula   The overall amount A in an account is given by the formula $A = P\left(1 + \dfrac{r}{n}\right)^{n \cdot t}$ where P is the principal, r is the annual interest rate written in decimal form, t is the length of time in years, and n is the number of times interest is compounded per year.

Concave Down When a parabola opens down, it is defined as concave down.

Concave Up When a parabola opens up, it is defined as concave up.

Concentration The concentration of a solution refers to the strength of the solution. The higher the concentration, the stronger the solution. Concentration of a component in a solution is determined by the formula:

$$\text{Concentration} = \frac{\text{Amount of component}}{\text{Amount of solution}}.$$

Conditional Equation An equation that is true for some values of the variable but not for others.

Conic Sections Curves formed when a plane intersects a pair of right circular cones.

Conjugate Axis In a hyperbola, an invisible line segment that passes through the center and lies perpendicular to the transverse axis.

Conjugates Two expressions of the form $u + v$ and $u - v$ are conjugates of each other. The conjugate of a binomial is obtained by changing the sign of the second term in the binomial.

Consecutive Even Integers Even integers that appear next to each other in an ordered list of all even integers. For example, 6, 8, and 10 are consecutive even integers.

Consecutive Integers Integers that appear next to each other in an ordered list of all integers. For example, 3, 4, and 5 are consecutive integers.

Consecutive Odd Integers Odd integers that appear next to each other in an ordered list of all odd integers. For example, 5, 7, and 9 are consecutive odd integers.

Consistent System A system of equations or inequalities that has at least one solution.

Constant A value that never changes.

Constant of Variation A fixed non-zero number, k, that relates two quantities that vary directly or indirectly. It is the ratio of the two variables in direct variation $k = \frac{y}{x}$, and the product of the two variables in indirect variation ($k = y \cdot x$). Also called the *proportionality* constant.

Constant Term A term without any variables.

Continuous Compounding The process of earning interest continuously.

Continuous Compound Interest Formula Continuous compound interest can be calculated using the formula $A = Pe^{rt}$, where $A =$ total amount after t years; $P =$ principal (original investment); $r =$ interest rate per year, and $t =$ number of years.

Contradiction A contradiction (equation or inequality) is always false, regardless of the value of the variable, if any.

Converge If $|r| < 1$, then the infinite geometric series has a finite sum and is said to converge.

Coordinate Axes Two perpendicular real number lines that divide the coordinate plane into four quadrants. The horizontal coordinate axis is called the *x-axis* and the vertical coordinate axis is called the *y-axis*. The point where the axes intersect is called the origin.

Coordinate Plane A plane represented by a rectangular coordinate system. It is also known as the Cartesian plane or the *xy*-plane.

Coordinates The term *coordinates* refers to the *x*- and *y*-coordinates of a point or ordered pair. The coordinates provide the position of the point in the coordinate plane.

Counting Numbers The set of natural numbers, or counting numbers, denoted by the symbol $\mathbb{N}$, is the set $\mathbb{N} = \{1, 2, 3, \dots \}$.

Cramer's Rule A method for solving systems of linear equations by using determinants.

Cross Multiplying A method used to clear an equation of fractions when each side consists of a single simple fraction. Multiplying the numerator of each side by the denominator of the other side and equating the two products obtained.

Cross Products The products $P \cdot S$ and $R \cdot Q$ are the cross products of the proportion $\frac{P}{Q} = \frac{R}{S}$. When a proportion is true, the cross products are equal.

Cube An exponential expression that consists of a base with an exponent of 3.

Cube Root The cube root of a real number a is a real number b that, when cubed, results in a. That is, $b = \sqrt[3]{a}$, if $b^3 = a$.

Cube Root Function $f(x) = \sqrt[3]{x}$. The domain of the cube root function is the set of all real numbers $\mathbb{R}$, or using interval notation, $(-\infty, \infty)$.

Daily Happening once each day (365 times each year), as with compound interest. See *Annually*.

Decimal Number (or Decimal) A number that contains a decimal point. The position of each digit in a decimal number determines its place value. Digits appearing in front of the decimal point form the whole-number part of the number, and digits appearing behind the decimal point form the decimal part (or fractional part) of the number.

Decimal Part (or Fractional Part) Digits appearing behind the decimal point form the *decimal part* (or *fractional part*) of the number.

Decimal Place A position of a digit to the right of the decimal point.

Decimal Point Separates the whole-number part and fractional part of a decimal number.

Degenerate Conic Section A curve formed when a plane intersects the vertex of a pair of right circular cones.

Degree of a Monomial The sum of the exponents on the variables in the monomial.

Degree of a Polynomial The degree of a polynomial is the largest degree of its terms. For example, the degree of $10x^5 - 3x^4 + x^3 - 2x^2 + 8x + 9$, is 5, the largest degree of any term in the polynomial.

Demand Equation An equation that relates quantity sold (demanded) to the price per item. When solved for x, it indicates the quantity that will be demanded for a given price.

Denominator The expression below the fraction bar in a fraction or quotient.

Dependent Equations Two equations that describe the same solution set. In two variables, the equations describe the same line. In three variables, the equations describe the same plane.

Dependent System A system of equations is dependent if an equation in the system can be obtained from one or more other equations in the system.

Dependent Variable A variable whose value depends on the value(s) of other variables. These are also called *output variables* because they represent the value that we get *out* after inputting values for the independent variables.

Descending Order When terms of a polynomial are placed in order from largest degree to smallest.

Determinant A value computed for a square matrix that can be used to solve systems of linear equations.

Devine Formula A popular formula in the health care profession that uses a person's height to compute his or her ideal body weight. The formula is $w = 110 + 5.06(h - 60)$ for men and $w = 100.1 + 5.06(h - 60)$ for women, where w is ideal body weight, in pounds, and h is height, in inches.

Diameter The diameter of a circle is the distance across the circle and through its center. The length of the diameter is two times the length of the radius.

Difference The result of subtracting two numbers. The number being subtracted is called the *subtrahend*, and the number being subtracted from is called the *minuend*.

Difference of Two Squares The result, $A^2 - B^2$, from multiplying the sum and difference of two terms (or conjugates).

Direct Variation For an equation of the form $y = kx$, we say that y varies directly with x, or y is proportional to x. The constant k is called the constant of variation or the proportionality constant.

Directly Proportional A relationship between two variables where one variable is a constant multiple of the other. See *Direct Variation*.

Directrix For a parabola, an imaginary line perpendicular to the axis of symmetry lying on the opposite side of the vertex as the focus, but the same distance away from the vertex.

Discriminant Given a quadratic equation $ax^2 + bx + c = 0, a \neq 0$, the expression $D = b^2 - 4ac$ is called the *discriminant*. If $D > 0$, then the quadratic equation has two real solutions. If $D < 0$, then the quadratic equation has two non-real solutions. If $D = 0$, then the quadratic equation has exactly one real solution.

Distance Formula The distance $d(A, B)$ between two points (x_1, y_1) and (x_2, y_2) is given by $d(A, B) = \sqrt{(x_2 - x_1)^2 + (y_2 - y_1)^2}$.

Diverge If $|r| \geq 1$, then an infinite geometric series does not have a finite sum and the series is said to diverge.

Divide The process of separating a quantity into groups of equal size.

Dividend The quantity that you divide into in a division problem. This is also the numerator of a fraction.

Divide Out To simplify a fraction, we can divide out common factors by dividing both the numerator and denominator by the common factors.

Division The process of separating a quantity into groups of equal size. The number being divided is called the *dividend*, whereas the number that divides is the *divisor*. The result of division is the *quotient*. If a and b represent two numbers and $b \neq 0$, then $a \div b = \dfrac{a}{b} = a \cdot \dfrac{1}{b}$.

Divisor The number being divided by in a division problem. This is called the denominator of a fraction.

Domain The domain of a relation is the set of all possible values for the independent variable. These are the first coordinates in a set of ordered pairs and are also known as *input values*.

Domain of a Rational Function The set of all real numbers except for those values that cause the denominator to equal zero.

Double-Negative Rule If a is a real number, then $-(-a) = a$.

Element An individual number of a set or other object included in a set.

Elimination Method A method for solving a system of linear equations in two variables that involves adding two equations together in a way that will eliminate one of the variables.

Ellipse The set of all points in a plane, the sum of whose distances from two fixed points is a positive constant. The two fixed points F_1 and F_2, and are called the *foci*.

Empty Set A set that contains no elements.

Endpoint A point that represents the end of a graph in one direction.

Equal Sign The symbol for equality ($=$).

Equation A statement that two quantities are equal. An equation is a *numeric equation* if it contains only numeric expressions and no variables. An equation is an *algebraic equation* if it contains one or more variables.

Equation in One Variable An equation that contains a single variable.

Equation in Two Variables An equation that contains two distinct variables.

Equivalent Equations Two or more equations with the same solution set.

Equivalent Fractions Two fractions that represent the same value.

Equivalent Functions Two functions that represent the same set of ordered pairs.

Equivalent Inequalities Two inequalities that have the same solution set. For example, the equivalent inequalities $4x > 12, 2x > 6$, and $x > 3$ all have the same solution set: $\{x \mid x > 3\}$, or $(3, \infty)$ in interval notation.

Equivalent Matrices Two or more matrices that have the same corresponding entries.

Equivalent Numeric Expressions Two numeric expressions are equivalent if they simplify to the same value. For example, $\dfrac{2}{4}$ is equivalent to $\dfrac{1}{2}$ and $3(4)$ is equivalent to 12.

Evaluate a Formula To find the value of an isolated variable by substituting values for all other variables in the formula.

Evaluate a Function To determine the value of a function for a given value of the independent variable by substituting the value for the independent variable into the function definition and simplifying the resulting numeric expression using the order of operations.

Evaluate Algebraic Expressions To substitute given values for the variables and simplify the resulting numeric expression using the order of operations.

Evaluate an Exponential Expression To multiply out the expression to a single numeric value.

Expansion by Minor A technique used to evaluate determinants of 3×3 and larger square matrices.

Exponent Also called *power*, it is a superscripted number that tells how many times a base expression is multiplied by itself.

Exponential Decay A model that describes the exponential decay of a population, quantity, or amount A, after a certain time, t, is $A(t) = A_0 e^{kt}$, where $A_0 = A(0)$ is the initial quantity and $k < 0$ is a constant called the *relative decay constant*.

Exponential Expression A constant or algebraic expression that is raised to a power. The constant or algebraic expression makes up the base, and the power is the exponent on the base.

Exponential Form Involving the use of exponents. For example, the expression x^3 is in exponential form (as opposed to $x \cdot x \cdot x$).

Exponential Function An exponential function is a function of the form $f(x) = b^x$, where x is any real number and $b > 0$ such that $b \neq 1$. The constant, b, is called the base of the exponential function.

Exponential Growth Model A model that describes the population, quantity, or amount A, after a certain time, t, is $A(t) = A_0 e^{kt}$, where $A_0 = A(0)$ is the initial quantity and $k > 0$ is a constant called the *relative growth rate*.

Extraneous Solutions A solution to a new equation that arises during the solving process but is not a solution to the original equation.

Factor In multiplication, each number or expression being multiplied. For example, in the expression $2 \cdot 3$, the factors are 2 and 3. In the expression $x(y + 1)$, the factors are x and $y + 1$.

Factor a Polynomial To write a polynomial as an equivalent expression that is the product of simpler polynomials.

Factor Over the Integers To write a polynomial as a product of polynomial factors where all coefficients in both the original polynomial and its factors are integers.

Factor Theorem For a polynomial function $f(x)$, (1). If $f(c) = 0$, then $x - c$ is a factor of $f(x)$. (2) If $x - c$ is a factor of $f(x)$, then $f(c) = 0$.

Factored Completely A whole number is *factored completely* if it is written as the product of all prime numbers. A polynomial is *factored completely* if it is written as the product of all prime polynomials.

Factored Form When a polynomial is written as an equivalent expression that is the product of polynomials.

Factorial of a Non-Negative Integer The factorial of a non-negative integer n, denoted as $n!$, is the product of all positive integers less than or equal to n. Thus, $n! = n(n - 1) \cdot \ldots \cdot 3 \cdot 2 \cdot 1$. Note that $0! = 1$.

Factoring Writing an expression as the product of two or more simpler expressions.

Factoring by Grouping A method of factoring where terms of the original polynomial are grouped together to look for common factors.

Feasible Domain The set of values for the independent variable that make sense, or are feasible, in the context of the application.

Feasible Solution A solution that satisfies the conditions of an application problem and makes sense within the context of the problem.

Fibonacci Number A of number that is part of the Fibonacci Sequence.

Fibonacci Sequence The Fibonacci sequence is defined recursively by $a_n = a_{n-1} + a_{n-2}$, where $a_1 = 1$ and $a_2 = 1$.

Finite Sequence A function whose domain is the finite set $\{1, 2, 3, \ldots, n\}$, where n is a natural number.

Finite Series A series with a fixed number of terms.

Finite Set A set with a fixed, or finite, number of elements.

First-Degree Equation A polynomial equation in which the largest exponent on a variable is 1.

Focus of a Parabola The point on the axis of symmetry such that all points on the parabola are equidistant from the point and the directrix.

FOIL Method FOIL is the acronym for a method of multiplying binomials; that is, to multiply the two First terms, the two Outside terms, the two Inside terms, and the two Last terms, then add the resulting products together.

Formula An equation that describes the relationship between two or more variables.

Fraction A number of the form $\frac{a}{b}$, where a and b are integers and $b \neq 0$. a is called the numerator, and b is called the denominator

Fraction Bar A horizontal line used to separate the numerator (or top expression) and the denominator (or bottom expression) in a fraction.

Function A special type of relation in which each value in the domain corresponds to exactly one value in the range.

Function Notation Notation used to name a function and identify the independent variable. An example of function notation is $f(x)$, read "f of x." For $y = f(x)$, f is the name of the function that relates the independent variable x to the dependent variable y.

General Equation for Percents Percent $\cdot$ Base $=$ Amount

General Form of the Equation of a Circle The general form of the equation of a circle is $Ax^2 + By^2 + Cx + Dy + E = 0$, where A, B, C, D, and E are real numbers and $A = B \neq 0$.

Geometric Series The sum of the terms of a geometric sequence.

Geometric Sequence A geometric sequence is a sequence of the form $a_1, a_1 r, a_1 r^2, a_1 r^3, a_1 r^4, \ldots$, where a_1 is the first term of the sequence and r is the common ratio such that $r = a_2/a_1 = a_3/a_2 = \ldots = a_{n+1}/a_n$ for all $n \geq 1$. The general term, or nth term, of a geometric sequence has the form $a_n = a_1 r^{n-1}$.

Graph A diagram or figure that shows relationships between quantities.

Graph an Ordered Pair To place a point at the ordered pair's location on the coordinate plane.

Graph a Number To place a point, or solid circle, ($\bullet$) at the number's location on the number line.

Graph of a Function The graph of a function is the graph of all ordered pairs that belong to the function. The point (a, b) lies on the graph of a function if and only if $f(a) = b$.

Graph of a System of Linear Inequalities in Two Variables The intersection of the graphs of each inequality in the system.

Graph of an Equation in Two Variables A visual display of all ordered pair solutions to the equation.

Greatest Common Factor (GCF) The largest factor common to all terms in a list. The GCF can Include both numeric factors, such as 2, and variable factors, such as x.

Greatest Common Factor (GCF) of a Group of Integers The largest integer that divides evenly into each integer in the group.

Greatest Common Factor (GCF) of a Group of Monomials The monomial with the largest coefficient and highest degree that divides each monomial evenly.

Greatest Common Factor (GCF) of a Polynomial The expression with the largest coefficient and highest degree that divides each term of the polynomial evenly.

Grouping Method When the terms are grouped together to be multiplied or divided.

Grouping Symbols Mathematical symbols used to group operations so they are treated as a single quantity. When simplifying, all operations within grouping symbols must be done first. Some examples of grouping symbols are parentheses ( ), brackets [ ], and braces { }. Operators such as absolute value | |; fraction bars, $-$; and radicals, $\sqrt{}$ are also treated as grouping symbols.

Half-Life The amount of time required for a quantity to fall to half its value as measured at the beginning of the time period.

Half-Plane One half of the coordinate plane, separated from the other half by a boundary line.

Hang Time Length of time that an object, such as a punted football or jumping athlete, remains in the air.

Horizontal Compression The squeezing of the graph towards the y-axis.

Horizontal Line A line with slope 0 passing through the point $(0, b)$. The equation of a horizontal line is given by $y = b$.

Horizontal Line Test   If every horizontal line intersects the graph of a function f at most once, then f is one-to-one.

Horizontal Stretch   The stretching of a graph away from the y-axis.

Hyperbolas   The set of all points in a plane, the difference of whose distances from two fixed points is a positive constant. The two fixed points, F_1 and F_2 are called the foci.

Hypotenuse   The longest side of a right triangle. It is the side opposite the right angle.

Identity   An equation or inequality that is always true, regardless of the value of the variable, if any.

Identity Property of Addition   If a is a real number, then $a + 0 = a$ and $0 + a = a$.

Identity Property of Multiplication   If a is a real number, then $a \cdot 1 = a$ and $1 \cdot a = a$.

Imaginary Number   An imaginary number, bi, is a complex number whose real part is equal to zero.

Imaginary Part of a Complex Number   For a complex number in standard or $a + bi$ form, the number b is called the *imaginary part of the complex number.*

Imaginary Unit i   The imaginary unit i is defined as $i = \sqrt{-1}$. Also, $i^2 = -1$.

Improper Fraction   An improper fraction is a fraction in which the absolute value of the numerator is greater than or equal to the absolute value of the denominator. The absolute value of an improper fraction is greater than or equal to 1.

Inclusive   Means to include the extremes as well as everything between. For example, the set of numbers between 4 and 7, *inclusive*, includes 4 and 7 along with all numbers in between them.

Inconsistent System of Equations   A system of equations without a solution. When solving an inconsistent system, a contradiction (false statement) will result.

Inconsistent System of Inequalities   A system of inequalities with no solution.

Independent System   A system of two equations in two variables is independent if there is exactly one solution to the system or no solution. In these cases, both equations give new information about the variables. If the two lines in the system intersect or are parallel, then the system is independent.

Independent Variable   A variable for which we can select values. These are also called *input variables* because they represent the set of values that we can *input* into a relation.

Indeterminate   A quantity whose value cannot be determined. For example, if a zero dividend is divided by a divisor of 0, the result is indeterminate. That is, $0 \div 0$ is indeterminate.

Index of a Radical Expression   In the radical expression $\sqrt[n]{a}$, n is called the index, and indicates the type of root.

Inequality   A statement that two quantities are not strictly equal to each other.

Inequality Symbols   Symbols used to indicate two quantities are not equal. For example, the symbol $>$ means greater than; the symbol $<$ means less than.

Infinite Set   A set with an unlimited number of elements.

Infinite Sequence   A function whose domain is the set of all natural numbers.

Infinite Series   A series with an infinite number of terms.

Integer   An element of the set $\mathbb{Z} = \{ \ldots, -4, -3, -2, -1, 0, 1, 2, 3, 4, \ldots \}$.

Intercept   A point where a graph crosses or touches a coordinate axis.

Interest   When we borrow money, we pay an extra amount of money called *interest* for this privilege. Similarly, if we

lend money to someone else, the payment we receive for this generosity is also called *interest*.

Interest Rate   The amount of interest paid or earned is a percent of the principal over a given time interval. The percent is called the *interest rate*.

Interior Angle   An angle inside a figure.

Intersecting Lines   Two lines that cross or touch at a single point on the same plane.

Intersection   For any two sets A and B, the intersection of A and B is given by $A \cap B$ and represents the elements that are in both set A and set B.

Interval Notation   An interval is a set of numbers that lie between a given lower and upper bound. The bounds indicate the end points of the interval. For example, the interval $[3, 9]$ is the set of all values between 3 and 9, including 3 and 9. The interval $(4, 11)$ is the set of all values between 4 and 11, but not including 4 or 11. The interval $(4, \infty)$ is the set of all values greater than 4.

Inverse Function   Let f be a one-to-one function with domain A and range B. Then f^{-1} is the inverse function of f with domain B and range A. Furthermore, if $f(a) = b$, then $f^{-1}(b) = a$.

Inverse Variation   For equations of the form $y = \dfrac{k}{x}$ or $y = k\dfrac{1}{x}$, we say that y varies inversely with x, or y is inversely proportional to x. The constant k is called the *constant of variation* or the *proportionality constant*.

Irrational Number   An irrational number is a real number that is not a rational number, so it cannot be written as the quotient $\dfrac{p}{q}$ of two integers p and q with $q \neq 0$. The decimal form of an irrational number will not terminate and will not repeat.

Isolate a Quantity   A quantity in an equation is isolated if it is by itself on one side of the equation and is the only occurrence of its type in the equation.

Isolated Radical Expression   A radical expression is an isolated radical expression when it stands alone on one side of the equal sign.

Joint Variation   A variable is directly proportional to the product of two or more other variables, such as $y = kxz$.

Leading Coefficient of a Polynomial in One Variable   For a polynomial in one variable written in standard form, the coefficient of the first term (the term with the highest degree).

Least Common Denominator (LCD)   The smallest multiple of all denominators in an algebraic expression or equation.

Least Common Multiple   The smallest multiple that is evenly divisible by all the numbers or expressions in a list of numbers or expressions.

Legs   The two sides of a right triangle that form the right angle.

Like Radicals   Two radicals that have the same index and the same radicand.

Like Terms   Two terms are like terms if they have variable factors that are exactly the same.

Line   A graph of an equation that can be written in the form $Ax + By = C$ (A and B cannot both be zero). Lines are straight and continue forever in both directions.

Linear Coefficient   The linear coefficient is the coefficient of the linear term in a polynomial.

Linear Equation in One Variable   An equation that can be written in the form $ax + b = c$, where x is a variable and a, b, and c are any numbers such that $a \neq 0$. A linear equation is also called a *first-degree equation* because the exponent on the variable is understood to be 1.

Linear Equation in Three Variables   An equation that can be written in the form $Ax + By + Cz = D$, where A, B, C, and D are real numbers, and $A, B,$ and C are not all equal to 0.

Linear Equation in Two Variables   An equation that can be written in the form $Ax + By = C$, where $A, B,$ and C are real numbers, and A and B are not both equal to 0.

Linear Function   A function of the form $f(x) = ax + b$, where a and b are real numbers and $a \neq 0$.

Linear Inequality in One Variable   An inequality that can be written in the form $ax + b < c$, where $a, b,$ and c are real numbers and $a \neq 0$. Note, the $<$ could also be replaced by $>, \leq,$ or $\geq$.

Linear Inequality in Two Variables   An inequality that can be written in the form $Ax + By < C$, where $A, B,$ and C are real numbers and A and B are not both equal to zero. Note, the $<$ could also be replaced by $>, \leq,$ or $\geq$.

Logarithm Property of Equality   If a logarithmic equation can be written in the form $\log_b u = \log_b v$, then $u = v$. Furthermore, if $u = v$, then $\log_b u = \log_b v$, for $b \neq 1$.

Logarithmic Form   The equation $y = \log_b x$ is said to be in logarithmic form.

Logarithmic Function   For $x > 0, b > 0$, and $b \neq 1$, the logarithmic function with base b is defined by $y = \log_b x$ if and only if $x = b^y$.

Logistic Model   A growth model of the form $P(t) = \dfrac{C}{1 + Be^{kt}}$, where $B, C,$ and k are constants with $C > 0$ and $k < 0$. For such models, population growth decreases as the size of the population increases and there is a limiting size of the population. The graph of a logistic model has a characteristic 'S' shape.

Long Division   A process for dividing two numbers that uses a sequence of steps involving division, multiplication, subtraction, and dropping down the next digit. The steps are repeated until the remainder can no longer be divided.

Long Division Symbol   The symbol $\overline{)}$ used to separate the dividend from the divisor in a long division problem.

Long Run Average Cost (LRAC)   The total long run cost divided by the number of units produced.

Long-Range Navigation   The precise location of the ship can be determined by finding the intersection of the two hyperbolas. This system of locating ships is known as long-range navigation or LORAN.

Lower Bound   The lower bound of an interval is the largest value that is less than or equal to all values in the interval. No values in the interval can be less than the lower bound. For example, the interval [6, 9] has a lower bound of 6. No value in the interval is less than 6.

Lower Limit of Summation   The value that indicates the number of the first term of a sequence to be included in a series.

Lowest Terms   A fraction is written in lowest terms, or simplest form, if the numerator and denominator have no common factors other than 1.

Main Fraction Bar   The fraction bar that separates the numerator and denominator of a complex fraction.

Major Axis   A line segment that passes through the foci of an ellipse and is the longer of the two axes of symmetry.

Markup   An amount added by a merchant to the cost of an item to obtain the selling price of the item.

Mathematical Expression   A statement containing symbols (including numbers) and/or mathematical operations (such as addition or subtraction).

Mathematical Model   Uses the language of mathematics to describe a problem. Typically, the model is an equation that describes a relationship within an application.

Mathematical Symbol   A character that represents a mathematical relation or operation.

Matrices   The plural of *matrix*.

Matrix   A rectangular array of numbers arranged in rows and columns.

Maximum   The largest value in the range of the function. It occurs at the highest point on the graph of the function.

Mean   A measure of central tendency found by adding a list of data values and then dividing the sum by the number of values. Also called the average.

Median   The middle value in a list of ordered data. If there is an even number of data values, the median is the average of the two middle values. The median is a measure of central tendency that is not greatly affected by a few extreme data values.

Member   Each object in a set is called an element or a member of the set.

Method of Relating the Bases   If an exponential equation can be written in the form $b^u = b^v$, then $u = v$.

Midpoint   The point that lies exactly halfway between the endpoints of the line segment.

Minimum   The smallest value in the range of the function. It occurs at the lowest point on the graph of the function.

Minor   Related to an entry in a determinant, a smaller determinant that results from crossing out the row and column in the larger determinant that contains the entry.

Minor Axis   The minor axis of an ellipse is the shorter of the two axes of symmetry.

Minor Fraction   The fractions within the numerator and/or the denominator of a complex fraction.

Minor Rational Expressions   The rational expressions within the numerator and denominator of a complex rational expression. Also called *minor fractions*.

Minuend   The number being subtracted from.

Mixed Number   A mixed number is a number of the form $a\dfrac{b}{c}$, where $a, b,$ and c are nonzero whole numbers, and $b < c$. The notation $a\dfrac{b}{c}$ means $a + \dfrac{b}{c}$, but the $+$ symbol is not written. For example, $2\dfrac{3}{4}$ and $5\dfrac{7}{8}$ are mixed numbers.

Mixture   A quantity formed by combining two or more other quantities.

Mode   A measure of central tendency, the mode of a list of data is the data value that appears most frequently. A list of data may have no mode, one mode, or more than one mode.

Model   A representation used to illustrate something that cannot be seen easily (for example, a DNA model). See also *Mathematical Model*.

Monomial   A simplified term in which all variables are raised to non-negative integer powers and no variables appear in any denominator.

Monthly   Happening once each month (12 times each year), as with compound interest. See *Annually*.

Multiple   The product of two whole number factors. A number that can be divided by a second number without a remainder is said to be a *multiple* of the second number.

Multiplication   The simplified process of writing repeated addition. Multiplication symbols include a multiplication sign ($\times$), an asterisk, a multiplication dot ($\cdot$), or parentheses.

Multiplication Property of Equality   If $a, b,$ and c are real numbers or algebraic expressions with $c \neq 0$ then $a = b$ and $a \cdot c = b \cdot c$ are equivalent equations.

Multiplication Property of Inequality   Let a, b, and c be real numbers. If $a < b$ and $c > 0$, then $ac < bc$ and $\dfrac{a}{c} < \dfrac{b}{c}$. Multiplying or dividing both sides of an inequality by a positive number c results in an equivalent inequality.

If $a < b$ and $c < 0$, then $ac > bc$ and $\dfrac{a}{c} > \dfrac{b}{c}$. Multiplying or dividing both sides of an inequality by a negative number c, and switching the direction of the inequality, results in an equivalent inequality.

Multiplication Property of Zero   If a is a real number, then $a \cdot 0 = 0$ and $0 \cdot a = 0$.

Multiplicative Identity   The number 1 is called the *multiplicative identity* because if any number is multiplied by 1, the *product* is the same identical number.

Multiplicative Inverse   The reciprocal of a number is also called the multiplicative inverse.

Natural Base   The irrational number that is equal to $2.718281828 \ldots$ The natural base appears very frequently in real-life applications and is denoted by e.

Natural Exponential Function   The natural exponential function is the exponential function with base e and is defined as $f(x) = e^x$.

Natural Logarithm   The base e logarithm, which is abbreviated as $\ln x$ instead of $\log_e x$.

Natural Logarithmic Function   For $x > 0$, the natural logarithmic function is defined by $y = \ln x$ if and only if $x = e^y$.

Natural Numbers   Also called *counting numbers*, the elements of the set $\mathbb{N} = \{1, 2, 3, 4, 5, \ldots \}$.

Negative Exponent   An exponent, or power, whose value is less than zero.

Negative Number   A number less than zero.

Negative-Power Rule   To remove a negative exponent, switch the location of the base (numerator or denominator) and change to a positive exponent: $a^{-n} = 1/a^n$ and $1/a^{-n} = a^n \ (a \neq 0)$.

Negative Slope   The slope of a line that slants downward, or falls, from left to right.

Negative Square Root   A negative real number b is the negative square root of a non-negative real number a, denoted $b = -\sqrt{a}$, if $b^2 = a$.

Nested Grouping Symbols   Grouping symbols within grouping symbols.

Newton's Law of Cooling   Newton's law of cooling states that the temperature of an object changes at a rate proportional to the difference between its temperature and that of its surroundings. The temperature T of an object at any time t is given by $T(t) = S + (T_0 - S)e^{kt}$, where T_0 is the original temperature of the object, S is the constant temperature of the surroundings, and k is the cooling constant.

Non-Negative   A real number, x, is non-negative if it is 0 or larger. That is, $x \geq 0$.

Non-Simplified Expression   An expression that can be written in a simpler form by combining like terms, removing grouping symbols, and so forth.

Non-Strict Inequality   An Inequality that contains one or both of the following Inequality symbols: $\leq$, $\geq$. The possibility of equality is included in a non-strict Inequality.

Nonzero   A real number is nonzero if it is not equal to 0. All positive and negative numbers are nonzero.

nth Partial Sum   The sum of the first n terms of a series.

Null Set   A set that contains no elements. Also called the *empty set*. Denoted by $\varnothing$ or $\{ \ \}$.

Null Symbol   The null symbol, $\varnothing$, denotes an empty set.

Numerator   The expression above the fraction bar in a fraction or quotient.

Numeric Expression   A combination of numbers and arithmetic operations.

Numerical Coefficient   See *Coefficient*.

One-to-One Functions   A function f is one-to-one if for any values $a \neq b$ in the domain of f, $f(a) \neq f(b)$. Alternatively, a function f is one-to-one if for any two range values, $f(u)$ and $f(v)$, $f(u) = f(v)$ implies that $u = v$.

Operation   In algebra, an operation is a mathematical action that is performed by following specific rules. Addition, subtraction, multiplication, and division are all operations.

Opposites   Two numbers are opposites if they are located the same distance away from 0 on the number line but lie on opposite sides of 0. The sum of two opposites equals zero.

Opposite of an Expression   Changing the sign of each term in the expression results in the opposite of the expression.

Opposite Polynomials   Two polynomials are opposite polynomials if they add to zero. To find the opposite of a polynomial, change the sign of each term.

Opposite Reciprocals   Two numbers are said to be opposite reciprocals if their product equals -1. This occurs when the two numbers have opposite signs and their absolute values are reciprocals.

Order of Operations   A set of rules for evaluating expressions with more than one operation. (1) Evaluate operations within parentheses (or other grouping symbols) first, starting with the innermost set and working out; (2) Work from left to right and evaluate any exponential expressions as they occur; (3) Work from left to right and perform any multiplication or division operations as they occur; and (4) Work from left to right and perform any addition or subtraction operations as they occur.

Ordered Pair (x, y)   A pair of real number values that corresponds to a point in the coordinate plane.

Ordered-Pair Notation   With ordered-pair notation, we indicate the solutions to a dependent system by expressing one coordinate in terms of the other in an ordered pair.

Ordered Pair Solution   An ordered pair solution to an equation in two variables, written as (x, y), gives values for the variables that, when substituted in the equation, result in a true statement.

Ordered Triple (x, y, z)   A set of three values that gives the location of a point in three dimensions.

Ordered-Triple Solution   An *ordered-triple solution* to an equation in three variables, written as (x, y, z) gives values for the variables that, when substituted in the equation, result in a true statement.

Order of Two Real Numbers   Refers to whether one real number is less than, greater than, or equal to a second real number.

Ordinary Annuity   Equal deposits that are made at the end of a compound period.

Origin (Real Number Line)   The point 0 is called the origin of a real number line. Numbers located to the left of the origin are negative numbers, and numbers located to the right of the origin are positive numbers.

Origin (Rectangular Coordinate System)   The point at which the x- and y-axes intersect in the rectangular coordinate system. The origin is represented by the ordered pair $(0, 0)$.

Original Price   The price of merchandise before any discount is subtracted or any tax is added. In applications of the percent equation involving percent increase or percent decrease, the *original price* is the base.

Overall Amount for Loan or Investment Whether borrowing or investing, the overall amount of money involved results from adding the principal and the interest. That is, Overall Amount = Principal + Interest, or $A = P + I$.

Paired Data Data represented as ordered pairs.

Parabola A parabola is the set of all points in a plane equidistant from a fixed point F and a fixed line D. The fixed point is called the *focus*, and the fixed line is called the *directrix*.

Parallel Lines Two lines that lie in the same plane but never cross or touch. Parallel lines have the same slope but different y-intercepts.

Partial Sum The sum of the first n terms of a series is called the nth partial sum of the series and is denoted as S_n.

Percent "Per hundred" or "out of 100." A percent is denoted by the percent symbol, %.

Percent Proportion $\dfrac{\text{Amount}}{\text{Base}} = \dfrac{\text{Percent Number}}{100}$. The amount is the part compared to the whole, the base is the number after *of*, and the percent number is the number before the percent symbol, %.

Percent Symbol See *Percent*

Perfect 4th Power A quantity that can be written as the 4th power of another quantity of the same type.

Perfect Cube A quantity that can be written as the cube of another quantity of the same type.

Perfect Square A quantity that can be written as the square of another quantity (including decimals that can be written in fraction form).

Perfect Square Trinomial A trinomial that can be written as the square of a binomial sum, $(A + B)^2$, or the square of a binomial difference, $(A - B)^2$.

Perimeter The distance around a figure. For a polygon, the perimeter is the sum of the lengths of the sides of the polygon.

Perimeter of a Rectangle If a rectangle has length l and width w, then its perimeter P is given by the formula $P = 2l + 2w$.

Perimeter of a Square If a square has sides of length s, then its perimeter P is given by the formula $P = 4s$.

Perimeter of a Triangle If a triangle has sides of length a, b, and c, then its perimeter P is given by the formula $P = a + b + c$.

Periods Groups of three digits separated by commas that are used to express whole numbers with many digits. The first five periods from right to left are *ones*, *thousands*, *millions*, *billions*, and *trillions*.

Perpendicular Lines Two lines are perpendicular if they meet at a right angle (90°). The product of the slopes will equal -1, so perpendicular lines have opposite-reciprocal slopes.

Periodic Compound Interest Formula Periodic compound interest can be calculated using the formula $A = P\left(1 + \dfrac{r}{n}\right)^{nt}$, where $A =$ total amount after t years; $P =$ principal (original investment); $r =$ interest rate per year (as a decimal), $n =$ number of times interest is compounded per year; and $t =$ number of years.

Place-Value Chart A chart that indicates the place value position of a decimal number.

Plot See *Graph a number* and *Graph an ordered pair*.

Plus or Minus Symbol The symbol $\pm$ is read as "plus or minus" and can be used to indicate two values, one positive and one negative. For example, $x = \pm 6$ is read as "x equals plus or minus six" and means $x = 6$ or $x = -6$. The symbol can also be used to indicate addition and subtraction.

For example $x = 5 \pm 3$ is read "x equals five plus or minus three." It means $x = 5 + 3 = 8$ or $x = 5 - 3 = 2$.

Point A location on a number line, on a plane, or in space. Usually represented by a dot.

Point-Slope Form Given the slope m of a line and a point (x_1, y_1) on the line, the point-slope form of the equation of the line is given by $y - y_1 = m(x - x_1)$.

Polygon A closed plane figure made up of connected line segments. Each line segment is called a *side* of the polygon.

Polynomial A monomial or a finite sum of monomials.

Polynomial Equation A polynomial equation in standard form is written as $P(x) = 0$, where $P(x)$ is a simplified polynomial in descending order. A polynomial equation results when two polynomials are equal to each other.

Polynomial Function A function of the form $P(x) = a_n x^n + a_{n-1} x^{n-1} + \cdots + a_1 x + a_0$, where each coefficient, a_i is a real number and n is a whole number.

Polynomial Inequality An inequality that can be written as $P(x) < 0, P(x) > 0, P(x) \leq 0, P(x) \geq 0$, where $P(x)$ is a polynomial function.

Polynomial in One Variable A polynomial that contains a single variable.

Polynomial in Several Variables A polynomial containing two or more variables.

Polynomial Long Division A process for dividing two polynomials that uses a sequence of steps involving division, multiplication, subtraction, and dropping down the next term. The steps are repeated until the remainder can no longer be divided.

Positive Numbers Numbers located to the right of the origin.

Positive Slope The slope of a line that slants upward, or rises, from left to right.

Power Also called *exponent*, it is a superscripted number that tells how many times a base number is multiplied by itself.

Power Principle of Equality If A and B represent algebraic expressions and n is a positive real number, then any solution to the equation $A = B$ is also a solution to the equation $A^n = B^n$.

Power-to-Power Rule When an exponential expression is raised to a power, multiply the exponents, that is, $(a^m)^n = a^{m \cdot n}$.

Positive Square Root A non-negative real number b is the *positive square root* of a non-negative real number a, denoted as $b = \sqrt{a}$, if $b^2 = a$. Also called *principal square root*.

Present Value If investors want to know how much money to invest now in order to reach a certain investment goal in the future, this amount of money, P, is known as the present value of A dollars.

Present Value Formula Present value can be calculated using the formula $P = A(1 + r/n)^{-nt}$, where $P =$ principal (original investment), $A =$ total amount after t years, $r =$ interest rate per year, $n =$ number of times interest is compounded per year, and $t =$ number of years.

Prime Factorization The expression of a composite number as the product of its prime factors. Every composite number has a unique prime factorization.

Prime Number A whole number greater than 1 whose only whole number factors are 1 and itself.

Prime Polynomial A polynomial whose only factors over the integers are 1 and itself.

Principal The amount of money that is borrowed or invested.

Principal *n*th Roots If a and b are real numbers and n is an integer such that $n \geq 2$, then b is the principal *n*th root of a, denoted as $b = \sqrt[n]{a}$, if $b^n = a$. Note that b must be non-negative when n is even.

Principal Square Root A non-negative real number b is the principal square root of a non-negative real number a, denoted as $b = \sqrt{a}$, if $b^2 = a$.

Product The result of multiplying two real numbers.

Product of Complex Conjugates The product of complex conjugates is equal to the sum of the squares of the real and imaginary parts. That is, $(a + bi)(a - bi) = a^2 + b^2$.

Product of Functions If f and g represent functions, then the product of f and g is $(f \cdot g)(x) = f(x) \cdot g(x)$.

Product Rule for Exponents When multiplying exponential expressions with the same base, add the exponents and keep the common base. That is, $a^m \cdot a^n = a^{m+n}$.

Product Rule for Radicals If $\sqrt[n]{a}$ and $\sqrt[n]{b}$ are real numbers, then $\sqrt[n]{a} \sqrt[n]{b} = \sqrt[n]{ab}$.

Product Rule for Square Roots If $\sqrt{a}$ and $\sqrt{b}$ are real numbers, then $\sqrt{a \cdot b} = \sqrt{a} \cdot \sqrt{b}$.

Product-to-Power Rule When raising a product to a power, raise each factor of the base to the common exponent. That is $(ab)^n = a^n b^n$.

Proper Fraction A proper fraction is a fraction in which the absolute value of the denominator is greater than the absolute value of the numerator. The absolute value of a proper fraction is less than 1.

Proper Subset If every element of a set A is also an element of set B, but A and B are not equal sets, then set A is a proper subset or strict subset of set, denoted as $A \subset B$.

Property of Equivalent Fractions If a, b, and c are numbers, then $\frac{a}{b} = \frac{a \cdot c}{b \cdot c}$ and $\frac{a}{b} = \frac{a \div c}{b \div c}$ as long as b and c are not equal to 0.

Proportion An equation stating that two ratios (or rates) are equal, such as $\frac{a}{b} = \frac{c}{d}$, where $b \neq 0$ and $d \neq 0$.

Proportionality Constant See *Constant of variation*.

Purchase Price The price of merchandise before the sales tax is added.

Pythagorean Theorem For right triangles, the sum of the squares of the lengths of the legs of the triangle equals the square of the length of the hypotenuse. That is, $a^2 + b^2 = c^2$.

Quadrants The four regions formed by the x- and y-axes in a rectangular coordinate system.

Quadratic Equation (in One Variable) A second-degree polynomial equation that can be written in the form $ax^2 + bx + c = 0$, where a, b, c and are real numbers and $a \neq 0$. Quadratic equations in this form are said to be in standard form.

Quadratic Formula The solutions to the quadratic equation $ax^2 + bx + c = 0, a \neq 0$, are given by the formula $x = \dfrac{-b \pm \sqrt{b^2 - 4ac}}{2a}$.

Quadratic Function A second-degree polynomial function of the form $f(x) = ax^2 + bx + c$, where a, b, c and are real numbers and $a \neq 0$. Every quadratic function has a U-shaped graph called a *parabola*.

Quadratic in Form An expression is considered to be quadratic in form when the variable part of the first term can be written as the square of the variable factor in the middle term by using an appropriate substitution.

Quadratic Trinomials Second-degree trinomials of the form $ax^2 + bx + c$.

Quarterly Happening once each quarter (four times each year). See *Annually*.

Quotient The result of division.

Quotient Rule for Exponents When dividing exponential expressions with the same nonzero base, subtract the denominator exponent from the numerator exponent and keep the common base. That is, $\dfrac{a^m}{a^n} = a^{m-n} (a \neq 0)$.

Quotient Rule for Radicals If $\sqrt[n]{a}$ and $\sqrt[n]{b}$ are real numbers and $b \neq 0$, then $\sqrt[n]{\dfrac{a}{b}} = \dfrac{\sqrt[n]{a}}{\sqrt[n]{b}}$.

Quotient Rule for Square Roots If $\sqrt{a}$ and $\sqrt{b}$ are real numbers and $b \neq 0$, then $\sqrt{a/b} = \sqrt{a}/\sqrt{b}$.

Quotient-to-Power Rule When raising a quotient to a power, raise both the numerator and denominator to the common exponent. $\left(\dfrac{a}{b}\right)^n = \dfrac{a^n}{b^n}$.

Radical Equation An equation that contains at least one radical expression with a variable in the radicand.

Radical Expression An expression that contains a radical sign ($\sqrt{}$).

Radical Function A function that contains one or more radical expressions.

Radical Sign ($\sqrt{}$) The symbol used to indicate roots of real numbers.

Radicand The expression beneath the radical sign.

Radius (of a Circle) The fixed distance, r, from the center of the circle, (h, k), to any point that lies on the circle. The length of the radius is half the length of the diameter.

Range The set of all values for the dependent variable. These are the second coordinates in the set of ordered pairs that define a relation and are also called *output values*.

Rate A special type of ratio in which quantities of different types are compared.

Rate of Change The ratio of the vertical change to the horizontal change when moving from one point on the graph to another. The slope of the line connecting the two points is equal to the average rate of change.

Rate of Work The number of jobs that can be completed in a given unit of time. If one job can be completed in t units of time, then the rate of work is given by $1/t$.

Ratio A comparison of two quantities, usually in the form of a quotient. There are three common notations used to express ratios: colon (e.g., 7:12); *to* (e.g., 7 to 12); and a fraction (quotient) (e.g., $\dfrac{7}{12}$).

Rational Equation A statement in which two rational expressions are set equal to each other.

Rational Exponent of the Form $a^{\frac{1}{n}}$ If n is an integer such that $n \geq 2$ and if $\sqrt[n]{a}$ is a real number, then $a^{\frac{1}{n}} = \sqrt[n]{a}$.

Rational Exponent of the Form $a^{\frac{m}{n}}$ If $\frac{m}{n}$ is a rational number in lowest terms, m and n are integers such that $n \geq 2$, and $\sqrt[n]{a}$ is a real number, then $a^{\frac{m}{n}} = \left(\sqrt[n]{a}\right)^m = \sqrt[n]{a^m}$.

Rational Expression An expression that can be written as the quotient $\dfrac{P}{Q}$ of two polynomials P and Q with $Q \neq 0$.

Rational Function A rational function is any function of the form $R(x) = \dfrac{P(x)}{Q(x)}$, where $P(x)$ and $Q(x)$ are polynomial functions with $Q(x) \neq 0$.

Rational Inequalities An inequality that can be written as $R(x) < 0, R(x) > 0, R(x) \leq 0, R(x) \geq 0$, where $R(x) = \dfrac{P(x)}{Q(x)}$ is a rational function.

Rational Number A number that can be written as a fraction $\frac{p}{q}$, where p and q are integers and $q \neq 0$.

Rationalizing a Denominator The process of removing radicals so that the denominator contains only a rational number.

Real Number A real number is any number that is either rational or irrational. Combining the set of rational numbers with the set of irrational numbers forms the set of real numbers, represented by $\mathbb{R}$.

Real Number Line A graph that represents the set of all real numbers, also known simply as the *number line*.

Real Part of a Complex Number For a complex number in the standard form $a + bi$ the number a is called the *real part of the complex number*.

Reciprocals Two nonzero numbers whose product is 1. Also called *multiplicative inverses*.

Rectangle A four-sided polygon in which the sides meet at right angles. Its opposite sides are equal in length.

Rectangular Coordinate System A system that consists of two number lines, one horizontal and one vertical. The horizontal number line is the *x*-axis and the vertical number line is the *y*-axis. The two axes intersect at a point called the *origin*. The plane represented by this system is called the *coordinate plane*. The axes divide the plane into four regions called *quadrants*. Also called the *Cartesian coordinate system*.

Rectangular Solid A solid that consists of six rectangular surfaces called its faces. A line segment where two faces touch (or intersect) is an edge of the solid. A corner point where three faces touch (or intersect) is a *vertex*. Opposite faces have the exact same shape and size.

Recursive Sequence A sequence in which each term is defined using one or more of its previous terms.

Regular Polygon A polygon whose sides are all of equal length and whose angles are all of equal measure.

Reference Rectangle A rectangle whose midpoints of each side are the vertices of a hyperbola or the endpoints of a conjugate axis.

Relation A correspondence between two sets of numbers that can be represented by a set of ordered pairs.

Relative Growth Rate See *Exponential Growth*.

Relative Decay Constant See *Exponential Decay*.

Remainder The amount that is left over after division, when the divisor can no longer divide into the dividend wholly.

Remainder Theorem If a polynomial $f(x)$ is divided by $x - c$, then the remainder is $f(c)$.

Restricted Value A restricted value for an algebraic expression is a value that makes the expression undefined.

Restricted Values for Rational Expressions Any value that causes the denominator of a rational expression to equal zero is a restricted value for that rational expression.

Revenue The dollar amount received from selling x items at a price of p dollars per item.

Right Angle An angle measuring 90°.

Right Triangle A triangle with a 90° angle.

Rise The vertical change between two points (x_1, y_1) and (x_2, y_2). We compute this by finding the difference of the *y*-coordinates: $y_2 - y_1$.

Root If c is a real number such that $f(c) = 0$ then c is called a root of the function f. Also called a *zero* of the function.

Roster Method A method of displaying the elements of a set in a list.

Repeating Decimal A decimal number in which a fixed number of digits form a pattern that repeats without end. A repeating decimal is indicated by placing a bar over the pattern of digits that repeat or by writing three dots (. . .), called an *ellipsis*, behind the established pattern. A repeating decimal can always be written as a fraction.

Row-Echelon Form A matrix form in which there are 1's along the diagonal from the upper left to lower right, and 0's below the diagonal.

Row Operation An operation on a row of a matrix that results in an equivalent matrix. Applying any of the following operations to a matrix will result in an equivalent matrix. (1) Interchange any two rows. (2) Multiply the elements of a row by the same nonzero constant. (3) Add a multiple of one row to another row.

Run The horizontal change between two points (x_1, y_1) and (x_2, y_2). We compute this by finding the difference of the *x*-coordinates: $x_2 - x_1$.

Sales Tax An amount charged by a state or local government when a purchase is made. Typically, the sales tax is a percent of the purchase price.

Sales Tax Rate If sales tax is charged as a percent of the price of the merchandise purchased, this percent is called the *tax rate*.

Satisfy A value satisfies an equation or inequality if, when substituted for the variable, the result is a true statement.

Scale An instrument for weighing or measuring.

Scalene Triangle A triangle with three different side lengths and three different angle measures.

Scatter Plots Graphs that display data as a set of points.

Scientific Notation A number is written in scientific notation if it has the form $a \times 10^n$, where a is a real number, such that $1 \leq |a| < 10$ and n is an integer.

Semiannually Happening twice each year. See *Annually*.

Series The sum of the terms of a sequence.

Set A collection of numbers or objects.

Set-Builder Notation A mathematical notation for defining a set by stating the properties that its members must satisfy.

Similar Triangles Triangles that have the same shape but not necessarily the same size. For similar triangles, corresponding angles are equal, and corresponding sides have lengths that are proportional.

Simple Interest Found by the formula $I = Prt$, where I is the interest earned or paid, P is the principal (money invested or borrowed), r is the interest rate per time period (in decimal form), and t is the number of time periods that the money is invested or borrowed.

Simplest Form A fraction is in simplest form, or lowest terms, when the numerator and denominator have no common factors other than 1.

Simplified In the simplest form possible.

Simplified Exponential Expressions An exponential expression is simplified if (1) there are no parentheses or other grouping symbols, (2) there are no zero or negative exponents, (3) there are no powers raised to powers, and (4) each base occurs only once.

Simplified Polynomial A polynomial is a simplified polynomial if all of its terms are simplified and none of its terms are like terms.

Simplified Term If a term contains a single constant factor (possibly a simplified fraction) and if none of the variable factors can be combined using the rules for exponents, then it is a simplified term.

Simplify a Complex Rational Expression To rewrite a rational expression in the form P/Q, $Q \neq 0$, where P and Q are polynomials with no common factors.

Slope The ratio of the vertical change in y, or rise, to the horizontal change in x, or run.

Slope 0 The slope of a horizontal line.

Slope Formula Given two points, (x_1, y_1) and (x_2, y_2), on the graph of a line, the slope m of the line is given by

$$m = \frac{y_2 - y_1}{x_2 - x_1}, \text{ where } x_1 \neq x_2.$$

Slope-Intercept Form A linear equation in two variables of the form $y = mx + b$ or $f(x) = mx + b$ is written in slope-intercept form, where m is the slope of the line and b is the y-intercept.

Solution A value that, when substituted for the variable, makes the equation or inequality true.

Solution Region To graph a system of linear inequalities, graph each inequality in the system and find the region they all have in common, if any. This common region is called the solution region.

Solution Set The set of all solutions to a given equation or inequality.

Solution to a System of Linear Equations in Three Variables An ordered triple that, when substituted for the variables, make all equations in the system true.

Solution to a System of Linear Equations in Two Variables An ordered pair that, when substituted for the variables, makes all equations in the system true.

Solution to a System of Linear Inequalities in Two Variables An ordered pair that, when substituted for the variables, makes all inequalities in the system true.

Solution to an Equation in Two Variables An ordered pair of values that, when substituted for the variables, makes the equation true.

Solve To solve an equation or inequality means to find its solution set, or the set of all values that make the equation or inequality true.

Solve a Formula for a Given Variable To isolate the variable on one side of the equation.

Solve an Equation To find every value that, when substituted for the variable, makes the equation true.

Sphere A sphere consists of all points in three dimensions that are located at an equal distance from a fixed point called the *center*. A *radius* extends from the center to the surface of a sphere. A *diameter* extends from one side of the sphere to the other, through the center.

Square A rectangle with four equal sides.

Square Matrix A matrix with an equal number of rows and columns.

Square Root For $a > 0$, $\sqrt{a}$ is the positive or principal square root of a. That is, $\sqrt{a} = b$ only if $b^2 = a$ and $b > 0$. Also, $\sqrt{0} = 0$.

Square Root Function A function involving a square root. The basic square root function is $f(x) = \sqrt{x}$.

Square Root Property If u is an algebraic expression and k is a real number, then $u^2 = k$ is equivalent to $u = -\sqrt{k}$ or $u = \sqrt{k}$. Equivalently, if $u^2 = k$ then $u = \pm\sqrt{k}$.

Squaring Property of Equality If A and B represent algebraic expressions, then any solution to the equation $A = B$ is also a solution to the equation $A^2 = B^2$.

Standard Form of a Complex Number A complex number written as $a + bi$, where a and b are real numbers and i is the imaginary unit.

Standard Form of a Quadratic Function A quadratic function is in standard form if it is written as $f(x) = a(x - h)^2 + k$. The graph is a parabola with vertex (h, k). The parabola "opens up" if $a > 0$ or "opens down" if $a < 0$. Also called the *Vertex Form*.

Standard Form of the Equation of a Circle The standard form of the equation of a circle with center (h, k) and radius r is $(x - h)^2 + (y - k)^2 = r^2$.

Straight-Line Depreciation An accounting method used for deducting losses in the value of equipment in equal amounts each year as it ages.

Strict Inequalities An inequality containing one or more of the following inequality symbols: $<$, $>$, $\neq$. There is no possibility of equality in a strict inequality.

Subscript Notation Use of subscripts to distinguish between different terms or quantities.

Subset If every element of set A is also an element of set B, then set A is a subset of set B, denoted as $A \subseteq B$.

Substitution A process in which a value or algebraic expression is substituted for a variable, or vice versa.

Substitution Method Involves solving one of two equations for one variable, substituting the resulting expression into the other equation, and then solving for the remaining variable.

Subtract To take one number away from another number to find the difference of the two numbers.

Subtraction Property of Equality See *Addition Property of Equality*.

Subtrahend The number being subtracted.

Sum The result of adding two real numbers.

Summation Notation A convenient way to express the sum of a series of terms using a short-hand notation. Also called *sigma notation*.

Supplementary Angles Two angles with measures that add to 180°. Also called *supplements*.

Supplements See *Supplementary Angles*.

Surface Area The total area of the surface on all sides of a solid. Measured in square units.

Surface Area of a Cube Found by the formula $SA = 6s^2$, where s is the side length.

Surface Area of a Rectangular Solid Found by the formula $SA = 2lw + 2lh + 2wh$, where l is the length, w is the width, and h is the height.

Surface Area of a Regular Pyramid Found by the formula

$$SA = B + \frac{1}{2}Pl,$$ where B is the area of the base, P is the

perimeter of the base, and l is the slant height.

Surface Area of a Right Circular Cylinder Found by the formula $SA = 2\pi r^2 + 2\pi rh$, where r is the radius and h is the height.

Surface Area of a Right Cone Found by the formula $SA = \pi rs + \pi r^2$, where r is the radius and $s = \sqrt{r^2 + h^2}$ is the slant height.

Surface Area of a Sphere Found by the formula $SA = 4\pi r^2$, where r is the radius.

Symmetric A two-dimensional graph is symmetric if it has mirror images (reflections) on opposite sides of a dividing line (called an *axis of symmetry*).

Synthetic Division A procedure for dividing a polynomial by a binomial of the form $x - c$. The procedure uses only the coefficients of the terms.

System of Linear Equations in Three Variables A collection of two or more linear equations in three variables considered together.

System of Linear Equations in Two Variables A collection of two or more linear equations in two variables considered together.

System of Linear Inequalities in Two Variables A collection of two or more linear inequalities in two variables considered together.

Term A constant, a variable, or the product of a constant and one or more variables raised to powers.

Terminating Decimal A decimal that ends, or has a finite number of digits after the decimal point.

Terms of the Polynomial The monomials that make up a polynomial.

Test Point A point used to find the half-plane that contains the ordered pair solutions to a linear inequality in two variables.

Total Revenue Found by multiplying the number of units sold by the price per unit, $R = xp$.

Translation When every point on a graph is shifted the same distance in the same direction.

Transverse Axis of a Hyperbola An imaginary line segment whose endpoints are the vertices of the hyperbola.

Trapezoid A four-sided polygon with one pair of parallel sides called bases.

Triangle A three-sided polygon with the measures of the angles adding to 180°.

Trinomial A simplified polynomial with three terms.

Undefined Slope The slope of a vertical line, because the slope has a denominator of 0 and division by 0 is undefined.

Union For any two sets A and B, the union of A and B is given by $A \cup B$ and represents the elements that are in set A or in set B.

Unit Rate A rate that has a denominator of 1 unit.

Unit Distance The distance from 0 to 1 represents the unit distance (the length of one unit) for a number line.

Upper Bound The smallest number that is greater than or equal to all values in an interval. For example, the interval $[6, 9]$ has an upper bound of 9.

Upper Limit of Summation The value that indicates the number of the last term of a sequence to be included in a series.

Variable A symbol (usually a letter) that represents a value that can change. For example, the variable x could represent 1 in one situation and 100 in another.

Variable Term A term that contains one or more variables.

Variation Equation An equation used in variation problems to describe how one quantity changes with respect to one or more additional quantities.

Varies Directly See *Direct Variation.*

Varies Inversely See *Inverse Variation*

Velocity The rate of change of distance with respect to time. The absolute value of velocity is speed.

Vertex of a Parabola The lowest point (if the parabola opens *up*) or the highest point (if the parabola opens *down*) on the graph of the parabola.

Vertex Form of a Quadratic Function See *Standard Form of a Quadratic Function.*

Vertex Formula for a Parabola Given a quadratic function of the form $f(x) = ax^2 + bx + c, a \neq 0$, the vertex of the parabola is given by $(h, k) = \left(-\dfrac{b}{2a}, f\left(-\dfrac{b}{2a} \right) \right)$.

Vertex of a Triangle A corner point of a triangle. The plural of vertex is *vertices*. A triangle has three vertices.

Vertical Compression If $0 < a < 1$, then the graph of $y = af(x)$ is a vertical compression of the graph of $y = f(x)$.

Vertical Line A line perpendicular to the *x*-axis that passes through point $(a, 0)$. The equation of the line is given by $x = a$. Vertical lines have undefined slope.

Vertical Line Test If a vertical line intersects the graph of a relation at more than one point, then the relation is not a function. If every vertical line intersects the graph of a relation at no more than one point, then the relation is a function.

Vertical Stretch If $a > 1$, the graph of $y = af(x)$ is a vertical stretch of the graph of $y = f(x)$.

Vertices The plural of *vertex.*

Vertices of an Ellipse The two points lying farthest apart on the ellipse. The vertices are on the major axis and are the points at which the ellipse makes its sharpest turns.

Vertices of a Hyperbola The two points on opposite branches of the hyperbola that are closest together. The vertices are the endpoints of the transverse axis and are the points at which the hyperbola makes its sharpest turns.

Volume The amount of space within a solid. Typically measured in cubic units.

Volume of a Cube Found by the formula $V = s^3$, where s is the side length.

Volume of a Rectangular Solid Found by the formula $V = lwh$, where l is the length, w is the width, and h is the height.

Volume of a Regular Pyramid Found by the formula $V = \dfrac{1}{3}Bh$, where B is the area of the base and h is the height.

Volume of a Right Circular Cylinder Found by the formula $V = \pi r^2 h$, where r is the radius and h is the height.

Volume of a Right Cone Found by the formula $V = \dfrac{1}{3}\pi r^2 h$, where r is the radius and h is the height.

Volume of a Sphere Found by the formula $V = \dfrac{4}{3}\pi r^3$, where r is the radius.

Whole Number An element of the set $\mathbb{W} = \{0, 1, 2, 3, 4, 5, \ldots\}$.

x-Axis The horizontal axis in the rectangular coordinate system.

x-Coordinate The first number in an ordered pair. Also called the *abscissa.*

x-Intercept The x-coordinate of a point where a graph crosses or touches the x-axis. (The y-coordinate is 0.)

y-Axis The vertical axis in the rectangular coordinate system.

y-Coordinate The second number in an ordered pair. Also called the *ordinate.*

y-Intercept The y-coordinate of a point where a graph crosses or touches the y-axis. (The x-coordinate is 0.)

Zero of a Function If c is a real number such that $f(c) = 0$, then c is called a zero, or root, of the function f.

Zero-Power Rule A nonzero base raised to the 0 power equals 1. That is, $a^0 = 1(a \neq 0)$.

Zero Product Property If A and B are real numbers or algebraic expressions and $A \cdot B = 0$, then $A = 0$ or $B = 0$.

Zero Slope See *Slope 0.*

Answers

R.1 Exercises

1. $\in$ **3.** $\notin$ **5.** $\in$ **7.** $\{3, 4, 5, 6, 7, 8, 9\}$ **9.** $\{21, 22, 23, \dots\}$ **11.** integer, rational number, real number
13. irrational number, real number **15.** integer, rational number, real number **17.** rational number, real number
19. natural number, whole number, integer, rational number, real number **21.** rational number, real number

23. a. 1 **b.** $0, 1$ **c.** $-17, 0, 1$ **d.** $-17, -\dfrac{25}{29}, 0, 0.331, 1$ **e.** $-\sqrt{5}, \dfrac{\pi}{2}$ **25.**

27. **29.** $>$ **31.** $<$ **33.** $>$ **35.** $>$ **37.** true **39.** true **41.** 18

43. $\dfrac{3}{8}$ **45.** 0.374

R.2 Exercises

1. 3 **3.** -2 **5.** $-\dfrac{11}{3}$ **7.** -2.3 **9.** 13 **11.** $\dfrac{4}{15}$ **13.** -3 **15.** 11 **17.** $\dfrac{40}{21}$ **19.** -2.6 **21.** 72 **23.** 153

25. $-\dfrac{5}{2}$ **27.** -48.6 **29.** $\dfrac{1}{12}$ **31.** -9 **33.** -25 **35.** 0 **37.** $-\dfrac{27}{20}$ **39.** 0.15 **41.** 512 **43.** -64 **45.** $\dfrac{81}{16}$
47. 0.00032 **49.** 13 **51.** 0.8 **53.** 6 **55.** -10 **57.** $c + ab$ **59.** $(4 + a) + 11$ **61.** $5x + 40$ **63.** $3x - 18$
65. $-7m + 7n - 35$ **67.** inverse property of addition **69.** identity property of multiplication **71.** commutative

property of addition **73.** -108 **75.** -0.1 **77.** 9 **79.** -8 **81.** 1223 **83.** 5 **85.** 1 **87.** $\dfrac{40}{7}$ **89.** $\dfrac{3}{2}$

91. -35 **93.** -11

R.3 Exercises

1. 225 **3.** -9 **5.** -3 **7.** 1.825 **9.** -7 **11.** 91 **13.** 7 **15.** $\dfrac{26}{5}$ **17.** 44 **19.** -1 **21.** $-5z$
23. $13ab$ **25.** $-10a + 30$ **27.** $4x^2 - 2xy - 3y^2$ **29.** $-1.35c^2$ **31.** $9t - 25$ **33.** $19x - 8$ **35.** $3x^2 - 2$
37. $22x^2 + 18x + 8$ **39.** $x + 10$ **41.** $5x + 9$ **43.** $2(x + 4)$ **45.** $3x - 9x$

1.1 Exercises

1. yes **3.** yes **5.** yes **7.** no **9.** $\{3\}$ **11.** $\{27\}$ **13.** $\{13\}$ **15.** $\{8\}$ **17.** $\{-9\}$ **19.** $\dfrac{5}{6}$ **21.** $\{3\}$

23. $\{4.8\}$ **25.** $\{-3\}$ **27.** $\{10\}$ **29.** $\{-1\}$ **31.** $\{0\}$ **33.** $\{-4\}$ **35.** $\{-2\}$ **37.** $\left\{\dfrac{17}{30}\right\}$ **39.** $\{-1\}$

41. $\left\{-\dfrac{8}{15}\right\}$ **43.** $\left\{\dfrac{2}{29}\right\}$ **45.** $\{1\}$ **47.** $\{4\}$ **49.** $\{2.24\}$ **51.** $\{8\}$ **53.** $\{-2\}$ **55.** contradiction, $\varnothing$

57. Identity, $\mathbb{R}$ **59.** Identity, $\mathbb{R}$ **61.** $x + 13 = 38$ **63.** $\dfrac{x}{4} - 5 = 31$ **65.** $\dfrac{x + 2.7}{6} = \dfrac{x}{9}$ **67.** $\dfrac{x}{2} = x - 19$

69. $h \approx 57.71$ inches   **71.** $N = 22$ sliders   **73.** Hannah: \$16,000, Taylor: \$19,000   **75.** *The Wizard of Oz*: 101 minutes, *Gone with the Wind*: 238 minutes.   **77.** Australian Open titles: 2, U.S. Open titles: 5, Wimbledon titles: 7   **79.** 2.8 lbs
81. Women's singles: 18, Women's doubles: 31, Mixed doubles: 10   **83.** \$75.80   **85.** 57, 59   **87.** 217, 219, 221, 223

1.2 Exercises

1. yes   **3.** no   **5.** no   **7.** (number line from −5 to 5)   **9.** (number line from −5 to 5)

11. (number line from −5 to 5)   **13.** (number line from −5 to 5)   **15.** $(-\infty, 1]$ **17.** $(-5.4, \infty)$ **19.** $[-6, 11]$

21. $\left[\dfrac{3}{2}, \dfrac{7}{3}\right)$ **23.** $(-\infty, 0)$ **25.** (number line from −5 to 5), $(-\infty, 2)$ **27.** (number line from −5 to 5), $(2, \infty)$

29. (number line from −5 to 5), $[2, \infty)$ **31.** $\left(-\dfrac{3}{2}, \infty\right)$, (number line from −5 to 5)   **33.** (number line from −10 to 0),

$(-\infty, -2)$ **35.** (number line from −3 to 7), $(-\infty, 5)$ **37.** $(-\infty, 0)$ **39.** $[-31, \infty)$ **41.** $(-\infty, 0)$

43. $(1, \infty)$ **45.** $(4, \infty)$ **47.** $\left(-\infty, -\dfrac{145}{46}\right]$ **49.** $(-2, \infty)$ **51.** $(-\infty, -5.5]$ **53.** $\{\}$ or $\varnothing$ **55.** $\{\}$ or $\varnothing$

57. $[-7, 3)$ **59.** $[-3, 1)$ **61.** $\left(-\dfrac{3}{2}, 3\right)$ **63.** $(-1, 2]$ **65.** up to 9 gigabytes   **67.** 105 guests

69. more than 250 miles   **71.** greater than \$148,850 and less than or equal to \$226,850

1.3 Exercises

1. $\{-2, 0, 6\}$ **3.** $\{-6, -4, -2, 0, 1, 2, 4, 5, 6, 9, 15\}$ **5.** $\{x \mid 2 \le x < 9\}$ **7.** $\{x \mid x \le -2\}$ **9.** $\{x \mid x \le -3 \text{ or } x \ge 5\}$
11. $\{x \mid \text{any real number}\}$ **13.** $(-8, 3)$ **15.** $[-3, 0) \cup (0, \infty)$ **17.** $(1, 4]$ **19.** $(-15, 2)$ **21.** $(-\infty, -3] \cup (4, \infty)$
23. $(-\infty, 0] \cup (6, \infty)$ **25.** $(-\infty, \infty)$ **27.** $(-\infty, -6)$ **29.** $(-\infty, -2]$

1.4 Exercises

1. $\{-3, 3\}$ **3.** $\left\{-\dfrac{5}{3}, -\dfrac{2}{3}\right\}$ **5.** $\left\{0, \dfrac{16}{3}\right\}$ **7.** $\left\{-\dfrac{3}{7}\right\}$ **9.** $\{\}$ or $\varnothing$ **11.** $\left\{-7, \dfrac{13}{3}\right\}$ **13.** $\left\{-\dfrac{1}{10}, \dfrac{7}{10}\right\}$ **15.** $\left\{0, \dfrac{3}{2}\right\}$

17. $\left[-\dfrac{1}{3}, \dfrac{1}{3}\right]$ **19.** $\left(-2, \dfrac{7}{2}\right)$ **21.** $\left\{\dfrac{7}{3}\right\}$ **23.** $(-\infty, -3) \cup (3, \infty)$ **25.** $\left(-\infty, \dfrac{9}{4}\right] \cup \left[\dfrac{15}{4}, \infty\right)$

27. $\left(-\infty, \dfrac{4}{9}\right) \cup \left(\dfrac{4}{9}, \infty\right)$ **29.** $\left[-1, \dfrac{5}{2}\right]$ **31.** $(-\infty, -8] \cup [-2, \infty)$ **33.** $(-\infty, \infty)$

1.5 Exercises

1. $w = \dfrac{A}{l}$ **3.** $h = \dfrac{3V}{\pi r^2}$ **5.** $r = \dfrac{E}{I} - R$ or $r = \dfrac{E - RI}{I}$ **7.** $h = \dfrac{A - 2\pi r^2}{2\pi r}$ **9.** $x = \sigma z + \mu$ **11.** 9 m by 18 m

13. 2100 square yards   **15.** 18 ft   **17.** 7 in.   **19.** 72 cubic inches   **21.** \$2143.05   **23.** 50°, 50°, 80°   **25.** 50°
27. soda: 60, popcorn: 32, candy: 18   **29.** \$2500 at 8.5% and \$1500 at 10%   **31.** 30 mph and 50 mph   **33.** 4 hours

35. 0.5 hour   **37.** 21.6%   **39.** 20 gallons   **41.** $11\dfrac{1}{4}$ pounds of the \$9.00 brand and $18\dfrac{3}{4}$ pounds of the \$5.00 brand

43. 28 ft to 68 ft

CHAPTER 2

2.1 Exercises

1.
Quadrant II

3.
Quadrant I

5.
x-axis

7. a. yes **b.** no **c.** yes **d.** yes **9. a.** no **b.** yes **c.** yes

d. yes **11. a.** -6 **b.** 6 **13.**

15.

17.

19.

21. *x*-intercept: 1, *y*-intercept: $\dfrac{3}{2}$ **23.** *x*-intercepts: -3 and 5, *y*-intercept: 5 **25.** *x*-intercepts: -5 and 3, *y*-intercept: 5

2.2 Exercises

1. independent: *x*, dependent: *y* **3.** independent: *b* and *h*, dependent: *A* **5.** independent: *x*, dependent: *y*

7. domain: $\{0, 1, 4\}$, range: $\{-2, -1, 0, 1, 2\}$ **9.** domain: $(-\infty, \infty)$ range: $[3, \infty)$ **11.** yes **13.** yes **15.** yes

17. yes **19.** no **21.** yes **23.** no **25.** yes **27.** no **29. a.** independent: percent with access to clean water,
dependent: percent undernourished **b.** $\{(87, 12), (99, 6), (99, 5), (86, 16), (91, 11)\}$ **c.** domain: $\{87, 99, 86, 91\}$,
range: $\{12, 6, 5, 16, 11\}$ **d.** No, it fails the vertical line test. **31. a.** Yes, it passes the vertical line test. **b.** independent:
year, dependent: price per gallon **c.** \$3.60 **d.** about 2017

2.3 Exercises

1. $f(x) = |2x - 5|$ **3.** $f(x) = -\sqrt{x} + 1$ **5.** $f(x) = -\dfrac{3}{4}x + 3$ **7.** $f(x) = 2x^2 - 5$ **9.** $f(x) = 2x^2 + 3$

11. 13 **13.** 1 **15.** 5 **17.** 54 **19.** **21.** **23.** **25. a.** \$20,000

b. 5 years **c.** domain: $[0, 5]$, range: $[0, 20,000]$ **d.** \$4000 **e.** \$4000 **27. a.** January, February, March, April, June,
August, September, October, and November **b.** May, July, and December **c.** December **d.** about \$13,900 **e.** about \$15,700
f. about \$150 per month **29.** first graph **31.**

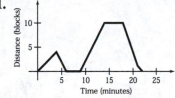

33. a. 442.225 m, the initial height of the object or the height of the building **b.** 319.725 m, the height of the object after
5 seconds **c.** -47.775 m, not possible, the object can't have a negative height **d.** 0 m, when the object hits the ground
e. domain: $[0, 9.5]$, range: $[0, 442.225]$ **f.**

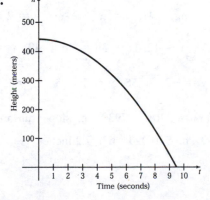

35. a. 34,515, the amount of money owed after 0 monthly payments   **b.** 20,709, the amount of money owed after 24 monthly payments   **c.** -6903, not possible, you can't owe a negative amount of money   **d.** $13,806   **e.** 60 months
f. domain: $[0, 60]$, range: $[0, 34,515]$ **g.**

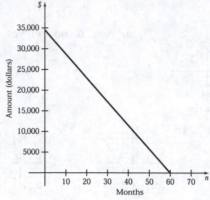

2.4 Exercises

1.   **3.**   **5.**   **7.**   **9.**   **11.**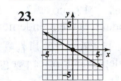

13.   **15.**   **17.**   **19.**   **21.** 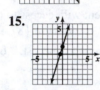   **23.**

25. 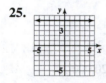   **27.**   **29.**

2.5 Exercises

1. $-\dfrac{8}{7}$ **3.** $m = \dfrac{3}{4}$ **5.** $m = -\dfrac{6}{5}$ **7.** $m = \dfrac{1}{2}$ **9.** $m = \dfrac{4}{3}$ **11.** $m = 0$ **13.** 0   **15.** slope: 4, y-intercept: -12

17. $m = -\dfrac{6}{5}, b = -\dfrac{4}{5}$ **19.** slope: $-\dfrac{5}{7}$, y-intercept: $\dfrac{12}{7}$ **21.** $m = 0, b = 10$ **23.**   **25.**

27.   **29.** slope: $\dfrac{1}{3}$, y-intercept: 3,   **31.**   **33.**

35. perpendicular   **37.** parallel   **39.** intersecting   **41.** $y = x - 2, x - y = 2$ **43.** $y = \dfrac{7}{9}x - \dfrac{8}{7}, 49x - 63y = 72$

45. $y = 3x + 18, 3x - y = -18$ **47.** $y = -\dfrac{4}{3}x + 12, 4x + 3y = 36$ **49.** $y = -x + 3, x + y = 3$

51. $y = \dfrac{2}{15}x + \dfrac{16}{15}, 2x - 15y = -16$ **53.** $y = -2$ **55.** $y = 1$ **57.** $y = -4x - 5, 4x + y = -5$

59. $y = \dfrac{1}{4}x + \dfrac{7}{2}, x - 4y = -14$ **61. a.** $T(x) = 240x + 7930$ **b.** slope: tuition and fees increase $240 per year, y-intercept: tuition and fees were $7930 in 2009   **c.** $10,090   **63.** 7.2 inches

2.6 Exercises

1. a. no **b.** no **c.** yes **3. a.** no **b.** yes **c.** no **5. a.** no **b.** yes **c.** no **7.**   **9.**

11.   **13.** 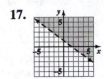 **15.**   **17.**  **19.**

21. a. $450c + 300s \le 90{,}000$ **b.** 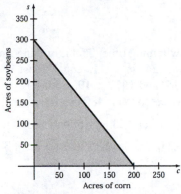 **c.** no $(93{,}750 > 90{,}000)$

23. a. $0.10d + 0.25q > 12.00$ **b.** 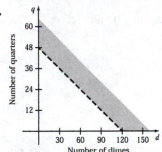 **c.** yes $(14.75 > 12.00)$

CHAPTER 3

3.1 Exercises

1. a. yes **b.** no **3. a.** no **b.** yes **5.** 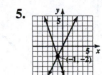 **7.** 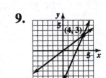 **9.** 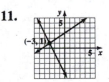 **11.**

13.   **15.** **17.** $(1, 3)$ **19.** $(1, 3)$ **21.** $\left(-\dfrac{1}{2}, -4\right)$ **23.** $\left(-\dfrac{5}{3}, \dfrac{26}{9}\right)$ **25.** $(-6, 6)$

27. $\left(\dfrac{5}{2}, \dfrac{1}{3}\right)$ **29.** $\left(\dfrac{2}{3}, \dfrac{1}{5}\right)$ **31.** $(6, 2)$ **33.** $(-2, -4)$ **35.** $(-2, 2)$ **37.** $(-1, 3)$ **39.** $(6, 9)$ **41.** $(-4, 7)$

43. $(1, -1)$ **45.** $\left(\dfrac{1}{2}, -1\right)$ **47.** inconsistent, $\varnothing$ **49.** $\{(x, y) \mid y = 2x + 6\}$ or $(x, 2x + 6)$ **51.** $\left(x, \dfrac{1}{4}x - \dfrac{1}{2}\right)$ or

$\left\{(x, y) \mid y = \dfrac{1}{4}x - \dfrac{1}{2}\right\}$ **53.** $\left(x, \dfrac{5}{7}x - 9\right)$ or $\left\{(x, y) \mid y = \dfrac{5}{7}x - 9\right\}$ **55.** inconsistent, $\{\,\}$ or $\varnothing$

57. Clayton: 12 years: Josh: 14 years   **59.** 40 nickels, 10 dimes   **61.** Missouri River: 2540 mi, Mississippi River: 2340 mi
63. engineering: $62,600; social science: $38,000   **65.** 2009: ≈ $0.77 trillion or $770,000,000,000

3.2 Exercises

1. yes: $(-1, 1, -2)$, no: $(1, -1, 2)$ **3.** $(-1, 2, 3)$ **5.** $\left(\dfrac{1}{4}, -\dfrac{1}{2}, \dfrac{3}{8}\right)$ **7.** $(19, -7, -15)$ **9.** $(0, -1, 1)$ **11.** $(-2, 3, 1)$

13. $\left(-1, \dfrac{2}{3}, \dfrac{1}{3}\right)$ **15.** no solution   **17.** $(-3y + 2z + 4, y, z)$ **19.** $(x, 8 - x, 5 - x)$ **21.** long swordsmen: 20,
spearmen: 20, crossbowmen: 35   **23.** cheese pizza: 10, supreme pizza: 10, pepperoni pizza: 15   **25.** one-point free throws: 207,
two-point field goals: 126, three-point field goals: 87   **27.** 6 townhouses, 13 houses, 9 condos   **29.** tuition: $7600,
materials: $1250, room/board: $9000

3.3 Exercises

1. Yessica: 9 mph, Zoe: 25 mph   **3.** 450 minutes   **5.** swimming time: 0.5 hr, running time: 1.0 hr   **7.** plane speed: 210
miles/hr, wind speed: 14 miles/hr   **9.** width: 75 cm, length: 124 cm   **11.** equal sides: 30 cm long, short side: 20 cm long
13. 40°, 40°, 100°   **15.** angle 1: 25°, angle 2: 65°   **17.** 15° and 165°   **19.** $x = 50°, y = 90°, z = 40°$
21. 1 glass of milk: 11 carbs, 1 snack bar: 13 carbs   **23.** cheeseburger: 480 calories, french fries: 395 calories

25. children: 85, adults: 194   **27.** $46\dfrac{2}{3}$ oz of Irish Breakfast tea and $53\dfrac{1}{3}$ oz of buttered rum tea

29. 18-carat gold: 140 ounces, 12-carat gold: 210 ounces   **31.** 1.5 oz of 15% and 3 oz of 30%   **33.** $200 at 6%, $1,400 at 8%,
and $600 at 9%

3.4 Exercises

1. a. yes **b.** no **c.** no   **3. a.** yes **b.** no **c.** yes   **5. a.** no **b.** no **c.** yes   **7.**   **9.**

11. 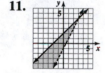   **13.**   **15.** 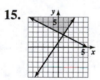   **17.** {} or ∅   **19.** 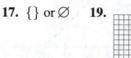   **21.**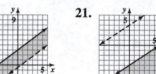

23. 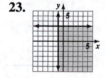   **25.**   **27.**   **29.** 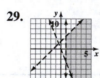   **31.** 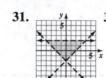   **33.**

35. a. 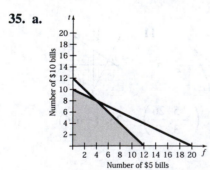   **b.** yes **c.** yes   **37. a.** 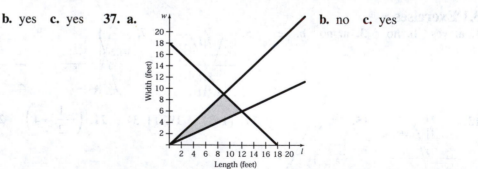   **b.** no **c.** yes

CHAPTER 4

4.1 Exercises

1. 7^6 **3.** y^{11} **5.** w^6 **7.** $30a^{10}$ **9.** $20a^7b^9$ **11.** $-15x^6yz^3$ **13.** b^4 **15.** $(-10)^3 = -1000$ **17.** $4q^{12}$

19. x^2y^6z **21.** $-\dfrac{x^2y^4}{4}$ **23.** 1   **25.** -1 **27.** 1   **29.** $\dfrac{1}{y^3}$ **31.** $\dfrac{6}{w^5}$ **33.** $\dfrac{1}{64}$ **35.** $\dfrac{1}{2^3} + \dfrac{1}{4} = \dfrac{3}{8}$ **37.** z^5

39. $(-3)^4 = 81$ **41.** $-12t^3$ **43.** $\dfrac{2^5}{9^2} = \dfrac{32}{81}$ **45.** x^{24} **47.** y^{14} **49.** $(-2)^{12}$ **51.** $243x^5$ **53.** $256m^{12}$

55. $64a^6b^{12}$ **57.** $16x^{12}y^8z^4$ **59.** $\dfrac{t^3}{4^3} = \dfrac{t^3}{64}$ **61.** $\dfrac{m^{20}}{n^{35}}$ **63.** $\dfrac{-64x^6}{y^{15}}$ **65.** $\dfrac{q^{36}}{p^{12}}$ **67.** $\dfrac{x^6}{216y^{15}}$ **69.** $\dfrac{-3y^4}{x^4}$ **71.** $\dfrac{q^{56}}{p^{16}}$

73. $\dfrac{49x^{12}}{y^{34}}$ **75.** $\dfrac{9a^{27}}{b^{23}}$ **77.** $-\dfrac{448a^4m^{21}}{b^6n^{15}}$ **79.** 8.9×10^{10} **81.** 4.06×10^{-7} **83.** 2.54×10^4 **85.** 5×10^{-9}

87. 0.00000297 **89.** 0.0000000008 **91.** $5{,}878{,}000{,}000{,}000$ **93.** 0.000053 **95.** 8×10^{11} **97.** 5×10^5

99. 4.26×10^2

4.2 Exercises

1. degree: 4, coefficient: 6   **3.** degree: 1, coefficient: $\dfrac{3}{8}$ **5.** degree: 0, coefficient: -13 **7.** degree: 4, coefficient: 4

9. 8   **11.** 6   **13.** $-\dfrac{2}{3}m^7 + m^5 - 9m^3 - 5m + 8$, degree: 7, leading coefficient: $-\dfrac{2}{3}$ **15.** $5x^2 - 9x + 4$, degree: 2, leading

coefficient: 5   **17.** $-3x + 9$, degree: 1, leading coefficient: -3 **19.** 5   **21.** 124   **23.** 7   **25.** $8x^3 - 3x^2 - 5x - 15$

27. $3w^3 + 4w^2 - 2w - 3$ **29.** $10a^2 - 9a - 7$ **31.** $3a^3 - 3a^2b + 10b^3$ **33.** $0.4m^3n^2 + 1.2m^2n^3 - 4.4mn$

35. $12y^3 + 5y^2 - 8y + 21$ **37.** $7m^3 - m^2 - 2m + 11$ **39.** $2.3x^2 - 3.2x + 2.9$ **41.** $-3x - 2xy + 2y$

43. $\dfrac{3a}{5} + \dfrac{3b}{8}$ **45.** $2x^3 - 14x^2 + 2x - 13$ **47.** $x^3 + 4x - 20$ **49.** $x^3 - 18x^2 - 4x - 4$

4.3 Exercises

1. $30x^7$ **3.** $8w^5$ **5.** $-\dfrac{3}{10}a^7b^9$ **7.** $-32x^2 + 72x$ **9.** $-x^4 - 2x^3 - 3x^2$ **11.** $3x^3y + 3x^2y^2$ **13.** $6.8a^5b^4 + 23.8a^4b^2$

15. $x^2 + 7x + 10$ **17.** $y^2 - \dfrac{11}{15}y - \dfrac{4}{5}$ **19.** $x^2 + 13x + 30$ **21.** $6x^2 + 19x + 10$ **23.** $3m^4 - 14m^2n^2 + 8n^4$

25. $x^2 - \dfrac{x}{6} - \dfrac{1}{6}$ **27.** $6x^3 - 4x^2 - 9x + 6$ **29.** $0.6a^3 - 0.08a^2b^2 + 9ab - 1.2b^3$ **31.** $y^2 + 10y + 25$

33. $100x^4 + 220x^2y + 121y^2$ **35.** $x^2 - 16x + 64$ **37.** $49p^2 - 28pq + 4q^2$ **39.** $x^2 - 100$ **41.** $9m^4 - 4n^2$

43. $x^3 + x^2 - 7x + 20$ **45.** $12a^3 - 17a^2b + 3ab^2 + 2b^3$ **47.** $2x^3 + 5x^2y - 5xy^2 + y^3$ **49.** $12x^4 - 33x^3 - 135x^2$

51. $x^3 + 6x^2 + 11x + 6$ **53.** $x^3 + 12x^2 + 48x + 64$ **55.** $-6x^4 + 6x^3y - 2x^2y + 2xy^2$ **57.** $6x^3 + 17x^2 - 26x + 8$

59. $9x^2 + 12xy + 4y^2 - 25$ **61.** $2x^2 + 11x - 40$ **63.** $0.42x^2 - 0.62x - 8.64$

4.4 Exercises

1. $6x^8$ **3.** $\dfrac{1}{6}$ **5.** $2x - 5$ **7.** $8y^5 - 4y^3 + 1$ **9.** $5x^5 + 3x^3 - 2x^2 + 8$ **11.** $2a - 7 - \dfrac{3}{a}$ **13.** $4x^2 + \dfrac{xy}{2} - y^2$

15. $x + 12$ **17.** $x^2 - 5x + 7$ **19.** $x^2 - 3x + 6 - \dfrac{34}{x + 5}$ **21.** $x + 5$ **23.** $3x^2 - 11x + 15 - \dfrac{3}{x^2 - 2}$

25. $4m^2 - 8m + 21 - \dfrac{49}{m + 2}$ **27.** $3x^3 + x$ **29.** $4x^2 - 8x + 21 - \dfrac{49}{x + 2}$ **31.** $x + 4 + \dfrac{3}{x - 5}$ **33.** $x^2 + 3x - 3$

35. $x^3 + 2x^2 + 9x + 18 - \dfrac{5}{x - 2}$ **37.** $15x^2 + 20x - 35$ **39.** $3x - 1 + \dfrac{3}{x - 2}, x \neq 2$ **41.** $2x^2 + 3x - 4$ **43.** 8

45. -10 **47.** $(x - 8)(x - 3)$ **49.** $(x + 1)$ is not a factor.

CHAPTER 5

5.1 Exercises

1. x^2 **3.** a^3b **5.** $6x^5$ **7.** 5   **9.** $18mn^4$ **11.** $4(5x + 2)$ **13.** $4xy(3x + 1)$ **15.** $4x^2(x^2 - 5x + 3)$

17. $4x^2y^2z^4(4yz^2 - 3x^2y^2 + 5xz)$ **19.** $-8x^2(x - 6)$ **21.** $-2(2x^2 + 8x - 9)$ **23.** $(x + 7)(11 + 3a)$

25. $(9y - 7)(x^2 + 11)$ **27.** $(x - 8)(x^2 + 3)$ **29.** $(m^2 + 3)(4m^3 - 3)$ **31.** $(x + 3)(y + 5)$

33. $(3x + 2)(3y + 4)$ **35.** $(x + 3)(5x + 4y)$ **37.** $(x^2 + 4)(x + 9)$ **39.** $(x^2 + 1)(x + 8)$

5.2 Exercises

1. $(x + 3)(x + 5)$ **3.** $(z + 4)(z + 12)$ **5.** $(y + 6)(y - 3)$ **7.** $(z - 8)(z + 4)$ **9.** $(p - 22)(p - 2)$
11. prime **13.** $(n + 8)(n + 30)$ **15.** prime **17.** $(y - 28)(y - 14)$ **19.** $8(x - 3)(x + 4)$
21. $2y^2(y + 5)(y + 12)$ **23.** $m^3(m - 12)(m - 5)$ **25.** $-(z - 8)(z + 12)$ **27.** $-(h - 6)(h + 16)$
29. $(x - 18y)(x - 2y)$ **31.** $6xy^2(x - 8y)(x - 4y)$ **33.** $(x + 3)(5x + 1)$ **35.** $(3w + 2)(7w + 2)$
37. $(y + 1)(5y - 8)$ **39.** $(p - 3)(6p - 1)$ **41.** prime **43.** $(3z - 4)(6z + 1)$ **45.** $(2y + 3)(11y - 10)$
47. $(3m - 2n)(4m - 3n)$ **49.** $(6m + 7n)(6m + 7n)$ or $(6m + 7n)^2$ **51.** $(3x + 1)(3x + 5)$ **53.** $(3m - 8)(4m - 5)$
55. $(2x - 3)(4x + 3)$ **57.** $(2m + 7)(11m + 2)$ **59.** $(3a - 1)(5a + 7)$ **61.** $(2p - 9q)(6p + q)$
63. prime **65.** $4(2x + 3)(3x - 5)$ **67.** $x^2(x + 5)(5x - 3)$ **69.** $3x^2y^2(2x + 5y)(6x - 5y)$
71. $-(4m + 3)(5m - 2)$ **73.** $-2w(2x + 7)(4x + 3)$ **75.** $(x^3 - 5)(x^3 + 4)$ **77.** $(x + y - 10)(5x + 5y + 12)$

5.3 Exercises

1. $(y - 4)(y + 4)$ **3.** $(x - y)(x + y)$ **5.** $(x^5 - 2)(x^5 + 2)$ **7.** $(3y - 1)(3y + 1)$ **9.** $(2x - 7y)(2x + 7y)$
11. $\left(2xy^2 + \frac{3}{2}z^4\right)\left(2xy^2 - \frac{3}{2}z^4\right)$ **13.** $3(2x - 3)(2x + 3)$ **15.** $5(4x^2 + 9)$ **17.** $3y(x - 2)(x + 2)$

19. $(x^2 + 1)(x + 1)(x - 1)$ **21.** $(x^2 + y^2)(x + y)(x - y)$ **23.** $5(x^4 + 1)(x^2 + 1)(x + 1)(x - 1)$
25. $(x - 1)^2$ **27.** $(2x + 7)^2$ **29.** prime **31.** $(m - 7n)^2$ **33.** $x(5y + 7)^2$ **35.** $(z^3 + 5)^2$
37. $(y - 5)(y^2 + 5y + 25)$ **39.** $(2x - 1)(4x^2 + 2x + 1)$ **41.** $(2z + 5)(4z^2 - 10z + 25)$
43. $(3x - y)(9x^2 + 3xy + y^2)$ **45.** $(xy - 3)(x^2y^2 + 3xy + 9)$ **47.** $2ab^2(a - 2b)(a^2 + 2ab + 4b^2)$
49. $3(x - 2)(x + 2)$ **51.** $(a + 2b)(a^2 - 2ab + 4b^2)$ **53.** $(x + 3)(x - 2)$ **55.** $5(z - 3)^2$ **57.** $(2x - 5)^2$
59. $x(5x - 1)(x + 3)$ **61.** prime **63.** $(3x + 2)^2$ **65.** $-3(p - 2q)(2p + q)$ **67.** $9y(1 + 2y)$ **69.** $(5m + 4n)^2$
71. $(x - y + 2)(x + y)$ **73.** $(x - 5)(x + 4)$ **75.** $3y(2y + 1)(8y - 3)$ **77.** $(p - 1)(p - 9)$

5.4 Exercises

1. $\{-1, 2\}$ **3.** $\left\{-3, -\frac{1}{2}\right\}$ **5.** $\left\{-\frac{5}{3}, -1, \frac{3}{2}\right\}$ **7.** $\{-2, -1\}$ **9.** $\left\{-\frac{1}{2}, 3\right\}$ **11.** $\{-2, -1, 0\}$ **13.** $\{-7, -2\}$

15. $\{-4, 7\}$ **17.** $\{-3, 3\}$ **19.** $\{-5, 5\}$ **21.** $\{-1, 0\}$ **23.** $\{-2, 3\}$ **25.** $\left\{-\frac{4}{3}\right\}$ **27.** $\left\{-\frac{5}{3}, 2\right\}$ **29.** $\{-3, 8\}$

31. $\{2, 10\}$ **33.** $\left\{-\frac{2}{3}, 4\right\}$ **35.** $\left\{-\frac{10}{3}, \frac{5}{2}\right\}$ **37.** $\{0, 7\}$ **39.** $\{-8, 0\}$ **41.** $\{-2, 0, 1\}$ **43.** $\left\{-2, -\frac{3}{7}, 2\right\}$

45. $\left\{-\frac{5}{2}, -\frac{1}{3}, \frac{5}{2}\right\}$ **47.** $\{1, 2, 3\}$ **49.** $\left\{-2, -\frac{5}{3}, 5\right\}$ **51.** $\{-1, 1, 2\}$ **53.** $\{-1, 4\}$ **55.** $\{-2, 3\}$

57. $\{-3, 3\}$ **59.** $\{-1\}$ **61.** $\{0, 3, 4\}$ **63.** $\{-3, -2, 2, 3\}$ **65.** 8 seconds **67.** 1.25 seconds **69.** 5 days
71. \$7 **73.** 130 feet **75.** 48 feet, 55 feet, 73 feet **77.** noon **79.** The width of the frame is 4 inches.
81. The sail has a base of 8 yards and a height of 12 yards.

CHAPTER 6

6.1 Exercises

1. $\{x \mid x \neq -2\}$ **3.** $\left\{x \mid x \neq -\frac{2}{5}\right\}$ **5.** $\{x \mid x \neq -2, 5\}$ **7.** $\left\{x \mid x \neq -4, \frac{2}{3}\right\}$ **9.** $\mathbb{R}$

11. $h(2) = 7; h(3) = 4; h(-3) = -\frac{1}{2}$ **13.** $R(1) = -1; R(0) = -\frac{1}{2}; R(-2)$ is undefined.

15. $f(-2) = 0; f(3) = -10; f(5)$ is undefined. **17.** $\frac{3}{5}$ **19.** $\frac{4}{9n^2}$ **21.** $\frac{n}{n - 2}$ **23.** $\frac{z - 5}{z + 7}$ **25.** simplest form

27. $\frac{2w - 1}{w + 3}$ **29.** $\frac{1}{x - 4}$ **31.** $\frac{x - 2}{2x^2 + 1}$ **33.** $\frac{x - 8y}{x - 4y}$ **35.** $\frac{1}{x + 2}$ **37.** -1 **39.** $-\frac{w + 2}{w + 8}$ **41.** $-\frac{1}{2}$

43. $\frac{1}{6x + 7}$ **45.** $\frac{a^2 + 2a + 4}{a + 2}$ **47.** simplest form **49.** $x - 2$

6.2 Exercises

1. $\dfrac{5b}{a}$ **3.** $\dfrac{z}{y^2}$ **5.** $\dfrac{5x}{21}$ **7.** $\dfrac{3}{5}$ **9.** $\dfrac{5(x-7)(x+1)}{16x(x+2)}$ **11.** $\dfrac{1}{(x+2)(x-1)}$ **13.** $\dfrac{3r(r+4)(r-4)}{r+2}$ **15.** $\dfrac{x+1}{x-1}$

17. $\dfrac{(x+1)(3x-2)}{(3x+2)(x-1)}$ **19.** 1 **21.** $\dfrac{3(x-7)(x-5)}{4x(x+5)}$ **23.** $\dfrac{5(x+2)(x-3)}{3(x-5)(x-2)}$ **25.** $\dfrac{x-2}{x+1}$ **27.** $\dfrac{x-2}{2(x+2)(x+1)}$

29. $-\dfrac{5x+1}{2x+1}$ **31.** 1 **33.** $\dfrac{x+3y}{3x-y}$ **35.** $\dfrac{y}{x^2+xy+y^2}$ **37.** 3 **39.** $\dfrac{(x+y)(x+3y)}{-2y^2}$ **41.** $\dfrac{5xy}{2}$ **43.** $\dfrac{x^2}{21(x-1)}$

45. $\dfrac{1}{3x}$ **47.** $\dfrac{(a-4)(a-3)}{2(a+3)}$ **49.** $(x+3)(x-1)$ **51.** $\dfrac{(2x+3)(3x+2)}{(x+2)(x+5)}$ **53.** $\dfrac{(x+2)(x+3)}{(x^2+3x+9)(x^2+2x+4)}$

55. $\dfrac{3(h+4)(h-2)}{(7h-2)(6h+1)}$ **57.** $\dfrac{(x-y)}{2}$ **59.** $\dfrac{(3x-y)(x+4y)}{3}$ **61.** $\dfrac{15x^3y}{16}$ **63.** $\dfrac{x+4}{x-2}$ **65.** $\dfrac{x+4}{x-2}$

67. $\dfrac{(3x-5)(y-4)}{(y+1)(2x-3)}$ **69.** $\dfrac{3x+8}{5(x+2)}$

6.3 Exercises

1. $\dfrac{6x}{7}$ **3.** $\dfrac{a^2+b^2+3}{5c^2}$ **5.** $\dfrac{x}{3a^3}$ **7.** $\dfrac{x}{8y^3+2}$ **9.** $x+5$ **11.** $\dfrac{5x-2}{5x-1}$ **13.** $\dfrac{x+3}{x-3}$ **15.** $\dfrac{1}{x+2}$ **17.** $\dfrac{2}{x+4}$

19. $18x^5$ **21.** $14a^3b^5c^3$ **23.** $12x(2x-1)(3x+1)$ **25.** $-(x-3)(x+2)$ **27.** $-(x-3)(x+1)(x+3)$
29. $(y-3)(y+3)^2$ **31.** $(2x+1)(x-3)(3x-2)$ **33.** $6x^2(3x-2)(x+5)(2x+1)$

35. $-(x-2)(x-1)(x+1)(x+2)$ **37.** $\dfrac{10x^3}{15x^4}$ **39.** $\dfrac{14a^3b^4}{21a^5b^3c^3}$ **41.** $\dfrac{7(x+3)(x-2)}{(x+2)(x-2)(x+3)(x-3)}$

43. $\dfrac{-5x(x+2)(x+1)}{(x^2-4)(x^2+x)}$ **45.** $\dfrac{(2x+1)(x+1)}{(x-1)(x+1)(x+2)}$ **47.** $\dfrac{x+2}{2x^2}$ **49.** $\dfrac{2(5x^2+6x+7)}{(x+2)(2x+1)}$

51. $\dfrac{3x+1}{x(x+1)}$ **53.** $\dfrac{11x^2-5x-10}{10x(x-2)}$ **55.** $\dfrac{x^2+4x+10}{2x(x+5)}$ **57.** $\dfrac{4x-6}{x-2}$ or $\dfrac{2(2x-3)}{x-2}$ **59.** $\dfrac{2x+1}{2x-5}$ **61.** $\dfrac{6}{x-3}$

63. $\dfrac{3(2x+3)}{(x-2)(x+2)}$ **65.** $\dfrac{2x-1}{2x+1}$ **67.** $\dfrac{x^2}{(x-2)(x-1)(x+1)}$ **69.** $\dfrac{5}{x+2}$ **71.** $\dfrac{5x^2-34x}{(x+6)(x-2)(x-6)}$

73. $\dfrac{x-3}{x-1}$ **75.** $\dfrac{3}{(x-1)(x+1)}$ **77.** $\dfrac{(2x+3)(x-15)}{(x+6)^2(2x-9)}$ **79.** $\dfrac{2}{x+2}$ **81.** $\dfrac{8x+6y}{x^2-y^2}$ or $\dfrac{2(4x+3y)}{(x+y)(x-y)}$

6.4 Exercises

1. 6 **3.** $\dfrac{5}{(x-3)(x-1)}$ **5.** $-\dfrac{x}{2x-3}$ or $\dfrac{x}{3-2x}$ **7.** $\dfrac{2(x+2)}{3(2x-1)}$ **9.** $\dfrac{3}{(x+1)(x+4)}$ **11.** $-\dfrac{3x-13}{5x-18}$ or $\dfrac{3x-13}{18-5x}$

13. $\dfrac{7x+1}{7x-1}$ **15.** $\dfrac{1}{(x-6)(x+5)}$ **17.** $\dfrac{x-4}{4x}$ **19.** $\dfrac{3}{x-5}$ **21.** $\dfrac{x-1}{x-7}$ **23.** $\dfrac{6x}{x^2+9}$ **25.** $\dfrac{6}{x-y}$

27. $\dfrac{7n}{n-7}$ **29.** $\dfrac{(x+2)(x+5)}{(x+1)(x-5)}$

6.5 Exercises

1. No, because $\dfrac{\sqrt{x^2-4x+1}}{x-5}$ is not a rational expression. **3.** yes **5.** $\left\{\dfrac{15}{2}\right\}$ **7.** $\{12\}$ **9.** $\left\{\dfrac{5}{21}\right\}$ **11.** $\{-15,-4\}$

13. $\{-12\}$ **15.** $\{\ \}$ or $\varnothing$ **17.** $\left\{w \mid w \neq -\dfrac{3}{2}, \dfrac{3}{2}\right\}$ **19.** $\{12\}$ **21.** $\{2\}$ **23.** $x=-\dfrac{4}{3}$ **25.** $x=\dfrac{1}{2}$

27. 127.5 minutes   **29.** He will need 2020 more pounds of corn seed.   **31.** 12 ohms and 36 ohms   **33.** 3 mph
35. 1.5 mph   **37.** It will take 18 minutes for both copy machines to copy the final exams.   **39.** 6 hours   **41.** 450 min

6.6 Exercises

1. a. $y = 3.5x$ **b.** $y = 24.5$ **3. a.** $y = 24x^2$ **b.** $y = 600$ **5. a.** $p = \dfrac{5}{8}n^3$ **b.** $p = 625$

7. She will be 40 feet below the surface.   **9.** 900 gallons could be pumped through the pipe.   **11. a.** $y = \dfrac{600}{x}$ **b.** $y = 20$

13. a. $n = \dfrac{518.4}{m^2}$ **b.** $p = 32.4$ **15.** The car will be 9 years old.   **17.** The satellite will weigh approximately 79.01 kg.

19. a. $y = \dfrac{15x}{z}$ **b.** $y = 20$ **21. a.** $P = 4.8 \text{ nm}^2$ **b.** $P = 67.2$ **23.** $5568   **25.** 45 kilograms per square meter

27. 18.75 ft or less

CHAPTER 7

7.1 Exercises

1. 10   **3.** $\dfrac{3}{11}$ **5.** not a real number   **7.** 7.416   **9.** 2.121   **11.** 3.550   **13.** 27   **15.** $8|x|$ **17.** $(2 - 9x)^2$

19. $9x^2$ **21.** $-11|x^5|$ **23.** $|x + 8|$ **25.** 7   **27.** $-\dfrac{3}{4}$ **29.** w^7 **31.** $-5z^2$ **33.** 4   **35.** $\dfrac{2}{5}$

37. $|x^5|$ **39.** 3.476   **41.** 2.187   **43.** 4.620   **45.** -7.786

7.2 Exercises

1. 9   **3.** not a real number   **5.** 2   **7.** 8   **9.** 2   **11.** 3   **13.** $\{x \mid x \geq -4\}$ or $[-4, \infty)$ **15.** $\{x \mid x \leq 3\}$ or $(-\infty, 3]$

17. $\left\{x \mid x \leq \dfrac{1}{3}\right\}$ or $\left(-\infty, \dfrac{1}{3}\right]$ **19.** $\mathbb{R}$ or $(-\infty, \infty)$ **21.** $\left\{x \mid x \geq -\dfrac{5}{2}\right\}$ or $\left[-\dfrac{2}{5}, \infty\right)$ **23.** $\left\{x \mid x \geq \dfrac{3}{32}\right\}$ or $\left[\dfrac{3}{32}, \infty\right)$

25. $\mathbb{R}$ or $(-\infty, \infty)$

27.   It looks like the graph of the square root function, but it is shifted two units to the right.

29.   It looks like the graph of the square root function, but it is shifted five units to the left.

31.   It looks like the graph of the square root function, but it is stretched by a factor of two.

33.   It looks like the graph of the cube root function, but it is shifted one unit to the left.

35. It looks like the graph of the cube root function, but it is shifted two units up.

37. It looks like the graph of the cube root function, but stretched by a factor of 3.

7.3 Exercises

1. $\sqrt{36} = 6$ **3.** $-\sqrt[4]{81} = -3$ **5.** $\sqrt[3]{27x^3} = 3x$ **7.** $\sqrt[5]{7xy}$ **9.** $(\sqrt{100})^3 = 1000$ **11.** $(\sqrt[4]{-16})^3$ is not a real number. **13.** 16 **15.** $\sqrt[3]{(-3ab)^7}$ or $(\sqrt[3]{-3ab})^7$ **17.** $10^{\frac{1}{2}}$ **19.** $\left(\dfrac{2x}{y}\right)^{\frac{1}{5}}$ **21.** $x^{\frac{4}{7}}$ **23.** $(2xy)^{\frac{4}{9}}$ **25.** $\dfrac{1}{25^{\frac{1}{2}}} = \dfrac{1}{5}$ **27.** $\left(\dfrac{8}{27}\right)^{\frac{2}{3}} = \dfrac{4}{9}$ **29.** $x^{\frac{5}{6}}$ **31.** $36^{\frac{1}{2}} = 6$ **33.** $\dfrac{q}{p^{\frac{11}{10}}}$ **35.** $\dfrac{2x^2}{y^{\frac{5}{2}}}$ **37.** $\dfrac{8}{27x^{\frac{7}{16}}y^{\frac{1}{2}}}$ **39.** $\sqrt[20]{x^{13}}$ **41.** $\sqrt{x}$ **43.** $\sqrt{3}$ **45.** $\sqrt[4]{35}$ **47.** $5x$ **49.** $6\sqrt{2}$ **51.** $-3\sqrt{7}$ **53.** $-5x^2$ **55.** $28xy\sqrt[3]{2xy^2}$ **57.** $xy^2z^2\sqrt[4]{xz}$ **59.** $6\sqrt{5}$ **61.** $6x^2y^2\sqrt{10y}$ **63.** $30x^2y$ **65.** $\dfrac{7}{9}$ **67.** $\dfrac{x\sqrt[3]{11}}{2}$ **69.** $\dfrac{x\sqrt[4]{21x}}{2}$ **71.** 3 **73.** $2x^5\sqrt{10}$ **75.** $-2m\sqrt[3]{3}$

7.4 Exercises

1. $13\sqrt{7}$ **3.** $3\sqrt{2} + 5\sqrt{6}$ **5.** $-6\sqrt[3]{6x}$ **7.** $-8\sqrt{3x} + 3\sqrt{x}$ **9.** $\dfrac{5}{8}\sqrt{x}$ **11.** $6\sqrt{5} + 6\sqrt{x}$ **13.** $7xy\sqrt[3]{2} - 5x\sqrt[3]{2}$ **15.** $-\sqrt{5}$ **17.** $12\sqrt{3} - 2\sqrt{6}$ **19.** $5\sqrt{3} + 6\sqrt{5}$ **21.** $\sqrt[3]{9}$ **23.** $11\sqrt[3]{2} + 6$ **25.** $3x\sqrt[4]{x}$ **27.** $5t\sqrt{t} + 6t$ **29.** $2a^2b\sqrt{ab}$ **31.** $\dfrac{23\sqrt{2}}{4y}$ **33.** $-4a\sqrt[4]{a^3}$ **35.** $2x\sqrt{5} + \sqrt{15}$ **37.** $2m\sqrt{3m} - 6m\sqrt{15}$ **39.** $29 - 13\sqrt{3}$ **41.** $x + 10\sqrt{x} + 24$ **43.** $12 - 3\sqrt{y} - 4\sqrt{x} + \sqrt{xy}$ **45.** $7z - 9$ **47.** $9 - x$ **49.** $4x + 20\sqrt{x} + 25$ **51.** $\dfrac{2\sqrt{6}}{3}$ **53.** $\dfrac{3\sqrt{2x}}{2x}$ **55.** $\dfrac{\sqrt[3]{18x}}{2x}$ **57.** $\dfrac{4\sqrt{3x}}{3x^2y^2}$ **59.** $10 + 5\sqrt{3}$ **61.** $\dfrac{-3\sqrt{x} + 3}{x - 1}$ **63.** $\dfrac{m + 3\sqrt{m} - 28}{m - 49}$

7.5 Exercises

1. $\{-6\}$ **3.** $\{4\}$ **5.** $\{-13\}$ **7.** $\{-8\}$ **9.** $\{13\}$ **11.** $\{\ \}$ or $\varnothing$ **13.** $\{4\}$ **15.** $\{4\}$ **17.** $\left\{\dfrac{5}{4}\right\}$ **19.** $\{2, 4\}$ **21.** $\left\{-\dfrac{7}{2}\right\}$ **23.** $\left\{-9, \dfrac{3}{2}\right\}$ **25.** $\{30\}$ **27.** $\{38\}$ **29.** $\{1\}$ **31.** $\{64\}$ **33.** $\{5\}$ **35.** $\{5\}$ **37.** $\left\{-\dfrac{7}{2}, 2\right\}$ **39.** $L = \dfrac{8T^2}{\pi^2}$ **41.** $h = \dfrac{3V}{\pi r^3}$ **43.** $r = \dfrac{A^2}{P^2} - 1$ or $r = \dfrac{A^2 - P^2}{P^2}$ **45. a.** 10th grade **b.** 36 words **47. a.** 2.12 m² **b.** 64.5 kg **49. a.** 0.96 sec **b.** 5.4 ft

7.6 Exercises

1. i **3.** i **5.** $-1 + i$ **7.** -1 **9.** $10 - 11i$ **11.** $7 - 3i$ **13.** 4 **15.** $-2\sqrt{7} + 9i$ **17.** $-1 + i$ **19.** $6 + 8i$ **21.** $20 - 9i$ **23.** $7 + 17i$ **25.** $\dfrac{33}{8} - \dfrac{1}{2}i$ **27.** $-45 + 28i$ **29.** $1 - 2\sqrt{2}i$ **31.** 29 **33.** $\dfrac{37}{4}$

35. 16   **37.** 116   **39.** 37   **41.** $\dfrac{2}{25} - \dfrac{11}{25}i$ **43.** $\dfrac{3}{4} + \dfrac{3}{4}i$ **45.** $-\dfrac{3}{5} - \dfrac{2}{5}i$ **47.** $-\dfrac{1}{7} + \dfrac{4\sqrt{3}}{7}i$ **49.** $8i\sqrt{5}$

51. $3 - 7i$ **53.** -8 **55.** $-2 - i\sqrt{5}$ **57.** $1 + \dfrac{\sqrt{2}}{2}i$

CHAPTER 8

8.1 Exercises

1. $\{0, 8\}$ **3.** $\{4, 6\}$ **5.** $\left\{-\dfrac{3}{4}, \dfrac{5}{2}\right\}$ **7.** $\{-16, 16\}$ **9.** $\left\{-\dfrac{5}{8}, \dfrac{5}{8}\right\}$ **11.** $\{-11, 11\}$ **13.** $\{-2\sqrt{7}, 2\sqrt{7}\}$

15. $\{-18, 32\}$ **17.** $\{-3, -3\sqrt{6}, -3 + 3\sqrt{6}\}$ **19.** $\{-\sqrt{5}, \sqrt{5}\}$ **21.** $\left\{-\dfrac{\sqrt{15}}{5}, \dfrac{\sqrt{15}}{5}\right\}$ **23.** $\{-7, 1\}$

25. $\{11 - 5\sqrt{7}, 11 + 5\sqrt{7}\}$ **27.** $\{-2i, 2i\}$ **29.** $\{-2i\sqrt{2}, 2i\sqrt{2}\}$ **31.** $\{1 - i\sqrt{5}, 1 + i\sqrt{5}\}$

33. $\{2 - 4i\sqrt{2}, 2 + 4i\sqrt{2}\}$ **35.** $\{15 - 12i, 15 + 12i\}$ **37.** $x^2 + 8x + 16 = (x + 4)^2$ **39.** $z^2 + 3z + \dfrac{9}{4} = \left(z + \dfrac{3}{2}\right)^2$

41. $\{-6, 4\}$ **43.** $\{7 - 2\sqrt{10}, 7 + 2\sqrt{10}\}$ **45.** $\left\{-\dfrac{7}{2} - \dfrac{\sqrt{7}}{2}i, -\dfrac{7}{2} + \dfrac{\sqrt{7}}{2}i\right\}$ **47.** $\{-1 - \sqrt{2}, -1 + \sqrt{2}\}$

49. $\{-2 - \sqrt{3}, -2 + \sqrt{3}\}$ **51.** $\left\{\dfrac{3 - \sqrt{65}}{4}, \dfrac{3 + \sqrt{65}}{4}\right\}$ **53.** $\{6 - 2i, 6 + 2i\}$ **55.** $\{2 - 3i, 2 + 3i\}$

57. $\left\{-\dfrac{4}{5} - \dfrac{2}{5}i, -\dfrac{4}{5} + \dfrac{2}{5}i\right\}$ **59.** $\left\{\dfrac{9 - \sqrt{65}}{8}, \dfrac{9 + \sqrt{65}}{8}\right\}$ **61.** $\{5 - 4\sqrt{2}, 5 + 4\sqrt{2}\}$ **63.** $\left\{-\dfrac{1}{3}, 3\right\}$

65. $\{4 - 3\sqrt{2}, 4 + 3\sqrt{2}\}$ **67.** $\left\{\dfrac{2}{3} - \dfrac{\sqrt{7}}{3}, \dfrac{2}{3} + \dfrac{\sqrt{7}}{3}\right\}$ **69.** $\left\{-2 - \dfrac{\sqrt{6}}{2}, -2 + \dfrac{\sqrt{6}}{2}\right\}$ **71.** $\{4 - \sqrt{5}, 4 + \sqrt{5}\}$

73. $\left\{\dfrac{1}{3}\right\}$ **75.** $\left\{\dfrac{2}{3} - \dfrac{\sqrt{10}}{3}, \dfrac{2}{3} + \dfrac{\sqrt{10}}{3}\right\}$ **77.** $\{2 - 5i, 2 + 5i\}$ **79.** $\left\{\dfrac{1}{8} - \dfrac{\sqrt{127}}{8}i, \dfrac{1}{8} + \dfrac{\sqrt{127}}{8}i\right\}$

81. $\left\{\dfrac{2}{3} - \dfrac{1}{3}i, \dfrac{2}{3} + \dfrac{1}{3}i\right\}$ **83.** $D = 5$; two real solutions   **85.** $D = 41$; two real solutions   **87.** $D = 0$; one real solution

89. $\{-2, -\sqrt{2}, \sqrt{2}, 2\}$ **91.** $\left\{\dfrac{1}{8}, 8\right\}$ **93.** $\left\{-2, \dfrac{1}{2}\right\}$ **95.** $\{81\}$ **97.** $\left\{0, \dfrac{4}{13}\right\}$

8.2 Exercises

1. up, wider   **3.** up, narrower   **5. a.** $(2, 4)$ **b.** $x = 2$ **c.** 0   **d.** 0 and 4   **e.** domain: $(-\infty, \infty)$ and range: $(-\infty, 4]$
7. a. $(2, 0)$ **b.** $x = 2$ **c.** 4   **d.** 2   **e.** domain: $(-\infty, \infty)$ and range: $[0, \infty)$ **9.** 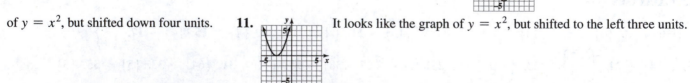 It looks like the graph

of $y = x^2$, but shifted down four units.   **11.** It looks like the graph of $y = x^2$, but shifted to the left three units.

13. It looks like the graph of $y = x^2$, but shifted to the right one unit and down two units.   **15. a.** $(2, -4)$

b. up   **c.** $x = 2$ **d.** 0 and 4   **e.** 0   **f.** 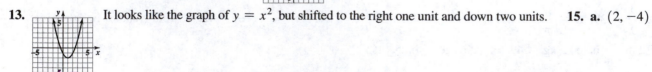   **g.** domain: $(-\infty, \infty)$ and range: $[-4, \infty)$

17. a. $(-1, -9)$ **b.** down **c.** $x = -1$ **d.** none **e.** -10 **f.** **g.** domain: $(-\infty, \infty)$ and range: $(-\infty, -9]$

19. a. $(4, 2)$ **b.** down **c.** $x = 4$ **d.** $4 - 2\sqrt{2} \approx 1.1716$ and $4 + 2\sqrt{2} \approx 6.8284$ **e.** -2 **f.**

g. domain: $(-\infty, \infty)$ and range: $(-\infty, 2]$ **21.** $f(x) = (x + 4)^2 - 7, (-4, -7)$ **23.** $f(x) = -2(x + 3)^2 + 8, (-3, 8)$

25. $f(x) = (x + 5)^2 - 25, (-5, -25)$ **27.** $f(x) = (x + 3)^2 - 16$ **a.** $(-3, -16)$ **b.** up **c.** $x = -3$

d. -7 and 1 **e.** -7 **f.** **g.** domain: $(-\infty, \infty)$ and range: $[-16, \infty)$ **29.** $f(x) = 4\left(x - \dfrac{7}{8}\right)^2 + \dfrac{79}{16}$

a. $\left(\dfrac{7}{8}, \dfrac{79}{16}\right)$ **b.** up **c.** $x = \dfrac{7}{8}$ **d.** none **e.** 8 **f.** **g.** domain: $(-\infty, \infty)$ and range: $\left[\dfrac{79}{16}, \infty\right)$

31. $f(x) = \dfrac{1}{4}(x - 4)^2 - 3$ **a.** $(4, -3)$ **b.** up **c.** $x = 4$ **d.** $4 - 2\sqrt{3} \approx 0.5359$ and $4 + 2\sqrt{3} \approx 7.4641$

e. 1 **f.** **g.** domain: $(-\infty, \infty)$ and range: $[-3, \infty)$ **33.** $\left(-\dfrac{5}{2}, -\dfrac{37}{4}\right)$ **35.** $\left(\dfrac{3}{2}, 9\right)$ **37. a.** $(2, -64)$

b. up **c.** $x = 2$ **d.** -6 and 10 **e.** -60 **f.** **g.** domain: $(-\infty, \infty)$ and range: $[-64, \infty)$

39. a. $(1, -5)$ **b.** down **c.** $x = 1$ **d.** none **e.** -6 **f.** **g.** domain: $(-\infty, \infty)$ and range: $(-\infty, -5]$

41. a. $\left(\dfrac{7}{6}, \dfrac{109}{12}\right)$ **b.** down **c.** $x = \dfrac{7}{6}$ **d.** $\dfrac{7}{6} - \dfrac{\sqrt{109}}{6} \approx -0.5734$ and $\dfrac{7}{6} + \dfrac{\sqrt{109}}{6} \approx 2.9067$ **e.** 5 **f.**

g. domain: $(-\infty, \infty)$ and range: $\left(-\infty, \dfrac{109}{12}\right]$

8.3 Exercises

1. $-\dfrac{4}{3}$ **3.** 1, 3, and 5, or 3, 5, and 7 **5.** 4 sec **7.** $\dfrac{10}{3}$ cm by 9 cm **9.** length: 10.05 inches, width: 6.55 inches

11. 3.76 ft **13.** 5 mph **15.** 120 min **17.** 3.5 sec, 61.025 m **19.** 3.5 sec, 201 ft **21. a.** \$71.60

b. 360, \$12,960 million **c.** \$36 **23. a.** $R(x) = -\dfrac{1}{40}x^2 + 8000x$ **b.** $P(x) = -\dfrac{1}{40}x^2 + 4000x - 20,000$

c. 80,000, \$159,980,000 **d.** \$6000 **25.** 1.2, \$11, \$7.20 **27.** 625 ft by 1250 ft, 781,250 ft^2 **29.** $\dfrac{1}{2}$, \$1.50

31. a. \$64, \$16 **b.** 22, \$0 **c.** no

8.4 Exercises

1. 5 **3.** $\sqrt{74}$ **5.** $\dfrac{\sqrt{829}}{6}$ **7.** 8 **9.** $(4, 8)$ **11.** $\left(-\dfrac{3}{2}, 1\right)$ **13.** $\left(-\dfrac{3}{2}, \dfrac{3}{4}\right)$ **15.** $(1, -1)$ **17.** $x^2 + y^2 = 1$

19. $(x + 2)^2 + (y - 3)^2 = 16$ **21.** $(x - 1)^2 + (y + 4)^2 = \dfrac{9}{16}$ **23.** $(x - 3)^2 + y^2 = 2$

25. $(h, k) = (0, 0), r = 3,$ 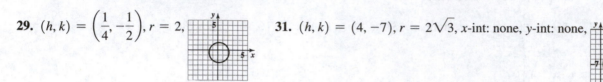 **27.** $(h, k) = (1, -5), r = 4,$

29. $(h, k) = \left(\dfrac{1}{4}, -\dfrac{1}{2}\right), r = 2,$ **31.** $(h, k) = (4, -7), r = 2\sqrt{3},$ x-int: none, y-int: none,

33. $(x + 3)^2 + y^2 = 4, (h, k) = (-3, 0), r = 2, x\text{-int: } x = -5, 1, y\text{-int: none,}$

35. $(x + 1)^2 + (y - 2)^2 = 4, (h, k) = (-1, 2), r = 2, x\text{-int: } x = -1, y\text{-int: } y = 2 \pm \sqrt{3},$

37. $(x - 2)^2 + (y - 4)^2 = 1, (h, k) = (2, 4), r = 1, x\text{-int: none, } y\text{-int: none,}$

39. $\left(x - \dfrac{3}{2}\right)^2 + \left(y - \dfrac{1}{2}\right)^2 = 3, (h, k) = \left(\dfrac{3}{2}, \dfrac{1}{2}\right), r = \sqrt{3}, x\text{-int: } x = \dfrac{3 \pm \sqrt{11}}{2}, y\text{-int: } y = \dfrac{1 \pm \sqrt{3}}{2},$

41. $(x - 1)^2 + (y + 2)^2 = 4, (h, k) = (1, -2), r = 2, x\text{-int: } x = 1, y\text{-int: } y = -2 \pm \sqrt{3},$

43. $\left(x - \dfrac{1}{4}\right)^2 + \left(y - \dfrac{1}{3}\right)^2 = 4, (h, k) = \left(\dfrac{1}{4}, \dfrac{1}{3}\right), r = 2, x\text{-int: } x = \dfrac{3 \pm 4\sqrt{35}}{12}, y\text{-int: } y = \dfrac{4 \pm 9\sqrt{7}}{12},$

8.5 Exercises

1. $(-\infty, -3] \cup [1, \infty)$ **3.** $\left[-\dfrac{2}{3}, 0\right]$ **5.** $[-1, 1]$ **7.** $\left(-\dfrac{1}{3}, 1\right)$ **9.** $(-\infty, 1]$ **11.** $[-3, 1)$ **13.** $(-\infty, 0) \cup (1, \infty)$

15. $(-1, 1]$ **17.** $(-4, -2]$ **19.** $\left(-3, -\dfrac{1}{2}\right] \cup (2, \infty)$ **21.** $(-3, 5)$

CHAPTER 9

9.1 Exercises

1.   **3.**   **5.** 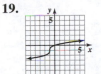   **7.**   **9.**   **11.**

13.   **15.**   **17.**   **19.**   **21.**   **23.**

25.   **27.**   **29.**   **31.**   **33.** 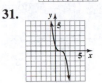   **35.**

37.   **39.** (–1, 3), (1, 3), (–2, 0)   **41.**   **43.**   **45.** (–2, 1), (2, 1), (–4, 0)   **47.**

49.   **51.** (–4, –1), (1, 0), (–3, –4)   **53.** (–2, 4), (–3, 2), (3, 2), (2, –4)   **55.** (–3, –2), (3, –2), (–2, –5)

9.2 Exercises

1. $(f \circ g)(x) = \dfrac{x + 7}{x + 1}$ **3.** $(f \circ h)(x) = 3\sqrt{x + 3} + 1$ **5.** $(h \circ f)(x) = \sqrt{3x + 4}$ **7.** $(f \circ f)(x) = 9x + 4$

9. 7   **11.** $\dfrac{2}{5}$ **13.** 2   **15.** –5   **17. a.** –4   **b.** 0   **c.** 2   **d.** –1   **19.** $(-\infty, \infty), (-\infty, \infty)$

21. $[0, \infty), (-\infty, \infty)$ **23.** $(-\infty, 1) \cup (1, 2) \cup (2, \infty), (-\infty, -1) \cup \left(-1, \dfrac{1}{2}\right) \cup \left(\dfrac{1}{2}, \infty\right)$ **25.** yes   **27.** yes   **29.** yes

31. no   **33.** yes   **35.** yes   **37.** no   **39.** yes   **41.–43.** $(f \circ g)(x) = (g \circ f)(x) = x$

45. 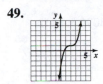   , domain: $(-\infty, 5]$, range: $[-1, \infty)$ **47.** 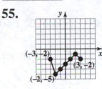   , domain: $(-\infty, \infty)$, range: $(-\infty, \infty)$

49.   , domain: $[-4, 3]$, range: $[-4, 4]$

51. $f^{-1}(x) = 3x + 15$, domain f = range $f^{-1} = (-\infty, \infty)$, range f = domain $f^{-1} = (-\infty, \infty)$

53. $f^{-1}(x) = \dfrac{x^3 + 3}{2}$, domain f = range $f^{-1} = (-\infty, \infty)$, range f = domain $f^{-1} = (-\infty, \infty)$

55. $f^{-1}(x) = \sqrt{-x - 2}$, domain f = range $f^{-1} = [0, \infty)$, range f = domain $f^{-1} = (-\infty, -2]$

57. $f^{-1}(x) = \dfrac{3}{x}$, domain f = range f^{-1} = range f = domain $f^{-1} = (-\infty, 0) \cup (0, \infty)$

59. $f^{-1}(x) = \dfrac{7x + 1}{5x + 8}$, domain f = range $f^{-1} = \left(-\infty, \dfrac{7}{5}\right) \cup \left(\dfrac{7}{5}, \infty\right)$, range f = domain $f^{-1} = \left(-\infty, -\dfrac{8}{5}\right) \cup \left(-\dfrac{8}{5}, \infty\right)$

9.3 Exercises

1.   **3.**   **5.**   **7.** $f(x) = 2^x$ **9.** $f(x) = \left(\dfrac{3}{4}\right)^x$ **11.** $f(x) = \left(\dfrac{1}{16}\right)^x$

13. , domain: $(-\infty, \infty)$, range: $(0, \infty)$, y-intercept: $\dfrac{1}{2}$, $y = 0$ **15.** , domain: $(-\infty, \infty)$,

range: $(-\infty, 0)$, y-intercept: -9, $y = 0$ **17.** , domain: $(-\infty, \infty)$, range: $(0, \infty)$, y-intercept: $\dfrac{1}{2}$, $y = 0$

19. , domain: $(-\infty, \infty)$, range: $(1, \infty)$, y-intercept: 2, $y = 1$ **21.** $\{4\}$ **23.** $\left\{\dfrac{1}{4}\right\}$ **25.** $\left\{-\dfrac{3}{5}\right\}$

27. $\left\{\dfrac{1}{3}\right\}$ **29.** $\{-1, 3\}$ **31. a.** \$300,000 **b.** \$265,640 **c.** \$163,299 **33. a.** 2560 bacteria **b.** 4096 bacteria

c. 8,589,934,592 bacteria **35.** \$12,752.18 **37.** Investment B **39.** \$469,068.04

9.4 Exercises

1. 20.085537 **3.** 1.395612 **5.** 32.725881 **7.** , domain: $(-\infty, \infty)$, range: $(-1, \infty)$, $y = -1$

9. , domain: $(-\infty, \infty)$, range: $(-\infty, -1)$, $y = -1$ **11.** $\{-2\}$ **13.** $\left\{\dfrac{1}{4}\right\}$ **15.** $\{-3, 4\}$

17. $\{-1, 1, 2\}$ **19.** \$6798.89, \$20,942.06 **21.** \$332,570.82 **23.** (a) has the lower present value.
25. a. $P(t) = 10e^{0.25t}$ **b.** 4034 bacteria **27. a.** 8 rabbits **b.** 478,993 rabbits

9.5 Exercises

1. $\log_3 9 = 2$ **3.** $\log_2 \dfrac{1}{8} = -3$ **5.** $\log_{1/3} 27 = t$ **7.** $5^0 = 1$ **9.** $\sqrt{2}^6 = 8$ **11.** $a^3 = (x - 1)$ **13.** $\dfrac{1}{2}$ **15.** 4

17. $-\dfrac{1}{3}$ **19.** 11 **21.** 0 **23.** 1 **25.** 0 **27.** $\log 1000 = 3$ **29.** $\ln 2 = k$ **31.** $\ln Z = 10$ **33.** $10^6 = 1,000,000$

35. $e^4 = Z$ **37.** -3 **39.** $\dfrac{2}{3}$ **41.** 49 **43.** 4 **45.** , $(0, \infty)$, $x = 0$ **47.** , $(1, \infty)$, $x = 1$

49. 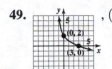 , $(-1, \infty), x = -1$ **51.** 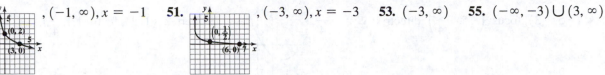 , $(-3, \infty), x = -3$ **53.** $(-3, \infty)$ **55.** $(-\infty, -3) \cup (3, \infty)$

57. $(-\infty, -5) \cup (8, \infty)$

9.6 Exercises

1. $\log_4 x + \log_4 y$ **3.** $2 + \log P$ **5.** $\log 9 - \log t$ **7.** $\log_2 M - 5$ **9.** $-9 \log_3 w$ **11.** $2 \log_7 x + 3 \log_7 y$

13. $\frac{1}{2} \log x - 1 - 3 \log y$ **15.** $1 + \frac{1}{2} \log_2 x + \frac{1}{2} \log_2 y$ **17.** $\frac{1}{5} + \frac{1}{5} \ln z - \frac{1}{2} \ln (x - 1)$ **19.** $-\frac{4}{3} \ln (x + 1) - 2 \ln (x - 1)$

21. $\log_b (AC)$ **23.** $\log_8 \left(x^2 \sqrt[3]{y} \right)$ **25.** 4   **27.** $\log_5 (x^2 - 4)$ **29.** $\log \left(\frac{x + 3}{x + 1} \right)$ **31.** $\{5\}$ **33.** $\{1\}$ **35.** $\{2\}$

37. 2.8362   **39.** -2.6572 **41.** $\log_3 (xw^2)$ **43.** $\log_2 (xyw)$ **45.** $\{5\}$ **47.** $\{36\}$

9.7 Exercises

1. $\{1.4650\}$ **3.** $\{-1, 3\}$ **5.** $\{12.7438\}$ **7.** $\{4.4841\}$ **9.** $\{0.3334\}$ **11.** $\{0.6931\}$ **13.** $\{-0.2055\}$

15. $\{-3.1192\}$ **17.** $\{7\}$ **19.** $\{2, 3\}$ **21.** $\left\{ -\frac{99}{5} \right\}$ **23.** $\left\{ \frac{1}{3} \right\}$ **25.** $\{6\}$ **27.** $\{9\}$ **29.** $\left\{ -1, \frac{1}{3} \right\}$ **31.** $\{\}$ or $\varnothing$

9.8 Exercises

1. about 7 years 9 months   **3.** 17.33%   **5.** 17.4 years   **7.** 6.50%   **9.** 4147 bacteria   **11.** 66 days

13. 200,119 people   **15.** 64.84%   **17. a.** 6000 families   **b.** 2000 families   **c.** 1988   **19. a.** 61,500   **b.** 40

c. $-\dfrac{\ln \left(\frac{43}{2} \right)}{6}$ **d.** 2009   **21.** about 11:52 PM

CHAPTER 10

10.1 Exercises

1. ellipse   **3.** circle   **5.** vertex: $(0, 0)$, focus: $(0, 4)$, directrix: $y = -4$, 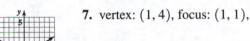   **7.** vertex: $(1, 4)$, focus: $(1, 1)$,

directrix: $y = 7$, 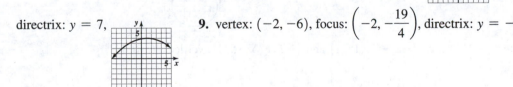   **9.** vertex: $(-2, -6)$, focus: $\left(-2, -\frac{19}{4} \right)$, directrix: $y = -\frac{29}{4}$,

11. vertex: $(0, 0)$, focus: $(-2, 0)$, directrix: $x = 2$,   **13.** vertex: $(2, 5)$, focus: $(1, 5)$, directrix: $x = 3$,

15. vertex: $(-3, -4)$, focus: $\left(-\frac{3}{4}, -4 \right)$, directrix: $x = -\frac{21}{4}$,   **17.** $y^2 = 8x$ **19.** $(x - 3)^2 = -8(y + 3)$

21. $(y + 2)^2 = -\left(x + \frac{11}{4} \right)$ **23.** $(x - 1)^2 = -12(y - 2)$ **25.** $(y - 4)^2 = 8(x + 2)$ and $(y - 4)^2 = -8(x - 2)$

27. vertex: $(-5, -3)$, focus: $\left(-5, -\dfrac{7}{4}\right)$, directrix: $y = -\dfrac{17}{4}$, **29.** vertex: $(1, 2)$, focus: $\left(1, \dfrac{7}{4}\right)$,

directrix: $y = \dfrac{9}{4}$, **31.** 10 cm **33.** $\dfrac{121}{16}$ ft **35.** 14.4 m

10.2 Exercises

1. center: $(0, 0)$, foci: $\left(\sqrt{12}, 0\right)$ and $\left(-\sqrt{12}, 0\right)$, vertices $(4, 0)$ and $(-4, 0)$, **3.** center: $(0, 0)$, foci: $\left(\sqrt{5}, 0\right)$

and $\left(-\sqrt{5}, 0\right)$, vertices $(3, 0)$ and $(-3, 0)$, **5.** center: $(2, 4)$, foci: $\left(2 + \sqrt{11}, 4\right)$ and $\left(2 - \sqrt{11}, 4\right)$,

vertices $(8, 4)$ and $(-4, 4)$, **7.** center: $(5, -1)$, foci: $(9, -1)$ and $(1, -1)$, vertices $(0, -1)$ and $(10, -1)$,

 9. $\dfrac{x^2}{49} + \dfrac{y^2}{24} = 1$ **11.** $\dfrac{(x + 6)^2}{16} + \dfrac{(y - 5)^2}{25} = 1$ **13.** $\dfrac{(x + 1)^2}{7} + \dfrac{(y - 4)^2}{16} = 1$

15. $\dfrac{(x - 4)^2}{4} + \dfrac{(y - 1)^2}{53} = 1$ **17.** $\dfrac{(x + 4)^2}{16} + \dfrac{y^2}{36} = 1$ **19.** $\dfrac{x^2}{16} + \dfrac{y^2}{12} = 1$ **21.** $\dfrac{(x + 2)^2}{20} + \dfrac{(y - 1)^2}{16} = 1$,

center: $(-2, 1)$, foci: $(-4, 1)$ and $(0, 1)$, vertices: $\left(-2 - 2\sqrt{5}, 1\right)$ and $\left(-2 + 2\sqrt{5}, 1\right)$,

23. $\dfrac{(x + 2)^2}{16} + \dfrac{(y - 1)^2}{1} = 1$, center: $(-2, 1)$ foci: $\left(-2 - \sqrt{15}, 1\right)$ and $\left(-2 + \sqrt{15}, 1\right)$, vertices: $(-6, 1)$

and $(2, 1)$, **25.** $\dfrac{x^2}{225} + \dfrac{y^2}{256} = 1$ **27.** $\dfrac{x^2}{81} + \dfrac{y^2}{441} = 1$ **29.** 44.80 in. **31.** about 39.69 ft

10.3 Exercises

1. center: $(0, 0)$, transverse axis: $y = 0$, vertices: $(-4, 0)$ and $(4, 0)$, foci: $(-5, 0)$ and $(5, 0)$, asymptotes: $y = \pm\dfrac{3}{4}x$,

3. center: $(2, 4)$, transverse axis: $x = 2$, vertices: $(2, -1)$ and $(2, 9)$, foci: $\left(2, 4 - \sqrt{61}\right)$ and $\left(2, 4 + \sqrt{61}\right)$, asymptotes: $y - 4 = \pm\frac{5}{6}(x - 2)$,   **5.** center: $(0, 0)$, transverse axis: $y = 0$, vertices: $\left(-\sqrt{5}, 0\right)$ and $\left(\sqrt{5}, 0\right)$,

foci: $(-5, 0)$ and $(5, 0)$, asymptotes: $y = \pm 2x$,   **7.** $\frac{x^2}{9} - \frac{y^2}{16} = 1$,

9. $\frac{(x - 4)^2}{1} - \frac{(y + 4)^2}{3} = 1$,   **11.** $\frac{(y - 6)^2}{4} - \frac{(x - 9)^2}{5} = 1$,

13. $\frac{y^2}{20} - \frac{x^2}{16} = 1$,   **15.** $\frac{(x - 2)^2}{4} - \frac{(y - 1)^2}{4} = 1$, center: $(2, 1)$, vertices: $(0, 1)$ and $(4, 1)$,

foci: $\left(2 - 2\sqrt{2}, 1\right)$ and $\left(2 + 2\sqrt{2}, 1\right)$, endpoints of conjugate axis: $(2, -1)$ and $(2, 3)$, asymptotes: $y - 1 = \pm(x - 2)$,

    17. $\frac{(y - 6)^2}{9} - (x + 2)^2 = 1$, center: $(-2, 6)$, vertices: $(-2, 3)$ and $(-2, 9)$, foci: $\left(-2, 6 - \sqrt{10}\right)$

and $\left(-2, 6 + \sqrt{10}\right)$, endpoints of conjugate axis: $(-3, 6)$ and $(-1, 6)$, asymptotes: $y - 6 = \pm 3(x + 2)$,

19. $\frac{(x - 6)^2}{25} - \frac{(y + 8)^2}{144} = 1$, center: $(6, -8)$, vertices: $(1, -8)$ and $(11, -8)$, foci: $(-7, -8)$ and $(19, -8)$, endpoints of

conjugate axis: $(6, -20)$ and $(6, 4)$, asymptotes: $y + 8 = \pm\frac{12}{5}(x - 6)$,   **21.** $\frac{y^2}{49} - \frac{x^2}{15} = 1$

23. a. $\frac{x^2}{900} - \frac{y^2}{2700} = 1$ **b.** 159 m

CHAPTER 11

11.1 Exercises

1. $4, 7, 10, 13$ **3.** $1, \frac{8}{5}, 2, \frac{16}{7}$ **5.** $30, 120, 600, 3600$ **7.** $-5, 10, -15, 20$ **9.** $-\frac{1}{42}, \frac{1}{56}, -\frac{1}{72}, \frac{1}{90}$ **11.** $7, 10, 13, 16$

13. $6, \frac{3}{2}, \frac{1}{6}, \frac{1}{96}$ **15.** $3, 0, 3, 6, 15, 36$ **17.** $a_n = 2n - 3, a_{11} = 19$ **19.** $a_n = n(n + 5), a_7 = 84$

21. $a_n = \frac{2}{5^n}, a_6 = \frac{2}{15{,}625}$ **23.** $a_1 = -6, a_n = (-1)^n(n + 2)!$ for $n \geq 2, a_5 = -5040$ **25.** 20 **27.** $\frac{11}{32}$ **29.** 12

31. 45 **33.** 196 **35.** 271 **37.** 27 **39.** $\sum_{i=1}^{10} 5i$ **41.** $\sum_{i=1}^{9} \frac{i + 3}{i + 4}$ **43.** $\sum_{i=1}^{6} \frac{(-1)^i}{9i}$ **45.** $\sum_{i=1}^{n+1} \frac{7^{i-1}}{(i - 1)!}$

11.2 Exercises

1. yes, 6   **3.** yes, $\dfrac{3}{2}$ **5.** yes, 2   **7.** $a_n = 5n - 3, a_{10} = 47$ **9.** $a_n = \dfrac{3}{2}n, a_{50} = 75$ **11.** $a_n = 2n - 1, a_{17} = 33$

13. 71   **15.** 97   **17.** $\dfrac{25}{2}$ **19.** 3240   **21.** 2375   **23.** -1002 **25.** no solution   **27.** 10,100   **29.** 147

31. $3185   **33.** $200, $65,000   **35.** 1.3 sec, 312 in.

11.3 Exercises

1. 8, 16, 32, 64, 128   **3.** $25, -5, 1, -\dfrac{1}{5}, \dfrac{1}{25}$ **5.** $-4, 8, -16, 32, -64$ **7.** yes, -1 **9.** yes, $-\dfrac{5}{3}$ **11.** $a_n = 3 \cdot 2^{n-1}$

13. $a_n = \dfrac{1}{5}\left(-\dfrac{2}{3}\right)^{n-1}$ **15.** $-\dfrac{951}{256}$ **17.** 78,732   **19.** $-91,749$ **21.** ≈ 50.31 **23.** $\dfrac{85}{128}$ **25.** converges, $\dfrac{2401}{6}$

27. diverges   **29.** $6561   **31.** $\approx \$1,760,290.38$ **33.** 200 ft   **35.** $\dfrac{322}{990}$ **37.** $\approx \$30,200.99$

11.4 Exercises

1. $m^6 + 6m^5n + 15m^4n^2 + 20m^3n^3 + 15m^2n^4 + 6mn^5 + n^6$

3. $x^7 - 7x^6y + 21x^5y^2 - 35x^4y^3 + 35x^3y^4 - 21x^2y^5 + 7xy^6 - y^7$

5. $64x^6 + 576x^5y + 2160x^4y^2 + 4320x^3y^3 + 4860x^2y^4 + 2916xy^5 + 729y^6$ **7.** 7   **9.** 1

11. $x^7 + 14x^6 + 84x^5 + 280x^4 + 560x^3 + 672x^2 + 448x + 128$

13. $1024x^5 + 1280x^4 + 640x^3 + 160x^2 + 20x + 1$

15. $3125x^5 - 9375x^4y + 11{,}250x^3y^2 - 6750x^2y^3 + 2025xy^4 - 243y^5$ **17.** $x^2 - 4\sqrt{2}x^{\frac{3}{2}} + 12x - 8\sqrt{2}x^{\frac{1}{2}} + 4$

19. $70a^4b^4$ **21.** $1{,}088{,}640x^4$ **23.** 126,720   **25.** 19,779,228

APPENDICES

Appendix B Exercises

1. $\begin{bmatrix} 2 & 3 & | & 5 \\ 5 & -2 & | & 1 \end{bmatrix}$ **3.** $\begin{bmatrix} \dfrac{1}{2} & -\dfrac{1}{3} & | & 0 \\ \dfrac{2}{7} & \dfrac{5}{9} & | & 2 \end{bmatrix}$ **5.** $\begin{bmatrix} 7 & -2 & 5 & | & 3 \\ 0 & 5 & 2 & | & -2 \\ 3 & 0 & -2 & | & 1 \end{bmatrix}$ **7.** $(0, -2)$ **9.** $(2, -1)$ **11.** $(5, -2)$

13. $\left(\dfrac{1}{2}, \dfrac{1}{3}\right)$ **15.** $\{\}$ or $\varnothing$ **17.** $(4, 3)$ **19.** $(1, -2, -1)$ **21.** $(1, 3, 2)$ **23.** $(4, 3, 4)$ **25.** $(0, 2, 0)$

27. $\left(-2, \dfrac{1}{2}, -3\right)$ **29.** $(-3, 0, 1)$

Appendix C Exercises

1. 18   **3.** 0   **5.** $(1, 2)$ **7.** $(-2, -4)$ **9.** $\left(\dfrac{1}{2}, \dfrac{1}{3}\right)$ **11.** $\{(x, y) \mid y = 3x - 5\}$ **13.** 164   **15.** -39

17. $(2, 1, -1)$ **19.** $(-3, 2, -4)$ **21.** $\{\}$ or $\varnothing$

Index